# An Advanced Introduction to Calculus-Based Physics (Mechanics)

Chris McMullen, Ph.D.
Northwestern State University of Louisiana

An Advanced Introduction to Calculus-Based Physics (Mechanics)

Copyright © 2012 Chris McMullen, Ph.D.

All rights are reserved.  This includes the right to reproduce any portion of this book in any form.

Chris McMullen, Ph.D.
Physics Instructor
Department of Mathematics and Physical Sciences
Northwestern State University of Louisiana
http://mathematics.nsula.edu/faculty/
mcmullenc@nsula.edu

CreateSpace

FIRST EDITION, May, 2012

Textbooks / science / physics / mechanics
Nonfiction / science / physics / mechanics

ISBN: 1463644086

EAN-13: 978-1463644086

# Contents

| | |
|---|---|
| Introduction | 5 |
| How to Succeed in Physics | 7 |
| Suggestions for Reading This Book | 9 |
| Which Material Is Practical, Which Is Abstract? | 10 |
| Units, Dimensions, and Significant Figures | 12 |
| What Is Physics? | 18 |
| 1 Single-Component Motion | 19 |
|     1.1 Basic 1D Motion Quantities | 20 |
|     1.2 Net and Average Values in 1D | 23 |
|     1.3 Instantaneous Values in 1D | 27 |
|     1.4 Calculus Applied to 1D Motion | 30 |
|     1.5 Single-Component Motion Graphs | 33 |
|     1.6 Uniform Acceleration in 1D | 43 |
|     1.7 Vertical Free Fall | 52 |
|     1.8 Non-Uniform Acceleration in 1D | 58 |
|     1.9 Multiple Moving Objects in 1D | 62 |
| 2 Multi-Component Motion | 76 |
|     2.1 Scalars and Vectors | 77 |
|     2.2 Vector Addition and Subtraction | 87 |
|     2.3 Vector Motion Quantities | 96 |
|     2.4 Polar and Cylindrical Coordinates | 108 |
|     2.5 Spherical Coordinates | 113 |
|     2.6 Projectile Motion | 117 |
|     2.7 Multi-Dimensional Motion | 129 |
| 3 Newton's Laws of Motion | 140 |
|     3.1 Newton's Laws | 142 |
|     3.2 Applying Newton's Second Law | 150 |
|     3.3 Mass and Weight | 154 |
|     3.4 Tension Forces | 158 |
|     3.5 Normal and Friction Forces | 165 |
|     3.6 Hooke's Law | 172 |
|     3.7 Mutual Surface Forces | 180 |
|     3.8 Air Resistance | 185 |
|     3.9 Gravitational Forces | 190 |
| 4 Uniform Circular Motion | 214 |
|     4.1 Basic Rotational Quantities | 214 |
|     4.2 Centripetal Acceleration | 218 |
|     4.3 Centripetal Forces | 222 |
|     4.4 Swinging in a Circle | 225 |

| | |
|---|---:|
| 4.5 Rounding a Turn | 231 |
| 4.6 Satellite Motion | 236 |
| **5 Work and Energy** | **246** |
| 5.1 The Scalar Product | 247 |
| 5.2 Work and Power | 250 |
| 5.3 Potential and Kinetic Energy | 262 |
| 5.4 Conservation of Energy | 269 |
| 5.5 Conservative Force Fields | 281 |
| 5.6 Nonconservative Work | 289 |
| **6 Systems of Objects** | **306** |
| 6.1 Center of Mass | 306 |
| 6.2 Momentum and Impulse | 320 |
| 6.3 Conservation of Momentum | 324 |
| 6.4 Motion of the Center of Mass | 329 |
| 6.5 Collisions in 1D | 335 |
| 6.6 Collisions in 2D | 344 |
| 6.7 The Ballistic Pendulum | 352 |
| 6.8 Scattering | 355 |
| 6.9 Rocket Motion | 360 |
| **7 Rotation** | **378** |
| 7.1 Angular Acceleration | 380 |
| 7.2 Moment of Inertia | 388 |
| 7.3 The Vector Product | 405 |
| 7.4 Torque | 420 |
| 7.5 Summing the Torques | 424 |
| 7.6 Static Equilibrium | 435 |
| 7.7 Rotational Kinetic Energy | 442 |
| 7.8 Angular Momentum | 451 |
| 7.9 Conservation of Angular Momentum | 455 |
| Appendix A: Propagation of Errors | 479 |
| Appendix B: Linear Regression | 484 |
| Hints | 489 |
| Selected Answers | 545 |
| Index | 565 |
| About the Author | 572 |

# Introduction

I was fortunate to teach a unique calculus-based physics course for several years. Two features made this course very unique: The audience included many advanced students, and the first semester of the course had a Calculus II corequisite. The norm for calculus-based physics at colleges and universities is a Calculus I corequisite. This means that my students were fluent in differentiation and basic integration at the beginning of the course, instead of just learning these skills during the course.

The unique corequisite of Calculus II proved to be meaningful in several ways. For one, the students were eager for me to incorporate a high level of calculus in my lectures from Day 1. They thrived on the higher-level math skills, both in the lectures and on the homework. Once I realized this, I applied calculus as often as possible in my lectures and examples, and also embedded it as a standard part of the homework, quiz, and exam problems. Standard physics textbooks, however, expect the course to have a Calculus I corequisite, which means that I had to create much of my own material in order to fully utilize their calculus skills during the first semester.

One of the major advantages of having a Calculus II corequisite on the first-semester calculus-based physics course has to do with the transition to the second semester. Traditionally, calculus plays a minor role in the first semester of calculus-based physics, where students are just learning calculus along with the course, but suddenly becomes a major component of the second semester of calculus-based physics – especially, in the mathematical techniques of electricity and magnetism. The sudden change in the level of mathematics makes this abrupt transition very challenging. When students have a Calculus II corequisite during the first semester, however, and get the opportunity to apply their calculus skills as a regular part of the lectures and coursework, most of the students are already comfortable with this level of mathematics when the second semester begins. This makes for a much smoother transition. For example, if the students perform rigorous center of mass and moment of inertia integrals in cylindrical and spherical coordinates, using the methods of trig substitution learned in Calculus II, these skills carry right into the electric field integrals that are standard in electricity and magnetism.

I find, also, that students have a better opportunity to fully grasp the precise meaning of the concepts when there is a higher math prerequisite. One of my top students called my approach "mathematically-motivated physics concepts" – and she was right on the button of exactly what I strive to do. I emphasize that students are most likely to make mistakes – or be totally in the dark – trying to reason concepts out using their intuition, but can usually use a precise definition, a mathematical equation, or a physical law to reason concepts out correctly. After a student answers a conceptual question incorrectly during a class discussion, I often tell the class to let the equations be their guide. The equations convey the concepts most precisely when we express them with more eloquent mathematics. Therefore, by requiring the students to have higher mathematics skills to get into the course and by expressing the mathematics at as high a level as possible, they have the best opportunity to learn the concepts most precisely, and to understand more subtle points.

Having taught such a unique calculus-based physics course, I saw how it could potentially be beneficial to students learning calculus-based physics to have a text that incorporated a higher level of calculus in the beginning and which emphasized the connections between the mathematics and the concepts. This could be useful for advanced students who have a strong background in mathematics. It could also serve as a useful reference for physics majors and engineers who have gone beyond the first year of physics, but who would like to review the fundamentals as they explore more advanced fields of physics.

Since there are very few first-year physics courses that place a Calculus II corequisite on first-semester physics, I do not expect this text to be adopted as a required textbook. There are, however, many very bright students taking or reviewing physics who may benefit from a text that incorporates a higher level of calculus, emphasizes connections between formal mathematics and precise concepts, and was written not for a general audience, but specifically for advanced students. For these reasons, I have prepared this text for independent use by bright, motivated students who have already completed Calculus I, who may be learning physics for the first time or who are reviewing the fundamentals. No prerequisite physics knowledge is assumed of the reader.

The following features of this book are intended to help students who are using this book independently to improve their understanding of physics. There are separate sections of hints and answers at the back of the book for the end-of-chapter exercises. It is highly recommended that you try to solve a variety of problems to assess your understanding of the material. The hints section, which you won't find in standard textbooks, is specifically designed to help independent learners who might get stuck on a problem. I hope that you appreciate that 100% of the conceptual questions and problems have hints. Also, 100% of the conceptual questions have answers, since it's very important to understand the concepts well if you want to become adept at solving the problems. Some of the problems have answers so that you can gain some confidence by reproducing the same answers, but even those problems that don't have answers do have hints to help you see if you are solving them correctly. The examples were selected based on their instructiveness in elucidating important concepts or illustrating how to carry out important problem-solving strategies; quality was favored over quantity. Simple plug-and-chug examples and problems are scarce, since the audience for this book is advanced, independent students. Several notes highlight common mistakes to help you learn from mistakes that many other students have made without having to suffer through them yourself.

May you develop a passion for physics – whether you love it or hate it (or love to hate it or hate to love it), may you be passionate toward it (after all, there's just a simple minus sign difference between the two). But please don't be indifferent about the subject. ☺

Chris McMullen, Ph.D.

# How to Succeed in Physics

Students often think of physics as being very mathematical, and it is, but it is much more than just math. Physics is also very conceptual. The mathematics and concepts are inherently intertwined. If you subtract the concepts from physics, all that would be left is pure mathematics. Similarly, you can't completely subtract the mathematics from physics. Surely, you can try, and you can do this to a large extent – and this is done in conceptual physics courses. But you can't completely remove the mathematics from physics because the math and concepts go together. The math is a quantitative, structured, logical expression of the theory of physics, which is based on experimental measurement. The equations provide much insight into the concepts, and one's conceptual understanding guides the application of the equations to solve problems and make predictions. You can't truly have one – i.e. pure math or pure concepts – without the other in a proper description of the physical world.

Students who are fluent in calculus have the mathematical potential to excel in physics, yet many of the students who excel in calculus struggle with physics. The reason is associated with the challenging conceptual aspect of the course. Many of the concepts in physics are very counterintuitive to most students, despite the fact that everyone has several years of first-hand experience with motion. The student's conceptual understanding is a large factor in successful problem-solving in physics. Students who do not fully grasp the concepts have trouble setting up the problem correctly. Unfortunately, if you make a mistake setting up the problem, no matter how good your math is, your solution will be incorrect. The concepts even play pivotal roles in applying the techniques of mathematics to carry out the solutions. One must also be able to reason out the mathematical results conceptually in order to answer follow-up questions at the completion of solving problems.

Obviously, since physics does have a strong mathematical component, you need to master the mathematical techniques – including new techniques learned in physics, as well as reviewing any areas of math that you are rusty in, such as trigonometry or integration. Most students with a strong background who make a concerted effort find that the math is reasonably straightforward to learn – i.e. there is a clear method behind the mathematical techniques. What many students struggle with is understanding the physics concepts, and seeing how the concepts relate to the math.

Therefore, in order to succeed in physics, you must dedicate time toward mastering the physics concepts, and spend some time studying how the concepts relate to the mathematics in the problem-solving strategies. Many students neglect this very important aspect of the course, and it shows in both their mathematical solutions and conceptual reasoning. It's easy to be overwhelmed with homework that involves mostly carrying out mathematical steps, but if you find time to keep up with the concepts and also keep up with the problem-solving, you should find this to be a great help in your endeavors to learn physics well.

With this in mind, and from experience working with students of differing abilities, learning styles, and levels, the following strategy is recommended for learning calculus-based physics successfully:

- Give regular thought to the concepts that you learn. Your understanding of the concepts develops incrementally in stages. Also, your retention of the concepts improves with repetition – i.e. each separate occasion in which you consider the concept.

- Keep an ongoing list of concepts to review – like the meaning of acceleration or the definition of inertia. Write the term, what it means, precisely, and also the gist of what it means. Also, write the distinction between similar terms – like mass and weight, or speed and velocity. Review these definitions, concepts, and laws regularly.
- When you solve problems, think about the concepts involved and how they relate to the problem and its solution. Also, think about how the concepts relate to the equations used.
- Identify the main problem-solving strategies and strive to master these strategies and their application to solve problems. You do this by studying the general steps of the strategy, reviewing examples and problems that you have solved, and mainly you master the strategies from ample practice solving a variety of physics problems. Practice, practice, practice. Practice makes permanent though, but not necessarily perfect, so seek help if you get an incorrect answer or if you don't fully understand a problem that you are solving.
- If you are taking a course, actively participate – or, if you are very introverted, actively think your way through the material as it is presented. At the end of the class, you don't just want to have several pages of notes, but you want to gain some understanding of the material.
- As you read any physics textbook, actively think your way through the text. You don't learn by reading math, but by thinking through the logic and trying to fill in the steps. Think about the relationship between the concepts and mathematics when studying examples or derivations.
- Keep an ongoing list of starting equations to review. This includes useful formulas from mathematics, like the quadratic equation or the volume of a sphere. You don't merely want to know the equations, but understand what they represent conceptually and how to apply them.
- Show your work clearly and completely. Someone else should be able to see the logic of your solution and to easily understand how you solved the problem. Organize your ideas, and realize that you can write sentences to supplement your math. Your goal is to convince your instructor that you come to class well-prepared and understand what you are doing. Also, don't underestimate the importance of including correct units with your answers and expressing your results with appropriate use of significant figures.
- If you don't understand something, or get an answer wrong, seek help. If you don't understand something on homework, you want to understand it before a quiz. If you make a mistake on a quiz problem, you want to learn from your mistake before an exam. Not only may you understand something better after receiving help, but you may become more confident, too.
- But you don't want to rely on too much help. Try it yourself first. When you get help, just look for a hint to point you in the right direction, and subsequent hints as needed. The more you strive to learn on your own, the better you will learn physics.

# Suggestions for Reading This Book

Some sections of this textbook are largely geared toward main problem-solving strategies in physics, some sections are geared toward elucidating important concepts, and some sections express physics on more formal, general, and/or abstract terms. Some sections will be easier to digest, some will be much more challenging to grasp. Generally, students catch onto more practical problem-solving strategies and more tangible concepts readily, and struggle with material that is more general, formal, or abstract.

One goal of this book is to help you learn how to solve a wide variety of physics problems with a minimum number of starting equations and problem-solving strategies. Since problem-solving in physics is inherently very conceptual, there are also many sections and subsections that are geared toward elucidating the underlying concepts and making connections between the equations and the concepts. The audience for this book is advanced, independent learners. As such, some sections express ideas and mathematics on much more abstract terms.

If you're learning physics for the first time, you may wish to focus more on the fundamental concepts and practical problem-solving, and skip some of the abstract material the first time through the book. However, if you're reviewing first-year physics concepts, or if you are among the few students who have such a knack for physics that you are able to breeze your way through the material, you may benefit from the challenge of trying to grasp the more abstract and advanced material in this book. The more advanced and abstract material may be especially useful to bridge the gap between a traditional first-year physics course and a course in classical mechanics, for students who are studying physics beyond the first year.

My main point is that you may not wish to read straight through this textbook. If you're struggling with some abstract material, instead of being frustrated by it you could move onto some more practical material and try to develop more confidence in your physics skills before tackling more abstract concepts. The book is ordered by topic, not necessarily by easiness.

If you're new to physics, or if you're currently enrolled in a first-year physics course, you may wish to replace the subscript $T$ for tangential in Chapter 1 with an $x$. The use of the adjective 'tangential' in Chapter 1 allows the same concepts to be used in a more general sense, such as gaining speed while traveling in a circle. However, many beginning students feel much more comfortable with $x, v_x, a_x$, and $t$ than $s_T, v_T, a_T$, and $t$. So if you see a $v_T$ in Chapter 1, feel free to think of it as $v_x$. However, looking ahead to Chapter 7, it will eventually become important to understand this tangential notation; there the $T$'s and $x$'s will no longer be interchangeable.

The table on the next two pages highlights which sections are more practical and which sections are more abstract in order to help you navigate your way through this textbook.

# Which Material Is Practical, Which Is Abstract?

Here is a guide to which sections of this textbook are more tangible and more geared toward problem-solving, and which sections of this textbook discuss things in more abstract terms. Students tend to grasp the more tangible and practical material more readily, and struggle with the more abstract material. There may be times when you want to focus more on the practical material. Hopefully, you will also devote some time attempting to grasp the more abstract material, too.

Chapter 1: The notation, involving tangential quantities, is abstract and general. You might prefer to replace all of the subscript $T$'s for tangential with $x$'s. (In Chapter 7, this distinction will be important.)
- Practical:
    - Strive to understand the distinction between net displacement and total distance traveled, and between average speed and average velocity.
    - Master the 1DUA strategy, including its application to vertical free fall problems.
    - The algebra of the multiple-moving objects strategy will be useful later in the course.
- Abstract:
    - Counting the degrees of freedom and number of dimensions may not be intuitive.
    - The distinction between average versus instantaneous values is abstract.
    - Most students struggle trying interpret motion graphs, but this is a very useful skill.

Chapter 2: Sections 2.1-2.2 and 2.6 are highly practical, while Sections 2.3-2.5 are very abstract.
- Practical:
    - The strategies for vector addition and projectile motion are essential.
- Abstract:
    - Beginners may want to skip vector motion quantities and other coordinate systems.

Chapter 3: This chapter is very fundamental to physics, and is mostly practical.
- Practical:
    - If you want to master physics, dedicate yourself to this chapter, which is very practical.
- Abstract:
    - Mutual surface forces and air resistance are the more challenging sections.

Chapter 4: Although some of the concepts may seem abstract, this chapter will be very useful.
- Practical:
    - Master this chapter, as it is a necessary prerequisite to learning rotation.
- Abstract:
    - Every section is practical, but you will find some abstract discussion mixed in.

Chapter 5: The concept of energy is abstract, while conservation of energy is highly practical.
- Practical:

- o The scalar product is easy to grasp and apply.
- o It is worth studying the work integrals. Although the integral is expressed on abstract terms, mainly you will just need to use the results of the integrals given in Section 5.2.
- o You will need to learn the different types of potential and kinetic energies, and will want to master the strategy for how to conserve energy. Conservation of energy is one of the most useful and widely applicable techniques of physics.
- Abstract:
  - o The distinction between conservative and nonconservative forces and the work integrals of Sec. 5.5 are much more abstract.

Chapter 6: Master the strategies for solving different types of collision problems.
- Practical:
  - o Finding the center of mass for a discrete system is straightforward.
  - o Being able to apply conservation of momentum to solve collision problems is useful.
- Abstract:
  - o The center of mass integral is abstract, yet it's a mathematical calculation – so if you work at it, you can make the strategy seem more practical. It's worth tackling, as a useful prelude to learning how to perform other abstract integrals in physics.
  - o Understanding the connection between the net external force and when momentum is conserved for a system is challenging for many students to grasp fully.
  - o The motion of the center of mass is an abstract notion.
  - o Beginners may wish to skip scattering and rocket motion.

Chapter 7: Rotation is always a challenging chapter for students, yet it's very practical.
- Practical:
  - o The strategy for uniform angular acceleration is very similar to the 1DUA strategy.
  - o There is a straightforward method for computing the vector product.
  - o The equation for torque is practical and reasonably straightforward; the main hurdle is for students to motivate themselves to be interested in the concept.
  - o Summing the torques is very similar to Newton's second law, and is just as practical when it comes to solving rotation problems.
  - o Conservation of energy in rotation and conservation of angular momentum are very useful problem-solving strategies.
- Abstract:
  - o Students have trouble motivating themselves to want to understand moment of inertia, but it can be understood well and is involved in many formulas in rotation.
  - o The moment of inertia integral is expressed on abstract terms, but the series of substitutions can be understood in practical terms through effort and practice.
  - o Very advanced students may want to challenge themselves with the tensor notation in Sec. 7.3.

# Units, Dimensions, and Significant Figures

Following are some basic things that you should know before you begin your studies of physics. This includes units, significant figures, and dimensionality. These are important for a number of reasons. Understanding dimensionality can help you check if a formula is viable, and so is something you should do whenever you derive an equation – if you have time to check and correct a solution on a quiz or exam problem, for example, this may save you some valuable points. Bear in mind that checking the dimensionality only shows if the formula is viable – so it can tell you that an equation is incorrect, but it can't guarantee that it's right (since you can write down a variety of equations that have the same dimensions, but aren't all the same). It's important to state your units after each quantitative answer because, for example, there is a significant difference between a foot and a meter, and a huge difference between a meter and a kilometer. Expressing your results with appropriate use of significant figures shows how many of the digits are actually relevant – it tells you the precision with which you know the calculated result. Also, if you ignore units and significant figures in your work, you're going to lose points for every question you answer all year long – which becomes a significant number of points at the end of the course. After this is a discussion of what mathematical prerequisites are expected for this text.

If you're not already, you want to quickly acquaint yourself with the SI system of units and the common metric prefixes. The SI units of length, time, and mass are the meter (m), (s), and (kg), respectively. The strange thing is that the kilogram includes a metric prefix, whereas the meter and second do not. It's a common mistake for students to work with grams instead of kilograms. All other units in first semester physics can be expressed in terms of kilograms, meters, and seconds.[1] For example, you will learn that a Newton (N) is equivalent to a kg·m/s$^2$. Therefore, if you are given a distance in feet, a time in hours, or a mass in grams, for example, it will generally be necessary to convert these units to SI units before you plug them into any formulas. Sometimes students think that if they are given centimeters, and see that they are solving for a distance, that they don't need to convert to meters, but that the final result will simply come out in meters. However, sometimes this backfires. If the formula involves $g$, for example, and you use 9.81 for this constant, its SI units are m/s$^2$. So if you are using an equation where a distance is in centimeters, but $g$ is in m/s$^2$, you have a units mismatch. It's tricky when meters are hiding in a constant like this. When in doubt, convert everything to SI units before you plug the numbers into equations and you can avoid such complications.

---

[1] Technically, you also need to include the SI unit for amount of substance, or mole number, which is the mole (mol). When you study electricity, optics, and thermal physics, you will also need to include the Ampère, the candela, and the Kelvin with the list of base SI units. In cgs units, we use the centimeter, gram, and second.

Just look at NASA's expensive mistakes for not paying attention to units and you can see that units are actually quite important. Sometimes you will need to convert between units, and you should know how to do it if you don't already – or maybe you did once, but now you need a quick review. Obviously, you have to look up the conversion factor – in your textbook, or, well, this is the Information Age, but make sure that you obtain reliable information from you-know-where. All that you really have to do is multiply by one. It's that simple. The 'one' is actually a fraction where the numerator equals the denominator. For example, there are 3600 s in 1 hr, so the fraction 3600 s/1 hr is equal to one. So is the reciprocal of this fraction, 1 hr/3600 s. So the important points are to (1) make sure that the fraction is actually equal to one and (2) determine whether or not your fraction is inverted. This second part is easy – only one will actually cancel the units.

---

**Example**. Given that 1 in. = 2.54 cm, what is the height of a 4.00-ft. tall gorilla in centimeters?

In this case, the fraction that equals one is either 1 in./2.54 cm or 2.54 cm/1 in. Of course, you also need to convert 4.00 ft. to 48.0 in. Now think about this: Which fraction should you multiply 48.0 in. by? If you do it correctly, the inches will cancel, but if your fraction is inverted, you'll get in.²/cm, which is definitely not what you want. Observe the cancellation of units with the correct fraction:

$$48.0 \text{ in.} \frac{2.54 \text{ cm}}{1 \text{ in.}} = 122 \text{ cm}$$

---

Sometimes you need to multiply by one more than once. For example, to convert from mph to km/s, you have to separately convert the units of distance and time (unless you look up a direct conversion, which you're not likely to be given on a test).

There is a tricky part to unit conversions: This is when units are raised to a power. For example, there are 10 mm in 1 cm, but there are 100 mm² in 1 cm². If you don't see this, draw a $10 \times 10$ grid and count the squares. Here is a reliable way to treat these powers: Raise the entire conversion factor to the power. For example, write 1 cm² as $(10 \text{ mm})^2$, which equals 100 mm².

If you need or want more practice with units or unit conversions, there are a few practice exercises coming at the end of this chapter with answers at the back of the book. If that's not enough for you, borrow some textbooks from the library and read the examples and practice the odd-numbered problems (so you can check your answers) to your heart's content.

It will be handy to be familiar with the common metric prefixes. The ones that we use most frequently in first-semester physics are kilo (k), centi (c), and milli (m), but over the course of the first year you are apt to come across micro (μ), nano (n), pico (p), mega (M), and giga (G). You can easily find less common ones in a standard textbook – or, again, you shouldn't have trouble finding basic information this day and age. Each prefix corresponds to a particular power of 10, according to the table below. So a cm equals 0.01 m, a MN (megaNewton) equates to $10^6$ N, and 20 g is equal to 0.020 kg in SI units.

| p | n | μ | m | c | k | M | G |
|---|---|---|---|---|---|---|---|
| $10^{-12}$ | $10^{-9}$ | $10^{-6}$ | $10^{-3}$ | $10^{-2}$ | $10^3$ | $10^6$ | $10^9$ |

> **Example.** A monkey charges you one million 'millicents' for a bunch of bananas. How much does this cost in dollars?
> Since the prefix milli- means $10^{-3}$, a millicent would apparently be a thousandth of a penny, or 0.001 ¢. So a million millicents equals $10^6 10^{-3}$ ¢ $= 1,000$ ¢. Divide by 100 to convert from pennies to dollars, and you'll see that the monkey is charging you $10 for a bunch of bananas.

All relationships among physical quantities – i.e. the formulas – must satisfy dimensionality. Dimensionality recognizes that there are different types of physical quantities, and you can't add apples to oranges. For example, it does not make any sense to try to add 3 m to 2 s because one is a length and the other is a time. However, you can add 3 m to 2 mi. since both are distances – provided that you first convert them to the same units. Dimensionality helps you keep track of what you can or can't add together, and it also helps ensure that the dimensions of one side of an equation match the dimensions on the other side of the equation – since you also can't set apples equal to oranges.

For each different type of SI unit, there is a corresponding physical dimension. For example, units of meters, seconds, and kilograms correspond to dimensions of length, time, and mass. So the dimensions of speed are length divided by time. This is traditionally expressed as $[v] = L/T$, where the brackets [ ] mean to look just at the dimensions (and not the numerical value). However, you can make perfectly equivalent dimensional comparisons just by looking at the SI units – i.e. there is no need to work with dimensions of length, time, and mass, when working with units of meters, seconds, and kilograms will do just as well. However, if you're taking a physics class, you might be expected to go with traditional dimensionality. In this text, though, you will see units analysis in lieu of dimensional analysis. It's easy to swap out meters with dimensions of length, seconds with dimensions of time, and so on, if you want to go back and forth between the two methods.

To check for unit consistency, just plug in SI units for each quantity. The main rules are: (1) You can't add incompatible units, like meters plus seconds, (2) the units on each side of the equation must match, and (3) the argument of a function must be unitless (or expressed in radians, in the case of trig functions). Radians are sort of strange, in that they often get swept under the rug in the context of meters (a point we will explore when we next come across it in the text). If you're trying to determine what the SI units of a particular physical quantity are, you just need to find any equation involving that physical quantity, plug in the SI units for everything else, and simply solve for its units algebraically. You'll see examples of solving for units in this textbook, so we won't bother with it now.

> **Example.** What values of the unitless exponents $c$ and $d$ would make the following equation dimensionally viable: $s^2 = \frac{1}{4} a^c t^d$, where $s$ is a distance, $t$ is a time, and $a$ is an acceleration, which has SI units of m/s²?
> Plug in the SI units for each quantity in the equation, and ignore the numerical coefficient (since the ½ doesn't affect the units): $(m)^2 = (m/s^2)^c (s)^d$. The only way to get the meters to match on both sides of the equation is if $c = 2$. Plugging this in, $m^2 = (m^2/s^4)(s)^d$. The meters cancel, and what remains is $s^4 = s^d$, which shows that $d = 4$.

When you receive your homework, quizzes, exams, and lab reports back in physics, you quickly learn that units and significant figures are worth points – and even if only a point or two per question, if you think about how many questions you answer in one semester, that's a healthy number of points. The number of significant figures that are appropriate for your answer shows how good the number is. A length of 2.00 m means that the length is good to about $\pm 1$ cm. Compare this to 2 m, which could be off by a whole meter, or 2.0000 m, which is good to about $\pm 0.1$ mm. The best rule for determining the number of figures that are significant is described in Appendix A, but requires knowing the uncertainties of all the given values. When these uncertainties are not available, there are a couple of simple rules for determining how many significant figures are appropriate for your answer. Bear in mind that you should keep all of the digits on your calculator for intermediate calculations, but should round to the appropriate decimal position to express your answer. It may also be necessary to use a metric prefix or scientific notation in order to express the number of significant figures correctly. For example, 1,242 g could easily be expressed with three significant figures using a metric prefix: 1.24 kg.

Here are the two rules for determining how many significant figures to keep when the uncertainties of the given values are not available. (1) If you multiply or divide two numbers, the result should have the same number of significant figures as whichever given quantity had the fewest number of significant figures. (2) If you add or subtract two numbers, the result should have the same precision as the least precise given quantity.

**Example.** Express $\Delta y$ with the appropriate number of significant figures, where $\Delta y = v_{y0}t + \frac{a_y t^2}{2}$ and given that $v_{y0} = 4.2$ m/s, $t = 0.634$ s, and $a_y = -9.81$ m/s$^2$.

Plug these numbers into the equation and apply the two rules for significant figures:[2]

$$\Delta y = v_{y0}t + \frac{a_y t^2}{2} = (4.2 \text{ m/s})(0.634 \text{ s}) + \frac{(-9.81 \text{ m/s}^2)(0.634 \text{ s})^2}{2}$$
$$\Delta s_y = 2.6[628] \text{ m} - 1.97[159418] \text{ m}$$

The brackets [ ] have been used to show which digits are not significant. The first term has two significant figures, and so would be expressed as 2.7 m, because 4.2 m/s has only two significant figures; and the last terms has three significant figures, and so would be expressed as $-1.97$ m, because both numbers in the product have three significant figures. On our calculator, we will use all of the digits to perform the subtraction, but the result will be expressed with a precision of $\pm 0.1$ m, since this is the precision of the least precise term (the first term, in this case). Our final answer is $\Delta y = 0.7$ m.

Finally, let's discuss the mathematical prerequisites for this text. Since this is a calculus-based textbook, it should be reasonable to assume that the reader recalls enough algebra to solve for one or more unknowns using standard algebraic techniques, remembers the basic trig functions and relationships for special triangles (like the 30°-60°-90° triangle), knows some common geometric formulas (like the area or circumference of a circle), can apply derivatives to polynomial and trigonometric expressions, and can perform similar anti-derivatives and definite integrals.

---

[2] Always write the equation in symbols before you plug in numbers so that your instructor can follow your logic.

However, you will be reminded of things like the quadratic equation when we first come across it in the text, and you should learn some elements of algebraic strategy as part of this text. In physics, the algebra may look different, especially since we tend to have several symbols in every equation, rather than just one or two. So you will learn methods for how to apply algebra to physics problem-solving strategies, and also some pointers for performing the algebra more efficiently. Regarding geometry, you should be able to easily look up any formulas that you might have forgotten – in case, for example, you need to figure out the surface area of a sphere. Yet, you will also learn how to apply calculus to derive such formulas – this will be taught in the text.

The text will teach calculus skills beyond Calculus I as they are needed, and will remind the reader of less-frequently needed Calculus I skills, too. If you don't know how to integrate via trigonometric substitution, or if you've never seen a double integral before, don't worry, when we first have occasion to use these techniques in this textbook, they will be taught to you. (If you want to study these things in more detail than they are described in this text, however, you might want a handy calculus book to supplement your understanding or to offer more practice.) As long as you have made it successfully through Calculus I, you should be in good shape.

I strongly recommend that you think your way through the examples. When you attempt to solve the practice problems, work with symbols and carry out the calculus and algebra as far as you can, and only use the calculator at the end. Try to estimate the answers to each calculation as you use your calculator. It will help strengthen your math skills and remind you of useful things like the cosine of thirty degrees. If you're a student in my class, you know you won't be able to use the calculator on quizzes or exams, so this will be good practice. If you become adept at solving problems without a calculator, then when you do use a calculator, you will have increased your capacity to think your way through the numbers and catch potential calculator mistakes,[3] rather than write the numbers down blindly.

## Practice Problems

The hints and answers to all of the practice problems in this section can be found, separately, toward the back of the book.

### Unit Conversions

1. A monkey walks 2.4 miles. Using 1 mile = 1760 yards and 1 in. = 2.54 cm (instead of looking up a direct conversion), determine how far the monkey walks in kilometers.
2. (A) If you had one 'kilosec' to finish a test, how much time would remain in minutes? (B) How many hours would there be in a 'milliyear'? (C) How many seconds would there be in a 'microday'?
3. (A) A rectangle has an area of 63 mm$^2$. Using 1 in. = 2.54 cm, determine the area of the rectangle in square feet. (B) A sphere has a volume of 5.7 cm$^3$. What is its volume in cubic meters?
4. A monkey drives 70 mph. Using 1 mile = 1760 yards and 1 in. = 2.54 cm (instead of looking up a direct conversion), express this speed in meters per second.

---

[3] What I really mean is user-mistakes typing the calculations. You should try to divide 1 by $2\pi$ on your calculator. It's a good exercise, as most students enter this incorrectly. If you get a number larger than 1, you made the common mistake. Figure out what you did wrong and you will have learned a valuable lesson.

5. A monkey has an acceleration of 2.3 m/s². Using 1 mile = 1760 yards and 1 in. = 2.54 cm (instead of looking up a direct conversion), express this in miles per hour-squared.

## Dimensions

6. Given that $v = A\omega \sin(\omega t)$, where $v$ has dimensions of speed and $t$ is a time, determine the dimensions of $\omega$ (omega) and $A$.

7. Given that $\mu = \sqrt{2\lambda\kappa}$, where $\mu$ (mu) has dimensions of speed and $\lambda$ (lambda) has dimensions of length, determine the dimensions of $\kappa$ (kappa).

8. Given that $\rho = \sigma\delta + \nu\delta^2/2$, where $\rho$ (rho) is a distance and $\delta$ (delta) is a time, determine the dimensions of $\sigma$ (sigma) and $\nu$ (nu).

## Significant Figures

9. Given that $L = 2.4$ m and $m = 7.38$ kg, express $\rho = m/L^3$ in SI units with appropriate use of significant figures.

10. Given that $L = 14.631$ m and $W = 8.54$ m, express $P = 2L + 2W$ in SI units with appropriate use of significant figures.

11. Given that $a = 17.529$ m and $b = 4.63$ m, express $c = a^2b + b^2a$ in SI units with appropriate use of significant figures.

# What Is Physics?

Surely, you wouldn't enroll in a course or read a textbook without already knowing what the subject is, right? But every semester I ask students what physics is on the first day of class, and it's not an easy question to answer – and they hope I don't ask this seemingly simple question on the final exam, too. Following is an idea of what physics is, so if someone asks you this, you can say more than just, "Hard!"

Like any other science, theoretical physics developed by applying the scientific method to a host of experimental observations. It involves measurement, partly to continually test the current theory and partly to explore new areas of physics; it involves mathematics, which developed to model the experimental data and to make predictions; and it involves conceptual analysis, which goes hand-in-hand with the quantitative analysis. But these qualities do not define physics, as they apply to any field of science. Science is a systematic, measurement-based study of the natural, physical universe. Physics is just one of many fields of science, yet physics lies at the core of every field of science.

Science can be divided into two branches – the physical sciences and biological sciences. The distinction is made by whether or not the content includes non-living or living things, respectively. Examples of physical sciences include physics, astronomy, chemistry, and geology. Some biological sciences are biology, zoology, anatomy, and genetics.

Physics is distinguished from other fields of science in that the subject of physics is the fundamental properties of and interactions between various forms of matter and radiation (such as electromagnetic, thermal, or nuclear radiation). Physics is often referred to as the fundamental science. The reason for this is that the principles of physics are involved at a fundamental level in every field of science. For example, all biological processes inherently involve interactions – at a fundamental, microscopic level – between biological components (like cells) and forms of radiation. Chemistry is often referred to as the central science, as it connects the fundamental science – physics – to the applied sciences. In the case of biology, for example, underlying all biological reactions are chemical reactions between elements, and ultimately these chemical reactions involve interactions between elementary particles (that make up the atoms that make up the molecules that make up the cells of living things) and particles of radiation (like photons or gluons).

Traditionally, the study of physics begins with classical mechanics, which is the area of physics with which we naturally have the most intuitive experience. The motion of macroscopic objects – such as marbles, humans, or moons – involves interactions like gravity, friction, tension, and air resistance. Classical mechanics serves as the basis for mechanical engineering. Other fields of classical physics include electromagnetism and thermodynamics. In classical physics, energy is continuous and the speeds involved are small compared to the speed of light. Physics also includes areas of modern physics, such as quantum mechanics – where energy comes in discrete packets called quanta – and relativity – where we find that the physics of relative motion deviates considerably from our intuitive experience with classical mechanics when dealing with relative speeds that are comparable to the speed of light. The scope of this textbook is limited to classical mechanics.

# 1 Single-Component Motion

**One-dimensional (1D) motion**: Single-component motion could correspond to an object that is just moving in one direction – say, north. It could also correspond to an object moving back and forth along a line – like east and west, or straight up and down. These are obvious cases of 1D motion, but 1D motion does not necessarily mean that the object is moving in a straight line. If the motion of an object is curved, the motion is considered to be 1D if the object only has a single degree of freedom. For example, if you bend a wire coat hanger into the shape of a curve – such as a parabola or an ellipse – and a bead slides along the wire, the bead's motion is effectively 1D because the bead experiences only a single degree of freedom – it can only move forward or backward along the predetermined shape of the wire.

**Degrees of freedom**: The number of degrees of freedom available to a moving object corresponds to the number of independent variable coordinates needed to describe the object's motion. The motion of an object is said to be 1D if its motion can be fully described with a single variable coordinate.

**Conceptual Examples.** Suppose that a monkey is running through a circular corridor. The corridor restricts the freedom of the monkey – i.e. the monkey may only move forward or backward through the corridor (or he may be at rest – not move at all). The motion of the monkey can be described with a single coordinate: You just need to know the tangential displacement of the monkey from some reference mark in the corridor; the sign of this tangential displacement will describe which direction the monkey is displaced from the reference mark.

You might want to think of the circle itself as being two-dimensional (2D), but if you are only concerned with the circumference of the circle (and not its interior), then it is actually a 1D object. If you are used to describing the circle in terms of the Cartesian coordinates $x$ and $y$, then what you are thinking is that the equation of the circle – which is $x^2 + y^2 = R_0^2$ if the circle is centered about the origin, where the radius of the circle, $R_0$, is constant – involves two coordinates. However, these two coordinates, $x$ and $y$, are not independent: If you know one, you can compute the other – e.g. given $x$, the other coordinate $y$ is found to be $y = \pm\sqrt{R_0^2 - x^2}$. If instead you use a different coordinate system (namely, polar coordinates, which we will learn in the next chapter), you can express the equation of the circle using a single variable (in particular, the equation $r = R_0 = const.$ defines a circle of radius $R_0$; and since one polar coordinate, $r$, is constant, the remaining coordinate, $\theta$, describes the position of every point on the circle).

If the monkey now runs around a gym, rather than a restricted corridor, his motion becomes 2D, as he gains a degree of freedom – i.e. he can move along any combination of north/south and east/west to travel in any direction in the horizontal plane. He may choose to run in a circle even in the gym, but he has the freedom to travel along any 2D path. In this case, two independent variable coordinates – which may be $x$ and $y$, but could also be $r$ and $\theta$, are required to describe the monkey's motion.

# 1 Single-Component Motion

## 1.1 Basic 1D Motion Quantities

**Time**: We use the symbol $t$ to represent a precise instant in time, and $\Delta t$ to represent a time interval: $\Delta t = t - t_0$. The SI unit for time is the second (s). So if you are given a time in some other units, like hours or milliseconds, it will generally (but not always) be convenient (and sometimes necessary) to convert the time to seconds.

> **Notation.** In the textbook, symbols for variables and constants appear italicized, whereas units do not. In this way, you can distinguish between mass, $m$, and the SI unit of length, which is the meter, m. In writing, however, it is difficult to make this distinction, but confusion is still generally avoided since units tend to follow numbers. That is, you might say that an object is 1.4 m tall, but you probably wouldn't say that it has 1.4 masses.

> **Notation.** We use a subscript nought (or naught), which is a zero, 0, to designate an initial value of a quantity.[4] For example, $t_0$ represents initial time and $v_0$ represents initial speed. Unlike an exponent, a subscript is not a mathematical operation. Rather, a subscript is a means of identification. That is, the two times $t_1$ and $t_2$ can be distinguished by giving them different subscripts.

> **Notation.** The Greek symbol $\Delta$, which is a capital delta, represents the change in a quantity. The change in time, $\Delta t$, corresponds to a time interval. You can also speak of change in mass, $\Delta m$, change in height, $\Delta h$, and a myriad of other quantities. The change in the quantity literally means the final value of the quantity minus its initial value. For example, $\Delta h$ means final height minus initial height: $\Delta h = h - h_0$. While $\Delta$ represents a finite change, in physics we often work with infinitesimal changes, for which we will use the symbol $d$, since this symbol is used to express differential quantities in calculus. So, for example, $dt$ represents an infinitesimal time interval, whereas $\Delta t$ represents a finite time interval.

> **Note.** It is a good habit to convert quantities to SI units before you plug them into any formulas.

**Direction in 1D**: The direction of motion can be completely described by using plus and minus signs in 1D, since the only possible directions are forward and backward. Once you set up a coordinate system to establish the location of the origin (where the coordinate is zero) and which way is forward, the position, velocity, and acceleration of the object will be completely specified by plus and minus signs.

Sometimes, it will be convenient (or necessary) to distinguish between positive and negative quantities in 1D motion – e.g. if you want to know whether the object is moving forward or backward. In this case, we will work with tangential displacement, tangential velocity, and tangential acceleration. These quantities may be positive or negative (or zero), and the signs will have precise meanings. Other times, we will just care about the absolute values in 1D. In this case, we will work with arc length, speed, and the magnitude of acceleration. These quantities cannot be negative.

---

[4] One convention is to use a subscript nought for initial quantities, and no subscript for final quantities – as in $x_0$ and $x$. Another popular convention is to use subscripts $i$ and $f$ for initial and final quantities – as in $x_i$ and $x_f$.

**Tangential displacement**: The position of an object in 1D motion is specified by the tangential displacement, $s_T$. The tangential displacement of the object equals the length of arc between the origin and the object's instantaneous position, and its sign indicates whether the object is displaced in the positive or negative direction from the origin. The tangential displacement is negative if the object is on the negative axis. The SI unit for displacement – and all other distances – is the meter (m).

**Tangential velocity**: The velocity of an object in 1D motion is given by its tangential velocity, $v_T$. The absolute value of the tangential velocity indicates how fast the object is instantaneously moving, and the sign of the tangential velocity reveals whether the object is moving forward or backward. The tangential velocity is negative if the object is moving in the negative direction (even if the object has a positive tangential displacement). The SI units for tangential velocity are m/s.

> **Note**. If you want to know whether an object is moving forward or backward at some specified time, in 1D, solve for the tangential velocity, $v_T$, and examine its sign.

**Tangential acceleration**: The acceleration of an object in 1D is given by its tangential acceleration, $a_T$. The tangential acceleration provides an instantaneous measure of how the tangential velocity is changing in time. There are two ways that the tangential acceleration of an object may be negative: (1) The object could be slowing down while heading in the positive direction, and (2) the object could be speeding up while heading in the negative direction. The SI units for tangential acceleration are m/s$^2$.

> **Signs**. The sign of the tangential acceleration, $a_T$, by itself does not necessarily tell you whether an object (in 1D) is speeding up or slowing down. For example, if an object is slowing down, its tangential acceleration can still be positive – if it is heading in the negative direction while slowing down. However, the sign of the tangential acceleration, $a_T$, can be combined with the sign of the tangential velocity, $v_T$, in order to determine whether or not the object is speeding up or slowing down.

> **Note**. Conceptually, it may be helpful to think of the units of acceleration, m/s$^2$, as (m/s)/s. That is, when the numerical value of acceleration describes how the tangential velocity is changing, these units, (m/s)/s, tell you how many m/s of tangential velocity the object is instantaneously gaining or losing per second.
>
> In future chapters we will learn that an object can have acceleration even if its speed is not changing. In this case, the object has acceleration because its direction is changing. In that case, when the numerical value of acceleration describes how the direction of the velocity is changing, these units, (m/s)/s, tell you how many m/s of speed the object is instantaneously being pulled away from its original direction of velocity. For example, the moon travels with a roughly constant speed (relative to the earth), but is accelerating because it changes direction; its acceleration describes how far its curved path is deviating from the straight-line motion associated with a line that is tangent to its path. That is, the moon is constantly being pulled away from its straight-line tangent by the earth's gravitational field. We will return to this point when we discuss uniform circular motion in a practice problem in Chapter 4.

**Arc length**: The total distance that an object has traveled equals the arc length, $s$, that has been traversed along the path of motion. Unlike the tangential displacement, arc length is nonnegative. The arc length does not equal the absolute value of the tangential displacement in 1D, as we will see when we explore the distinction between total distance traveled and net 1D displacement.

**Speed**: If you want to know how fast an object is moving at a given instant, but do not care which way it is headed, then you want to compute the object's speed, $v$. Unlike tangential velocity, speed is nonnegative. The speed of an object equals the absolute value of its tangential velocity. The SI units for speed are m/s.

> **Important Distinction.** In 1D, tangential velocity, $v_T$, describes both how fast and which way an object is moving, and therefore may be positive or negative (or zero). However, the similar quantity, speed, $v$, describes only how fast, and not which way; and is therefore nonnegative. So if a monkey is traveling 5 m/s to the south and the positive direction is taken to be north, its tangential velocity is $-5$ m/s, while its speed is 5 m/s.

**Magnitude of acceleration**: If you want to know the instantaneous acceleration of an object, but do not care what its sign is, then you should work with the magnitude of the object's acceleration, $a$. Unlike tangential acceleration, the magnitude of the acceleration is nonnegative. In 1D, the magnitude of the acceleration equals the absolute value of the tangential acceleration.[5] The SI units for acceleration are m/s$^2$.

> **Signs.** The magnitude of the acceleration, $a$, is always positive, so it does not help you to distinguish between an object that is speeding up or slowing down.

> **Important Distinction.** There are two sets of physical quantities that are useful for describing 1D motion: One set includes the directional quantities $s_T$, $v_T$, and $a_T$ – the tangential displacement, tangential velocity, and tangential acceleration – and the other set includes the directionless quantities $s$, $v$, and $a$ – the arc length, speed, and magnitude of acceleration.[6] While we restrict our attention to 1D motion, whether or not there is a subscript $T$ will designate whether or not the quantity may be negative, and whether or not the sign of the quantity provides directional information.

---

[5] Actually, as we will learn in Chapters 4 and 7, if the path is curved (even a 1D curved path), there are two components of acceleration – tangential and centripetal. For a curved path, the magnitude of the acceleration is not equal to the absolute value of the tangential acceleration, but combines the tangential and centripetal components together with the Pythagorean theorem. We will disregard the centripetal component of acceleration in Chapter 1, where we focus on 1D motion.

[6] It is more common to use $x$, $v_x$, and $a_x$ for 1D quantities that may be negative, but $s_T$, $v_T$, and $a_T$ are more general in that tangential quantities naturally apply to curved paths, whereas the $x$-component of motion is generally linear since $(x, y, z)$ are rectangular coordinates. We will see these same tangential quantities in later chapters, and the tangential role that they play in that sense is the same as the tangential variables used here.

## 1.2 Net and Average Values in 1D

**Net 1D displacement**: The net 1D displacement of an object is a directed distance that extends from the object's initial position to its final position. Defined as such, the net 1D displacement equates to the change in the object's tangential displacement, $\Delta s_T = s_T - s_{T0}$. The net 1D displacement may be positive or negative (or zero), depending upon whether the object is displaced forward or backward.

**Total distance traveled**: The arc length grows as an object changes position, even if the object retraces part or all of its path, such that the change in arc length, $\Delta s$, equals the total distance traveled by the object during the corresponding time interval. The total distance traveled, $\Delta s$, can be found by summing the absolute values of the positive and negative tangential displacements, $\Delta s_{T,i}$, of the object:

$$\Delta s = \sum_{i=1}^{N} |\Delta s_{T,i}|$$

In practice, this means first determining all of the instants for which the object changes direction, and then computing the displacement separately for each duration of time that it moved exclusively forward or exclusively backward. Unlike the net 1D displacement, the total distance traveled is nonnegative.

> **Notation.** The summation symbol, capital sigma ($\Sigma$), is shorthand for adding multiple terms together. For example, $\sum_{i=1}^{N} x_i$ reads, "the sum as $i$ goes from 1 to $N$ of $x_i$," and equals $x_1 + x_2 + \cdots + x_N$. This notation is also used in Chapter 6, Chapter 7, and Appendix B.

**Path-dependence**: The definition of net 1D displacement, $\Delta s_T$, depends only on the initial and final positions of the object, and does not depend upon how the object got there. In contrast, the actual path taken is critical toward calculating the total distance traveled, $\Delta s$. The net 1D displacement, $\Delta s_T$, is said to be path-independent, whereas the total distance traveled is path-dependent.

> **Important Distinction.** The net 1D displacement, $\Delta s_T$, asks how far (and which way) the final position lies from the initial position, and is therefore path-independent, whereas the similar quantity, the total distance traveled, $\Delta s$, asks how much distance was traversed all together, and therefore is very much path-dependent. The total distance traveled, $\Delta s$, will be greater than the magnitude of the net 1D displacement, $\Delta s_T$, only if the object changes direction one or more times during its motion.

> **Conceptual Example.** If a monkey walks 12 m to the west and then walks 5 m to the east, using a coordinate system in which east is positive, the net 1D displacement of the monkey is $-7$ m, whereas the total distance traveled is 17 m.

**Mean-value theorem**: The average value of a function, $f(x)$, over an interval $a \leq x \leq b$ can be found from the mean-value theorem of calculus:

## 1 Single-Component Motion

$$\overline{f(x)} = \frac{\int_{x=a}^{b} f(x)dx}{\int_{x=a}^{b} dx}$$

The denominator is often written as $b - a$. However, the above form generalizes to account for any weighting factors (distribution functions). Namely, the weighted average is

$$\overline{f(x)} = \frac{\int_{x=a}^{b} f(x)w(x)dx}{\int_{x=a}^{b} w(x)dx}$$

**Notation.** A bar over a quantity indicates that it is an average value. For example, $\overline{a_T}$ represents the average value of the tangential acceleration.

**Notation.** Hopefully, by now you have had enough math that you are used to seeing functions expressed in terms of their arguments, like a function of $x$, like $f(x)$, or a function of time, as in $g(t)$. You definitely don't want to look at $v(t)$ and mistakenly read it as a product of two variables.

**Example.** Determine the average value of $\sin \theta$ from $\theta = 0$ to $\theta = \pi$ rad.
    Apply the mean-value theorem, noting that the argument of the function is $\theta$, which plays the role of the variable $x$ in the definition above.

$$\overline{\sin \theta} = \frac{\int_{\theta=0}^{\pi} \sin \theta \, d\theta}{\int_{\theta=0}^{\pi} d\theta} = \frac{[-\cos \theta]_{\theta=0}^{\pi}}{\pi} = \frac{-\cos \pi + \cos 0}{\pi} = \frac{2}{\pi}$$

What would the average value of the cosine function be over one complete cycle? **Hint:** You should be able to see the answer without doing any math.[7]

**Average speed:** Thinking of speed as a function of time, $v(t)$, we can find its average value over some time interval by applying the mean-value theorem:

$$\overline{v(t)} = \frac{\int_{t=t_0}^{t} v(t)dt}{\int_{t=t_0}^{t} dt} = \frac{\Delta s}{\Delta t} = \frac{\text{total distance traveled}}{\text{elapsed time}}$$

If you want to find the average speed of an object, you should first compute the total distance traveled and divide by the total time. The average speed of an object is nonnegative.

**Note.** If you are asked to find average speed, you should definitely begin with its definition. If instead you try to go with your intuition, most likely you will be completely wrong.

---
[7] It's zero!

**Conceptual Example.** A monkey gets in a bananamobile and drives 10.00 m/s for 2.0 min. and then drives 20.00 m/s for 8.000 hr. Since the monkey spent a mere 2.0 min. driving 10.00 m/s, but the vast majority of his time, 8.000 hr., driving 20.00 m/s, it would be totally crazy to think that the average speed is 15.00 m/s – which would be the arithmetic mean, or the kind of average you are probably most accustomed to from math class – right? That is, you have to consider how much time the monkey spends driving each speed – you can't just naïvely add the initial and final speeds and divide by two. Keep this in mind when you consider the following example, and in the future. If you apply the definition, you should be able to see that the average speed of the monkey is 19.96 m/s.

**Example.** A monkey drives to work with an average speed of 40 m/s because he is running late, but drives home at a more leisurely average speed of 20 m/s. What is the monkey's average speed for the roundtrip?

First, you should be able to show conceptually that you expect the average speed for the roundtrip to lie in the range 20 m/s $< \bar{v} <$ 30 m/s.[8] If not, study the previous example. In order to solve the problem mathematically, apply the definition of average speed. You may be wondering how we can find the total distance traveled and the elapsed time, since all we have are two speeds – no distances or times are given.[9] What do you do in algebra when you don't know something? That's right, you just write a symbol for it. This will definitely happen again in physics, so when it does, don't get stuck, just write a symbol for it and then proceed to solve the problem. Let's denote the distance between the monkey's home and his work as $s$. In this case, the total distance traveled for the roundtrip is $2s$. The roundtrip time equals $t_1 + t_2$, where the time spent driving to work is $t_1 = s/40$ and the time spent driving home is $t_2 = s/20$.[10] Plugging this into the definition of average speed,[11]

$$\bar{v} = \frac{\Delta s}{\Delta t} = \frac{2s}{\frac{s}{40} + \frac{s}{20}} = \frac{2}{\frac{1}{40} + \frac{1}{20}} = \frac{2}{\frac{1+2}{40}} = \frac{80}{3} \text{ m/s}$$

**Notation.** Units are important. However, it may be cumbersome to drag units throughout a solution – particularly, when you solve rather involved problems – and it may also be confusing to mix symbols and units. As such, most of the examples in this text will suppress units in the calculations, as in $t_2 = s/20$, where the units of 20 m/s must be implied. This avoids confusion between seconds and arc length. It's especially important to include correct units at the end of a problem, and convert given units (usually, to SI units) so that they are compatible before you plug numbers into an equation.

---

[8] The symbol $\bar{v}$ is no different from $\overline{v(t)}$, except that the argument is implicit rather than explicit.
[9] In case you may be wondering, you are not 'allowed' to make a distance up. This problem can be solved in general terms – i.e. without knowing what the distance is.
[10] Just apply the definition of average speed (but when you do, make sure that you are, in fact, working with an average or a constant speed – if instead it's an initial or final speed, for example, this formula does not apply!), in order to see that time equals distance traveled divided by the average speed. You can avoid making simple algebraic mistakes with this simple equation if you analyze the units carefully when you complete the algebra.
[11] Don't just read math, think your way through it. You learn much be trying to fill in the steps. For one, you learn how to solve problems by learning how to think ahead a couple of steps. In this particular example, doing so would also provide a good review of how to add and divide fractions.

## 1 Single-Component Motion

**Average tangential velocity**: The definition of the average 1D velocity is analogous to the definition of average speed, and can similarly be found by applying the mean-value theorem:

$$\overline{v_T(t)} = \frac{\int_{t=t_0}^{t} v_T(t)dt}{\int_{t=t_0}^{t} dt} = \frac{\Delta s_T}{\Delta t} = \frac{\text{net 1D displacement}}{\text{elapsed time}} \quad \text{(in 1D)}$$

If you want to find the average 1D velocity of an object, you should first compute the net 1D displacement of the object and divide by the total time. In contrast to average speed, average 1D velocity may be negative.

**Important Distinction.** Average speed and average 1D velocity are similar quantities, but have significantly different definitions. Average speed involves the total distance traveled, and is nonnegative, whereas average 1D velocity involves the net 1D displacement, and may be negative.

**Constant speed**: If an object travels with constant speed, it's trivial to compute its average speed.[12] The same equation that defines average speed also applies to an object that travels with constant speed:

$$v = \frac{\Delta s}{\Delta t} \quad \text{(only if } v = const.\text{)}$$

**Conceptual Example.** If a monkey travels 5.0 m/s to the west for 12 s and then travels 10/3 m/s to the east for 18 s, using a coordinate system in which east is positive, the average speed of the monkey is 4.0 m/s, since the total distance traveled is 120 m, whereas the average 1D velocity of the monkey is 0 because there is no net displacement.[13]

**Uniform tangential velocity**: If an object travels with constant (or uniform) tangential velocity,

$$v_T = \frac{\Delta s_T}{\Delta t} \quad \text{(only if } v_T = const.\text{)}$$

**Note.** In the vast majority of physics problems, speed and velocity are changing. If you use the equation for constant speed (or velocity) for an object that is changing speed (or velocity), not only will you get the wrong answer, but if you are enrolled in a physics course, you probably won't receive any partial credit for such a major conceptual mistake.

---

[12] However, if an object travels with one constant speed for a period of time, and then travels with a different constant speed for another period of time, the overall average speed is not so obvious, as you can see by reviewing the previous example. Also, while it's trivial to find the average speed of an object that has constant speed, it may not be so easy to find its average 1D velocity.

[13] Once again, you will learn more if you take the time to think through the content of this text. In this case, you should think about where these numbers came from until you can reason it out. Physics is definitely a thinking-man's (or woman's) subject.

> **Important Distinction.** Uniform speed means that speed is constant, whereas uniform velocity requires both speed and direction to be constant. If an object is changing direction, its velocity is definitely not uniform; if an object has constant speed, its velocity may not be uniform.

**Average tangential acceleration**: The definition of the average acceleration in 1D is analogous to the definition of average 1D velocity:

$$\overline{a_T(t)} = \frac{\int_{t=t_0}^{t} a_T(t)dt}{\int_{t=t_0}^{t} dt} = \frac{\Delta v_T}{\Delta t} = \frac{v_T - v_{T0}}{\Delta t} = \frac{\text{change in velocity}}{\text{elapsed time}} \quad \text{(in 1D)}$$

If you want to find the average acceleration of an object, you should first compute the change in the object's velocity and divide by the total time. Average acceleration may be negative.[14]

> **Important Distinction.** It's very important to note whether or not the adjective 'average' is used to describe speed, velocity, or acceleration. For example, if you're looking for acceleration, or some specific kind of acceleration besides average acceleration, finding the change in velocity and dividing by the elapsed time will generally be incorrect (only if acceleration happens to be uniform – which is not always the case – will it coincidentally be correct). The equation above applies specifically to average acceleration, but certainly not to acceleration in general.

## 1.3 Instantaneous Values in 1D

**Measurement of speed**: Suppose that a monkey is driving a bananamobile and you wish to measure its speed. You might use a stopwatch and tape measure to make this measurement. You could use the stopwatch to determine the time that elapses after the bananamobile passes a tree until it reaches a boulder, for example, and then you could measure the distance between these objects. Dividing the distance traveled by elapsed time, what you get is the average speed. If the speed of the bananamobile changes significantly between the tree and the boulder – i.e. if the speed is appreciably non-uniform – then this average speed is of limited use. It would be foolish to measure the average speed and think of this as 'the' speed of the bananamobile. If the speed of the bananamobile is not constant, then there will be an infinite number of instantaneous speeds over the duration of the measurement. You might like to know the initial and final speeds, or perhaps a table of speeds at various intervals over the course of the trip, in order to obtain a good quantitative measure of precisely how the speed varies as a function of time. The important point is that the distance traveled over the elapsed time will not be worth calculating – unless you have a question that specifically asks you to find the average speed, or unless the speed happens to be constant.

---

[14] In principle, you could also compute the average magnitude of acceleration, and distinguish between this and average tangential acceleration. However, unlike average 1D velocity and average speed, it is conventional to present only average (tangential) acceleration. It would be straightforward to calculate, though.

You could get a better measure of speed by working with a smaller time interval, and hence a smaller distance. The speed can't vary as much over a small time interval as it can over a large interval. In principle, you could measure several short time intervals and corresponding differences to obtain a set of average speeds that correspond to different segments of the trip. As you make the time intervals shorter and shorter, theoretically and average speed becomes closer and closer to an instantaneous speed.[15] In this way, we can think of the instantaneous speed – i.e. how fast the bananamobile is traveling at any particular moment – as the limiting case of an average speed:

$$v = \lim_{\Delta t \to 0} \frac{\Delta s}{\Delta t}$$

Instantaneous speed is nonnegative.

**Measurement of velocity in 1D**: Analogous to instantaneous speed, instantaneous velocity is defined as a limiting case of the average 1D velocity as

$$v_T = \lim_{\Delta t \to 0} \frac{\Delta s_T}{\Delta t} \quad \text{(in 1D)}$$

Instantaneous velocity may be negative.

**Note.** The limit as $\Delta t$ approaches zero in the definition of instantaneous speed and instantaneous velocity is both conceptually and mathematically crucial. If speed or velocity is changing and you disregard the limit in the formulas above, this is a major conceptual mistake. The limit conceptually reminds you that the average speed and average 1D velocity only become approximately equal to their instantaneous values for very short time intervals. Mathematically, it should tell you that you need to treat the time-dependence by taking a derivative – and that simply dividing distance by time will generally not get you an initial, final, or instantaneous speed (and similarly for instantaneous velocity).

**Important Distinction.** The difference between instantaneous speed and instantaneous velocity in 1D is that you must measure the direction of the displacement – which may be positive or negative for a given 1D coordinate system – when measuring instantaneous velocity.

**Derivatives**: Recall from calculus that the derivative is defined in terms of a limit. In particular, a derivative of the function $y(x)$ with respect to $x$ is defined by:

$$\frac{dy}{dx} = \lim_{h \to 0} \frac{y(x+h) - y(x)}{h}$$

---

[15] On the other side, experimentally, there comes a limit where the relative error in measuring the average speed increases with a smaller time interval. However, we can put a couple of closely-spaced photogates out there, or find some other means to improve the precision of the measurements so that the time intervals are reasonably small without too much experimental error. The focus of this discussion is not on the experiment itself, but on the theoretical definition of instantaneous speed.

> **Example**. Apply the definition of a derivative in terms of a limit to find the derivative of $x^2$ with respect to $x$.
>
> According to the definition of the derivative,[16]
>
> $$\frac{d}{dx}x^2 = \lim_{h \to 0} \frac{(x+h)^2 - x^2}{h} = \lim_{h \to 0} \frac{2xh + h^2}{h} = 2x$$

**Instantaneous tangential velocity**: Recall from our conceptual discussion of measuring velocity that the instantaneous value of tangential velocity relates to a limiting case of average 1D velocity. This limit can be cast in a form that looks more like the usual mathematical definition of a derivative in calculus:

$$v_T = \lim_{\Delta t \to 0} \frac{\Delta s_T}{\Delta t} = \lim_{\Delta t \to 0} \frac{s_T(t + \Delta t) - s_T(t)}{\Delta t} \quad \text{(in 1D)}$$

By comparison, you can see that $\Delta t$ plays the role of $h$, $s_T$ plays the role of $y$, and $t$ plays the role of $x$. Therefore, the derivative $dy/dx$ corresponds to $ds_T/dt$. That is, the instantaneous tangential velocity, $v_T$, equals the derivative of tangential displacement with respect to time:

$$v_T = \frac{ds_T}{dt} \quad \text{(in 1D)}$$

You should recognize this result from calculus, except that you probably expressed this with different symbols in calculus. The instantaneous tangential velocity may be negative and has units of m/s.

**Instantaneous tangential acceleration**: Analogously, the instantaneous acceleration, $a_T$, equates to a derivative of the instantaneous tangential velocity with respect to time:

$$a_T = \frac{dv_T}{dt} = \frac{d^2 s_T}{dt^2} \quad \text{(in 1D)}$$

The instantaneous tangential acceleration may be negative and has units of m/s$^2$.

**Instantaneous speed**: Similar to the instantaneous tangential velocity, the instantaneous speed, $v$, equals a derivative of arc length with respect to time:

$$v = \frac{ds}{dt}$$

The instantaneous speed is nonnegative and has units of m/s.

---

[16] You generally won't need to apply limits to take derivatives in physics. The purpose of this exercise is just to remind you of the connection between limits and derivatives so that you can see how the conceptual interpretation of instantaneous motion quantities like velocity and acceleration relate to the derivatives that you know from Calculus I. Henceforth, to take a derivative, just apply the techniques that you learned in calculus.

> **Important Distinction.** The instantaneous speed, $v$, tells you how fast an object is moving at a given moment, while the instantaneous velocity, $v_T$, in 1D tells you both how fast and which way an object is moving. In 1D, the direction of motion is revealed by the sign of the tangential velocity.

**Instantaneous magnitude of acceleration in 1D**: The absolute value of a derivative of the instantaneous speed with respect to time equals the magnitude of the acceleration, $a$, in 1D:

$$a = \left|\frac{dv}{dt}\right| = \left|\frac{d^2s}{dt^2}\right| \quad \text{(in 1D)}$$

The instantaneous magnitude of acceleration is nonnegative and has units of $m/s^2$.

> **Important Distinction.** In 1D, the distinction between the magnitude of the acceleration and the tangential acceleration is that tangential acceleration describes how the tangential velocity is changing. The magnitude of the acceleration does not quite tell you how the speed is changing in 1D: Since the magnitude of the acceleration is nonnegative, it does not tell you if the object is speeding up or slowing down. The tangential acceleration also does not directly tell you whether the object is speeding up or slowing down, but information combined from both the tangential acceleration and the tangential velocity can be combined to determine this (as described in a previous note).

## 1.4 Calculus Applied to 1D Motion[17]

**Relationship between tangential motion quantities**: The tangential motion quantities are related to one another via subsequent derivatives with respect to time:

$$v_T = \frac{ds_T}{dt} \quad , \quad a_T = \frac{dv_T}{dt} = \frac{d^2s_T}{dt^2} \quad \text{(in 1D)}$$

These relations can also be expressed as definite integrals:[18]

$$\Delta s_T = \int_{t=t_0}^{t} v_T dt \quad , \quad \Delta v_T = v_T - v_{T0} = \int_{t=t_0}^{t} a_T dt \quad \text{(in 1D)}$$

---

[17] The general principles are discussed in this section. Note that examples of how to apply these calculus techniques to specific functions are given in Sec. 1.8 on non-uniform acceleration.

[18] In calculus, it is common to express these as indefinite integrals with a constant of integration, and to later determine the constant of integration by applying boundary conditions. In physics, we can conceptually build the boundary conditions into the problem by instead performing definite integrals. For example, the initial tangential velocity, $v_{T0}$, is essentially a constant of integration that corresponds to the initial time, $t_0$.

It is important to note that the definite integral of tangential acceleration does not equal the tangential velocity, but equals the change in tangential velocity.

**Potential Mistake.** If you wish to evaluate a derivative with respect to time – such as tangential velocity or tangential acceleration – at a particular instant, you must first take the derivative and then plug in the time. If you know that $s_T = 4$ m at $t = 3$ s, for example, it would be absurd to think that the derivative of tangential displacement with respect to time is zero because 4 m is a constant. Instead, you must express $s_T$ as a function of time, then take a derivative with respect to time, and finally plug in $t = 3$ s. Furthermore, it would be a major conceptual mistake to think that the tangential velocity is 4/3 m/s: In physics, tangential velocity usually does not equal tangential displacement divided by time because tangential velocity often is not constant (even if $v_T$ is constant, you must factor in $s_{T0}$).

**Potential Mistake.** If you are integrating over time – e.g. to find tangential displacement or tangential velocity – you can't pull any symbols out of the integral unless you know that they are constant. For example, $\int_{t=t_0}^{t} a_T dt$ will usually not be equal to $a_T \Delta t$. Instead, if the symbol is not a constant, you must express it as a function of time in order to perform the integral.

**Common Mistake.** The definite integral of tangential acceleration over time does not equal tangential velocity, but instead equals the change in tangential velocity. It is a common mistake for students to forget to account for the initial tangential velocity when computing the final tangential velocity from an integral.

**Relationship between directionless motion quantities in 1D**: Like the tangential motion quantities, the set of directionless 1D motion quantities are related to one another via subsequent derivatives with respect to time:

$$v = \frac{ds}{dt} \quad , \quad a = \left|\frac{dv}{dt}\right| = \left|\frac{d^2 s}{dt^2}\right| \quad \text{(in 1D)}$$

In integral form, these relations are:

$$\Delta s = \int_{t=t_0}^{t} v \, dt \quad , \quad |\Delta v| = |v - v_0| = \int_{t=t_0}^{t} a \, dt \quad \text{(in 1D)}$$

It is very important to note that these integrands involve speed and the magnitude of acceleration, and not tangential velocity and tangential acceleration.

**Important Distinction.** In 1D, the integral of tangential velocity equals the net 1D displacement, whereas the integral of speed equals the total distance traveled. Similarly, the integral of tangential acceleration differs from the integral of the magnitude of acceleration.

**Relationship between tangential and directionless motion quantities in 1D**: If you want to relate tangential motion quantities to one another, you simply apply calculus, and similarly if you want to relate directionless 1D motion quantities to one another. However, you must be careful if you want to relate a tangential motion quantity to a directionless motion quantity. The instantaneous speed equals the absolute value of the instantaneous tangential velocity, and similarly for the 1D accelerations, but the relationship between the tangential displacement and the arc length is not as obvious (see the definitions of the total distance traveled and the net 1D displacement):

$$v = |v_T| \quad , \quad a = |a_T| \quad \text{(in 1D)}$$

> **Important Distinction.** The instantaneous speed equals the absolute value of the instantaneous tangential velocity in 1D, but the same is not generally true regarding the average speed and average 1D velocity, nor does the analogous relationship generally hold regarding the tangential displacement and the arc length (or the total distance traveled and the net 1D displacement).

**Net 1D displacement integral**: The net 1D displacement integral, $\Delta s_T = \int_{t=t_0}^{t} v_T dt$, is path-independent. That is, if the moving object changes direction, you do not need to split the integral into pieces to account for each change in direction. The definite integral very naturally provides the net 1D displacement for the entire trip, regardless of whether or not the object changes direction.

**Total distance traveled integral**: The total distance traveled integral, $\Delta s = \int_{t=t_0}^{t} v dt$, is path-dependent. In this case, the fact that $v = |v_T|$, in 1D, causes a change in direction to change the outcome of the integral. Therefore, if you know tangential velocity, $v_T$, as a function of time, in order to compute the total distance traveled, you must first find all of the turning points of the motion, then integrate $|v_T|$ over the corresponding time intervals, and finally sum these distances. You can find an example of this, and other principles discussed in this section, in Sec. 1.8.

> **Important Distinction.** If you know tangential velocity as a function of time, $v_T(t)$,[19] and integrate it over time, the result is the net 1D displacement, $\Delta s_T$. Given $v_T(t)$, if instead you want the total distance traveled, $\Delta s$, in 1D, you must break the integral up into separate distances – corresponding to each time that the object changes direction – and add the distances together (each distance being positive). In the case of net 1D displacement, $\Delta s_T$, you may skip the 'middle' man and integrate directly from the initial position to the final position.

> **Note.** If you are looking for helpful examples on how to apply calculus to 1D motion problems, see Sec. 1.8, which is dedicated to non-uniform acceleration.

---

[19] Hopefully, you remember from math or have realized by now that the notation $f(x)$ means that $f$ is a function of the variable $x$, and does not mean to multiply $f$ by $x$.

## 1.5 Single-Component Motion Graphs

**1D motion data**: In the real-world, sometimes you have experimental data, but have not yet established a mathematical relationship for it – i.e. you may not know the equation of the curve, but may just have a set of data points that more or less follow some unknown smooth curve. You can apply the techniques from this section, regarding how to analyze motion graphs, to motion data, even if you don't have an algebraic relationship for the data.

**Slope of the tangent line**: Conceptually, recall from calculus that the derivative of $y$ with respect to $x$, $dy/dx$, represents the slope of the curve $y = y(x)$ evaluated at $x$, which is the same as the slope of the line that is tangent to the curve at the point $(x, y)$. Therefore, given a plot of data corresponding to the relationship $y = y(x)$, you can numerically evaluate the derivative $dy/dx$ at a specified point $(x, y)$ by finding the slope of the tangent line at point $(x, y)$.

> **Note.** In order to numerically find the slope of a tangent line at a specified point, first draw the tangent line – this line should match the slope of the curve at the desired point. Next, choose two points on the tangent line from which to determine its slope. These points should lie on the tangent line, but not on the curve itself. Also, these two points should be far away, not close together, in order to reduce relative error in interpolating the values needed to compute the slope. Interpolate to read the $x$- and $y$-coordinates of the two points on the tangent line. Lastly, compute the rise and run in order to determine the slope of the tangent line: $m = \frac{rise}{run} = \frac{\Delta y}{\Delta x} = \frac{y - y_0}{x - x_0}$.

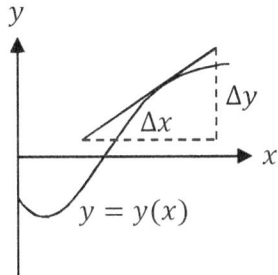

**Area under the curve**: Conceptually, you should also have learned in calculus that the definite integral $\int_{x=x_0}^{x} y(x) dx$ represents the area between the curve and the $x$-axis over the interval $(x_0, x)$. Thus, if you are given a plot of data for $y = y(x)$, you can numerically compute the definite integral $\int_{x=x_0}^{x} y(x) dx$ by finding the area between the curve and the $x$-axis over the interval $(x_0, x)$. Remember from calculus that area may be positive or negative, depending upon whether or not the curve lies above or below the $x$-axis.

**Note**. You can divide the area between the curve and the horizontal axis into shapes like rectangles and triangles that approximately equal the area. You can then compute the total area as the sum of the areas of the rectangles and triangles. Recall that the area of a triangle is one-half of its base times its height: $A_\triangle = \frac{1}{2}bh$. Also, recall from calculus that the area is negative for a region where the curve lies below the horizontal axis and positive if it lies above the horizontal axis.

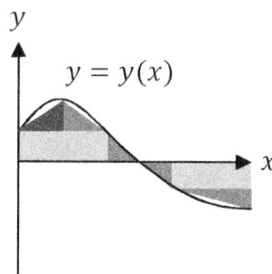

**Position graphs in 1D**: Given a graph of tangential displacement, $s_T$, in 1D, as a function of time, $t$:[20]
- The vertical intercept represents the initial position, $s_{T0}$.
- The net 1D displacement simply equals the final $s_T$-value minus $s_{T0}$: $\Delta s_T = s_T - s_{T0}$.
- To find the total distance traveled, you have to sum all of the individual forward and backward displacements – for each time the object changes direction: $\Delta s = \sum_{i=1}^{N}|\Delta s_{Ti}|$.
- You can determine the tangential velocity, $v_T$, at any instant from the slope of a tangent line.
- The speed equals the absolute value of the tangential velocity in 1D: $v = |v_T|$.
- If you want to know which way an object is moving, check the sign of $v_T$.

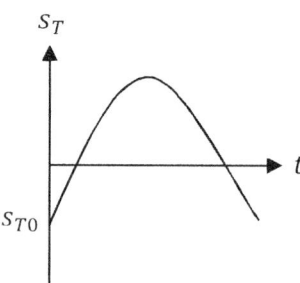

**Important Distinction**. Maximum velocity refers to the greatest positive slope (or least negative slope, if none are positive) for a 1D graph of tangential displacement as a function of time, whereas maximum speed means where the slope is steepest (regardless of sign). Similarly, if the minimum velocity is negative, the minimum speed may actually be zero, as in the example that follows.

---

[20] It is common to work with $x$, $v_x$, and $a_x$ in 1D – although, as pointed out previously, the tangential notation is a little more general, naturally allowing for curved 1D paths (like circular motion). In this case, note that $t$ is on the horizontal axis, and that $x$ is on the vertical axis. So if you refer to the graph and say, "the $x$-axis," you better realize that this is actually the vertical, and not the horizontal, axis for such a plot.

**Conceptual Example**. Which points in the graph below correspond to the minimum tangential velocity and to the minimum speed?[21]

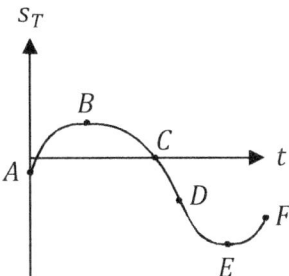

The instantaneous tangential velocity equals the slope of the tangent line at any point on the curve. The slope is minimum – in this case, most negative – at point $D$. Therefore, the tangential velocity is minimum at point $D$. Speed is nonnegative, however. The slope of the tangent line is zero at points $B$ and $E$, where the tangent is horizontal. The speed is therefore minimum – in this case, zero – at points $B$ and $E$. (If this were instead a graph of tangential velocity or tangential acceleration, we would have to use a different method to find tangential velocity and speed – other than the slope.)

**Example**. Determine the maximum speed, maximum tangential velocity, average speed, and average 1D velocity for the object that traveled in 1D according to the position graph illustrated below.

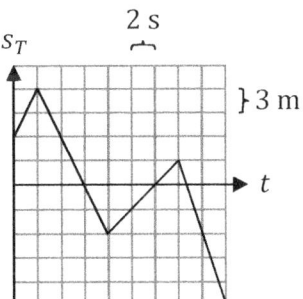

Since the tangential velocity equals the derivative of the tangential displacement with respect to time, $v_T = ds_T/dt$, which corresponds to the slope of the tangent line, and since speed equals the absolute value of the tangential velocity in 1D, $v = |v_T|$, the speed is maximum where the slope is steepest. By inspecting the slopes, it can be seen that the final slope is steepest – i.e. over the interval $14.0 \text{ s} \leq t \leq 18.0 \text{ s}$. The maximum speed is given by the absolute value of this steepest slope:

---

[21] Note that, physically, objects do have initial and final velocities, which must be physically meaningful. In calculus, technically, the derivative is ill-defined at the endpoints. This is due to a discontinuous jump, or a lack of smoothness, which is not really the case here. The curve extends beyond the endpoints shown; we have simply graphed a particular time interval, which is just a section of the complete curve. Furthermore, note that the initial and final speeds may be nonzero. We often start and end a problem while an object is in motion.

## 1 Single-Component Motion

$$v_{max} = \left|\frac{rise}{run}\right|_{max} = \left|\frac{-15.0 \text{ m} - 3.0 \text{ m}}{18.0 \text{ s} - 14.0 \text{ s}}\right| = \left|\frac{-18.0 \text{ m}}{4.0 \text{ s}}\right| = 4.5 \text{ m/s}$$

In contrast to the maximum speed, the tangential velocity is maximum where the slope is greatest – which means the most positive slope (or, if none of the slopes are positive, it means the least negative slope). By inspecting the slopes, it can be seen that the initial slope – over the interval $0 \leq t \leq 2.0$ s – is the steepest positive slope. Therefore, the maximum velocity is:[22]

$$v_{T,max} = \left(\frac{rise}{run}\right)_{max} = \frac{12.0 \text{ m} - 6.0 \text{ m}}{2.0 \text{ s} - 0} = \frac{6.0 \text{ m}}{2.0 \text{ s}} = 3.0 \text{ m/s}$$

The definition of average speed is the total distance traveled divided by the elapsed time: $\bar{v} = \Delta s/\Delta t$. Looking at the graph, the object first traveled 6.0 m forward, then 18.0 m backward, then 9.0 m forward, and finally 18.0 m backward. The total distance traveled equals the sum of the absolute values of these displacements: $\Delta s = 6.0 \text{ m} + 18.0 \text{ m} + 9.0 \text{ m} + 18.0 \text{ m} = 51.0 \text{ m}$. The total time is $\Delta t = 18.0$ s. The average speed equals $\bar{v} = \Delta s/\Delta t = 51.0 \text{ m}/18.0 \text{ s} = 2.83 \text{ m/s}$.

The average 1D velocity is defined as the net 1D displacement divided by the elapsed time: $\overline{v_T} = \Delta s_T/\Delta t$. The net 1D displacement can be found by subtracting the initial tangential displacement from the final tangential displacement for a 1D graph: $\Delta s_T = s_T - s_{T0} = -15.0 \text{ m} - 6.0 \text{ m} = -21.0 \text{ m}$. The average 1D velocity is therefore $\overline{v_T} = \Delta s_T/\Delta t = -21.0 \text{ m}/18.0 \text{ s} = -1.17 \text{ m/s}$.

**Velocity graphs in 1D**: Given a graph of tangential velocity, $v_T$, in 1D, as a function of time, $t$:
- The vertical intercept represents the initial tangential velocity, $v_{T0}$.
- The net 1D displacement, $\Delta s_T$, equals the area between the curve and the $t$-axis, keeping in mind that area[23] is negative when the curve lies below the $t$-axis.
- The total distance traveled, $\Delta s$, equals the sum of the absolute values of the areas between the curve and the $t$-axis.
- You can determine the tangential velocity, $v_T$, at any instant just by reading off the value from the vertical axis.
- The speed equals the absolute value of the tangential velocity in 1D: $v = |v_T|$.
- If you want to know which way an object is moving, check the sign of $v_T$.
- You can determine the tangential acceleration, $a_T$, at any instant from the slope of a tangent line. Recall that $a_T$ may be negative, but the magnitude of the acceleration in 1D is $a = |a_T|$.

---

[22] That trailing zero is a significant figure. It declares that there might be a little uncertainty on the order of $\pm 0.1$ m/s from reading the graph (which generally involves interpolation error). Contrast this with 3 m/s, which could be off by a whole $\pm 1$ m/s. Surely, our calculation is better than that! You can find a discussion of the rules for determining how many significant figures to keep in the section on How to Succeed in Physics.

[23] What we normally think of as area has dimensions of length-squared, or SI units of $m^2$. However, when we refer to the 'area' between a velocity curve and the $t$-axis, such 'area' actually has dimensions of length, or SI units of m. This is why we are able to interpret the 'area' as net 1D displacement or total distance traveled – in order to do so, it must have units of m. Similarly, the 'area' between an acceleration curve and the $t$-axis has dimensions of length/time, or SI units of m/s, which is consistent with its interpretation as the change in velocity.

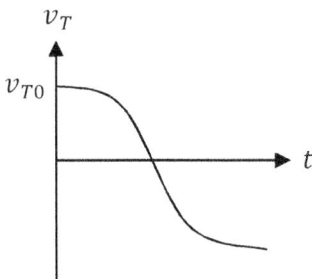

**Example.** What are the minimum and maximum tangential velocities and speeds for the graph of 1D motion below?

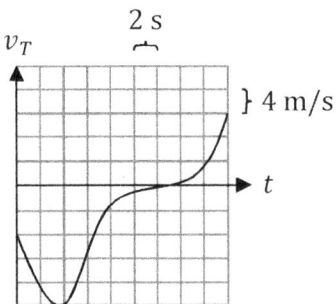

Since the vertical axis represents the tangential velocity for this graph, we can simply read off the vertical coordinates to obtain instantaneous values of tangential velocity. The minimum tangential velocity is $v_{T,min} = -20.0$ m/s and the maximum tangential velocity is $v_{T,max} = 12.0$ m/s. For speed, we are concerned only with how fast, and not which way. Speed is the absolute value of tangential velocity in 1D: $v = |v_T|$. The minimum speed for the graph above is $v_{min} = 0$. The maximum speed is $v_{max} = 20.0$ m/s.

**Example.** Determine the initial and final tangential accelerations for the graph in the previous example.
Tangential acceleration equals the derivative of the tangential velocity with respect to time, $a_T = dv_T/dt$, which equates to the slope of the tangent line for a plot of tangential velocity as a function of time. Tangent lines at the initial and final points on the curve are illustrated below.

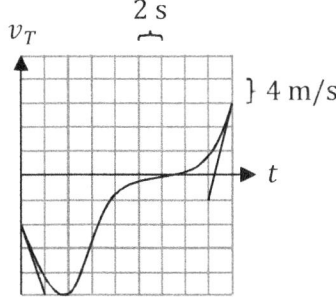

The initial and final values of the tangential acceleration equal the initial and final slopes:

$$a_{T0} = \left(\frac{rise}{run}\right)_0 = \frac{-20.0 \text{ m/s} - (-8.0 \text{ m/s})}{2.0 \text{ s} - 0} = \frac{-12.0 \text{ m/s}}{2.0 \text{ s}} = -6.0 \text{ m/s}^2$$

$$a_T = \left(\frac{rise}{run}\right) = \frac{12.0 \text{ m/s} - (-4.0 \text{ m/s})}{18.0 \text{ s} - 16.0 \text{ s}} = \frac{16.0 \text{ m/s}}{2.0 \text{ s}} = 8.0 \text{ m/s}^2$$

**Example.** Determine the net 1D displacement and total distance traveled in the previous two examples.

The net 1D displacement equals the definite integral of the tangential velocity over time, $\Delta s_T = \int_{t=t_0}^{t} v_T dt$, which corresponds to the area between the curve and the horizontal axis for a plot of velocity as a function of time. This area is approximated by the triangles and rectangles illustrated in the graph below.[24] Note that the area where the curve lies below the horizontal axis is negative.

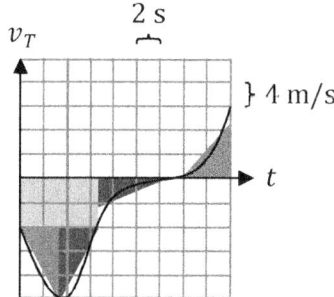

The net 1D displacement is approximately equal to the sum of the areas (keeping in mind that some of these areas are negative) of the shapes that approximate the region between the curve and the horizontal axis:[25]

$$\Delta s_T \approx \sum_{i=1}^{N} A_i$$

$$\approx -(6.7 \text{ s})(8.0 \text{ m/s}) - \frac{(3.3 \text{ s})(12.2 \text{ m/s})}{2} - \frac{(3.4 \text{ s})(12.2 \text{ m/s})}{2} - \frac{(6.1 \text{ s})(4.2 \text{ m/s})}{2} + \frac{(4.8 \text{ s})(8.5 \text{ m/s})}{2}$$

$$\Delta s_T \approx -87 \text{ m}$$

The difference between the total distance traveled and the net 1D displacement, for a graph of tangential velocity as a function of time, is that the areas are all positive for total distance traveled. Adding the absolute values of the areas, the total distance traveled is found to be $\Delta s \approx 128$ m.

---

[24] There is a little error in approximating the area of the curve with a set of triangles and squares. In those cases where you do not know the equation of the curve – which can be the case when working with new experimental data, for example – the integral will be numerical, rather than algebraic.

[25] The usual rules for significant figures would give an additional significant figure, but the error involved in approximating the area with triangles and rectangles suggests that a result quoted to the nearest 1 m/s is more appropriate (see Appendix A).

**Acceleration graphs in 1D**: Given a graph of tangential acceleration, $a_T$, in 1D, as a function of time, $t$:
- The vertical intercept represents the initial tangential acceleration, $a_{T0}$.
- The change in tangential velocity, $\Delta v_T$, equals the area between the curve and the $t$-axis, keeping in mind that area[26] is negative when the curve lies below the $t$-axis.
- In order to determine the tangential velocity, $v_T$, at a given instant, you must solve for it in terms of the initial tangential velocity and the area up to that moment: $v_T = v_{T0} + area$.
- The speed equals the absolute value of the tangential velocity in 1D: $v = |v_T|$.
- If you want to know which way an object is moving, check the sign of $v_T$.
- You can determine the tangential acceleration, $a_T$, at any instant just by reading off the value from the vertical axis.
- The magnitude of the acceleration equals the absolute value of the tangential acceleration in 1D: $a = |a_T|$.

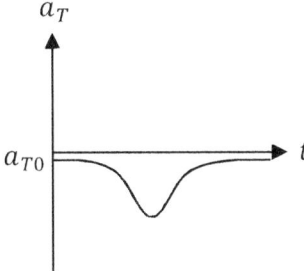

**Example**. Determine the tangential velocity at $t = 8.0$ s for the object that traveled in 1D according to the tangential acceleration graph illustrated below, given that its initial tangential velocity was $v_{T0} = -20.0$ m/s.

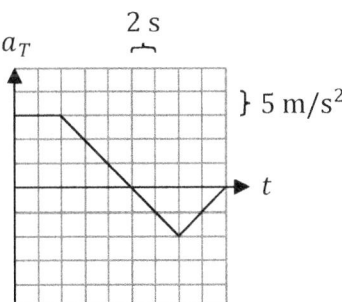

---

[26] What we normally think of as area has dimensions of length-squared, or SI units of m². However, when we refer to the 'area' between a velocity curve and the $t$-axis, such 'area' actually has dimensions of length, or SI units of m. This is why we are able to interpret the 'area' as net 1D displacement or total distance traveled – in order to do so, it must have units of m. Similarly, the 'area' between an acceleration curve and the $t$-axis has dimensions of length/time, or SI units of m/s, which is consistent with its interpretation as the change in velocity.

In 1D, the change in tangential velocity equals the definite integral of the tangential acceleration over time, $\Delta v_T = \int_{t=t_0}^{t} a_T dt$, which equates to the area between the curve and the horizontal axis for a graph of tangential acceleration as a function of time. In order to find the instantaneous tangential velocity at $t = 8.0$ s, we find the area corresponding to the interval $0 \le t \le 8.0$ s. This area is divided into rectangles and a triangle in the illustration below.

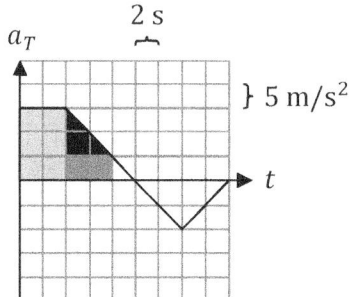

The change in tangential velocity equals the sum of the areas (keeping in mind that areas may be negative) of the shapes that approximate the region between the curve and the horizontal axis:

$$\Delta v_T = \sum_{i=1}^{N} A_i = (4.0 \text{ s})(15.0 \text{ m/s}^2) + \frac{(4.0 \text{ s})(10.0 \text{ m/s}^2)}{2} + (4.0 \text{ s})(5.0 \text{ m/s}^2)$$

$$\Delta v_T = 100 \text{ m/s}$$

It is important to realize that this is the change in tangential velocity, and not the answer to the question. We can solve for the tangential velocity at $t = 8.0$ s from the change in tangential velocity over the interval $0 \le t \le 8.0$ s and the initial tangential velocity since $\Delta v_T = v_T - v_{T0}$:

$$v_T = v_{T0} + \Delta v_T = -20.0 \text{ m/s} + 100.0 \text{ m/s} = 80.0 \text{ m/s}$$

**Common Mistake.** The area between the tangential acceleration curve and the time axis does not equal tangential velocity, but instead equals the change in tangential velocity. It is a common mistake for students to forget to account for the initial tangential velocity when computing the final tangential velocity from a plot of tangential acceleration as a function of time. Conceptually, the initial tangential velocity equates to the constant of integration going from tangential acceleration to tangential velocity.

**Potential Mistakes.** How you analyze a graph depends on what type of graph you have. Students sometimes remember finding slope to get tangential velocity, and then apply the same reasoning to all problems, regardless of what type of graph is given. However, how you determine tangential velocity depends very much on what type of graph you are analyzing: For a graph of $s_T(t)$, tangential velocity is indeed the slope of the tangent, but for a graph of $v_T(t)$, you just read tangential velocity directly, and for a graph of $a_T(t)$, the change in tangential velocity equals the area. You need to learn how information about distance, velocity, and acceleration can be determined from each type of graph.

**Hint.** Looking at the units can help you determine whether or not the slope or area will give you what you are looking for. If you take the units of the vertical axis and divide by the units of the horizontal axis, this will give you the units of the slope. For example, given a plot of $s_T$ as a function of $t$, the units of the slope are m/s, which corresponds to the units of tangential velocity. If you multiply the units of the two axes together, this gives you the units of the area. For example, given a plot of $v_T$ as a function of $t$, the area has units of (m/s)s = m, which is the unit for tangential displacement.

**Curve sketching**: Suppose that you are given a plot of the tangential displacement, $s_T$, of an object as a function of time and wish to sketch its tangential velocity, $v_T$, as a function of time. Since tangential velocity equals the derivative of tangential displacement with respect to time, $v_T = ds_T/dt$, the desired graph equates to a sketch of the slopes of the tangent lines. If you want to make a precise plot, you can draw tangents at several points and calculate the slope of each tangent. However, if you just wish to make a sketch, it will suffice to label whether the slope is very steep and positive (++), positive but not too steep (+), zero (0), negative but not too steep (−), or very steep and negative (−−). The slopes of the tangential displacement graph become the values for the tangential velocity graph.

This same method applies whenever the quantity that you are sketching equals the derivative of the quantity that you are given with respect to the variable plotted on the horizontal axis − so it would also work to make a sketch of tangential acceleration, $a_T$, as a function of time given a plot of tangential velocity, $v_T$, as a function of time.

**Conceptual Example.** Make a sketch of tangential velocity as a function of time corresponding to the 1D position graph illustrated below.

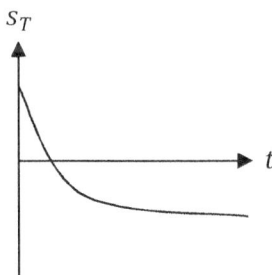

The tangential velocity equals the derivative of tangential displacement with respect to time, $v_T = ds_T/dt$, which equates to the slope of the tangent line. Therefore, we need to sketch the slope of the tangential displacement as a function of time. The initial slope of tangential displacement is steep and negative (−−). As time progresses, the slope of tangential displacement becomes less negative (−) and approaches (but doesn't quite reach) zero (0). These slopes serve as the values of the tangential velocity. A plot of tangential velocity as a function of time must then begin negative and grow toward zero. Examining the plot of tangential displacement more closely, you can see that the slope is fairly constant in the beginning and also levels out toward the end, but there is an interval in between where the slope changes more rapidly. These details should also be reflected in the sketch. A sketch of the tangential velocity as a function of time is shown on the following page.

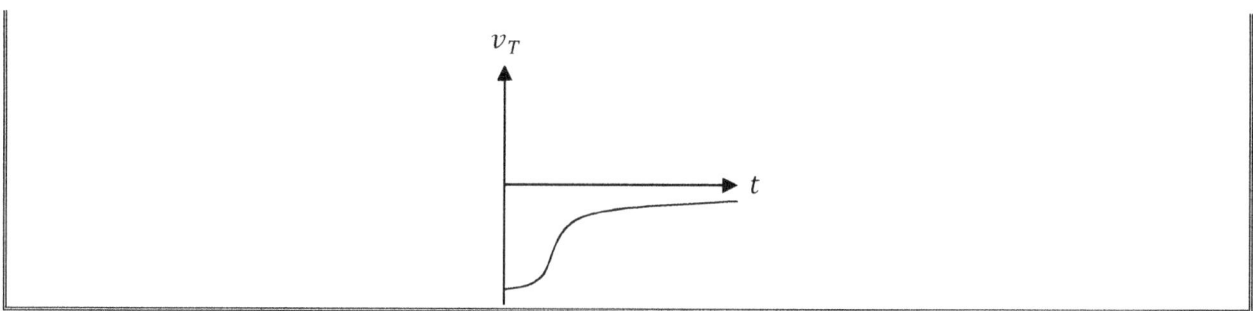

Now imagine that you are given a plot of the tangential acceleration, $a_T$, of an object as a function of time and wish to sketch its tangential velocity, $v_T$, as a function of time. This time, tangential acceleration equals the derivative of tangential velocity with respect to time, $a_T = dv_T/dt$. Note that this is effectively backwards compared to the previous case (where we discussed how to plot tangential velocity given a plot of tangential displacement). In this case, the values of the tangential acceleration graph serve as the slopes of the tangent lines for the tangential velocity graph. You can make a rough sketch by labeling whether the tangential acceleration values (not the slopes!) are large and positive (++), small and positive (+), zero (0), small and negative (−), and large and negative (−−). These correspond to slopes of tangential velocity that are very steep and positive (/), positive but not too steep (╱), zero (−), negative but not too steep (╲), and very steep and negative (\). Draw a smooth curve for tangential velocity that matches this set of slopes; if you are given an initial tangential velocity, $v_{T0}$, start your curve at this value.[27]

This same method applies whenever the quantity that you are sketching equals the integral of the quantity that you are given with respect to the variable plotted on the horizontal axis – so it would also work to make a sketch of tangential displacement, $s_T$, as a function of time given a plot of tangential velocity, $v_T$, as a function of time.

**Conceptual Example.** Make a sketch of tangential displacement as a function of time corresponding to the 1D tangential velocity graph illustrated below, assuming that the initial tangential displacement is positive.

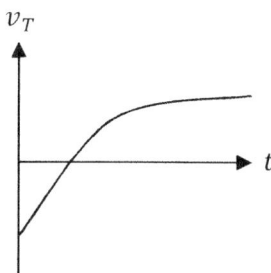

---

[27] If you want to make a precise plot, since the change in tangential velocity equals the integral of tangential acceleration as a function of time, $\Delta v_T = \int_{t=t_0}^{t} a_T dt$, in 1D, you could divide the curve up into incremental areas, and use each area to determine how the tangential velocity has changed since its previous value. This area technique is not efficient, however, for just sketching the key behavior.

The tangential velocity equals the derivative of tangential displacement with respect to time, $v_T = ds_T/dt$, which means, in this case, that the given values (not slopes) of the tangential velocity yield the slopes of the tangent lines of tangential displacement as a function of time. Being careful to look at the values of the given tangential velocity curve (instead of the slopes), we see that the tangential velocity begins negative $(--)$ and steadily increases to a positive value, until leveling out toward a final positive value $(++)$. This means that the slope of the tangential displacement curve must begin steep and negative $(\backslash)$ and steadily increase to a steep and positive slope $(/)$, where its slope levels out toward the end. A sketch of the tangential displacement as a function of time is illustrated below.

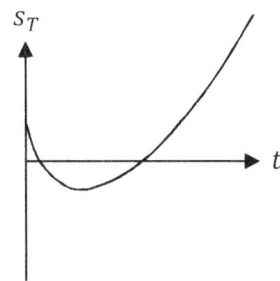

Note that the horizontal slope of the tangential displacement curve corresponds to the point of zero tangential velocity.

**Important Distinction.** If you are given a pot of $s_T$ as a function of $t$ and wish to sketch $v_T$ as a function of $t$, or if you are given a pot of $v_T$ as a function of $t$ and wish to sketch $a_T$ as a function of $t$, such that the graph that you wish to sketch is the derivative of the graph given, then you should use the slopes of the given plot as the values that you wish to sketch. If instead you are given a pot of $v_T$ as a function of $t$ and wish to sketch $s_T$ as a function of $t$, or if you are given a pot of $a_T$ as a function of $t$ and wish to sketch $v_T$ as a function of $t$, such that the graph that you wish to sketch is the anti-derivative of the graph given, then you should use the values of the given plot as the slopes of the curve that you wish to sketch; in this case, you must piece the slopes together to form a smooth curve with the desired initial value.

## 1.6 Uniform Acceleration in 1D

**Uniform acceleration**: Acceleration is termed uniform if it is constant. Thus, uniform acceleration means that the velocity changes at a constant rate with respect to time. The concept of uniform acceleration is important in physics because it corresponds to a large class of problems – e.g. an object freely falling near the surface of the earth experiences approximately uniform acceleration (provided that the change in altitude is small compared to earth's radius). It is therefore convenient to be familiar with the equations of uniform acceleration. However, it is also important to realize that these equations do not apply to all problems – and to recognize problems where the acceleration may be non-uniform.

**Conceptual meaning of uniform acceleration in 1D**: A uniformly accelerated object with 1D motion gains or loses tangential velocity at a constant rate. This is easy to see if you think of the units, $m/s^2$, as $(m/s)/s$. For example, a car that has a uniform acceleration of $4\ m/s^2$ gains $4\ m/s$ of tangential velocity every second, while a truck with a uniform acceleration of $-20\ m/s^2$ loses $20\ m/s$ of tangential velocity every second.

> **Note**. An object that gains $4\ m/s$ of tangential velocity each second is not necessarily gaining speed. Similarly, an object that loses $4\ m/s$ of tangential velocity each second is not necessarily losing speed. For example, if an object has an initial tangential velocity of $-12\ m/s$ and gains $4\ m/s$ of tangential velocity, after one second its tangential velocity will be $-8\ m/s$, which is a decrease in speed. It's important to remember that tangential acceleration describes directly how tangential velocity changes in time, and that speed is only determined after removing the direction.

> **Conceptual Example**. A car drives east, accelerating from rest at a constant rate of $5\ m/s^2$ for $6\ s$. Conceptually, we can reason that the car's speed will be $5\ m/s$ after $1\ s$, $10\ m/s$ after $2\ s$, $15\ m/s$ after $3\ s$, and so on. The car's final speed will be $30\ m/s$.

**The sign of uniform tangential acceleration**: There are two ways that the tangential acceleration, $a_T$, can be negative: The object can be losing speed while moving forward, or it can be gaining speed while moving backward. Similarly, there are two ways that the tangential acceleration can be positive: The object can be gaining speed while moving forward, or it can be losing speed while moving backward.

> **Common Mistake**. It's common for students to think that a negative tangential acceleration implies that an object is losing speed, or that a positive tangential acceleration means that speed is increasing. However, you must consider both the direction of the tangential velocity and the sign of the tangential acceleration to determine whether the speed is increasing or decreasing. Remember, the sign of the tangential acceleration relates directly to how tangential velocity changes, and that speed is determined, in 1D, by taking the absolute value of the tangential velocity.

> **Conceptual Example**. A train is traveling south with an initial speed of $24\ m/s$. It then experiences a uniform acceleration of $8\ m/s^2$ for $4\ s$, where the direction of the acceleration is to the north. If we setup our coordinate system with the positive tangential direction to the south, then the initial tangential velocity is $24\ m/s$ and the tangential acceleration is $-8\ m/s^2$.[28] The train loses $8\ m/s$ of tangential velocity – but not necessarily speed – each second. The train's speed will be $16\ m/s$ after $1\ s$, $8\ m/s$ after $2\ s$, and it will momentarily be at rest after $3\ s$. So the train does lose speed for the first three seconds. However, assuming that the acceleration remains constant, the train's will be moving $8\ m/s$ to the north after $4\ s$, $16\ m/s$ after $5\ s$, and so on – gaining speed to the north.

---

[28] If instead we choose the positive tangential direction to be north, the initial tangential velocity is $-24\ m/s$ and the tangential acceleration is $8\ m/s^2$. Either way, these two quantities differ by a relative minus sign. The physics remains unchanged, as it can't be affected by our choice of coordinates: The train still loses speed to the south until momentarily coming to rest, and then gains speed to the north.

**Mathematical definition of uniform acceleration in 1D**: Tangential acceleration equals a derivative of tangential velocity with respect to time. In the case of 1D uniform acceleration (1DUA), this means that the tangential acceleration is a constant, which we may call $a_{T0}$:

$$a_{T0} = \frac{dv_T}{dt} \quad \text{(for 1DUA)}$$

This means that, in 1D, the tangential velocity changes at a constant rate if acceleration is uniform.

> **Notation.** We often use the subscript zero – called nought, literally meaning the digit zero, or naught, which means nothing – to refer to an initial value. For example, when an object accelerates for a period of time, we use $v_{T0}$ to distinguish the initial tangential velocity from the final tangential velocity. However, we sometimes also use the subscript nought (0) to designate a quantity that is constant. There should be no confusion because if a quantity is constant, its initial and final values are the same.

**Velocity equation for 1D uniform acceleration**: We can derive an equation for the time-dependence of the tangential velocity in 1D uniform acceleration by integrating over the acceleration. The correct procedure is to split the differentials in order to separate variables. In the previous equation, the tangential acceleration is constant, so the two variables are tangential velocity and time. We separate variables by algebraically placing time and tangential velocity on opposite sides of the equation:

$$dv_T = a_{T0} dt \quad \text{(for 1DUA)}$$

Now we may integrate both sides:[29]

$$\int_{v_T = v_{T0}}^{v_T} dv_T = \int_{t = t_0}^{t} a_{T0} dt \quad \text{(for 1DUA)}$$

The left integral trivially equals the change in the tangential velocity. The tangential acceleration can be pulled out of the right integral because it is constant:[30]

$$v_T - v_{T0} = a_{T0}(t - t_0) \quad \text{(for 1DUA)}$$

Notice how this equation involves initial tangential velocity, $v_{T0}$. We obtained this naturally by properly separating variables and integrating both sides – a fundamental technique for solving a class of simple differential equations. It's a common mistake for students to neglect $v_{T0}$ when integrating.

---

[29] To reiterate an important distinction between conventional approaches in math and physics, it is common to perform indefinite integrals in math, and later solve for the constants of integration by applying boundary conditions, whereas in physics we can conceptually identify the significance of the constants by building them into definite integrals. The initial tangential velocity, $v_{T0}$, corresponds to the initial time, $t_0$.

[30] Think twice before you pull *any* symbol out of an integral in physics: If it's not constant, you're making a major conceptual (and mathematical) mistake!

# 1 Single-Component Motion

**Displacement formula for 1D uniform acceleration**: We can express the tangential velocity in terms of the tangential displacement using calculus – i.e. $v_T = ds_T/dt$ – and apply the same procedure to derive an equation for the time-dependence of tangential displacement in 1D uniform acceleration:

$$\frac{ds_T}{dt} - v_{T0} = a_{T0}(t - t_0) \quad \text{(for 1DUA)}$$

We can separate variables – i.e. $s_T$ and $t$ – by moving $v_{T0}$ to the right and multiplying by $dt$:

$$ds_T = v_{T0}dt + a_{T0}(t - t_0)dt \quad \text{(for 1DUA)}$$

Now we may integrate both sides:

$$\int_{s_T=s_{T0}}^{s_T} ds_T = \int_{t=t_0}^{t} v_{T0}dt + \int_{t=t_0}^{t} a_{T0}(t - t_0)dt \quad \text{(for 1DUA)}$$

The three quantities with a subscript nought are treated as constants during the integration:[31]

$$\Delta s_T = v_{T0}(t - t_0) + \frac{a_{T0}(t - t_0)^2}{2} \quad \text{(for 1DUA)}$$

**Time-independent formula for 1D uniform acceleration**: We can derive a convenient time-independent formula for 1D uniform acceleration by combining the two previous results – i.e. the equation for tangential velocity and the equation for the net 1D displacement. Let us eliminate time by solving for it in the tangential velocity equation:

$$t = t_0 + \frac{v_T - v_{T0}}{a_{T0}} \quad \text{(for 1DUA)}$$

Substituting this expression in for time in the net 1D displacement equation,

$$\Delta s_T = v_{T0}\frac{v_T - v_{T0}}{a_{T0}} + \frac{a_{T0}}{2}\left(\frac{v_T - v_{T0}}{a_{T0}}\right)^2 \quad \text{(for 1DUA)}$$

This simplifies to[32]

$$v_T^2 - v_{T0}^2 = 2a_{T0}\Delta s_T \quad \text{(for 1DUA)}$$

---

[31] You may write the net 1D displacement, $\Delta s_T$, as $s_T - s_{T0}$ if you prefer. It is a little simpler to work with $\Delta s_T$, rather than two separate quantities, when using the equations of 1D uniform acceleration. Since $s_T$ and $s_{T0}$ always appear in the form $s_T - s_{T0}$ in these equations, we choose to use $\Delta s_T$ instead of $s_T - s_{T0}$.

[32] You become a better problem-solver when you can look a few steps ahead – to see where you're going. So, occasionally we will skip a few steps to give you practice. Pull out a sheet of paper and try to fill in the steps.

**Equations of 1D uniform acceleration**: We have derived three equations of 1D uniform acceleration:

$$v_T - v_{T0} = a_{T0}\Delta t$$
$$\Delta s_T = v_{T0}\Delta t + \frac{a_{T0}\Delta t^2}{2} \quad \text{(for 1DUA)}$$
$$v_T^2 - v_{T0}^2 = 2a_{T0}\Delta s_T$$

> **Notation**. We have expressed the equations of 1D uniform acceleration in general terms by using the notation for tangential variables. These equations apply to any object that experiences uniform tangential acceleration, even if the path is curved – like a circle.[33] However, if you apply these equations to linear uniform acceleration, you may find it convenient to work with $x$ (or $y$). In this case, the equations will look like this:[34]
>
> $$v_x - v_{x0} = a_{x0}\Delta t$$
> $$\Delta x = v_{x0}\Delta t + \frac{a_{x0}\Delta t^2}{2} \quad \text{(for linear 1DUA[35])}$$
> $$v_x^2 - v_{x0}^2 = 2a_{x0}\Delta x$$

**Conceptual meaning of the equations of 1D uniform acceleration**: Our conceptual interpretation of uniform tangential acceleration is described by the equation $v_T - v_{T0} = a_{T0}\Delta t$, which mathematically states that when tangential acceleration is constant, the object gains or loses tangential velocity at a constant rate. You may think of the displacement formula for 1D uniform acceleration, $\Delta s_T = v_{T0}\Delta t + \frac{a_{T0}\Delta t^2}{2}$, the following way: The first term, $v_{T0}\Delta t$, represents the displacement that an object would have if it traveled with constant tangential velocity, while the second term, $\frac{a_{T0}\Delta t^2}{2}$, represents the difference in displacement caused by the fact that the tangential velocity is actually changing. In Chapter 5, we will see that the remaining equation, $v_T^2 - v_{T0}^2 = 2a_{T0}\Delta s_T$, corresponds to a special case of conservation of energy.[36]

---

[33] However, we will learn in later chapters that an object traveling along a curved path also has other kinds of acceleration, and that the overall acceleration is non-uniform – even if the tangential component of acceleration is.
[34] This notation, in terms of $x$ (or $y$), is what you'll see in almost every textbook in the chapter on uniform acceleration. However, if you look ahead to the chapter on rotation (not the chapter on uniform circular motion), you will find a discussion of uniform angular acceleration and uniform tangential acceleration – for which this more general tangential notation is applied. It is also common to write $a_x$ instead of $a_{x0}$, but it's important to distinguish between what is constant and what is not – an important consideration before you pull a quantity out of an integral. Compare the derivations of this section to the examples in Sec. 1.8.
[35] A car that is driving in a circle with constant tangential acceleration (but not constant overall acceleration) experiences 1D uniform tangential acceleration (but not uniform acceleration, which is overall acceleration). The point is that 1D uniform acceleration need not be linear. We can distinguish between linear and tangential 1DUA.
[36] Specifically, for an object falling vertically through a uniform gravitational field, conservation of energy would lead to the equation $v^2 - v_0^2 = 2gh$.

**Interpreting the signs for 1D motion**: It's important to keep in mind that the tangential quantities – $\Delta s_T$, $v_T$, $v_{T0}$, and $a_{T0}$ – (or the linear quantities – $\Delta x$, $v_x$, $v_{x0}$, and $a_{x0}$) may all be positive or negative (or zero). Here is what the signs mean, conceptually:
- The net 1D displacement, $\Delta s_T$, is positive if the final position is ahead of the initial position. Looking at your coordinate system, is the final position more positive than the initial position (or, if neither coordinate is positive, is the final coordinate less negative)? For the sign of $\Delta s_T$, just compare where the final position is relative to the initial position.
- The tangential velocity, $v_T$, (initial or final) is positive if the object is moving in the positive direction and negative if moving in the negative direction. For the sign of $v_T$, just look at which way the object is moving (relative to your coordinate system).
- The tangential acceleration, $a_{T0}$, is positive if the object is moving in the positive direction and gaining speed or moving in the negative direction and losing speed, and is negative if the object is moving in the positive direction and losing speed or moving in the negative direction and gaining speed. For the sign of $a_{T0}$, you must look at both which way the object is moving and what is happening to its speed.

---

## Problem-Solving Strategy for 1D Uniform Acceleration

0. The first step (or in this case, the zeroth step) toward solving any problem in physics is first to identify the concepts involved and the relationships among them. In this case, make sure that the acceleration is uniform and that the problem is 1D before applying this problem-solving strategy. If you use this strategy to solve a problem where acceleration is not constant or to a problem that is not 1D, you will be making a major conceptual/strategic mistake.
1. Draw and label a diagram. Don't work in the dark: A labeled diagram is like using a flashlight to see what's going on. Label the initial ($i$) and final ($f$) positions on your diagram, since the equations of 1D uniform acceleration involve initial and final quantities. Also draw and label the positive direction.
2. List your knowns and identify the desired unknown(s). You should generally know 3 of the 5 following symbols: $\Delta s_T$, $v_T$, $v_{T0}$, $a_{T0}$, and $\Delta t$. If you think you only know 2, re-read the problem carefully. If you think you know 4, it's likely that you don't actually know one of them.
3. Look at the three equations of 1D uniform acceleration. See if you can plug the quantities you know into one of these equations and directly solve for the unknown that you are looking for. If so, use this equation to solve for the unknown.
4. Sometimes you can't solve for the unknown using a single equation. When this happens, plug the information that you know into two of the equations. You may find that you have two equations in two unknowns – a type of problem that you should know how to solve from algebra. If so, use these equations to solve for the unknowns – e.g. solve for an unknown in one equation and substitute this expression in to the other equation.
5. Some problems give you information that you can express mathematically – like "when the final speed equals twice the initial speed." Use this extra equation to solve a system of equations. Generally, you are ready to proceed with the algebra when you have $N$ equations in $N$ unknowns.

**Note**. Only 2 of the 3 equations for 1D uniform acceleration are algebraically independent.[37] Therefore, we need to have at most 2 unknowns of the 5 quantities used in the equations: $\Delta s_T$, $v_T$, $v_{T0}$, $a_{T0}$, and $\Delta t$. This way, we would be solving a system of 2 algebraically independent equations in 2 unknowns. The exception to this rule is when you can express an additional relationship between these 5 symbols that is algebraically independent from the equations of 1D uniform acceleration. For example, if you are trying to find when the tangential velocity equals one-half its initial value, you could use the equation $v_T = v_{T0}/2$, allowing you to solve 3 equations in 3 unknowns.

**Math Note**. In principle, you can solve a system of $N$ equations in $N$ unknowns. For $M \geq 1$, you generally need more information in order to solve a system of $N$ equations in $N + M$ unknowns; and generally if you have $N + M$ equations in $N$ unknowns, the superfluous information may lead to a contradiction. It is worth counting knowns and unknowns before you begin the algebra so that you know whether or not the algebra can potentially lead to an answer. If you have too much information, you should reconsider what you think you know — probably, you don't really know something that you think you know — and if you have too little information, don't waste your time on the math until you reason out what else you know — or what other equation you could use.[38]

**Signs**. The signs are critical when you apply the equations of 1D uniform acceleration (and many other equations in physics). For example, if you throw a banana and want to know when it will reach the ground, whether the initial velocity is upward or downward affects the time that the banana is in the air. The equations aren't psychic: You have to use the correct signs so that the equations give you a longer time when you throw the banana upward and a shorter time when you throw the banana downward. You can review the rules for determining the signs on the previous page.

**Signs**. The signs are also important when you're doing the math. For example, if you solve for the final tangential velocity, and find it to be $v_T = \sqrt{16}$ m/s, you need to realize that there are two possible solutions: $v_T = 4$ m/s and $v_T = -4$ m/s. You must consider both positive and negative roots. You may be accustomed to ignoring the negative squareroot for some math courses, but either root (or both) may be physically meaningful in physics. For example, if you throw a banana straight upward and ask when it will be 5 m above the ground, you may find that there are two possible answers: A positive or negative tangential velocity distinguishes between the banana still on its way up from the banana on its way back down. If you use the wrong root — positive or negative — and subsequently plug this value into another equation, you should expect to get the wrong answer.

---

[37] In terms of calculus, however, only one is independent. Thinking about how to use the equations to solve problems, it's the algebraic strategy that's important here: We're trying to figure out how many knowns we need before we will be able to solve for an unknown.

[38] There are occasional exceptions to this rule. Technically, you should count the number of independent combinations of unknowns. For example, given the two equations $3xz - 2y = 8$ and $6y - xz = 16$, you can actually solve for $y$ (but not for $x$ or $z$). Define $u \equiv xz$ to rewrite these 2 equations in terms of just 2 unknowns.

# 1 Single-Component Motion

**Potential Mistake.** Students sometimes want to make the initial or final tangential velocity equal to zero when it shouldn't be. Remember that the equations of 1D uniform acceleration only apply while the object is uniformly accelerating. If you throw a banana up into the air, when we speak of the initial or final tangential velocity, we really mean just after leaving your hand and just before striking the ground. So if a question asks you when a banana reaches the ground – the final tangential velocity is not zero (since it's gaining speed as it falls downward), and the question really implies for you to find the time just before impact.

**Potential Mistake.** Suppose that you are solving a multipart problem, and found the final tangential velocity in part (A) of the problem. You should be careful before you use this answer in a subsequent part of the problem – like part (B). If the final point that you are working with is different in part (B) than it was in part (A), it's no longer the final tangential velocity. For example, if you throw a banana straight upward and solve for the time it takes to strike the ground in part (A), and part (B) asks you for the maximum height of the banana above the ground, you can't plug the complete trip time that you solved for in (A) into the equations of 1D uniform acceleration to solve part (B).

**Hint.** Sometimes you can perform a quick estimate to check some of your answers. At the completion of a problem, if you know $v_T$, $v_{T0}$, $a_{T0}$, and $\Delta t$, you can check for consistency using the conceptual interpretation of uniform tangential acceleration. For example, if a car uniformly accelerates to the west from 20 m/s to 40 m/s in a time of 5 s, it must gain 4 m/s of speed each second, so its tangential acceleration better be 4 m/s$^2$.

**Math Note.** You may need to use the quadratic equation when applying the equations of 1D uniform acceleration. For example, if you know the net 1D displacement, initial tangential velocity, and the tangential acceleration, you can solve for the time interval, $\Delta t$, using the equation $\Delta s_T = v_{T0}\Delta t + \frac{a_{T0}\Delta t^2}{2}$. Recall that you must first express the quadratic in standard form – i.e. move all of the terms to the same side of the equation, as in $ax^2 + bx + c = 0$. The solution to the standard form of the quadratic equation is:

$$x = \frac{-b \pm \sqrt{b^2 - 4ac}}{2a}$$

You must consider both signs: Often, one sign is physically relevant and the other is not. You should usually expect to obtain a positive time interval if solving for time – otherwise, you need a time machine! An imaginary time pertains to something that's not physically possible – or signifies that you made a mistake somewhere (either in your assumptions or your algebra). For example, if you throw a banana upward and ask when it will be 100 m above the ground, unless its initial speed is very large (faster than a human can throw), the math will give you an imaginary time to tell you that the banana will never reach such a great height.

**Example**. A monkey driving 36 m/s begins to decelerate uniformly when he sees a banana 72 m away. What deceleration is required for the monkey to stop the car prior to reaching the banana?

First, draw and label a diagram, showing the initial and final positions and setting up a coordinate system.

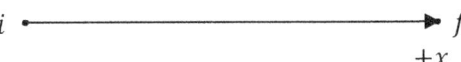

Next, list which quantities you know and identify the desired unknown:

$$\Delta x = 72 \text{ m} \quad , \quad v_{x0} = 36 \text{ m/s} \quad , \quad v_x = 0 \quad , \quad a_{x0} = ?$$

Note that the final tangential velocity is zero when the car comes to a stop. We don't know time, and we're not looking for time, so the time-independent equation of 1D uniform acceleration is handy:

$$v_x^2 - v_{x0}^2 = 2a_{x0}\Delta x$$

$$a_{x0} = \frac{v_x^2 - v_{x0}^2}{2\Delta x} = \frac{0^2 - (36 \text{ m/s})^2}{2(72 \text{ m})} = -9.0 \text{ m/s}^2$$

**Example**. A monkey runs around a 50-m diameter circle with an initial speed of 2.0 m/s and with a uniform tangential acceleration of 0.10 m/s². The monkey completes one lap. How long does it take for the monkey to complete the lap?

As usual, begin with a labeled diagram.

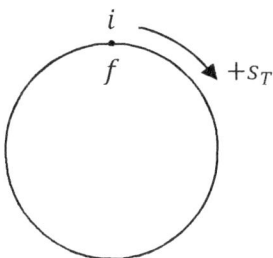

Tabulate the knowns and identify the desired unknown:

$$D = 50 \text{ m} \quad , \quad v_{T0} = 2.0 \text{ m/s} \quad , \quad a_{T0} = 0.10 \text{ m/s}^2 \quad , \quad \Delta t = ?$$

This only includes 2 of the 5 quantities in the equations for 1D uniform acceleration, but we can obtain a third known by relating the net 1D displacement[39] to the diameter:

---

[39] In the next chapter, we will see that, in 2D and 3D motion, the net displacement is defined in such a way that it would be zero for this trip. However, the net 1D displacement is nonzero – it doesn't realize that the circle is cyclic, but acts as if you could go infinitely far in one direction without returning to the starting position.

$$\Delta s_T = \pi D = 50\,\pi \text{ m}$$

We can solve for the elapsed time by applying the quadratic equation:[40]

$$\Delta s_T = v_{T0}\Delta t + \frac{a_{T0}\Delta t^2}{2}$$
$$50\,\pi = 2\,\Delta t + \frac{0.1\,\Delta t^2}{2}$$
$$0.05\,\Delta t^2 + 2\Delta t - 50\,\pi = 0$$
$$\Delta t = \frac{-2 \pm \sqrt{2^2 - 4(0.05)(-50\,\pi)}}{2(0.05)} = 40\text{ s}$$

## 1.7 Vertical Free Fall

**Free fall**: An object is said to be freely falling if the only force acting on it equals the object's weight – i.e. the force of gravity exerted on the object. In practice, we treat many falling objects as if they are in free fall provided that the effects of air resistance are negligible. If air resistance is negligible, the object falls as if it is in a vacuum.

**Note**. An object may be in a state of free fall even if it is moving up – i.e. away from the surface of a planet (or moon). It is 'falling' in the sense that it loses speed on the way up, and eventually will fall back down (provided that its speed does not exceed escape speed).

**Vacuum**: A volume that is completely void of matter is called a vacuum. Therefore, an object that falls through a vacuum does not experience a force of air resistance.[41]

**Note**. In this textbook, neglect air resistance unless stated otherwise. If a monkey is falling near the surface of the earth, assume that the monkey is falling through a vacuum unless a problem or question explicitly asks you to consider the effects of air resistance.[42] Also assume that the change in altitude is small compared to the radius of the planet unless it is clearly not a uniform gravity problem.

---

[40] This is just a reminder that we will sometimes suppress the units and even significant figures in the middle of a calculation (but not in the answers) in order to avoid clutter and potential confusion between symbols and units.

[41] Observe that there may be – and generally is – gravity in a vacuum. Space is a near vacuum, and there is definitely gravity pulling on objects in space – from the myriad astronomical bodies that all have gravitational fields. When we study Newton's law of universal gravitation, this should become clear (if it isn't already). When you see video of an astronaut and various objects floating in a ship, it's not because there is any lack of gravity in space – you could get the same result in an elevator by cutting the cable, but you would definitely concede that there was gravity in the elevator when it struck the ground. We will return to this point in a later chapter.

[42] If you are wondering how the monkey will breathe, you can assume that there is air for biological purposes, but neglect it for the physics (or just give the poor little monkey an oxygen tank).

**Gravitational acceleration**: The acceleration that an object experiences in a state of free fall is termed gravitational acceleration ($g$).

**Earth's gravity**: The value of standard gravity near the surface of the earth is $9.81$ m/s$^2$. Gravitational acceleration actually varies slightly over the surface of the earth, depending upon the altitude and the distance to the axis of earth's rotation. We will treat the dependence on altitude when we consider Newton's law of universal gravitation. However, unless the altitude is clearly an important aspect of the problem, use the value above for earth's gravity.

> **Note**. In this textbook, assume that all objects are near the surface of the earth unless stated otherwise.

**Moon's gravity**: The value of gravitational acceleration near the surface of the moon is approximately one-sixth that of the earth: $g_m \approx g_e/6$.

**Mass-independence of gravitational acceleration**: In free fall, all objects fall with the same gravitational acceleration regardless of their mass and shape. You should only concern yourself with mass and shape if the problem clearly states that you must account for the effects of air resistance. We will see why mass does not affect an object's free fall acceleration when we investigate Newton's laws of motion.

> **Conceptual Examples**. A monkey drops a 0.5-kg[43] banana and a 5-kg bunch of bananas – both from rest – simultaneously from the same height. Which object reaches the ground first?[44] Neither – they both accelerate with the same gravitational acceleration. Even if you considered the effects of air resistance, there probably wouldn't be a significant difference for bananas, especially from a reasonably low height.
> 
> Now the monkey drops a feather and a bunch of bananas, both from rest, simultaneously from the same height. Which of these objects reaches the ground first? Neither! Remember, we neglect any effects air resistance unless stated otherwise, and the question made no mention about air resistance. If you find this difficult to believe, you can easily find a demonstration of a feather and a coin falling through a vacuum on the internet.

**Uniform gravitational fields**: An object that is freely falling through a gravitational field experiences approximately uniform acceleration provided that the change in altitude is not significant compared to the radius of the planet (or other astronomical body). If the motion is also vertical – i.e. straight up or down – then the equations of 1D uniform acceleration apply. We will treat the more general case of projectile motion in the following chapter.

**Sign of gravitational acceleration**: Gravitational acceleration ($g$) is a magnitude of acceleration, and so is always positive. The vertical component of acceleration, $a_{y0}$, however, may be negative.

---

[43] A note of mathematical English: The hyphen serves to join the 0.5 and the kg together to form an adjective, as in the high-flying monkey. In contrast, without the hyphen, as in, "the monkey has a mass of 0.5 kg," the separate 0.5 and kg form a noun, like a monkey that was flying high.

[44] It would be senseless to ask which falls with greater speed, since the speed of both objects steadily increases. We could ask which falls with greater acceleration, however.

**Sign of vertical component of acceleration in free fall**: If you setup your coordinate system so that $+y$ is upward, an object in free fall has a vertical component of acceleration equal to $a_{y0} = -g$.[45] On the way up, this minus sign tells you that the object is moving forward, but losing speed. On the way down, it tells you that the object is gaining speed, but moving backward. Either way, $a_{y0}$ is negative. Conceptually, the acceleration is downward due to the gravitational pull, and downward acceleration is negative if you choose $+y$ to be upward.

**Common Mistake**. Intuitively, most students want the vertical component of acceleration to be negative on the way up, but positive on the way down – but it's negative in both cases. Remember, on the way back down, although the object gains speed, it's moving backward, which makes $a_{y0}$ negative. Also, the acceleration wouldn't be uniform if it changed sign on the way back down, would it? If you review our derivation of the equations of 1D uniform acceleration, you will see that we assumed that $a_{T0}$ was constant in both sign and magnitude throughout the trip.

**Conceptual Example**. A monkey throws a banana straight upward. What is the banana's tangential acceleration when it reaches its maximum height? It's $-9.81$ m/s$^2$, if you choose $+y$ to be upward.[46] Remember, the acceleration is uniform. It wouldn't be constant if it were nonzero going up, but suddenly zero at the top. Following is an argument to help convince you that acceleration is not zero at the top. Recall that acceleration describes instantaneously how velocity changes in time. The velocity of the banana is zero at the top. If its velocity were zero at the top and, as you might intuitively like to be the case, its acceleration were zero at the top – would the banana ever come back down? Chances are, except in cartoons, if you throw a banana straight upward, it doesn't stop when it reaches the top of its trajectory!

**Signs**. If you choose $+y$ to be upward for an object in vertical free fall, the vertical component of acceleration, $a_{y0}$, will be negative regardless of the object's motion, the vertical component of its velocity, $v_{y0}$ or $v_y$, will be negative if the object is moving downward, and the vertical net displacement, $\Delta y$, will be negative if the final position is beneath the initial position.

**Conceptual meaning of gravitational acceleration**: The value of gravitational acceleration, $g$, tells you how many m/s the vertical component of velocity loses each second. If you throw an object straight upward, it tells you how many m/s of speed the object loses each second going up, and how many m/s of speed it gains each second coming back down. For example, near the surface of the earth, an object loses 9.81 m/s of speed each second on the way up, and gains 9.81 m/s of speed each second on its way back down. If the object has an initial speed of 20 m/s, it would reach the top of its trajectory in just over 2 s (when it runs out of speed), and would return to its starting point in just over 4 s.

---

[45] Now that we are talking about vertical free fall, and not general 1D motion, instead of $T$ for tangential, we use $y$, since we will use the same symbol to represent the vertical component of projectile motion in the next chapter.
[46] Remember, all questions, problems, and discussions imply that the objects are near earth's surface, that air resistance effects are negligible, and that the altitude change is small unless stated otherwise. Also neglect earth's rotation unless stated otherwise.

**Conceptual Examples.** If a monkey drops a banana from rest over a deep well, after 1 s its speed will be about 10 m/s, after 2 s its speed will be about 20 m/s, after 3 s its speed will be about 30 m/s, etc. If a banana travels straight upward with an initial speed of 40 m/s, after 1 s its speed will be about 30 m/s, after 2 s its speed will be about 20 m/s, after 3 s its speed will be about 10 m/s, etc. The banana will reach its maximum height after about 4 s, and then it will gain speed from rest just like the banana falling down the well. It will return to its starting position with a roundtrip time of about 8 s.

**Note.** Pay attention to whether a problem asks you to find gravitational acceleration, $g$, or the vertical component of acceleration, $a_{y0}$. Remember that $a_{y0}$ is negative for an object in free fall if $+y$ is upward, but $g$ is positive because it is the magnitude of the free fall acceleration.

**Potential Mistake.** Remember that $a_{y0} = -g$ for an object in free fall where $+y$ is upward. This does not apply to objects that are not in free fall. So if a car is accelerating down a highway, its acceleration will probably not equal $-9.81$ m/s$^2$ (unless it drives over the edge of a cliff!). Similarly, in later chapters, if a banana is in the air, but connected to a box via a rope and pulley, $a_{y0}$ will probably not be $-9.81$ m/s$^2$ because the object is not in free fall (unless the rope gets cut). Also, watch out for problems where an object is not near earth's surface – or not even on the earth – in which case gravitational acceleration will have a different value.

**Hints.** You can check some of your answers using the conceptual interpretation of acceleration. If you know $v_{y0}$, $v_y$, $a_{y0}$, and $\Delta t$, for example, you can check that the amount of tangential velocity gained or lost each second agrees with these values. For example, if you know that $v_{y0} = 25$ m/s and $v_y = -15$ m/s, and the object is near the surface of the earth, $\Delta t$ better be about 4 s because it would take about 2.5 s for the object to reach its maximum height and it would spend about 1.5 s more coming partway back down.

You can also check that your values and signs make sense conceptually. For example, if a monkey throws a banana downward, its final tangential velocity better be negative. As another example, if a monkey throws a banana upward with an initial speed of 18 m/s and you compute the speed of the banana when it is 10-m above the ground, the speed better not be greater than 18 m/s. You can get some funny numbers by entering equations incorrectly in the calculator, using the wrong mode of a calculator, or making silly algebra or sign mistakes, for example, but sometimes you can catch a mistake by checking that your answers are reasonable.

**Example.** A chimpanzee visits another planet. When she gets there, she drops a banana from rest from the top of a cliff. The banana descends 162 m in a time of 9.0 s. What is the value of gravitational acceleration on this planet?

The first step is to draw a diagram. Label the initial and final positions on the diagram, which in this case correspond to the top and bottom of the cliff. Also, setup a coordinate system. The illustration on the following page shows $+y$ upward.

Write down the known quantities and identify the desired unknown:

$$\Delta y = -162 \text{ m} \quad , \quad v_{y0} = 0 \quad , \quad \Delta t = 9.0 \text{ s} \quad , \quad a_{y0} = ?$$

Note that the net vertical displacement, $\Delta y$, is negative because the final position lies below the initial position (since $+y$ is upward). We can solve for the tangential acceleration with a single equation:[47]

$$\Delta y = v_{y0}\Delta t + \frac{a_{y0}\Delta t^2}{2} = \frac{a_{y0}\Delta t^2}{2}$$

$$a_{y0} = \frac{2\Delta y}{\Delta t^2} = \frac{2(-162 \text{ m})}{(9.0 \text{ s})^2} = -4.0 \text{ m/s}^2$$

The problem asks for gravitational acceleration, which equals the absolute value of the vertical component of the acceleration: $g = |a_{y0}| = 4.0 \text{ m/s}^2$.

**Example.** A sloth jumps straight upward and has a hangtime of 1.0 s. How high does the sloth jump?

As usual, we begin with a labeled diagram. The trick here is that we either need to put the initial position at the top or the final position at the top – if instead you work with the roundtrip, you won't be able to solve for the maximum height. It turns out to be advantageous to put the initial, rather than the final, position at the top: Looking at the three equations of 1D uniform acceleration that we're working with, we see that one equation has $v_{y0}$, but not $v_y$. So, given the choice, it's better to make $v_{y0} = 0$.

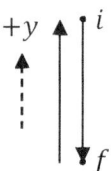

---

[47] Trust the equations and problem-solving strategies. You can solve problems and answer questions correctly in physics if you let problem-solving strategies, laws, and equations serve as your guide. As a teacher, it's a hair-pulling experience to see how many gifted students want to abandon the things you teach and instead make up crazy equations – namely, for a problem like this, students like to take the distance and divide by either the time or the time-squared, neither of which provides a correct answer (but, hopefully, you can see that the former yields the average speed; and while the latter has the right units, it's certainly not the tangential acceleration). If you're enrolled in a physics course, realize that you're more likely to earn partial credit (and actually solve the problem correctly) if you apply the equations and strategies that you've learned in the course, than if you just do something simple with the numbers without any reason for it.

However, if you put the final position at the top, you can still solve the problem with the usual strategy – it just makes the algebra a little less efficient. Next, we tabulate the knowns and the desired unknown:

$$v_{y0} = 0 \quad , \quad a_{y0} = -9.81 \text{ m/s}^2 \quad , \quad \Delta t = 0.50 \text{ s} \quad , \quad \Delta y = ?$$

Note that it would be a major conceptual mistake to set the time interval equal to 1.0 s: Looking at how we defined the initial and final positions, the corresponding time interval is only one-half of the hangtime. One equation allows us to solve for the net vertical displacement directly:

$$\Delta y = v_{y0} \Delta t + \frac{a_{y0} \Delta t^2}{2} = 0 + \frac{(-9.81 \text{ m/s}^2)(0.50 \text{ s})^2}{2} = -1.2 \text{ m}$$

The minus sign is important because the final position is below the initial position. However, the question asked how high the sloth jumped, not what its net vertical displacement was for the trip down (which is only negative because of how we choose the $+y$ direction): $H = |\Delta y| = 1.2$ m.

**Note.** For a symmetric free fall trajectory – i.e. where the beginning and ending points have the same height – the time that the object spends rising equals the time that it spends falling. This means that you can divide the entire trip time by two and apply it to just the trip up or just the trip down. However, it would be a major conceptual mistake to do this if the beginning and ending points do not have the same height.

**Example.** A monkey sitting in a tree throws a banana straight upward. The banana strikes the ground 7.0 s later with a final speed of 40 m/s. From what height was the banana thrown?
   We begin with a labeled diagram. Note that this time the initial position is higher than the final position, and we are working with the complete trip.

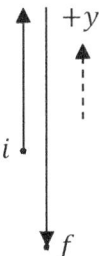

The next step is to list the known and desired unknown quantities:

$$v_y = -40 \text{ m/s} \quad , \quad a_{y0} = -9.81 \text{ m/s}^2 \quad , \quad \Delta t = 7.0 \text{ s} \quad , \quad \Delta y = ?$$

Note that the final tangential velocity is negative because the banana is moving downward in the final position, and the coordinate system designates upward as the positive direction. This time, we must solve two equations in two unknowns in order to obtain the answer:

$$v_y - v_{y0} = a_{y0}\Delta t \quad , \quad \Delta y = v_{y0}\Delta t + \frac{a_{y0}\Delta t^2}{2}$$

$$-40 - v_{y0} = (-9.81)(7.0) \quad , \quad \Delta y = 7v_{y0} + \frac{(-9.81)(7.0)^2}{2}$$

$$v_{y0} = 28.7 \quad , \quad \Delta y = 7v_{y0} - 240$$

$$\Delta y = -39 \text{ m}$$

Therefore, the banana was thrown from a height of $H = |\Delta y| = 39$ m (the top of a rather tall tree).[48]

**Important Distinction.** The net vertical displacement, $\Delta y$, is path-independent. So if a monkey throws a banana straight up into the air from a tree and the banana lands on the ground, you can skip the middleman to solve for, say, the time that the banana spends in the air. That is, it is completely unnecessary to split the problem into two pieces, and work separately with the trip up and the trip down.[49] It is much more efficient to work with the initial position just after the banana is thrown and the final position just before it lands on the ground. On the other hand, if you want to find the total distance traveled, $\Delta s$, then you must separately determine the upward and downward net displacements and add their absolute values.

## 1.8 Non-Uniform Acceleration in 1D

**Equations of motion from tangential displacement as a function of time**: Given an equation for the 1D tangential displacement, $s_T$, expressed in terms of time, $t$, you can derive equations for the time-dependence of the tangential velocity, $v_T$, and tangential acceleration, $a_T$, by taking successive derivatives with respect to time:

$$v_T = \frac{ds_T}{dt} \quad , \quad a_T = \frac{dv_T}{dt} = \frac{d^2 s_T}{dt^2} \quad \text{(in 1D)}$$

**Math Note.** If you want to evaluate the instantaneous tangential velocity or instantaneous tangential acceleration at a particular time, you must apply the derivative(s) before plugging in the numerical value for time.

---

[48] In this case, we were able to solve for one unknown first and then use it to solve for the other unknown. However, solutions to two equations and two unknowns will not always be this trivial. Usually, you will need to solve for one unknown algebraically in terms of the other, and then substitute this expression into the other equation. Then you can solve for one of the unknowns. Or you can apply a different, but equivalent, technique, like establishing equal and opposite coefficients, using Cramer's rule, or even Gaussian elimination.

[49] However, if you want to know the maximum height of the banana relative to the ground or the speed of the banana when it is halfway down, for example, then you must work with a portion of the trip.

**Equations of motion from tangential velocity as a function of time**: Given an equation for the 1D tangential velocity, $v_T$, expressed in terms of time, $t$, you can derive an equation for the time-dependence of the tangential acceleration, $a_T$, by taking a derivative with respect to time, and an equation for the time-dependence of the net 1D displacement, $\Delta s_T$, by integrating over time:

$$a_T = \frac{dv_T}{dt} \quad , \quad \Delta s_T = \int_{t=t_0}^{t} v_T dt \quad \text{(in 1D)}$$

**Math Note.** You must substitute an equation for the tangential velocity, $v_T$, in terms of time, $t$, before you can perform the integration to find the net 1D displacement, $\Delta s_T$. It would be a major conceptual mistake to pull $v_T$ out of the integral if $v_T$ is time-dependent.

**Equations of motion from tangential acceleration as a function of time**: Given an equation for the 1D tangential acceleration, $a_T$, expressed in terms of time, $t$, you can derive equations for the time-dependence of the tangential velocity, $v_T$, and tangential displacement, $s_T$, by integrating over time:

$$\Delta v_T = v_T - v_{T0} = \int_{t=t_0}^{t} a_T dt \quad , \quad \Delta s_T = \int_{t=t_0}^{t} v_T dt \quad \text{(in 1D)}$$

**Common Mistake.** The integral of tangential acceleration over time yields the change in tangential velocity, not the tangential velocity. It is a common mistake for students to forget to factor the initial tangential velocity into their math when integrating tangential acceleration.

**Math Note.** You must substitute an equation for the tangential acceleration, $a_T$, in terms of time, $t$, before you can perform the integration to find the change in tangential velocity, $\Delta v_T$. It would be a major conceptual mistake to pull $a_T$ out of the integral if $a_T$ is time-dependent. Furthermore, before you can integrate to find the net 1D displacement, $\Delta s_T$, you must algebraically solve for $v_T$ – by adding $v_{T0}$ to both sides – and substitute this expression for time in for $v_T$: The tangential velocity, $v_T$, must be expressed in terms of time, $t$, before you can integrate – you can't plug a numerical value for $v_T$ into the integrand.

**Potential Mistake.** If you are given $s_T$, $v_T$, or $a_T$ – or their linear equivalents, such as $x$, $v_x$, or $a_x$, for linear motion – as a function of time, you should instinctively know that you can relate the tangential displacement, tangential velocity, and tangential acceleration via calculus. If instead you attempt to use the equations of 1D uniform acceleration when the tangential acceleration is time-dependent, this would be a major conceptual mistake. In this case, you cannot apply the problem-solving strategy for 1D uniform acceleration, but must instead apply calculus, conceptual reasoning, and other problem-solving skills. This is summarized in the problem-solving strategy that follows.

## Problem-Solving Strategy for 1D Non-Uniform Acceleration

0. This strategy is useful for relating tangential displacement, tangential velocity, and tangential acceleration in 1D motion problems where the acceleration may be non-uniform.
1. If you are given tangential displacement, $s_T$, tangential velocity, $v_T$, or tangential acceleration, $a_T$, – or their linear equivalents, such as $x$, $v_x$, and $a_x$ – as a function of time, derive equations for the time-dependence of the other tangential quantities (i.e. of the set $s_T$, $v_T$, and $a_T$) by applying calculus. Remember, the definite integral of the tangential acceleration over time yields the change in tangential velocity – not the tangential velocity by itself. For other cases where the acceleration is non-uniform, but you are not given the time-dependence of one of these three quantities, see Step 4.
2. You should now have 3 time-dependent expressions – one for $s_T$, $v_T$, and $a_T$. Count how many unknowns you have. You generally need to set up $N$ equations in $N$ unknowns. If you have more unknowns than you have equations, you need to determine what other equations apply to the problem. Restrict yourself to the quantities $\Delta s_T$, $v_{T0}$, $v_T$, $a_{T0}$, $a_T$, and $\Delta t$, if possible.
3. Once you have $N$ equations in $N$ unknowns, solve the system algebraically. The unknown that you are ultimately after should be related to one or more of the quantities listed in Step 2.
4. If you are given $v_T$ as a function of $s_T$ – instead of as a function of $t$ – you can apply the chain rule to relate $v_T$ to other tangential quantities; then you can go onto Steps 2 and 3. Here is another possibility: Once you learn Newton's second law and conservation of energy in later chapters, you will see that the mathematical formulation of these laws are second- and first-order, respectively, differential equations. These differential equations can be solved (in principle) to derive time-dependent expressions for non-uniform tangential acceleration for general systems. Then you can proceed to Step 1.

**Example.** A monkey runs around a circle according to the equation $s_T(t) = 8t^{3/2} - 4t$, where SI units have been suppressed to reduce clutter.[50] The monkey begins running at $t = 0$. Determine the initial tangential velocity and find the tangential acceleration of the monkey at $t = 9.0$ s.

Equations for the tangential velocity and tangential acceleration can be derived by applying successive derivatives with respect to time to the given equation for tangential displacement:

$$v_T = \frac{ds_T}{dt} = \frac{d}{dt}(8t^{3/2} - 4t) = 12t^{1/2} - 4$$

$$a_T = \frac{dv_T}{dt} = \frac{d}{dt}(12t^{1/2} - 4) = 6t^{-1/2}$$

The initial tangential velocity, $v_{T0}$, is the tangential velocity evaluated at the initial time, which in this case is $t = 0$: $v_{T0} = v_T(0) = -4.0$ m/s. The tangential acceleration at the desired time can be found by plugging in the specified value of time: $a_T(t = 9) = 6(9)^{-1/2} = 2.0$ m/s$^2$.

---

[50] Also, assume that the given coefficients are good to at least as many significant figures as the specified time, 9.0 s. When it may help to avoid confusion, we will omit the units and trailing zeroes after decimals (i.e. the .0's) in equations that have numerical values and variables mixed together.

**Example**. A monkey runs along a straight line[51] according to the equation $v_x(t) = 20\sin(5t)$, where SI units have been suppressed to reduce clutter. What is the monkey's speed when $\Delta x = 2.0$ m, assuming that the motion starts at $t = 0$?

We can derive an expression for the net displacement, $\Delta x$, of the monkey as a function of time by integrating the given equation for tangential velocity over time:

$$\Delta x = \int_{t=t_0}^{t} v_x dt = \int_{t=0}^{t} 20\sin(5t)\,dt = -4[\cos(5t)]_{t=0}^{t} = 4 - 4\cos(5t)$$

Note the importance of the lower limit: If you plug in $t = 0$ to the equation for $\Delta x$, you will see that the initial net displacement is zero, which makes sense because the monkey hasn't moved yet. If you had forgotten to evaluate the antiderivative at the lower limit of the integral, you would have missed this significant term (namely, the 4). Now we can plug in the given net displacement and solve the system of two equations in two unknowns:[52]

$$2 = 4 - 4\cos(5t)$$
$$4\cos(5t) = 2$$
$$\cos(5t) = \tfrac{1}{2}$$
$$\sin(5t) = \pm\sqrt{3}/2$$
$$v_x(t) = 20\sin(5t) = 20(\pm\sqrt{3}/2) = \pm 10\sqrt{3} \text{ m/s}$$

The monkey's speed is thus $10\sqrt{3}$ m/s.

---

**Example**. A monkey climbs a vertical rope according to the equation $a_y(t) = 4t^3 - 9t^2$ with an initial tangential velocity of 3.0 m/s, where SI units have been suppressed to reduce clutter. Determine the tangential velocity of the monkey at $t = 2.0$ s, assuming that the motion begins at $t = 0$.

The definite integral of the tangential acceleration over time equals the change in the monkey's tangential velocity:

$$v_y - v_{y0} = \int_{t=t_0}^{t} a_y dt = \int_{t=0}^{t}(4t^3 - 9t^2)dt = t^4 - 3t^3$$
$$v_y = v_{y0} + t^4 - 3t^3$$
$$v_y(t=2) = 3 + 2^4 - 3(2)^3 = 3 + 16 - 24 = -5.0 \text{ m/s}$$

---

[51] The monkey's path corresponds to a straight line in that he moves only forward and backward along the $x$-axis. If you are visualizing the sine function and wondering why it's not 2D motion, it's because the sine function is expressing the 1D relationship between position (rather, its derivative – tangential velocity) and time. That is, the sine function involves $x$ and $t$, not $x$ and $y$.

[52] Observe that we don't need to apply the inverse cosine to both sides to solve for time. Looking ahead, we see that we only need to know $\sin(5t)$, and not the time itself. We solved this problem efficiently by applying the trigonometric relationship $\sin^2\theta + \cos^2\theta = 1$.

# 1 Single-Component Motion

**Chain rule**: You may recall from calculus that the chain rule can be applied to take a derivative of a function of one variable with respect to a different variable. For example, given $f(u)$, the chain rule may be applied to take a derivative of $f$ with respect to $x$:

$$\frac{df}{dx} = \frac{df}{du}\frac{du}{dx}$$

**Example.** Express the derivative $df/dx$, where $f(u) = 3\sin u$ and $u = 4x^3$, in terms of $x$.
Apply the chain rule to find this derivative:

$$\frac{df}{dx} = \frac{df}{du}\frac{du}{dx} = \left[\frac{d}{du}3\sin u\right]\left[\frac{d}{dx}4x^3\right] = (3\cos u)12x^2 = 36x^2\cos(4x^3)$$

**Applying the chain rule to 1D non-uniform acceleration**: Suppose that you are given an equation for the tangential velocity, $v_T$, but instead of being expressed as a function of time, $t$, it is expressed as a function of the tangential displacement, $s_T$. If you want to find the tangential acceleration, $a_T$, for example, you really need to take a derivative of $v_T$ with respect to $t$, but you don't know $v_T$ as a function of time. In this case, you must apply the chain rule in order to relate $dv_T/dt$ to $dv_T/ds_T$, as in the following example.

**Example.** A monkey runs along a straight line according to the equation $v_x(x) = 4x^{3/2}$, where SI units have been suppressed to reduce clutter. What is the monkey's acceleration when $x = \frac{1}{2}$ m?
Since we need to take a derivative of $v_x$ with respect to $t$, but know $v_x$ as a function of $x$, we must apply the chain rule:

$$a_x = \frac{dv_x}{dt} = \frac{dv_x}{dx}\frac{dx}{dt} = v_x\frac{dv_x}{dx} = 4x^{3/2}\frac{d}{dx}\left(4x^{3/2}\right) = 24x^2$$
$$a_x(x = \tfrac{1}{2}) = 24(\tfrac{1}{2})^2 = 6.0 \text{ m/s}^2$$

## 1.9 Multiple Moving Objects in 1D

**Distinguishing between multiple objects**: If there are two or more objects moving in the same problem, include subscripts on all variables that may be different. For example, if two objects travel with different initial tangential velocities, use subscripts to distinguish between the initial tangential velocities of objects 1 and 2: $v_{1,T0}$ and $v_{2,T0}$. If $v_{1,T0}$ and $v_{2,T0}$ are different, but instead you simply wrote $v_{T0}$ for both objects, once both $v_{T0}$'s appeared in the same equation, you might cancel quantities that are not actually equal. The subscripts serve as a reminder of which quantities may be different.

**Common variables and constants**: You don't need to include subscripts for each object for quantities that are exactly the same for both objects. For example, if two objects travel for the same time interval, you may just write $\Delta t$. In this case, it is not necessary to distinguish between $\Delta t_1$ and $\Delta t_2$.

> **Note**. It's important to correctly reason out which quantities are definitely the same and which quantities may be different for multiple objects moving in the same problem. For example, if you know that two objects start and finish in the same position, then you know that both have the same net displacement, $\Delta s_T$, but if they start and finish in different positions, then you must distinguish between $\Delta s_{1T}$ and $\Delta s_{2T}$. Similarly, if one object has a headstart, you will need to distinguish between $\Delta t_1$ and $\Delta t_2$, but if they both start and finish at the same time, omit the subscript from the time interval and work with $\Delta t$.

**Separate equations of motion**: When solving a problem where two or more objects are moving, write down (at least) one equation of motion for each object using subscripts appropriately. The needed equation of motion, in 1D, is usually the net 1D displacement, $\Delta s_T$, as a function of the time interval, $\Delta t$, because multiple-moving-object problems tend to involve relating the tangential displacements and elapsed times for the various objects. That is, very often one object catches another, in which case you know that they meet at the same place at the same time (different than the same net displacement if they start in different positions, and different from the same time interval if one has a headstart).

**Constraints**: The equations of constraint relate information from one object's equations of motion to the equations of motion for other objects. For example, if object 1 has a 3-s headstart over object 2, one equation of constraint is $\Delta t_1 = \Delta t_2 + 3$. Similarly, if the total distance traveled is 150 m and if object 1 travels in the positive $x$-direction while object 2 travels oppositely, then you know that $\Delta x_1 - \Delta x_2 = 150$ m. The equations of constraint connect the otherwise separate equations of motion.

---

### Problem-Solving Strategy for Multiple Moving Objects in 1D

0. Apply this strategy to solve problems with multiple objects moving simultaneously in 1D, where you want to relate tangential displacement and time for various objects.[53]
1. Draw and label a diagram, including initial and final positions and a coordinate system. Identify the known quantities. Use subscripts to distinguish between symbols that may be different for each object.
2. Write (at least) one equation of motion for each object, using subscripts. Usually, the equation of motion that you want is $\Delta s_T$ as a function of $\Delta t$. You should know the form of the equation of motion if the object travels with either constant tangential velocity or uniform tangential acceleration. If the tangential acceleration is non-uniform, apply the strategy from the previous section.
3. Write the equation(s) of constraint that relate information from the objects' equations of motion. Usually, equations of constraint relate $\Delta s_{1T}$ to $\Delta s_{2T}$ or $\Delta t_1$ to $\Delta t_2$.
4. If you are looking for an extreme value of an unknown, one of the equations of constraint will come from applying calculus to solve for the critical points.
5. Identify the knowns and count the unknowns. Once you have $N$ equations in $N$ unknowns, apply algebra to solve for the unknowns. If there is a calculus constraint, you will need to express the function to be minimized or maximized in terms of one unknown only before you apply the derivative.

---

[53] In later chapters, we will consider problems where multiple objects are moving simultaneously, but where we are interested in relating quantities – like force and acceleration – other than tangential displacement and time.

**Example**. Two monkeys in bumper cars are heading toward one another. When the monkeys are separated by a distance of 50 m, one monkey is instantaneously moving 0.60 m/s and uniformly accelerating at a rate[54] of 0.20 m/s² and the other monkey is instantaneously moving 0.40 m/s and uniformly accelerating at a rate of 0.60 m/s². Where do the monkeys meet?

Since the equations of 1D uniform acceleration involve tangential quantities, which may be positive or negative (or zero), you need to be mindful that some quantities may be negative and reason through the signs carefully. It is necessary to draw a diagram and setup a coordinate system in order to determine the signs. There are two initial positions and one final position in the illustration below.

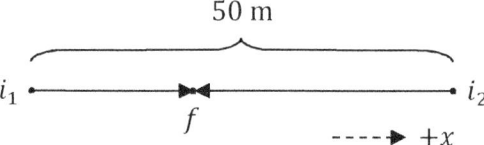

Identify the knowns and the unknown that we're looking for, using subscripts for quantities that might not be the same for each monkey:

$$s = 50 \text{ m} \quad , \quad v_{1,x0} = 0.60 \text{ m/s} \quad , \quad v_{2,x0} = -0.40 \text{ m/s}$$
$$a_{1,x0} = 0.20 \text{ m/s}^2 \quad , \quad a_{2,x0} = -0.60 \text{ m/s}^2 \quad , \quad \Delta x_1 = ?$$

Note that the total distance traveled, 50 m, does not have a subscript because it does not relate just to one monkey, but the two monkeys collectively. Also, observe that some of these quantities are negative; all of the signs are dictated by the choice of the $+x$-direction. Next, express the net 1D displacement of each monkey in terms of time. Apply the equations of 1D uniform acceleration since each monkey moves with constant 1D acceleration, using subscripts:

$$\Delta x_1 = v_{1,x0} \Delta t + \frac{a_{1,x0} \Delta t^2}{2} = 0.6 \Delta t + 0.1 \Delta t^2$$
$$\Delta x_2 = v_{2,x0} \Delta t + \frac{a_{2,x0} \Delta t^2}{2} = -0.4 \Delta t - 0.3 \Delta t^2$$

We have 3 unknowns in these 2 equations, but can write a third equation by applying the constraint:

$$\Delta x_1 - \Delta x_2 = s = 50 \text{ m}$$

---

[54] There are many different kinds of rates in physics. So if you just say rate, the word is ambiguous. It's not ambiguous here, of course, since the rate is clearly specified to be an acceleration. The point is that rate in general can refer to speed, acceleration, or a host of other quantities (like $dm/dt$, the rate at which mass is ejected from a rocket). However, if you say something like a rate of 0.20 m/s², where you declare the rate to be a particular value with units, the units limit what kind of rate this could be. In this example, the rate 0.20 m/s² has the units of acceleration, so it definitely couldn't be a speed.

Note the relative sign that reflects that monkey 2 moves in the negative $x$-direction. This system of equations can be solved by direct substitution:

$$0.6\Delta t + 0.1\Delta t^2 + 0.4\Delta t + 0.3\Delta t^2 = 50$$
$$0.4\Delta t^2 + \Delta t - 50 = 0$$
$$\Delta t = \frac{-1 \pm \sqrt{1^2 - 4(0.4)(-50)}}{2(0.4)} = \frac{-1 \pm 9}{0.8} = 10 \text{ s}$$

We can determine where they meet by substituting the time into the net displacement formulas:

$$\Delta x_1 = 0.6(10) + 0.1(10)^2 = 16 \text{ m}$$
$$\Delta x_2 = -0.4(10) - 0.3(10)^2 = -34 \text{ m}$$

They meet 16 m from the initial position of the monkey with the acceleration of 0.20 m/s². We can check for self-consistency by noting that the absolute values of the net 1D displacements total 50 m.

**Example.** A chimpanzee tags an orangutan and declares, "You're it!" The chimpanzee then runs at a constant speed of 4.0 m/s in a straight line. The orangutan gives chase 3.0 s later, uniformly accelerating from rest at a rate of 0.50 m/s². When does the orangutan catch the chimpanzee?

As usual, begin with a labeled diagram, remembering to identify the positive direction.

$$i \longrightarrow f$$
$$+x$$

In this problem, the chimpanzee and orangutan both start and finish in the same place, but one has a headstart. We next identify the knowns and desired unknown, using a subscript 1 for the chimpanzee and 2 for the orangutan:

$$v_{1,x0} = 4.0 \text{ m/s} \quad , \quad v_{2,x0} = 0 \quad , \quad t_0 = 3.0 \text{ s} \quad , \quad a_{1,x0} = 0 \quad , \quad a_{2,x0} = 0.50 \text{ m/s}^2 \quad , \quad \Delta t_1 = ?$$

Of course, $t_0 = 3.0$ s refers only to the headstart, and not to the elapsed time for either primate. The chimpanzee travels with uniform velocity, while the orangutan travels with 1D uniform acceleration:

$$\Delta x = v_{1,x0}\Delta t_1 = 4\Delta t_1$$
$$\Delta x = v_{2,x0}\Delta t_2 + \frac{a_{2,x0}\Delta t_2^2}{2} = 0.25\Delta t_2^2$$

The net 1D displacement is the same for each, and the different elapsed times are constrained by the headstart time:

$$\Delta t_1 = \Delta t_2 + t_0 = \Delta t_2 + 3$$

Since the chimpanzee – primate 1 – begins running 3 s sooner than the orangutan – primate 2 – the chimpanzee's trip time must be greater than the orangutan's trip time. We now have 3 equations in 3 unknowns, which are readily solved through substitution:

$$\Delta x = 4\Delta t_1 = 0.25\Delta t_2^2$$
$$4(\Delta t_2 + 3) = 0.25\Delta t_2^2$$
$$0.25\Delta t_2^2 - 4\Delta t_2 - 12 = 0$$
$$\Delta t_2 = \frac{-(-4) \pm \sqrt{(-4)^2 - 4(0.25)(-12)}}{2(0.25)} = \frac{4 \pm 2\sqrt{7}}{0.5} = 19 \text{ s}$$

The orangutan catches the chimpanzee approximately 19 s after he begins to chase him, or about 22 s after being tagged 'It.' It is a good idea to substitute the answer into both net 1D displacement equations to check that both net 1D displacements are the same.

**Extreme values of unknowns**: Suppose that a monkey climbs a rope vertically upwards with a constant speed and that you wish to throw a banana straight upward to the monkey.[55] If you know how fast the monkey is climbing and where the monkey is when you throw the banana, you can determine the minimum speed that the banana needs to reach the monkey. If you throw the banana faster than the minimum speed, it will still reach the monkey. The minimum speed is the critical speed for which the banana will reach the monkey.

If you wish to find the extreme value of a variable, you can do so by applying calculus. Recall that the critical values of a function $f$ of the variable $x$ can be determined by setting a derivative of $f$ with respect to $x$ equal to zero: $\frac{df}{dx} = 0$.[56] The solutions to the equation $\frac{df}{dx} = 0$ yield the values of $x$, which we shall denote by $x_c$, for which the function $f(x)$ has a relative minimum, relative maximum, or a point of inflection.

Evaluate the second derivative, $\frac{d^2f}{dx^2}$, at each value of $x = x_c$ that corresponds to a critical point in order to determine whether $f(x_c)$ is a relative minimum, relative maximum, or a point of inflection:[57]

- If $\frac{df}{dx}\big|_{x=x_c} = 0$ and $\frac{d^2f}{dx^2}\big|_{x=x_c} < 0$, $f(x_c)$ is a relative maximum.
- If $\frac{df}{dx}\big|_{x=x_c} = 0$ and $\frac{d^2f}{dx^2}\big|_{x=x_c} > 0$, $f(x_c)$ is a relative minimum.
- If $\frac{df}{dx}\big|_{x=x_c} = 0$ and $\frac{d^2f}{dx^2}\big|_{x=x_c} = 0$, $f(x_c)$ is a point of inflection.

If you are looking for the absolute minimum or absolute maximum, you must also remember to check the endpoints of the interval (even if the endpoints correspond to $x = \pm\infty$) because $f$ could have its most extreme value (which may not even be finite) at the endpoints.

---

[55] Notice that it says 'to' the monkey, not 'at' the monkey. We're trying to feed the monkey, not hurt the poor guy.
[56] Conceptually, from calculus, we know that if $f(x)$ is a smooth and continuous (i.e. differentiable) function of $x$, it will have extrema where its slope, $df/dx$, is zero (i.e. where the curve is instantaneously horizontal).
[57] Conceptually, from calculus, whether the region around the critical point is concave upward or concave downward – determined by the sign of $d^2f/dx^2$ – tells us what type of extremum we have at the critical point.

Returning to the climbing monkey, strategically you could solve for the banana's initial vertical velocity, $v_{y0}$, as a function of time, $t$, and set a derivative of $v_{y0}$ with respect to $t$ equal to zero in order to determine which value of $t$ yields the minimum value of $v_{y0}$ needed for the banana to reach the monkey.[58] In this case, the equation $\frac{dv_{y0}}{dt} = 0$ will be needed in addition to the other equations – i.e. the equation of motion of each object and any equation(s) of constraint – in order to make $N$ equations in $N$ unknowns. Essentially, $\frac{dv_{y0}}{dt} = 0$ is a calculus-based constraint to supplement the algebraic constraints.

**Minimum headstart**: If object 1 begins moving a time $t_0$ sooner than object 2 begins to move, one equation of constraint will be $t_1 = t_2 + t_0$. If you are trying to solve for the minimum time $t_0$ that object 2 can wait before moving and still catch object 1, you will need to take a derivative of $t_0$ with respect to some other variable and set it equal to zero in order to solve for its minimum value.[59] Similarly, If object 1 has a headstart in distance, $x_0$, compared to object 2, and object 2 will catch object 1, one equation of constraint will be $\Delta x_2 = \Delta x_1 + x_0$. In this case, object 2 must travel further than object 1 in order to make up for the headstart.

---

**Example.** A monkey is driving with an initial velocity of 36 m/s to the west. The car ahead of him is traveling with a constant velocity of 20 m/s to the west. The monkey uniformly decelerates[60] at a constant rate of 0.50 m/s² in order to avoid bumping into the car in front. What minimum separation must they have when the monkey begins to decelerate in order to avoid a collision?

We establish our choice of coordinates in the diagram below.

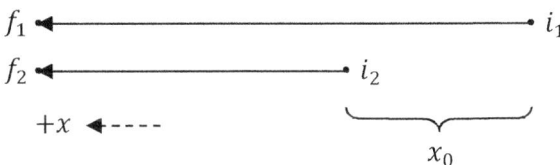

Note that the monkey, object 1, and car ahead, object 2, are both traveling in the $+x$-direction in this diagram. The knowns and desired unknown include:

---

[58] Those students who have an excellent grasp of the concepts can find an alternative to applying the calculus for many of the extreme-value motion problems. The non-calculus alternative, which is conceptually trickier, will be left as an exercise to the reader (see the Practice Problems).

[59] Observe that $t_0$ and $v_{y0}$ are being treated as variables (not constants) here.

[60] If you're wondering why the deceleration isn't negative in the statement of the problem, it's because you can't determine the sign of the acceleration until you setup a coordinate system: If you establish that $+x$ is to the east, the monkey's acceleration would actually be positive, since losing speed in the negative direction equates to positive acceleration. It's not the most intuitive coordinate system to use – and it's easy to make other sign mistakes if you decide to be a rebel and adopt such a coordinate system; plus, your instructor may dock points if your unconventional method and work is not very clear, or if there is any inconsistency – but the physics is independent of which way we choose to orient the $+x$-axis. That is, the relative signs will be the same regardless of this choice, but the individual signs very much depend upon which way is $+x$.

# 1 Single-Component Motion

$$v_{1,x0} = 36 \text{ m/s} \quad , \quad v_{2,x0} = 20 \text{ m/s} \quad , \quad a_{1,x0} = -0.50 \text{ m/s}^2 \quad , \quad a_{2,x0} = 0 \quad , \quad x_{0,min} = ?$$

The monkey's acceleration is negative because he is losing speed while heading in the $+x$-direction. The monkey decelerates uniformly in 1D and the car ahead travels with uniform velocity:[61]

$$\Delta x_1 = v_{1,x0} t + \frac{a_{1,x0} t^2}{2} = 36t - 0.25t^2$$
$$\Delta x_2 = v_{2,x0} t = 20t$$

The algebraic equation of constraint is

$$\Delta x_1 = \Delta x_2 + x_0$$

and the extremum condition is[62]

$$\frac{dx_0}{dt} = 0$$

It is convenient to first combine equations to express $x_0$ as a function of $t$, and then apply the derivative:[63]

$$\Delta x_1 = \Delta x_2 + x_0$$
$$36t - 0.25t^2 = 20t + x_0$$
$$x_0 = 16t - 0.25t^2$$
$$\left.\frac{dx_0}{dt}\right|_{t=t_c} = 16 - 0.5t_c = 0 \quad \Rightarrow \quad t_c = 32 \text{ s}$$
$$x_{0,min} = x_0(t = 32) = 16(32) - 0.25(32)^2 = 256 \text{ m}$$

It is a good idea to check that both net 1D displacements agree with the equation of constraint.[64]

---

[61] The elapsed time, $\Delta t$, is a time interval: $\Delta t = t - t_0$. If you choose to start your stopwatch at $t_0 = 0$, which is the choice of origin for your time coordinate, then $\Delta t = t$. This is particularly aesthetic in this example, where we will be taking a derivative with respect to time: $df/d(\Delta t)$ is a rather silly – and unnecessary – notation.

[62] Mathematically, we also wish to ensure that the second derivative, $d^2 x_0/dt^2$, is positive when evaluated at the critical point such that $x_0$ is indeed a relative minimum, and not a relative maximum or point of inflection. Furthermore, we desire to find the absolute minimum value of $x_0$, which means that we also need to check the endpoints. However, you may be able to conceptually reason that the solution corresponds to the absolute minimum, in which case this extra math would be superfluous. Your solutions to such problems should either show the math described in this footnote, or should include an equivalent conceptual argument.

[63] The symbol $\Rightarrow$ reads, "...implies that..."

[64] You can also check your answer for consistency if you solve it both by applying the calculus-based minimization technique and the conceptual, non-calculus approach described in a prior footnote. See the Practice Problems (and Hints to the Chapter 1 Practice Problems toward the back of the book).

## Conceptual Questions

You will receive the most benefit if you first try the problems yourself, then consult the hints, and finally check your answer after working it out again. The hints and answers to all of the conceptual questions can be found, separately, toward the back of the book.

1. For each activity listed, indicate whether the motion is effectively 1D, 2D, or 3D: A racecar completes thirty laps around a track, an ant crawls along the outside of a cylindrical trash can, a bowling ball rolls along an alley, a hiker walks along a narrow trail, a bumblebee flies through the air, a passenger rides in an elevator, a mountain climber climbs a steep cliff, a bead slides long a curved wire, a whale swims underwater, a guard paces back and forth.

2. A monkey releases a ball from rest from a height of one meter above the ground. After each bounce, the ball returns to half of its previous maximum height. When the ball strikes the ground for the fourth time, what are its total distance traveled and net displacement?

3. A monkey paces back and forth along a length of 20 m with a constant speed of 2.0 m/s. After walking west three times and east two times, what are the monkey's average speed and average tangential velocity?

4. A monkey drives a car to the north as follows: Beginning from rest, he accelerates from zero to 20 m/s; he then travels with a constant speed of 20 m/s; and lastly he decelerates from 20 m/s to rest. The entire trip takes one minute. What is the monkey's average tangential acceleration for the trip?

5. Describe how you can figure out when an object changes direction by examining (A) a plot of tangential displacement as a function of time, (B) a plot of tangential velocity as a function of time, and (C) a plot of tangential acceleration as a function of time.

6. Given the plot of tangential displacement as a function of time below, make sketches of the tangential velocity and tangential acceleration as functions of time.

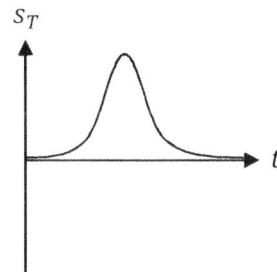

7. Given the plot of tangential velocity as a function of time below, make sketches of the tangential displacement and tangential acceleration as functions of time (assuming $s_{T0} = 0$).

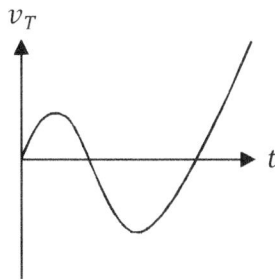

8. Given the plot of tangential acceleration as a function of time below, make sketches of the tangential displacement and tangential velocity as functions of time (assuming $s_{T0} < 0$ and $v_{T0} = 0$).

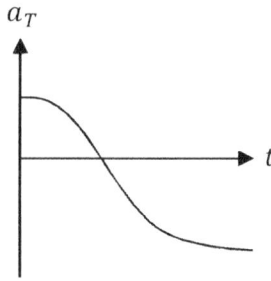

9. Given a plot of $v_T$ as a function of $s_T$, describe how you could make a graph of $a_T$ as a function of $s_T$. (Observe that you are given no direct information about time.)

10. A train with an initial velocity of 56 m/s to the north has a uniform deceleration of $-8.0$ m/s$^2$. Without using any equations: (A) What is the train's velocity after 4.0 s? (B) When will the train run out of speed?

11. A monkey throws a banana straight upward with an initial speed of 35 m/s. Without using any equations: (A) Approximately when does the banana run out of speed? (B) What is the approximate velocity of the banana 5.0 s after release? (C) Approximately when does the banana return to the monkey? (D) What is the approximate speed of the banana when the monkey catches it?

12. Make a sketches of 1D uniform acceleration where $s_{T0} > 0$, $v_{T0} < 0$, and $a_{T0} < 0$. Make all three plots – i.e. plot $s_T$, $v_T$, and $a_T$ as functions of time.

## Practice Problems

You will receive the most benefit if you first try the problems yourself, then consult the hints, and finally check your answer after working it out again. The hints to all of the practice problems and the answers to selected practice problems can be found, separately, toward the back of the book.

**Hint.** Look at units to help determine which physical quantity a numerical value corresponds to. For example, the number 3.5 m/s$^2$ must be an acceleration, whereas 3.5 m/s must be some type of speed or velocity (like initial, final, or average tangential speed or velocity).

**Note.** Many, but not all, of the problems in this textbook can be solved by applying one of the main problem-solving strategies (described step-by-step in the text). It is worth studying those strategies.

### Total, net, and average values

1. A monkey bicycles 60 km in 3.0 hr and then bicycles 60 km in 5.0 hr. (A) Determine the average speed of the monkey for the entire trip. (B) Conceptually, explain why the answer does not equal the arithmetic mean of the two speeds (16.0 km/hr), and justify whether it should be larger or smaller than this value.

2. A monkey bicycles 40 km in 5.0 hr, 20 km in 4.0 hr, and then 30 km in 3.0 hr. (A) Determine the average speed of the monkey for the entire trip. (B) Conceptually, explain why the answer does not equal the arithmetic mean of the two speeds (7.7 km/hr), and justify whether it should be larger or smaller than this value.

3. A monkey travels 350 m north with a constant speed of 2.0 m/s, 150 m south with a constant speed of 4.0 m/s, and 250 m north with a constant speed of 3.0 m/s. (A) What is the monkey's net displacement? (B) What is the monkey's total distance traveled? (C) What is the monkey's average speed? (D) What is the monkey's average velocity?

4. A monkey runs 120 m east at *some* constant speed, then runs 180 m west at a constant speed of 6.0 m/s. The whole trip takes one minute. (A) How fast did the monkey run to the east? (B) What is the net displacement of the monkey? (C) What is the average speed of the monkey? (D) What is the average velocity of the monkey?

5. A monkey walks from his dorm to class with a constant speed of 3.0 m/s. The monkey realizes that he forgot to bring his backpack, and so runs back to his dorm with a constant speed of 6.0 m/s. The monkey then returns to class on skateboard with a constant speed of 9.0 m/s. Each time, the monkey follows the same route. What is the average speed of the monkey for the entire trip?

6. A monkey is driving a bananamobile north with a speed of 10 km/hr, accelerates up to a top speed of 50 km/hr, slows down to 20 km/hr, and speeds up to 40 km/hr. The whole trip takes 5.0 min. What is the average acceleration of the bananamobile for the entire trip?

## Motion graphs

7. The position of a chimpanzee is plotted below. (A) What is the chimpanzee's net displacement? (B) What is the chimpanzee's average speed? (C) What is the velocity of the chimpanzee at $t = 12$ s? (D) What is the initial speed of the chimpanzee? (E) Where is the chimpanzee at $t = 8.0$ s? (F) For what time interval(s) is the chimpanzee heading in the negative $s_T$-direction? (G) At what time(s) is the chimpanzee momentarily at rest? (H) What is the average acceleration of the chimpanzee? (I) What is the maximum speed of the chimpanzee? (J) Sketch the chimpanzee's velocity as a function of time.

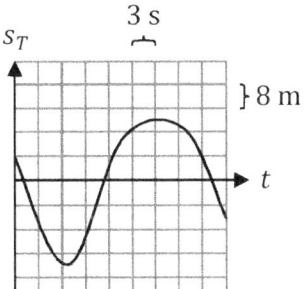

8. The velocity of a gorilla is plotted on the following page. (A) What is the total distance traveled by the gorilla? (B) What is the gorilla's average velocity? (C) What is the velocity of the gorilla at $t = 12$ s? (D) What is the initial acceleration of the gorilla? (E) Where is the gorilla at $t = 8.0$ s? (F) What is the gorilla's acceleration at $t = 8.0$ s? (G) For what time interval(s) is the gorilla heading in the negative $s_T$-direction? (H) At what time(s) is the gorilla momentarily at rest? (I) What is the average acceleration of the gorilla? (J) What is the maximum speed of the gorilla? (K) Sketch the gorilla's acceleration as a function of time. (L) Sketch the gorilla's position as a function of time.

# 1 Single-Component Motion

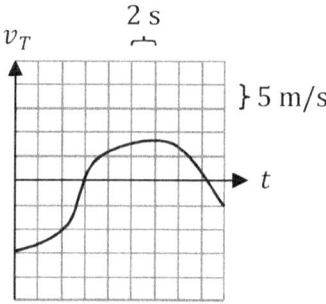

9. The acceleration of an orangutan is plotted below. The initial velocity of the orangutan is 15 m/s. (A) What is the velocity of the orangutan at $t = 12$ s? (B) What is the initial acceleration of the orangutan? (C) What is the speed of the orangutan at $t = 8.0$ s? (D) What is the orangutan's acceleration at $t = 8.0$ s? (E) For what time interval(s) is the orangutan heading in the negative $s_T$-direction? (F) At what time(s) is the orangutan momentarily at rest? (G) What is the average acceleration of the orangutan? (H) What is the maximum speed of the orangutan? (I) Sketch the orangutan's velocity as a function of time.

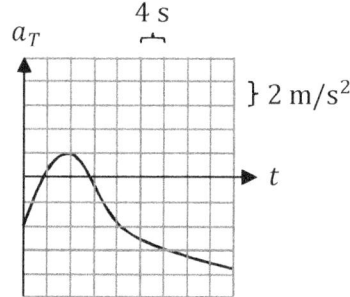

**One-dimensional uniform acceleration**

10. A gorilla uniformly accelerates a car from rest to 80 m/s in 5.0 s. (A) What is the acceleration of the car? (B) How far does the car travel during this time?

11. A chimpanzee riding a bicycle 15 m/s uniformly accelerates at a rate of 3.5 m/s$^2$ as the bicycle travels 200 m. (A) How long does this take? (B) What is the final speed of the bicycle?

12. A monkey driving a bananamobile 40 m/s spots a baby monkey crossing the street 100 m ahead. The monkey uniformly decelerates, stopping just in time. For how much time does the monkey decelerate?

13. A princess runs upstairs to rescue a knight from a high tower. As she tires, she uniformly decelerates at a rate of 0.20 m/s$^2$. Five seconds later, her speed is 3.60 m/s. During these five seconds, how many 20-cm tall stairs did she climb?

14. A monkey slides down a 40-m long incline from rest, reaching the bottom in four seconds. What is his acceleration?

15. A monkey driving a bananamobile spots a banana split 200 m ahead on the highway. The monkey uniformly decelerates for 5.0 s, stopping just in time to avoid squashing the banana split. Find (A) the initial velocity and (B) acceleration of the bananamobile.

16. A monkey makes a wish (for more bananas, of course!) as she drops a banana into a wishing well. She hears the splash of the banana two and one-half seconds later. (Neglect the time it takes for the sound to travel to her ears.) (A) How deep is the well? (B) What was the final speed of the banana?

17. A monkey leaps upward from a trampoline with an initial speed of 30 m/s. How high does the monkey leap?

18. On the moon, a monkey feels like Super Monkey. If the monkey jumps straight upward on the moon with the same initial speed with which he could jump straight upward on earth, (A) how much higher (i.e. by what factor) can the monkey jump and (B) how much longer will the monkey be in the air?

19. A monkey ties a firecracker to his physics textbook and launches the textbook straight upward with an initial speed of 40 m/s.[65] The firecracker explodes six seconds after launch. (A) How high is the textbook above the ground when the firecracker explodes? (B) What is the velocity of the textbook just before the firecracker explodes? (C) What is the maximum height of the textbook above the ground?

20. You are kidnapped by little green men and transported to Planet Fyzx. Unfortunately, they give you physics lab equipment to make measurements and complete a quiz. (A) You drop your watch from the top of a 100-m high cliff. (Time flies!) Your watch strikes the ground 5.0 seconds later. What is the acceleration of gravity near the surface of Planet Fyzx? (Use this value in any subsequent calculations.) (B) Then you throw a pig downward with an initial speed of 20 m/s. (Pigs fly!) What is the velocity of the pig just before it strikes the ground? (C) Now you throw a cow upward, and it returns to you 6.0 seconds later. (Cows fly!) Find the (i) height, (ii) velocity, and (iii) acceleration of the cow at the top of its trajectory.

21. A skunk is thrown straight upward from a bridge. The skunk is moving at a rate of 50 m/s just before it splashes into the lake below 8.0 seconds after being thrown. (A) How high is the bridge above the water? (B) What is the initial speed of the skunk? (C) What is the average velocity of the skunk for the whole stinking trip? (D) What is the average speed of the skunk for the whole stinking trip?

22. You lean out of your second-story dorm window (10 meters above the ground) and throw a ball upward with an initial speed of 30 m/s. Unfortunately, while the ball is returning downward somebody in a room above leans out the window (35 meters above the ground) and intercepts it. (A) How much time elapsed after the ball was thrown before it was intercepted? (B) What was the maximum height of the ball above the ground?

23. To illustrate that the pen is mightier than the sword, a brave knight throws his sword high up into the air. He patiently waits for its return, holding a pen up. After the sword is 10 m above the ground on the way up, four seconds elapse before the sword is moving 15 m/s on the way down. (A) Where is the sword when it is moving 15 m/s on the way down? (B) What are (i) the average velocity and (ii) the average speed of the sword for this part of the trip (from 10 m above the ground on the way up until it is moving 15 m/s on the way down)?

24. A chimpanzee throws a banana vertically upward at a speed of 30 m/s. How fast is the banana moving when it is halfway up?

25. A monkey on the ground throws a 3.0-kg monkey wrench vertically upward. The wrench is moving 15 m/s when it reaches 50% of its maximum altitude. (A) How long the wrench is in the air (from when it is thrown to when it returns to the monkey)? (B) How high does the wrench rise? (C) How fast was the wrench thrown?

---

[65] Professional stunt monkey. Do not attempt.

## 1 Single-Component Motion

### Multiple moving objects and non-uniform acceleration

26. Two monkeys jog around a circular track with a circumference of one kilometer. One monkey jogs clockwise at a rate of 5.0 m/s, while his sister jogs counterclockwise at a rate of 8.0 m/s. They begin from the same position. (A) How much time passes before they first meet? (B) How many laps has the sister completed when her brother completes his $15^{\text{th}}$ lap? (C) After one minute, how much further has the sister traveled compared to her brother?

27. The court jester is jousting with a pencil as his lance. He stands 200 m from his opponent. From rest, he uniformly accelerates at a rate of 2.5 m/s$^2$ as he approaches his fierce opponent, who gallops at a constant rate of 10 m/s. (A) Where will they meet? (B) When will they meet? (C) What is the speed of the court jester when they meet?

28. A monkey steals his uncle's banana and runs at a constant rate of 10 m/s. Two seconds later his uncle chases the monkey, accelerating at a constant rate of 5.0 m/s$^2$. (A) When does the uncle catch his nephew? (B) How far do the monkeys run? (C) What is the final speed of the uncle?

29. A monkey at the top of a 100-m cliff drops a banana at the same time that a monkey on the ground below throws a banana vertically upward with a speed of 40 m/s. (A) When will the bananas collide? (B) At what height will the bananas collide?

30. A chimpanzee throws an acid-filled flask vertically upward at a rate of 40 m/s. Three seconds later the chimpanzee throws a base-filled test tube vertically upward at a rate of 50 m/s. Find the height and time of the collision between the acid-filled flask and the base-filled test tube.

31. A monkey steals his uncle's prized golden banana and uniformly accelerates from rest with an acceleration of 0.25 m/s$^2$. After a short delay, the monkey's uncle pursues the monkey with a constant speed of 3.0 m/s. Apply calculus to determine the maximum amount of time that the monkey's uncle can wait and still catch the monkey.

32. Solve the previous problem without using calculus. There is a conceptual constraint that can be reasoned, which is equivalent to the calculus constraint.

33. A monkey steals the king's crown and makes a mad dash for it. A royal guard chases the monkey. The monkey has a 20-m head-start as he runs 2.0 m/s when he sees the royal guard pursuing at a constant rate of 5.0 m/s. The monkey then uniformly accelerates. Apply calculus to determine the minimum acceleration of the monkey that allows the monkey to escape with the king's crown.

34. Frustrated with physics, you throw your notebook into a trash can at the top of a hill. It tips over and rolls from rest with a uniform acceleration of 0.50 m/s$^2$. Then you realize that your homework was in your notebook, which you still need to turn in. So you chase the stinking homework assignment at a constant rate of 4.0 m/s after the trash can has rolled a distance $x_0$ down the incline. Apply calculus to determine the minimum value that $x_0$ can be if you are to catch the trash can before it lands in a pile of manure at the bottom of this very high hill.

35. SI units have been suppressed for convenience, but (of course) you must express your answers with SI units. Also, in this problem, it is implied that all of the numbers are good to two significant figures (in order to reduce clutter and potential confusion from all of the 0's). (A) Given $x(t) = 3t^3 - 2t^2 + 4$, find $v_{x0}$ and $a_x(3)$. (B) Given $x(t) = \frac{4}{t^3}$, find $a_x(2)$ and $\overline{v_x}$ from $t = 1.0$ s to $t = 4.0$ s. (C) Given $v_T(t) = 2t + 3$, find $a_T(4)$ and $\Delta s_T(5)$. (D) Given $a_y(t) = \sqrt{t}$ and $v_{y0} = 2$, find $v_y(4)$ and $\Delta y(4)$. (E) Given $x(t) = 2t^{5/2} - 8t^{3/2}$, in 1D, when is the velocity zero? (F) Given $v_x(t) = \frac{3}{t^2} - \frac{2}{t}$, what is the minimum value of $v_x$?

36. A monkey on roller skates has a tangential velocity given by $v_T(t) = \alpha(\beta - t)^2 - \gamma$, where alpha ($\alpha$), beta ($\beta$), and gamma ($\gamma$) are constants. In their appropriate SI units, $\alpha = 1/2$ and $\beta = 1/\alpha$. The monkey skates for 8.0 seconds (from $t = 0$) with an initial tangential velocity of $-10$ m/s. (A) What are the SI units of $\alpha$ and $\beta$? (B) Solve for $\gamma$. (C) What is the maximum speed of the monkey? (D) What is the tangential acceleration of the monkey after skating for 4.0 seconds? (E) How far does the monkey travel before changing direction? (F) What is the average tangential velocity of the monkey? (G) What is the average speed of the monkey?

37. In 1D, a monkey on roller skates has an $x$-component of acceleration given by $a_x(t) = \beta\sqrt{t}$, where $\beta = 15$ (this is the Greek letter beta) in SI units. The monkey skates for 4.0 s (from $t = 0$) with an initial $x$-component of velocity of $-40$ m/s. (A) What are the SI units of $\beta$? (B) What is the final $x$-component of velocity of the monkey? (C) What is the average $x$-component of velocity of the monkey?

38. A monkey is *so* lazy that instead of actually running around, he just imagines running around in 1D with the following $y$-component of velocity: $v_y(t) = 3(2 - t)^2$ where SI units have been suppressed for convenience (and where the constants are good to two significant figures). The monkey imagines his motion to begin at $t = 0$ and end at $t = 6.0$ seconds. (A) What $y$-component of acceleration does the monkey imagine having at $t = 4.0$ seconds? (B) What is the net displacement for the monkey's imagined trip? (C) At what time(s), if any, does the monkey imagine running 27 m/s?

39. A monkey runs with a tangential velocity of $v_T(t) = \dfrac{t^2}{2} + \dfrac{7}{12}$ after a black cat sneaks up on him and meows. The motion begins at $t = 0$. SI units have been suppressed for convenience and the constants are good to two significant figures. (A) What is the net tangential displacement of the monkey after 4.0 seconds? (B) What is the tangential acceleration of the monkey at $t = 13$ seconds?

40. A homing pigeon takes off to deliver a homework assignment to a physics student along a 1D path. Its acceleration is $a_x = 3t^2$, where SI units have been suppressed. Derive the following equations:

$$\Delta x = v_{x0} t + \frac{at^2}{12}$$

$$v_x = v_{x0} + \frac{at}{3}$$

$$\overline{v_x} = \frac{3v_{x0} + v_x}{4}$$

# 2 Multi-Component Motion

**Two-dimensional (2D) motion**: The motion of an object is considered to be 2D motion if it has two degrees of freedom. An object that is always moving in the same plane, and which is not moving along a predetermined course, experiences 2D motion. Driving a bumper car is 2D because you have the freedom to drive in any direction (except for the restriction of the bounding wall), but driving along a country highway is very much 1D because your only choices are to drive forward or backward along the predetermined route – even if the highway is curved. An object's motion may also be 2D even if its path of motion does not lie in a plane: The motion is 2D if the path of motion is restricted to lie on a curved surface. For example, while the earth is 3D, its surface is 2D. Traveling along the surface, you may change your latitude and longitude, but your altitude is fixed by the altitude of the surface.[66] Since your motion can be described using two independent coordinates – latitude and longitude – the motion is 2D. If you simply walk along the equator, then your motion is 1D since it can be described with a single independent coordinate – just longitude.[67]

**Three-dimensional (3D) motion**: An object is experiencing 3D motion if three independent coordinates are required to describe its motion.[68] You can experience fully 3D motion by climbing on a jungle gym – in which case you may climb upward/downward, east/west, or north/south, or any combination of these (by climbing, say, diagonally) – or by swimming underwater (whereas swimming along the surface is, at most, 2D). An object has three degrees of freedom when its motion is 3D.[69]

---

[66] Mountains and other forms of terrain do not make the motion 3D, they just make the surface less ellipsoidal. If you are on the surface, you are constrained to have whatever altitude that surface point has. In order to make this motion 3D, you must have the freedom of climbing or flying above the surface or digging below it.

[67] Visualizing the earth with its poles at the top and bottom, circles of longitude are vertical circles of the same size passing through both poles, while circles of latitude are horizontal circles of varying size parallel to the equator. If you walk along the equator, which is a circle of latitude, you must specify your longitude to know where you are.

[68] Things suddenly become more complicated if you think about their size or the fact that most objects are composed of many other smaller objects. For example, a ball may be moving in a straight line while spinning: Does not the spinning represent a higher-dimensional motion than the 1D path of the ball? Each part of the ball (except those infinitesimal parts on the axis of rotation itself) travels in a circle. That circular path is 1D, but the ball is also moving in a straight line, which means that each part of the ball actually travels along a sine wave (the combined circular and linear motions). Each part of the ball is not constrained to travel along the sine wave, for the path of the sine wave can easily be reshaped if the ball simply changes its angular speed. Hence, you would say that each part of the ball experiences 2D motion. Rather than worry about such complications, let's just restrict ourselves to what the center of mass of the object does. In this case, such a ball's motion would be 1D.

[69] It's interesting to consider what 'freedom' means in this context. A bead sliding along a parabolic wire experiences a single degree of freedom – forward or backward along the parabola. The bead's motion is 1D. However, if you throw a ball off a roof and neglect the rotation of the earth, the fact that gravity varies with altitude, and air resistance, the ball will travel along a parabola, but we would say that the ball's motion is 2D. Strictly speaking, gravity constrains the motion of the ball just as much as the hanger constrains the bead.

## 2.1 Scalars and Vectors

**Types of physical quantities**: Some physical quantities that we measure have a definite direction, while others do not. As an illustration of this important difference, consider the distinction between mass and weight. We will describe what these two terms mean in much more detail in a later chapter, but for now consider the following definitions:
- Weight – the gravitational pull exerted on an object.
- Mass – a measure of the resistance of an object to be accelerated.

Based on these definitions, the weight of an object has a definite direction, whereas the mass of an object is directionless. If you release a banana from your hand, it will definitely fall downward toward the center of the planet. Weight consistently pulls in the same direction. Instead, if you are trying to accelerate a banana from rest, you will find it equally difficult to change its velocity by pushing it to the north as to the east.[70] There is no preferred direction to associate with mass. Here, mass is an example of a scalar and weight is an example of a vector.

**Magnitude**: When we make a measurement of a physical quantity – like time, velocity, mass, or weight – we always record a magnitude, which includes a numerical value and units. Conceptually, the magnitude is a measure of how much[71] of the quantity there is. When measuring a physical quantity that has a direction, when we speak of the magnitude of that quantity, we are referring to the part of the measurement that excludes the direction. For example, consider a banana that has a weight of 0.9 N downward. The weight includes both the magnitude, which is just 0.9 N, and the direction, which is downward. Similarly, if the velocity of a car is 25 m/s northeast, the magnitude of its velocity is just 25 m/s. Here, you can see that speed is the magnitude of the velocity vector. If you measure a scalar quantity, it only has a magnitude. For example, for a box of bananas with a mass of 32 kg, the magnitude of its mass is the same 32 kg; in this case, there is no direction to exclude in order to obtain the magnitude.

**Note**. Don't let problems that ask you for the magnitude of a quantity confuse you. For example, a problem might ask you to compute the magnitude of the acceleration of a banana, or the magnitude of the weight of a monkey. This simply means how much, without including direction. Sometimes students remember finding the magnitude of the net displacement in one problem, and remember that they ultimately found a distance for their answer, and then when another problem asks them to find the magnitude of the net force, they start thinking about distance – but instead they should be finding the net force vector and determining what its magnitude is (i.e. once you know how much force the net force has and in what direction it points, the magnitude is just the 'how much' part).

---

[70] It's implied that all other things are equal. If there is a brick just to the north of the banana, that's an external influence. The mass of the banana is an internal quantity, independent of external influences.

[71] Realize that 'how much' may apply to any kind of measurement – it's not necessarily a volume or a mass, but could be a distance, time, voltage, pressure, or any other type of measurement. If you measure the water pressure for your faucet, for example, you are asking 'how much' force per unit area of water comes out.

## 2 Multi-Component Motion

**Direction**: In addition to a magnitude, many physical quantities also have a direction. When we speak of a car that has a velocity of 18 m/s to the south, the direction of this quantity is specified as south. We can also quantify the direction. In 2D, the direction of a vector can be established by setting up an $xy$ (Cartesian) coordinate system and specifying the direction counterclockwise from the $+x$-axis. In 3D, a single angle would be ambiguous, and it is instead necessary to specify two angles (which we will define precisely when we reach the section on spherical coordinates).

> **Note**. If a problem asks you to find the direction of a vector, it usually means to find the angle that the vector makes with the $+x$-axis, measured in a counterclockwise sense (or if you have a 3D problem, then it means to find the pair of angles comprising the spherical coordinate system).[72]

**Scalars**: A scalar is a quantity that has a magnitude, but which does not have a direction. For example, speed has only a magnitude because it means only how fast, but not which way. Mass is another example of a scalar, as it does not include a direction. We usually think of distance as a scalar, but displacement (aka directed distance) is actually a vector: Unlike distance, when we measure displacement we note both the distance and direction together.[73]

**Mathematics of scalars**: Scalar quantities can be added together, subtracted from one another, multiplied together, and divided by one another according to the rules of ordinary arithmetic. You might think that this is a trivial point to make, but the reason it's important is that the mathematics of vector quantities do not follow the rules of the ordinary arithmetic of numbers, as we will learn.

**Vectors**: A vector is a quantity that includes a direction in addition to a magnitude.[74] For example, velocity is a vector because it means how fast and which way combined. Weight is another example of a vector because it always pulls an object along the gravitational field lines. When you speak of a vector, you must bear in mind that it has both magnitude and direction. If you want to speak only of its magnitude, then explicitly state something like, "The magnitude of velocity is 40 m/s." It would be senseless to write, "The velocity is 40 m/s," but, "The velocity is 40 m/s to the north," or, "The velocity is 40 m/s at an angle of 30°," are fine (where in the last example, there must be an established coordinate system, which defines the direction as this angle counterclockwise from the $+x$-axis).

---

[72] When restricting ourselves to 2D, we usually work with $x$- and $y$-coordinates.

[73] You might be inclined to classify time as a scalar, and we probably think of it this way very often, but if you think about it, time actually proceeds in a particular direction – if you were going backward in time instead of forward, you would definitely know it. On the other hand, you don't really have a choice – it seems that you can only go forward through time. It's not so clear-cut in first-year physics. In more advanced courses, we define a vector based on how its components transform, which would resolve the issue much more readily. For example, in special relativity, we can make a four-vector out of space and time coordinates. A time vector can then be obtained as a special case of a spacetime vector by setting all of the space coordinates to zero. So if you can have a vector in 1D space, you can also define one in 1D time, in principle.

[74] I once told a class of students that anything that has direction is a vector. One clever student instantly asked me if an angle was therefore a vector, since angle itself is a measure of direction. From this statement, you might similarly wonder if a paper with the word 'north' on it is a vector. But if he had asked me whether a compass was a vector, I would immediately have replied that the compass reading is a unit vector.

**Notation**. In textbooks, the distinction between a scalar symbol and a vector symbol is that a scalar symbol appears in italics while a vector symbol appears in boldface. When writing on the board or on paper, it's difficult to distinguish between italics and boldface, so in this case an arrow is placed over a vector symbol in order to distinguish it from a scalar symbol. In this text, we use italics for scalar symbols and both write vector symbols in boldface and place arrows above them. For example, the symbol for mass – a scalar quantity – appears as $m$, while the symbol for weight – a vector quantity – appears as $\vec{\mathbf{W}}$. The magnitude of a vector quantity is a scalar, and so is often represented with the same symbol without the arrow. If we write $W$ for weight, we mean just its magnitude, and not its direction.

**Notation**. The magnitude of a vector can be explicitly declared by placing two vertical double lines on both sides of the vector. For example, $\|\vec{\mathbf{F}}\|$ represents just the magnitude of the force vector $\vec{\mathbf{F}}$.[75] Very often, we write the same symbol without the arrow over it as a simpler means of identifying just the magnitude. For example, $F$ also refers to the magnitude of the force vector $\vec{\mathbf{F}}$. That is, $F = \vec{\mathbf{F}}$. However, there are times when the same symbol without the arrow over it already has a precise meaning, which may differ from the magnitude of the vector; such is the case, for example, with the distinction between average velocity and average speed.

**Exceptions**. When working with net and average quantities, the double vertical lines are necessary to avoid confusion between the magnitude of the vector and a different scalar quantity. For example, the average velocity $\|\vec{\bar{\mathbf{v}}}\|$ is distinctly different from the average speed, $\bar{v}$: Generally, $\bar{v} \neq \|\vec{\bar{\mathbf{v}}}\|$, so it would be incorrect to write $\bar{v}$ to mean the magnitude of the average velocity. Similarly, the arc length, $s$, is distinctly different from the magnitude of the net displacement, which we will denote by $\|\Delta\vec{\mathbf{r}}\|$ in this chapter; we do not use the notation $\Delta r$ to represent net displacement.[76] When we discuss work in a later chapter, we will also work with displacement, $\vec{\mathbf{s}}$, and we will see that $s \neq \|\vec{\mathbf{s}}\|$, in general.

**Math Note**. Be careful not to abuse the notation of vectors (as many students do, even when I point this out with examples). Namely, be careful not to set vectors equal to scalars when you write equations. For example, it is a conceptual mistake to write an equation like $\vec{\mathbf{v}} = 40$ m/s, since the left-hand-side of the equation includes direction, while the right-hand-side is a scalar. It would be okay to write $v = 40$ m/s, though, since $v$ is a scalar, or to write $\vec{\mathbf{v}} = 40$ m/s to the northeast or $\vec{\mathbf{v}} = 40$ m/s @ 45°, since direction is included on both sides of the equation.

**Important Distinction**. A vector quantity has direction, whereas a scalar quantity does not.

---

[75] The quantity $\|\vec{\mathbf{A}}\|$ reads, "the magnitude of $\vec{\mathbf{A}}$." It does not read, "the absolute value of $\vec{\mathbf{A}}$." Of course, if we mean to take the absolute value of a quantity, we declare that with single vertical lines on both sides, as in $|x|$. Magnitudes and absolute values do not mean the same thing, and the absolute value of a vector does not even make sense unless you are working strictly in 1D, where the direction of a vector quantity is described by its sign (cf. tangential velocity, which is a 1D quantity, and the more general term, velocity, which may be 2D or 3D).
[76] You can see that we are using a different notation in 2D and 3D motion than we used in the previous chapter. We will define quantities like velocity and speed, including their notation, precisely when we introduce them.

## 2 Multi-Component Motion

**Examples of scalars and vectors**: Following are some examples to illustrate the difference between physical quantities that are scalars and vectors:

- The total distance traveled, $\Delta s$, is a scalar. It is a measure of how far an object has traveled all together over the course of its trip; this quantity does not have a direction. The net displacement, $\Delta \vec{r}$, is a vector that extends from the initial position to the final position, and so has a direction.[77]
- Similarly, the arc length (aka distance traveled), $s$, is a scalar, while the displacement, $\vec{s}$, is a vector. The distinction between the distance traveled and displacement from the total distance traveled and the net displacement is that these may be instantaneous values – i.e. they are not necessarily the endpoints of a complete trip.
- The speed, $v$, of an object is a scalar because it provides a measure only of how fast an object is moving instantaneously, whereas the velocity, $\vec{v}$, is a vector that combines speed with direction.
- The mass, $m$, of an object is a scalar: The more mass an object has, the more net force is required to change its velocity (i.e. to accelerate it), regardless of which direction the net force is exerted; therefore, mass does not have a particular direction to associate with it. Mass is related to a similar, but significantly different quantity in nature, called weight. Weight, $\vec{W}$, is a vector because the force of gravity clearly pulls in a definite direction.

**Visualizing vectors**: A vector can be represented graphically by drawing an arrow (not to be confused with the arrow that we draw on top of the symbol for the vector): The length of the arrow visually represents the magnitude and the arrow points along its direction. Some vectors are represented visually in the diagram below.

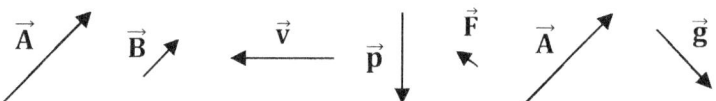

**Translation of a vector**: Conceptually, it is important to realize that a vector can be translated and still be exactly the same vector. The translation of a vector means to move the vector horizontally or vertically (or a combination of both), but to preserve its magnitude and direction while doing so. So if you are drawing arrows to represent vectors, you may move the arrow around and it will still be the same vector as long as the arrow has the same length and orientation. That is, the tail of the arrow does not have a definite location. Observe that the concept of a vector makes sense even if you don't setup a coordinate system.[78] That is, the tail of a vector is not constrained to lie at a particular place.

---

[77] Note that the total distance traveled, $\Delta s$, is a scalar, but although it does not have a direction, it very much depends upon the path taken. The net displacement, $\Delta \vec{r}$, is a vector, but although it has a direction, it is a path-independent quantity. Students sometimes want to associate direction with path-dependence, but it's backwards.

[78] The position coordinates of an object, however, are critically linked to how the coordinate system is setup. If you move the origin or rotate the coordinate system, you definitely get different values of the Cartesian coordinates $(x, y, z)$. A vector that extends from the origin of a coordinate system to the object's instantaneous position, such as the position vector, $\vec{r}$, therefore serves as an exception to the vector translation rule: You can't move $\vec{r}$ (which we will learn about in Sec. 2.3) without changing $\vec{r}$. However, we will see that the net displacement, $\Delta \vec{r}$, is independent of the setup of the coordinate system.

**Note**. When dealing with two or more vectors visually, it is sometimes conceptually very significant to realize that the vectors may be moved around, provided that you preserve their magnitude and direction. For example, a problem may draw two vectors joined at their tails, but if you wish to add them (a method described in the following section), you need to move one to instead join the two vectors tip-to-tail.[79]

**Mathematics of vectors**: The rules for the arithmetic of vectors[80] are considerably different from the rules for the arithmetic of ordinary numbers.[81] For example, as we will learn in the next section, if you wish to add two vectors, it would be a major conceptual mistake to simply add together their magnitudes, as it would be a complete disregard for the fact that vectors have direction. As it turns out, there are actually three useful ways to multiply vectors (which we will explore in later chapters).[82]

**Projection of a vector**: A vector can be projected onto another object, such as a line or a plane. You can visualize this as a stream of parallel rays of light shining at the vector: In this case, the shadow left on the object represents the vector's projection. The projection of the vector depends upon the angle from which the light is shone and the geometry of the object onto which it is projected.[83]

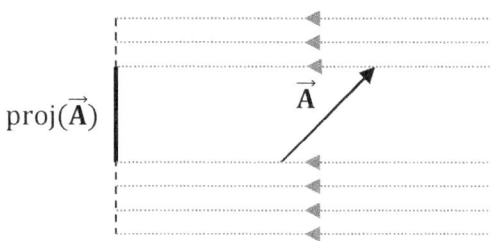

**Components**: The projection of a vector onto a coordinate axis yields a component of the vector. A vector in 3D space can be resolved into three components – one for each coordinate axis. These axes need not be the usual Cartesian coordinates – $x$, $y$, and $z$. In addition to Cartesian components, a vector may be resolved into components in other coordinate systems, such as spherical coordinates or cylindrical coordinates (to be discussed in a later section). Another conceptually useful choice is to work with tangential – along the direction of motion – and normal – perpendicular to the direction of motion – components.

---

[79] This point is usually counterintuitive to students, who – without remembering otherwise – imagine, incorrectly, that vectors' tails must be fixed in position. Perhaps this is from graphing curves in math class: For example, if you move a straight line on a graph, you change its equation (its $y$-intercept changes).

[80] The arithmetic of vectors is actually called vector algebra. The rules of the vector algebra – essentially, the analogs for commutativity, associativity, distribution, and additive inverses, and such – form the basis for how to add, subtract, multiply, and find additive inverses.

[81] By analogy, you might want to say that the arithmetic taught in grade school is essentially scalar algebra. However, it's not quite the same – e.g. there is such a thing as a pseudoscalar, creating a subtle distinction.

[82] The three useful ways to multiply two vectors together include the scalar product (which we use to compute work, for example), the vector product (for computing torque, for example), and the outer product (very important for the completeness relation in linear algebra) – which is easiest to understand in matrix form.

[83] If you are curious about this, you may have an interest in the subject of projective geometry.

## 2 Multi-Component Motion

**Cartesian components**: The projection of a vector, $\vec{A}$, onto the $x$-axis yields the $x$-component of the vector, $A_x$, onto the $y$-axis yields the $y$-component of the vector, $A_y$, and onto the $z$-axis yields the $z$-component of the vector, $A_z$. A vector $\vec{A}$ lying in the $xy$ plane has just two nonzero components, which are $A_x$ and $A_y$ in Cartesian coordinates.

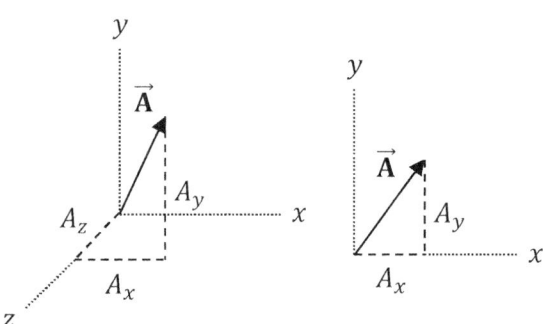

---

**Notation**. The components of a vector are denoted through the use of subscripts. As examples, $A_y$ represents the $y$-component of vector $\vec{A}$, and $A_T$ represents the tangential component of vector $\vec{A}$. Remember, a subscript (unlike an exponent) is not a mathematical operation, but a bookkeeping device – it's a note that helps distinguish similar quantities from one another.

---

**Math Note**. The components of a vector, like $A_x$ and $A_y$, may be positive or negative, depending upon which quadrant a 2D vector (or which octant a 3D vector) lies in. However, the magnitude of a vector is always nonnegative. That is, $A \geq 0$, whereas the signs of $A_x$ and $A_y$ are not restricted.

---

**Tangential and normal components**: For a vector that relates to a path (such as velocity or acceleration) such that its magnitude and direction generally vary at each point along the path, it is often convenient to work with tangential and normal components. Adjoining the tail of the vector to the point on the path where it is instantaneously defined (keeping in mind that the vector will generally look different at other points on the path), the vector can be resolved into tangential and normal components by drawing a tangent line and also a line perpendicular to the tangent to serve as the tangential ($T$) and normal ($N$) axes.[84] In 3D, there is a whole plane that is perpendicular to the tangent line, and so it is necessary to draw two normal axes, where both normals and the tangent are mutually perpendicular. The convenience of doing this is that the tangential and normal components of acceleration have a very natural interpretation, conceptually, as we will see later, and mathematically it is also practical to treat these components separately. Tangential and normal components are particularly common in circular motion. For circular motion the tangential and normal directions will relate to the 2D polar coordinates that we use (see Sec. 2.4). These concepts should seem abstract now, and a few students generally shudder at the thought of polar coordinates, but these things actually make life simpler in circular motion, and things will clear up when we proceed to apply these ideas.

---

[84] In mathematics, the word 'normal' means perpendicular to a curve or surface.

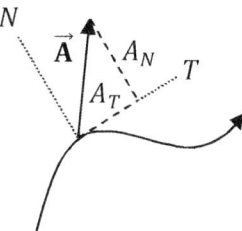

**Relating components to magnitude and direction**: In 2D, the magnitude and components of a vector form a right triangle, with the magnitude lying on the hypotenuse.[85]

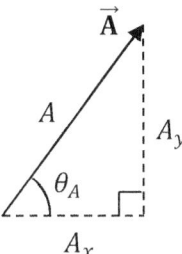

Given the magnitude, $A$, and direction, $\theta_A$, of a vector, $\vec{A}$, lying in the $xy$ plane, the Cartesian components, $A_x$ and $A_y$, can be computed by applying trigonometry:

$$A_x = A \cos \theta_A \quad , \quad A_y = A \sin \theta_A \quad \text{(in 2D)}$$

If instead you are given the Cartesian components, $A_x$ and $A_y$, the magnitude, $A$, can be found by applying the Pythagorean theorem and the direction, $\theta_A$, of the vector, $\vec{A}$, can be determined from an inverse tangent:

$$A = \sqrt{A_x^2 + A_y^2} \quad , \quad \theta_A = \tan^{-1}\left(\frac{A_y}{A_x}\right) \quad \text{(in 2D)}$$

Similar relationships relate the tangential and normal components, $A_T$ and $A_N$, of a vector, $\vec{A}$, to its magnitude, $A$, and direction, $\theta_A$, in 2D. For example, $A = \sqrt{A_T^2 + A_N^2}$.

The magnitude, $A$, and direction, $\theta_A$, of a vector lying in the $xy$ plane are related to the Cartesian components, $A_x$ and $A_y$, in the same way that 2D polar coordinates are related to Cartesian (or rectangular) coordinates. In 3D, the direction of a vector is specified by two angles. We will postpone the mathematics of 3D vectors until we introduce the cylindrical and spherical coordinate systems in Sec.'s 2.4 and 2.5.

---

[85] When we draw this diagram, students are tempted to interpret the sides as distances, from their experience with geometry and trigonometry. However, these may be speeds, forces, or other quantities, remembering that velocity, force, acceleration, and many other physical quantities are vectors.

## 2 Multi-Component Motion

**Math Note.** When applying an inverse trig function, such as $\cos^{-1}$, there are two possible answers, mathematically – corresponding to two possible quadrants. Physically, however, there is only one viable answer, corresponding to the quadrant in which the vector actually lies. You must consider both possibilities when doing the math, and then select the one correct answer when applying it to physics.

**Math Note.** Most calculators (including the popular TI-83 and other TI models) only provide inverse trig functions in two of the four possible quadrants. For example, you will only get an inverse tangent in Quadrants I or IV on most calculators. However, the vector sometimes lies in Quadrants II or III. You need to draw the vector to decide if the calculator angle is the one you need, and draw the two right triangles to determine the alternate angle, when necessary.

**Hint.** When applying an inverse tangent, there is a very simple rule for determining whether or not the calculator angle is the one you need, and, if not, how to find the alternate angle: Simply add 180° to your calculator's answer if the $x$-component of the vector is negative.[86]

**Math Note.** Recall that the angle of a vector is measured counterclockwise from the $+x$-axis. Also, recall that you may add (or subtract) multiples of $360°$ to an angle without, effectively, changing the angle. For example, $-60°$ is the same angle as $300°$.

**Math Note.** In the previous chapter, we worked exclusively with 1D motion, in which case the direction of a vector was simply given by its sign. However, in 2D and 3D, the direction is specified by one or two angles. Thus, in 2D and 3D, you can't simply take the absolute value of the vector quantity to determine the magnitude. Rather, you must use the Pythagorean theorem in order to compute the magnitude of a vector from its components.[87]

**Conceptual Example.** How is it possible for two different vectors lying in the $xy$ plane to have an $x$-component that is equal to one-half of the magnitude of the vector?

There are two ways to draw such a vector – it could lie in Quadrant I or IV.[88] That is, the $y$-component of the vector could be positive or negative. These two vectors are illustrated in the following diagram.

---

[86] This same rule does not apply to the other inverse trig functions, like inverse sine.

[87] Strictly speaking, you can't even take the absolute value of a vector in 1D to get the magnitude. In the previous chapter, we did not actually work with vector quantities, but with tangential components of vectors. Components of a vector may be positive and negative. The absolute value of a vector does not make sense, but the magnitude of a vector does. If you want to find the magnitude of a vector, apply the Pythagorean theorem.

[88] The $x$-component of the vector can't be negative and still be equal to half of the magnitude: Since the magnitude of a vector can't be negative, the condition $A_x = A/2$ requires $A_x$ to be nonnegative. There are Quadrant II and III vectors where $A_x = -A/2$, for which $A_x$ is negative, but if you read the question carefully, you will see that it does not allow for such a minus sign. If you change the wording to 'minus one-half of the magnitude of the vector,' then the two vectors would instead lie in Quadrants II and III.

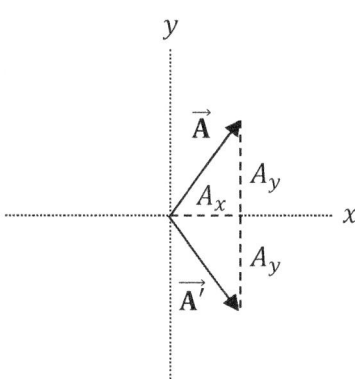

We could also answer this question mathematically. For example, we can use the Pythagorean theorem, $A^2 = A_x^2 + A_y^2$, to solve for the $y$-component of the vector. The condition of the problem is that $A_x = A/2$. Substituting this into the Pythagorean theorem and solving for the $y$-component, we find that $A_y = \pm\sqrt{A^2 - A_x^2} = \pm\sqrt{A^2 - (A/2)^2} = \pm\sqrt{A^2 - A^2/4} = \pm A\sqrt{3}/2$. The two possible values of $A_y$ correspond to the two possible roots.

Alternatively, we could solve for the angle of the vector using an inverse cosine: $\theta_A = \cos^{-1}\left(\frac{A_x}{A}\right) = \cos^{-1}\left(\frac{1}{2}\right) = 60°$ or $300°$. These are the two possible directions of the vector, which lie in Quadrants I and IV.

**Example**. A vector lying in the $xy$ plane has a magnitude of 30 N and a direction of $150°$. Determine the Cartesian components of the vector.

The $x$- and $y$-components of the vector can be found through trigonometry:

$$A_x = A\cos\theta_A = 30\cos 150° = -15\sqrt{3}\text{ N}$$
$$A_y = A\sin\theta_A = 30\sin 150° = 15\text{ N}$$

**Unit vectors**: A unit vector is a vector that has a magnitude equal to one mathematical unit.[89] Unit vectors serve a useful directional purpose.

**Cartesian unit vectors**: The Cartesian unit vectors, $\hat{\mathbf{i}}, \hat{\mathbf{j}}$, and $\hat{\mathbf{k}}$ (called "i-hat," "j-hat," and "k-hat;" the 'hat' is called the caret), point along the $x$-, $y$-, and $z$-axes, respectively. Any vector, $\vec{\mathbf{A}}$, can be expressed in terms of its Cartesian (or rectangular) components ($A_x, A_y$, and $A_z$) and Cartesian unit vectors as $\vec{\mathbf{A}} = A_x\hat{\mathbf{i}} + A_y\hat{\mathbf{j}} + A_z\hat{\mathbf{k}}$. This representation of a vector shows that the zigzag path formed by traveling $A_x$ units along the $x$-axis, $A_y$ units along the $y$-axis, and $A_z$ units along the $z$-axis is equivalent to the straight-line path from the tail of the vector to its tip – i.e. the components provide information that is equivalent to the magnitude and direction (which includes two angles for a vector in 3D space).

---

[89] Despite its name, a unit vector is actually unitless, so it's a mathematical unit, not a physical unit. For example, both a force vector and components of force may be expressed in Newtons, which does not permit the unit vectors to have any physical units. The magnitude of a unit vector is one, not one physical unit (like 1 N).

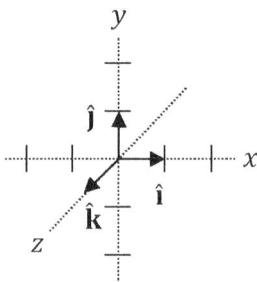

**Note**. The Cartesian unit vectors – $\hat{\imath}$, $\hat{\jmath}$, and $\hat{k}$ – are constants; they always have a magnitude of one unit, and each always points in the same direction – e.g. $\hat{k}$ always points along the $+z$-axis. Other unit vectors – such as those of 2D polar, cylindrical, and spherical coordinate systems – are not constants, however. This is important when applying calculus to vectors. For example, when taking a derivative of a vector with respect to time, the Cartesian unit vectors may be treated as constants, while other vectors must be treated as variables (applying the product and chain rules of differentiation).

**Example.** Determine the magnitude and direction of the vector $\vec{A} = 4\hat{\imath} - 4\sqrt{3}\hat{\jmath}$.
The components of the vector are the coefficients of the unit vectors: $A_x = 4$ and $A_y = -4\sqrt{3}$. As the components are mutually perpendicular, the magnitude is given by the Pythagorean theorem, $A = \sqrt{A_x^2 + A_y^2} = \sqrt{16 + 48} = 8$, and the direction can be found from trig, $\theta = \tan^{-1}(A_y/A_x) = \tan^{-1}(-\sqrt{3}) = 300°$. This vector lies in the $xy$ plane and is directed $300°$ counterclockwise from the positive $x$-axis (looking from the $+z$-direction).

**Tangential and normal unit vectors**: The tangential and normal unit vectors, $\hat{t}$ and $\hat{n}$, point along the tangential and normal directions, respectively – i.e. tangent and perpendicular to the path of motion. In 3D, there are two normal vectors, $\hat{n}_1$ and $\hat{n}_2$, which are mutually perpendicular to one another and to the unit tangent vector, $\hat{t}$. Any vector, $\vec{A}$, can be expressed in terms of tangential and normal unit vectors as $\vec{A} = A_T\hat{t} + A_{n_1}\hat{n}_1 + A_{n_2}\hat{n}_2$. For a curved path, $\hat{t}$, $\hat{n}_1$, and $\hat{n}_2$ are variables – not constants like the Cartesian unit vectors – because their directions change as the path curves.[90]

**Null vector**: The sum of a vector, $\vec{A}$, and its additive inverse, $-\vec{A}$, equals the null vector, which has zero magnitude. Since the null vector has no magnitude, it does not make sense to ascribe a direction to it. Hence, the null vector is traditionally expressed as a scalar zero, as in the vector subtraction $\vec{A} - \vec{A} = 0$.

---

[90] In Chapter 1, when we worked with tangential displacement, tangential velocity, and tangential acceleration, these quantities were actually the tangential components of the displacement, velocity, and acceleration vectors. The components of a vector can be positive or negative, and in 1D the sign of these components provided a measure of the direction of the corresponding vectors. In 2D and 3D motion, it is necessary to use the Pythagorean theorem to find the magnitude from the components and to apply trigonometry to find the direction – one or two angles, depending upon whether we are working in 2D or 3D – of a vector. Compare to 1D, where we simply took the absolute value to find the magnitude, and the sign indicated the direction.

## 2.2 Vector Addition and Subtraction

**Vector addition**: Two or more vectors, $\vec{A}_1, \vec{A}_2, \cdots, \vec{A}_N$, added together equal a resultant vector, $\vec{R}$. The resultant vector, $\vec{R}$, is equivalent to the sum $\vec{A}_1 + \vec{A}_2 + \cdots + \vec{A}_N$, which we express as vector addition: $\vec{R} = \vec{A}_1 + \vec{A}_2 + \cdots + \vec{A}_N$. The resultant vector, $\vec{R}$, is equivalent to the set of vectors, $\vec{A}_1, \vec{A}_2, \cdots, \vec{A}_N$, acting together in the sense illustrated by the following examples.

---

**Conceptual Examples.** A monkey walks 50 m north and then walks 30 m southeast. The resultant vector — which is the net displacement, $\Delta\vec{r}$, in this case — is equivalent to the two displacements, $\vec{s}_1$ and $\vec{s}_2$, added together — i.e. $\Delta\vec{r} = \vec{s}_1 + \vec{s}_2$ — in the sense that the monkey could walk from the same initial position and reach the same final position by traveling directly along the resultant, $\Delta\vec{r}$, as by walking first along $\vec{s}_1$ and then along $\vec{s}_2$.

Two monkeys lasso a calf. One monkey pulls the calf with a force of 300 N to the east and the other monkey pulls the calf with a force of 200 N to the southeast. The resultant vector, $\vec{F}_{net}$, is equivalent to these two forces, $\vec{F}_1$ and $\vec{F}_2$, working together — i.e. $\vec{F}_{net} = \vec{F}_1 + \vec{F}_2$ — in the sense that $\vec{F}_{net}$ acting alone would result in the same acceleration of the calf as the combination of $\vec{F}_1$ and $\vec{F}_2$.

A monkey rows a boat with a speed of 2 m/s directly across a river with a river current of 0.4 m/s. If the monkey angled the boat straight across and attempted to row directly across the river, the current would cause the boat to drift at an angle — partly across and partly downriver. Therefore, the monkey must have angled his boat and attempted to row at an angle partly upriver, in order for the river current to compensate and take the monkey across in a straight line. The resultant velocity, $\vec{v}$, is equivalent to the monkey's rowing velocity, $\vec{v}_r$, which is angled partly upriver, and the river current's velocity, $\vec{v}_c$: $\vec{v} = \vec{v}_r + \vec{v}_c$.

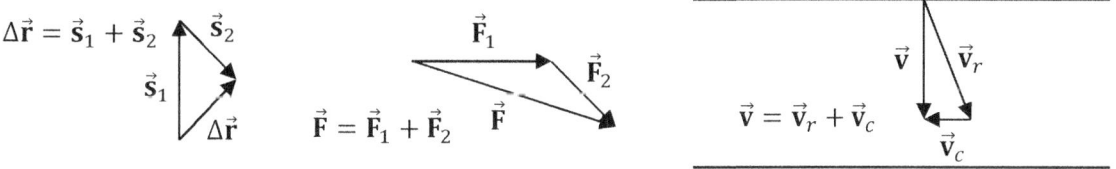

---

**Resultant vector**: The resultant vector, which we call $\vec{R}$ here, but will generally have whichever symbol best suits the nature of the problem (e.g. $\vec{F}_{net}$ would be better suited for the resultant of adding together force vectors), is the sum of two or more vectors added together: $\vec{R} = \sum_{i=1}^{N} \vec{A}_i$ (summation notation is defined in Sec. 1.2), where $N$ is the number of vectors, $\vec{A}_1, \vec{A}_2, \cdots, \vec{A}_N$, added together to make $\vec{R}$. This means that $\vec{R}$ is equivalent to the set of vectors, $\vec{A}_1, \vec{A}_2, \cdots, \vec{A}_N$, working together.

**Graphical vector addition**: The resultant of $N$ vectors added together, $\vec{R} = \sum_{i=1}^{N} \vec{A}_i$, can be depicted visually by connecting the vectors, $\vec{A}_1, \vec{A}_2, \cdots, \vec{A}_N$, tip-to-tail — i.e. such that the tip of one connects to the tail of another. The resultant vector, $\vec{R}$, then extends from the tail of the first to the tip of the last, such that it begins and ends at the same points as the chain of vectors, $\vec{A}_1, \vec{A}_2, \cdots, \vec{A}_N$.

## 2 Multi-Component Motion

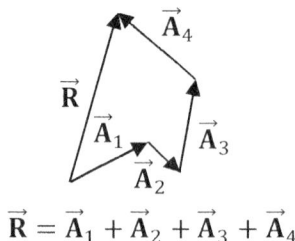

$$\vec{R} = \vec{A}_1 + \vec{A}_2 + \vec{A}_3 + \vec{A}_4$$

In principle, the graphical method of vector addition can be applied to determine the magnitude and direction of the resultant vector given the magnitudes and directions of the given vectors. If the given vectors are drawn to scale and joined tip-to-tail, the resultant vector's magnitude and direction will come out to the same scale. However, it is generally more convenient – and involves no drawing error – to apply trigonometry to add vectors. We will describe the trigonometric method in the problem-solving strategy for vector addition.

**Note.** The vectors must be joined tip-to-tail in order to obtain a correct visual representation of the resultant vector. Problems sometimes draw vectors joined tail-to-tail. In this case, if you wish to sketch the resultant vector (which is useful for predicting in which quadrant the resultant will lie in), you need to translate one of the vectors in order to join them tip-to-tail before you can draw the resultant vector.

**Note.** The resultant vector may have a smaller magnitude than one (or all) of the given vectors; it is not necessarily the longest arrow of the arrangement of vectors in the tip-to-tail diagram. However, the resultant is always easy to identify if you remember that it joins the tail of the first vector to the tip of the last vector in the addition. Examine the tip-to-tail arrangements on the previous page.

**Conceptual Example.** Consider a monkey that first walks 200 m at a direction of 30° and then walks 400 m at a direction of 60°. These trips can be represented as displacement vectors, $\vec{s}_1$ and $\vec{s}_2$. The monkey could achieve the same end result by traveling along a single vector, $\Delta\vec{r}$, which is the net displacement of the monkey. The net displacement, $\Delta\vec{r}$, is a vector that extends from the initial position to the final position. Thus, the net displacement, $\Delta\vec{r}$, is the vector sum of the individual displacements: $\Delta\vec{r} = \vec{s}_1 + \vec{s}_2$. However, the total distance traveled, $\Delta s$, is the sum of the distances traveled in each trip. The total distance traveled, $\Delta s$, is the scalar addition of the two distances: $\Delta s = s_1 + s_2$, which is the sum of the two magnitudes. The total distance traveled, $\Delta s$, equals 600 m, while the magnitude of the net displacement is clearly smaller: 200 m $< \|\Delta\vec{r}\| <$ 600 m.

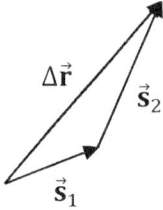

**Conceptual Example**. How can you draw three vectors with equal magnitudes, for which the resultant is zero? You can't do this if you first draw one vector and then draw an equal and opposite vector because although the first two vectors cancel each other out, you are forced to draw a third vector that makes the sum nonzero. Instead, you should think about what it means for the resultant vector to be zero: It means that the tip of the last vector must join to the tail of the first vector when the three vectors are joined tip-to-tail, which suggests drawing all three vectors along edges of an equilateral triangle.

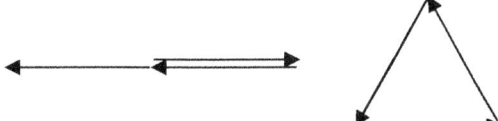

**Parallelogram for the addition of two vectors**: When two vectors, $\vec{A}_1$ and $\vec{A}_2$, are joined tip-to-tail to form a resultant vector, $\vec{R} = \vec{A}_1 + \vec{A}_2$, it does not matter which vector, $\vec{A}_1$ or $\vec{A}_2$, you draw first. That is, whether you join the tip of $\vec{A}_1$ to the tail of $\vec{A}_2$, or the tip of $\vec{A}_2$ to the tail of $\vec{A}_1$, does not affect the resultant vector, $\vec{R}$. You can draw it both ways in the same parallelogram (a quadrilateral – a four-sided polygon – with two sets of parallel edges), as illustrated below. Either way, the resultant vector, $\vec{R}$, lies along the diagonal of the parallelogram.

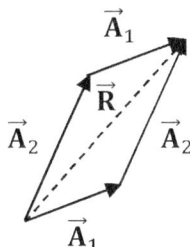

**Parallelepiped for 3D addition of three vectors**. If you extend the previous parallelogram discussion to the addition of three vectors, $\vec{R} = \vec{A}_1 + \vec{A}_2 + \vec{A}_3$, to the case where all three vectors, $\vec{A}_1, \vec{A}_2$, and $\vec{A}_3$, do not lie in the same plane, the resultant vector, $\vec{R}$, lies along the body diagonal of a parallelepiped[91] (a polyhedron with 6 parallelogram-shaped sides, 8 vertices, and 12 edges, where there are 3 sets of 4 parallel edges), as illustrated below, that is formed by drawing the different combinations of joining the three vectors, $\vec{A}_1, \vec{A}_2$, and $\vec{A}_3$, tip-to-tail.

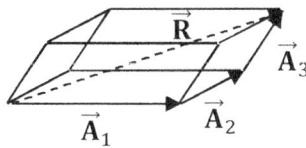

---

[91] If you pronounce it like it looks, the last two syllables sound like 'pie-pid.' There are actually two acccepted pronunciations, though, so if you hear someone end it with, 'pip-id,' that's also correct.

## 2 Multi-Component Motion

**Principle of vector addition**: Here, we visually illustrate the underlying principle for how to add vectors together trigonometrically. Study the diagram below, which shows two vectors, $\vec{A}_1$ and $\vec{A}_2$, joined tip-to-tail to form a resultant vector, $\vec{R} = \vec{A}_1 + \vec{A}_2$.

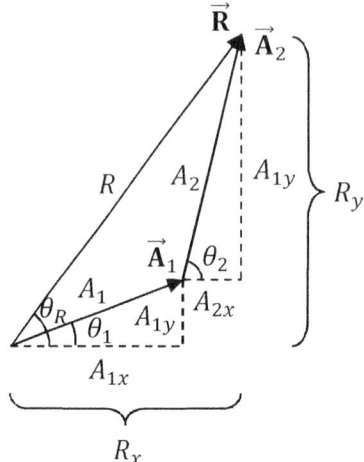

If you examine the Cartesian coordinates of all three vectors, you will see that:

$$R_x = A_{1x} + A_{2x} \quad , \quad R_y = A_{1y} + A_{2y} \quad \text{(for 2 vectors added together in 2D)}$$

This is the underlying principle for how to add vectors trigonometrically: Add the respective Cartesian components of the given vectors together to find the corresponding Cartesian components of the resultant vector. Although the diagram above shows all three vectors lying in Quadrant 1 (if, traditionally, you draw $+x$ to the right and $+y$ upward), this result applies in general, even if one or more vectors lies in other quadrants. In other quadrants, it is important to keep in mind that one or more components of the vectors are negative.

This result also generalizes readily to the case of adding together three or more vectors: Consider $N$ vectors added together: $\vec{R} = \sum_{i=1}^{N} \vec{A}_i$, where the vectors, $\vec{A}_1, \vec{A}_2, \cdots, \vec{A}_N$, may not all lie in the $xy$ plane. You could add the first two vectors together, to obtain $A_{12x} = A_{1x} + A_{2x}$, $A_{12y} = A_{1y} + A_{2y}$, and $A_{12z} = A_{1z} + A_{2z}$, and then you could add this intermediate resultant vector to the third vector to obtain $A_{123x} = A_{12x} + A_{3x} = A_{1x} + A_{2x} + A_{3x}$, and similarly for $y$- and $z$-components. If you continue this to $N$ vectors, you will find that the same underlying principle of vector addition applies:

$$R_x = \sum_{i=1}^{N} A_{ix} \quad , \quad R_y = \sum_{i=1}^{N} A_{iy} \quad , \quad R_z = \sum_{i=1}^{N} A_{iz}$$

> **Math Note.** If you are adding vectors together in 3D, you will need to find the $x$-, $y$-, and $z$-components using the prescription given in Sec. 2.5. For example, generally $A_x \neq A \cos \theta_A$ in 3D because a 3D vector is defined by two angles. The relation $A_x = A \cos \theta_A$ only holds for a vector lying in the $xy$ plane.

## Problem-Solving Strategy for Adding Vectors

0. If you have the magnitudes and directions of two or more given vectors, $\vec{A}_1, \vec{A}_2, \cdots, \vec{A}_N$, you can apply this strategy to determine the magnitude and direction of the resultant, $\vec{R} = \sum_{i=1}^{N} \vec{A}_i$.

1. First draw and label a sketch of the given vectors, $\vec{A}_1, \vec{A}_2, \cdots, \vec{A}_N$, drawn tip-to-tail. Also draw and label the resultant, $\vec{R}$, which extends from the tail of the first vector, $\vec{A}_1$, to the tip of the last vector, $\vec{A}_N$. Use the sketch to check the signs of all of the components and to check that the result lies in the correct quadrant (or octant in 3D).

2. Determine the Cartesian components of each given vector. If all of the vectors lie in the $xy$ plane, you can compute the $x$- and $y$-components of the given vectors using the following formulas:

$$A_{ix} = A_i \cos \theta_i \quad , \quad A_{iy} = A_i \sin \theta_i \quad \text{(in 2D)}$$

However, if one or more vectors do not lie in the $xy$ plane, you must use two angles (discussed in Sec. 2.5) to determine the components of the given vectors:

$$A_{ix} = A_i \cos \varphi_i \sin \theta_i \quad , \quad A_{iy} = A_i \sin \varphi_i \sin \theta_i \quad , \quad A_{iz} = A_i \cos \theta_i$$

3. The Cartesian components of the resultant vector, $R_x$, $R_y$, and $R_z$, can now be determined by adding together the respective components of the given vectors:

$$R_x = \sum_{i=1}^{N} A_{ix} \quad , \quad R_y = \sum_{i=1}^{N} A_{iy} \quad , \quad R_z = \sum_{i=1}^{N} A_{iz}$$

4. Apply the Pythagorean theorem to determine the magnitude of the resultant vector from its Cartesian components:[92]

$$R^2 = R_x^2 + R_y^2 + R_z^2$$

Apply trigonometry to determine the direction of the resultant. Solve for the polar angle, $\theta$, if the resultant lies in the $xy$ plane; otherwise, solve for the azimuthal and polar angles, $\varphi$ and $\theta$, respectively, of spherical coordinates (see Sec. 2.5):

$$\theta = \tan^{-1} \left( \frac{R_y}{R_x} \right) \quad \text{(in 2D)}$$

$$\varphi = \tan^{-1} \left( \frac{R_y}{R_x} \right) \quad , \quad \theta = \cos^{-1} \left( \frac{R_z}{R} \right) \quad \text{(in 3D)}$$

---

[92] This is actually a generalization of the Pythagorean theorem to find the long diagonal of a 3D rectangular box (called a cuboid – which is to a cube as a rectangle is to a square). We will discuss this, also, in Sec. 2.5.

**Important Distinction.** Vector addition is very different from scalar addition. When adding two physical quantities, it's critical to first determine whether they are vectors or scalars, and then add them appropriately. It would be a major conceptual mistake to add two magnitudes together for quantities that are vectors – direction must be taken into account by working with components, according to the strategy for adding vectors.[93]

**Law of cosines:** Recall from trigonometry that the sides of any triangle – even if it is not a right triangle – can be related through the law of cosines: $c^2 = a^2 + b^2 - 2ab\cos\theta$, where $c$ is the side opposite to the angle $\theta$. Note that if the angle $\theta$ is obtuse – i.e. $\theta > 90°$ – then $\cos\theta$ is negative, in which case $-2ab\cos\theta$ is positive (since there are two minus signs in this case): This should make sense, conceptually, since $c^2$ should be less than $a^2 + b^2$ if $\theta$ is acute, equal to $a^2 + b^2$ if $\theta$ is right, and greater than $a^2 + b^2$ if $\theta$ is obtuse. Note that $\theta = 90°$ reproduces the Pythagorean theorem.[94]

If you draw two vectors tip-to-tail, the two vectors and their resultant make a triangle. The magnitudes of these vectors could be related to one another by the law of cosines. However, the strategy that we have learned for adding vectors is much more practical than applying the law of cosines for two reasons: (1) Vector addition is much more efficient than the law of cosines when adding together three or more vectors, as is often the case in physics; and (2) vector addition is also much more straightforward for 3D problems than trying to apply the law of cosines for a general 3D problem.

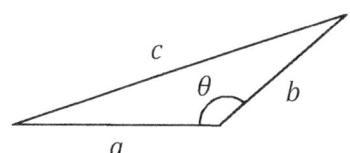

**Math Note.** Students who are fluent in the law of cosines sometimes try to apply it to physics problems, but it's not a good idea since many problems in physics – especially, in subsequent chapters – involve three or more vectors added together. In this case, applying the law of cosines repeatedly is not nearly as efficient as the strategy for adding vectors outlined on the previous page. Furthermore, the law of cosines is inconvenient for general 3D problems, compared to adding components of vectors together.

**Example.** A monkey walks 200 m at a direction of 120° and then walks 100 m at a direction of 240°. Find the magnitude and direction of the monkey's net displacement.

The problem is visually depicted by the diagram below.

---

[93] For example, in second-semester physics, which covers electricity and magnetism, the superposition of electric fields must be carried out by treating the electric fields as vectors. There are always a few students who mistakenly try to just add their magnitudes together. However, there are also students who try to add electric potentials together as if electric potential were a vector – when electric potential is, in fact, a scalar. It's very important to first determine the nature of what you're adding, and then add the quantities appropriately.

[94] In this sense, the law of cosines is also a generalization of the Pythagorean theorem – a generalization to include triangles that are not right triangles.

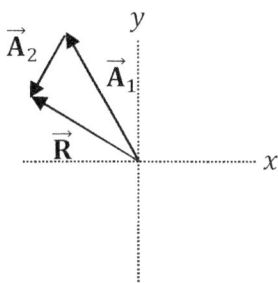

First, draw a sketch to help determine the signs of the components, and to check that the final answer comes out in the right quadrant. Draw the given vectors tip-to-tail in the sketch, then draw the resultant from the tail of the first to the tip of the last.[95]

Next, find the Cartesian components of the given vectors. In the sketch above, you can see that $A_{1x}$, $A_{2x}$, and $A_{2y}$ are negative, while $A_{1y}$ is positive, which must agree with the math:

$$A_{1x} = A_1 \cos \theta_1 = 200 \cos 120° = -100 \text{ m} \quad , \quad A_{1y} = A_1 \sin \theta_1 = 200 \sin 120° = 100\sqrt{3} \text{ m}$$
$$A_{2x} = A_2 \cos \theta_2 = 100 \cos 240° = -50 \text{ m} \quad , \quad A_{2y} = A_2 \sin \theta_2 = 100 \sin 240° = -50\sqrt{3} \text{ m}$$

Now add the respective components together to find the Cartesian components of the resultant vector:

$$R_x = A_{1x} + A_{2x} = -100 \text{ m} - 50 \text{ m} = -150 \text{ m}$$
$$R_y = A_{1y} + A_{2y} = 100\sqrt{3} \text{ m} - 50\sqrt{3} \text{ m} = 50\sqrt{3} \text{ m}$$

Lastly, apply the Pythagorean theorem and trigonometry to determine the magnitude and direction of the resultant vector, checking that the answers agree with the sketch:[96]

$$R = \sqrt{R_x^2 + R_y^2} = \sqrt{(-150)^2 + (50\sqrt{3})^2} = 50\sqrt{3^2 + (\sqrt{3})^2} = 100\sqrt{3} \text{ m}$$
$$\theta = \tan^{-1}\left(\frac{R_y}{R_x}\right) = \tan^{-1}\left(\frac{50\sqrt{3}}{-150}\right) = \tan^{-1}\left(-\frac{1}{\sqrt{3}}\right) = 150°$$

---

[95] Don't make assumptions about the angles. For example, this sketch looks like a right triangle, but is it? Don't assume it is. If you can prove that it is, showing your work/reasoning completely, that's different – but even so, your instructor probably wants to see evidence that you have learned the general strategy. Round numbers were used so that you can follow the math readily without need of a calculator, so the triangle might be right. If so, just ignore it – vector addition usually doesn't involve a right triangle, so you need to learn the general strategy.
[96] Try this on your calculator, then study the sketch above.

**Additive inverse**: The vector that could be added to a given vector, $\vec{A}$, to produce the null vector defines the additive inverse, $-\vec{A}$, of a vector: $\vec{A} + (-\vec{A}) = 0$. In order for a vector, $\vec{A}$, and its additive inverse, $-\vec{A}$, to add up to zero when added tip-to-tail, the additive inverse, $-\vec{A}$, must have the same magnitude as $\vec{A}$, but opposite direction. You obtain the additive inverse by simply negating the vector.

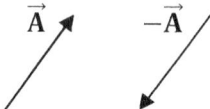

Given the magnitude, $A$, and direction, $\theta$, of a vector, $\vec{A}$, its additive inverse, $-\vec{A}$, can be constructed by preserving the magnitude, $\|-\vec{A}\| = A$, and adding (or subtracting[97]) 180° to its direction, $\theta \to \theta + 180°$. This has the effect of negating both components: $A_x \to -A_x$ and $A_y \to -A_y$.

**Vector subtraction**: The subtraction of one vector, $\vec{A}$, from another vector, $\vec{B}$, to make a third vector, $\vec{C} = \vec{B} - \vec{A}$, can be expressed as the sum of $\vec{B}$ and $-\vec{A}$, where $-\vec{A}$ is the additive inverse of $\vec{A}$: That is, $\vec{C} = \vec{B} + (-\vec{A})$. In this way, any vector subtraction problem can be expressed as a vector addition problem.

---

### Problem-Solving Strategy for Subtracting Vectors

0. Given the magnitudes and directions of two vectors, $\vec{A}$ and $\vec{B}$, apply this strategy to determine the magnitude and direction of $\vec{C} = \vec{B} - \vec{A}$.
1. Follow the same steps of the problem-solving strategy for adding vectors, with the following changes.
2. In the sketch, first negate $\vec{A}$ by drawing a vector in the opposite direction as $\vec{A}$. This negated vector is its additive inverse, $-\vec{A}$. Then add the additive inverse, $-\vec{A}$, and $\vec{B}$ tip-to-tail. The remaining vector, $\vec{C}$, extends from the tail of $-\vec{A}$ (realizing that its tail is opposite to where it was on the original vector, $\vec{A}$) to the tip of $\vec{B}$.
3. Either add 180° to $\theta_A$ and apply the remaining steps as usual, or leave $\theta_A$ unchanged and subtract components instead of adding components – i.e. $C_x = B_x - A_x$, $C_y = B_y - A_y$, and $C_z = B_z - A_z$.

---

**Math Note**. When you subtract two vectors, either add 180° to the angle of one vector and then proceed normally, or leave the angle unchanged and subtract components, rather than add them. You must choose one or the other – if instead you do both, you will cancel yourself out and effectively add, instead of subtract, the given vectors.

---

[97] You may add or subtract multiples of 360° to an angle without changing its direction. For example, 450° is the same direction as 90°. Observe that there is a 360° difference between adding 180° and subtracting 180°. For example, starting with 200°, adding 180° results in 380°, while subtracting 180° results in 20°; and 380° and 20° correspond to the same direction. Therefore, it doesn't make a difference if you choose to subtract, instead of add, 180° to the angle to negate a vector.

**Conceptual Example.** Graphically subtract the two vectors below according to $\vec{C} = \vec{B} - \vec{A}$.

First, negate $\vec{A}$ by flipping it to obtain $-\vec{A}$, and then join $-\vec{A}$ to $\vec{B}$ to obtain $\vec{C}$, which extends from the tail of $-\vec{A}$ to the tip of $\vec{B}$.

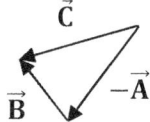

**Note.** If you're asked to subtract two vectors graphically, don't connect them tip-to-tail (without first negating the appropriate vector), as that would instead add the two vectors. Some students guess that the rule for subtracting two vectors might be to join them tail-to-tail or tip-to-tip, but you don't have to guess in physics – it's easy enough to work it out with a method that you know will work. If you join them tail-to-tail, you will have a dilemma – from which tail to which tail should the third vector point? You won't have this problem if you use the method of the previous example (highly recommended).

**Magnification of a vector:** If you wish to multiply a scalar quantity, $c$, with a vector quantity, $\vec{A}$, the new vector, $c\vec{A}$, has the same direction as $\vec{A}$, and the magnitude is multiplied by $c$: $\|c\vec{A}\| = c\|\vec{A}\|$.

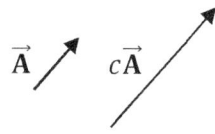

### Problem-Solving Strategy for Adding/Subtracting Magnified Vectors

0. Given the magnitudes and directions of $N$ vectors, $\vec{A}_1, \vec{A}_2, \cdots, \vec{A}_N$, apply this strategy to determine the magnitude and direction of $\vec{B} = \sum_{i=1}^{N} c_i \vec{A}_i$, where $c_1, c_2, \cdots, c_N$ are scalars (which may be negative).
1. Follow the same steps of the problem-solving strategy for adding vectors, with the following changes.
2. To make the sketch, first rescale the magnitudes of the given vectors, $\vec{A}_1, \vec{A}_2, \cdots, \vec{A}_N$, by the scaling factors, $c_1, c_2, \cdots, c_N$, negating any vector, $\vec{A}_i$, for which the corresponding scalar, $c_i$, is negative.
3. The simplest method is to leave the magnitudes and directions of the given vectors unchanged when you work out the math, but to use the scaling factors when you work out the components:

$$B_x = \sum_{i=1}^{N} c_i A_{ix} \quad , \quad B_y = \sum_{i=1}^{N} c_i A_{iy} \quad , \quad B_z = \sum_{i=1}^{N} c_i A_{iz}$$

**Example.** Given $\vec{A}(u) = 3u\hat{\imath} - 2\hat{\jmath}$ and $\vec{B}(u) = u^2\hat{\imath} - 3u\hat{\jmath}$, where $u$ is a dimensionless[98] variable, find

$$\vec{C}(u) = 3\vec{A} - 2\frac{d\vec{B}}{du}$$

Note that we're given the components of $\vec{A}$ and $\vec{B}$ (as functions of $u$), which means that we will skip the step where we find the components of the given vectors from their magnitudes and directions. Vector addition and subtraction are rather simple when expressed in terms of unit vectors. For this reason, many students work out all of their vector addition and subtraction solutions in terms of unit vectors, even when the problems give magnitudes and directions. When taking the derivative, the Cartesian unit vectors are constant (but, remember, other kinds of unit vectors generally are not constants). We can simply substitute the expressions for the given vectors into the equation for $\vec{C}$:

$$\vec{C}(u) = 3(3u\hat{\imath} - 2\hat{\jmath}) - 2\frac{d}{du}(u^2\hat{\imath} - 3u\hat{\jmath}) = 9u\hat{\imath} - 6\hat{\jmath} - 4u\hat{\imath} + 6\hat{\jmath} = 5u\hat{\imath}$$

Since the $y$-component of $\vec{C}$ vanished, its magnitude is simply $5u$ and its direction is $\hat{\imath}$ (or $0°$).

## 2.3 Vector Motion Quantities

**Position vector:** The position vector, $\vec{r}$, is a vector that extends from the origin of a coordinate system to the instantaneous position of a particle: The position vector, $\vec{r}$, is therefore unique, having coordinates $(x, y, z)$ for its Cartesian components:

$$\vec{r} = x\hat{\imath} + y\hat{\jmath} + z\hat{k}$$

The position vector of a particle indicates where the particle is relative to the origin of a coordinate system at any given time.

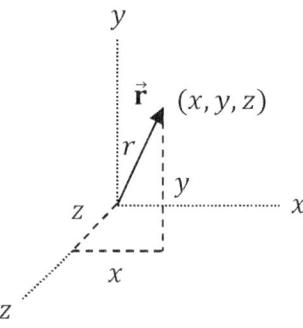

---

[98] Therefore, the variable $u$ has no units.

> **Note**. It is generally incorrect to use Cartesian coordinates, $x$, $y$, and $z$, for the Cartesian components, $A_x$, $A_y$, and $A_z$, of a vector, $\vec{\mathbf{A}}$. The position vector, $\vec{\mathbf{r}}$, is the only vector defined precisely this way.

**Net displacement**: The initial position vector, $\vec{\mathbf{r}}_0$, extends from the origin to the starting position $(x_0, y_0, z_0)$ of a particle, while the position vector, $\vec{\mathbf{r}}$, extends from the origin to the instantaneous position $(x, y, z)$ of the particle at time $t$. The vector subtraction of these position vectors, $\Delta\vec{\mathbf{r}} = \vec{\mathbf{r}} - \vec{\mathbf{r}}_0$, equals the net displacement of the particle – a vector that extends from the initial position to the instantaneous position. The net displacement can be expressed in Cartesian coordinates as

$$\Delta\vec{\mathbf{r}} = (x - x_0)\hat{\mathbf{i}} + (y - y_0)\hat{\mathbf{j}} + (z - z_0)\hat{\mathbf{k}} = \Delta x\hat{\mathbf{i}} + \Delta y\hat{\mathbf{j}} + \Delta z\hat{\mathbf{k}}$$

since $\vec{\mathbf{r}} = x\hat{\mathbf{i}} + y\hat{\mathbf{j}} + z\hat{\mathbf{k}}$ and $\vec{\mathbf{r}}_0 = x_0\hat{\mathbf{i}} + y_0\hat{\mathbf{j}} + z_0\hat{\mathbf{k}}$. The magnitude of the net displacement is

$$\|\Delta\vec{\mathbf{r}}\| = \sqrt{(x-x_0)^2 + (y-y_0)^2 + (z-z_0)^2} = \sqrt{(\Delta x)^2 + (\Delta y)^2 + (\Delta z)^2}$$

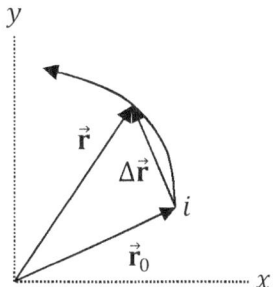

> **Note**. The initial position vector, $\vec{\mathbf{r}}_0$, and the net displacement, $\Delta\vec{\mathbf{r}}$, join tip-to-tail to form the position vector, $\vec{\mathbf{r}}$. Observe that $\vec{\mathbf{r}}$ is the resultant: $\vec{\mathbf{r}} = \vec{\mathbf{r}}_0 + \Delta\vec{\mathbf{r}}$. Solving for the net displacement, $\Delta\vec{\mathbf{r}} = \vec{\mathbf{r}} - \vec{\mathbf{r}}_0$, which is why the symbol $\Delta\vec{\mathbf{r}}$ is used to represent net displacement. Net displacement, $\Delta\vec{\mathbf{r}}$, is a vector representation of the change in an object's position.

**Differential displacement vector**: The net displacement, $\Delta\vec{\mathbf{r}}$, can also be represented on the differential scale. In the limit that the net displacement becomes infinitesimal, $\Delta x \to dx$, $\Delta y \to dy$, and $\Delta z \to dz$. In Cartesian coordinates, the differential displacement is[99]

$$d\vec{\mathbf{s}} = \hat{\mathbf{i}}dx + \hat{\mathbf{j}}dy + \hat{\mathbf{k}}dz$$

**Differential arc length**: The differential arc length, $ds$, is obtained from the distance formula in 3D space in the limit that $\Delta x \to dx$, $\Delta y \to dy$, and $\Delta z \to dz$. In Cartesian coordinates,

---

[99] The quantities $d\vec{\mathbf{s}}$ and $d\vec{\mathbf{r}}$, which are derived from different perspectives, turn out to be the same. Some texts use $d\vec{\mathbf{s}}$, while others use $d\vec{\mathbf{r}}$, but both are the same. We explain the origin of these two different symbols, $d\vec{\mathbf{s}}$ and $d\vec{\mathbf{r}}$, which represent the same quantity, in the next footnote. In this text, we will favor $d\vec{\mathbf{s}}$ over $d\vec{\mathbf{r}}$.

## 2 Multi-Component Motion

$$ds = \sqrt{dx^2 + dy^2 + dz^2}$$

**Relationship between differential displacement and differential arc length**: The differential displacement, $d\vec{s}$, is a vector quantity corresponding to an infinitesimal change in position, and the differential arc length, $ds$, is a scalar quantity corresponding to an infinitesimal distance traveled. At the differential level, the differential displacement, $d\vec{s}$, has a magnitude equal to the differential arc length, $ds$, and a direction tangent to the path: $d\vec{s} = \hat{t} ds$, where $\hat{t}$ is a unit tangent. This definition of the differential displacement, in terms of the differential arc length, leads to the same definition of net displacement that we obtained previously by looking at an infinitesimal change in position – namely, $d\vec{s} = \hat{i} dx + \hat{j} dy + \hat{k} dz$.[100] Observe that the differential arc length equals the magnitude of the differential displacement vector: $ds = \|d\vec{s}\|$.

---

**Note**. Although the differential arc length equals the magnitude of the differential displacement vector, $ds = \|d\vec{s}\|$, it is important to realize that this is true only at the infinitesimal level. A finite distance traveled is generally greater than the magnitude of the corresponding displacement (they are only equal for an object that does not change direction – i.e. forward motion in a straight line only).

---

**Net displacement integral**: The net displacement, $\Delta\vec{r}$, can be found at the finite level by integrating the differential displacement:

$$\Delta\vec{r} = \int_i^f d\vec{s} = \int_{x=x_0}^{x} \hat{i} dx + \int_{y=y_0}^{y} \hat{j} dy + \int_{z=z_0}^{z} \hat{k} dz = (x - x_0)\hat{i} + (y - y_0)\hat{j} + (z - z_0)\hat{k}$$

As expected, this integral agrees with our previous definition of net displacement in terms of the change in position. Integrating the differential displacement, $d\vec{s}$, which is a vector quantity, is very easy in Cartesian coordinates, and leads to a path-independent result – i.e. it depends only on the endpoints.[101]

**Total distance traveled integral**: The total distance traveled, $\Delta s$, (or change in arc length) can be obtained at the finite level by integrating over the differential arc length:

---

[100] Defining differential displacement as the infinitesimal limit of the net displacement, $\Delta\vec{r}$, leads to $d\vec{s} = \hat{i} dx + \hat{j} dy + \hat{k} dz$, while defining the differential displacement in terms of the differential arc length, $d\vec{s} = \hat{t} ds$, leads to the same relation. The former definition would suggest calling the differential displacement $d\vec{r}$, while the latter suggests $d\vec{s}$. Once again, both quantities, $d\vec{r}$ and $d\vec{s}$, turn out to be the same.

[101] We would obtain the exact same result by integrating $\hat{t} ds$, since $d\vec{s} = \hat{t} ds$, but – as we will see shortly – integrating over $\hat{t} ds$ along the path can be quite a challenge compared to the trivial integration that we performed over $\hat{i} dx + \hat{j} dy + \hat{k} dz$. In order to integrate over $\hat{t} ds$, you have to express the unit tangent and the differential arc length each in terms of the same coordinates (such as $x$, $y$, and $z$), and integrate over the actual path (which provides a relationship between the coordinates, which wasn't necessary when integrating over $\hat{i} dx + \hat{j} dy + \hat{k} dz$). Observe that $\hat{t}$ is not a constant for a curved path, and therefore cannot be pulled out of the integral like the Caretesian unit vectors. The form $d\vec{s} = \hat{i} dx + \hat{j} dy + \hat{k} dz$ is much more convenient than $d\vec{s} = \hat{t} ds$ for this integral.

$$\Delta s = \int_i^f ds = \int_i^f \sqrt{dx^2 + dy^2 + dz^2}$$

In practice, this is applied by factoring – e.g. for a curve in the $xy$ plane, $\Delta s = \int_{x=x_0}^{x} \sqrt{1 + (dy/dx)^2}\, dx$. Observe that the scalar quantity, $ds$, is much more difficult to integrate over than the vector quantity, $d\vec{s}$. Also, the integral over $ds$ is path-dependent, whereas the integral over $d\vec{s}$ is path-independent. One little arrow makes a very significant difference!

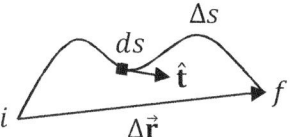

**Important Distinction.** The differential arc length is the magnitude of the differential displacement, $ds = \|d\vec{s}\|$, but the magnitude of the finite displacement vector is always less than or equal to the corresponding finite arc length: $s \geq \|\vec{s}\|$. That is, the arc length is not, in general, the magnitude of the displacement vector. Whether you speak of the differential quantities, $ds$ and $\|d\vec{s}\|$, or the finite quantities, $s$ and $\|\vec{s}\|$, makes a significant difference.

**Conceptual Example.** A monkey walks halfway around a circle with a radius of 25 m, from the eastern most point to the western most point on the circle. What are the total distance traveled and net displacement for the monkey's trip?

The trip is illustrated below, where a coordinate system has been setup.

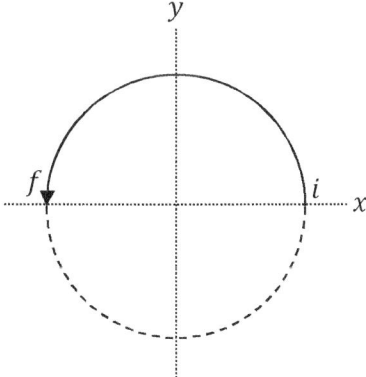

The total distance traveled (change in arc length), $\Delta s$, in this case equals half the circumference: $\Delta s = C/2 = \pi R_0 = 25\pi$ m. The net displacement is a vector that extends from the initial position to the final position. In this example, it has a magnitude equal to the diameter and the direction is along the $-x$-axis: $\Delta \vec{r} = -2R_0 \hat{\imath} = -50$ m $\hat{\imath}$. Note that the components of the net displacement are $\Delta x = -50$ m and $\Delta y = 0$. This problem was simple enough that no integration was necessary.

**Example.** A monkey strolls along a parabolic arc given by the equation $y = 3x^2$ from $x = -1.00$ m to $x = 2.00$ m, where $y$ has the same SI units as $x$.[102] Determine the total distance traveled and net displacement for the monkey's stroll.

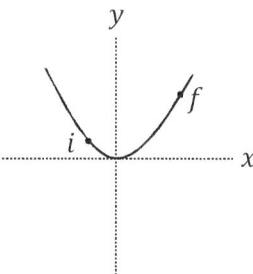

The total distance traveled can be computed by integrating the differential arc length. It is convenient to factor out a $dx$ from $ds = \sqrt{dx^2 + dy^2}$. Also, the resulting derivative, $dy/dx$, can be evaluated for the equation of the curve, $y = 3x^2$: $dy/dx = 6x$. Putting this all together:

$$\Delta s = \int_i^f ds = \int_i^f \sqrt{dx^2 + dy^2} = \int_{x=x_0}^{x} \sqrt{1 + (dy/dx)^2}\, dx \quad \text{(in 2D)}$$

$$\Delta s = \int_{x=-1}^{2} \sqrt{1 + 36x^2}\, dx$$

This integral can be performed by making the substitution $x \equiv (\tan\theta)/6$. The reason for this substitution is that $1 + 36x^2 = 1 + \tan^2\theta = \sec^2\theta$, which collapses the two terms inside the squareroot down to a single term. Since the integrand will then be expressed in terms of $\theta$, we must also express the differential in terms of $\theta$: $dx = (\sec^2\theta)d\theta/6$. With these substitutions, the integral becomes:[103]

$$\Delta s = \frac{1}{6}\int_{x=-1}^{2} \sec^3\theta\, d\theta$$

This integral can be performed by parts, identifying $u \equiv \sec\theta$ and $dv \equiv \sec^2\theta\, d\theta$. In this case, $du = \sec\theta \tan\theta\, d\theta$ and $v = \tan\theta$. Plugging this into the rule for integrating by parts, the total distance traveled becomes:

---

[102] SI units were suppressed in the equation to avoid potentially confusing clutter. With SI units, the equation must read $y = 3 \text{ m}^{-1} x^2$. The general problem is that the same symbol may be used for a different quantity and units – e.g. mass, $m$, could easily be confused with the meter, m. So by leaving the units out of the equation, we may focus exclusively on the symbols. At the end of the calculation, careful consideration of the units is important.
[103] Note that we haven't changed the limits yet. You could write the limits in terms of $\theta$, or, as we will do here, you can rewrite the antiderivative in terms of $x$ before evaluating it at the limits.

$$6\Delta s = \int_{v=v_0}^{v} u\,dv = [uv]_i^f - \int_{u=u_0}^{u} v\,du = [\sec\theta \tan\theta]_{x=-1}^{2} - \int_{x=-1}^{2} \sec\theta \tan^2\theta\, d\theta$$

$$6\Delta s = [\sec\theta \tan\theta]_{x=-1}^{2} - \int_{x=-1}^{2} \sec\theta\,(\sec^2\theta - 1)d\theta$$

$$6\Delta s = [\sec\theta \tan\theta]_{x=-1}^{2} - \int_{x=-1}^{2} \sec^3\theta\, d\theta + \int_{x=-1}^{2} \sec\theta\, d\theta$$

$$6\Delta s = [\sec\theta \tan\theta]_{x=-1}^{2} - 6\Delta s + [\ln|\sec\theta + \tan\theta|]_{x=-1}^{2}$$

$$\Delta s = \frac{[\sec\theta \tan\theta + \ln|\sec\theta + \tan\theta|]_{x=-1}^{2}}{12}$$

Recall that, at one point, we had $6\Delta s = \int_{x=-1}^{2} \sec^3\theta\, d\theta$: In the algebra above, we brought this term to the left-hand-side, making $6\Delta s + 6\Delta s$, and then divided both sides by 12.[104] We can now return to the original variable, $x$, by writing $\tan\theta = 6x$ and $\sec\theta = \sqrt{1 + \tan^2\theta} = \sqrt{1 + 36x^2}$:

$$\Delta s = \frac{\left[6x\sqrt{1 + 36x^2} + \ln\left|6x + \sqrt{1 + 36x^2}\right|\right]_{x=-1}^{2}}{12}$$

$$\Delta s = \frac{12\sqrt{145} + 6\sqrt{37} + \ln\left|12 + \sqrt{145}\right| - \ln\left|-6 + \sqrt{37}\right|}{12} = 15.6 \text{ m}$$

The net displacement is much simpler to calculate. Since the net displacement is path-independent, only the coordinates of the initial and final positions matter. Use the equation $y = 3x^2$ to determine the initial and final $y$-coordinates: $y(-1) = 3$ and $y(2) = 12$. The net displacement is

$$\Delta\vec{r} = \int_i^f d\vec{s} = \int_{x=x_0}^{x} \hat{\imath}\,dx + \int_{y=y_0}^{y} \hat{\jmath}\,dy = (x - x_0)\hat{\imath} + (y - y_0)\hat{\jmath} = [2 - (-1)]\hat{\imath} + (12 - 3)\hat{\jmath}$$

$$\Delta\vec{r} = 3\hat{\imath} + 9\hat{\jmath}$$

Observe that the magnitude of the net displacement, $\|\Delta\vec{r}\| = \sqrt{(\Delta x)^2 + (\Delta y)^2} = \sqrt{3^2 + 9^2} = 9.49$ m, is smaller than the total distance traveled, $\Delta s = 15.6$ m.

---

[104] One possible way to perform an integral of the form $\int_i^f f(x)dx$ is to integrate it by parts. In order to do this, you need to separate $f(x)dx$ into two pieces: $f(x)dx = u\,dv$, where $u$ includes part of $f(x)$ and $dv$ includes the remaining factor of $f(x)$ as well as $dx$. In this example, we separated $\int_i^f \sec^3\theta\, d\theta$ into $u \equiv \sec\theta$ and $dv \equiv \sec^2\theta\, d\theta$. The prescription for integrating by parts originates from the product rule: That is, in terms of implicit differentials, $d(uv) = u\,dv + v\,du$. In terms of integrals, bringing one term to the other side: $\int_i^f u\,dv = [uv]_i^f - \int_i^f v\,du$. If you can't find the anti-derivative of $u(v)$ with respect to $v$, but you can find a way to split the integrand so that you could find the anti-derivative of $v(u)$ with respect to $u$, integrating by parts may be useful. For example, $\int_{\theta=0}^{\theta} \theta \sin\theta\, d\theta$ can be split into $u = \theta$ and $dv = \sin\theta\, d\theta$, for which $du = d\theta$ and $v = -\cos\theta$. Integration by parts then gives $\int_{\theta=0}^{\theta} \theta \sin\theta\, d\theta = [-\theta \cos\theta]_{\theta=0}^{\theta} + \int_{\theta=0}^{\theta} \cos\theta\, d\theta$, which is now easy to integrate.

## 2 Multi-Component Motion

**Velocity vector**: The derivative of the position vector, $\vec{r}$, with respect to time, $t$, equals the velocity vector, $\vec{v}$:

$$\vec{v} = \frac{d\vec{r}}{dt}$$

Conceptually, the velocity, $\vec{v}$, provides a measure of both how fast and which way an object is instantaneously moving. Velocity is the instantaneous rate at which the position of an object is changing in time.

> **Math Note.** When taking a derivative of the position vector, $\vec{r}$, with respect to time, $t$, to obtain velocity, $\vec{v}$, you must consider whether or not the unit vectors in the expression are constants. The Cartesian unit vectors – $\hat{\imath}$, $\hat{\jmath}$, and $\hat{k}$ – are constants, whereas most other unit vectors (like polar, spherical, tangential, and normal unit vectors) are usually not constant. You must apply the product rule for differentiation when the position vector is expressed in terms of unit vectors that are not constants. The same holds for derivatives of other vectors (with respect to any variable), such as taking a derivative of velocity, $\vec{v}$, with respect to time, $t$, to find acceleration, $\vec{a}$.

**Speed**: A derivative of the arc length, $s$, with respect to time, $t$, equals the speed, $v$:

$$v = \frac{ds}{dt}$$

The speed is a nonnegative quantity because arc length only increases[105] with increasing time: $v \geq 0$. While speed, $v$, is defined as the instantaneous time rate of change of arc length (total distance traveled), velocity, $\vec{v}$, is defined as the instantaneous time rate of change of the position vector. Yet, from these two distinctly different definitions, speed turns out to equal the magnitude of the velocity: $v = \|\vec{v}\| = \sqrt{v_x^2 + v_y^2 + v_z^2}$.

**Proof that speed is the magnitude of velocity**: From the definition of velocity,[106]

$$\vec{v} = \frac{d\vec{r}}{dt} = \frac{d}{dt}(x\hat{\imath} + y\hat{\jmath} + z\hat{k}) = \hat{\imath}\frac{dx}{dt} + \hat{\jmath}\frac{dy}{dt} + \hat{k}\frac{dz}{dt} = v_x\hat{\imath} + v_y\hat{\jmath} + v_z\hat{k}$$

This vector equation incorporates three separate equations – one for each component:

---

[105] You can shorten the magnitude of your net displacement, $\|\Delta\vec{r}\|$, by retracing your footsteps, but not your total distance traveled, $\Delta s$. If you run clockwise around a circle, and then counterclockwise, for example, your net tangential displacement, $\Delta s_T$, (see Chapter 1) shrinks when you retrace your footsteps, but your total distance traveled (change in arc length), $\Delta s$, increases regardless of which way you move.

[106] Since the differential displacement is $d\vec{s} = \hat{\imath}dx + \hat{\jmath}dy + \hat{k}dz$, observe that velocity, $\vec{v}$, both equals $\vec{v} = \frac{d}{dt}\vec{r}$ and $\vec{v} = \frac{d}{dt}\vec{s}$. It's another way to see that $d\vec{r}$ is the same as $d\vec{s}$.

$$v_x = \frac{dx}{dt} \quad , \quad v_y = \frac{dy}{dt} \quad , \quad v_z = \frac{dz}{dt}$$

The magnitude of the velocity vector is found by applying the 3D generalization of the Pythagorean theorem:

$$\|\vec{v}\| = \sqrt{v_x^2 + v_y^2 + v_z^2} = \sqrt{\left(\frac{dx}{dt}\right)^2 + \left(\frac{dy}{dt}\right)^2 + \left(\frac{dz}{dt}\right)^2} = \frac{\sqrt{dx^2 + dy^2 + dz^2}}{dt} = \frac{ds}{dt} = v$$

**Average velocity**: Applying the mean-value theorem from calculus, the average velocity, $\vec{\bar{v}}$, equals the net displacement, $\Delta\vec{r}$, divided by the elapsed time, $\Delta t$:

$$\vec{\bar{v}} = \frac{\int_{t=t_0}^{t} \vec{v}(t)dt}{\int_{t=t_0}^{t} dt} = \frac{\Delta\vec{r}}{\Delta t} = \frac{\text{net displacement}}{\text{elapsed time}}$$

If you need to determine the average velocity of an object, you should first determine its net displacement and then divide by the corresponding time interval.

**Average speed**: Similarly, the mean-value theorem from calculus can be applied to find the average speed, $\bar{v}$, from the total distance traveled, $\Delta s$, and the elapsed time, $\Delta t$:

$$\overline{v(t)} = \frac{\int_{t=t_0}^{t} v(t)dt}{\int_{t=t_0}^{t} dt} = \frac{\Delta s}{\Delta t} = \frac{\text{total distance traveled}}{\text{elapsed time}}$$

In order to find the average speed of an object, first determine the total distance traveled and then divide by the corresponding time interval.

**Important Distinction.** Average speed and average velocity are similar quantities, but have significantly different definitions. Average speed involves the total distance traveled, and is a nonnegative scalar, whereas average velocity involves the net displacement, and is a vector.

**Conceptual Example.** A monkey circumnavigates the earth with a constant speed of 35 m/s, returning to his starting position. What are the average speed and average velocity of the monkey relative to the monkey's starting position?[107]

---

[107] If the problem instead stated, "relative to the center of the earth," then you would have to account for the earth's rotation. If you sit down on your couch in New York City, for example, you experience a net displacement – and therefore, also, a nonzero average velocity – as the earth rotates (though the net displacement is zero for multiplies of 24 hours). Similarly, relative to the sun the earth sweeps out an astronomical net displacement (except for multiples of 1 year, for which the net displacement is zero). Neglect the rotation and revolution of the earth (and motion of the sun through the galaxy, relative motion of the galaxy, etc.) unless stated otherwise.

## 2 Multi-Component Motion

> You should be thinking about finding the total distance traveled and the elapsed time because that's how average speed is defined, but you don't actually need to look up the circumference of the earth in this case. Since the speed is constant, it doesn't change, which makes it very easy to average: The average speed is simply $\bar{v} = 35$ m/s.
>
> However, the velocity is not constant. The direction of the monkey's velocity changes as he circumnavigates the globe; if he had traveled in a straight line, he would have gone off on a tangent out into space! If he thought he was moving in a straight line the whole time, he was actually traveling in a circle – the circumference of the earth. It's very easy to calculate the net displacement of the monkey: $\Delta \vec{\mathbf{r}} = 0$ because he finished where he started. The average velocity is therefore zero: $\bar{\vec{\mathbf{v}}} = 0$.

**Uniform velocity**: An object's velocity is said to be uniform if its velocity is constant – i.e. if its speed is constant and it travels along a straight line. In the case of constant velocity, the average velocity of the object is simply the velocity:

$$\vec{\mathbf{v}} = \frac{\Delta \vec{\mathbf{r}}}{\Delta t} \quad \text{(only if } \vec{\mathbf{v}} = const.\text{)}$$

**Constant speed**: Similarly, if an object travels with constant speed, it's really easy to average:

$$v = \frac{\Delta s}{\Delta t} \quad \text{(only if } v = const.\text{)}$$

> **Note**. In the vast majority of physics problems, speed and velocity are changing. It would be a major conceptual mistake to use the equation for constant speed (or velocity) for an object that is changing speed (or velocity).

> **Important Distinction**. Uniform speed means that speed is constant, whereas uniform velocity requires both speed and direction to be constant. If an object is changing direction, its velocity is definitely not uniform; if an object has constant speed, its velocity may not be uniform.

**Acceleration vector**: The derivative of the velocity, $\vec{\mathbf{v}}$, with respect to time, $t$, equals the acceleration vector, $\vec{\mathbf{a}}$:

$$\vec{\mathbf{a}} = \frac{d\vec{\mathbf{v}}}{dt} = \frac{d^2 \vec{\mathbf{r}}}{dt^2}$$

Conceptually, the acceleration, $\vec{\mathbf{a}}$, provides a measure of the instantaneous rate at which the velocity of an object is changing in time. An object can have acceleration by changing its speed or by changing its direction of motion (or both).

**Average acceleration**: Applying the mean-value theorem from calculus, the average acceleration, $\bar{\vec{\mathbf{a}}}$, equals the change in velocity, $\Delta \vec{\mathbf{v}}$, divided by the elapsed time, $\Delta t$:

$$\vec{\bar{a}} = \frac{\int_{t=t_0}^{t} \vec{v}(t)dt}{\int_{t=t_0}^{t} dt} = \frac{\Delta \vec{v}}{\Delta t} = \frac{\vec{v} - \vec{v}_0}{\Delta t} = \frac{\text{change in velocity}}{\text{elapsed time}}$$

If you need to determine the average acceleration of an object, you should first determine its change in velocity – which is a vector subtraction – and then divide by the corresponding time interval.

**Important Distinction.** The adjective 'average' makes a significant difference. For example, if you're looking for acceleration, tangential acceleration, or centripetal acceleration, it would be a major conceptual mistake to find the change in velocity and divide by the elapsed time (only if acceleration happens to be uniform – which is not always the case – will it coincidentally be correct). The equation above applies specifically to average acceleration, but certainly not to acceleration in general.

**Conceptual Example.** A monkey gets into his car and accelerates from rest to 30 m/s in 6.0 s, then drives with constant velocity to the north for 20 s. The monkey then uniformly decelerates for 10 s until coming to rest. What is the monkey's average acceleration for this trip? It's zero: The initial and final velocities are the same (both zero), so the change in velocity is zero. To help see how the average acceleration can be zero, consider that the velocity increased during the first 6.0 s, but decreased during the last 10 s – that is, the monkey's acceleration was both positive and negative during the trip. Note that the monkey's average velocity is not zero, though, because he does have a nonzero net displacement; nor is the average speed zero because there is a nonzero total distance traveled.

**Tangential and normal components of motion**: When a particle travels along a curved path, it is often convenient to work with tangential and normal components for two reasons: For one, as we will see shortly, these components have a natural conceptual interpretation; and for another, when we study uniform circular motion we will learn a useful formula for the component of acceleration associated with the normal component, which makes the mathematics also convenient in terms of these components.

The velocity vector is naturally a tangential vector – i.e. the direction of the velocity is tangential to the path. Therefore, the velocity, $\vec{v}$, can be expressed as the tangential component of velocity, $v_T$, times a unit tangent, $\hat{t}$:

$$\vec{v} = v_T \hat{t}$$

As in Chapter 1, the tangential velocity can be negative. The speed, $v$, which is the magnitude of the velocity, $v = \|\vec{v}\|$, equals the absolute value of the tangential velocity, $v = |v_T|$, and is therefore nonnegative.[108]

---

[108] The tangential component of velocity is not to be confused with the transverse component of velocity, which is useful in the context of orbits. The transverse component is perpendicular to the position vector, $\vec{r}$, whereas the tangential velocity is along the tangent, $\hat{t}$; the transverse direction is tangential for a circular orbit, but not in general. When applying Kepler's second law to describe how the speed of a planet or comet, for example, varies with its distance from the sun, it is useful to use the transverse component in conservation of angular momentum.

In general, the acceleration vector, $\vec{a}$, has both tangential and normal components, $a_T$ and $a_c$ (the normal component designated as 'centripetal,' meaning inward, with a subscript $c$), which together – as we will see, precisely, momentarily – describe the instantaneous rate at which velocity (both speed and direction) changes in time. When we study forces in relation to curved paths, we will often therefore resolve them into tangential and normal components in order to relate them to the tangential and normal components of the acceleration.

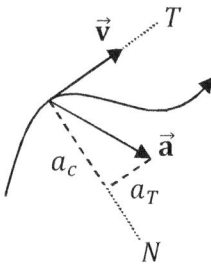

**Tangential acceleration**: The tangential component of acceleration, $a_T$, describes the instantaneous rate of change of the tangential velocity, $v_T$.

$$a_T = \frac{dv_T}{dt}$$

Since the speed is related to the tangential component of velocity by $v = |v_T|$, in essence (i.e. aside from the signs) the tangential component of acceleration provides a measure of how the speed changes in time.

> **Note**. It is important to realize that $a_T$ is only the tangential component of acceleration, not the magnitude of the acceleration vector, $\vec{a}$; the tangential component of acceleration, $a_T$, describes how the tangential component of velocity, $v_T$, changes in time.

**Centripetal acceleration**: The centripetal component of acceleration, $a_c$, is a normal component of acceleration directed along the inward normal, $\hat{n}_i$.[109] The centripetal component of acceleration describes how the direction of the velocity changes in time. We will explore centripetal acceleration further in Chapter 4, where the direction of velocity will change, but the speed will not.

**Acceleration in terms of tangential and normal components**: In general, the acceleration vector will have one tangential component, $a_T$, and one normal component – the centripetal acceleration, $a_c$, which is directed inward:

---

[109] In vector calculus, you learn about the radius of curvature vector. At any point on a curve, you can find the circle that best matches the curvature – analogous to finding a line that best matches the slope, which we call a tangent line. The radius of the circle that matches the curvature at any given point defines the direction of $\hat{n}_i$.

$$\vec{a} = a_T\hat{t} + a_c\hat{n}_i$$

The magnitude of the acceleration, $a = \|\vec{a}\|$, is related to the tangential and normal components by the Pythagorean theorem:

$$a = \|\vec{a}\| = \sqrt{a_T^2 + a_c^2}$$

For 1D motion, the acceleration vector reduces to $\vec{a} = a_T\hat{t}$, which we studied in Chapter 1, and for circular motion it becomes $\vec{a} = a_c\hat{n}_i$, which we will explore in detail in Chapter 4.

**Conceptual Example.** A monkey drives a bananamobile with a constant speed of 4.0 m/s, yet he claims that his acceleration is not zero. Is this possible? Yes! If the monkey is changing direction – i.e. traveling along a curved path, rather than a straight line – then he will have acceleration even though his speed is constant. Remember, acceleration describes how velocity – which is a combination of speed and direction changes in time. If either speed or direction change, the monkey is accelerating. In this case, the monkey's tangential acceleration is zero, but his centripetal acceleration (and therefore his overall acceleration) is not zero, provided that he is changing direction.

**Important Distinction.** Since there are several different types of acceleration, it's very important to pay attention to which adjective is used, if any. Whether you are asked for tangential acceleration, $a_T$, centripetal acceleration, $a_c$, or the magnitude of the acceleration, $a = \|\vec{a}\|$, makes a significant difference. If you are just asked for the acceleration, strictly speaking, this means to find the acceleration vector, $\vec{a}$, although textbooks often really mean just to find the magnitude, $a = \|\vec{a}\|$. There is also gravitational acceleration, $g$, and later we will learn about angular acceleration, $\alpha$.

**Relationship between vector motion quantities**: The vector motion quantities are related to one another via subsequent derivatives with respect to time:

$$\vec{v} = \frac{d\vec{r}}{dt} \quad , \quad \vec{a} = \frac{d\vec{v}}{dt} = \frac{d^2\vec{r}}{dt^2}$$

These relations can also be expressed as definite integrals:

$$\Delta\vec{r} = \int_{t=t_0}^{t} \vec{v}\, dt \quad , \quad \Delta\vec{v} = \vec{v} - \vec{v}_0 = \int_{t=t_0}^{t} \vec{a}\, dt$$

**Relationship between scalar motion quantities**: The total distance traveled, $\Delta s$, can be computed directly by integration:

## 2 Multi-Component Motion

$$\Delta s = \int_i^f ds = \int_i^f \sqrt{dx^2 + dy^2 + dz^2}$$

The speed, $v$, is most easily computed when the components of the velocity vector are known:

$$v = \|\vec{\mathbf{v}}\| = \sqrt{v_x^2 + v_y^2 + v_z^2}$$

The speed and arc length are also related directly through calculus:

$$v = \frac{ds}{dt} \quad , \quad \Delta s = \int_{t=t_0}^{t} v\,dt$$

**Note.** If you are looking for help regarding how to apply calculus to 2D and 3D motion problems, see Sec. 2.7.

### 2.4 Polar and Cylindrical Coordinates

**2D Polar coordinates**: It's much easier to work with 2D polar coordinates than Cartesian coordinates to solve the equations of motion for an object traveling along a circular arc or to find the center of mass of a slice of pie, for example. In these cases, instead of working with horizontal and vertical coordinates, $x$ and $y$, it is much simpler to work with a radial coordinate $r$ and a corresponding angle $\theta$.

**2D Radial coordinate**: The radial coordinate, $r$, in 2D polar coordinates equals the distance from the origin to a given point $(x, y)$ in the $xy$ plane. The radial coordinate is strictly nonnegative: $r \geq 0$, unlike the Cartesian coordinates, $x$ and $y$, which can be negative.

**2D Polar angle**: The 2D polar angle, $\theta$, measures the angle between the positive $x$-axis and the radial distance, $r$, in a counterclockwise sense. Every point in the $xy$ plane can be mapped by $r \geq 0$ and $0 \leq \theta \leq 2\pi$.

**Math Note.** The direction comes from $\theta$, such that $r$ is nonnegative.[110]

---

[110] It would be redundant to allow $0 \leq \theta \leq 2\pi$ and to allow $r$ to be positive or negative. We could instead restrict $\theta$ to lie in the range $0 \leq \theta \leq \pi$ and allow $r$ to be positive or negative, but the convention is to restrict $r$ to be nonnegative. This is consistent with the magnitude of a vector, which must be nonnegative.

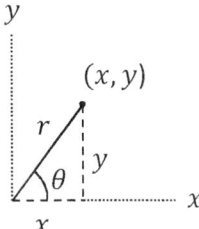

**Relating 2D polar and Cartesian coordinates**: The 2D polar coordinates are related to the Cartesian coordinates by

$$x = r\cos\theta \quad , \quad y = r\sin\theta \quad \text{(in 2D)}$$

The inverse relations are

$$r = \sqrt{x^2 + y^2} \quad , \quad \theta = \tan^{-1}\left(\frac{y}{x}\right) \quad \text{(in 2D)}$$

**Note**. The relationship between the 2D polar coordinate $(r, \theta)$ and the Cartesian coordinate $(x, y)$ has the same form as the relationship between the magnitude and direction $(A, \theta_A)$ of a vector, $\vec{A}$, and its Cartesian components $(A_x, A_y)$ in 2D. Similarly, when applying the inverse tangent to determine the 2D polar angle, it is necessary to examine the signs of $x$ and $y$ in order to establish in which quadrant the angle lies.

**Conceptual Example**. What is the equation for a circle centered at the origin in 2D polar coordinates? Simply set $r = const.$ to obtain the equation for a circle: $r = R_0$ corresponds to a circle of radius $R_0$ (since $\theta$ is a free parameter by virtue of its absence from the equation). In terms of Cartesian coordinates, this becomes $x^2 + y^2 = R_0^2$.

**Notation**. A lowercase $r$ designates the 2D polar coordinate – or, in 3D, the spherical radial coordinate; the lowercase $r$ is, in general, a variable, and often does not correspond to a radius of sorts. Henceforth, we will use the symbol $R_0$ to represent the radius of a circle, in order to avoid possible confusion with $r$; the radius $R_0$ is a constant.

**Conceptual Example**. Describe the curve given by the equation $r = c\theta$ in 2D, where $c$ is a constant. As $\theta$ grows, so does $r$. This curve will look like a spiral. Note that this curve has significance even for angles $> 2\pi$. Although adding multiples of $2\pi$ radians to an angle reproduces the same direction, the value of $r$ will continue to grow as $\theta$ increases.

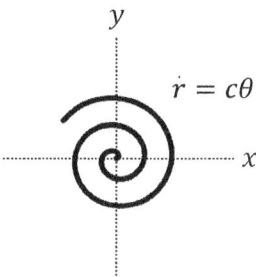

**2D Polar unit vectors**: In 2D polar coordinates, the unit vectors include the radial unit vector, $\hat{r}$, which is directed outward from the origin along a direction specified by $\theta$, and the tangential unit vector, $\hat{\theta}$, which is tangential to the unit circle in a counterclockwise sense. In 2D polar coordinates, the tangential unit vector, $\hat{\theta}$, is always 90° counterclockwise compared to the radial unit vector, $\hat{r}$.

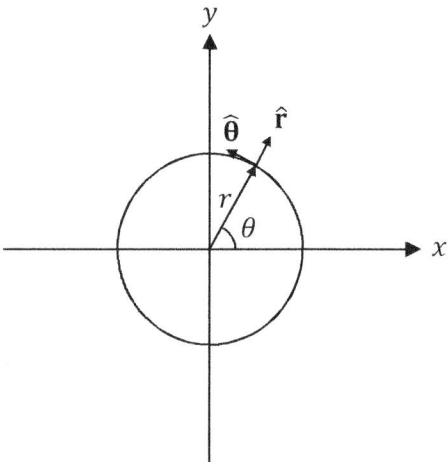

**Important Distinction.** The Cartesian unit vectors are constant, but polar, spherical, and cylindrical unit vectors are functions of angular coordinates. When differentiating or integrating vector quantities expressed in polar, spherical, or cylindrical coordinates, it is important to remember that the non-Cartesian unit vectors are not constant – i.e. you may not pull them out of the derivative or integral.

**Position vector in 2D polar coordinates**: The position vector can be expressed in polar coordinates as $\vec{r} = r\hat{r}$, which says to go outward from the origin a distance $r$ at a direction specified by the radial unit vector $\hat{r}$. The position vector, $\vec{r}$, is the same whether it is expressed in 2D polar coordinates or Cartesian coordinates: $\vec{r} = r\hat{r} = x\hat{i} + y\hat{j} = \hat{i}\,r\cos\theta + \hat{j}\,r\sin\theta$. Comparing the second and fourth expressions here, it can be seen that $\hat{r} = \hat{i}\cos\theta + \hat{j}\sin\theta$. The tangential unit vector is related to the radial unit vector by adding 90° to $\theta$, $\hat{\theta} = \hat{i}\cos(\theta + 90°) + \hat{j}\sin(\theta + 90°) = -\hat{i}\sin\theta + \hat{j}\cos\theta$.[111]

---

[111] Recall the trig identities $\cos(x + y) = \cos x \cos y - \sin x \sin y$ and $\sin(x + y) = \sin x \cos y + \sin y \cos x$. Another way to relate the polar unit vectors is to note that $\frac{d\hat{r}}{d\theta} = \hat{\theta}$ and $\frac{d\hat{\theta}}{d\theta} = -\hat{r}$.

$$\hat{\mathbf{r}} = \hat{\mathbf{i}}\cos\theta + \hat{\mathbf{j}}\sin\theta \quad \text{(in 2D)}$$
$$\hat{\boldsymbol{\theta}} = -\hat{\mathbf{i}}\sin\theta + \hat{\mathbf{j}}\cos\theta \quad \text{(in 2D)}$$

**Differential arc length in 2D polar coordinates**: At the differential level, the radial arc length, $dr$, and circular arc length, $rd\theta$, give the more general equation for arc length as

$$ds = \sqrt{dr^2 + (rd\theta)^2} \quad \text{(in 2D)}$$

**Circular arc length**: For a circular arc, $r = R_0 = const.$, so $dr = 0$ and $ds$ simplifies to $ds = R_0 d\theta$. In this case, the total distance traveled equals $\Delta s = \int_i^f ds = \int_{\theta=\theta_0}^{\theta} R_0 d\theta = R_0 \Delta\theta$.

> **Math Note**. In the formula for arc length, $\theta$ must be expressed in radians. As a general rule, if you see $\theta$ in a formula and it's not the argument of a trig function, it needs to be in radians.[112]

**Tangential and normal components for circular motion**: For an object traveling in a circle centered about the origin in the $xy$ plane, the radial unit vector, $\hat{\mathbf{r}}$, is a unit normal and $\hat{\boldsymbol{\theta}}$ is a unit tangent. However, the inward normal (associated with centripetal acceleration) equals $-\hat{\mathbf{r}}$. Thus, 2D polar coordinates are naturally suited for dividing circular motion into tangential and normal components.

**Cylindrical coordinates**: Adding the third Cartesian coordinate, $z$, to 2D polar coordinates makes (3D) cylindrical coordinates: $r_c$, $\theta$, and $z$.

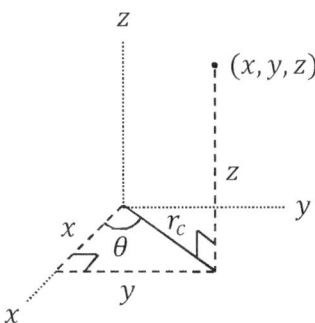

> **Note**. Illustrations of 3D coordinate systems can have no more than two axes lying in the plane of the plane, which means that one axis must point out of the page. In the illustration above, the $x$ axis is perpendicular to the plane of the page – we view the diagram with a perspective so that the projection of the $x$-axis onto the $yz$ plane appears to come at an angle, but it is really straight out of the page.[113]

---

[112] School is fun when students do their physics work mostly in degrees and their math work mostly in radians. A student might forget to switch the calculator mode from degrees to radians before taking a calculus quiz. Oops!

[113] It is conventional to draw right-handed coordinate systems. If you point the fingers of your right hand toward $+x$, curl them toward $+y$ (you may need to rotate your wrist, making sure that if you extend your fingers they still point along $+x$), and your extended thumb points along $+z$, then the coordinate system is right-handed; if instead your thumb points along $-z$, then the coordinate system is left-handed.

**Cylindrical radial coordinate**: The cylindrical radial coordinate, $r_c$, extends perpendicularly from the $z$-axis to the point $(x, y, z)$. After projecting the point $(x, y, z)$ along the $\pm z$-axis onto the $xy$ plane, the radial coordinate, $r_c$, and polar coordinate, $\theta$, are just like those of 2D polar coordinates.

> **Notation**. The importance of the subscript $c$ on the cylindrical radial coordinate, $r_c$, is to distinguish it from the spherical radial coordinate, $r$, which extends from the origin – instead of from the $z$-axis – to the point $(x, y, z)$. The symbols $r$ and $r_c$ cannot be used interchangeably.

**Relating cylindrical and Cartesian coordinates**: The cylindrical coordinates are related to the Cartesian coordinates by

$$x = r_c \cos\theta \quad , \quad y = r_c \sin\theta \quad , \quad z = z \quad \text{(cylindrical)}$$

The inverse relations are

$$r_c = \sqrt{x^2 + y^2} \quad , \quad \theta = \tan^{-1}\left(\frac{y}{x}\right) \quad , \quad z = z \quad \text{(cylindrical)}$$

**Unit vectors in cylindrical coordinates**: The cylindrical unit vectors are $\hat{\mathbf{r}}_c$, $\hat{\boldsymbol{\theta}}$, and $\hat{\mathbf{k}}$. The first two unit vectors, $\hat{\mathbf{r}}_c$ and $\hat{\boldsymbol{\theta}}$, lie in a plane that is parallel to the $xy$ plane and are defined just like the 2D polar unit vectors; the third unit vector, $\hat{\mathbf{k}}$, is the same as the third unit vector from Cartesian coordinates.

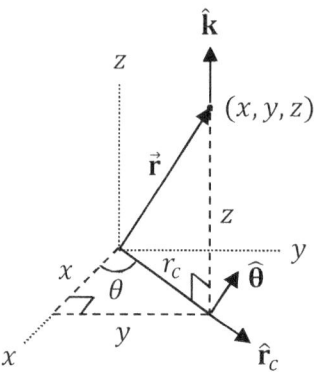

> **Math Note**. The polar unit vectors, $\hat{\mathbf{r}}_c$ and $\hat{\boldsymbol{\theta}}$, of cylindrical coordinates are generally variables (since they depend upon the angle $\theta$), whereas $\hat{\mathbf{k}}$ is a constant. It is important to remember this when applying calculus to vector expressions in cylindrical coordinates.

**Position vector in cylindrical coordinates**: In cylindrical coordinates, the position vector, $\vec{\mathbf{r}}$, has two components: $\vec{\mathbf{r}} = r_c \hat{\mathbf{r}}_c + z \hat{\mathbf{k}}$. The cylindrical unit vectors bear the same relationship with the Cartesian unit vectors as do the 2D polar unit vectors:

$$\hat{\mathbf{r}}_c = \hat{\mathbf{i}}\cos\theta + \hat{\mathbf{j}}\sin\theta \quad \text{(cylindrical)}$$
$$\hat{\boldsymbol{\theta}} = -\hat{\mathbf{i}}\sin\theta + \hat{\mathbf{j}}\cos\theta \quad \text{(cylindrical)}$$

**Differential arc length in cylindrical coordinates**: At the differential level, the radial arc length, $dr_c$, circular arc length, $r_c d\theta$, and vertical arc length, $dz$, give the more general equation for arc length as

$$ds = \sqrt{dr_c^2 + (r_c d\theta)^2 + dz^2} \quad \text{(cylindrical)}$$

## 2.5 Spherical Coordinates

**Spherical coordinates**: Any point in 3D space can be located by specifying its distance from the origin, $r$, and two angles, which we call $\varphi$ and $\theta$.

**Radial coordinate**: In spherical coordinates, the radial coordinate, $r$, is the distance from the origin to a given point $(x, y, z)$.

**Important Distinction.** The radial coordinate, $r$, of spherical coordinates is the distance from the origin to a given point $(x, y, z)$, while the radial cylindrical coordinate, $r_c$, of cylindrical coordinates is the perpendicular distance from the $z$-axis to a given point $(x, y, z)$. In general, $r_c \leq r$.

**Azimuthal angle**: The azimuthal angle, $\varphi$, is measured after first projecting the radial distance, $r$, onto the $xy$ plane: The azimuthal angle, $\varphi$, is the angle between the $+x$-axis and this projection, measured in a counterclockwise sense when viewed from the $+z$-direction.[114]

**Polar angle**: The polar angle, $\theta$, of spherical coordinates measures the angle between the $z$-axis and the radial distance, $r$.[115]

---

[114] The lowercase Greek letter $\varphi$ may be pronounced with either a long 'i' sound – as in 'pie' – or a long 'e' sound – as in 'fee.' Well, you probably can't get away with that if you speak Greek: The two common pronunciations ultimately result from the difference between traditionalists who try to imagine how the ancients would pronounce words in their original Latin and Greek and modernists who Americanize the words that we borrow from other languages. For example, if you pronounce 'cacti' as 'kak-tie,' you are being inconsistent if you say 'fee' instead of 'fie.' Perhaps the ultimate example of Americanizing words from other languages belongs to the word 'alumnus.' At graduation, a single male is an alumnus, a single female is an alumna, a group of males are alumni, and a group of females are alumnae. In the original tongue, you would say 'alumni' with a long 'e' (alum-nee) and 'alumnae' with a long 'i' (alum-nigh); but in modern English, we have swapped the pronunciations of these plural forms. Nevertheless, 'phi' is pronounced both 'fee' and 'fie' so often presently that it would be absurd to claim that either way is correct or better.

[115] Unfortunately, it is conventional to call the polar angle $\theta$ and the azimuthal angle $\varphi$ in physics, but to call the polar angle $\varphi$ and the azimuthal angle $\theta$ in math. It's really fun when you're taking physics and math exams on the same day, and both involve problems in spherical coordinates.

## 2 Multi-Component Motion

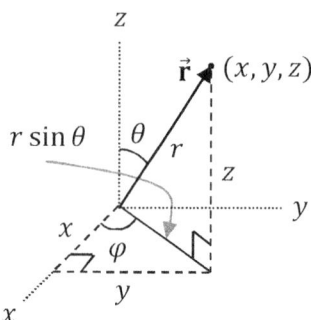

**Angular ranges**: The convention is to let the azimuthal angle, $\varphi$, vary over the range $0 \leq \theta \leq 2\pi$ and the polar angle, $\theta$, vary over the range $0 \leq \theta \leq \pi$.

**Note.** Observe that this choice covers every point $(x, y, z)$ in 3D space; it would be redundant to let both $\theta$ and $\varphi$ vary from $0 \leq \theta, \varphi \leq 2\pi$. To see this, imagine a fixed value for the polar angle, $\theta$, such as $45°$. Varying the radial coordinate, $r$, for this choice of $\theta$ produces a semi-infinite ray. Also varying the azimuthal angle, $\varphi$, this semi-infinite rays sweeps out a cone. If you now vary the polar angle, $\theta$, over the range $0 \leq \theta \leq \pi$, the cone will sweep out all of 3D space. Thus, $0 \leq \theta \leq 2\pi$ would be redundant.

**Longitude**: Setting both $r = R_0 = const.$ and $\varphi = const.$, while varying the polar angle, $\theta$, sweeps out a vertical circle called a longitude. The set of possible longitudes is a set of great circles – called 'great' because they all have the same radius. All of the longitudes pass through the two points $z = \pm R_0$, called the poles.

**Latitude**: Setting both $r = R_0 = const.$ and $\theta = const.$, while varying the azimuthal angle, $\varphi$, sweeps out a horizontal circle called a latitude. Unlike the longitudes, the latitudes come in different sizes, all with radii less than or equal to $R_0$. The largest latitude is called the equator and lies in the $xy$ plane.

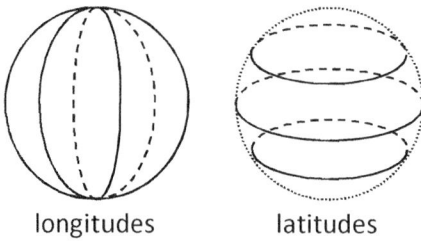

longitudes        latitudes

**Relating spherical and Cartesian coordinates**: The spherical coordinates are related to the Cartesian coordinates by

$$x = r \cos\varphi \sin\theta \quad , \quad y = r \sin\varphi \sin\theta \quad , \quad z = r \cos\theta \quad \text{(spherical)}$$

The inverse relations are

$$r = \sqrt{x^2 + y^2 + z^2} \quad , \quad \theta = \cos^{-1}\left(\frac{z}{r}\right) \quad , \quad \varphi = \tan^{-1}\left(\frac{y}{x}\right) \quad \text{(spherical)}$$

**Relating the radial coordinate to the cylindrical radial coordinate**: The radial coordinate, $r$, of spherical coordinates is related to the cylindrical radial coordinate, $r_c$, of cylindrical coordinates by

$$r = \sqrt{r_c^2 + z^2}$$

**Unit vectors in spherical coordinates**: The spherical unit vectors are $\hat{r}$, $\hat{\theta}$, and $\hat{\varphi}$. The radial unit vector, $\hat{r}$, is directed outward from the origin along the radial coordinate, $r$. The polar unit vector, $\hat{\theta}$, of spherical coordinates is tangent to a longitude passing through the end of the radial coordinate, with a downward sense. The azimuthal unit vector, $\hat{\varphi}$, is tangent to a latitude passing through the end of the radial coordinate, with a counterclockwise sense when viewed from the $+z$-direction; equivalently, the radial coordinate can first be projected onto the $xy$ plane before drawing $\hat{\varphi}$, since a vector may be translated in space.

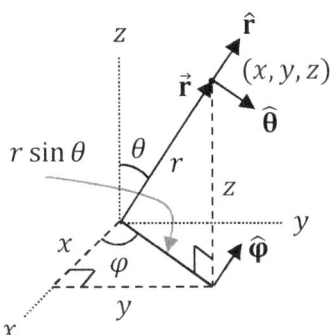

**Position vector in spherical coordinates**: In spherical coordinates, the position vector is simply $\vec{r} = r\hat{r}$, which says to go a distance $r$ outward from the origin along $\hat{r}$, which is specified by two angles – the polar angle, $\theta$, and the azimuthal angle, $\varphi$. Comparing the position vector in spherical and Cartesian coordinates, $\vec{r} = r\hat{r} = x\hat{i} + y\hat{j} + z\hat{k} = \hat{i}\,r\cos\varphi\sin\theta + \hat{j}\,r\sin\varphi\sin\theta + \hat{k}\,r\cos\theta$, the radial unit vector is found to be $\hat{r} = \hat{i}\cos\varphi\sin\theta + \hat{j}\sin\varphi\sin\theta + \hat{k}\cos\theta$. As we did in 2D polar coordinates, we could then add $90°$ to $\theta$ to obtain the polar unit vector, $\hat{\theta}$, from $\hat{r}$, applying the trigonometric identities for adding angles to the argument of trig functions. The result is $\hat{\theta} = \hat{i}\cos\varphi\cos\theta + \hat{j}\sin\varphi\cos\theta - \hat{k}\sin\theta$. In order to obtain the azimuthal unit vector, $\hat{\varphi}$, we first need to project $\hat{r}$ onto the $xy$ plane. We thus set $\theta = 90°$ in $\hat{r}$ and then add $90°$ to $\varphi$, applying the same trigonometric identities. This leads to the expression $\hat{\varphi} = -\hat{i}\sin\varphi + \hat{j}\cos\varphi$.

$$\hat{r} = \hat{i}\cos\varphi\sin\theta + \hat{j}\sin\varphi\sin\theta + \hat{k}\cos\theta \quad \text{(spherical)}$$
$$\hat{\theta} = \hat{i}\cos\varphi\cos\theta + \hat{j}\sin\varphi\cos\theta - \hat{k}\sin\theta \quad \text{(spherical)}$$
$$\hat{\varphi} = -\hat{i}\sin\varphi + \hat{j}\cos\varphi \quad \text{(spherical)}$$

**Differential arc length in spherical coordinates**: At the differential level, the radial arc length, $dr$, longitudinal arc length, $rd\theta$, and latitudinal arc length, $r\sin\theta\, d\varphi$, give the more general equation for arc length as

$$ds = \sqrt{dr^2 + (rd\theta)^2 + (r\sin\theta\, d\varphi)^2} \quad \text{(spherical)}$$

**Generalizing the Pythagorean theorem to 3D**: The equation $r = \sqrt{x^2 + y^2 + z^2}$ is essentially a generalization of the Pythagorean theorem to 3D. A cuboid – a 3D rectangular box – with dimensions $a \times b \times c$ has a long diagonal equal to $r = \sqrt{a^2 + b^2 + c^2}$. One way to see this is to first use the (ordinary) Pythagorean theorem to find the length of the diagonal of the $a \times b$ face, which is $\sqrt{a^2 + b^2}$. Noting that the face $a \times b$ is perpendicular to the edge $c$, the face diagonal $\sqrt{a^2 + b^2}$ makes a 90° angle with edge $c$. Using the (ordinary) Pythagorean theorem to combine $\sqrt{a^2 + b^2}$ and $c$ leads to the equation $r = \sqrt{a^2 + b^2 + c^2}$ for the long diagonal.

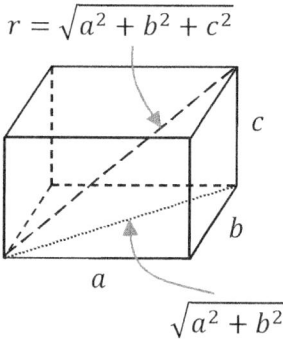

**Magnitude of a vector in 3D space**: The magnitude of a vector is related to its Cartesian components by the 3D generalization of the Pythagorean theorem:

$$A^2 = A_x^2 + A_y^2 + A_z^2$$

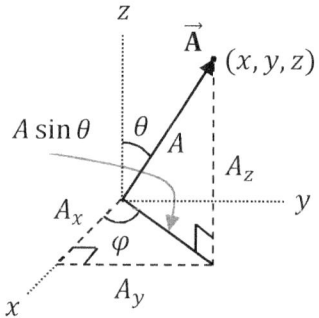

**Direction of a vector in 3D space**: It is necessary to specify two angles in order to unambiguously establish the direction of a vector in 3D space. The natural choice is to use the azimuthal and polar angles, $\varphi$ and $\theta$, from spherical coordinates. The azimuthal angle, $\varphi$, and polar angle, $\theta$, can be computed from the magnitude, $A$, of a vector, $\vec{A}$, using trigonometry:

$$\varphi = \tan^{-1}\left(\frac{A_y}{A_x}\right) \quad , \quad \theta = \cos^{-1}\left(\frac{A_z}{A}\right)$$

In spherical coordinates, the vector $\vec{A}$ can be expressed as $\vec{A} = A\hat{r}$, which is equivalent to the Cartesian form, $\vec{A} = A_x\hat{i} + A_y\hat{j} + A_z\hat{k}$. That is, going $A$ units outward along $\hat{r}$ is equivalent to going $A_x$ units along $\hat{i}$, $A_y$ units along $\hat{j}$, and $A_z$ units along $\hat{k}$.

---

**Example**. Determine the magnitude and the two angles that specify the direction of the vector $\vec{A} = 4\hat{i} - 4\sqrt{3}\hat{j} - 8\hat{k}$.

The magnitude is found by applying the 3D generalization of the Pythagorean theorem:

$$A = \sqrt{A_x^2 + A_y^2 + A_z^2} = \sqrt{(4)^2 + \left(-4\sqrt{3}\right)^2 + (-8)^2} = 8\sqrt{2}$$

The polar and azimuthal angles are

$$\theta = \cos^{-1}\left(\frac{A_z}{A}\right) = \cos^{-1}\left(\frac{-8}{8\sqrt{2}}\right) = 135°$$

$$\varphi = \tan^{-1}\left(\frac{A_y}{A_x}\right) = \tan^{-1}\left(\frac{-4\sqrt{3}}{4}\right) = -60°$$

---

## 2.6 Projectile Motion

**Projectile**: An object that is given an initial velocity (by being shot or thrown, for example) and travels through air or space is termed a projectile.

**Assumptions of projectile motion**: In this text, the following assumptions apply to projectile motion except where explicitly stated otherwise:
- The effects of air resistance are negligible.
- The projectile is in free fall – i.e. the net force acting on it is the force of gravity (i.e. its weight).
- Gravitational acceleration is approximately uniform – i.e. the variation in altitude over the course of the projectile's trajectory is small compared to the radius of the planet.
- The effects of the planet's rotation (and revolution about its star, etc.) are negligible.
- We will also assume that the projectile is near earth's surface unless stated otherwise.

## 2 Multi-Component Motion

**Equations of projectile motion**: We will setup our coordinate system with $+y$ vertically upward and $+x$ horizontal, such that the projectile's trajectory lies in the $xy$ plane. Since the projectile is assumed to be in a state of free fall, its acceleration will be $\vec{a} = -g\hat{j}$.[116] The change in velocity can then be found by integrating the acceleration over time:[117]

$$\Delta \vec{v} = \vec{v} - \vec{v}_0 = \int_{t=t_0}^{t} \vec{a}\, dt = -\int_{t=t_0}^{t} g\hat{j}\, dt = -g\Delta t \hat{j} \quad \text{(for 2DUA)}$$
$$\vec{v} = \vec{v}_0 - g\Delta t \hat{j} \quad \text{(for 2DUA)}$$

We were able to pull $g$ out of the integral because we have assumed gravitational acceleration to be approximately uniform. The net displacement can now be found by integrating velocity over time:

$$\Delta \vec{r} = \int_{t=t_0}^{t} \vec{v}\, dt = \int_{t=t_0}^{t} (\vec{v}_0 - g\Delta t \hat{j})\, dt = \vec{v}_0 \Delta t - \frac{g\Delta t^2}{2}\hat{j} \quad \text{(for 2DUA)}$$

Note that the initial velocity, $\vec{v}_0$, is a constant – i.e. the progression of time can't change what the velocity was initially.

Each of these vector equations is really two equations in one – one equation for each component of motion. The velocity equals $\vec{v} = v_x \hat{i} + v_y \hat{j}$ in general (in 2D), which means that $v_x \hat{i} + v_y \hat{j} = v_{x0}\hat{i} + v_{y0}\hat{j} - g\Delta t \hat{j}$. Since $\hat{i}$ and $\hat{j}$ are perpendicular, and therefore independent directions, we conclude that

$$v_x = v_{x0} \quad , \quad v_y = v_{y0} - g\Delta t \quad \text{(for 2DUA)}$$

This is consistent with $a_x = 0$ and $a_y = -g$, which are the components of the equation $\vec{a} = -g\hat{j} = a_x \hat{i} + a_y \hat{j}$. The components of the net displacement, $\Delta \vec{r} = \Delta x \hat{i} + \Delta y \hat{j} = v_{x0}\Delta t \hat{i} + v_{y0}\Delta t \hat{j} - \frac{g\Delta t^2}{2}\hat{j}$, are

$$\Delta x = v_{x0}\Delta t \quad , \quad \Delta y = v_{y0}\Delta t - \frac{g\Delta t^2}{2} \quad \text{(for 2DUA)}$$

From our work with 1DUA (see Sec.'s 1.6-1.7), we know that $v_y = v_{y0} - g\Delta t$ and $\Delta y = v_{y0}\Delta t - \frac{g\Delta t^2}{2}$ combine to make $v_y^2 - v_{y0}^2 = -2g\Delta y$. There are thus four useful equations for solving projectile motion problems (cf. three useful equations for solving 1DUA problems), which include one $x$-equation and three $y$-equations.

---

[116] Recall that gravitational acceleration, $g$, which is the magnitude of the free-fall acceleration, is always positive, whereas the components of the acceleration vector may be negative.

[117] 2DUA stands for two-dimensional uniform acceleration. The acceleration itself is purely vertical; the 2D refers to the trajectory (i.e. an object can reach any point in 2D space by adjustment of its initial velocity or launch angle, as opposed to 1DUA).

$$\Delta x = v_{x0}\Delta t \quad , \quad \begin{array}{l} \Delta y = v_{y0}\Delta t - \dfrac{g\Delta t^2}{2} \\ v_y = v_{y0} - g\Delta t \\ v_y^2 - v_{y0}^2 = -2g\Delta y \end{array} \quad \text{(for 2DUA)}$$

**Math Note.** In these equations, $x$ is horizontal and $y$ is vertical. Furthermore, the choice $a_y = -g$ (where $g$ itself is positive) assumes that $+y$ is upward.

**Note.** These equations apply only while the projectile is in free fall. These equations do not apply after the projectile motion strikes the ground or other object, nor do they apply before the projectile leaves the hand, gun, or other object that launches it. Thus, the initial velocity, $\vec{v}_0$, may refer to the instant just after the projectile is launched, but not before; similarly, the final velocity, $\vec{v}$, may refer to the instant just before impact, but not after.

**Conceptual interpretation of projectile motion**: The motion of a projectile has two components – a horizontal component ($x$) and a vertical component ($y$). Horizontally, there is no acceleration (i.e. $a_x = 0$) because there are no net forces pushing or pulling horizontally on the projectile to overcome its inertia (definition to come shortly). Vertically, there is a downward acceleration (i.e. $a_y = -g$) because there is a gravitational force (equal to the projectile's weight) pulling downward on the projectile. The horizontal component of motion is governed by inertia: The horizontal component of velocity is constant (i.e. $v_x = v_{x0}$ at any time). The vertical component of motion is governed by 2D uniform acceleration with $a_y = -g$. The horizontal and vertical components of the motion act independently of one another, and the trajectory of a projectile is a combined effect of these two components.

If a monkey throws a banana upward at an angle (but not straight upward), the banana has both horizontal and vertical components of velocity, $v_{x0}$ and $v_{y0}$, initially. The horizontal component of velocity, $v_x$, remains constant – i.e. $v_x = v_{x0}$ – throughout its trajectory. The vertical component of velocity, $v_y$, decreases with uniform acceleration: On the way up, $v_y$ decreases until reaching zero at the top; on the way down, $v_y$ continues to decrease in the sense that it grows more and more negative.

The two components together form the instantaneous speed of the projectile: $v = \sqrt{v_x^2 + v_y^2}$.

The horizontal and vertical components of the velocity of a projectile are illustrated in the figure below. Observe that $v_x$ is the same at every point on the trajectory, while $v_y$ diminishes on the way up, is zero at the top, and becomes more and more negative on the way down.

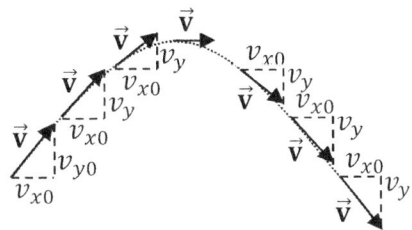

## 2 Multi-Component Motion

**Inertia**: According to Newton's first law, which we will explore in more detail in Chapter 3, all objects have a property termed inertia, which is a natural tendency to maintain constant momentum (mass times velocity). For an object with constant mass, constant momentum is synonymous with constant velocity – which is synonymous with no acceleration. Thus, an object with constant mass has a natural tendency to maintain constant velocity – or has a tendency to not accelerate. If the object is at rest, its velocity is zero, and so the object has a natural tendency to stay at rest; but if the object is already moving, its velocity is nonzero, and so the object has a natural tendency to keep moving with constant speed in a straight line. A 'natural' tendency is what the object does if the net external force acting on it is zero. If instead there is a net force pushing or pulling on the object, it will not act according to its natural tendency. A net force overcomes the object's inertia, causing it to accelerate.

Observe that no force is needed to keep an object moving. Rather, a net force is required to change an object's velocity. In the absence of a net force, an object would keep its velocity forever. We will return to this point in Chapter 3 when we consider Newton's laws of motion.

You do need to exert a net force on an object in order to launch it, but once it is launched, that force no longer affects the motion; once it is launched, horizontally it acts according solely to its sense of inertia, and vertically its inertia is overcome by gravity.

---

**Conceptual Example.** One monkey throws a banana horizontally at the same time that another monkey releases a banana from rest. Both bananas are released from the same height above horizontal ground. Which banana strikes the ground first?[118]

The way to answer this question is to realize that gravity governs the vertical component of motion of both bananas the same way – i.e. the vertical components of motion of both bananas obey the equations of 1DUA – and that the horizontal component of motion is independent of the vertical component of motion. Since $v_{y0} = 0$ for each, both bananas strike the ground at the same time. The net displacement and final speed of the bananas differ, but the elapsed time is the same for each.

---

**Conceptual Example.** A monkey is standing on the top of a train that is traveling horizontally in a straight line to the north with a constant speed of $50 \text{ m/s}$. There is a letter X just under the monkey's feet. The monkey throws a banana straight up into the air – relative to the train. The banana leaves the monkey's hand just as the banana and the letter X pass over the letter Y on the tracks beneath the train. Where does the banana land?

The banana is moving horizontally with a speed of $50 \text{ m/s}$ before the monkey throws it. The banana has inertia – i.e. the banana's natural tendency is to keep moving $50 \text{ m/s}$, until a net force acts on it to change its velocity. The monkey imparts a force on the banana to give it an upward component of velocity, and gravity pulls the banana downward, decelerating it on the way up and accelerating it back down (both are downward accelerations); but nothing acts horizontally (since we neglect air resistance unless otherwise stated) to overcome its inertia horizontally. Thus, the banana's horizontal component of velocity remains a constant $50 \text{ m/s}$ throughout its motion, and it lands on the X.

---

[118] If the monkey fires the banana with a rocket launcher, instead of throwing it with his hand, and the banana has escape velocity, it will never strike the ground (since we neglect air resistance unless stated otherwise, burning up in the atmosphere is not an option). Let's assume that the banana travels a small distance compared to the circumference of the earth.

If you intuitively expected the banana to land on the Y, back on the tracks, instead of on the X on top of the train (the distinction is illustrated below), then consider the following argument. Imagine that your friend is driving at a constant speed of 70 mph in a straight line. If you toss your cell phone straight up inside the car (close to, but not reaching the top of the car), consider how far the car travels during this time interval. That is, if you expect the cell phone's tendency to be to land back on the highway where it was when you let go of it, it should smack the back window before it reaches the top of its arc! But surely it lands back in your hand if you throw it straight upward relative to the car, and doesn't smack the back window. For a more extreme example, imagine that you are traveling a few hundred miles per hour in an airplane (that is flying horizontally with constant speed) and toss your cell phone upward – again, it returns to you, and does not hit a passenger sitting behind you.

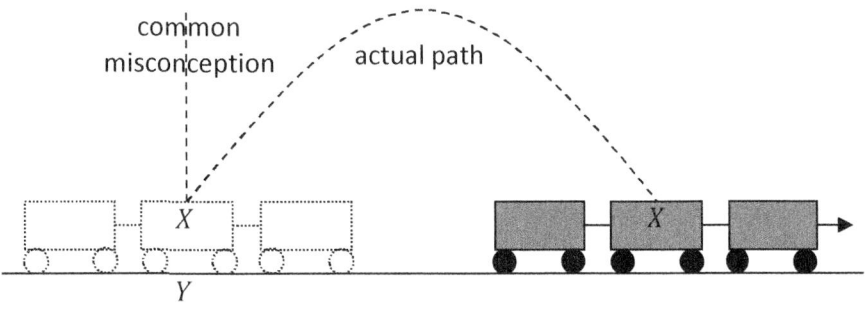

**Conceptual Example.** A monkey is flying an airplane horizontally with constant speed. The monkey wishes to drop a crate full of bananas (from rest, relative to the plane) off for the inhabitants below. Which path below best illustrates the path that the crate of bananas will follow?

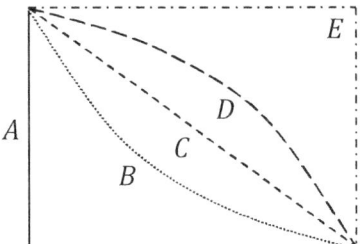

The crate has inertia and so maintains a constant horizontal component of velocity, $v_x = v_{x0}$, equal to the speed of the plane throughout. Gravity accelerates the crate downward, steadily increasing its vertical component of velocity, $v_y$, as it descends. That is, $v_y$ begins at zero and grows steadily, while $v_x$ is constant throughout. Thus, the crate travels more horizontally than vertically in the beginning, and begins to travel more vertically in comparison the further it falls. The result is the parabolic trajectory labeled $D$ in the diagram above.

The same parabolic path would result if a monkey threw a banana horizontally from the top of a cliff. Releasing a banana from 'rest' relative to a horizontally moving object results in the same trajectory as if the banana had been thrown horizontally with the same initial speed (as the horizontally moving object) by a monkey that was at rest relative to the ground.

2 Multi-Component Motion

## Problem-Solving Strategy for Projectile Motion

0. Apply this strategy for a projectile that is freely falling through a uniform gravitational field.
1. Draw and label a diagram. Include the initial ($i$) and final ($f$) positions on your diagram, and also set up your $xy$ coordinate system with $x$ horizontal and $y$ vertical.
2. Resolve the initial velocity into components (if the projectile is launched relative to a moving source – e.g. relative to a train – see the last paragraphs of Sec. 2.6 before you begin):

$$v_{x0} = v_0 \cos \theta_0 \quad , \quad v_{y0} = v_0 \sin \theta_0$$

If you know the initial speed, $v_0$, or the initial angle, $\theta_0$, plug these into the equations for $v_{x0}$ and $v_{y0}$. If you know both $v_0$ and $\theta_0$, obtain numerical values of $v_{x0}$ and $v_{y0}$ now.
3. List your knowns and identify the desired unknown(s). You should generally know 4 of the 7 following symbols: $\Delta x$, $\Delta y$, $v_{x0}$, $v_{y0}$, $v_y$, $g$, and $\Delta t$. If you think you know 5 of these, it's likely that you don't actually know one of them. If you think you only know 3 of these, re-read the problem carefully. If you know $v_0$, but not $\theta_0$ – or if you know $\theta_0$, but not $v_0$ – then the two equations from Step 2 give you the equivalent of one unknown from the list.
4. Look at the four equations of projectile motion. Think about which unknown (from the list of 7) you wish to solve for, and then think about how to combine two or more equations together to solve for that unknown. You will generally need one equation from $x$ and one equation from $y$ (at a minimum).
5. Some problems give you information that you can express mathematically – like "when the final speed equals twice the initial speed." Express this extra information in an equation. In this case, you will probably have fewer than 4 knowns from the list when you are working on Step 3.
6. Try to set up $N$ equations in $N$ unknowns, at which point you are ready to proceed with the algebra.
7. Once you have solved for one or more of the three unknowns from the list of 7 – $\Delta x$, $\Delta y$, $v_{x0}$, $v_{y0}$, $v_y$, $a_y$, and $\Delta t$ – you can find other quantities that you may be seeking. Here are some examples:

- To obtain the final speed, $v$, first find $v_y$ and $v_{x0}$, since $v = \|\vec{v}\| = \sqrt{v_x^2 + v_y^2}$ and $v_x = v_{x0}$.
- To find the magnitude of the net displacement, note that $\|\Delta \vec{r}\| = \sqrt{\Delta x^2 + \Delta y^2}$.
- To find the direction of the final velocity, recall that velocity is a vector: $\theta = \tan^{-1}\left(\frac{v_y}{v_{x0}}\right)$.
- To find the average velocity, apply its definition, $\bar{\vec{v}} = \Delta \vec{r}/\Delta t$, and note that $\Delta \vec{r} = \Delta x \hat{\imath} + \Delta y \hat{\jmath}$.

**Hint.** If an object is thrown horizontally, along the $+x$-axis, then you know the angle of the initial velocity, $\theta_0$: It's zero! In this case, $v_{y0} = v_0 \sin 0° = 0$ and $v_{x0} = v_0 \cos 0° = v_0$. This should make sense: If an object is thrown horizontally, then initially there is no vertical component of velocity – it's entirely horizontal. Only for a horizontal launch does $v_{x0} = v_0$; this is certainly not the case in general.

**Potential Mistake.** If you know how fast, $v_0$, and which way, $\theta_0$, a projectile is moving initially, you must apply trig to solve for $v_{x0}$ and $v_{y0}$. Many students intuitively want to plug the value of $v_0$ in for either $v_{x0}$ or $v_{y0}$, but that's incorrect unless the projectile is launched either horizontally or vertically (i.e. $\theta_0 = 0°, 90°,$ or $180°$).

**Interpreting the signs for projectile motion**: It's important to keep in mind that the components of the net displacement and velocity – $\Delta x$, $\Delta y$, $v_{x0}$, $v_{y0}$, $v_y$ – may all be positive or negative (or zero). Here is what the signs mean, conceptually:[119]

- The vertical displacement, $\Delta y$, is positive if the final position is above the initial position.
- The vertical component of velocity – initial, $v_{y0}$, or final, $v_y$ – is positive if the projectile is moving upward and negative if it is moving downward; and it (but not $v_{x0}$) is zero if the projectile is instantaneously moving horizontally. The launch angle, $\theta_0$, has the same sign as $v_y$.
- The vertical component of acceleration, $a_y = -g$, is negative because gravity is pulling downward. The projectile is losing speed when it is heading upward and gaining speed on the way down.
- If the horizontal $+x$-axis is along the direction of motion, then $\Delta x$ and $v_{x0}$ are positive. You probably wouldn't choose to point $+x$ in the opposite direction for a typical projectile motion problem, but if you have a problem with two objects moving in opposite directions, you won't have a choice but to have $\Delta x$ and $v_{x0}$ negative for one of the objects.

**Signs.** The signs are critical when you apply the equations of projectile motion. Apply the sign conventions above to use the equations correctly. Also, consider the signs carefully when doing the math. For example, when taking a squareroot to find $v_y$, you need the negative root if the projectile is moving downward. Also, when solving for $t$ in a quadratic equation, you must consider the physical meaning behind both possible answers (and both times may be positive).

**Conceptual Example.** A monkey throws a banana with an initial speed of 30 m/s at an angle of 60° above the horizontal. What are the acceleration and speed of the banana at the top of its trajectory?

The acceleration of the banana is uniform throughout – from the moment just after it leaves the monkey's hand until the moment just before impact. The acceleration is no different at the top of the trajectory than it is anywhere else. The horizontal component of the banana's acceleration is zero, $a_x = 0$, and the vertical component is $a_y = -g = -9.81$ m/s$^2$. The acceleration vector is therefore $\vec{a} = -g\hat{j} = -9.81\hat{j}$ m/s$^2$, which has a magnitude of $a = \|\vec{a}\| = \sqrt{a_x^2 + a_y^2} = 9.81$ m/s$^2$.

The vertical component of the banana's velocity is zero, $v_y = 0$, at the top, but the speed of the banana is not zero, $v \neq 0$, at the top because it still has a horizontal component of velocity, $v_x$. The horizontal component of velocity is constant throughout the trajectory: $v_x = v_{x0} = v_0 \cos\theta_0 = 30\cos 60° = 15$ m/s. The speed of the banana at the top is therefore $v = \|\vec{v}\| = \sqrt{v_x^2 + v_y^2} = 15$ m/s.

If your intuition (incorrectly) told you that the speed should be zero at the top of the trajectory, consider the following argument. Visualize a baseball player hitting a line drive that peaks just over the third baseman's head. If the speed of the baseball were zero at the peak of its trajectory, it would then drop straight down instead of continuing along a parabolic arc. It would seem rather silly to see a baseball flying 100 mph and suddenly fall straight down when it got to the top of its trajectory. Rather, the baseball has inertia, so it continues to move rapidly horizontally even when it reaches the top.

---

[119] For projectile motion problems, we are setting our coordinate system up with $+y$ vertical and $+x$ horizontal.

**Conceptual Example.** A monkey throws a banana upward – with some specified speed – at a specified angle above the horizontal. You could then apply the projectile motion strategy to determine how long it takes for the banana to reach a given height. When you do this, you may find that there are two positive times. Both times are physically significant: The smaller time corresponds to when the banana is moving upward, and the larger time corresponds to when the banana is falling back down.

If the given height is greater than the maximum height of the banana, when you proceed to solve for the time, you will get an imaginary number. An imaginary solution could have significance: It could be the equation's way of telling you that the banana will never get there. However, very often when a student gets an imaginary time, it's because of a conceptual mistake setting up the problem or an algebraic mistake. It's totally incorrect to just 'ignore' a negative sign in a squareroot; if this happens, you either need to hunt down your conceptual or algebra mistake (most likely), or you could, in principle, have a trick question that's asking you about something that can't physically happen (but standard textbook questions generally don't include such tricks).

**Example.** A monkey throws a banana with an initial velocity of 10 m/s at an angle of 30° below the horizontal from the roof of a building. The banana leaves the monkey's hand at a height of 15 m above horizontal ground. How far does the banana travel horizontally before landing on the ground?

Drawing and labeling a diagram really helps to visualize the problem and work through the signs correctly; it can also help to avoid making mistakes on multi-part problems – e.g. to help you see whether a quantity used in part (A) also applies to part (B).

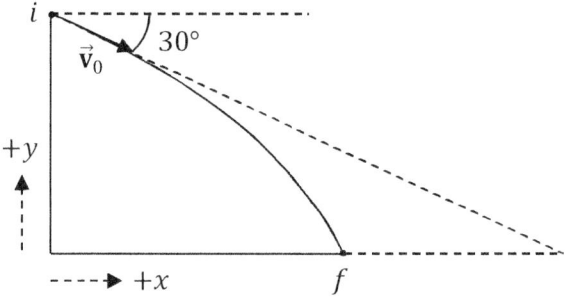

Before doing any other math for a projectile motion problem, first use trigonometry to resolve the initial velocity into components (even if you don't have numbers for $v_0$ and $\theta_0$ yet, you will still need to make use of these equations):

$$v_{x0} = v_0 \cos \theta_0 = 10 \cos(-30°) = 5\sqrt{3} \text{ m/s} \quad , \quad v_{y0} = v_0 \sin \theta_0 = 10 \sin(-30°) = -5 \text{ m/s}$$

Note the importance of the minus sign: $\theta_0$ and $v_{y0}$ are negative because the banana was thrown below the horizontal. If you wish to solve for the time it takes the banana to reach the ground, for example, surely the elapsed time will be less if the banana is thrown downward rather than upward; the signs are crucial for the equations to yield meaningful answers. List the knowns and desired unknown:

$$v_{x0} = 5\sqrt{3} \text{ m/s} \quad , \quad v_{y0} = -5 \text{ m/s} \quad , \quad g = 9.81 \text{ m/s}^2 \quad , \quad \Delta y = -15 \text{ m} \quad , \quad \Delta x = ?$$

The vertical displacement of the banana, $\Delta y$, is negative because it finished at a lower height than it started. The equation for the unknown, $\Delta x = v_{x0}\Delta t$, involves time, which we do not yet know, which suggests using a $y$-equation to first find time:

$$\Delta y = v_{y0}\Delta t - \frac{g\Delta t^2}{2}$$
$$-15 = -5\Delta t - \frac{9.81\Delta t^2}{2}$$
$$4.905\Delta t^2 + 5\Delta t - 15 = 0$$
$$\Delta t = \frac{-5 \pm \sqrt{5^2 - 4(-15)4.905}}{2(4.905)} = 1.3 \text{ s}$$
$$\Delta x = v_{x0}\Delta t = (5\sqrt{3})(1.3) = 11 \text{ m}$$

**Potential Mistake.** Given that a projectile is launched at an angle of $30°$ below the horizontal and descends 15 m vertically, for example, students sometimes proceed to make a right triangle from this information and proceed to use it to solve for the horizontal side. However, this would be a major conceptual mistake because the projectile takes a curved path – i.e. it does not travel along a straight line. Thus, it's not really a triangle – one 'side' is actually a curve. Compare the two diagrams on the previous page: One shows the actual curved path, while the other shows the triangle that would have been made had the path instead been straight. Note that the horizontal distance traveled, $\Delta x$, is shorter than the leg of the triangle as a result of the curvature of the path. Trust the laws, starting equations, and problem-solving strategies because they work (if applied correctly) and if you make mistakes, you may receive partial credit for having a solution similar to the strategies you are supposed to be studying.

**Example.** A monkey wants to throw a banana horizontally from the roof of a building so that it reaches his uncle below. The banana needs to travel 20 m horizontally and descend 12 m vertically to reach the monkey's uncle. How fast does the monkey need to throw the banana?

As usual, we begin with a labeled diagram that shows our coordinate system.

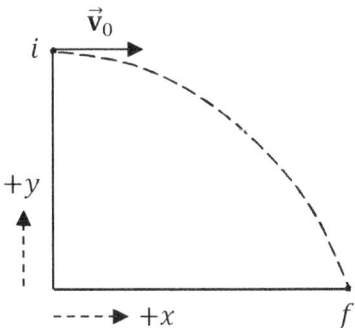

Next, use trigonometry to relate the horizontal and vertical components of the initial velocity to the initial speed and launch angle. Since the banana is thrown horizontally, note that the launch angle is $0°$:
$$v_{x0} = v_0 \cos\theta_0 = v_0 \cos(0°) = v_0 \quad , \quad v_{y0} = v_0 \sin\theta_0 = v_0 \sin(0°) = 0$$

Tabulate the known quantities and the desired unknown:

$$v_{y0} = 0 \quad , \quad g = 9.81 \text{m/s}^2 \quad , \quad \Delta y = -12 \text{ m} \quad , \quad \Delta x = 20 \text{ m} \quad , \quad v_{x0} = v_0 = ?$$

The equation for $v_{x0}$ requires first using the $y$-information to solve for time:

$$\Delta y = v_{y0}\Delta t - \frac{g\Delta t^2}{2}$$

$$-12 = 0 - \frac{9.81 \Delta t^2}{2}$$

$$\Delta t = \sqrt{\frac{24}{9.81}} = 1.6 \text{ s}$$

$$v_{x0} = v_0 = \frac{\Delta x}{\Delta t} = \frac{20}{1.6} = 13 \text{ m/s}$$

Remember, $v_{x0}$ is generally not equal to $v_0$; this is only true for a horizontal launch.

**Path-independence of projectile motion**: The equations of 2DUA – i.e. the four equations of projectile motion – depend only upon the specified initial and final positions, and do not require breaking the motion up into smaller segments. For example, if a banana is thrown upward at some angle and lands on the ground below, you can relate the variables $\Delta x$, $\Delta y$, $v_{x0}$, $v_{y0}$, $v_y$, $g$, and $\Delta t$ directly for the entire trip – it is completely unnecessary to first work with the trip up and separately work with the trip down (unless you are asked to find the total distance traveled or maximum height, for example).[120]

**Trajectory of projectile motion**: The $\Delta x$ and $\Delta y$ equations can be combined by eliminating time in order to obtain an equation for the trajectory of projectile motion:

$$\Delta x = v_{x0}\Delta t \quad , \quad \Delta y = v_{y0}\Delta t - \frac{g\Delta t^2}{2} \quad \text{(for 2DUA)}$$

$$\Delta y = \frac{v_{y0}\Delta x}{v_{x0}} - \frac{g\Delta x^2}{2v_{x0}^2} \quad \text{(for 2DUA)}$$

$$\Delta y = \Delta x \tan\theta_0 - \frac{g\Delta x^2}{2v_0^2 \cos^2\theta_0} \quad \text{(for 2DUA)}$$

This shows that a projectile traveling through a uniform gravitational field travels along a parabola.[121]

---

[120] The math 'knows' whether the banana first went up by the sign of $v_{y0}$: If $v_{y0}$ is positive, the result is a longer time interval, while a negative $v_{y0}$ results in a shorter time interval.

[121] The trajectory becomes an ellipse if you account for the dependence of gravitational acceleration with altitude, but the difference is minute unless $\Delta y$ is significant compared to the dimensions of the planet.

**Horizontal range**: Consider a projectile that is launched from and lands back on horizontal level ground such that $\Delta y = 0$. This is approximately the case for throwing a javelin or hitting a baseball, for example. Setting $\Delta y = 0$ in the equation for the trajectory of a projectile gives the horizontal range, $R = \Delta x$, of such a projectile:[122]

$$0 = R \tan \theta_0 - \frac{gR^2}{2v_0^2 \cos^2 \theta_0} \quad \text{(for 2DUA if } \Delta y = 0\text{)}$$

$$R = \frac{2v_0^2 \sin \theta_0 \cos \theta_0}{g} \quad \text{(for 2DUA if } \Delta y = 0\text{)}$$

$$R = \frac{v_0^2 \sin 2\theta_0}{g} \quad \text{(for 2DUA if } \Delta y = 0\text{)}$$

where the trig identity $\sin 2\theta = 2 \sin \theta \cos \theta$ has been applied.

The launch angle that maximizes the horizontal range, $\theta_0^{max}$, can be found by taking a derivative of the horizontal range with respect to $\theta_0$ and setting it equal to zero:[123]

$$\frac{dR}{d\theta_0} = \frac{2v_0^2 \cos 2\theta_0^{max}}{g} = 0$$

$$\cos 2\theta_0^{max} = 0$$

$$\theta_0^{max} = \frac{\cos^{-1} 0}{2} = \frac{\pi}{4} \text{ rad}$$

Therefore, in the absence of air resistance, a launch angle of 45° maximizes the horizontal range of the projectile.

**Relative velocity**: Suppose that two observers are in relative motion with one another: In particular, observer $A$ is traveling with instantaneous velocity $\vec{v}_{AB}$ relative to observer $B$. If observer $A$ measures the velocity of an object relative to himself, $\vec{v}_{oA}$, the velocity of the same object relative to observer $B$ is obtained through vector addition:

$$\vec{v}_{oB} = \vec{v}_{oA} + \vec{v}_{AB}$$

The equation for relative velocity can be understood by drawing position vectors from each observer to the moving object in coordinate systems for each observer, as illustrated in the following figure. In the figure, $O_A$ and $O_B$ represent observers located at the origins of two coordinate systems, $(x_A, y_A)$ and $(x_B, y_B)$, where observer $O_A$ and the corresponding coordinate system are moving with velocity $\vec{v}_{AB}$ relative to observer $O_B$ and his/her corresponding coordinate system.

---

[122] Some instructors do not permit students to use the range equation or trajectory of a parabola as starting equations, as these are derived equations. All of the projectile motion problems can be solved by applying the projectile motion strategy and starting with the four 2DUA equations. If you apply the range equation, beware that it does not apply to all problems, but only those few problems for which $\Delta y = 0$.

[123] It is indeed a maximum as the second derivative is negative when evaluated at the resulting angle.

## 2 Multi-Component Motion

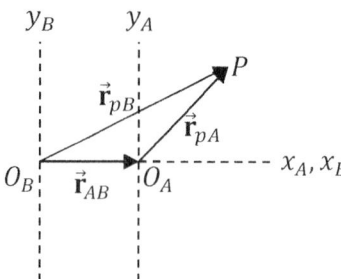

Here, $\vec{r}_{AB}$ gives the position of observer $A$ relative to observer $B$, $\vec{r}_{pA}$ gives the position of the object at point $P$ relative to observer $A$, and $\vec{r}_{pB}$ gives the position of the object at point $P$ relative to observer $B$. Observer $A$ is moving relative to observer $B$, and the object at point $P$ is moving relative to observer $A$. Observe that these three position vectors are arranged tip-to-tail:

$$\vec{r}_{pB} = \vec{r}_{pA} + \vec{r}_{AB}$$

The relative velocity equation is then obtained by taking derivatives with respect to time.
If the relative velocity, $\vec{v}_{AB}$, is constant, then $\vec{r}_{AB} = \vec{v}_{AB}\Delta t$:

$$\vec{r}_{pB} = \vec{r}_{pA} + \vec{v}_{AB}\Delta t \quad (\text{if } \vec{v}_{AB} = \text{const.})$$

This is the Galilean transformation equation.[124] Taking a derivative with respect to time, we obtain the same relative velocity equation, $\vec{v}_{pB} = \vec{v}_{pA} + \vec{v}_{AB}$. Taking a second derivative with respect to time, we find that the relative accelerations are equal when the relative velocity is constant:

$$\vec{a}_{pB} = \vec{a}_{pA} \quad (\text{if } \vec{v}_{AB} = \text{const.})$$

**Example.** A monkey stands on the top of a train that is traveling horizontally with a constant speed of 20 m/s. The monkey, facing the direction of motion of the train, throws a banana with an initial velocity of 20 m/s at an angle of 60° above the horizontal relative to the train. What are the magnitude and direction of the initial velocity relative to the ground?
With $x$ along the direction that the train is moving, we find, for the monkey (m) and ground (g):

$v_{pm,x0} = v_{m0}\cos\theta_{m0} = 20\cos 60° = 10$ m/s , $v_{pm,y0} = v_{m0}\sin\theta_{m0} = 20\sin 60° = 10\sqrt{3}$ m/s

$v_{pg,x0} = v_{pm,x0} + v_{mg,x} = 10 + 20 = 30$ m/s , $v_{pg,y0} = v_{pm,y0} + v_{mg,y} = 10\sqrt{3} + 0 = 10\sqrt{3}$ m/s

$v_{pg,0} = \sqrt{v_{pg,x0}^2 + v_{pg,y0}^2} = \sqrt{(30)^2 + (10\sqrt{3})^2} = 20\sqrt{3}$ m/s , $\theta_{pg,0} = \tan^{-1}(v_{pg,y0}/v_{pg,x0}) = 30°$

---

[124] The Galilean transformation applies to objects moving slowly compared to the speed of light. Einstein's theory of relativity must instead be applied for relative speeds comparable to the speed of light. The Galilean transformation equation is usually written from the perspective of observer $A$, who is moving relative to observer $B$, as $\vec{r}_A = \vec{r}_B - \vec{v}_r\Delta t$, and so usually features a minus, rather than a plus, sign.

**Launch from a moving source**: If a projectile is launched from a moving source, simply apply the relative velocity equation to determine the velocity of the projectile relative to the ground and then use the usual projectile motion strategy.

## 2.7 Multi-Dimensional Motion

**Equations of motion from position as a function of time**: Given an equation for the position vector, $\vec{r}$, expressed in terms of time, $t$, you can derive equations for the time-dependence of the velocity, $\vec{v}$, and acceleration, $\vec{a}$, by taking successive derivatives with respect to time:

$$\vec{v} = \frac{d\vec{r}}{dt} \quad , \quad \vec{a} = \frac{d\vec{v}}{dt} = \frac{d^2\vec{r}}{dt^2}$$

**Equations of motion from velocity as a function of time**: Given an equation for the velocity, $\vec{v}$, expressed in terms of time, $t$, you can derive an equation for the time-dependence of the acceleration, $\vec{a}$, by taking a derivative with respect to time, and an equation for the time-dependence of the net displacement, $\Delta \vec{r}$, by integrating over time:

$$\vec{a} = \frac{d\vec{v}}{dt} \quad , \quad \Delta \vec{r} = \int_{t=t_0}^{t} \vec{v}\, dt$$

**Equations of motion from acceleration as a function of time**: Given an equation for the acceleration, $\vec{a}$, expressed in terms of time, $t$, you can derive equations for the time-dependence of the velocity, $\vec{v}$, and net displacement, $\Delta \vec{r}$, by integrating over time:

$$\Delta \vec{v} = \vec{v} - \vec{v}_0 = \int_{t=t_0}^{t} \vec{a}\, dt \quad , \quad \Delta \vec{r} = \int_{t=t_0}^{t} \vec{v}\, dt$$

---

### Problem-Solving Strategy for Non-Uniform Acceleration

0. This strategy is useful for relating the position vector, velocity, and acceleration in motion problems where the acceleration may be non-uniform.
1. If you are given position, $\vec{r}$, velocity, $\vec{v}$, or acceleration, $\vec{a}$, as a function of time, derive equations for the time-dependence of the other motion vectors by applying calculus.
2. Count how many unknowns you have. You generally need to set up $N$ equations in $N$ unknowns. If you have more unknowns than you have equations, you need to determine what other equations apply to the problem.
3. Once you have $N$ equations in $N$ unknowns, solve the system algebraically.

**Example.** A monkey has an acceleration of $\vec{a} = 4t\hat{\imath} - 20\hat{\jmath}$ and an initial velocity of $\vec{v}_0 = 16\,\hat{\jmath}$, where SI units have been suppressed. The motion begins at $t = 0$. What is the monkey's speed at $t = 4.0$ s?

The instantaneous velocity can be found readily through a definite integral:

$$\vec{v} - \vec{v}_0 = \int\limits_{t=t_0}^{t} \vec{a}\, dt$$

$$\vec{v}(t=4) = \vec{v}_0 + \int\limits_{t=0}^{4} (4t\hat{\imath} - 20\hat{\jmath})\, dt = 16\,\hat{\jmath} + [2t^2\hat{\imath} - 20t\hat{\jmath}]_{t=0}^{4} = 32\hat{\imath} - 64\hat{\jmath}$$

The speed is the magnitude of the velocity vector:

$$v = \|\vec{v}\| = \sqrt{v_x^2 + v_y^2} = \sqrt{32^2 + (-64)^2} = 32\sqrt{5}\text{ m/s}$$

## Conceptual Questions

1. Indicate whether each physical quantity is a scalar or vector: volume, electric charge (such as the net charge of an ionized atom), the 'jerk' (defined as a derivative of acceleration with respect to time), density, magnetic field (as indicated by a compass reading), wavelength (as of light, which we interpret as color), temperature (the average kinetic energy of the molecules of a substance), wind readings (measurements a golfer would like to know before hitting his shot), the strong nuclear force (responsible for attracting protons and neutrons together in the nucleus), angular momentum (a measure that describes how a Frisbee is spinning, for example), memory (such as how much information you can store on a hard drive), relative star position (where to aim your telescope to view a star).

2. Indicate whether each physical quantity can or can't be negative, depending on the situation: $v$, $v_z$, $a_T$, $\|\vec{F}\|$, $\Delta s$, $\Delta s_T$, $F_y$, $t$.

3. Sketch the resultant of the two vectors illustrated below.

4. Sketch the subtraction of the right vector from the left vector illustrated below.

5. Sketch $-3\vec{A}$ for the vector $\vec{A}$ illustrated below.

6. Sketch $3\vec{A} - 2\vec{B}$ for the vectors $\vec{A}$ and $\vec{B}$ illustrated below.

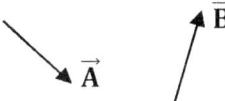

7. Given that $\vec{A}$ has a magnitude of 5.0 N and $\vec{B}$ has a magnitude of 8.0 N, what are the minimum and maximum possible values that the magnitude of the resultant, $\vec{R} = \vec{A} + \vec{B}$, could have?

8. The 4 vectors illustrated below have equal magnitudes of 5.0 m with their tips lying at 4 corners of a regular pentagon. Find the magnitude and direction of the resultant.

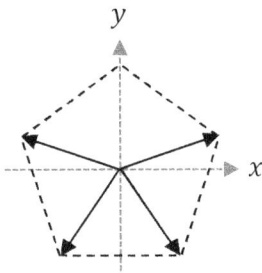

9. (A) Beginning near the equator, a monkey walks one mile south, one mile east, and one mile north. What are the monkey's net displacement and total distance traveled? (B) There are an infinite set of starting points from which the monkey could walk one mile south, one mile east, and one mile north, and have a net displacement of zero. One way is to start one mile south of the North Pole. Where are the possible starting points in the southern hemisphere for which this can happen?

10. A rocket orbits a planet with a constant speed of 15 km/s. What are the average speed and average velocity of the rocket for one complete orbit?

11. A monkey is riding a bicycle to the north with a speed of 12 m/s, and is riding to the south with a speed of 12 m/s two minutes later. What is the monkey's average acceleration over the course of these two minutes?

12. Whereas the velocity only has a tangential component, $\vec{v} = v_T \hat{t}$, the acceleration may have a component perpendicular to the velocity, $\vec{a} = a_T \hat{t} + a_c \hat{n}_i$ (i.e. a normal component along $\pm \hat{n}_i$, where $\hat{n}_i$ is the inward normal). Given that $\vec{a} = d\vec{v}/dt$, explain conceptually how it's possible for the acceleration to have a normal component when the velocity does not. Mathematically, what is the origin of the second term when taking a derivative?

13. Sketch the curve $r = 3\sin\theta$ in the $xy$ plane.

14. Sketch the curve $\tan\theta = 2r\cos\theta$ in the $xy$ plane.

15. In cylindrical coordinates, describe the shape that is defined by the equation $r_c = z/2$.

16. In spherical coordinates, describe the shape that is defined by the equation $\theta = \pi/3$ rad.

17. A monkey sails in a pirate ship with constant velocity. The monkey launches a cannonball straight upward relative to the ship. Where does the cannonball land? Explain. Also, describe the path of the cannonball relative to the ship and relative to the earth.

## 2 Multi-Component Motion

18. An airplane travels with a constant speed of 300 mph horizontally to the east, heading toward an island populated mostly by monkeys. The pilot wishes to drop a food package (from rest relative to the airplane) such that it lands on the island. Sketch the path that the food package will follow, and describe conceptually where the food package should be released relative to the island.[125]

19. The 'monkey gun' refers to the following classic physics problem: A hunter aims his rifle directly at a monkey that is hanging from a tree, while the monkey lets go at the same instant that the hunter pulls the trigger. Since that problem must surely upset animal rights' activists, let's change the problem so that the hunter throws a banana directly at the monkey's original position. Without knowing the initial speed of the banana or launch angle,[126] prove that the banana will be delivered right to the monkey.

20. A monkey hits golf balls from top of a cruise ship. Neglect air resistance and the height of the cruise ship. We know that a launch angle of 45° maximizes the range of the golf ball if the cruise ship is at rest. If the ship travels with constant velocity and the monkey aims away from the ship's velocity, how is the launch angle (relative to the ship) that maximizes the range of the golf ball (relative to the earth) affected, if at all? Explain.

21. A monkey has a motor boat that provides a constant boat speed of 15 m/s when traveling horizontally across still water. The monkey wishes to use the motor boat to travel across a river with constant width and a river current of 10 m/s. The monkey will aim his motor boat in a particular direction before he turns the motor on, but will not manually redirect the motor boat once the motor turns on (he will instead follow whatever course the river current provides). Sketch and describe how the monkey should aim his motor boat if he wants to reach the other side (A) in the least amount of time, (B) with the shortest net displacement, and (C) with the fastest boat speed relative to the ground.

22. Given a plot of $x$ as a function of $t$ in addition to a plot of $y$ as a function of $t$, describe how to determine (A) where the object is moving fastest and (B) where the object has the greatest acceleration.

23. Make sketches for a projectile with $x_0 > 0$, $y_0 > 0$, $v_{x0} > 0$, and $v_{y0} > 0$. Make all six plots – i.e. plot $x$, $v_x$, $a_x$, $y$, $v_y$, and $a_y$ as functions of time.

24. Sketch a frictionless hill that a monkey could slide down such that the monkey's tangential speed increases while the magnitude of the monkey's tangential acceleration decreases.

25. Sketch a frictionless hill that a monkey could slide down such that the monkey's tangential speed and the magnitude of the monkey's tangential acceleration both increase.

---

[125] Obviously, the airplane needs air, but neglect the effects of air resistance for the food package.
[126] You do need to assume that the banana will travel far enough horizontally to reach the monkey before he strikes the ground, but otherwise you don't need to know how fast or at what angle the banana is thrown.

## Practice Problems

### Components, vector addition and subtraction

1. A monkey pirate etches pictures of monkeys on large rocks. The monkeys' tails are vectors pointing to buried bananas. (A) One of the tail vectors is $\vec{A}$, with magnitude 300 m and direction 35°. What are the $x$- and $y$-components of this vector? (B) Another of the tail vectors is $\vec{B}$, with an $x$-component of 40 m and $y$-component of 200 m. What are the magnitude and direction of this vector? (C) To find buried bananas, you walk along $\vec{A}$, then walk along $\vec{B}$. This is equivalent to walking along $\vec{C}$, where $\vec{C} = \vec{A} + \vec{B}$. Find the magnitude and direction of $\vec{C}$.

2. A monkey vector, $\vec{M}$, has a magnitude of 8.0 N and a direction of 210°. A banana vector, $\vec{B}$, has a magnitude of 4.0 N and a direction of 330°. Use the strategy of vector addition to determine the magnitude and direction of their resultant.

3. A monkey is going bananas with monkey vectors. (A) The monkey finds that he can draw two different monkey vectors with a $y$-component of $-40$ m and a magnitude of 50 m. Explain why this is possible and find all possible directions for this monkey vector. (B) The monkey draws a monkey vector along the positive $x$-axis. Then he draws two more monkey vectors. All three monkey vectors have equal magnitude, yet the resultant is zero. Explain how this is possible and find the directions of all three monkey vectors.

4. Monkey vector $\vec{A}$ has a magnitude of 350 bananas and a direction of 63°. Monkey vector $\vec{B}$ has a magnitude of 200 bananas and a direction of 345°. Monkey vector $\vec{C} = \vec{B} - \vec{A}$. Find the magnitude and direction of $\vec{C}$.

5. In the diagram below, the *Star Wreck Monkeyprize* is being pulled by two tractor beams. Tractor beam $\vec{A}$ has a magnitude of 50 kN and tractor beam $\vec{B}$ has a magnitude of 70 kN. (A) Draw a sketch to illustrate the net force exerted on the *Monkeyprize*. Clearly label the magnitude and direction of this net force. (B) Find the magnitude and direction of the net force exerted on the *Monkeyprize*.

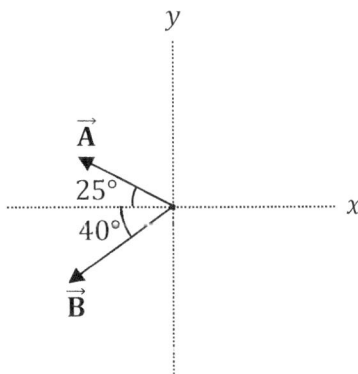

6. A monkey sails in a boat where the ocean current contributes a velocity 50 m/s at 160° and the wind contributes a velocity 30 m/s at 220° to the boat. What are the speed and direction of the boat relative to the ground?

7. Consider the monkey vector 🧍 and banana vector 🍌 illustrated below. The magnitude of each vector is given in meters per second. The flower vector 🌼 is defined by the following equation:

$$\vec{🌼} = 3\vec{🍌} - 4\vec{🧍}$$

Find the magnitude and direction of the flower vector 🌼.

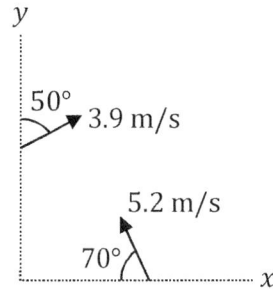

8. His Highness has defined the vectors below.

$$\vec{K}(x) = 3x^2\hat{i} + 4\hat{j} - 3\sin(2x)\hat{k} \quad , \quad \vec{Q}(x) = \frac{d\vec{K}}{dx}$$

$$\vec{P}(x) = \int_{x=0}^{x} \vec{K}(x)dx \quad , \quad \vec{R}(x) = (3x^2 - 5, 220°)$$

where $x$ is in radians. (A) Express $\vec{R}\left(\frac{\pi}{2}\right)$ in terms of $\hat{i}$ and $\hat{j}$. (B) Express the magnitude and direction of $\vec{P}(2\pi)$ in spherical coordinates. (C) Express $\vec{P} + 2\vec{Q}$ in terms of Cartesian unit vectors. (D) Express the magnitude and direction of $3\vec{Q}\left(\frac{\pi}{2}\right) - 2\vec{P}\left(\frac{\pi}{2}\right)$ in spherical coordinates. Note that the comma in the vector R separates the magnitude from the direction (the angle is a 3-digit number, not a 4-digit number – i.e. note the presence of the minus sign – but is instead a function, comma, a direction).

9. A monkey and his uncle love to dangle from tree limbs and discuss silly physics concepts. The monkey defines the silliness vector, $\vec{S}(\theta)$, to have a magnitude of $\sqrt{3}\sec\theta$ and a direction of $\theta$. His uncle defines the ludicrous vector, $\vec{L}(\theta)$, according to:

$$\vec{L}(\theta) = 2\vec{S}(\theta) - \sqrt{3}\frac{d}{d\theta}\vec{S}(\theta)$$

(A) Find the $x$- and $y$-components of $\vec{S}\left(\frac{\pi}{6}\right)$. (B) Find the magnitude and direction of $\vec{L}\left(\frac{\pi}{6}\right)$.

10. Consider the monkey vector $\vec{\text{🐵}}$ and banana vector $\vec{)}$ illustrated below. The magnitude of each vector is expressed as a function of $t$, where SI units have been suppressed for convenience. The flower vector $\vec{🌸}$ is defined by the following equation:

$$\vec{🌸}(t) = \sqrt{3}\,\vec{)} - 2\frac{d}{dt}\vec{\text{🐵}}$$

Find the magnitude and direction of the flower vector $\vec{🌸}$ at $t = 4.0$ seconds.

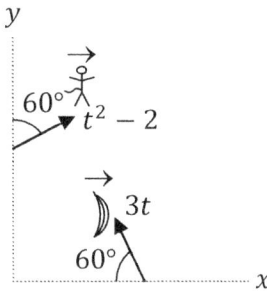

**Position, velocity, and acceleration vectors and scalars**

11. A monkey runs 4.0 m/s to the east for 25 s, 8.0 m/s to the north for 15 s, and then 6.0 m/s to the south for 10 s. Determine (A) the net displacement, (B) the total distance traveled, (C) the average speed, and (D) the average velocity of the monkey for the whole trip.

12. A monkey jogs around the track illustrated below. The bottom arc is semicircular with 50-m radius. The monkey completes the lap in 40 seconds. The monkey runs:
    (i) at a constant speed of 10 m/s from $A$ to $B$ along the semicircular arc
    (ii) at a different constant speed from $B$ to $A$ along the straight section.

(A) What is the total distance traveled by the monkey in trip (i)? (B) What is the net displacement of the monkey in trip (i)? (C) How fast did the monkey travel in trip (ii)? (D) What is the average speed of the monkey for the whole lap? (E) What is the magnitude of the average velocity of the monkey for the whole lap?

13. A monkey runs along the path $y(x) = \left(x - \frac{4}{9}\right)^{3/2}$ from $x = \frac{4}{9}$ to $x = \frac{8}{9}$, where SI units have been suppressed. Determine (A) the total distance traveled and (B) the net displacement of the monkey by formally integrating the appropriate differential element. Keep enough significant figures (i.e. assume that the given values are good to as many digits as needed) such that your two answers are different.

## 2 Multi-Component Motion

14. The position of a wild monkey is given by $\vec{r}(t) = \lambda t\hat{i} + (\beta t^2 + \gamma)\hat{j}$, where (lambda) $\lambda = 3.0$ m/s, (beta) $\beta = 2.0$ m/s$^2$, and (gamma) $\gamma = -4.0$ m. (A) How far from the origin is the monkey at $t = 4.0$ seconds? (B) What is the net displacement (magnitude and direction) of the monkey from $t = 2.0$ to $t = 4.0$ seconds? (C) What is the velocity (magnitude and direction) of the monkey at $t = 3.0$ seconds? (D) What is the acceleration (magnitude and direction) of the monkey at $t = 5.0$ seconds?

15. When you flush the toilet, the contents travel down a curved pipe with a velocity given by $\vec{v}(t) = (\beta t^2 - \rho)\hat{i} + \lambda\sqrt{t}\hat{j}$, where (beta) $\beta = 2.0$, (rho) $\rho = 3.0$, and (lambda) $\lambda = 4.0$. (A) What are the SI units of $\beta$, $\rho$, and $\lambda$? (B) Find the magnitude and direction of the acceleration at $t = 2.0$ seconds. (C) Find the average velocity for the first 4.0 seconds.

16. A stray cat has interrupted the jousting tournament. Several knights chase the cat through the arena. The cat's acceleration is given by $\vec{a}(t) = 3t\hat{i} + (t^2 - 2)\hat{j}$, where SI units have been suppressed. When $t = 2.0$ seconds, the cat's velocity is $5.0$ m/s at an angle of $40°$. (A) What is the cat's initial velocity (magnitude and direction)? (B) What is the cat's minimum acceleration? (C) What is the net displacement of the cat after running for four seconds?

### Coordinate systems, 3D vectors

17. Transform the 2D polar equation $\tan\theta = 2r\cos\theta$ into Cartesian coordinates. What kind of curve is this?

18. Transform the equation $(y - x)^2 = 1$ into 2D polar coordinates. Use trig substitutions to rewrite the expression in a simplified form using a single trig function.

19. Consider the curve described by the parametric equations $x = \cos z$ and $y = \sin z$.[127] Transform these into a pair of equations in cylindrical coordinates. What kind of curve is this?

20. Consider the surface described by $\theta = \pi/6$ in spherical coordinates. Transform this equation into Cartesian coordinates. What kind of surface is this?

21. Consider the surface described by $\varphi = \pi/6$ in spherical coordinates. Transform this equation into Cartesian coordinates. What kind of surface is this?

22. A monkey wishes to place a monkeystick in a 36 cm × 24 cm × 12 cm trunk. What is the maximum length of monkeystick that will fit into the trunk?

23. A monkey needs to express the vector $(2, 2\sqrt{3}, 4\sqrt{3})$ in spherical coordinates. Help him out.

24. Consider the vector function, $\vec{A}(x, y)$, where $x$ and $y$ are the arguments of the vector function, $\vec{A}$, and not its components, $A_x$ and $A_y$. That is, $A_x$ and $A_y$ are themselves functions of the coordinates $x$ and $y$. For example, $A_x$ could be $4xy^2$, or any other function of $x$ and $y$. Derive equations for the 2D polar components of $\vec{A}$, denoted by $A_r$ and $A_\theta$ in terms of the Cartesian components $A_x$ and $A_y$. Note that $A_\theta$ is not an angle. Show that the angle of vector $\vec{A}$, denoted as $\theta_A$, is, in general, not equal to the angle $\theta$ defined by $\tan^{-1}(y/x)$.[128]

---

[127] Note that a single equation in 3D space defines a surface – not a curve. For example, the equation $y = 2x + 3$ describes a plane in 3D space – not a line – since this equation holds for any value of $z$. In contrast, the pair of equations, $y = 2x + 3$ and $z = 4$, define a line, now that the value of $z$ has been fixed. The pair of equations, $y = 2x + 3$ and $z = 5x - 2$ also define a line – it's the intersection of these two planes.

[128] These are very important distinctions to understand when working with vector fields (e.g. gravitational field or electric field), which are commonly encountered in higher-level physics courses and in vector calculus.

## Projectile motion, relative velocity, multi-dimensional motion

25. A monkey launches a cannonball with an initial speed of 100 m/s at a 30° angle. What is the speed of the cannonball at the top of its trajectory?

26. Sick of studying, a physics student vomits from a window. The vomit has an initial speed of 20 m/s at an angle of 65° above the horizontal. The vomit lands 50 m away from the building. As usual, assume that air resistance effects are negligible. (A) How high is the window above the ground? (B) What is the maximum height of the vomit? (C) What is the speed of the vomit at the top of its trajectory? (D) What is the acceleration of the vomit at the top of its trajectory? (E) What is the magnitude of the net displacement of the vomit? (F) What is the direction of the net displacement of the vomit?

27. A fire-breathing dragon spits knights in smoldering armor out from the roof of a 200-m tall castle. (A) What will be the final velocity (implying both magnitude and direction in the answer – in contrast to final speed) of a knight spat out with a velocity of 30 m/s at an angle of 25° above the horizontal? (B) Repeat for a knight spat out with a velocity of 30 m/s at an angle 25° below the horizontal? (C) Repeat for a knight spat out with a horizontal velocity of 30 m/s. (D) Which knight strikes the ground with the greatest speed? Prove your answer symbolically in general terms.

28. In order to relieve some physics stress, you make a piñata that looks like Sir Isaac Newton, suspend it from a tree 10 m above the ground, and throw apples at it. You throw one apple from the roof of a building at a rate of 20 m/s and at an angle 30° below the horizontal and it strikes Newton right in the belly. Good aim! The horizontal distance between the building and Newton is $40\sqrt{3}$ m. (A) How tall is the building? (B) For how much time is the apple in the air? (C) Find the final speed of the apple.

29. Upon hearing you creatively express your physics frustrations, the ghost of Isaac Newton grabs you, takes you to the roof of a building, and throws you at a velocity of 40 m/s at an angle of 30° below the horizontal. You strike the ground below a distance $100\sqrt{3}$ away from the base of the building. How tall is the building?

30. A cat sharpens its nails on the furniture, then hides 15.0 m above the ground in a tree. When the cat's companion returns home from a long day at work and discovers the damage, the companion goes outside and throws a shoe at a speed of 20 m/s directly at the cat. The shoe is thrown from a height of 2.0 m above the ground and a distance of 10.0 m from the tree. At the instant the shoe is thrown, the cat falls from the tree (hoping to avoid the oncoming shoe). (A) At what angle is the shoe thrown? (B) What is the vertical position of the shoe when the horizontal position of the shoe matches the horizontal position of the cat? (C) What is the vertical position of the cat at this time? (D) Does the shoe strike the cat?

31. A young lad has prepared romantic physics poems in order to serenade his lover, but first he must get her attention by throwing an orange against her window. If he throws the orange at an angle of 50°, her window is a height of 10.0 m above the ground, he stands a distance of 20.0 m away from her wall, and he throws the orange from height of 2.0 m, how fast must he throw the orange in order for it to strike her window?

32. A plane is flying 250 km/hr at a constant altitude of 20 km. How far away from its target should the bomb be when it releases a pilot? (Yes, you read that correctly.)

33. A football kicks a quarterback with an initial speed of 50 m/s at an angle of 60° above the horizontal. (A) What is the maximum height of the quarterback? (B) What is the speed of the quarterback at the top of his trajectory? (C) How far away from the football does the quarterback land on the ground? (D) For how much time is the quarterback in the air?

34. A baseball on the top of a 20-m high building throws an outfielder at a speed of 30 m/s horizontally toward a catcher who is standing 50 m away from the base of the building. (A) How far must the catcher below run in order to catch the outfielder? (B) Must he run toward or away from the building?

35. A sore golf ball tees up a frustrated golfer at the bottom of a hill with a steady 20° incline. The golf ball then swings a golf club, striking the golfer, who flies off with an initial speed of 30 m/s at an angle of 55° relative to the horizontal (so 35° relative to the incline). How far up the incline has the golfer traveled when he lands?

36. A smoldering cannonball places a pirate inside a cannon on the rear of a ship that is sailing southbound at a rate of 25 m/s. The cannonball lights the cannon, launching the pirate southbound with an initial speed of 40 m/s. What angle, relative to the pirate ship, of the cannon produces maximum range of the pirate relative to the ocean?

37. One pirate ship heads 10 m/s east, while another approaches at 20 m/s west. (A) If the cannons launch the cannonballs at 80 m/s at 30° angles above the horizontal relative to the pirate ship, what is the minimum distance the pirate ships can be apart when a cannonball can be launched and reach the enemy ship? (B) What will be the maximum height of the cannonball in the air?

38. A monkey kicks a banana ball with an initial velocity $\vec{v}_0$ at an angle $\theta_0$ above the horizontal. Derive an equation for its maximum height, $H$, relative to its starting point in terms of $v_0$, $\theta_0$, and/or $g$. No other symbols may appear in your final expression.

39. A wild monkey has a uniform acceleration with components $a_x = 4.0$ m/s$^2$ and $a_y = 8.0$ m/s$^2$. The initial velocity of the monkey is $\vec{v}_0 = (10$ m/s$)\,\hat{\imath}$. Find the magnitude and direction of (A) the acceleration, (B) the velocity, and (C) the net displacement of the monkey at $t = 3.0$ s.

40. A monkey runs with a constant speed, $v_m$, on horizontal ground. Neglecting the height of the monkey, derive an equation for the launch angle (relative to the monkey) that maximizes the horizontal range of the javelin.

41. A monkey's son just marries his brother's friend's daughter. The monkey's son and his brother's friend's daughter head east on the roof of a horizontally moving train at a constant rate of 20 m/s. As the train passes the monkey's brother, the monkey's son throws a coconut straight up with an initial speed of 10 m/s relative to the monkey's son. Find each of the following relative to (i) the monkey's son and (ii) the monkey's brother: (A) the initial velocity (magnitude and direction) of the coconut; (B) the time the coconut is in the air; (C) the horizontal distance traveled by the coconut; (D) the velocity of the coconut at the top of its trajectory; (E) the acceleration of the coconut at the top of its trajectory.

42. A monkey rides a train that moves with a constant horizontal speed of 10 m/s. The monkey throws a coconut with an initial speed of 20 m/s relative to the train. At what angle relative to the horizontal should the monkey throw the coconut so that it appears to go straight up relative to the ground?

43. A bored physics student blindfolds a monkey, spins the monkey around several times, and then lets go. The monkey proceeds to run, but not in a straight line, since he is very dizzy. Thus, his motion is two-dimensional. The $x$- and $y$-coordinates of his position are plotted as functions of time in the graphs on the following page.

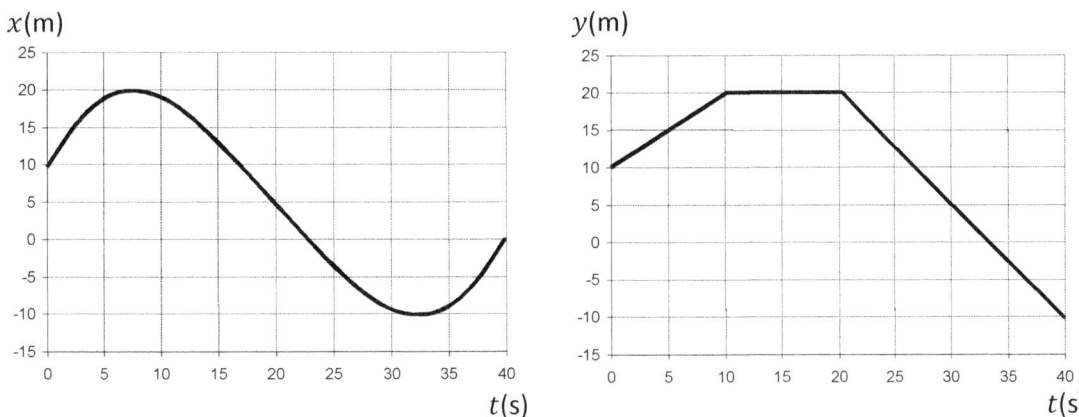

(A) What is the net displacement of the monkey? (B) What is the speed of the monkey at $t = 25$ seconds? (C) What is the direction of the monkey's velocity at $t = 25$ seconds? (D) When is the monkey is Quadrant III? (E) Sketch the monkey's velocity and acceleration. (F) Find the monkey's average speed for the trip.

# 3 Newton's Laws of Motion

**Ancient Greek physics**: The ancient Greeks made a plethora of brilliant contributions to many areas of academia, especially geometry and philosophy. The work of many of the most notable contributors – such as Ptolemy, Aristotle, Euclid, and Pythagoras – was held in such high esteem for several centuries that their few mistakes left very lasting impressions, not leading to accepted corrections in some instances until the Renaissance period. One of the most significant mistakes was concerning the 'natural' tendency of objects, which considerably slowed the development of theoretical physics.

The ancient Greeks believed that there was a clear division between terrestrial and celestial motion: We lived in an imperfect world on earth, while the heavenly bodies that we see in the sky were perceived to be perfect orbs that traveled in aesthetic geometric paths.[129] The moon was a natural dividing line between these two worlds, as the moon's appearance and orbit clearly do not match their perception of other heavenly bodies. Looking terrestrially, the Greeks observed that all objects that are set in motion tend to come to rest. This repeated observation – no doubt supported by experiment – led to the natural, but incorrect, conclusion that objects have a natural tendency to be at rest. A second important mistake that the ancient Greeks made with regard to motion was the perception that heavier objects tend to fall with a greater acceleration.

The ancient Greeks also made a very well-known mistake regarding the motion of objects in our solar system: They believed that the solar system was geocentric rather than heliocentric. Actually, one of the ancient Greeks, Hipparchus, proposed that the solar system could be explained more simply under the assumption that the earth orbited the sun. Unfortunately, Hipparchus lost faith in his argument when the following objections were raised: The ancient Greeks calculated the enormous speed that the earth would need to have in order to orbit the sun, and since they believed that terrestrial objects had a natural tendency to be at rest, it was argued that birds and other objects would not keep up with the earth, but would fall off (just like any objects that you accidentally place on the top of your car will get left behind when you start driving[130]); and furthermore, the earth is evidently extremely heavy,[131] and since they believed that all terrestrial objects required something to push them to keep them from coming to rest, it was unimaginable that something could exert enough force to continually power the earth through its orbit.[132]

---

[129] They first postulated that the sun and planets orbited the earth in perfect circles; when experimental data became too precise for this model to work, they revised the orbits to feature a deferent and epicycle(s) – a small circle holding the celestial body rotates as it revolves around a large circle called a deferent.

[130] The object may not wind up in your driveway, but may make a small part of the trip before falling off. If you understand why the object falls off in modern terms, you wouldn't have accepted this argument.

[131] Technically, in this sense the earth is extremely massive, but in ancient Greece the distinction between weight and mass had not yet developed as they did not have a proper understanding of the natural tendency of objects.

[132] In the minds of the ancient Greeks, the massive sun and other heavenly objects were subject to a different set of laws – celestial rather than terrestrial. So the same objection couldn't be applied to the sun.

Ancient Greeks are often criticized for the notable mistakes that they made in the fields of physics and astronomy. Some of the critics blame the ancient Greeks for not having based more of their scientific models on experiment, especially citing the following example: If the ancient Greeks had dropped a pebble and a boulder from the same height at the same time, it should have dispelled the myth that heavier objects fall with greater acceleration. However, before faulting the ancient Greeks too much, consider this counterexample: If the ancient Greeks had dropped objects like feathers, paper, and pencils, in this case they may have indeed observed that objects with different mass do not necessarily fall with the same acceleration. As it turns out, cross-sectional area plays a more significant role than mass for objects falling with air resistance, but you can see how the complications of air resistance may have made it difficult to deduce precise postulates regarding gravity.

Regarding the ancient Greek concept of the natural tendency of an object, this idea seemed to be supported by terrestrial experiments.[133] Also, ancient Greek astronomy very closely followed the scientific method: The most basic astronomical observations suggested the intuitive notion that the sun and planets orbit the earth – after all, that's what it looks like from our perspective. As their data improved, they developed and revised their model to fit the data, very much in the spirit of science – this is the origin of the idea of having epicycles in epicycles. So it is not entirely accurate to claim that ancient Greek science was not based on experimental measurements. However, this claim does have some merit, as many of the ancient Greeks who dabbled in astronomy and physics were philosophers at heart.

**Renaissance physics**: About two millennia later, the Renaissance period (15th and 16th centuries A.D.) challenged the predominant ways of thinking in Europe. Copernicus literally put his belief in the heliocentric model of the solar system before his own life. Galileo observed sunspots, which challenged the Greek notion that the heavens beyond the moon were perfect orbs, and then he observed a full Venus – a phase that would not be possible in the geocentric model. Galileo also observed that objects fall with the same acceleration regardless of their mass, made many careful and precise measurements pertaining to accelerated motion, developed the notion that a satellite (such as a moon orbiting a planet) is really just a projectile that has so much tangential speed that it doesn't fall to the ground, and almost developed the correct notion of the natural tendency of an object (instead, he wrote that objects have a natural tendency to travel in circles – obviously, he was still thinking about satellites in circular orbits when he wrote this). The challenge to the prevailing views of physics and the discoveries made by Galileo and others paved the way for Isaac Newton to develop a thorough treatise of the physical motion of objects (which he wrote in his famous text called the *Principia*).

**Newtonian physics**: Isaac Newton discovered that the same underlying principle of gravity could explain the motion of terrestrial objects falling near the surface of the earth as well as the motion of celestial bodies in the solar system, correcting the ancient Greek expectation that the realms of earth and outer space were vastly different in nature (imperfect versus perfect orbs) governed by different physical laws. Newton also formulated a precise definition of the natural tendency of an object that corrected the ancient Greek notion that objects tend to be at rest – as exemplified by celestial objects. Newton's laws of motion and gravity are fundamental principles that are central toward understanding physics.

---

[133] The ancient Greeks didn't observe objects coming to rest out in space, but the heavens were viewed as a perfect world subject to a different set of physical laws. Terrestrially, however, objects seemed to come to rest.

# 3 Newton's Laws of Motion

## 3.1 Newton's Laws

**Momentum**: The momentum, $\vec{\mathbf{p}}$, of an object equals the product of its mass, $m$, and its velocity, $\vec{\mathbf{v}}$:

$$\vec{\mathbf{p}} = m\vec{\mathbf{v}}$$

Momentum is a vector. The SI units for momentum are kg·m/s, which we will learn equate to a Newton times a second, N·s.

> **Hint**. It's easy to work out the SI units for any physical quantity. You just need to know an equation for the physical quantity: Plug the SI units for everything else into the equation, and then solve for the SI units of the quantity of interest. In the case of momentum, $[\vec{\mathbf{p}}]_u = [m]_u[\vec{\mathbf{v}}]_u = $ kg·m/s.

> **Notation**. The brackets, [ ], are generally employed to indicate the dimensions of a quantity. For example, $[\vec{\mathbf{p}}]$ refers to just the dimensions of momentum, which are the dimension of mass ($M$) times the dimensions of velocity; and the dimensions of velocity are those of length ($L$) divided by those of time ($T$). Therefore, $[\vec{\mathbf{p}}] = [m][\vec{\mathbf{v}}] = ML/T$. The dimensions of all physical quantities in mechanics can similarly be expressed in terms of $M$, $L$, and $T$.
> 
> Dimensionality is related to units, but the two terms are not interchangeable. The dimensions refer to the nature of the physical quantity, while the units refer to a standard measure that could be used to determine the magnitude of the quantity. There are usually multiple units in which a measurement may be made. For example, distance – which has the dimensions of length, $L$ – can be measured in meters (m), centimeters (cm), inches (in.), and many other units of distance. We will include a subscript $u$ with the brackets to distinguish between the dimensions and units of a physical quantity – e.g. $[m]_u$ refers to the unit of mass, which is the kilogram (kg) in SI units, while $[m]$ refers to the dimension of mass.

> **Note**. Equations for dimensions and units are not formulas. It would be a major conceptual mistake to let the relation between dimensions or units dictate what arithmetic you should do to calculate a quantity. For example, suppose you drop a marble from rest and measure its height above the ground and the time it takes to reach the ground, and you wish to use these measurements to calculate gravitational acceleration. If you looked solely at the units of acceleration, m/s², and took the height and divided by the square of the time, your answer would be incorrect: The 1DUA formula that relates the height and time is $\Delta y = v_{yo}t + \frac{1}{2}a_{yo}t^2$. It is a good idea to look at the units to check if a formula that you derive is dimensionally consistent, but it is never a good idea to look at the units to guide your arithmetic. Instead, let the concepts, strategies, laws, and equations be your guide.

**Inertia**: All objects have a natural tendency pertaining to their physical motion, which is termed inertia. Most precisely, an object's inertia is its natural tendency to maintain constant momentum. Since $\vec{\mathbf{p}} = m\vec{\mathbf{v}}$, a single object with constant mass has a natural tendency to maintain constant velocity; or, put another way, inertia is the natural tendency of an object to resist being accelerated.

Recall that velocity includes both speed and direction, such that a single object with constant mass has a natural tendency to maintain constant speed and travel in a straight line. As a special case, an object at rest has zero velocity, and therefore has a natural tendency to remain at rest. An object that is moving, however, does not have a natural tendency to come to rest, but instead has a natural tendency to maintain constant momentum.[134]

The ancient Greeks observed that objects at rest tend to stay at rest, and also observed that terrestrial objects that are set in motion tend to come to rest. However, it is the natural tendency of all objects – terrestrial or otherwise – to maintain constant momentum. Terrestrial objects come to rest because net external forces – generally, as the result of resistive forces like friction and air resistance – overcome their inertia. The motion of celestial objects evidently persists forever due to a lack of such resistive forces.

> **Important Distinction**. Objects at rest do have a natural tendency to be at rest, but moving objects do not have a natural tendency to come to rest. Many students find this idea to be counterintuitive, though they should come to refine their intuition after studying physics. All objects – moving or not – have a natural tendency to maintain constant momentum.

**Force**: A force is something, that if applied by itself, would cause an object to change its momentum (very often though, there are multiple forces acting on an object, and it is the resultant of these force vectors that determines whether or not an object will move according to its natural tendency). Every force can effectively be classified as either a push or a pull.

The SI unit of force is the Newton (N). As we will see from the formula for Newton's second law, the Newton is related to other SI units by: $1\text{ N} = 1\text{ kg·m/s}^2$. Thus, if you are working with forces in Newtons, or if you are calculating force and want to express it in Newtons, you need to ensure that all of the distances are in meters, the times are in seconds, and the masses are in kilograms.

> **Hint**. It is a good habit to convert all units to kilograms (kg), meters (m), and seconds (s) because all derived units, like the Newton (N), are related to these base SI units. If, for example, you work with masses in grams, force will not come out in Newtons – you first need to convert grams to kilograms.

**Common forces**: Following is a list of some of the types of forces that are commonly encountered in mechanics problems, with a brief description of each to help you determine what forces may be acting on an object. We will discuss each in more detail as we come across them in later sections.

---

[134] It is imprecise to say that an object in motion has a natural tendency to remain in motion, although it is generally stated as such in physical science courses. The problem with this statement is that there are many ways to be in motion, only one of which – uniform momentum – is the correct natural tendency. Objects do not have a tendency to maintain constant acceleration, but acceleration is a kind of motion. It is most precise to state that objects have a natural tendency to maintain constant momentum; when speaking of a single object with constant mass, this equates to maintain constant velocity. It is also imprecise to say that objects have a natural tendency to maintain constant speed, because they also tend to travel in a straight line – you could travel with constant speed in a curved path, which is not a natural motion. In physical science, just the basic idea is conveyed; but in physics, it is necessary to understand the precise definition of inertia, and to distinguish between different types of motion – so state 'maintain constant momentum' instead of ' stay in motion.'

- Every object in the universe is in a gravitational field and therefore experiences a gravitational force, which we call the object's weight, $\vec{\mathbf{W}}$. Weight acts along the gravitational field line. Near the surface of a planet, the gravitational field lines are directed radially inward, toward the center of the planet. On paper, we usually draw illustrations with gravity directed down toward the bottom of the page (except when noted otherwise, as in 'top view'). It is a good idea to label weight as $m\vec{\mathbf{g}}$, so that you don't confuse the magnitude of the weight, $W$, with work later.
- An object that is in contact with a surface experiences a normal force, $\vec{\mathbf{N}}$. The normal force is a support force that the surface exerts on the object, which generally limits or (inhibits) the effect of gravity's downward pull. In mathematics, the word 'normal' means perpendicular. If you remember this, it may help you draw the normal force correctly.
- An object that is in contact with a surface may also experience a friction force, $\vec{\mathbf{f}}$. The friction force acts tangential to the surface, directed against the motion (or, if the object is stationary, in a direction so as to oppose potential motion). Look at the direction of the velocity (or potential for velocity to result) to determine the direction of the friction force.
- Similarly, when air is not neglected, a force of air resistance, $\vec{\mathbf{f}}_a$, acts against the velocity.
- If you see a cord, string, rope, thread, cable, or similar material (but not a spring), there will be a tension, $\vec{\mathbf{T}}$, pulling on the object along the length of the cord, string, rope, etc.
- If an object is connected to a spring that is displaced from equilibrium, the spring exerts a restoring force, $\vec{\mathbf{F}}_r$, on the object, directed toward the equilibrium position of the spring.
- A mechanical (as opposed to, say, electromagnetic) push or pull that cannot be classified as one of the above forces will generally be labeled as $\vec{\mathbf{P}}$.

**Notation.** We will use $\vec{\mathbf{F}}$ to speak of force in general terms – i.e. without a specified type of force in mind. When we declare a specific type of force, we will use a symbol especially for it: $m\vec{\mathbf{g}}$ (or $\vec{\mathbf{W}}$) for weight, $\vec{\mathbf{N}}$ for normal force, $\vec{\mathbf{f}}$ for friction, $\vec{\mathbf{T}}$ for tension, $\vec{\mathbf{F}}_r$ for restoring force, and $\vec{\mathbf{P}}$ for other types of mechanical pushes or pulls.[135]

**Note.** If the SI unit of a quantity is not a Newton, it's definitely not a force! When trying to determine which forces may be acting on an object, it's important to consider only forces, and not other quantities like mass, velocity, or acceleration, which you can tell are not forces just by examining their units.

**Natural tendency**: Inertia is defined as the 'natural tendency' of an object. When we refer to an object's natural tendency, what we mean is what the object would tend to do if the net external force acting on it were zero. That is, if the vector sum of all of the external forces acting on an object equals zero, the object will travel with constant momentum.[136]

---

[135] The symbols are not all universal – e.g., $\vec{\mathbf{F}}_N$ is often used for normal force and $\vec{\mathbf{F}}_T$ is sometimes used for tension.
[136] It is imprecise to say that inertia is what an object would do if there were no forces acting on it. Actually, objects often travel according to their inertia when there are forces acting on them – namely, when the force vectors add up to zero. An object travels according to its natural tendency when all of the forces acting on it cancel out, not necessarily when there are no forces acting on it. As long as the net force acting on an object equals zero, it will travel with constant momentum.

**Newton's first law**: According to Newton's first law (aka the law of inertia), if the net external force acting on an object (or system of objects) equals zero the object will maintain constant momentum:

$$\sum \vec{F}_{ext} = 0 \quad \Rightarrow \quad \vec{p} = const.$$

If the net external force acting on a single object with constant mass equals zero, the object will maintain constant velocity – since $\vec{p} = m\vec{v}$ – which also means that the acceleration, $\vec{a}$, is zero.

> **Note**. Inertia is most generally a natural tendency to maintain constant momentum. In a typical physics course, the vast majority of problems involve an object with constant mass, for which inertia can be stated as a natural tendency to maintain constant velocity. Not all objects have constant mass, however. For example, the mass of a rocket is reduced if it ejects steam or burns fuel. For objects with variable mass, inertia is a tendency to maintain constant momentum, but not constant velocity.

**General form of Newton's second law**: If the net external force acting on object (or system of objects) is nonzero, its momentum changes according to Newton's second law:

$$\sum \vec{F}_{ext} = \frac{d\vec{p}}{dt}$$

> **Note**. The above equation is the most general form of Newton's second law.[137] There is a more common form that we will derive soon, which applies to most of the problems that we solve in first-year physics. However, the common form is not applicable to all problems, as we shall see. It is therefore sometimes necessary to apply the most general form of Newton's second law.

**Newton's first law as a special case of Newton's second law**: Strictly speaking, we only need two of Newton's laws of motion because Newton's first law is just a special case of Newton's second law. Looking at Newton's second law, you can see that if the net external force is zero – i.e. if $\sum \vec{F}_{ext} = 0$ – then $d\vec{p}/dt = 0$, which requires $\vec{p}$ to be a constant, in accordance with Newton's first law.

**Derivation of the common form of Newton's second law**: Let us apply Newton's second law to a single object. Since $\vec{p} = m\vec{v}$, we can express Newton's second law as[138]

$$\sum \vec{F}_{ext} = \frac{d(m\vec{v})}{dt} \quad \text{(single object)}$$

In general, it is necessary to apply the product rule to carry out the differentiation. However, if the mass of the object is constant, Newton's second law can be expressed as[139]

---

[137] We must be careful how we define our system in the most general variable-mass case (see Sec. 6.9).
[138] To see how to generalize the same relation to a system of objects, see Chapter 6.

## 3 Newton's Laws of Motion

$$\sum \vec{F}_{ext} = m \frac{d\vec{v}}{dt} = m\vec{a} \quad \text{(single object with constant mass)}$$

**Common form of Newton's second law**: Most of the problems in first-year physics involve working individually with objects that have constant mass. For these problems, it is more convenient to work with acceleration than it is to work with momentum, and so the special case of Newton's second law that we just derived is much more common than the general form:[140]

$$\sum \vec{F}_{ext} = m\vec{a} \quad \text{(single object with constant mass)}$$

The sum of the external forces represents vector addition. Recalling that the method of adding vectors involves working with components, Newton's second law is generally expressed in terms of components. We will describe how to apply Newton's second law to solve problems in the next section.

> **Note**. The common form of Newton's second law – in terms of acceleration – is convenient when treating objects with constant mass individually. We will see that it is more convenient to work with the most general form of Newton's second law – in terms of momentum – when treating a system of objects (see Chapter 6). It is necessary to use the general form of Newton's second law when working with objects that have variable mass, which is typical of rocket problems (discussed in Chapter 6).

**SI units for force and momentum**: From Newton's second law, we see that a Newton (N) must equal a kilogram (kg) times a meter per second-squared ($m/s^2$):

$$[\vec{F}]_u = [m]_u [\vec{a}]_u$$
$$1 \text{ N} = 1 \text{ kg·m/s}^2$$

The SI units for momentum may be expressed as N · s or kg·m/s, since both sets of units are equivalent:

$$[\vec{p}]_u = [m]_u [\vec{v}]_u = \text{kg·m/s} = \text{N·s}$$

**Inertial mass**: The mass of an object is most accurately defined as a measure of the object's inertia. In this sense, mass is a measure of the resistance of an object to accelerate. This definition of mass is consistent with the common form of Newton's second law: An object with greater mass requires a greater net external force in order to achieve a given acceleration. It should also agree with experience: It's easier to accelerate (i.e. to change the velocity) of a less massive object.

---

[139] For a system of objects, this equation can also be expressed in terms of the acceleration of the center of mass. However, for a system of objects, it is generally more convenient to work with momentum. See Chapter 6.

[140] In physical science courses, Newton's second law is sometimes written as $\vec{F} = m\vec{a}$, but the sum is important because there are generally several forces acting on an object, not just one. The summation symbol, $\Sigma$, reminds you that you must add vectors to apply Newton's second law. Writing Newton's second law as $F = ma$ is worse yet: You certainly can't add the magnitudes of the forces together, in general, to find the acceleration of an object. The forces generally act in different directions, and so you must add the forces as vectors.

> **Note**. If you have studied chemistry, you may be accustomed to thinking of mass in terms of the number of protons and neutrons that compose an object.[141] This is convenient when working with chemical reactions, but not for solving motion problems. In physics, it is much more convenient to think of mass as a measure of inertia. Furthermore, it is more accurate to think of mass as a measure of inertia than to express it in terms of proton and neutron masses: The reason is that the nuclear energy that binds protons and neutrons together in the nucleus is equivalent to an amount of mass determined by Einstein's famous equation, $E = mc^2$.[142] If you are used to thinking of mass in different terms, conceptually it will be well worthwhile to retrain yourself to think of mass as a measure of inertia.

**Newton's third law**: All forces come in pairs according to Newton's third law: When one object – call it object 1 – exerts a force on another object – call it object 2 – object 2 exerts a force back on object 1 that is equal in magnitude, but opposite in direction:[143]

$$\vec{\mathbf{F}}_{12} = -\vec{\mathbf{F}}_{21}$$

Apply Newton's third law when you want to relate the forces that two objects exert on one another. It is easiest to apply Newton's third law correctly if you identify the two objects and call them objects 1 and 2.

> **Note**. Newton's third law is sometimes stated as follows: For every action, there exists a reaction that is equal and opposite to the action. Most students who try to apply Newton's third law in terms of action and reaction make conceptual mistakes – they generally have trouble identifying the action and the corresponding reaction. Students who instead try to identify the two objects and think about the force that object 1 exerts on object 2 compared to the force that object 2 exerts on object 1 tend to make fewer conceptual mistakes than those who think in terms of action and reaction. However, the two statements of Newton's third law are the same: If the force that object 1 exerts on object 2 is the action, the reaction is the force that object 2 exerts on object 1. Students often find it counterintuitive that these forces should be equal and opposite – but if you study the conceptual examples, you might come to understand why this is and reconcile it with your experience.

**Internal forces**: Due to Newton's third law, the internal forces acting on a system always cancel out. This is why we only concern ourselves with the external forces acting on an object (or a system of objects) in Newton's first and second laws.

---

[141] The electrons have negligible mass compared to the protons and neutrons.

[142] For example, the most abundant isotope of helium is $^4_2$He, which has 2 protons and 2 neutrons. A single neutral $^4_2$He atom does not have a mass equal to $2m_p + 2m_n + 2m_e$. If you experimentally determine the atomic mass of $^4_2$He (which is different than the average atomic mass for helium given in a periodic table because it weights isotopes by natural abundance), it differs from $2m_p + 2m_n + 2m_e$ by precisely $E = mc^2$, where $E$ is the nuclear binding energy that holds the $^4_2$He nucleus together. The point is that the mass of an object does not actually equate to the sum of the masses of its constituents. However, mass is always a measure of inertia.

[143] We often say 'equal and opposite,' but two force vectors can't technically be equal if they are oppositely directed. When we say 'equal and opposite,' what we really mean is that only the magnitudes are equal.

**Conceptual Examples**. A monkey, who is riding in a train on level tracks, releases a banana from rest (relative to the train) directly above a small coin on the floor of the train. Where does the banana land?

The answer depends upon how the train is moving. If the train is traveling with constant velocity, the banana will land right on the coin because the banana has inertia. Relative to the train, the banana travels straight down. Relative to an observer standing on the ground outside and looking through a window, the banana's path would look like a parabola. That is, the path looks like 1DUA relative to the monkey, and like projectile motion relative to an observer outside of the train.

If the train is traveling with constant speed, but the track is curved, the banana will appear to be deflected to the right or left relative to the train. For example, if the train is turning to the left, the banana will appear to be deflected to the right. Actually, the banana has inertia and so travels forward when it is released, and it is the train that deflects to the left, causing the banana to miss to the right.

If the train travels in a straight line, but decelerates, then the banana will land ahead of the coin. When the banana is released, the horizontal component of the banana's velocity remains constant – with a value equal to the speed of the train when the banana was dropped – while the vertical component of the banana's velocity increases due to gravity. The banana is traveling faster, horizontally, than the train in this case because the train is slowing down, while the banana is not.

**Common Mistake**. Most students find many physics concepts – especially Newton's laws of motion – to seem counterintuitive at first. For this reason, students who answer questions based on their intuition generally make numerous mistakes. If instead you try to base your answers on physical laws, definitions, or equations, you are more likely to arrive at correct answers – and also you will be able to explain your answers (many instructors grade your reasoning more heavily than your answer). You don't have to guess at physics questions; rather, you should let the laws, definitions, and equations serve as your guide.

**Conceptual Example**. A monkey throws a bowling ball horizontally to the west. Why does the monkey feel pushed to the east when he throws the bowling ball?

Just as when a shooter fires a rifle, the monkey experiences recoil when he throws the bowling ball. The force of recoil pushing the monkey backward can be understood by considering Newton's third law: A consequence of the force that the monkey exerts on the bowling ball is that the bowling ball exerts a force back on the monkey that is equal in magnitude and opposite in direction.

**Conceptual Example**. A monkey wearing boots is stranded on horizontal frictionless ice. How can the monkey get home?[144]

First, observe that the monkey can't walk on frictionless ice – well, he can try, but he would look rather silly, probably doing somersaults instead. When there is friction, we walk by pushing one direction on the ground, being propelled in the opposite direction by Newton's third law; but the monkey can't push horizontally on the ground on frictionless ice. However, he can still make use of Newton's third law by throwing his boot – receiving a push in the direction opposite to his throw.

---

[144] Another good question is how the monkey got there in the first place. Maybe he was dropped in by helicopter (though we had better not neglect the presence of air for the helicopter).

**Note.** Notice that the force pairs associated with Newton's third law cancel out if you treat the system as a whole, but not if you apply Newton's second law to each object individually. In the previous example, when the monkey throws the boot, it would be incorrect to think that nothing happens because the two forces – i.e. the force of the throw and the equal, but opposite, reaction – cancel out. If you sum the forces for the system – thinking of the system as the monkey plus the boots – you will indeed find that the net external force is zero, but this means that the center of mass of the system will not accelerate. In this case Newton's second law tells you only what will happen to the system as a whole, not what may happen to the individual components of the system. If instead you apply Newton's second law to just one object – say, the boot – then the force of the throw is an external force for the boot, which accelerates the boot during the throw (once the monkey lets go of the boot, of course, the throw no longer accelerates the boot). The force that the boot exerts on the monkey does not contribute to the boot's acceleration – only forces acting on the boot affect the boot's acceleration. The force that the boot exerts on the monkey instead accelerates the monkey during the throw.[145]

It is a common mistake for students to get confused when applying Newton's second and third laws to the same problem. The force pairs only cancel out for the system as a whole. If you sum the forces for individual objects, only a single force is involved (because the two forces of the mutual force pair act on different objects).

**Conceptual Example.** A 60-kg monkey running 5 m/s to the east tackles a 20-kg monkey who was at rest prior to the collision. Determine the ratio of the forces that they exert on one another during the collision, and also determine the ratio of their accelerations during the collision.

According to Newton's third law, the force that the 60-kg monkey exerts on the 20-kg monkey are equal in magnitude, but opposite in direction. Thus, the ratio of the forces is $1:-1$. These contact forces accelerate each monkey according to Newton's second law: $\sum \vec{F}_{ext} = m\vec{a}$. Although the forces are equal in magnitude, the masses and accelerations are not: $m_1 \vec{a}_1 = -m_2 \vec{a}_2$ or $\vec{a}_1 : -\vec{a}_2 = m_2 : m_1 = 3.0$. That is, the less massive monkey accelerates three times as much as the more massive monkey during the collision, and in the opposite direction. This means that the 60-kg monkey slows down less from 5 m/s than the 20-kg monkey speeds up from rest.

**Note.** Students often find it counterintuitive that the contact forces should be equal in magnitude during a collision – especially, if a bug is struck by the windshield of a car. However, Newton's second law explains that although the forces are equal in magnitude, the less massive object will experience greater acceleration during the collision. The different accelerations and their relation to the mass ratio should agree with experience.

Also, note that Newton's third law does not tell you who gets hurt more during a collision, only that the forces will be equal. There are other factors involved in how much somebody gets hurt or how much something gets damaged – e.g. structural properties, biological processes, etc.

---

[145] Of course, the monkey can't 'swim' home, since we neglect air unless stated otherwise – you can't 'swim' by waving your arms in a vacuum (and it isn't so effective in air, either). Also, you can't 'fake out' Newton, so to speak: That is, if the monkey holds onto the boot at the end of his throw, he won't get home (but he might look silly doing somersaults, for example) – the action of holding onto the boot comes with a reaction from the boot.

## 3.2 Applying Newton's Second Law

**Practical form of Newton's second law**: Newton's second law involves the vector sum of the external forces, $\sum \vec{F}_{ext}$, acting on an object. The key step to adding vectors is to work with components. Therefore, when Newton's second law is applied to solve problems, it is generally expressed in terms of components. For a single object with constant mass, the practical form of Newton's second law is

$$\sum_{i=1}^{N} F_{ix} = ma_x \quad , \quad \sum_{i=1}^{N} F_{iy} = ma_y \quad , \quad \sum_{i=1}^{N} F_{iz} = ma_z \quad \text{(single object with constant mass)}$$

in Cartesian coordinates, where $\sum_{i=1}^{N} F_{ix}$, for example, represents the sum of the $x$-components of the forces acting on the object and $a_x$ is the $x$-component of acceleration. Newton's second law can also be expressed in component form in other coordinate systems, such as 2D polar coordinates. For an object traveling along a curved path, it is often convenient to work with tangential and normal components.

> **Note**. We will use the component form of Newton's second law – i.e. the equations above – so frequently that we will not remind the reader every time that this form of the law – in terms of acceleration – applies only if mass is constant. It is important to realize that the equations above hold for the respective components of forces and accelerations: It would be a major conceptual mistake to sum the magnitudes of the forces and set that equal to the mass times the magnitude of the acceleration.

**Equilibrium**: The net external force equals zero for an object (or system of objects) in equilibrium:[146]

$$\sum \vec{F}_{ext} = 0 \quad \text{(equilibrium)}$$

Thus, an object travels according to its natural tendency – i.e. its inertia – if it is in equilibrium. Equilibrium does not imply that there are no forces acting on the object; rather, there may be forces acting on the object, but all of the forces cancel out if the object is in equilibrium. There are two types of equilibrium – static and dynamic equilibrium.

**Static equilibrium**: An object is in static equilibrium if it is not moving.

**Dynamic equilibrium**: An object is in dynamic equilibrium if it is moving with constant momentum – which means, for an object with constant mass, traveling with constant velocity.

---

[146] The equation $\sum \vec{F}_{ext} = 0$ demands that the center of mass of the object (or system of objects) has zero acceleration (assuming that the total mass is constant). We must also ensure that the object is in rotational equilibrium by setting the net external torque equal to zero: $\sum \vec{\tau}_{ext} = 0$. We will reconsider this in Chapter 7.

**Free-body diagram**: It is useful to draw a free-body-diagram (FBD) in order to identify and visualize all of the forces acting on an object.[147] This helps to determine the components of the forces when applying Newton's second law. A FBD consists of a dot to represent the object in addition to arrows pulling on the dot to represent the forces acting on the object. It is also useful to label the arrows with their respective forces, label any angles that may be relevant for determining components of forces, and to setup a coordinate system. It is conventional to draw the arrows as pulls, rather than pushes.

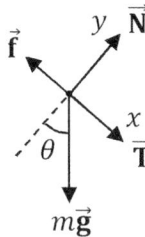

**Note**. Only include forces in the FBD: Don't include other quantities like velocity, since only forces are involved in $\sum \vec{F}_{ext}$. You shouldn't feel that you need to guess which forces are acting on an object: Review the bulleted list of common forces in the previous section, which describes when each of these forces is acting on an object. For example, if an object is in contact with a surface, you need a normal force, and if the surface is not frictionless, you also need a friction force; if an object is connected to a cord (or rope, string, etc.), you need a tension force. There are also rules for how to determine in what direction each force is acting. For example, normal force is always perpendicular to the surface, and friction is tangential to the surface, acting against the motion (or potential motion).

**Note**. When there are multiple objects in the problem, draw a separate FBD for each object (otherwise, you won't be able to solve for any internal forces, like tension in a connecting cord). Only draw forces acting directly on an object in its FBD. For example, if a spy is dangling from a helicopter by holding onto a rope, when you draw the FBD for the helicopter, include the helicopter's weight and the tension in the cord, but do not include the spy's weight in the helicopter's FBD – instead, the spy's weight will appear in the spy's FBD. In this example, the spy's weight does not directly act on the helicopter, but the effect of the spy's weight indirectly affects the helicopter through the tension in the rope. In general, the tension will not equal the spy's weight (especially, if the helicopter has vertical acceleration).

**Conceptual Example**. A monkey standing at the bottom of an incline gives a box of bananas a shove (and then lets go). The box of bananas slides first up the incline and then back down. Draw FBD's for the box of bananas as it travels both up and then back down the incline.

First, identify the forces acting on the box of bananas. The earth pulls the box of bananas straight down toward the center of mass of the earth: This is the weight, $m\vec{g}$, of the box of bananas. The box of bananas is in contact with a surface (the incline), which means that there is a normal force, $\vec{N}$, perpendicular to the incline, and friction, $\vec{f}$, which acts against the direction of motion.

---

[147] It is not merely useful, but required for virtually all physics courses.

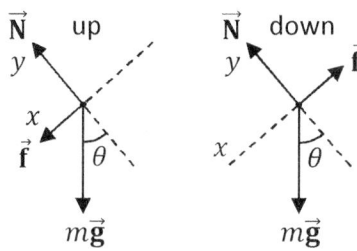

Note that the force of the monkey's shove does not appear in the FBD's because it does not affect the acceleration of the box of bananas as it slides up and down the incline (once the monkey lets go) because the box of bananas has inertia. That is, the monkey needed to exert a force to overcome the box of bananas in order to give it an initial velocity, but once the box of bananas is moving, it has a natural tendency to keep traveling up the incline. Only the forces that act on the box instantaneously contribute to the acceleration of an object; past forces do not (the monkey's shove only accelerated the box of bananas while the monkey was touching the box, which is how the box of bananas received its initial velocity). Observe the change in the direction of the friction force in the two cases.

**Types of forces to consider**: Consider the following forces when drawing a FBD for mechanics problems:
- Every object will have weight, $m\vec{g}$.[148] Near the surface of a planet, the object's weight pulls down toward the center of mass of the planet.[149] Weight pulls down toward the bottom of the page unless noted otherwise or if the drawing includes a picture of astronomical bodies.
- Is the object in contact with a surface? If so, draw a normal force, $\vec{N}$, perpendicular to the surface.
- Is the object in contact with a surface that is not declared to be frictionless? If so, draw a friction force, $\vec{f}$, tangential to the surface, in a direction that opposes the motion (or, if the object is stationary, in a direction that opposes potential motion).
- Is there a cord, string, rope, thread, cable, or similar material? If so, draw a tension, $\vec{T}$, along the cord, string, rope, etc.
- Is there a spring in the problem? If so, draw a restoring force, $\vec{F}_r$, which would return the spring to its equilibrium position.
- Is air resistance negligible? Neglect air resistance unless stated otherwise. If the effects of air resistance are included in the problem, the force of air resistance, $\vec{F}_a$, is opposite to the velocity.
- Study the problem (and any accompany diagrams) carefully to determine whether or not there are any other forces involved in the problem – e.g. a declared push or pull, $\vec{P}$, or a drive force to accelerate a car, $\vec{F}_D$.

---

[148] As we will explore in more detail in Sec. 3.9, even objects in free-fall orbits which appear to be weightless actually have weight. There is an exception, however: You could be at some magic position between the earth and the moon, for example, where their gravitational fields cancel out; but in general objects will have weight.
[149] The direction of gravitational field lines is not as immediately obvious when there are multiple sources – e.g. out in space where the earth pulls one way, the moon pulls another, the sun pulls another, etc. In such cases, you can find the magnitude and direction of the net gravitational field through vector addition.

**Note**. We will discuss these types of forces in more detail as we come across them in later sections.

## Problem-Solving Strategy for Applying Newton's Second Law (Common Form)

0. The common form of Newton's second law is useful for relating forces to acceleration, and is also useful for relating forces to one another for problems where there is no acceleration. The common form of Newton's second law applies to single objects with constant mass.[150] (For cases where mass is constant or for how to treat systems of objects, see Chapter 6).
1. Draw and label a FBD for each object, using appropriate subscripts, for which you are concerned with its motion.[151] Label any angles that will be useful for determining components of forces.
2. Setup a coordinate system (not necessarily Cartesian). It is wise to choose one coordinate along the direction of acceleration; the remaining coordinates must be chosen such that all of the independent coordinates are mutually perpendicular to one another. This choice ensures that (at least) two components of acceleration will be zero. If there is a pulley in the problem, it is useful to think of the pulley as bending a coordinate axis – see Sec. 3.4.
3. Sum the components of forces separately for each object for each coordinate axis. You will need to apply trig to resolve each force into components. It's conventional (and convenient) to work with small angles (i.e. $0° \leq \theta \leq 90°$) in FBD's, so we will abandon the convention from vector addition and not label all of our angles counterclockwise from the $+x$-axis.
4. If there is friction in the problem, you will need to apply the equation for friction (see Sec. 3.5).
5. If there is a spring in the problem, you will need to apply Hooke's law (see Sec. 3.6).
6. Count your knowns and unknowns. In algebra courses, you generally need $N$ equations in $N$ unknowns, but beware that in physics courses, you can often solve a system of $N$ equations with fewer than $N$ unknowns – the reason is that some variables, like mass, often cancel out after substitutions.

**Note**. You won't be able to solve for internal forces – like tension in a rope connecting two boxes – unless you draw FBD's and sum the components of forces separately for each object.

**Hint**. If a force lies exclusively on one coordinate axis, it will only appear in the sum of the components for that coordinate. In 2D, if you write a $\sin \theta$ in the sum of components for one coordinate, you will need to write a $\cos \theta$ in the corresponding coordinate's sum, and vice-versa.

---

[150] Newton's second law can also be applied to systems of objects where the total mass is constant. If the objects have different accelerations, then the common form of Newton's second law relates to the acceleration of the center of mass, as we will see in Chapter 6. Even though we may, in principle, apply Newton's second law to a system of objects, it is generally preferable to apply it to objects individually – one reason for this is that you won't be able to solve for internal forces within the system unless you treat the objects as separate entities.

[151] For example, if a monkey pulls a box of bananas with a specified force, and you are interested in the box's acceleration, you can sum the components of forces just for the box of bananas, where the monkey's pull will be one of the external forces acting on the box. On the other hand, if two boxes are connected by a cord, you generally need to draw a FBD for each box and sum the components of forces for both boxes.

> **Note.** You don't necessarily want to choose $x$ to be horizontal and $y$ to be vertical. For example, if an object slides down an incline, it is much more convenient to choose $x$ down the incline and $y$ perpendicular to the incline: This choice makes $a_y = 0$, whereas working with horizontal and vertical directions would involve two nonzero components of acceleration. In this case, it would be silly to rotate your FBD to make $x$ appear horizontal and $y$ appear vertical on your paper: All this extra work does is increase your chances of making a mistake; it's much easier just to rotate your head if you wish to look at your diagram that way.

> **Note.** You will find examples of how to apply Newton's second law throughout this chapter.

## 3.3 Mass and Weight

**Inertial and gravitational mass**: Mass, $m$, actually serves two roles. As we have already seen, the mass of an object is a measure of the object's inertia,[152] as a greater net external force is needed to achieve a given acceleration (i.e. to change an object's velocity by a specified amount) for a more massive object. Additionally, all massive objects create gravitational fields and experience a gravitational force – called weight – in the presence of other[153] gravitational fields. In both senses, mass is a scalar.[154] The SI unit for mass is the kilogram (kg).

**Weight**: The weight, $\vec{W}$, of an object is the gravitational force exerted on the object. Weight is a vector. The SI unit for weight is the Newton (N).

> **Important Distinction.** Mass is a scalar quantity that provides a measure of inertia, and also creates gravitational fields. Weight is a vector quantity equal to the gravitational force exerted on an object.

---

[152] Momentum is also a measure of inertia. Here is an example to illustrate the distinction. A 40-kg monkey running 8 m/s to the east collides head-on with a 60-kg monkey running 4 m/s to the west. They cling together after the collision. Which way do they travel after the collision? We will learn in Chapter 6 that it's momentum, and not mass, that determines the answer. Specifically, the less massive monkey has more momentum (recall that $\vec{p} = m\vec{v}$), so they travel to the east after the collision (with a speed of 0.8 m/s in order to conserve momentum). For a single object with constant mass, mass provides a useful measure of inertia: A greater net external force must be applied to stop a very massive moving object quickly compared to a less massive object with the same initial speed. Momentum is a more generally useful measure of inertia than mass, in that it accounts for velocity. Consider two boxes with equal mass, but different initial speeds: A greater net external force is required to bring the faster box to rest with the same stopping distance.

[153] Objects don't exert gravitational forces on themselves, so an object doesn't experience a gravitational force due to the gravitational field that it creates.

[154] A point-mass creates gravitational field lines that radiate inward toward it. Since these field lines are isotropic (the same in all directions), there is no preferred direction to associate with gravitational mass. A mass that's placed in a gravitational field created by another mass does experience a force in a particular direction, but that force is called weight, not mass. Mass is a scalar, weight is a vector.

**Note.** Mass and weight are fundamentally different terms with precise meanings. Although these two physical quantities turn out to be related through an equation, the terms are not interchangeable.

**Important Distinction.** The term 'gravity' refers to the gravitational field, $\vec{g}$, or to its magnitude, $g = \|\vec{g}\|$, which is called gravitational acceleration. The units of $g$ are m/s$^2$. If you want to refer to the force associated with gravity, $\vec{W}$, use the term 'weight' or call it gravitational force.

**Relationship between mass and weight**: The simplest way to relate mass to weight is to apply Newton's second law to an object in free fall. The only force acting on an object in free fall is the weight of the object:

$$\sum_{i=1}^{1} F_{iy} = ma_y$$
$$-W = ma_y \quad \text{(free fall)}$$
$$-W = -mg \quad \text{(true in general)}$$
$$W = mg$$

Recall that the acceleration of an object in free fall equals its gravitational acceleration. The result, $W = mg$, which relates the magnitude of the weight vector to the mass and gravitational acceleration, $g$, can also be expressed in vector form in terms of the gravitational field, $\vec{g}$, as:

$$\vec{W} = m\vec{g}$$

Although we derived the equation $\vec{W} = m\vec{g}$ by considering the special case of free fall, this equation actually holds in general. An object experiences the same force of gravity, $\vec{W} = m\vec{g}$, whether it is at rest near the surface of a planet or if it is falling near the surface of the planet; the difference is that an upward support force is exerted in one case to prevent the object from falling. If you hold an object in your hand, you can feel the pull of its weight: Try holding a more massive object and you will feel a greater pull. Release the object and you will see that its weight immediately accelerates the object downward as soon as you remove the support force.

> **Note**. It is fundamentally impossible to 'convert' a mass in kilograms to a weight in Newtons and vice-versa (and similarly for converting kilograms to pounds and so on). You can, however, compute the weight of an object in Newtons from its mass in kilograms by applying the formula $W = mg$.[155]

**Measuring weight**: Scales come in a variety of forms, but every scale that measures weight actually does so indirectly. A scale directly measures a scale force, $\vec{F}_s$, which in practice is either a normal force, $\vec{N}$, tension, $\vec{T}$, or restoring force, $\vec{F}_r$, depending upon how the scale works. When the scale is utilized according to its designed intentions, the magnitude of the scale force, $F_s = \|\vec{F}_s\|$, equals the magnitude of the weight, $W = \|\vec{W}\|$, of the object. We will see, later in this section, how a scale reading can differ from the weight of an object.

**Measuring mass**: Scales that measure normal force, tension, or restoring force provide an indirect means of measuring weight, and although many such scales indicate the mass of an object in grams, kilograms, or other unit of mass, such scales really provide a measure of weight and not mass. If you place an object on a scale near the surface of the earth and then bring the same object and scale to the surface of the moon (under the same conditions), if the two scale readings are significantly different, the scale is definitely measuring weight and not mass. Many scales expect to be used near the surface of the earth and so divide the measured weight by earth's gravitational acceleration in order to indicate the corresponding mass; such a scale would not provide an accurate reading for either mass or weight on the top of the mountain or on the moon, for example. Nonetheless, the most convenient means to determine mass is to use a scale designed to measure weight and apply the formula $W = mg$.

However, it is possible to measure mass directly. An inertial balance is one such instrument for achieving a direct measure of inertial mass. An inertial balance is a device that swings side-to-side, allowing for a direct measurement of how much an object resists being accelerated. An inertial balance would provide the same value for mass if the same object (in the same physical condition) was measured by the same inertial balance on the surface of the earth, atop a tall mountain, on the moon, or anywhere else (under the same laboratory conditions).

**Universality of mass**: The mass of an object is independent of its location. Whether you transport an object to the top of a tall mountain, to the moon, to another planet, or anywhere else in the universe, the object will offer the same resistance to acceleration (under the same laboratory conditions).

**Non-universality of weight**: The weight of an object is dependent upon its location, as it depends upon the value of gravitational acceleration. Gravitational acceleration and hence weight are greater near the surface of a planet than at the top of a mountain, have different values near the surfaces of different planets, and vary throughout the universe. If you want to weigh less, just climb a mountain – or if you really want to lose weight, go to the moon (but if you want to reduce your mass, that's a different issue).

---

[155] There are, however, conversion tables that do provide a 'conversion' from kilograms to Newtons, and vice-versa. Such a 'conversion' assumes that the object is near the surface of the earth and applies the formula $W = mg$ for you. It is conceptually important to realize that mass and weight are fundamentally different types of quantities with different meanings. Therefore, if you want to determine one from the other, apply the formula $W = mg$ rather than think of this as a conversion of sorts.

**Note**. The mass of an object would be the same everywhere in the universe because its resistance to accelerate is independent of its location, but the weight of an object varies from place to place because gravitational acceleration is position-dependent.

**Conceptual Example**. A monkey trains for a trip to the moon as follows. First the monkey puts on a spacesuit. The monkey jumps straight up in the air to see how high she can jump. She lifts weights to see how much weight she can lift. Then she punches a large punching bag that offers much resistance.

After her training, she takes a rocket to the moon. When she arrives on the moon, she trains some more. She finds that she can jump higher on the moon because there is less gravitational acceleration. For the same initial velocity, she can jump about 6 times higher because moon's gravity is about one-sixth that of earth's gravity. You shouldn't guess, though: Let the equation $v_y^2 - v_{y0}^2 = 2a_{y0}\Delta y$ serve as your guide. The monkey reaches the peak of her jump when $v_y = 0$, so for a given $v_{y0}$, $\Delta y$ must be 6 times greater on the moon because $a_{y0} \approx -g_e/6$.

The monkey also lifts weights on the moon. When she lifts weights on the earth, the heaviest weight she can carry is a brick with a 40-kg mass. She brought several of these 40-kg bricks to the moon, and finds that she can easily lift them on the moon. In fact, she can stack 6 of these bricks on top of one another and just barely lift the entire stack. The reason is that these bricks weigh less on the moon. Although their mass is still 40 kg on the moon, they weigh less because gravity is reduced by a factor of 6.

Finally, the monkey punches the same large punching bag that she used when training on earth. When she punches the bag with all of her might, she finds that the bag hurts her hand just as much as it did back on earth. This is because the punching bag has the same mass on the earth and moon, and therefore is equally resistant to acceleration on both the earth and moon.

**Apparent weight**: The apparent weight of an object equals what a scale would read if measuring the weight of the object, which may differ from the actual weight of the object. For example, in a spaceship in a free-fall orbit, objects have weight – by definition, since in free fall the only force acting on an object is its weight; it's the gravitational force that provides the acceleration – though they appear weightless. They not only appear weightless, floating in the spaceship if not fastened, but an astronaut inside such a spaceship actually feels weightless. You don't feel weightless standing on the ground – you don't feel weight, either: What you feel is the normal force that the ground exerts on you, preventing your weight from accelerating you downward. If you remove that normal force, you will feel weightless – but you will also fall downward with an acceleration equal to gravitational acceleration. The expression, "It's not the fall that kills you," is right: It's Newton's laws that kill you when you smack the ground below.

**Example**. A 60-kg monkey stands on a scale in an elevator. What does the scale read when the elevator accelerates upward with a constant rate of $3.0 \text{ m/s}^2$?

Solve for the scale reading to determine the apparent weight of the monkey by applying Newton's second law. First, draw and label a FBD and setup a coordinate system. The forces acting on the monkey include the monkey's weight, which is downward, and the force that the scale exerts on the monkey, which is upward. The scale really reads the force that the monkey exerts on the scale, which is equal and opposite to the force that the scale exerts on the monkey by Newton's third law.

$$\sum_{i=1}^{2} F_{iy} = ma_y$$
$$F_s - mg = ma_y$$
$$F_s = m(g + a_y) = 60(9.81 + 3.0) = 0.77 \text{ kN}$$

Since the $+y$-axis is upward and the elevator is accelerating upward, $a_y$ is positive; if the elevator had instead been accelerating downward, in this coordinate system $a_y$ would have been negative.

**Note**. Observe that the force that is accelerating the elevator does not appear in the FBD in the previous example because that force directly acts on the elevator, but only indirectly acts on the monkey. The forces that directly act on the monkey are only the weight of the monkey and the scale force. The force that accelerates the elevator ultimately accelerates the monkey upward, but is communicated to the monkey through the scale force.

## 3.4 Tension Forces

**Tension**: There is a tension force, $\vec{T}$, in a cord, string, rope, thread, cable, and other materials that can be pulled taut. The tension acts along the cord, string, rope, etc. Springs are different – we call the force associated with a stretched or compressed spring a restoring force, as springs have a different physical behavior than cords, strings, ropes, etc. We will explore springs in Sec. 3.6.

**Note**. Unlike weight and friction, there are no master equations for tension force. The equation for tension will vary from problem to problem. However, you can always solve for tension by applying the strategy for Newton's second law: When there is a cord, string, rope, etc. in a problem, tension will naturally appear in the sums of the components of the forces, and you can eliminate tension or solve for it as needed just by applying algebra.

**Example**. A monkey connects a 5.0-kg bunch of bananas to a string and holds the other end of the string so that the bananas are suspended vertically in the air. What is the tension in the string when the monkey accelerates the bananas downward at a rate of $2.5 \text{ m/s}^2$?

Begin with a labeled FBD with a choice of coordinates. The forces acting on the bananas include an upward tension force and the weight of the bananas pulling downward.

$$\sum_{i=1}^{2} F_{iy} = ma_y$$
$$T - mg = ma_y$$
$$T = m(g + a_y) = 5.0(9.81 - 2.5) = 37 \text{ N}$$

**Note.** Observe that the monkey's pull is not directly included in the FBD for the bananas in the previous example, but that the effect of the monkey's pull is incorporates through the tension force. Only the tension force and weight directly act on the bananas. Also, note that the bananas are not in free fall.

**Example.** A 4.0-kg bunch of bananas is suspended as illustrated below. The system is in static equilibrium. Determine the tension in each section of cord.

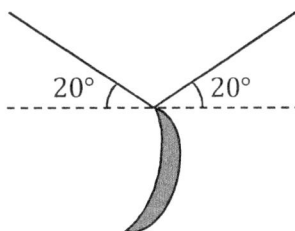

Begin by drawing a labeled FBD and setting up a coordinate system. There are two tension forces pulling on the bananas – one along each cord – and the bananas also have weight pulling down.

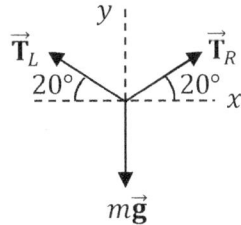

3 Newton's Laws of Motion

By symmetry, we can deduce that the two tension forces have equal magnitudes: $T_L = T_R \equiv T$. This FBD is 2D and so requires resolving the forces into $x$- and $y$-components. The weight is pulling along the $-y$-axis and so has only a $y$-component, while the tension forces have both $x$- and $y$-components. The acceleration of the banana is zero because it is in equilibrium:

$$\sum_{i=1}^{3} F_{ix} = ma_x \quad , \quad \sum_{i=1}^{3} F_{iy} = ma_y$$

$$T\cos 20° - T\cos 20° = 0 \quad , \quad T\sin 20° + T\sin 20° - mg = 0$$

$$2T\sin 20° = mg$$

$$T = \frac{mg}{2\sin 20°} = \frac{(4.0)(9.81)}{2\sin 20°} = 57 \text{ N}$$

**Tension force pairs**: Like all forces, tension forces come in pairs according to Newton's third law. When two objects are connected by a cord, tension forces that are equal in magnitude, but opposite in direction, pull both objects along the cord. Tension forces may be different in different cords or in different sections of the same cord, but come in force pairs that are equal in magnitude, but oppositely directed, in each section of cord according to Newton's third law.

**Conceptual Example**. A 500-kg gorilla and 50-kg calf are standing on horizontal frictionless ice. The gorilla lassoes the calf and pulls the (horizontal) rope with a force of 200 N. Compare the forces exerted on each animal and also compare their accelerations while the gorilla is pulling on the rope.

According to Newton's third law, the two tension forces are equal in magnitude, but opposite in direction: The gorilla exerts a 200-N force on the calf and the calf exerts a 200-N force on the gorilla. During the pull, however, because the forces are equal in magnitude, the calf's acceleration is 10 times greater than the gorilla's acceleration (this figure comes from the ratio of the masses, which can be seen by applying Newton's second law). The gorilla and calf accelerate towards one another. They will meet closer to the gorilla's starting position since the calf accelerates more during the pull. As the gorilla and calf approach one another, the tension in the cord decreases as the cord slackens, until the tension is reduced to zero; this occurs very quickly, such that the gorilla and calf spend most of the time sliding without any tension in the rope.

**Example**. A 1,000-kg gorilla in mid-air hangs desperately onto the free end of a rope, the other end of which is connected to a 5,000-kg helicopter that is traveling straight upward. The lifting force exerted on the helicopter is 90 kN. Determine the acceleration of the system and the tension in the connecting cord.

This problem requires two FBD's – one for the helicopter and one for the gorilla. The helicopter and gorilla will have the same acceleration (assuming that the rope does not stretch significantly). The lifting force, $\vec{F}_l$, pulls the helicopter upward, while the helicopter's weight and tension pull the helicopter downward. Similarly, tension pulls the gorilla upward, while the gorilla's weight pulls him downward. The magnitudes of the two tension forces are equal according to Newton's third law: $T_h = T_g \equiv T$.

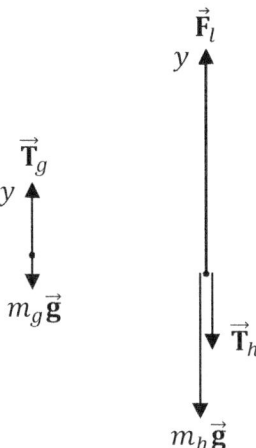

$$\sum_{i=1}^{2} F_{g,iy} = m_g a_y \quad , \quad \sum_{i=1}^{3} F_{h,iy} = m_h a_y$$

$$T - m_g g = m_g a_y \quad , \quad F_l - T - m_h g = m_h a_y$$

We have two equations and two unknowns, $a_y$ and $T$. By virtue of Newton's third law, the most efficient way to eliminate the unknown tension force is to add the two equations together:

$$F_l - (m_g + m_h)g = (m_g + m_h)a_y$$

$$a_y = \frac{F_l}{m_g + m_h} - g = \frac{90{,}000}{1{,}000 + 5{,}000} - 9.81 = 5.2 \text{ m/s}^2$$

The tension can be found by substituting the result for acceleration into one of the prior equations:

$$T = m_g(a_y + g) = 1{,}000(5.2 + 9.81) = 15 \text{ kN}$$

**Note**. When two or more forces pull on an object in the same direction, it is useful to draw the forces side-by-side in the FBD. This helps to visualize all of the forces, in order to help inadvertently forgetting to include one of the forces in the sums.

**Note**. It is crucial that subscripts be used on any symbols that are not equal – e.g. $m_g$ distinguishes the mass of the gorilla from the mass of the helicopter, $m_h$. Students who neglect the subscripts usually wind up canceling quantities that are not actually equal to one another. On the other hand, it is also important to conceptually (or mathematically) reason when two physical quantities are equal and use a common symbol for them – such as using $T$ for two tension forces acting on the same section of cord.

**Hint**. Work with symbols throughout your solution. This makes it easier for you to find and correct any mistakes in your solution, and easier for others (especially, your grader) to follow your logic.

# 3 Newton's Laws of Motion

**Accelerometer**: Note that a necklace dangling from a rearview mirror serves as an accelerometer – a device that measures acceleration. When the car travels with constant velocity, the necklace dangles vertically because the necklace has inertia. If the car slows down, the necklace leans forward due to its inertia; similarly, if the car speeds up, the necklace leans backward. If the car turns, the necklace leans outward – as the necklace has a natural tendency to travel in a straight line. A measurement of the angle that the necklace makes with the vertical can be applied to determine the acceleration of the car, as illustrated in the next example.

---

**Example**. A monkey drives a bananamobile with uniform acceleration in a straight line on a horizontal road. A banana is tied to the rearview mirror by a cord, and the cord makes an angle of 30° with the vertical. What is the acceleration of the bananamobile?

The bananamobile will have the same acceleration as the banana. The forces acting on the banana include tension along the cord and weight pulling downward. The banana is accelerating horizontally – the same direction as the car is accelerating – so it is convenient to choose the $x$-axis to be horizontal and the $y$-axis to be vertical: With this choice, $a_y = 0$. The FBD for the banana illustrated below features this coordinate system.

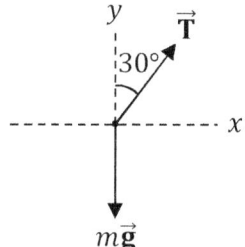

$$\sum_{i=1}^{2} F_{ix} = ma_x \quad , \quad \sum_{i=1}^{2} F_{iy} = ma_y$$
$$T \sin 30° = ma_x \quad , \quad T \cos 30° - mg = 0$$
$$T \sin 30° = ma_x \quad , \quad T \cos 30° = mg$$

The tension can be efficiently eliminated by dividing both equations:[156]

$$a_x = g \tan 30° = 9.81 \tan 30° = 5.7 \text{ m/s}^2$$

---

**Pulleys**: A cord (or string, cable, etc.) bends when it passes around a pulley. Thus, a pulley effectively serves to change the direction in which the tension forces at each end of a cord act.

---

[156] If $a = b$ and $c = d$, it must be true that $\frac{a}{c} = \frac{b}{d}$ – i.e. if the numerators and denominators are equal, the fractions must also be equal – as long as $c \neq 0$ (which implies $d \neq 0$). You better be careful to rearrange terms, if needed, to ensure that you don't divide by zero. For example, if you divide $T \sin 30° = ma_x$ by $T \cos 30° - mg = 0$ before moving $mg$ to the right, you are dividing both sides of the equation by zero, which is a big mistake.

**Frictionless pulleys**: Until we reach Chapter 7, we will assume that all of the pulleys are frictionless, such that the cords (or cables, strings, etc.) slide over the pulley as if the pulley were covered with a layer of ice. Under this assumption, the tension forces applied in each section of cord – one on each side of the pulley – will be equal in magnitude (but due to the pulley, only oppositely directed in terms of the shape of the cord). We will treat the more realistic case of a pulley that rotates with the cord in Chapter 7, where we will see that such rotation of the pulley is negligible provided that the mass of the pulley's wheel is small compared with any masses that are suspended from the pulley. In Chapter 7, we will also learn that the two sections of cord divided by the pulley actually involve different tension forces when the pulley is not assumed to be frictionless.

**Effect of pulleys on coordinates**: Conceptually, it is useful to view the pulley as bending one of the coordinate axes. That is, if you are drawing FBD's for two objects that are connected by a cord that passes over a pulley, think of one of the coordinate axes as being along the length of the cord, and think of the coordinate axis actually bending where the cord passes over the pulley. The reason for this choice of coordinates is that the component of acceleration of both objects along the cord will then be equal, which simplifies the notation. This is illustrated in the next example.

**Atwood's machine**: If you proceed to determine gravitational acceleration by dropping an object and measuring the time and height of descent, one of the experimental challenges lies in measuring the time interval with as little error as possible. The problem is that, near the surface of earth, an object gains about 10 m/s of speed each second as it falls, and so covers a large distance in little time even if released from rest.

One experimental method designed to improve upon this determination of gravitational acceleration involves the construction of Atwood's machine. Atwood's machine consists of two masses suspended from the ends of a cord that passes over a pulley. If the ratio of the masses is not large, the acceleration of the masses will be significantly smaller than gravitational acceleration. This smaller acceleration can be measured with less error, in principle, from which gravitational acceleration can be calculated.

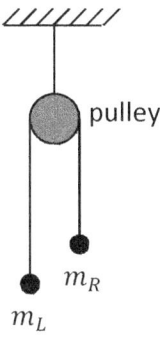

**Example.** A monkey makes Atwood's machine by suspending bananas of mass $m_L$ and $m_R > m_L$ from the two ends of a cord and passing the cord over a pulley. The cord slides over the pulley without friction. Derive equations for the acceleration of the masses and the tension in the connecting cord in Atwood's machine in terms of the suspended masses and the value gravitational acceleration.

The two FBD's include tension forces pulling each mass upward and weight pulling each mass downward. The mass $m_L$ will travel upward with the same magnitude of acceleration as the mass $m_R$ travels downward. Thus, it is convenient to choose the $+y$-axis to be up for $m_L$, and to bend around the pulley so as to be down for $m_R$. With this choice, $a_{Ly} = a_{Ry} \equiv a_y$. Since there is no friction between the cord and pulley, the magnitudes of the tension forces pulling on the two masses are equal: $T_L = T_R \equiv T$.

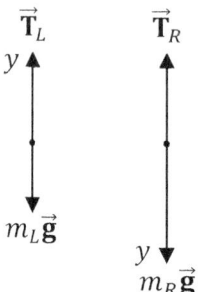

$$\sum_{i=1}^{2} F_{L,iy} = m_L a_y \quad , \quad \sum_{i=1}^{2} F_{R,iy} = m_R a_y$$
$$T - m_L g = m_L a_y \quad , \quad m_R g - T = m_R a_y$$
$$(m_R - m_L)g = (m_R + m_L) a_y$$
$$a_y = \frac{m_R - m_L}{m_R + m_L} g$$
$$T = m_L(g + a_y) = m_L g \left(1 + \frac{m_R - m_L}{m_R + m_L}\right) = \frac{2 m_R m_L g}{m_R + m_L}$$

It's a good idea to check that a derived equation provides reasonable results in limiting cases that are easy to verify. For example, in the limit that $m_L \approx m_R$, the acceleration of the system approaches zero. In fact, if the masses were equal, we would expect no acceleration. In the other extreme, in the limit that $m_L \ll m_R$, the acceleration approaches $g$. You could achieve this limit by replacing one of the bananas with a feather, in which case the other banana would practically be in free fall. We thus see that our result for the acceleration agrees with our expectations in two different extremes.

Now let's explore the tension in these limits. In the limit that $m_L \approx m_R$, the tension in the cord approaches $2m_L g$, and in the limit that $m_L \ll m_R$, the tension in the cord approaches zero.

**Hint**. The most efficient way to eliminate the unknown tension forces is to add equations together, as in the previous example.

**Note**. If you derive an equation, it's a good idea to check the results in limiting cases. The limits can't ensure that your formula is correct, but they can give you some confidence; and if the limits don't check out, then you know that the formula is incorrect.

## 3.5 Normal and Friction Forces

**Normal force**: A normal force, $\vec{N}$, acts on any object that is in contact with a surface. Normal force is a support force that a surface exerts on an object, and by Newton's third law the object also exerts a normal force on the surface. Normal force is always perpendicular to the surface.

> **Note**. Unlike weight and friction, but like tension, there are no master equations for normal force. The equation for normal force will vary from problem to problem. However, you can always solve for normal force by applying the strategy for Newton's second law: When an object is in contact with a surface in a problem, normal force will naturally appear in the sums of the components of the forces, and you can eliminate normal force or solve for it as needed just by applying algebra.
> 
> Only in the simplest possible cases will the normal force equal the object's weight or the object's weight times the cosine of an angle. It is totally incorrect (and naïve) to guess that the normal force will equal the object's weight times a trig function, even though this will turn out to be the case in some of the problems. In many problems, normal force will equal a significantly different expression. The only correct method for determining the normal force is to apply Newton's second law and solve for it.

> **Note**. Some students inadvertently say 'natural force' instead of 'normal force,' mistakenly thinking that 'natural' is sort of synonymous with 'normal.' However, this is totally incorrect because 'normal' signifies perpendicularity in mathematics, and 'natural' does not have a meaning synonymous with this.

> **Conceptual Example**. Would you feel less normal force sleeping on a bed made of concrete, wood, or rubber? Would you feel less normal force lying down or standing up?
> 
> In either case, the forces acting on you include your weight, which is downward, and normal force, which is upward (assuming that the bed is horizontal). The normal force must equal your weight, so long as your bed is not accelerating vertically. The same goes for standing up, so long as the ground is not accelerating vertically. Therefore, it does not matter what your bed is made of, nor whether you stand or lie down: In each case the normal force is the same.

> **Note**. Force by itself does not tell you about comfort or pain. For example, if your mother exerted the same force with which she would normally pat your check, but applied this force using the tip of a needle instead of her fingers, even though the forces would be the same, surely the pain would not. In the case of the needle, it is pressure – force per unit area – that makes the difference. Similarly, standing is less comfortable than lying down because the force is spread over less area, resulting in greater pressure.

> **Example**. On a cold winter day, a monkey steps outside and slides down an icy driveway. The driveway is inclined at an angle $\theta$, and the icy surface is virtually frictionless. Derive an equation for the acceleration of the monkey in terms of the angle of the incline and gravitational acceleration.

The forces acting on the monkey include the monkey's weight, which acts downward, and normal force, which is perpendicular to the incline. The monkey slides down the incline, so it is convenient to choose $+x$ to be down the incline and $+y$ to be along the normal force. This choice of coordinates makes $a_y = 0$. A labeled FBD with this coordinate system is illustrated below.

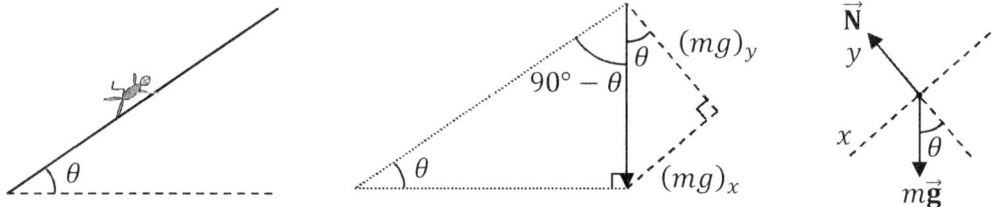

Application of Newton's second law yields:

$$\sum_{i=1}^{2} F_{ix} = ma_x \quad , \quad \sum_{i=1}^{2} F_{iy} = ma_y$$
$$mg \sin \theta = ma_x \quad , \quad N - mg \cos \theta = 0$$
$$a_x = g \sin \theta \quad , \quad N = mg \cos \theta$$

Note that the $x$-component of the weight involves $\sin \theta$, while the $y$-component of the weight involves $\cos \theta$, which is backwards compared to the convention that we had adopted in Chapter 2, where we first introduced vectors. The reason for the difference is that in Chapter 2 we defined the angles of the vectors to be counterclockwise from the $+x$-axis, whereas in the context of Newton's laws it is generally more convenient to work with smaller angles. In Chapter 2, the signs came naturally, while in the context of Newton's laws, one must look at the diagram and reason out the signs for each component. It is worth studying the diagrams and the components of weight above because many problems in physics involve an object on an incline.

Let us consider the limiting cases of the result, $a_x = g \sin \theta$. In the limit that $\theta$ approaches zero, the acceleration approaches zero; this corresponds to a horizontal surface, where gravity is canceled by the normal force. In the other extreme, where $\theta$ approaches 90°, the acceleration approaches gravitational acceleration; this corresponds to a vertical surface, in which case the monkey would be freely falling as the surface would not actually be in contact with the monkey.

**Note.** Study the incline diagram and the sums carefully in the previous example because many FBD's in physics involve an object on an incline.

**Example.** Two boxes of bananas are connected by a cord that passes over a pulley. One box is on frictionless horizontal ground, while the other is in midair, as illustrated on the next page. Derive equations for the acceleration of the boxes and the tension in the connecting cord in terms of the masses of the boxes, assuming that the cord slides over the pulley without friction.

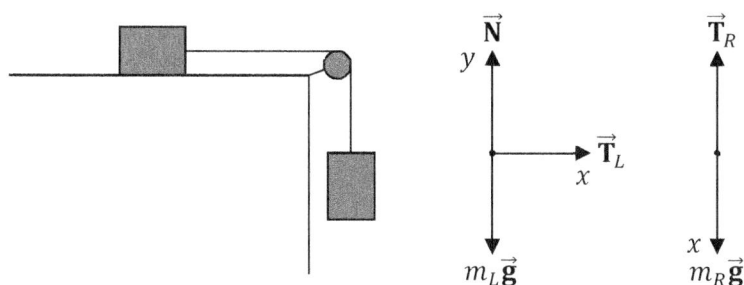

The left box experiences an upward normal force, downward weight, and a tension to the right, while the right box is pulled upward by tension and downward by its weight. The magnitudes of the two tension forces are equal because the cord slides over the pulley without friction: $T_L = T_R \equiv T$. The left box accelerates to the right as the right box accelerates downward. It is therefore convenient to think of the pulley as bending the $x$-axis downward, as illustrated by our choice of coordinates. With this choice, $a_{Lx} = a_{Rx} \equiv a_x$ and $a_{Ly} = a_{Ry} = 0$. Applying Newton's second law,

$$\sum_{i=1}^{3} F_{L,ix} = m_L a_x \quad , \quad \sum_{i=1}^{3} F_{L,iy} = m_L a_y \quad , \quad \sum_{i=1}^{2} F_{R,ix} = m_R a_x$$

$$T = m_L a_x \quad , \quad N - m_L g = 0 \quad , \quad m_R g - T = m_R a_x$$

$$m_R g = (m_L + m_R) a_x$$

$$a_x = \frac{m_R}{m_L + m_R} g$$

$$T = \frac{m_L m_R}{m_L + m_R} g$$

where we added two equations together to eliminate tension. It would be a good exercise to check that these results exhibit the expected physical behavior in limiting cases – e.g. in the limit that $m_R \gg m_L$.

**Friction force**: There is generally a friction force, $\vec{f}$, acting on each surface when two surfaces are in contact with one another and the surfaces are sliding against one another or if they have the potential to slide against one another (we will elaborate on this when we discuss static friction). The friction force acts tangential to the surface, in a direction that opposes the velocity of the surface (or the potential motion of the surface). The magnitude of the friction force, $f = \|\vec{f}\|$, is related to the magnitude of the normal force, $N = \|\vec{N}\|$, through the relation

$$f_s \leq \mu_s N \quad , \quad f_k = \mu_k N$$

where $\mu_s$ and $\mu_k$ represent the coefficients of static and kinetic friction, respectively, and $f_s$ and $f_k$ are the magnitudes of the corresponding friction forces. The formula $f_k = \mu_k N$ applies to a surface that is presently sliding against another surface, while the formula $f_s \leq \mu_s N$ applies to a surface that is presently stationary relative to another surface.

> **Note.** When you apply Newton's second law to a problem that involves friction, you will need to use the equation that relates the magnitude of the friction force to the magnitude of the normal force.

**Coefficient of friction**: The coefficient of friction – $\mu_s$ in the case of static friction and $\mu_k$ for kinetic friction – is the ratio of the magnitude of the friction force (the maximum amount possible in the case of static friction) to the normal force, and is a property of the materials of the two surfaces. For example, the coefficient of friction is different for cork sliding against aluminum compared to cork sliding against plastic or to cork sliding against cork. The coefficient of friction, $\mu_s$ or $\mu_k$, is dimensionless – i.e. it has no units. This can be seen from the equations $f_s \leq \mu_s N$ and $f_k = \mu_k N$.

**Static friction**: The force of static friction, $\vec{f}_s$, applies when two surfaces are in contact with one another, but there is (instantaneously) no relative motion between them. Conceptually, the inequality $f_s \leq \mu_s N$ relates to the fact that friction can only resist motion, and hence cannot contribute toward motion. That is, the force of static friction will have a maximum magnitude of $\mu_s N$ in its effort to prevent two surfaces from sliding against one another. If less force than $\mu_s N$ is needed to prevent such sliding, then $f_s$ equal only that amount which is required to prevent sliding from occurring.

> **Conceptual Examples.** Consider a 10-kg box of bananas resting on a horizontal surface, where the coefficient of friction between the box and surface is 0.50. Weight pulls the box of bananas downward, but it does not accelerate downward because the surface counteracts its weight with an upward force. These two forces cancel out. Thus, the friction force, in this case $\vec{f}_s$, is zero. Since no other forces acting on the box have a tangential component, if there were a friction force, it would cause acceleration. Friction can't cause acceleration – it can only limit acceleration or cause deceleration. Thus, the force of static friction equals only what is needed in this case, which is nothing at all. No other forces are contributing toward tangential acceleration, so no friction is induced.
>
> Imagine that a monkey now pushes horizontally on the box with a force of 30 N. Application of Newton's second law to this simple problem reveals that the normal force, in this case, equals the weight of the box, $N = mg = (10)(9.81) = 98$ N. The maximum possible magnitude of the force of static friction is $f_s^{max} = \mu_s N = (0.50)(98) = 49$ N. However, the friction force only needs to have a magnitude of $f_s = 30$ N in order to counteract the monkey's push. Therefore, the force of static friction will be only $f_s = 30$ N, and not its maximum possible value of $f_s^{max} = 49$ N.
>
> If the monkey now exerts a horizontal force of 60 N on the box of bananas, the fore of static friction will equal its maximum possible value of $f_s^{max} = 49$ N. The net horizontal component of force will then be 11 N, giving rise to an acceleration of 1.1 m/s² according to Newton's second law.
>
> If instead the monkey exerts the 60-N force at an angle of 45° above the horizontal, the horizontal component of the monkey's push will be $60\cos 45° = 42$ N. In this case, the friction force will only equal $f_s = 42$ N, and the box of bananas will remain at rest.

**Kinetic friction**: The force of kinetic friction, $\vec{f}_k$, applies when two surfaces are in contact with one another and there is (instantaneously) relative motion between them. In this case, the magnitude of the friction force equals $f_k = \mu_k N$. Conceptually, this is an equality – rather than inequality – because there is motion to resist (in the form of deceleration, if possible; otherwise, to limit any acceleration).

**Static and kinetic coefficients of friction**: Experimentally, the coefficients of static and kinetic friction, $\mu_s$ and $\mu_k$, respectively, are approximately equal. If either is slightly larger, it must be the coefficient of static friction:[157]

$$\mu_s \gtrsim \mu_k$$

Following is an argument for why $\mu_s$ may be slightly greater than $\mu_k$. If an object is at rest, as in the case of static friction, a net external force must overcome the object's inertia in order to accelerate it. If instead the object is already moving, the object's inertia is something that helps, rather than something that must be overcome. This is the common, intuitive expectation for $\mu_s > \mu_k$.

The most straightforward experimental procedure for determining $\mu_s$ and $\mu_k$ is consistent with this argument. In order to determine $\mu_s$, you experimentally determine the minimum force required to accelerate an object from rest, in which case the applied force is overcoming the object's inertia. In order to determine $\mu_k$, you set the object in motion and experimentally determine the force required to keep the object in motion. This time, inertia is on your side, and you are only overcoming friction.

However, not all physicists agree with the argument that $\mu_s$ should be larger than $\mu_k$, perhaps most notably Nobel Prize recipient Richard Feynman.[158] The coefficient of kinetic friction, $\mu_k$, applies for any kind of motion, not necessarily constant velocity. If you set the object in motion and apply a net force that causes acceleration or deceleration, and measure both the net force and the acceleration, you can also determine $\mu_k$ experimentally. In this case, you are overcoming the object's inertia. Inertia is a natural tendency to have constant momentum – it's not a preference to be at rest. Similarly, Einstein found that motion is relative – there is no absolute reference frame. Nature shows no preference for being at rest compared to having constant momentum – so from this perspective it would be strange for the coefficient of friction to be different in the static and kinetic cases.

Experimentally, precise determination of $\mu_s$ and $\mu_k$ has its own challenges: There are obvious sources of error in the case of kinetic friction – e.g. the surfaces may not be perfectly uniform, such that the coefficient of friction may actually vary with position as the object slides. If, in fact, $\mu_s$ is greater than $\mu_k$, the difference is evidently slight.

**Independence of area**: The coefficient of friction, $\mu_s$ or $\mu_k$, depends only upon the materials of the two surfaces, and does not depend upon the area of contact between the two surfaces.[159]

> **Note**. The coefficient of friction – and hence the friction force – does not depend upon the area of contact between the two surfaces. Most students find this to be counterintuitive. However, you can try this out by performing a demonstration like that described in the next conceptual example in order to see it for yourself.

---

[157] The symbol $\gtrsim$ means that two quantities are approximately equal, but one may be greater than the other.
[158] See the *Feynman Lectures on Physics*.
[159] The coefficient of friction does depend upon the nature of the contact between the two surfaces, which is why it depends upon the materials. If two materials effectively interact over a greater percentage of area, microscopically, you would expect this to affect the coefficient of friction. Reducing the area of contact, macroscopically, however, does not affect the coefficient of friction.

**Conceptual Example.** Imagine a rectangular block with the following dimensions: 12 cm × 6 cm × 1 cm. If the block slides down an incline on its largest side – i.e. the one that's 12 cm × 6 cm – it will have the same acceleration as it would if it slid down the same incline on either of its smaller sides – i.e. the 12 cm × 1 cm side or the 6 cm × 1 cm side (assuming that the block is uniformly smooth throughout each of its surfaces). You can conduct a simple experiment to test this out.

**Conceptual Example.** Imagine a rectangular block sliding down an incline. If a second, identical block were glued to the top of this block (effectively doubling the mass of the block), the acceleration down the incline would be the same. The coefficient of friction is the same because the same two surfaces are in contact. The normal force is doubled because the mass is effectively doubled. The friction force therefore also doubles. The component of the block's weight down the incline also doubles. However, twice the mass also means twice the inertia. If you apply Newton's second law, you will see that mass cancels out, and so the acceleration remains unchanged.

**Example.** A monkey pulls a box of bananas at an angle $\theta$ above the horizontal, dragging the box horizontally along the ground with constant velocity. Derive an equation for the magnitude of the monkey's pull, $P$, in terms of $m$, $g$, $\mu_k$, and $\theta$. What value of $\theta$ minimizes the monkey's pull?

The forces acting on the box include the monkey's pull, the weight of the box pulling downward, the normal force pushing upward, and friction pulling horizontally against the motion. These forces are illustrated in the FBD below.

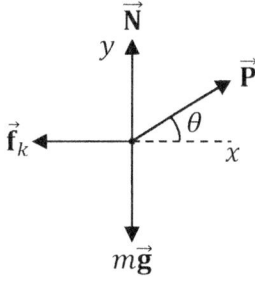

Since the velocity is constant, $a_x = 0$ and $a_y = 0$:

$$\sum_{i=1}^{4} F_{ix} = ma_x \quad , \quad \sum_{i=1}^{4} F_{iy} = ma_y$$
$$P\cos\theta - f_k = 0 \quad , \quad P\sin\theta + N - mg = 0$$
$$N = mg - P\sin\theta$$
$$f_k = \mu_k N = \mu_k mg - \mu_k P\sin\theta$$
$$P\cos\theta - \mu_k mg + \mu_k P\sin\theta = 0$$
$$P = \frac{\mu_k mg}{\cos\theta + \mu_k \sin\theta}$$

Setting $dP/d\theta$ equal to zero, one finds that the minimum effort pull corresponds to $\theta = \tan^{-1}\mu_k$.

**Example.** A monkey exerts a force, $\vec{P}$, at an angle $\theta$ above the horizontal in order to hold a physics textbook against the wall, as illustrated below. Derive an equation for the magnitude of the monkey's push, $P$, needed to support the physics textbook in terms of $m$, $g$, $\mu_s$, and $\theta$.

The forces acting on the physics textbook include the monkey's push, the weight of the physics textbook pulling downward, normal force perpendicular to the wall, and friction, which acts along the wall. These forces are illustrated in the FBD below.

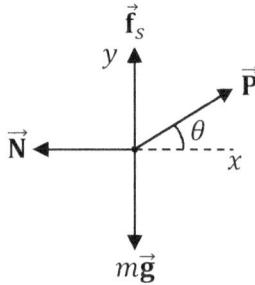

There is actually a range of values for $P$ that will solve the problem, since $f_s \leq \mu_s N$. The minimum value of $P$ needed to support the physics textbook corresponds to an upward friction force equal to $\mu_s N$. The physics textbook is supported if $a_x = 0$ and $a_y = 0$:

$$\sum_{i=1}^{4} F_{ix} = ma_x \quad , \quad \sum_{i=1}^{4} F_{iy} = ma_y$$

$$P\cos\theta - N = 0 \quad , \quad P\sin\theta + f_s - mg = 0$$

$$N = P\cos\theta$$

$$f_s = \mu_s N = \mu_s P \cos\theta$$

$$P\sin\theta + \mu_s P \cos\theta - mg = 0$$

$$P = \frac{mg}{\sin\theta + \mu_s \cos\theta}$$

## 3.6 Hooke's Law

**Springs**: A spring has a natural equilibrium position.[160] If a spring is stretched from equilibrium, it exerts a compressing force, and if it is compressed from equilibrium, it exerts a stretching force, in an effort to restore equilibrium.

**Restoring force**: When a spring is not in its natural equilibrium position, it exerts a restoring force, $\vec{F}_r$, in an effort to restore equilibrium. When the spring is stretched from equilibrium, the restoring force acts to compress the spring, and when the spring is compressed from equilibrium, the restoring force acts to stretch the spring. The direction of the restoring force is toward the natural equilibrium position.

> **Note**. A spring exerts a restoring force, $\vec{F}_r$, when it is not in its natural equilibrium position. Although there is a tension force, $\vec{T}$, in a string, rope, cable, and other similar materials, we call the force associated with a spring a restoring force, $\vec{F}_r$, and not a tension force, $\vec{T}$. The reason is that Hooke's law (described below) generally applies to springs, but not in general to strings, ropes, cables, etc.[161]

**Applied force**: A force that is exerted to cause a spring to stretch or compress from equilibrium is definitely not a restoring force – restoring forces cause a spring to stretch or compress toward equilibrium, not away from it. An applied force is a force that is exerted on a spring by another object or by a person. A restoring force is a force that is exerted by the spring, not on the spring. For example, if a mass is suspended from a vertical spring, its weight is an applied force that causes the spring to stretch from its natural length; in response, the spring exerts a restoring force in an effort to compress back to its natural length. These are not equal and opposite force pairs from Newton's third law;[162] however, these two forces will be equal in the new (i.e. vertical, which is not 'natural') equilibrium position. In fact, the weight of the object is constant, while the restoring force varies with position – this is easy to see, for example, if the object oscillates vertically about the equilibrium position.

---

[160] The equilibrium position actually varies, depending upon the circumstances. For example, the length of a spring that is in equilibrium while suspended vertically is a little longer than its length when it is in equilibrium and suspended horizontally (all else being equal). Similarly, if a mass is suspended from a vertical spring, its new equilibrium position is stretched compared to its previous equilibrium position. The 'natural' equilibrium position corresponds to a spring lying horizontally without any forces acting on it that have horizontal components.

[161] Assume that strings, ropes, cables, etc. do not stretch unless specifically noted otherwise. If a string is elastic, like a rubber band, its stretching could obey a force law similar to Hooke's law. A spring definitely may stretch or compress; assume all springs to obey Hooke's law unless a clear exception is made.

[162] The object's weight is the force that the planet exerts on it; its equal and opposite force is the force the object exerts on the planet. The spring exerts a restoring force on the object, and the object exerts an equal and opposite force on the spring; but the force that the object exerts on the spring does not equal the object's weight, in general – that's only true when the object is at the new (vertical) equilibrium position. If the object oscillates about this new equilibrium position, the restoring force will constantly be changing, whereas the object's weight will be constant (well, it may change slightly as gravity varies with altitude – negligibly over a small change in altitude). Note also that in the problems, we generally apply Newton's second law to objects connected to a spring, but not to the spring itself. In this example, the object's weight and the restoring force would be relevant.

**Hooke's law**: A system obeys Hooke's law if the restoring force, $\vec{F}_r$, is linearly proportional to the negative of the displacement from equilibrium, $\Delta\vec{r}$:[163]

$$\vec{F}_r = -k\Delta\vec{r} = -(\vec{r} - \vec{r}_e) \quad \text{(Hooke's law)}$$

The proportionality constant, $k$, is called the spring constant in the case of a spring. Hooke's law applies to a variety of other systems that exhibit oscillatory behavior besides springs, such as the motion of molecules in many chemical systems.

The significance of the minus sign is that the spring (or other system) is exerting the force in an effort to restore equilibrium. The negative sign means that the restoring force will act to stretch the spring if the spring is compressed from equilibrium, and will act to compress the spring if it is stretched from equilibrium.

It is convenient to setup a coordinate system such that the $x$-axis extends along the direction that the spring would stretch, such that the restoring force has only an $x$-component:

$$F_{rx} = -k\Delta x = -k(x - x_e) \quad \text{(Hooke's law in 1D)}$$

Furthermore, if the origin is placed at the equilibrium position, then $x_e$ will be zero and $F_{rx} = -kx$.

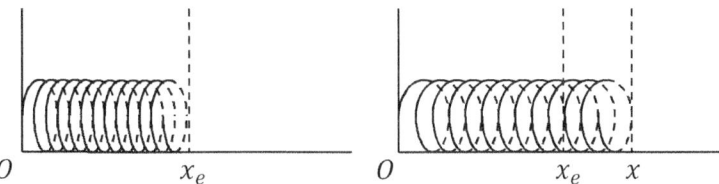

> **Hint.** Draw the actual direction of the restoring force, $\vec{F}_r$, in the FBD and realize that the magnitude of the restoring force is $k|x - x_e|$ (with a good choice of coordinates) when applying Newton's second law. Be careful not to include one sign (plus or minus) associated with the direction of the restoring force in the FBD in addition to the minus sign from the formula, which equates to using the minus sign of Hooke's law twice – a common mistake.

> **Hint.** Whenever possible, setup your coordinate system with the origin at equilibrium and with the $x$-axis along the direction that the spring would stretch. This will make $x_e$ equal to zero.

**Spring constant**: The spring constant, $k$, is the proportionality constant in Hooke's law. Conceptually, it provides a measure of how stiff the spring is – a spring that requires more applied force to stretch it (some specified amount) has a higher spring constant. The SI units of the spring constant are N/m, which equates to $kg/s^2$.

---

[163] The subscript ($e$) designates the equilibrium position, which is not, in general, the initial position. In fact, if the spring is oscillating about equilibrium, the equilibrium position is probably not the position from which the oscillatory motion was initiated (though it could be, as long as it were given an initial velocity from equilibrium).

**Equilibrium positions**: The spring has a natural length in its natural equilibrium position, when it lies horizontally and there are no external forces acting on it that have horizontal components. If a constant force is applied to the spring, the spring will have a new equilibrium position. For example, if the spring is suspended vertically, its new equilibrium position will correspond to a length that is a little longer than the spring's natural length. In either case, if the spring is displaced from equilibrium, it will oscillate about its equilibrium position.

When applying Hooke's law, it is convenient to work with $F_{rx} = -k(x - x_e)$, where $x_e$ corresponds to the current equilibrium position, and not necessarily the natural equilibrium position. Furthermore, it is convenient to place the origin at the current equilibrium position, such that $F_{rx} = -kx$. Regardless of where you put the origin, the restoring force will be the same – since the restoring force only depends upon the displacement from equilibrium.

It is completely unnecessary – and often inconvenient – to work with the natural equilibrium position when the current equilibrium position is not at the natural position. If you work with $F_{rx} = -k(x - x_e)$, where $x_e$ corresponds to the natural equilibrium position, you will still get the same value for the restoring force as you would if $x_e$ were to correspond to the current equilibrium position. Both $x$ and $x_e$ are defined differently in the two cases, such that the difference, $x - x_e$, is the same in the two scenarios.

---

**Hint**. Don't worry about the natural equilibrium position; just concern yourself with the current equilibrium position. (The exception is if the problem specifically gives you information pertaining to the natural equilibrium position, or asks questions specifically about the natural equilibrium position.) The math is generally simplest if you choose the origin to be at the current equilibrium position and setup your coordinate system so that $x$ measures how much the spring is stretched from the current equilibrium position (of course, $x$ will be negative if the spring is instead compressed).

---

**Conceptual Example**. One end of a horizontal spring is fixed to a vertical wall, while a mass is connected to its free end. The system is displaced from equilibrium and released from rest, and the mass oscillates back and forth. Where, precisely, are the magnitude of the acceleration and velocity greatest?

The forces acting on the mass include its weight, normal force, and the restoring force, as illustrated in the FBD below.

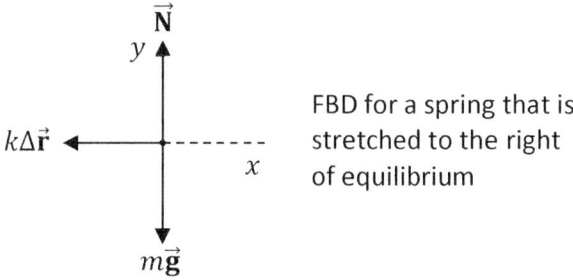

FBD for a spring that is stretched to the right of equilibrium

The weight and normal force cancel out, so the net external force equals the restoring force. Thus, the restoring force is proportional to the acceleration, according to Newton's second law. The restoring force is proportional to the negative of the displacement from equilibrium, according to Hooke's law.

Like the restoring force, the acceleration is proportional to the negative of the displacement from equilibrium. This shows that the magnitude of the mass's acceleration is greatest when the mass is furthest from equilibrium.

The speed (the magnitude of the velocity) is zero at the turning points. Therefore, the speed is not greatest where the magnitude of the acceleration is greatest. Rather, the speed is greatest where the acceleration is zero – i.e. at equilibrium. The mass gains speed all of the way to equilibrium, then loses speed until reaching the turning point, continuing this sequence as it oscillates back and forth.

**Example.** One end of a horizontal 4.0 N/cm spring is fixed to a vertical wall, while a 2.0-kg box of banana-shaped chocolates is connected to its free end. The coefficient of friction between the box of bananas and the ground is 0.25. What is the acceleration of the box of banana-shaped chocolates when the spring is stretched 5.0 cm from, and returning to, equilibrium?

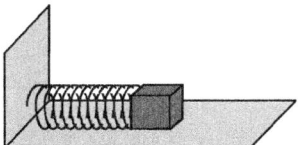

Weight, $m\vec{g}$, pulls downward, while normal force, $\vec{N}$, pushes upward. Since the spring is stretched from equilibrium, it exerts a restoring force, $k\Delta\vec{r}$, to the left (toward equilibrium) in the diagram above. Since the box of banana-shaped chocolates is sliding to the left in the diagram above, the friction force, $\vec{f}$, acts to the right. The FBD for this scenario is shown below, with a coordinate system setup with the origin at equilibrium and the $x$-axis extending to the right.

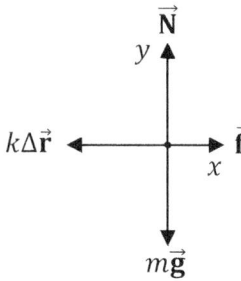

Summing the components of the forces acting on the banana-shaped chocolates,

$$\sum_{i=1}^{4} F_{ix} = ma_x \quad , \quad \sum_{i=1}^{4} F_{iy} = ma_y$$
$$f - kx = ma_x \quad , \quad N - mg = 0$$
$$a_x = \mu g - \frac{kx}{m} = (0.25)(9.81) - \frac{(400)(0.05)}{2.0} = -7.5 \text{ m/s}^2$$

In this problem, the minus sign represents that the box is gaining the negative $x$-direction.

**Note**. Neglect the mass of the spring itself (meaning that we will assume that the mass of the spring is very small compared to the suspended mass) unless stated otherwise.

**Note**. Make sure that your units are consistent. If, for example, you express $x - x_e$ in centimeters, while using the value of $g$ in m/s$^2$, you have a units mismatch. It's a good habit to convert all units to meters, kilograms, and seconds.

**Example**. A 60 N/cm spring has a length of 8.0 cm when lying on the (horizontal) floor. The spring is then suspended vertically from the ceiling, with a 4.0 kg load hanging from its free end. How long will the spring be when it is in the new equilibrium position?

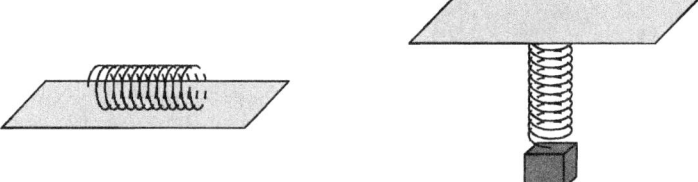

This problem involves two different equilibrium positions. Let us place the origin at equilibrium when the spring is horizontal, with the $x$-axis corresponding to the direction that the spring would stretch; and affix this coordinate system to the spring so that in the vertical position, the position $x$ of the new equilibrium position will equal the distance that the spring has stretched to reach the new equilibrium position.

When the spring is vertical, the weight of the load, $m\vec{g}$, pulls downward. When the spring is in the new equilibrium position, there will be a restoring force, $k\Delta\vec{r}$, pulling the load upward, trying to return the spring to its natural equilibrium position.

The acceleration of the load is zero in the new equilibrium position:

$$\sum_{i=1}^{2} F_{ix} = ma_x$$
$$kx - mg = 0$$

$$x = \frac{mg}{k} = \frac{(4.0)(9.81)}{6000} = 0.65 \text{ cm}$$

Thus, the length of the spring is $8.0 + 0.65 = 8.7$ cm in the new equilibrium position.

**Note**. Observe that 1 N/cm = 100 N/m; since the cm is in the denominator, the 0.01 conversion factor in the denominator equates to a factor of 100 in the numerator.

**Note**. If the same problem involves two different equilibrium positions for the same spring, there will be a nonzero restoring force in one of the two equilibrium positions. For problems that involve only a single equilibrium position, the restoring force will be zero at equilibrium (even if it's not the natural equilibrium position) for a wise choice of coordinates.

**Two bodies interacting via Hooke's law**: Up until now, we have considered problems where one end of the spring has been fixed. Suppose now that a horizontal spring is connected to two boxes with masses $m_L$ and $m_R$, where friction between the boxes and horizontal surface is negligible.

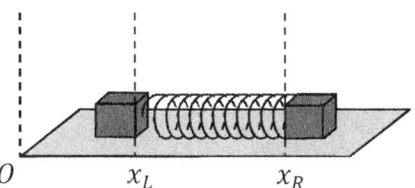

Let us setup our coordinate system with the $x$-axis oriented along the axis of the spring, as illustrated above, where the coordinates $x_L$ and $x_R$ represent the positions of the two ends of the spring relative to the origin. The actual length of the spring will then equal $x_R - x_L$, which may be less or greater than the natural length of the spring, which we denote by $L$.

The spring exerts restoring forces $\pm k \Delta \vec{r}$ on the two boxes, which are equal in magnitude, but opposite in direction, according to Newton's third law. The FBD's for the two boxes also include weights and normal forces, as illustrated below.[164]

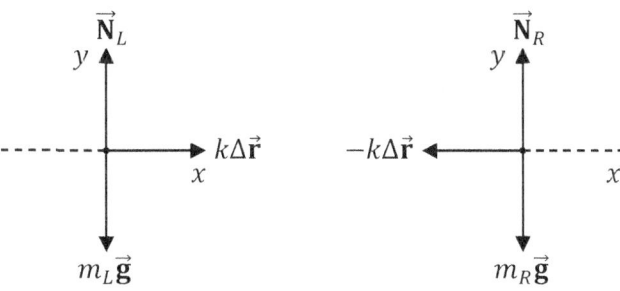

---
[164] We don't draw a FBD for the spring, or apply Newton's second law to the spring itself, when the mass of the spring is neglected compared to the masses of the boxes – an assumption we make unless stated otherwise.

Notice that the amount that the spring is stretched (or compressed, if this quantity is negative) is given by the difference between the actual length of the spring, $x_R - x_L$, and its natural length, $L$:

$$\|\Delta \vec{r}\| = \Delta x = x_R - x_L - L \quad \text{(two-body Hooke's law in 1D)}$$

Applying Newton's second law to the two boxes,

$$\sum_{i=1}^{3} F_{L,ix} = m_L a_{Lx} \quad , \quad \sum_{i=1}^{3} F_{L,iy} = m_L a_{Ly} \quad , \quad \sum_{i=1}^{3} F_{R,ix} = m_R a_{Rx} \quad , \quad \sum_{i=1}^{3} F_{R,iy} = m_R a_{Ry}$$

$$k(x_R - x_L - L) = m_L a_{Lx} \quad , \quad N_L - m_L g = 0 \quad , \quad -k(x_R - x_L - L) = m_R a_{Rx} \quad , \quad N_R - m_R g = 0$$

The ratio of the accelerations of the two boxes is inversely proportional to the ratio of their masses, and is negative as they accelerate in opposite directions:[165]

$$\frac{a_L}{a_R} = -\frac{m_R}{m_L} \quad \text{(two-body Hooke's law in 1D)}$$

With a little algebra, we can combine the two $x$-component equations together in a useful way:

$$m_R k(x_R - x_L - L) = m_R m_L a_{Lx} \quad , \quad -m_L k(x_R - x_L - L) = m_L m_R a_{Rx}$$
$$-(m_R + m_L)k(x_R - x_L - L) = m_R m_L (a_{Rx} - a_{Lx}) \quad \text{(two-body Hooke's law in 1D)}$$
$$-kx = \frac{m_R m_L}{m_R + m_L}(a_{Rx} - a_{Lx}) = \mu a$$

where we have defined $x \equiv \Delta x$ to simplify the notation and $\mu$ is the standard notation for the reduced[166] mass (not to be confused with the coefficient of friction) of the two-body system,

$$\mu = \frac{m_R m_L}{m_R + m_L} \quad \text{or} \quad \frac{1}{\mu} = \frac{1}{m_R} + \frac{1}{m_L} \quad \text{(two-body Hooke's law in 1D)}$$

Since $x = x_R - x_L - L$ and because the natural length of the spring, $L$, is a constant, it follows that

$$a = \frac{d^2 x}{dt^2} = \frac{d^2}{dt^2}(x_R - x_L - L) = \frac{d^2 x_R}{dt^2} - \frac{d^2 x_L}{dt^2} = a_{Rx} - a_{Lx} \quad \text{(two-body Hooke's law in 1D)}$$

which was used to derive the equation $-kx = \mu a$.

---

[165] The two boxes may accelerate towards one another, as suggested in the FBD's, or away from one another. If $x_R - x_L - L > 0$, they accelerate towards each other, if $x_R - x_L - L < 0$, they accelerate away from each other, and if $x_R - x_L - L = 0$, their accelerations are instantaneously zero. Note that these inequalities do not say which direction they boxes are instantaneously moving, but only which way that they are instantaneously accelerating.

[166] It is called the 'reduced' mass because $\mu$ is smaller than both $m_R$ and $m_L$. You should be able to convince yourself that if $x$ and $y$ are both positive, the quantity $u$ defined as $u = x + y$ is greater than both $x$ and $y$, and the quantity $v$ defined as $\frac{1}{v} = \frac{1}{x} + \frac{1}{y}$ is smaller than both $x$ and $y$.

The important result, $-kx = \mu a$, shows that the two-body Hooke's law problem can be treated as a single-body Hooke's law problem with one end of the spring fixed, provided that you use the reduced mass, $\mu$, as the mass of the system in Newton's second law.

> **Hint**. If you see a two-body Hooke's law problem, you should solve the problem just like any other Hooke's law problem, except for using $\mu$ for the mass when you apply Newton's second law.

**Simple harmonic motion**: If the net external force acting on a system has the form of Hooke's law (including the minus sign and direct proportionality with the displacement from equilibrium),[167] then the system will oscillate with simple harmonic motion if displaced from equilibrium.[168] Consider, for example, a horizontal spring with one end fixed to a vertical wall and a box connected to its free end, where there is no friction between the box and the ground. If the box is displaced from equilibrium, it will oscillate back and forth with simple harmonic motion (illustrated below).

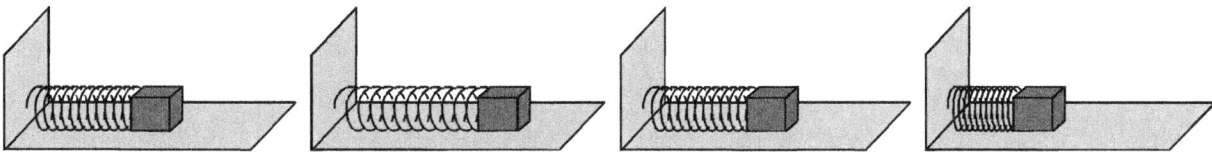

Let us consider the motion of the box for one complete cycle. For definiteness, let us assume that the box is displaced to the right of the equilibrium position and released from rest. We will setup our coordinate system with the origin at equilibrium and the $x$-axis extending to the right. The initial position of the box is then given by $x_m$, which will be the maximum displacement of the box from the equilibrium position.

- Initially, the position of the box, $x$, is a maximum, $x(0) = x_m$, the velocity is zero, $v_x(0) = 0$, and the acceleration is most negative, $a_x(0) = -kx_m/m \equiv -a_m$, using Newton's second law.
- As the box gains speed toward equilibrium, the acceleration becomes less negative.
- When the box reaches equilibrium, $x_e = 0$, the velocity is its most negative, $v_{xe} = -v_m$, and the acceleration is zero, $a_{xe} = 0$.
- The box does not stop at equilibrium because it has inertia and its speed is maximum at equilibrium. After passing equilibrium, the acceleration becomes positive – corresponding to losing speed in the negative direction.
- The acceleration is most positive, $+a_m$, when it reaches $-x_m$, where the speed is again zero.
- From this turning point, the box heads in the positive $x$-direction, gains speed, and its acceleration becomes less negative as the box approaches equilibrium from $-x$.
- At equilibrium, $x_e = 0$, this time the velocity is its most positive, $v_{xe} = +v_m$, and the acceleration is again zero, $a_{xe} = 0$.
- The box loses speed returning to its initial position, $x_m$, and initial acceleration, $-a_m$, temporarily having zero speed at the turning point. The cycle then repeats itself.

---

[167] This doesn't mean that the restoring force has to be the only force acting on the system. Rather, other forces may act on the system, so long as they cancel out in the sums of the components of the forces.
[168] This is the case for reasonable displacements – i.e. if the spring is not stretched beyond the elastic limit.

> **Note**. Simple harmonic motion is non-uniform acceleration.[169] It would be a major conceptual mistake to apply the equations of 1DUA to a spring (or any other system that obeys Hooke's law).

**Applications of Hooke's law**: Simple harmonic motion applies to any system that obeys a force law that has the form of Hooke's law; and, conversely, any system that exhibits simple harmonic motion (not any type of oscillatory behavior, but specifically simple harmonic motion) can be described by Hooke's law. Hooke's law applies to many systems other than springs, such as oscillations of a diatomic molecule (which is a two-body oscillation).

## 3.7 Mutual Surface Forces

**Mutual force pairs**: According to Newton's third law, all forces come in pairs. That is, forces are mutual: One object can't exert a force on another object without simultaneously receiving a force that is equal in magnitude, but opposite in direction. It's impossible to push something harder than it pushes you back, to pull something harder than it pulls you back, or to hit someone harder than he/she hits you back. For example, if one boxer has his hands tied behind his back while another boxer punches him in the face, his face hits the fist with exactly the same force that the fist hits his face.[170]

> **Conceptual Example**. When a monkey pushes a box of bananas, the box of bananas exerts an equal and opposite force back on the monkey. Since these forces cancel out, how is the monkey able to cause the box of bananas to accelerate?
>   Newton's third law involves two forces that act on two different objects. If you apply Newton's second law to just the box of bananas, the monkey's push is an external force for the box of bananas, which causes the box of bananas to accelerate.
>   If you apply Newton's second law to the system of the monkey and box of bananas combined together, the monkey's push becomes an internal force and cancels out. The system accelerates because friction between the monkey and ground is external to the system, and so there is a net external force acting on the system.

---

[169] You can see this as follows: The restoring force varies with position according to Hooke's law, and the sum of the external forces equals the mass of the object connected to the spring times its acceleration, which means that the acceleration will vary with position. Thus, the acceleration is not uniform (i.e. constant). You can also see this conceptually since the object connected to a spring (or any other system that obeys Hooke's law) will reverse direction multiple times. (Note that it would be incorrect to state that acceleration is non-uniform if an object changes direction: If you throw a rock straight upward, its acceleration is uniform, but it comes back down; and if you throw a rock at an angle, it constantly changes direction. It is correct, however, to state that any object that reverses direction two or more times during its motion definitely has non-uniform acceleration.)

[170] Notice that Newton's third law states only that the forces are equal, but not whether the fist or face gets hurt more. Pain is more complicated than force. Newton's second law combined with Newton's third law allows you to compare the accelerations. But if you want to compare pain, you need look at physiology, not just physics.

If instead you apply Newton's second law to the system of the monkey plus the box of bananas plus the earth, then both the monkey's push and the friction force cancel out. This sum tells you that the center of mass of the earth, monkey, and box of bananas does not accelerate (relative to the center of the earth – not relative to the sun, of course). As the monkey and box of bananas accelerate one direction relative to the earth, they impart an angular acceleration to the earth, having a miniscule effect on the earth's rotational speed. The earth has an enormous mass (and hence inertia) compared to the monkey and box of bananas that its angular acceleration is nearly zero.

**Mutual normal forces**: When two objects are in contact with one another, they exert equal and opposite normal forces on one another. If you apply Newton's second law to one object individually, only one normal force (per surface that makes contact with another object) will appear in its FBD and sum of the forces.

**One object in contact with two surfaces**: It's important to realize that if an object is in contact with two surfaces, there will be two separate normal forces acting on it. When one object has another object lying above it, the downward normal force is often overlooked when applying Newton's second law, but is just as important as the upward normal force it receives from the object below it.

**Note**. If you see an object in contact with two surfaces, remember to include two normal forces when you draw and label its FBD.

**Conceptual Example**. Would you feel less, more, or the same amount of normal force while lying down in bed if you are under the covers compared to being coverless? (Assume the bed to be stationary.)

When you lie down on the bed coverless, the bed exerts an upward normal force, $\vec{N}_b$, which balances your weight, $m\vec{g}$. Three forces act on you when you are in the covers: Weight, $m\vec{g}$, pulling downward, an upward normal force from the bed, $\vec{N}_b$, and a downward normal force from the covers, $\vec{N}_c$. The magnitude of the net normal force must equal your weight, according to Newton's second law: $N_b - N_c = mg$. Therefore, the bed exerts a greater normal force on you when you are under the covers, as $N_b = mg + N_c > mg$. However, the magnitude of the net normal force is the same whether or not you are covered – either way, it equals your weight. The two normal forces, $\vec{N}_b$ and $\vec{N}_c$, partly cancel since they act in opposite directions, yet they do have a squeezing effect, with one pushing upward on you from the bottom with a force greater in magnitude than your weight, and another pushing downward on you from the top. Fortunately, typical bed covers are rather lightweight, but if you slept with a pile of bricks on top of you, it would certainly feel much more uncomfortable than having no bricks; you would feel no consolation that the net normal force is the same in each case.

**Example**. Determine all of the normal forces in the stack of boxes (at rest) illustrated below.

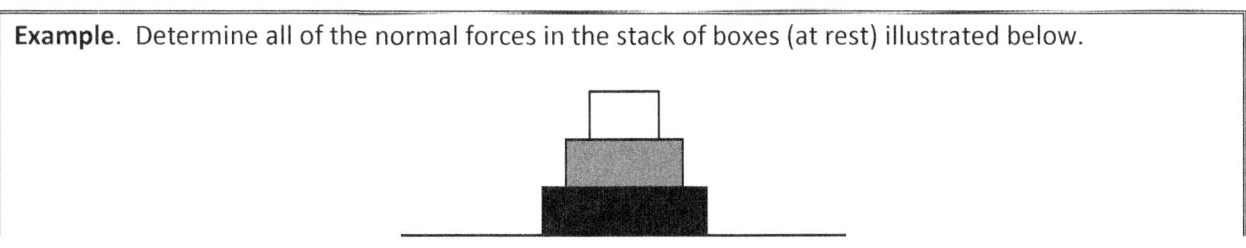

Each box experiences two normal forces and its own weight, except for the top box, which experiences just one normal force. The FBD's for the three boxes are illustrated below.

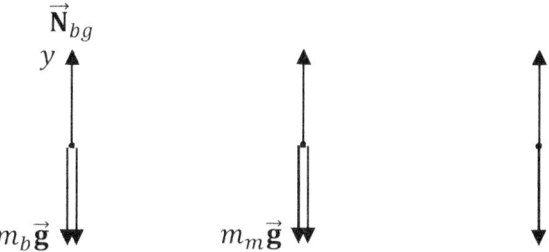

The subscripts $t$, $m$, and $b$ designate the top, middle, and bottom boxes, and the subscript $g$ refers to the ground. The notation $\vec{\mathbf{F}}_{12}$ represents the force exerted on object 1 by object 2, so, for example, $\vec{\mathbf{N}}_{mb}$ is the force exerted on the middle box by the bottom box. From Newton's third law, $N_{mb} = N_{bm}$ and $N_{tm} = N_{mt}$. Applying Newton's second law,

$$\sum_{i=1}^{3} F_{b,iy} = m_b a_y \quad , \quad \sum_{i=1}^{3} F_{m,iy} = m_m a_y \quad , \quad \sum_{i=1}^{2} F_{t,iy} = m_t a_y$$

$$N_{bg} - m_b g - N_{bm} = 0 \quad , \quad N_{mb} - m_m g - N_{mt} = 0 \quad , \quad N_{tm} - m_t g = 0$$

$$N_{tm} = N_{mt} = m_t g$$

$$N_{mb} = N_{bm} = (m_m + m_t)g$$

$$N_{bg} = (m_b + m_m + m_t)g$$

**Mutual friction forces**: When two objects slide against one another – or attempt to slide against one another – they exert equal and opposite friction forces on one another. Of course, an exception is made in problems where the surfaces are declared to be (approximately) frictionless.

**Common Mistake.** When two objects are in contact with one another, when drawing FBD's for both objects it's important to realize that there will be equal and opposite normal forces and equal and opposite friction forces acting on each object. It's a common mistake for students to leave one or more of these forces out of the FBD's.

**Conceptual Example.** A box of bananas lies on the horizontal bed in the back of a flatbed truck, as illustrated below. The truck is gaining speed in the forward direction. Draw separate FBD's for both the box of bananas and the truck.

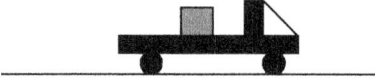

The forces acting on the box of bananas include a downward gravitational pull, $m_b \vec{g}$, an upward normal force, $\vec{N}_{bt}$, exerted by the truck, and a force of friction, $\vec{f}_{bt}$, directed along the motion of the truck.[171] The forces acting on the truck include its weight, $m_t \vec{g}$, an upward normal force from the ground, $\vec{N}_{tg}$, a downward normal force, $\vec{N}_{tb}$, from the box of bananas, a force of friction, $\vec{f}_{tb}$, exerted by the box of bananas, a drive force, $\vec{F}_D$, and other resistive forces, $\vec{F}_R$.[172]

These forces are illustrated below. Observe the mutual normal and friction forces between the truck and box of bananas, corresponding to Newton's third law.

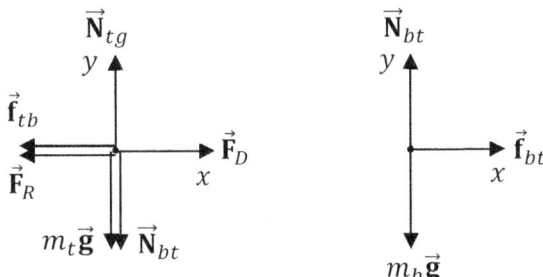

**Example.** In the figure below, the large gray box is being pushed horizontally to the right by an external force, $\vec{P}$. If the magnitude of $\vec{P}$ is large enough, the small black box can actually be supported against gravity through friction between the gray and black boxes. Neglecting friction between the gray box and the horizontal — but not, of course, between the gray and black boxes — derive an equation for the minimum value of $P$ required to prevent the small black box from falling in terms of the masses of the two boxes and the coefficient of friction.

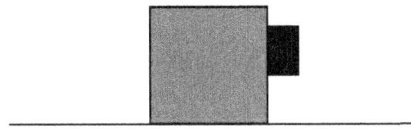

---

[171] It's a mistake to think that the box of bananas is moving to the right, and therefore friction acts to the left: Relative to the truck, the box of bananas is either stationary or sliding to the left (in the diagram, right is the forward direction and left is backward). The box of bananas has inertia – friction is what makes the box of bananas gain speed. If the force of friction isn't great enough to give the box of bananas as much acceleration as the truck, then the box of bananas will slide backwards relative to the truck.

[172] The drive force is ultimately the result of the truck's engine burning fuel, and is exerted by the ground on the tires. The friction force between the ground and tires creates a torque, which results in angular acceleration of the tires; this is how the truck gains speed. The rotation of the tires has rotational inertia, in addition to the inertia of the truck. In Chapter 7 we will see that we need to sum both the forces and the torques separately to account for the effects of the rotational inertia of the tires. For now, we are just exploring mutual friction and normal forces. The resistive force includes all other (i.e. besides friction between the box of bananas and the bed, which has been accounted for separately) frictional effects, such as air resistance. Notice that in the case of rolling, friction from the ground actually provides a forward torque, and so friction between the ground and tires does not have the same resistive effect that it has on sliding. We will explore this more fully in Chapter 7.

There are two pairs of internal forces exerted between the gray and black boxes – horizontal normal forces, $\vec{\mathbf{N}}_{LR}$ and $\vec{\mathbf{N}}_{RL}$, and vertical friction forces, $\vec{\mathbf{f}}_{LR}$ and $\vec{\mathbf{f}}_{RL}$, where the subscripts $L$ and $R$ designate the left and right boxes. The weights of the two boxes pull downward, normal force from the ground pushes the gray box upward, and the external push, $\vec{\mathbf{P}}$, acts horizontally to the right on the left box.[173] These forces are illustrated in the FBD's below.

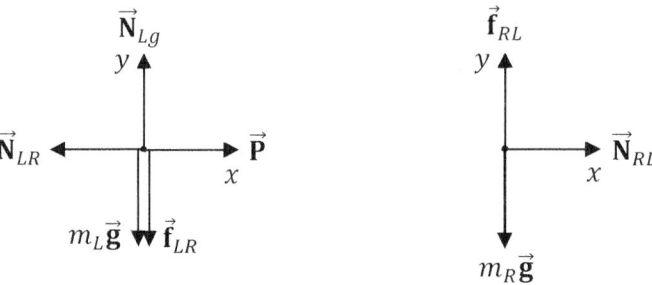

Applying Newton's second law to each box individually,

$$\sum_{i=1}^{5} F_{L,ix} = m_L a_x \quad , \quad \sum_{i=1}^{5} F_{L,iy} = m_L a_y \quad , \quad \sum_{i=1}^{3} F_{R,ix} = m_R a_x \quad , \quad \sum_{i=1}^{3} F_{R,iy} = m_R a_y$$
$$P - N_{LR} = m_L a \quad , \quad N_{Lg} - m_L g - f_{LR} = 0 \quad , \quad N_{RL} = m_R a \quad , \quad f_{RL} - m_R g = 0$$

From Newton's third law, we know that $N_{LR} = N_{RL}$ and $f_{LR} = f_{RL}$.[174] Expressing $P$ in terms of $\mu$, $m_L$, $m_R$, and $g$ is a matter of combining equations together with substitutions, including $f_{LR} = \mu N_{LR}$:

$$f_{LR} = m_R g = \mu N_{LR}$$
$$N_{LR} = \frac{m_R g}{\mu}$$
$$a = \frac{N_{LR}}{m_R} = \frac{g}{\mu}$$
$$P = N_{LR} + m_L a$$
$$P = \frac{m_R g}{\mu} + \frac{m_L g}{\mu}$$
$$P = \frac{m_R + m_L}{\mu} g$$

Conceptually, it should make sense that a smaller coefficient of friction requires a greater push, $\vec{\mathbf{P}}$.

---

[173] The external push, $\vec{\mathbf{P}}$, acts directly on the gray box, and so is only included in the FBD and sum of the x-components for the gray box. Its effect on the black box is indirectly communicated through the horizontal normal forces.

[174] Notice that $\vec{\mathbf{N}}_{LR} = -\vec{\mathbf{N}}_{RL}$, but $N_{LR} = N_{RL}$. Magnitudes of vectors can't be negative, but vectors and components can; signs are directional.

## 3.8 Air Resistance

**Fluid**: The term fluid is used to describe both liquids and gases.[175]

**Fluid resistance**: An object experiences a resistive force when it travels through a fluid.

**Air drag**: When the fluid is the air, the force is called the force of air resistance or air drag.

**Buoyancy**: An object submerged in a fluid also experiences an upward buoyant force that depends on the object's density (as well as what fraction of the object is submerged in the fluid) according to Archimedes' principle: Archimedes found that the magnitude of the buoyant force equals the magnitude of the weight of the displaced fluid.[176]

The buoyant force results from the fact that the pressure of a fluid varies with depth according to the formula $P = \rho_f g h$, where $\rho_f$ is the density (mass per unit volume) of the fluid and $h$ is the standard symbol for depth (not height!). Pressure is greatest at the bottom of the fluid, where depth is greatest (i.e. where $h$ is greatest, but where height is zero). Thus, if an object is submerged in a fluid, greater pressure is exerted on its bottom than on its top, and as a result of this difference in pressure there is a net upward force, which we refer to as buoyancy.

Buoyancy has a negligible effect on the acceleration of an object traveling through a fluid as long as the object is significantly denser than the fluid. If the object is not very dense, however, the effect of buoyancy can be significant. As an extreme example, consider a balloon filled with helium that rises upward due to the buoyant force (i.e. helium is less dense than air and so rises upward due to its buoyancy).

**Apparent weight in a fluid**: Objects appear to weigh less when submerged in a fluid due to the upward buoyant force. Since the buoyant force is a generally constant for an object traveling through a fluid, its effect can be accounted for by using the object's apparent weight instead of its actual weight and otherwise ignoring buoyancy.[177]

---

**Note**. In Volume 1 of this textbook, we will neglect the buoyant force unless stated otherwise.

---

**Force of fluid resistance**: When an object travels through a fluid, the force of fluid resistance, $\vec{F}_f$, (aka retarding force) depends upon a number of factors, including:

---

[175] If you can remember that fluids 'flow,' this may help you recall the meaning of the word 'fluid.'
[176] It's the weight of the amount of fluid that is displaced by the object's submersion, and not, in general, the weight of the object (however, the object's weight equals the weight of the displaced fluid if the object is floating due to its buoyancy, like a boat – but in that case, the object is only partly submerged). If you place an object in a fluid, such as a beaker of water, the fluid level rises by an amount equal to the volume of the object. If you weigh this displaced fluid, it will equal $\rho_f V g$, where $\rho_f$ is the density of the fluid.
[177] Observe that helium – and any other gas that is less dense than air (it would be a conceptual mistake to say 'lighter' than air) – has a negative apparent weight when submerged in air.

- the object's velocity. The faster the object is moving, the greater the magnitude of the force of fluid resistance; and the force of fluid resistance acts to oppose the object's velocity.
- the object's cross-sectional area – i.e. how large the object looks when viewed from the direction in which it is heading. For example, a basketball has a considerably larger cross-sectional area than a golf ball, and a sheet of paper falling vertically has a much larger cross-sectional area when it is oriented horizontally rather than vertically.
- how aerodynamic the object's shape is. For example, if an umbrella is falling vertically, the force of fluid resistance will be much greater if it is oriented upright rather than up-side-down: When it falls upright, its concave surface traps the air molecules, but when it falls up-side-down, its convex surface pushes air to the side, out of its way. Aerodynamic design is an important element to consider when making fuel-efficient cars, especially since a single car will be used for hundreds of thousands of miles and there are millions of cars – getting a few more miles per gallon really adds up.
- the density of the fluid. A more dense fluid generally provides greater resistance to motion.
- other factors to a much lesser extent, in general – the predominant factors are generally the object's velocity and shape.

The factors that affect the force of fluid resistance are generally constant factors except for the object's velocity. While the force of fluid resistance can, in general, have a complicated dependence upon the object's shape, fluid density, velocity, and other factors, in can often be reasonably approximated in the following form:

$$\vec{F}_f = -bv^c \hat{v} = -bv^{c-1}\vec{v}$$

where $b$ is a positive constant[178] that depends upon the constant factors of fluid resistance, like shape, $c$ is a constant exponent, and $\hat{v}$ is a unit vector pointing in the direction of the object's velocity. Note that the object's velocity can be expressed in terms of its speed and direction as $\vec{v} = v\hat{v}$. Also notice that the dimensions of the coefficient $b$ depend upon the power $c$.

**Force of air resistance**: When the fluid is air, we will use a subscript $a$ to designate the force of air resistance (aka wind resistance or air drag), $\vec{F}_a$. Two common situations arise in the case of air resistance: $c = 1$ (linear drag) and $c = 2$ (quadratic drag).

**Linear drag**: The case $c = 1$ corresponds to linear drag (aka Stokes' law of resistance or viscous resistance), and provides a good approximation for rather small objects with low speeds. In the case of linear drag, $\vec{F}_a = -bv\hat{v} = -b\vec{v}$.

**Quadratic drag**: The case $c = 2$ corresponds to quadratic drag (aka Newton's law of resistance), and provides a good approximation for larger objects with higher speeds (but less than the speed of sound, which is about 340 m/s in air at STP). This applies to pitching a baseball, driving a racecar, or launching a rocket, for example. $\vec{F}_a = -bv^2\hat{v} = -bv\vec{v}$.

---

[178] It is tempting to call this constant, $b$, the drag coefficient, but you will soon see that there is another coefficient with this name, which is related to, but not equal to, $b$.

In the case of quadratic drag, the coefficient $b$ is given by the expression,

$$b = \frac{\rho A D}{2}$$

where $\rho$ is the density of the fluid, $A$ is the cross-sectional area of the object, and $D$ is the (dimensionless) drag coefficient, which accounts for the object's shape.

**Resistive automobile forces**: An automobile experiences a force of air resistance as well as other resistive forces, which include road resistance (though friction plays a different role in the case of rolling compared to sliding, which we will investigate more thoroughly in Chapter 7) and internal frictional forces involving the drive shaft and axles, for example. We will use $\vec{\mathbf{F}}_R$ to speak of the complete resistive force in the case of driving, which includes $\vec{\mathbf{F}}_a$.

> **Conceptual Example.** A monkey throws a banana straight upward with an initial speed of 10 m/s, and catches it at the same height on its return down. Accounting for the effects of air resistance, how does the final speed of the banana (just before being caught) compare to its initial speed?
> 
> As a reference, it is useful to first establish what would happen in a perfect vacuum: Neglecting the air, the final speed would be the same as the initial speed. Now adding the effects of air resistance, it should be clear that the maximum height of the banana will be reduced. At the top of the arc, the banana's speed will temporarily be zero.[179] Next, consider the banana's descent from the reduced height: The banana can't reach a final speed of 10 m/s from this reduced height because more height was needed to reach this speed in the case of a perfect vacuum. Therefore, the final speed of the banana is less than 10 m/s.

**Equations of motion with fluid resistance**: The equations for position, velocity, and acceleration as functions of time – called the equations of motion – for an object traveling through a fluid can be derived by applying Newton's second law and solving the resulting differential equation (since acceleration is a derivative of velocity with respect to time and velocity is a derivative of the position vector with respect to time).

> **Note.** The acceleration of an object traveling through a fluid is generally non-uniform, in which case the 1DUA and 2DUA equations do not apply.

**Terminal velocity**: Fluid resistance has the tendency to reduce the velocity of an object until its acceleration is zero. When the acceleration becomes zero, the object is said to travel with terminal velocity, $\vec{\mathbf{v}}_t$ – i.e. with constant speed, called terminal speed, $v_t$, in a straight line. For example, an object falling through the air from a great height gains speed until reaching its terminal speed; and if a parachute is then opened, the object loses speed until reaching a lower terminal speed.

---

[179] The speed is only zero because the banana was thrown straight upward. If instead the banana were thrown upward at an angle, its speed would not be zero at the top of its trajectory – only the vertical component of its velocity would be zero.

3 Newton's Laws of Motion

**Hint**. Draw the actual direction of the force of fluid resistance, $\vec{\mathbf{F}}_f$, in the FBD and realize that the magnitude of the force of fluid resistance is $bv^c$ when applying Newton's second law. Be careful not to include one sign (plus or minus) associated with the direction of the force of fluid resistance in the FBD as well as the minus sign from the formula, $\vec{\mathbf{F}}_f = -bv^c\hat{\mathbf{v}}$ – which is a common mistake. For linear drag ($c = 1$), this force will have components of the form $F_{fy} = \pm bv_y$, where $v_y$ itself may be negative; again, you must choose the overall sign carefully so as to make the math and FBD consistent.

## Problem-Solving Strategy for Motion with Fluid Resistance

0. Apply this strategy to problems where an object travels through a fluid, including problems that state that the effects of air resistance are to be accounted for.
1. Follow the strategy for applying Newton's second law, including drawing the FBD, setting up a coordinate systems, and summing the components of the forces. Include the force of fluid resistance in the FBD and sums, which is directed opposite to the velocity.
2. Include the equation for fluid resistance when summing the components of the forces.
3. If you are just looking for terminal speed, you can obtain an equation for it by setting the acceleration equal to zero. If you need the equations of motion, go onto Steps 4 and 5.
4. Note that Newton's second law is a second-order differential equation for position as a function of time (since acceleration is the second derivative of position with respect to time). Separate variables and integrate to obtain the equations of motion.[180] In physics, it is generally more convenient to perform definite integrals with the constants built into the math, as the limits have a natural physical interpretation (such as initial time, initial position, or initial component of velocity).
5. Count your knowns and unknowns. You generally need $N$ equations in $N$ unknowns.

**Example**. A monkey releases a banana-shaped cereal flake from rest from the top of a skyscraper. Assuming that the force of air resistance corresponds approximately to linear drag for the entire descent, derive the equations of motion for the banana-shaped cereal flake.

Begin by drawing a FBD for the banana-shaped cereal flake.

---

[180] Not every differential equation is separable; only the simpler problems can be solved this way. The problems in this book involving differential equations can be solved by the technique of separation of variables and illustrated in the following example (and, sometimes, using the chain rule, as we will see in later material). If you take a course on differential equations, you will learn a set of techniques for solving a variety of differential equations.

In the previous illustration, we chose upward to correspond to the $+y$-axis. For linear drag, $c = 1$ and $\vec{F}_a = bv\hat{j}$ (since the force of air resistance is upward, it's along $+\hat{j}$, which is opposite to the velocity, which is along $-\hat{j}$). In terms of components, however, $F_{ay} = -bv_y$ because $v_y$ itself is negative – which makes $F_{ay}$ positive. Assuming that the skyscraper is relatively short compared to earth's radius, we will neglect the variation of gravity with altitude. Summing the $y$-components of the forces:

$$\sum_{i=1}^{2} F_{iy} = ma_y$$

$$-mg - bv_y = m\frac{dv_y}{dt}$$

Next, we separate variables by algebraically placing $t$ and $v_y$ on opposite sides of the equation:

$$dt = -m\frac{dv_y}{bv_y + mg}$$

After separating the two variables, we integrate both sides of the equation. When $t$ varies from the initial time, $t_0$, to the final time, $t$, the $y$-component of the velocity varies from $v_{y0} = 0$ to $v_y$:

$$\int_{t=t_0}^{t} dt = -m \int_{v_y=0}^{v_y} \frac{dv_y}{bv_y + mg}$$

$$\Delta t = -\frac{m}{b} \ln\left(\frac{bv_y + mg}{mg}\right)$$

where we used the rule for the difference of logs, $\ln(y) - \ln(x) = \ln(y/x)$. We can solve for $v_y$ by isolating the logarithm and then exponentiating both sides:

$$\frac{bv_y + mg}{mg} = e^{-\frac{b\Delta t}{m}}$$

$$v_y = -\frac{mg}{b}\left(1 - e^{-\frac{b\Delta t}{m}}\right)$$

As expected, the speed, $v = |v_y|$ in 1D, grows more and more slowly, asymptotically approaching a terminal speed, $v_t = \frac{mg}{b}$. A final integration yields position as a function of time:

$$\Delta y = \int_{y=y_0}^{y} dy = \int_{t=t_0}^{t} v_y dt = -\int_{t=t_0}^{t} \frac{mg}{b}\left(1 - e^{-\frac{b\Delta t}{m}}\right) dt = -\frac{mg}{b}\Delta t - \frac{m^2 g}{b^2}\left(1 - e^{-\frac{b\Delta t}{m}}\right)$$

The lower limit makes a contribution to the definite integral on the right-hand side since $e^0 = 1$.

3 Newton's Laws of Motion

**Example**. If all we wanted in the previous example was the terminal speed, $v_t$, we could have found it without any integration. We could have simply set the acceleration equal to zero after applying Newton's second law:

$$-mg - bv_{yt} = ma_y = 0$$
$$v_{yt} = -\frac{mg}{b}$$
$$v_t = |v_{yt}| = \frac{mg}{b}$$

**Note**. One component of the force of fluid resistance can reverse direction in the middle of a problem. For example, if you throw a banana straight upward, the force of air resistance will act downward as the banana rises, and upward as the ball descends. When accounting for the effects of fluid resistance, it is necessary to split the motion into two parts when the direction of the force of fluid resistance reverses direction during a problem. It's a common mistake for students to forget about this in their solutions.

## 3.9 Gravitational Forces

**Newton's law of universal gravitation**: Any two massive objects experience an attractive gravitational force that follows Newton's law of universal gravitation. Two point-masses experience a force of gravitational attraction according to

$$\vec{\mathbf{F}}_{21} = -G\frac{m_1 m_2}{R^2}\hat{\mathbf{R}}_{12} \quad \text{(point-masses)}$$

where the proportionality constant, $G$, is called the gravitational constant, $\hat{\mathbf{R}}_{12}$, is a unit vector directed from object 1 to object 2, $R$ is the separation between the point-masses, and $\vec{\mathbf{F}}_{21}$ is the force exerted on object 2 by object 1. The direction of the gravitational force, $\vec{\mathbf{F}}_{21}$, is $-\hat{\mathbf{R}}_{12}$, meaning that object 1 pulls object 2 toward object 1 – that is, the force is attractive.

The two masses need not be point-masses for the above formula to apply. If they are spherically symmetric objects (like planets or stars), the same formula still applies; in this case, $R$ is the separation between the centers of the two massive objects.[181] If the masses are not spherically symmetric, or not even spherical, then you need to integrate to find the force; this integration technique will be covered in a subsequent volume of this textbook on electricity and magnetism.

Newton's law of universal gravitation is called an inverse-square law because it varies inversely with the square of the separation between the masses. For example, if the masses are placed twice as far apart, the force drops by a factor of four.

---

[181] If you look up the earth-sun distance, for example, the value quoted will actually be the average distance between their centers. Many students who don't realize this incorrectly add the radii of the sun and earth to the earth-sun distance when trying to find the force of attraction between them.

Note that Newton's third law is reflected in Newton's law of universal gravitation:

$$\vec{F}_{12} = -G\frac{m_1 m_2}{R^2}\hat{R}_{21} = G\frac{m_1 m_2}{R^2}\hat{R}_{12} = -\vec{F}_{21} \quad \text{(point-masses)}$$

For example, the moon exerts a force on the earth that is equal in magnitude, but opposite in direction, to the force that the earth exerts on the moon. Although the forces are equal, the force has a greater effect on the moon, so that the moon appears to orbit the earth, because the earth has much more mass (and therefore inertia) than the moon.[182]

**Note.** The symbol $R$ represents the separation between the centers of two masses, and very often does not serve as a radius. For example, if you want to compute the force that the earth exerts on Jupiter at a particular instant, you need to know the instantaneous distance between earth and Jupiter; in this case, $R$ is clearly not a radius.

**Note.** You may need to look up constants to solve problems when applying Newton's law of universal gravitation. For example, you will definitely need to know the value of the gravitational constant, $G$. You may also need to look up the mass of, or average separation between, astronomical bodies such as the sun, planets, or moons. Many instructors will require memorization of $G$, but usually do not require students to memorize the masses or sizes of astronomical bodies.

**Math Note.** When applying Newton's law of universal gravitation, you have to work with exponents. For example, the calculation of the magnitude of the gravitational force that the sun exerts on the earth looks like this:

$$F = G\frac{m_1 m_2}{R^2} = \left(6.67 \times 10^{-11} \, \frac{\text{N m}^2}{\text{kg}^2}\right) \frac{(5.97 \times 10^{24} \text{ kg})(1.99 \times 10^{30} \text{ kg})}{(1.50 \times 10^{11} \text{ m})^2}$$

Therefore, it is necessary to be familiar with the algebra of exponents, which we summarize below:

$$x^a x^b = x^{a+b} \quad , \quad \frac{x^a}{x^b} = x^{a-b} \quad , \quad (x^a)^b = x^{ab} \quad , \quad x^0 = 1$$

$$x^{-a} = \frac{1}{x^a} \quad , \quad x^a x^{-b} = x^{a-b} \quad , \quad \frac{x^a}{x^{-b}} = x^{a+b} \quad , \quad \frac{x^{-a}}{x^{-b}} = x^{-a+b} \quad , \quad \frac{x^{-a}}{x^b} = x^{-a-b}$$

<u>Some Examples</u>

$$(3 \times 10^{24})(8 \times 10^{22}) = 24 \times 10^{46} = 2.4 \times 10^{47}$$

$$(4 \times 10^{11})^2 = 16 \times 10^{22} = 1.6 \times 10^{23}$$

$$\frac{9 \times 10^{18}}{(6 \times 10^6)^2} = \frac{9 \times 10^{18}}{36 \times 10^{12}} = 0.25 \times 10^6 = 2.5 \times 10^5$$

---

[182] In the two-body problem (i.e. just the earth and moon), the earth and moon both orbit the center of mass of the system. However, the center of mass of the earth-moon system lies very close to the earth, so the moon has a significant orbit, whereas the earth's orbit around the center of mass is much, much smaller.

## 3 Newton's Laws of Motion

**Gravitational constant**: The proportionality constant in Newton's law of universal gravitation is a universal constant – i.e. it has the same value anywhere in the universe.[183] In SI units, its value is

$$G = 6.67 \times 10^{-11} \text{ N m}^2/\text{kg}^2$$

You don't need to memorize the units of $G$ (but you may need to memorize its numerical value); rather, if you know Newton's law of universal gravitation, you need only plug in the SI units of the other quantities to be able to solve for the units of $G$ (try it!).

> **Important Distinction.** Physics is often case-sensitive. For example, the lowercase italicized symbol $g$ represents gravitational acceleration,[184] whereas the uppercase italicized $G$ represents the gravitational constant. The two quantities have significantly different conceptual meaning, and do not even share the same SI units (N m$^2$/kg$^2$ does not reduce to m/s$^2$). The gravitational constant, $G$, is the constant of proportionality in Newton's law of universal gravitation, and has the same value throughout the universe; while gravitational acceleration, $g$, is the acceleration that an object in free fall would have at a particular point in space, and varies with location.

**The strength of gravity**: Gravity is actually the weakest of the four fundamental forces of nature.[185] The reason is that it takes an astronomical amount of mass to create a force on the order of Newtons because there is a factor of $6.67 \times 10^{-11}$ in the proportionality constant. If you place two 1-kg masses 1 m apart, the force of attraction will be a mere 0.0000000000667 N! Two 100-kg people standing 1 m apart experience a gravitational force of 0.000000667 N. Two loaded 10,000-kg semi trucks placed 10 m apart experience a gravitational force of 0.000667 N. The gravitational attraction between two terrestrial objects is insignificant compared to resistive forces like friction and air resistance. Thus, you don't notice that ordinary terrestrial objects – like cars, people, and bowling balls – attract one another gravitationally, and don't see them accelerate toward one another due to their gravitational attraction.

At least one of the masses must be 'astronomical' in order to achieve an appreciable gravitational force. So in practice, Newton's law of universal gravitation is useful for computing the force that the sun exerts on the earth or the force that the earth exerts on a monkey, for example. The mass of the earth, $5.97 \times 10^{24}$ kg, for example, easily overcomes the factor of $10^{-11}$ in $G$.

---

[183] In recent years, extrapolation of cosmological data has caused physicists to question the constancy of one or more of the fundamental constants, which include the gravitational constant, the speed of light, and Planck's constant. In this text, we will treat these three fundamental constants as actual constants – i.e. completely independent of position and time. If one of these constants has slowly evolved over billions of years, or varies spatially over the course of millions of light-years, for example, it won't have a significant effect on any of the exercises in this book.

[184] The non-italicized lowercase symbol g represents the unit gram. It's difficult to distinguish between a handwritten $g$ and g, but the distinction between gravitational acceleration and grams should be clear from context. For example, compare an object with mass of 40 g to the equation $ma_y = -mg$.

[185] The four fundamental forces are the gravitational attraction between two masses, the electromagnetic attraction or repulsion between electric charges, the strong nuclear force that binds protons and neutrons to one another in the nucleus, and the weak nuclear force responsible for the radioactive decays of unstable nuclei.

**Conceptual Example.** In principle, we can make the gravitational attraction between two masses greater by making $R$ smaller. In practice, however, it's quite an experimental challenge to separate two very massive objects by a very small distance. Imagine, for example, that we have two fully-loaded semi trucks. Why can't you make $R$ as small as, say, a millimeter?

Of course, you can park two semi trucks very close to one another, such that there is just a tiny distance separating their sides, but the $R$ that you would need to use in Newton's law of universal gravitation is the distance between their centers – not the separation between their nearest sides. If you want to make $R$ one millimeter for two semi trucks, you would have to build two overlapping trucks![186]

**Gravitational field:** Imagine an object of mass $m$ in the gravitational field of a single isolated planet (i.e. let's ignore all of the stars, moons, and other planets in the universe) of mass $m_p$. If the object is in a state of free fall, then by definition its acceleration is gravitational acceleration:

$$\sum_{i=1}^{1} \vec{F}_i = m\vec{a}$$

$$-G\frac{m_p m}{R^2}\hat{R} = m\vec{g} \quad \text{(point-masses)}$$

$$\vec{g} = -G\frac{m_p}{R^2}\hat{R} \quad \text{(single point-mass)}$$

This is the equation for the gravitational field created by a single pointlike mass or spherically symmetric massive object at a distance $R$ from its center of mass.[187]

**Important Distinction.** The distance $R$ in the equation for the gravitational field created by a spherically symmetric object is the distance from its center of mass to the point where you are trying to determine the gravitational field. In general, this $R$ is not the radius of an object; it would only be the radius of a planet if you are trying to find the gravitational field at the planet's surface. If you are trying to calculate gravitational field at the top of a mountain or out in space, $R$ will be greater than the planet's radius.

**Example.** Calculate the value of gravitational acceleration near earth's surface.

The mass and average radius of the earth are $5.97 \times 10^{24}$ kg and $6.37 \times 10^6$ m, respectively. Plugging these into the formula for the gravitational acceleration, assuming a perfect sphere, we find:[188]

$$g_e = G\frac{m_e}{R_e^2} = \left(6.67 \times 10^{-11} \frac{\text{N m}^2}{\text{kg}^2}\right)\frac{(5.97 \times 10^{24} \text{ kg})}{(6.37 \times 10^6 \text{ m})^2} = 9.81 \text{ m/s}^2$$

---

[186] Semi trucks are also neither pointlike at such close proximity, nor are they spherically symmetric.
[187] For a spherically symmetric object, this expression is only valid outside of the object. If you dig a tunnel through the surface of the earth, for example, this expression does not apply. We will learn how to calculate gravitational field inside of a planet in a subsequent volume of this text on electricity and magnetism (on Gauss's law).
[188] The rotation of the earth makes a small contribution to the effective value of $g$, which we are neglecting here.

**Vector fields**: Gravitational field is a vector field. A vector field is a vector whose magnitude and direction depend upon the location of the vector. Gravity is stronger nearer an astronomical body and weaker further away, and points in different directions at different points in space.

---

**Example.** Calculate the value of gravitational acceleration at an altitude equal to earth's average radius.

The quantity $R$ is measured from the center of mass of the earth to the point where we wish to compute the gravitational field. This distance equals the radius of the earth plus the altitude: $R = R_e + h = 2R_e$. We do not need to look up astronomical data to solve this problem, but can actually obtain a number directly from the value of $g$ at earth's surface (which denote by $g_e$):

$$g(R = 2R_e) = G\frac{m_e}{(2R_e)^2} = G\frac{m_e}{4R_e^2} = \frac{g_e}{4} = 2.45 \text{ m/s}^2$$

---

**Example.** What would gravitational acceleration be near the surface of a planet that has 3 times earth's mass and 2 times earth's radius?

Observe that we can solve such comparison problems without looking up astronomical data. Just take a ratio as follows:

$$\frac{g_p}{g_e} = \frac{G\dfrac{m_p}{R_p^2}}{G\dfrac{m_e}{R_e^2}} = \frac{m_p R_e^2}{m_e R_p^2} = (3)\left(\frac{1}{2}\right)^2 = \frac{3}{4}$$

$$g_p = 0.75 g_e = 7.36 \text{ m/s}^2$$

---

**Gravitational mass**: Recall that mass has two different meanings in physics: Mass is a measure of inertia, and mass is also the source of a gravitational field. All objects that have mass create gravitational fields. All objects with mass also experience a gravitational force – called weight – due to the net gravitational field created by all of the other massive objects in the universe.

---

**Important Distinction.** A massive object creates its own gravitational field, but experiences a gravitational force (weight) due to gravitational fields made by other masses. We use the formula $g = Gm_p/R^2$ to find the magnitude of the gravitational field that a spherically symmetric massive object creates, and the formula $W = mg$ to find the force that a mass $m$ experiences in the presence of an external gravitational field. Putting these together, $W = Gm_p m/R^2$ is the mutual force of attraction that one massive object exerts on another: One mass acts as the source – it makes the gravitational field – that the other experiences. Both masses are the source of a gravitational field, and both experience the same gravitational force (except in opposite directions) from one another. It's important to realize that this is a mutual force; a mass does not exert a force on itself.

---

**Test mass**: We can calculate gravitational field at any point in space – there doesn't even need to be a massive object there. Once you know the gravitational field at some given point in space, conceptually the value of gravitational field tells you how much weight a massive object would experience if placed at that point. We call such a hypothetical object a test mass, $m_t$. Then $W = m_t g$.

**Net gravitational field**: If you want to compute the net gravitational field created by multiple objects, use the principle of superposition. This means to find the gravitational field for each source, and add these vector fields using the strategy of vector addition.

## Problem-Solving Strategy for the Superposition of Gravitational Fields

0. If you are given $N$ point-masses (or spherically symmetric masses, where the field point is not within any of the objects), $m_1, m_2, \cdots, m_N$, and their coordinates, $(x_1, y_1, z_1), (x_2, y_2, z_2), \cdots, (x_N, y_N, z_N)$ you can apply this strategy to determine the magnitude and direction of the net gravitational field, $\vec{g}$, at any specified point, $(x, y, z)$, called the field point. This is called superposition (vector addition). The field point is the place where you are computing the gravitational field. The given masses are called sources.
1. First draw a sketch of the gravitational fields, $\vec{g}_1, \vec{g}_2, \cdots, \vec{g}_N$, that the sources make at the field point. Do this by visualizing a test-mass at the field point and drawing an arrow toward each source to represent the attractive gravitational field due to each source individually.
2. Setup a Cartesian coordinate system.
3. Use the distance formula $R_i = \sqrt{(x - x_i)^2 + (y - y_i)^2 + (z - z_i)^2}$ to determine the distance between each source and the field point. If the sources are spheres, measure $R_i$ from their centers.
4. Determine the magnitude of each gravitational field using the formula $g_i = Gm_i/R_i^2$.
5. Use trig to determine the direction of each gravitational field. If the field point and all of the sources lie in the same plane, the direction of each gravitational field can be specified by a single angle, $\theta_i$. If the configuration is instead three-dimensional, you will need to compute two angles, $\theta_i$ and $\varphi_i$ of spherical coordinates, for each vector. Check your sketch to make sure that the angles agree with the directions of the gravitational fields. In two dimensions, this means making sure that each $\theta_i$ agrees with the quadrant in which $\vec{g}_i$ lies. Recall that $\theta_i$ is measured counterclockwise from the $+x$-axis.
6. Now that you have the magnitude, $g_i$, and direction of each vector, apply the strategy for how to add vectors (see Chapter 2) in order to determine the magnitude and direction of the resultant.

**Common Mistake**. In general, the angle $\theta_i$ will not lie in the first quadrant when adding vectors. Use your gravitational field sketch to make sure that each angle, $\theta_i$, corresponds to the proper quadrant.

**Example**. In the diagram below, the planet on the left has a mass of $8.0 \times 10^{24}$ kg and the planet on the right has a mass of $4.0 \times 10^{24}$ kg. The separation between the centers of the planets is $6.0 \times 10^9$ m. Determine the magnitude and direction of the net gravitational field at the point marked $X$.

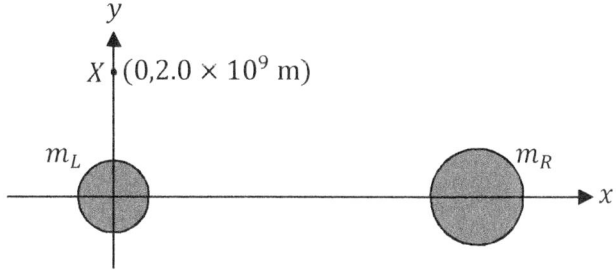

First, sketch the gravitational fields at the field point; each points toward the corresponding source. Setup a coordinate system.

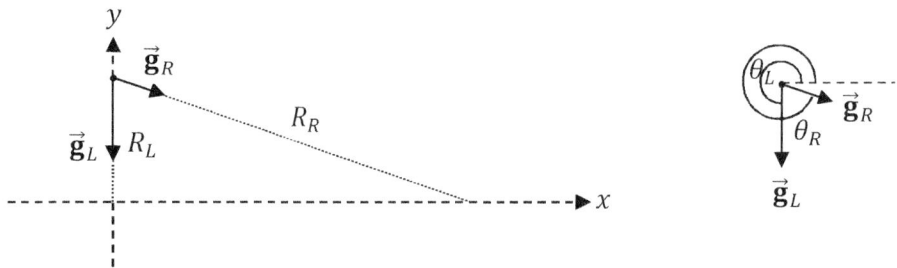

The distance $R_L$ equals $2.0 \times 10^9$ m, and the distance $R_R$ can be found from the distance formula:

$$R_R = \sqrt{(6.0 \times 10^9)^2 + (2.0 \times 10^9)^2} = 6.3 \times 10^9 \text{ m}$$

The magnitudes of the two gravitational fields are:

$$g_L = G\frac{m_L}{R_L^2} = \left(6.67 \times 10^{-11} \frac{\text{N m}^2}{\text{kg}^2}\right) \frac{(8.0 \times 10^{24} \text{ kg})}{(2.0 \times 10^9 \text{ m})^2} = 1.3 \times 10^{-4} \text{ m/s}^2$$

$$g_R = G\frac{m_R}{R_R^2} = \left(6.67 \times 10^{-11} \frac{\text{N m}^2}{\text{kg}^2}\right) \frac{(4.0 \times 10^{24} \text{ kg})}{(6.3 \times 10^9 \text{ m})^2} = 6.7 \times 10^{-6} \text{ m/s}^2$$

The direction of $\vec{\mathbf{g}}_L$ is clearly $\theta_L = 270°$. The reference angle for the direction of $\vec{\mathbf{g}}_R$ can be found by applying trig to the right triangle illustrated above:

$$\tan^{-1}\left(\frac{2.0 \times 10^9}{6.0 \times 10^9}\right) = 18°$$

Since $\vec{\mathbf{g}}_R$ lies in the fourth quadrant, this translates to $\theta_R = 360° - 18° = 342°$.

Now that we have the magnitudes and directions of the vectors $\vec{\mathbf{g}}_L$ and $\vec{\mathbf{g}}_R$, we can determine the magnitude and direction of the net gravitational field, $\vec{\mathbf{g}}$, by applying the strategy for adding vectors:

$$g_{Lx} = g_L \cos\theta_L = 1.3 \times 10^{-4} \cos 270° = 0$$
$$g_{Ly} = g_L \sin\theta_L = 1.3 \times 10^{-4} \sin 270° = -1.3 \times 10^{-4} \text{ m/s}^2$$
$$g_{Rx} = g_R \cos\theta_R = 6.7 \times 10^{-6} \cos 342° = 6.3 \times 10^{-6} \text{ m/s}^2$$
$$g_{Ry} = g_R \sin\theta_R = 6.7 \times 10^{-6} \sin 342° = -2.1 \times 10^{-6} \text{ m/s}^2$$
$$g_x = g_{Lx} + g_{Rx} = 0 + 6.3 \times 10^{-6} \text{ m/s}^2 = 6.3 \times 10^{-6} \text{ m/s}^2$$
$$g_y = g_{Ly} + g_{Ry} = -1.3 \times 10^{-4} \text{ m/s}^2 - 2.1 \times 10^{-6} \text{ m/s}^2 = -1.3 \times 10^{-4} \text{ m/s}^2$$
$$g = \sqrt{g_x^2 + g_y^2} = \sqrt{(6.3 \times 10^{-6})^2 + (-1.3 \times 10^{-4})^2} = 1.3 \times 10^{-4} \frac{\text{m}}{\text{s}^2}$$
$$\theta = \tan^{-1}(-1.3 \times 10^{-4}/6.3 \times 10^{-6}) = -87°$$

**Vertical free fall in a varying gravitational field**: Consider an object that is in vertical free fall in a region of space where the only significant gravitational field is due to a spherically symmetric planet. If the object makes a significant change in altitude, its acceleration will not be approximately constant; rather, it will vary with the inverse-square of its distance from the center of mass of the planet. Taking the $+y$-direction to be directly away from the planet's center,

$$\sum_{i=1}^{1} F_{iy} = ma_y$$

$$-G\frac{m_p m}{y^2} = ma_y \quad \text{(vertical free fall)}$$

$$-G\frac{m_p}{y^2} = \frac{dv_y}{dt} \quad \text{(vertical free fall)}$$

where we placed the origin at the planet's center so that $y$ is the distance from the freely falling object to the planet's center. This differential equation has three variables – $y$, $v_y$, and $t$. We can separate variables and integrate if we can rewrite it in terms of just two of these three variables. This can be achieved using the chain rule:

$$-G\frac{m_p}{y^2} = \frac{dv_y}{dy}\frac{dy}{dt} = v_y\frac{dv_y}{dy} \quad \text{(vertical free fall)}$$

Now we can separate variables – i.e. put all of the $y$'s on one side and all of the $v_y$'s on the other – and then integrate:

$$-\int_{y=y_0}^{y} \frac{Gm_p dy}{y^2} = \int_{v_y=v_{y0}}^{v_y} v_y dv_y \quad \text{(vertical free fall)}$$

$$Gm_p\left(\frac{1}{y} - \frac{1}{y_0}\right) = \frac{1}{2}(v_y^2 - v_{y0}^2) \quad \text{(vertical free fall)}$$

This equation of motion relates the final velocity to the final position in terms of the initial values. An equation for position as a function of time can similarly be found by solving for $v_y$, writing $v_y = dy/dt$, and integrating a second time.

> **Note**. The equations of 1DUA do not apply to a freely falling object that makes a significant change in altitude (compared to its distance from the center of the planet). Similarly, the equations of projectile motion do not feature 1DUA in the vertical component of motion when a projectile changes altitude significantly. Gravitational acceleration is only approximately uniform for objects that make relatively small changes in altitude (such that $\Delta R/R$ is a small fraction).

**Weightlessness**: A person in free fall experiences a feeling of weightlessness. Of course, a person does have weight in free fall: Recall that we defined an object to be in a state of free fall if the only force acting on the object was its weight. However, even though a person has weight in free fall, one feels weightless when in free fall. You actually feel your weight when you are at rest, but feel your apparent weight when you are accelerating. You may want to briefly review the concept of apparent weight that we discussed toward the end of Sec. 3.3. What you actually feel – and perceive as apparent weight – is the normal force acting on you.

When you stand on a horizontal scale at rest (which is the way a scale is designed to be used), the net force acting on you is zero. In this case, the normal force balances your weight. It's the normal force that you actually 'feel.' If you stand on an elevator that is accelerating vertically (as in the last example of Sec. 3.3), what you 'feel' – your apparent weight – is different from your actual weight: When the elevator accelerates upward, your apparent weight is greater than your actual weight, and when the elevator accelerates downward, your apparent weight is less than your actual weight. In the unfortunate circumstance that the cable of an elevator is cut, the passengers inside would feel weightless (until impact, anyway) – as does anyone in a state of free fall.

You can measure your normal force by standing on a scale in an elevator. If you do so, you will see that the normal force is greater if the elevator accelerates upward (which could mean moving upward and gaining speed, but could also mean moving downward while slowing down) and that the normal force is less if the elevator accelerates downward (either gaining speed while moving downward or losing speed while moving upward). This normal force equals your apparent weight, and differs from your actual weight. If the elevator cable were cut, the scale would read zero. It's not only that the scale reads zero, but you would also feel weightless, and everything in the elevator car would float just like the contents and astronauts of a spaceship (or space station) float when it's in a free-fall orbit.

Regardless of what the elevator is doing, you experience a gravitational force acting downward – equal to your actual weight. When the elevator is at rest, something (the scale, if you're standing on one; otherwise, the floor) exerts an upward normal force on you equal to your weight that prevents you from falling. It's this upward force that you feel, even though both your downward weight and upward normal force are acting on you (or rather, you feel squeezed from a combination of these forces). In the extreme case that the cable is cut, both the floor and scale are accelerating downward at the same rate that you are. As everything is falling downward, nothing below you is exerting an upward normal force on you. The only force acting on you is your actual weight pulling you downward (and hence you do not feel squeezed). All of the contents of the elevator are falling with the same acceleration, and so everything appears to float in the elevator car (relative to the elevator). Your digestive system would also be accelerating downward at the same rate, and therefore floating relative to you – whereas ordinarily your digestive system is being pulled downward by gravity, but is usually restricted because a normal force prevents it from falling. In this strange state, you would very likely be nauseous enough to vomit.

Astronauts similarly experience weightlessness when their spaceships (or space stations) are in free-fall orbits. When their ships eject steam (or other exhaust) to travel in a different orbit, however, they feel apparent weight instead – equal to the normal force. This apparent weight generally will be different from their actual weight (at that point in space); the apparent weight will be zero if the ship is in free fall; and the apparent weight will be the same as actual weight (at that point in space) if the ship travels with constant momentum. Note that you feel weightless even if the free-fall orbit is curved (see Sec. 4.6).

## Conceptual Questions

1. For each statement that follows, give an example that shows that the statement is incorrect or imprecise: "Objects have a natural tendency to be at rest," "Objects have a natural tendency to stay in motion," "Objects have a natural tendency to maintain constant speed," "Objects have a natural tendency to maintain constant velocity." What is a better statement that would be both correct and precise for any situation?

2. Can you feel inertia? Explain.

3. Indicate whether or not each of the following physical quantities is a type of force: velocity, tension, inertia, mass, weight, pressure, a quantity measured in foot-pounds, $\vec{N}$, $\vec{g}$, the push of a finger, the pull of a tractor beam.

4. Only two forces act on a 20-kg object. One force has a magnitude of 300 N and the other force has a magnitude of 200 N. What are the minimum and maximum possible magnitudes of acceleration of the object under these conditions? Explain.

5. A necklace dangles from the rearview mirror of a car. Describe the direction of the necklace's lean in each of the following situations: the car is at rest, the car gains speed in a straight line on a horizontal road, the car travels with constant speed in a circle, the car travels with constant speed in a straight line up a hill, the car travels with constant speed in a straight line on a horizontal road, the car travels with constant speed in a straight line down a hill, the car decelerates in a straight line on a horizontal road, the car travels in the air (but neglect the air) in a parabolic arc after going driving over a ramp.

6. A monkey gets in a car (at rest) and straps the seatbelt on. (A) Starting from rest on a horizontal road, the monkey presses his foot against the gas pedal. As he does so, describe the first (new) force that the monkey feels. (B) Now the monkey is traveling with constant speed in a straight line along a horizontal road. The monkey presses his foot against the brake pedal. As he does so, describe the first (new) force that the monkey feels.

7. Two small airplanes are flying with the same constant horizontal velocity. One plane is flying at a higher altitude than the other. The higher airplane is flying up-side-down. A monkey in the higher airplane wants to unbuckle his seatbelt and land in the lower airplane. Where should the higher airplane be relative to the lower airplane in order to achieve this? Explain.[189]

8. A car is driving with a constant velocity of 70 mph to the north along a straight, level highway. The windshield strikes a mosquito that was hovering above the highway just before impact. What can you say about the ratio of the forces that the car and mosquito exert on one another during the collision? What can you say about the ratio of their accelerations during the collision? Explain.

9. A monkey ties a cord to a banana, grabs the free end of the cord, and whirls the banana around in a horizontal circle (the cord is not horizontal, though). Sketch the motion from the top view, showing what happens when the monkey suddenly lets go of the cord.

10. At a miniature golf course, a golf ball rolling on a horizontal surface suddenly comes to a curved railing, which guides the ball along a quarter circle. The golf ball is heading north before reaching the guide, and heading west just after leaving the guide. Sketch the path of the golf ball to illustrate what happens to the golf ball after leaving the guide.

---

[189] Obviously, the airplanes need air, but neglect the effects of air resistance for the monkey.

11. An astronaut is in a satellite in a free-fall orbit (and therefore feels weightless). The astronaut needs to go outside of the ship to repair the ship. The astronaut goes outside in a suit, holding a wrench in one hand and a tie rope in the other. While making the repair, the astronaut accidentally loses his grip on the tie rope, while still holding the wrench. Unfortunately, during this accident, the astronaut begins to drift away from the ship. How can the astronaut return to the ship (without anyone else's assistance)? Explain.

12. What would happen if everyone on the earth suddenly started walking west at the same time? Explain. What happens when everyone stops? What happens when everyone goes back home?

13. In a classic physics question, a tablecloth covers a level table, and dinnerware is placed on top of the tablecloth. A monkey grabs a corner of the table cloth, quickly yanking the tablecloth off of the table, leaving the dinnerware still set on the table (i.e. not falling over or falling off the table). (A) Explain the physics behind this. (B) What happens if the monkey instead pulls the tablecloth slowly? Explain. (C) Consider the monkey's technique: Specifically, the results depend upon whether the corner is slightly above or slightly below the table height when the monkey yanks on it (and whether that corner pulls the tablecloth upward or downward as the monkey yanks on it). Discuss this.

14. In a classic physics question, a monkey places a block of wood on his head and then hits the block of wood with a hammer, and then the monkey places a block of metal between his head and the block of wood, and again hits the block of wood with a hammer. Why does the monkey's head feel less pain with the presence of the block of metal?

15. In a classic physics question, a banana is connected to the ceiling with one cord, while a second cord dangles from the banana. If a monkey pulls quickly on the free cord, the lower cord snaps, and the monkey is unable to reach the banana; but if the monkey pulls slowly on the free cord, the upper cord snaps and the banana falls down. Explain the physics concepts behind this experiment.

16. In a classic physics question, a monkey tries to lift himself up into the air by pulling upward on his bootstraps. Is this possible? Explain.

17. If the action is the weight of an object, identify the force that is the equal and opposite reaction associated with Newton's third law.

18. Two monkeys are each hanging from one end of a rope in Atwood's machine (and so both monkeys are above the ground). One monkey is initially 4.0 m above the (level) ground, while the other monkey is initially 3.0 m above the ground. Both monkeys have the same mass. Initially, both monkeys are at rest. The (initially) lower monkey 'climbs' upward by shortening the length of rope between the two monkeys by 1.0 m, while the other monkey simply hangs on. Neglecting friction between the cord and pulley, what are the heights of each monkey after the monkey finishes climbing? Explain.

19. On the moon, a monkey feels like Super Monkey. (A) On earth, the monkey can lift a box that weighs as much as 200 lbs. If we transport several of the same boxes from the earth to the moon, how many of these boxes can the monkey lift on the moon? Explain. (B) The monkey kicks a bowling ball near the surface of the earth, and kicks the same bowling ball near the surface of the moon under identical conditions (except, of course, for the moon's gravity). Compare how much the monkey's foot hurts in both cases. Explain. In light of this, would you advise the monkey to try to stop a locomotive by standing in front of the train and pushing against it?[190]

---

[190] For the extreme case, instead of comparing the earth and the moon, compare the earth and a non-spinning space station in a free fall orbit.

20. A 150-lb. monkey stands on a scale in an elevator. How does the scale reading compare to 150-lbs. during each of the following situations: the elevator is at rest, the elevator gains speed moving upward, the elevator travels upward with constant speed, the elevator gains speed moving downward, the elevator moves downward with constant speed, the elevator loses speed moving upward, the elevator loses speed going downward, the elevator cable is cut.

21. Draw a hill that a monkey could slide down for which the force of kinetic friction would increase as the monkey slides down (assuming a uniform coefficient of friction). Explain.

22. A box of bananas is at rest on an incline, with a cord connecting the upper end of the box to a wall. The cord is parallel to the incline. A range of tension forces is possible in the cord. Explain why there is not a single unique solution for the tension force in this case.

23. Three wooden blocks slide down an incline (with friction) from rest from the same height. All three blocks have identical surfaces. Block A has mass $m$ and surface area of contact $A$, block B has mass $2m$ and surface area of contact $A$, and block C has mass $m$ and surface area of contact $2A$. What ordering do you predict for the blocks when they reach the bottom of the incline. Explain.

24. Two boxes filled with sand slide down an incline (with friction) from rest from the same height. One box has no lid, allowing some sand to fall out. Compare the accelerations of the two boxes. Explain.[191]

25. A monkey wishes to launch a banana vertically by setting the banana on top of a spring and compressing the spring. Should the monkey use a spring with a larger or smaller spring constant if he wants to launch the banana as high as possible? Explain.

26. If a monkey digs a tunnel through the earth (and insulates it and creates a vacuum inside of it) and drops a banana through the tunnel, the acceleration of the freely-falling banana inside the tunnel does not have the form of an inverse-square law (which only applies outside of the earth), but is instead given by $\vec{g} = -(Gm_e r/R_e^3)\,\hat{r}$, where $R_e$ is earth's radius, $m_e$ is the earth's mass, and $\vec{r} = r\hat{r}$ is a vector that extends from the center of the earth to the instantaneous position of the banana. Does the force that the earth exerts on the banana inside the tunnel obey Hooke's law?

27. Will the effective spring constant of the combination be greater if two springs are connected in series (the end of one connects to the end of another, so that they are collinear) or if they are connected in parallel (the springs are side-by-side so that they both connect the same object in the same way). Explain.

28. In the illustration below, if the wedge is pushed to the right with sufficient force, $\vec{P}$, the block will actually slide up the frictionless incline. Explain this conceptually.

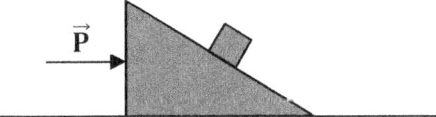

29. In the illustration below, if the large block is pushed to the right with sufficient force, $\vec{P}$, the small block will appear to defy gravity (i.e. it won't fall). Explain this conceptually.

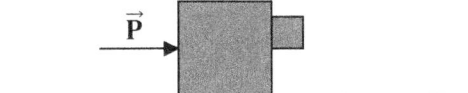

---

[191] Air may cause some of the sand to fall out, but neglect the effect of any air pushing against the boxes.

30. A monkey throws a banana straight upward and catches it at the same height at which it left his hand. Accounting for the effects of air resistance, how does the time that the banana spends traveling upward compare with the time that the banana spends traveling downward? Explain.

31. Accounting for the effects of air resistance, conceptually describe why an object dropped from the top of a skyscraper reaches a terminal speed.

32. An airplane is traveling horizontally with a constant speed of 100 m/s when a passenger jumps out with an unopened parachute. (A) Describe the horizontal and vertical components of the passenger's velocity and acceleration at each of the following moments: just after the passenger jumps out of the plane, as the passenger is falling but before he opens the parachute, a long time after falling but before opening the parachute, just after opening the parachute, a long time after opening the parachute. (B) There are two different terminal speeds in this problem. Explain. (C) At which moment(s) is it possible for the passenger's acceleration to be directed upward? Explain. What is the direction of the velocity at this (or these) time(s)?

33. There is a point between the earth and moon where the net gravitational field (neglecting other massive objects in the universe, such as the sun or other planets) is zero. (A) Is this point closer to the earth or moon? Explain. (B) If you proceed to solve for this position mathematically, you obtain two solutions, yet only one solution is physically viable. Explain how there are two mathematical solutions, and the conceptual meaning of the solution that is not physically viable.

34. If a moon is moved away from a planet so that its new orbit has four times its old orbital radius, by what factor does the gravitational force of attraction between the moon and planet change?

35. If the earth had only one-half of its present mass and one-half of its present diameter, what would earth's surface gravity be?

36. (A) In a non-spinning space station in a free-fall orbit, would an astronaut inside the space station perceive a bowling ball inside the space station to have weight? Would it be easy for the astronaut to 'lift' the bowling ball? Would it hurt less for the astronaut to 'punch' the bowling ball? Explain. (B) The moon is a satellite of the earth – it travels in a free-fall orbit around the earth. However, an astronaut near the surface of the moon does not feel weightless. Explain the distinction between the moon and the space station.

37. (A) Which exerts a greater force on the earth – the sun or the moon? What simple astronomical observation allows you to draw this conclusion without doing any calculations? (B) Which is more responsible for producing tides in earth's oceans – the sun or the moon? If your answer differs from part (A), explain why. (C) Draw a sketch of the earth, pretending that it is entirely covered with ocean (rather than just seven-tenths), illustrating the elongation of the ocean created by the tidal forces of your answer to part (B). (D) In one day, the body from part (B) remains in roughly the same position as the earth completes one rotation. In which position(s) in the diagram do earth's oceans experience low and high tides? How many low and high tides are there each day? (E) The moon and sun both produce significant tidal effects, although one is much more significant than the other. When the moon, sun, and earth are aligned, high tides are higher and low tides are lower than normal (these are called spring tides); when the moon, sun, and earth make a right triangle, high tides are lower and low tides are higher than normal (these are called neap tides). Draw illustrations of this. About how many times per month do each of these positions occur?

38. Express the units of the gravitational constant, $G$, in terms of kilograms, meters, and/or seconds only.

39. What must be the SI units of the constant $a$ in the equation $F = -ax^2$? First express your answer in terms of Newtons and/or meters only, and then express your answer in terms of kilograms, meters, and/or seconds only.

## Practice Problems

### Mass, weight, apparent weight, and weightlessness

1. A dummy deemed to be 180-lbs.[192] on earth is transported to the moon to partake in some free fall experiments. What are the dummy's mass and weight on the earth and moon (four answers in all)?
2. A dummy deemed to be 123-kg on earth is transported to another planet where he takes 4.0 seconds to strike the ground when dropped from a height of 100 m. What are the dummy's mass and weight on this planet (two answers in all)?
3. A monkey compares boxes of bananas on the earth and moon. (A) How much does a 45-kg box of bananas weigh on the earth? How much does it weigh on the moon? (B) A monkey could have a box of bananas that weighs 300-N on the earth or the moon. What would the mass of the box of bananas be in each case (two answers in all)?
4. A 50-kg monkey stands on a scale while riding a 1000-kg elevator. (A) What does the scale read (i) as the elevator travels upward at a constant rate of 4.0 m/s and (ii) as the elevator travels downward at a constant rate of 3.0 m/s$^2$? (B) As the elevator travels with uniform acceleration, the monkey weighs 800 N. What are the magnitude and direction of the acceleration? (C) In part (B), the monkey drops a quarter from a height of 2.0 m above the elevator floor. How long does it take to reach the floor? (D) The cable snaps when the elevator is 40 m above a giant spring while the elevator is at rest. What does the scale read as the elevator falls? (D) What is the speed of the elevator just before striking the spring?

### Applications of Newton's second law

5. A 25-g bullet traveling 300 m/s shoots into a physics textbook that is 3.5 cm wide, and exits the textbook traveling 40 m/s. Poor defenseless textbook! (A) What was the magnitude of the average deceleration of the bullet while it was penetrating through the textbook? (B) What was the magnitude of the average resistive force exerted on the bullet while it was penetrating through the textbook?

---

[192] Conceptually, it is improper to look up a 'conversion' from pounds to kilograms: You can't convert from apples to oranges – i.e. one is a unit of weight (or force), while the other is a unit of mass. Such 'conversion factors' only work for specific locations, like the surface of the earth. So if you convert pounds to kilograms for your solution to this problem, you won't understand how to solve a similar problem when you're told the weight of an object in pounds on some other planet or moon. The better technique is to look up a conversion factor from pounds to Newtons, since both are units of force. Finally, don't 'convert' from Newtons to kilograms – as this makes no sense; instead, use a formula to calculate the mass in kilograms from the weight in Newtons. It's a formula, not a conversion.

6. A 200-kg skydiver parachutes from an airplane. What is the acceleration of the skydiver when the force of air resistance is (A) zero, (B) 900 N, (C) 2500 N, (D) 1960 N. (E) Give the direction of the velocity and acceleration for each case. (F) What is the magnitude of the force of air resistance when the skydiver has reached his terminal velocity? (G) When the skydiver's speed is least, is the acceleration least or greatest? Explain.

7. You are kidnapped by little green monkeys who transport you to Planet Mnqy in a UFO. While you are being kidnapped, you hang onto a rope (in mid-air, but, as usual, neglect the air) connected to their UFO, as illustrated below. Assume that your weight is 200 N. Find the tension in the cord if: (A) The UFO accelerates upward at 15 m/s$^2$; (B) the UFO accelerates downward at a rate of 4.0 m/s$^2$; and (C) the UFO travels upward with a constant speed of 5.0 m/s.

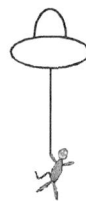

8. A physics spy, 00$\pi$, rescues a physics student from distractions (in mid-air) as illustrated below. The 1200-kg helicopter is pushed upward with a force of 20 kN.[193] The spy weighs 800 N, and the mass of the student is 60 kg. (A) Find the magnitude of the acceleration of the helicopter. (B) Find the tension in each rope.

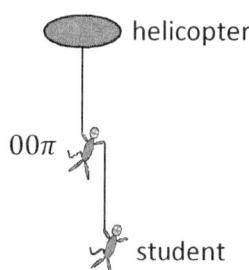

9. A 20-kg monkey is connected to a 40-kg treasure chest as shown below. The coefficient of kinetic friction between the chest and the ground is 0.25. The block is presently moving to the right. (A) Find the acceleration of the chest of bananas. (B) Find the magnitude of the acceleration of the monkey. (C) Find the tension in the cable.

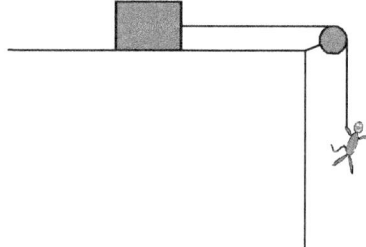

---

[193] Obviously, the helicopter requires air to fly. Nonetheless, use the given lift force, but neglect air resistance.

10. Lemur and Chimp found a box of lucky charms on their expedition in North Aperica. Unfortunately, both tripped and fell off a cliff when they were rigging a pulley system to get their treasure home. However, they are still hanging by the skin of their teeth, as illustrated below. Lemur, on the right, has a mass of 10 kg; Chimp, on the left, has a mass of 30 kg; and the box of lucky charms has a mass of 60 kg. The coefficient of friction between the box and ground is 0.12. The block is presently moving to the left. What are the magnitude and direction of the acceleration of the system?

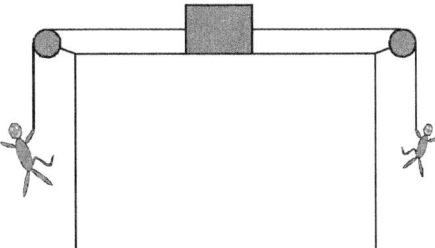

11. Three boxes of precious physics textbooks are connected by a cord. The left box has a mass of 500 kg, the middle box has a mass of 1000 kg, and the right box has a mass of 1500 kg. The leading block is pulled by a force $\vec{P}$, which has a magnitude of 2.0 kN and a direction 40° above the horizontal, as illustrated below. The system begins from rest. Neglecting friction, find (A) the magnitude of the acceleration of each box and (B) the tension in each cord.

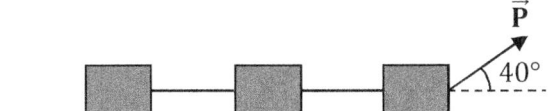

12. Two boxes of monkeys connected by a cord are pulled by a banana with a force of $800\sqrt{3}$ N as illustrated below. The left box of monkeys has a mass of $40\sqrt{3}$ kg and the right box of monkeys has a mass of $60\sqrt{3}$ kg. The coefficient of friction between the boxes of monkeys and the horizontal is $\sqrt{3}/3$. The system is initially at rest. (A) Find the magnitude of the acceleration of the system. (B) What is the tension in the connecting cord?

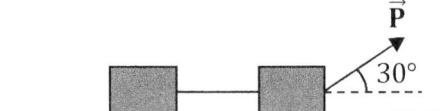

13. Two smiling students love physics so much that on Saturday, in the middle of Winter Break, they go outside to play with a $60\sqrt{3}$-kg box of physics textbooks. The smiling students pull on the box with the forces illustrated below. The box is initially at rest. The coefficient of friction between the box of physics textbooks and the horizontal is $\sqrt{3}/6$. Merrily find the magnitude and direction of the acceleration of the box of physics textbooks.

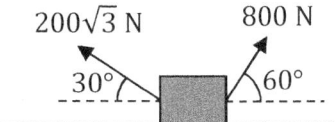

14. The monkey of mass $m$ hanging from the rope in the diagram on the following page is in static equilibrium. (A) Derive equations for the tension in each cord in terms of $m$, $\theta_1$, $\theta_2$, and/or $g$. (B) Simplify your expression in the limit that the two angles become equal.

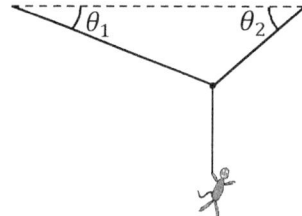

15. A monkey sits in a large saucer and slides down an incline of angle $\theta$ with respect to the horizontal. The coefficient of friction between the monkey and the ground is $\mu$. Solve for the acceleration of the monkey down the incline in terms of $\theta$, $\mu$, and/or $g$.

16. A Neanderthal built an escalator with a 200-kg passenger box, a very sturdy vine, a pulley, and a large rock as illustrated below. The coefficient of friction between the box and the ground is 0.60. The passenger box is presently moving up the incline. What minimum mass of the stone is needed to accelerate a 100-kg passenger (sitting in the passenger box) up the incline?

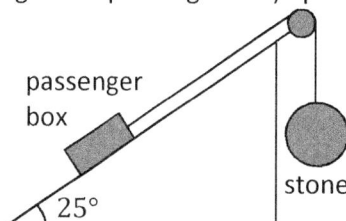

17. A 750-N Neanderthal slides <u>down</u> (even though there is a push up the incline, as illustrated below – how?) a dinosaur's back at a constant rate of 30 m/s as illustrated below. The coefficient of kinetic friction between the dinosaur and the Neanderthal is 0.30. What is the magnitude of the push, $\vec{P}$?

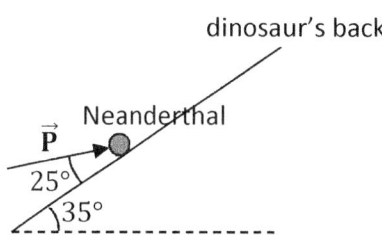

18. Lemur and Chimp find a 200-kg treasure chest filled with gold-plated bananas. The chest rests on an incline as illustrated below. The coefficient of friction between the chest and incline is 0.60. The incline makes a 35° angle with the horizontal. A cord parallel to the incline prevents the box from moving. (A) What are the possible tensions in the cord? (B) After Lemur cuts the cord, what is the magnitude of the acceleration of the treasure chest? (C) What is the speed of the treasure chest after sliding 30 m down the incline?

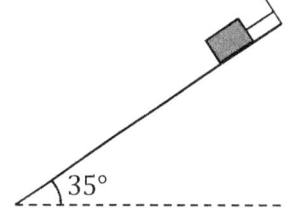

19. A 40-kg monkey is connected to a 20-kg box of bananas as illustrated below. The coefficient of friction between the box of bananas and the incline is 0.50. Find the two possible accelerations of the system. Explain why two are possible.

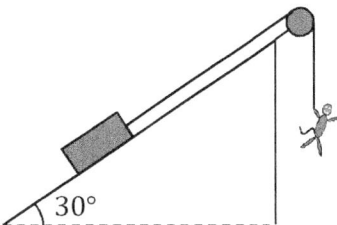

20. A 500-kg treasure chest rests on the deck of a pirate ship. The coefficients of friction between the treasure chest and the deck are 0.36 and 0.39. One is the coefficient of static friction, while the other is the coefficient of kinetic friction. The ship is sailing through rough weather. The crew is concerned that the treasure chest may slide and disturb the valuable contents. The crew is quarreling over the maximum angle that the deck can make with the horizontal before the treasure chest will begin to slide. Do the crew a favor and figure this out for them.

21. A monkey pulls two boxes of bananas up an incline with a force of 1500 N, as illustrated below. When he reaches the top, he will dump them into a volcano, hoping his great sacrifice will reap great rewards on his exam. The coefficient of friction between the boxes of textbooks and the incline is $\sqrt{3}/3$. (A) Find the magnitude of the acceleration of the boxes of physics textbooks. (B) Find the tension in the connecting cord. (C) If the connecting cord breaks, what will be the acceleration of each box?

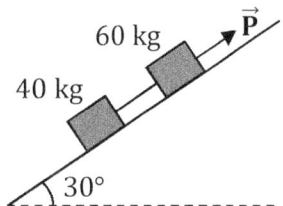

22. A 20-kg monkey is connected to a 40-kg monkey as illustrated below. The coefficient of friction is $\sqrt{3}/2$ between either monkey and the surface. The system is presently moving to the right. (A) Find the acceleration of the monkeys. Also, indicate the direction. (B) Find the tension in the connecting cord. (C) If the cord snaps, which monkey will have greater acceleration? Explain.

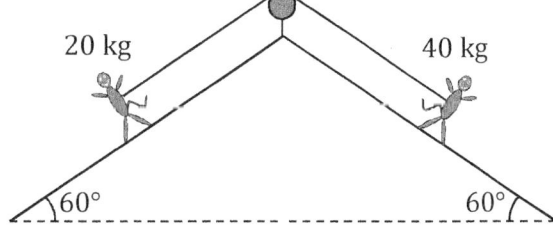

23. As illustrated on the following page, brave explorers Lemur and Chimp find themselves in a sticky situation on their quest for the Golden Banana. The coefficients of friction between Lemur and the incline are 0.60 and 0.70. In the position shown, 90-kg Lemur and 40-N Chimp are moving 5.0 m/s downward and Lemur is 80 m from the cliff. (A) Which coefficient is which, and which is needed to solve this problem? (B) Find the magnitude of the acceleration of Lemur and Chimp. (C) Find the tension in the rope. (D) What is the fate of Lemur and Chimp?

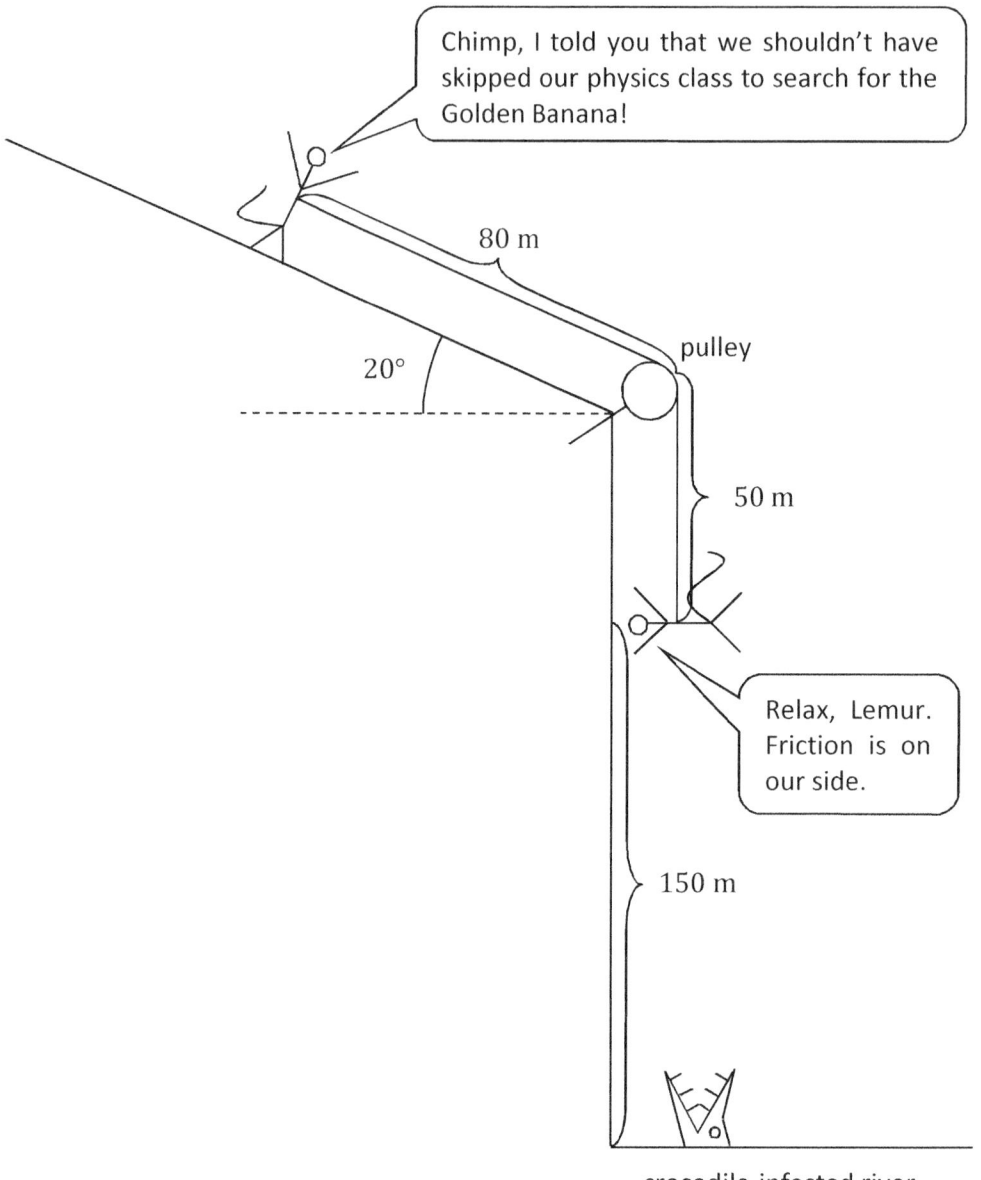

crocodile-infested river

24. Lemur and Chimp drive home on a level road. Lemur ties a coconut to a string and suspends it from the ceiling of the car. When the velocity of the car is constant, the pendulum is vertical. When the car accelerates, the pendulum makes an angle $\theta$ with the vertical. Thus, this pendulum functions as an accelerometer. Derive an equation for the magnitude of the acceleration of the car in terms of $\theta$ and $g$.

25. Lemur and Chimp find two more treasure chests – one 50-kg chest filled with copper-plated watermelons and a 100-kg chest filled with silver-plated coconuts. The coefficient of static friction between the chests is 0.40, but the ground is frictionless. As illustrated on the following page, Lemur and Chimp exert an 800-N horizontal force on the chest of watermelons. (A) What is the magnitude of the acceleration of the top chest? (B) What is the tension in the cord?

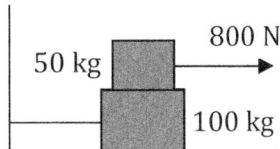

26. As illustrated below, a little green monkey pushes a treasure chest of coconuts of mass $m_1$ with force $\vec{P}$, yet a little purple monkey of mass $m_2$ does not fall. The ground is frictionless, but the coefficient of friction between you and the chest is $\mu$. (A) Derive an equation for the magnitude of the acceleration in terms of $m_1, m_1, \mu$, and/or $g$. (B) Derive an equation for $P$ in terms of $\mu$.

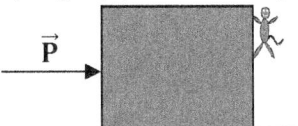

27. A monkey of mass $m_1$ sits on the frictionless wedge of mass $m_2$ in the diagram below, yet the monkey is stationary relative to the wedge. How can this be? Because the wedge is being pushed with a horizontal force $\vec{P}$. Derive an equation for $P$ in terms of $m_1, m_2, g$, and/or $\theta$, where $\theta$ is the angle of the wedge relative to the incline.

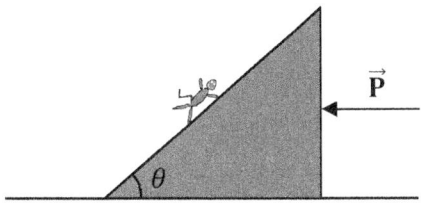

28. As illustrated below, 20-N/cm spring is stretched on a frictionless incline by fixing its upper end and connecting a 4.0-kg box of bananas to its lower end. The length between its coils is 18 cm in this stretched position, when the system is in static equilibrium. What is the natural length of the spring?

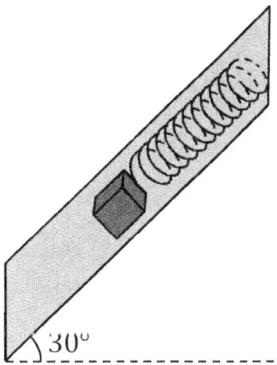

29. As illustrated on the following page, a 30-N/cm spring lies at the bottom of a frictionless incline, with its lower end fixed. A 15-kg box of bananas slides down the incline, compressing the spring. What is the acceleration of the box of bananas when the spring is compressed (A) 0 cm, (B) 20 cm, and (C) 40 cm?

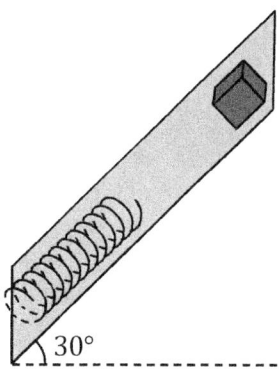

30. The coefficient of friction between an 8.0-kg box of bananas and a horizontal surface is 0.20. As illustrated below, the box of bananas is connected to two horizontal springs – a 50-N/cm spring to its left and a 20-N/cm to its right. The other ends of the springs are fixed. What is the initial acceleration of the box of bananas when it is released from rest at a position where the left spring is stretched 6.0 cm from its natural length and the right spring is compressed 4.0 cm from its natural length?

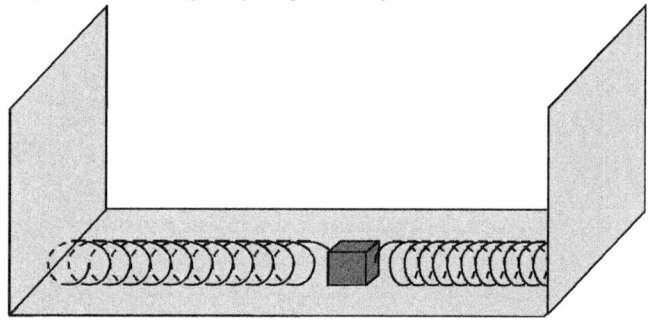

31. A 24-N/cm spring lies at the bottom of a frictionless incline, with its lower end fixed. As illustrated below, a 12-kg box of bananas is connected to the spring via a cord that passes over the pulley, such that the box of bananas dangles in midair (one section of the cord is parallel to the incline, the other hangs vertically). How much is the spring stretched at its current equilibrium position compared to the equilibrium position that it would have if the cord were cut?

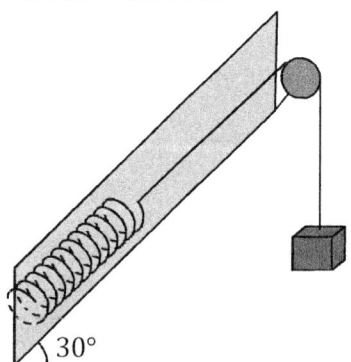

## Air resistance and using calculus with Newton's second law

32. A student finally finished completing her homework assignment. It only took 10,000,000 pages (double-sided). She fills a large crate with her homework, and attaches a rope to the center of one side. Now she is prepared to drag the crate to class by pulling along the rope with force $\vec{P}$ directed at an angle $\theta$ above the horizontal. See the diagram below. The coefficient of kinetic friction between the sidewalk and the crate is 0.40. (A) If she wants to pull the crate with constant velocity with minimum effort (i.e. minimum $P$), at what angle $\theta$ should she pull on the rope? (B) Explain why the answer isn't 0°.

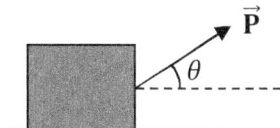

33. The entrance to Banana Cavern is blocked by a 200-kg rock. Lemur and Chimp summon their strength to push it out of the way. The net force exerted on the rock is: $\vec{F}(t) = \alpha t^2 \hat{i} + (\beta t - \lambda)\hat{j}$, where $\alpha = -3.0$, $\beta = 5.0$, and $\lambda = 15$ in their appropriate SI units. The initial velocity (at $t = 0$) of the rock is zero. (A) What are the SI units of alpha ($\alpha$), beta ($\beta$), and lambda ($\lambda$)? (B) What are the magnitude and direction of the initial acceleration? (C) At what time is the direction of the acceleration equal to 180°? (D) What is the magnitude of the momentum at $t = 3.0$ seconds?

34. The acceleration of a wild 20-kg monkey is $\vec{a}(t) = \delta t \hat{i} - \lambda \hat{j}$, where $\delta = 3.0$ and $\lambda = 4.0$ in SI units. The initial velocity (at $t = 0$) of the monkey is $\vec{v}_0(t) = \beta \hat{i}$, where $\beta = 5.0$ in SI units. (A) What are the SI units of delta ($\delta$), beta ($\beta$), and lambda ($\lambda$)? Find the magnitude and direction of: (B) the initial acceleration of the monkey; (C) the net force exerted on the monkey at $t = 3.0$ seconds; (D) the momentum of the monkey at $t = 3.0$ seconds; and (E) the net displacement of the monkey at $t = 3.0$ seconds.

35. A 60-kg dog paddles its way around a pool. The magnitude and direction of its velocity are
$$v(t) = 3t$$
$$\theta(t) = 2t$$
where $\theta$ has units of radians and $v$ has units of m/s when $t$ is expressed in seconds. (A) Find the magnitude and direction of the dog's momentum at $t = 3.0$ seconds. (B) Express the net force exerted on the dog in $\hat{i}$ and $\hat{j}$ notation. (C) At what time is the dog moving in the negative $y$-direction?

36. The momentum of a 700-g slice of banana-cream pie is:
$$\vec{p}(t) = \gamma t^{3/2} \hat{i} - (\beta t)^{1/2} \hat{j}$$
where $\gamma = 3.0$ and $\beta = 5.0$ in their respective SI units. (A) What are the SI units of gamma ($\gamma$) and beta ($\beta$)? (B) What is the initial velocity (at $t = 0$) of the slice of pie? (C) What is the magnitude of the net displacement of the pie for the first 4.0 seconds? (D) What is the direction of the net displacement of the pie for the first 4.0 seconds? (E) What is the net force exerted on the pie when $t = 2.0$ seconds? (F) At what time(s) does the net force exerted on the pie have a direction of 330°?

37. The position of a 2.0-kg flashlight is $\vec{r}(t) = r_0 \cos(\omega_0 t)\hat{i} + r_0 \sin(\omega_0 t)\hat{j}$, where $r_0 = 20$ m and $\omega_0 = 0.40$ rad/s (omega-nought). (A) What are the Cartesian coordinates of the flashlight's initial position (i.e. at $t = 0$)? (B) What are the magnitude and direction of the flashlight's initial velocity? (C) What are the magnitude and direction of the momentum of the flashlight when $t = 3.0$ seconds? (D) What are the magnitude and direction of the acceleration of the flashlight when $t = 3.0$ seconds? (E) What are the magnitude and direction of the net force exerted on the flashlight when $t = 3.0$ seconds? (F) At what time is flashlight accelerated along 135°?

38. The magnitudes and directions of two forces acting on a 600-g banana split are:
$$F_1 = 3t \quad , \quad \theta_1 = 110°$$
$$F_2 = \sqrt{2t} \quad , \quad \theta_2 = 240°$$
where the initial velocity (at $t = 0$) of the banana split is $3\hat{\imath} + 4\hat{\jmath}$. SI units have been suppressed for convenience. (A) What are the magnitude and direction of the net force exerted on the banana split when $t = 3.0$ seconds? (B) What is the initial momentum of the banana split? (C) What is the acceleration of the banana split when $t = 3.0$ seconds? (D) What is the velocity of the banana split when $t = 3.0$ seconds? (E) What is the average velocity of the banana split for the first 3.0 seconds?

39. Derive an equation for the terminal speed of an object falling with quadratic drag in terms of the mass of the object, gravitational acceleration, the density of the fluid, rho ($\rho$), the cross-sectional area of the object, $A$, and the drag coefficient, $D$.

40. An object slides along a horizontal surface with an initial speed $v_0$. Neglect friction, but account for the effects of air resistance. Derive equations for position and velocity as functions of time in the case of (A) linear drag and (B) quadratic drag.

41. Derive equations for position and velocity as functions of time if the net force acting on an object has the form $ct$, where $c$ is a constant and the object has initial velocity $\vec{v}_0$.

42. Derive equations for position and velocity as functions of time if the net force acting on an object has the form of Hooke's law, where $x_e = 0$ and the initial position is $x_m$ and the object starts from rest.

## Newton's law of universal gravitation

43. All monkeys know that the moon is not made of green cheese. It's really made of banana-coconut-cream pie. Mmmm delicious! (A) Use astronomical data to compute gravitational acceleration near the moon's surface. (B) What force does the earth exert on the moon? (C) What force does the moon exert on the earth?

44. Monkeys call our planet Monkey Earth (instead of Mother Earth). (A) Look up the mass and radius of the earth to compute the earth's gravitational acceleration near the surface of the earth. (B) At what altitude is the earth's gravity equal to half of what it is at the surface?

45. (A) A 30-kg monkey stands 5.0 m away from a 20-kg banana. According to Newton, what is the gravitational force of attraction between the monkey and the banana? (B) Two twin monkeys have identical mass and stand 10 m apart. What would the mass of each monkey need to be to make the gravitational force of attraction equal to one Newton?

46. As you begin to do your homework, a bright light suddenly appears in the classroom – first a tiny blinding ball of light, then it expands into a much dimmer rectangle. Hoping to avoid taking the quiz, you walk through this rectangle... and wind up surrounded by little green monkeys. Wouldn't you know it? They make you take a physics quiz!

Planet Ban, as they call it, which is a trillion light-years from earth, has a radius of 200 km. Gravitational acceleration near the surface of Ban is $4.0 \text{ m/s}^2$. Ban's moon, Ana, has a mass that is one-third of Ban's mass, and an orbital radius that is ten times Ban's radius. (A) What is Ban's mass? (B) What force does Ban exert on Ana?

47. Planet Mnqy has five times the mass of the earth and three times the diameter. You may not look up the mass and radius of the earth for this problem. (A) Calculate gravitational acceleration near the surface of Mnqy using only the value for the gravitational acceleration near the surface of the earth. (B)

A monkey weighs 100 N near the surface of Mnqy. Find the mass and weight of the monkey when he stands atop a mountain with an altitude equal to 10% of Mnqy's radius.

48. Five minutes after starting this problem, you are kidnapped by little green monkeys who transport you to Planet Mnqy in a giant flying banana. (Well, you can always hope.)

Mass of Mnqy:     $9.0 \times 10^{20}$ kg
Radius of Mnqy:   $2.0 \times 10^{5}$ m

You kiss the ground when you realize that Planet Mnqy is made of banana! (A) If you throw a rock into the air (yes, there is air, and, yes, you may neglect it, except as needed for biological purposes), what will be the rock's acceleration? (B) If you weigh 400 N on earth, what are your mass and weight on planet Mnqy? (C) Would it be easier for Supermonk to leap a tall building on the earth or planet Mnqy? Would it be easier for Supermonk to stop a locomotive on the earth or planet Mnqy? Explain in terms of physics concepts.

49. If a monkey puts a bunch of banana's near the earth, they fall toward the earth; if he puts the bananas near the moon, they fall toward the moon. However, there is a "magical" point where the bananas will neither fall toward the earth nor the moon. Where is this "magical" location?

50. A spherical moon with uniform density, radius $R_0$, and mass $m_p$ is made of banana. Monkeys mining the banana cut out a spherical cavity of radius $R_0/2$ such that one end of the cavity touches the moon's center and the opposite end touches the moon's surface. The monkeys take the mined banana back to their home planet. (A) What is the moon's mass now? (B) What force would be exerted on a test monkey of mass $m_t$ located a distance $3R_0$ from the moon's center on a line joining the moon's and cavity's centers? (C) What is the net gravitational field at the point $\left(\frac{R_0}{2}, \frac{R_0}{4}\right)$ in the diagram below?

**Note**: In part (C), according to Gauss's law,[194] the equation for the gravitational field inside of a spherically symmetric mass distribution is $\vec{g} = -(Gm_p r/R_p^3)\,\hat{r}$, where $R_p$ is radius of the planet/moon and $\vec{r} = r\hat{r}$ is a vector that extends from the center of the planet/moon to the instantaneous position of the banana.

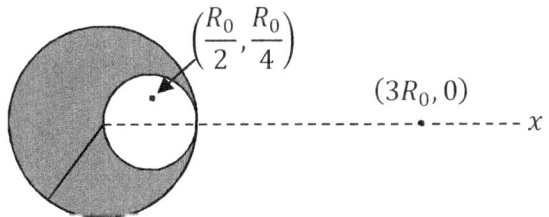

---

[194] Gauss's law will be covered in a subsequent volume of this text.

# 4 Uniform Circular Motion

**Uniform circular motion**: An object is said to travel with uniform circular motion (UCM) if it travels in a circle with constant speed. It is important to realize that any object that travels in a circle must be accelerating, even if its speed is constant. Recall that acceleration describes how velocity changes in time, and that velocity includes not only speed, but also direction. In order to travel in a circle, an object must be changing direction. An object traveling in a circle has a natural tendency – i.e. its inertia – to go off on a tangent. Therefore, there must be one or more forces with inward components giving the object a net inward force in order to cause the object to deviate from its tangential tendency – i.e. to change direction and travel in a circle rather than along its natural straight line.

If you travel in a circle with constant speed, you will indubitably experience your natural tendency – inertia – but will perceive it as a tendency to go outward from your perspective. If you consider your situation carefully, too, you will be able to identify one or more forces that are pushing you inward. If you remove these inward forces, you will suddenly go off on a tangent. This can be seen, for example, by whirling a necklace in a circle. The faster you whirl the necklace, the greater the tension. The necklace wants to go off on a tangent, so more tension is needed to make it instead travel in a circle. If you suddenly let go, the inward force is removed and the necklace immediately proceeds to move tangentially.

UCM occurs frequently and in a myriad of forms in nature, and also occurs in many diverse engineering applications. Many moons and planets, for example, travel in fairly circular orbits with relatively constant speeds. Rigid objects tend to rotate such that each part of the object travels in a circle – e.g. you can throw an object in the air so that it spins in addition to following an approximately parabolic arc. Objects at rest relative to the earth travel in circles as the earth spins on its axis. A great many man-made objects feature wheels, pulleys, and gears that rotate with constant angular speed. If you have experience with UCM from amusement park rides, consider how the concepts of this chapter relate to that experience.

## 4.1 Basic Rotational Quantities

**Angular position**: For an object traveling in a circle, its position can be specified completely in terms of its angular position, $\theta$, which measures how far counterclockwise the object is from the $+x$-axis. The angular position is the same as the polar angle in 2D polar coordinates, except that in circular motion it is convenient to remove the restriction on $\theta$ – i.e. $\theta$ may be negative or exceed $2\pi$ rad. In this way, the angular position, $\theta$, keeps track of how many times an object has traveled around a circle. The SI unit of angular position is the radian (rad). It can also be converted to revolutions (rev) and degrees (°):

$$1 \text{ rev} = 2\pi \text{ rad} = 360°$$

**Angular displacement**: The angular displacement, $\Delta\theta$, is simply the change in angular position: $\Delta\theta = \theta - \theta_0$. The net number of revolutions completed can be determined from $\Delta\theta$.

> **Hint**. The SI unit of $\theta$ is the radian, and some formulas – like the formula for arc length, $s = R_0\theta$ – require $\theta$ to be expressed in radians. However, if you are solving for the net number of revolutions completed, solve for $\Delta\theta$ and convert the answer to revolutions.

**Angular speed**: The magnitude of the instantaneous rate at which the angular position, $\theta$, of a particle changes in time equals the angular speed, $\omega$, of the particle:[195]

$$\omega = \left|\frac{d\theta}{dt}\right|$$

The SI units of angular speed are rad/s, which can also be converted to rev/s.

**Angular velocity**: The angular speed, $\omega$, is the magnitude of the angular velocity, $\vec{\omega}$. The direction of the angular velocity is perpendicular to the plane of rotation. We will define its direction more precisely in Chapter 7.

**Angular acceleration**: The magnitude of the instantaneous rate at which the angular speed, $\omega$, of a particle changes in time equals the magnitude of the angular acceleration, $\vec{\alpha}$, of the particle:[196]

$$\alpha = \|\vec{\alpha}\| = \left|\frac{d\omega}{dt}\right| = \left|\frac{d^2\theta}{dt^2}\right|$$

The SI units of angular acceleration are $\text{rad/s}^2$, which can also be converted to $\text{rev/s}^2$.

**Tangential displacement in circular motion**: The tangential displacement, $s_T$, differs from the arc length, $s$, in that it may be positive or negative, depending upon whether the object is displaced in a counterclockwise or clockwise sense, respectively; and also that an object may decrease $s_T$ by reversing direction, whereas $s$ is monotonically nondecreasing. The tangential displacement is related to the angular position by the 'arc length' formula:

$$s_T = R_0\theta \quad \text{(circular motion)}$$

Similarly, the net tangential displacement equals $\Delta s_T = R_0\Delta\theta$ in circular motion.

**Arc length in circular motion**: The arc length, $s$, equals the total distance traveled. The arc length is nonnegative and is a monotonically nondecreasing function of time. In general, $s \geq |s_T|$.

---

[195] In writing, it is often convenient to distinguish the lowercase Greek letter omega ($\omega$) from a double-u ($w$) by making it look like it's waving its left hand ($\omega$r); of course, its left hand appears on the right from our perspective.
[196] The lowercase Greek letter alpha ($\alpha$) is the first letter of the Greek alphabet.

**Tangential velocity in circular motion**: The tangential velocity, $v_T$, equals the derivative of tangential displacement, $s_T$, with respect to time:

$$v_T = \frac{ds_T}{dt}$$

In circular motion, since $s_T = R_0 \theta$ and $R_0$ is a constant, it follows that

$$|v_T| = R_0 \omega \quad \text{(circular motion)}$$

**Tangential speed in circular motion**: Since an object's velocity is inherently tangential, the object's speed (often called tangential speed in this context), $v$, equals the absolute value of the tangential component of the velocity: $v = |v_T|$. The tangential component of velocity can be positive or negative in order to distinguish between counterclockwise and clockwise motion, respectively, whereas speed cannot be negative.

**Tangential acceleration in circular motion**: The tangential acceleration, $a_T$, equals the derivative of tangential velocity, $v_T$, with respect to time:

$$a_T = \frac{dv_T}{dt}$$

In circular motion, since $v_T = R_0 \omega$ and $R_0$ is a constant, it follows that

$$a_T = R_0 \alpha \quad \text{(circular motion)}$$

Conceptually, the tangential component of acceleration, $a_T$, describes how the tangential velocity changes in time. Since the speed equals the absolute value of the tangential velocity, the tangential acceleration essentially describes (up to a possible minus sign) how the speed of an object changes.

**Centripetal acceleration in circular motion**: The acceleration also has a centripetal component, $a_c$. Conceptually, the centripetal acceleration describes how the direction of velocity changes in time. The word 'centripetal' means towards the center (or inward); we will understand why this adjective is used in Sec. 4.2, where we will also derive an equation for centripetal acceleration.

**Position vector in circular motion**: For motion in the $xy$ plane, the position vector can be expressed in 2D polar coordinates as $\vec{r} = r\hat{r}$: It has a magnitude $r$ and is directed from the origin to the instantaneous position of the particle – i.e. along $\hat{r}$. For circular motion with the origin at the center of the circle, this becomes $\vec{r} = R_0 \hat{r}$, since the particle is always the same distance $R_0$ from the origin.

**Velocity vector in circular motion**: The velocity, $\vec{v}$, can be found by differentiating the position vector with respect to time:

$$\vec{v} = \frac{d\vec{r}}{dt}$$

In circular motion, the velocity is tangential to the circle: $\vec{v} = v_T \hat{\theta}$.

**Acceleration vector in circular motion**: The acceleration, $\vec{a}$, can be found by differentiating the velocity with respect to time:

$$\vec{a} = \frac{d\vec{v}}{dt}$$

In circular motion, the acceleration has, in general, both tangential and centripetal components: $\vec{a} = a_T \hat{\theta} - a_c \hat{r}$. The minus sign represents that centripetal acceleration is inward, whereas $\hat{r}$ points outward. There is always a nonzero centripetal acceleration, $a_c$, in circular motion because the direction of the velocity is changing; the tangential component of acceleration, $a_T$, is also nonzero if the speed is changing.

---

**Important Distinction.** There are four distinct types of acceleration:[197] the acceleration vector, tangential acceleration, centripetal acceleration, and angular acceleration. The adjective, or lack thereof, is critical. By default, if we speak of the acceleration without an adjective, we mean either the acceleration vector, $\vec{a}$, or its magnitude, $a = \|\vec{a}\|$. The acceleration, $\vec{a}$, describes how the velocity changes, the tangential acceleration, $a_T$, essentially describes how the speed changes, the centripetal acceleration, $a_c$, describes how the direction changes, and the angular acceleration, $\alpha$, describes how the angular speed changes. It's important to know which type of acceleration is under consideration.

---

**Period**: The time it takes to complete one revolution is called the period, $T$. The SI unit of the period is the second (s).[198]

**Frequency**: The number of revolutions completed in a specified time interval is called the frequency, $f$. In SI units, the specified time interval is exactly one second, such that the SI units of frequency are rev/s, which is called a Hertz (Hz). That is, in SI units we measure frequency in Hz, where 1 Hz = 1 rev/s. Frequency and period share a reciprocal relationship:

$$f = \frac{1}{T}$$

**Angular frequency**: The angular frequency, $\omega_0$, is simply the frequency converted to radians per second:

---

[197] In addition, there is gravitational acceleration.
[198] We use $T$ for period much like we use $t$ for time because the period is a particular time. Physics is case-sensitive – i.e. the lowercase and uppercase 't' may not be interchanged, in general. The period refers specifically to the time it takes to go around exactly once, whereas the object may go just partly around or may complete more than a single revolution. Students who get these $t$'s mixed up make costly mistakes when solving problems.

# 4 Uniform Circular Motion

$$\omega_0 = 2\pi f = \frac{2\pi}{T}$$

The angular frequency, $\omega_0$, equals the angular speed, $\omega$, when $\omega$ is constant.

> **Note**. Period, frequency, and angular frequency are generally applied only to periodic motion – i.e. when the motion is repeated over even intervals of time. They apply to circular motion when the angular speed is constant. The other angular quantities apply even if the motion is not periodic.

> **Example**. A monkey spins a basketball on his fingertips with constant angular speed. The basketball completes 270 revolutions in one minute. What are the period and frequency of the motion?
> The frequency is 270 rev/min., which can be converted to SI units:
>
> $$f = 270 \frac{\text{rev}}{\text{min.}} \frac{1 \text{ min.}}{60 \text{ s}} = 4.5 \text{ Hz}$$
>
> The period is the reciprocal of the frequency:
>
> $$T = \frac{1}{f} = \frac{2}{9} \text{ s}$$

> **Example**. Determine the angular speed of the second hand of a clock. Which part of the second hand moves fastest?
> Since a second hand has constant angular speed, the angular speed is the same as the angular frequency. The period of the second hand is one minute:
>
> $$\omega = \frac{2\pi}{T} = \frac{2\pi}{60} = \frac{\pi}{30} \text{ rad/s}$$

## 4.2 Centripetal Acceleration

**Uniform circular motion**: For the remainder of this chapter, we will focus exclusively on uniform circular motion (UCM), which corresponds to an object that travels in a circle with constant speed. Although the speed is constant, the velocity is not because the object is constantly changing direction when it travels along a circle. Therefore, even though the speed is constant, the acceleration is not zero. The tangential acceleration and angular acceleration are zero, but the centripetal acceleration is not.

> **Note**. It's important to understand that the acceleration is not zero in UCM because the direction of velocity changes as an object travels in a circle. You definitely don't want to make the mistake of thinking that the acceleration is zero because the speed is constant.

**Position vector in UCM**: In UCM, it's convenient to setup a coordinate system such that the circle lies in the $xy$ plane with the origin at its center. The position vector, $\vec{r}$, is then given by $\vec{r} = R_0 \hat{r}$, where $R_0$ is the radius of the circle and the polar unit vector $\hat{r}$ points outward from the origin to the instantaneous position of the object that's traveling with UCM.

**Velocity vector in UCM**: The velocity, $\vec{v}$, is found by taking a derivative of the position vector, $\vec{r}$, with respect to time. The radius, $R_0$, is constant:

$$\vec{v} = \frac{d\vec{r}}{dt} = \frac{d(R_0 \hat{r})}{dt} = R_0 \frac{d\hat{r}}{dt} \quad \text{(UCM)}$$

Here, we need to take a derivative of the unit vector $\hat{r}$ with respect to time. We can do this by first expressing $\hat{r}$ in terms of the Cartesian unit vectors $\hat{i}$ and $\hat{j}$ (see Sec. 2.4):

$$\hat{r} = \hat{i} \cos \theta + \hat{j} \sin \theta \quad \text{(in 2D)}$$

We need to take a derivative of $\hat{r}$ with respect to time, but $\hat{r}$ is instead expressed in terms of $\theta$. Therefore, we apply the chain rule:[199]

$$\vec{v} = R_0 \frac{d\hat{r}}{dt} = R_0 \frac{d\hat{r}}{d\theta} \frac{d\theta}{dt} = R_0 \omega_0 (-\hat{i} \sin \theta + \hat{j} \cos \theta) \quad \text{(UCM)}$$
$$\vec{v} = R_0 \omega_0 \hat{\theta} = v_T \hat{\theta} \quad \text{(UCM)}$$

The speed, $v = |v_T|$, is constant and equal to

$$v = R_0 \omega_0 \quad \text{(UCM)}$$

and the velocity is tangential to the circle – i.e. along $\pm\hat{\theta}$.[200]

**Acceleration vector in UCM**: The acceleration, $\vec{a}$, is found by taking a derivative of the velocity, $\vec{v}$, with respect to time:

$$\vec{a} = \frac{d\vec{v}}{dt} = \frac{d(R_0 \omega_0 \hat{\theta})}{dt} = R_0 \omega_0 \frac{d\hat{\theta}}{dt} = R_0 \omega_0 \frac{d\hat{\theta}}{d\theta} \frac{d\theta}{dt} \quad \text{(UCM)}$$
$$\vec{a} = -R_0 \omega_0^2 \hat{r} = -\frac{v_T^2}{R_0} \hat{r} = -a_c \hat{r} \quad \text{(UCM)}$$

This vector describes how the direction of the velocity changes in time in UCM.

---

[199] Note that $\omega = d\theta/dt$; since $\omega$ is constant in UCM, it's the same as the angular frequency, $\omega_0$. Also note that $d\hat{r}/d\theta = \hat{\theta}$ in 2D polar coordinates, which is readily seen by inspecting the equations for these unit vectors in Sec. 2.4. Similarly, $d\hat{\theta}/d\theta = -\hat{r}$, which we will apply when we derive the equation for acceleration below.

[200] The magnitude of the velocity equals the speed, which is the absolute value of the tangential velocity: $v = |v_T|$. The direction is given by the sign of $v_T$ times $\hat{\theta}$ – i.e. motion may be clockwise or counterclockwise.

# 4 Uniform Circular Motion

**Magnitude of centripetal acceleration**: The magnitude of the acceleration vector equals the centripetal acceleration in UCM:

$$a_c = R_0 \omega_0^2 = \frac{v_T^2}{R_0} = \frac{v^2}{R_0} \quad \text{(UCM)}$$

The last step follows since the speed, $v$, equals the absolute value of the tangential velocity, $v = |v_T|$ (and the magnitude of the velocity $v = \|\vec{v}\|$).

> **Note**. The above formula is especially useful for solving UCM problems. There are many equations that apply to UCM, and different UCM problems require using different combinations of equations, but most solutions involve using the above formula. The above formula will also be involved in other contexts beyond this chapter. For example, when we apply conservation of energy to problems where an object travels in a circle, we will see that the formula above proves to be significant.

**Direction of centripetal acceleration**: The direction of the acceleration vector in UCM is $-\hat{r}$. Since $\hat{r}$ points outward, $-\hat{r}$ points inward. Therefore, the acceleration of an object traveling with UCM is inward (directed toward the center of the circle). The adjective 'centripetal' literally means center-seeking.

The direction of the acceleration in UCM can be understood conceptually, as follows. The object has inertia – a natural tendency to travel in a straight line with constant momentum. That is, the object wants to go off on a tangent according to its inertia, so there must be a force pulling it inward – called centripetal force, which we will describe more in the next section – which causes the object to change direction, and this centripetal force gives rise to centripetal acceleration through Newton's second law.

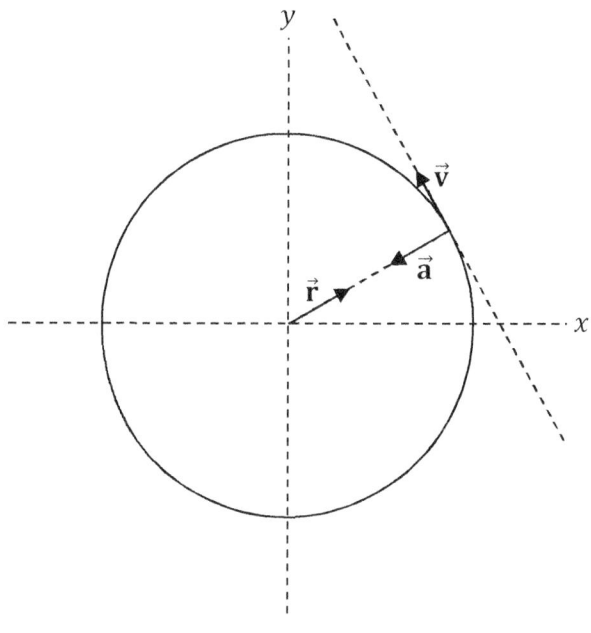

**Equations of UCM**: For an object traveling in a circle with constant speed, its acceleration is given by $a_c = v^2/R_0$. Its speed is related to its angular speed by $v = R_0\omega_0$. Since the speed is constant, it can also be expressed as the total distance traveled divided by the time interval, $v = \Delta s/\Delta t$. This must hold for any time interval, including the period, for which the total distance traveled is the circumference: $v = C/T = 2\pi R_0/T$. Similarly, the angular speed can be expressed as $\omega_0 = |\Delta\theta|/\Delta t$ or $\omega_0 = 2\pi/T$. In UCM, the angular speed is also equal to the angular frequency, which is related to the frequency by $\omega_0 = 2\pi f$. The frequency and period share an inverse relationship: $f = 1/T$. The total distance traveled is related to the angular displacement by $\Delta s = R_0|\Delta\theta|$.[201] It is often also useful to apply Newton's second law to UCM, which we will describe in the following sections. Any combination of these equations may be useful for solving UCM problems:

$$a_c = \frac{v^2}{R_0} \quad (\text{UCM})$$

$$v = \frac{\Delta s}{\Delta t} = \frac{2\pi R_0}{T} = R_0\omega_0 \quad , \quad \Delta s = R_0|\Delta\theta| \quad , \quad f = \frac{1}{T} \quad (\text{UCM})$$

$$\sum_{i=1}^{N} F_{i,in} = ma_c \quad , \quad \sum_{i=1}^{N} F_{i,T} = ma_T = 0 \quad , \quad \sum_{i=1}^{N} F_{i,z} = ma_z = 0 \quad (\text{UCM})$$

---

### Problem-Solving Strategy for Uniform Circular Motion

0. Apply this strategy to solve problems where an object travels in a circle with constant speed.
1. List the knowns and identify the desired unknown(s).
2. Determine whether or not it is necessary or useful to apply Newton's second law – i.e. determine whether or not it may be helpful to relate the forces to the acceleration (how to apply Newton's second law in the context of UCM is the topic of the remaining sections). Note that mass may eventually cancel out in the algebra, depending upon the situation and what the desired unknown is.
3. If you are applying Newton's second law, in the FBD, identify the center of the circle and choose one axis to point inward $(in)$. Next, identify the tangential direction $(T)$, not to be confused with tension. Lastly, identify the direction perpendicular to the plane of rotation $(z)$. Sum the inward, tangential, and $z$-components of the forces. Note that the inward component of acceleration is $a_c$, while the tangential and $z$-components of the acceleration are zero in UCM.
4. You will need to setup $N$ independent equations in $N$ unknowns, keeping in mind that sometimes an unknown mass may turn out to cancel in the algebra (meaning that it might seem like there are one too many unknowns, when actually there aren't). Solve the system algebraically.

---

**Hint**. Look at the units to help decipher which quantities are given in the problems. For example, the frequency, $f$, has SI units of Hz or rev/s, whereas the angular speed, $\omega_0$, has SI units of rad/s and the speed, $v$, has SI units of m/s. Both time, $\Delta t$, and period, $T$, share the same units, so it's important to determine whether or not any time given corresponds to precisely one revolution. It's similarly important to distinguish between the total distance traveled, $\Delta s$, and the circumference, $C$.

---

[201] In UCM, it's convenient to use $\Delta s$ and $v$ instead of $\Delta s_T$ and $v_T$ because the object doesn't change direction.

# 4 Uniform Circular Motion

> **Hint.** If you want to determine the number of revolutions completed, solve for $\Delta\theta$ and convert it from radians to revolutions. Note that it is possible to have $\Delta\theta > 2\pi$ or $\Delta\theta < 0$.

**Example.** A monkey runs in a circle with a diameter of 40 m with a constant speed of 5.0 m/s, completing 6.0 laps. Determine the monkey's acceleration, the monkey's angular speed, the period of revolution, the total distance traveled, and the total time that the monkey runs.

Note that the monkey completes 6.0 rev, which can be expressed as $12\pi$ rad. These questions can be solved simply by selecting the most efficient equation for each part:[202]

$$a_c = \frac{v^2}{R_0} = \frac{(5)^2}{20} = 1.25 \text{ m/s}^2$$

$$\omega_0 = \frac{v}{R_0} = \frac{5}{20} = 0.25 \text{ rad/s}$$

$$T = \frac{2\pi}{\omega_0} = \frac{2\pi}{0.25} = 8\pi \text{ s}$$

$$\Delta s = R_0 \Delta\theta = (20)(12\pi) = 240\pi \text{ m}$$

$$\Delta t = \frac{\Delta s}{v} = \frac{240\pi}{5} = 48\pi \text{ s}$$

## 4.3 Centripetal Forces

**Centripetal force**: In the previous section, we found that an object traveling with UCM experiences an acceleration toward the center of the circle. Therefore, from Newton's second law it follows that there must be one or more forces acting on the object that have an inward component. We call the net inward force a centripetal force. The centripetal force is not necessarily a single force, but is the sum of the inward components of the forces. There may not be any forces pointing straight inward, but there do need to be forces that have inward components so as to give a net force pointing toward the center.

**Centrifugal force**: Experience traveling in UCM intuitively suggests that the force should be outward rather than inward. The experience of feeling pushed outward when we travel with UCM is called centrifugal force – centrifugal meaning away from the center. Centrifugal force is considered to be a pseudoforce because from an inertial point of view UCM can be explained in terms of only centripetal forces and inertia. However, observers traveling with UCM feel as if they are being pushed outward, so this pseudoforce seems real to observers who are traveling in circles. The same pseudoforce can be explained without any outward forces by an observer who is not traveling in a circle. We will resolve this difference of perspective on the following page.

---

[202] If you setup your coordinate system (i.e. your choice of whether clockwise or counterclockwise is the positive direction) such that the object moves in the positive direction, then $\Delta\theta$ will be positive and the absolute value in $\Delta s = R_0 |\Delta\theta|$ will be superfluous. This will be the logical choice for now. We will explore the signs in circular motion in more detail in Chapter 7, where we will allow objects to change speed (i.e. non-uniform circular motion).

**Inertia**: An object traveling with UCM naturally wants to go off on a tangent according to its own inertia – i.e. it wants to travel in a straight line, not in a circle. Therefore, there must be a centripetal (not centrifugal) force acting on the object to cause it to continually change direction, rather than to fly off on a tangent.

Let us consider two different perspectives: A monkey traveling with UCM (relative to the ground), and chimpanzee who is not moving (relative to the ground). The monkey is an accelerating observer because he is changing direction, while the chimpanzee is an inertial observer because the chimpanzee is not changing his momentum.

The monkey feels like he is being pushed outward. This is a real perception that the monkey has, and the monkey actually feels a force that reinforces this perception. What the monkey really perceives results from his inertia: The monkey wants to fly off on a tangent. Because the monkey wants to fly off on a tangent, there must be a real centripetal force pushing the monkey inward. However, the monkey traveling with UCM perceives that he is being pushed outward. Remember Newton's third law: If there is a force pushing the monkey inward, the monkey pushes with an equal and opposite force outward.[203] The monkey perceives that he is pushing outward – in fact, if you remove the centripetal force, the monkey will suddenly go outward. This perception of wanting to outward when traveling in UCM gives the accelerating observer a perception of a centrifugal force.

The inertial observer – i.e. the chimpanzee – explains everything in terms of inertia and centripetal force. The monkey has inertia and wants to go off on a tangent, but a centripetal force pushes the monkey away from its natural tendency, causing the monkey to travel in a circle. If the centripetal force is removed, the monkey will travel in a straight line. See the difference? The chimpanzee says that the monkey wants to go off on a tangent, whereas the monkey feels this tangential tendency as an outward tendency. This is the difference in perspective between the accelerating and inertial observer. Both agree that the monkey won't travel in a circle if the centripetal force is removed, and so both must concede that there is a centripetal force acting in UCM.

**Important Distinction**. There is a real centripetal force acting on an object that travels in a circle (even if it's not UCM). The object traveling in the circle feels pushed outward, as if there were a centrifugal force, but from an inertial perspective the motion can be completely explained in terms of centripetal force and inertia.

**Conceptual Example**. A 40-kg monkey and a 120-kg monkey board an amusement park ride that spins in a circle. Who should sit on the outside?

The monkeys want to go off on a tangent according to their inertia. The outside of the car will push the monkeys inward, supplying the centripetal force. The outer monkey will be pushed inward by the car, and will push the inner monkey inward. By Newton's third law, the inner monkey will exert an equal and opposite force pushing the outer monkey outward. The outer monkey will therefore feel squeezed. Thus, the monkey with more mass – i.e. the 120-kg monkey – should sit on the outside, as he is better suited to handle the squeezing effect.

---

[203] When applying Newton's second law to the monkey, only the forces acting on the monkey are included in the diagram – the equal and opposite push outward is acting on something else (not the monkey), and therefore is not included in the sum of the forces acting on the monkey.

4 Uniform Circular Motion

> **Conceptual Example**. A monkey ties a banana to a string and whirls the banana in a horizontal circle.[204] In this case, the tension supplies the needed centripetal force.[205] The centripetal force is a real force that pulls the banana away from its natural tendency. This is easy to see when the monkey suddenly lets go of the string, in which case the banana instantaneously[206] goes off on a tangent.

**Sum of the inward components of the forces in UCM**: According to Newton's second law, for an object with constant mass, the net external force equals the object's mass times its acceleration: $\sum \vec{F}_{ext} = m\vec{a}$. In UCM, the acceleration is directed toward the center of the circle. Therefore, the sum of the inward components must equal mass times the centripetal acceleration:

$$\sum_{i=1}^{N} F_{i,in} = ma_c$$

The sum of the inward components is regarded as the centripetal force. That is, there is generally not a single force that is interpreted as 'the' centripetal force – it is, in general, a combined effect. There might not be any forces that are directed straight inward, but there will be at least one force with an inward component. Centripetal force is not a new kind of force, but the sum of the inward components of ordinary forces like tension, normal force, weight, etc.

> **Note**. Do not draw and label a centripetal force, $\vec{F}_c$, in your FBD. Rather, your FBD will consist of ordinary forces like tension, normal force, weight, friction, and other forces discussed in Chapter 3. At least one force will have an inward component, but there might not be any forces directed straight toward the center of the circle.

**Sum of the tangential components of the forces in UCM**: The tangential acceleration, $a_T$, is zero for an object traveling with UCM. Therefore, the sum of the tangential components of the forces acting on an object in UCM equals zero:

$$\sum_{i=1}^{N} F_{i,T} = ma_T = 0 \quad \text{(UCM)}$$

**Sum of the components of the forces perpendicular to the plane of rotation in UCM**: When an object travels with UCM, it does not move in the direction perpendicular to the plane of rotation – i.e. in the $z$-direction. Thus, the sum of the $z$-components of the forces acting on an object in UCM must also be zero:

---

[204] The banana travels in a horizontal circle, but the thread is not horizontal – rather, the thread sweeps out a cone, as we will see in the next section.
[205] The tension has an inward component, but does not point straight inward if the banana is in the air, for example, as we will learn in the next section. Thus, it would be a mistake to say that the tension *is* the centripetal force.
[206] The banana will start out heading along its tangent, but, of course, it will curve downward due to gravity's pull.

$$\sum_{i=1}^{N} F_{i,z} = ma_z = 0 \quad \text{(UCM)}$$

**Applying Newton's second law to UCM**: It is useful to apply Newton's second law to an object traveling with UCM when it is helpful to know how the forces are related to the centripetal acceleration (and therefore the speed) of the object. In order to apply Newton's second law in UCM, first draw a FBD. Draw and label the center ($C$) of the circle on the diagram. Draw and label a coordinate system with inward ($in$), tangential ($T$), and $z$-axes.[207] The inward axes points from the object's instantaneous position in the FBD toward the center of the circle, the tangential direction is the direction of the object's instantaneous direction in the FBD (which is tangent to the circle), and the $z$-direction is perpendicular to the plane of rotation.

**Note**. When you draw a FBD for an object in UCM, you should be thinking, "Where is the center of the circle?" This way, you can correctly draw and label the inward axis. It's necessary to identify the inward, tangential, and $z$-directions in order to be able to sum these respective components of the forces.

**Note**. A FBD for an object traveling with UCM often requires drawing a three-dimensional (3D) diagram. In this case, it is often convenient to draw the same diagram from two different perspectives in order to help visualize the forces acting on the object and what components they have.

**Note**. Examples of how to apply Newton's second law to UCM problems are provided in the remaining sections of this chapter.

## 4.4 Swinging in a Circle

**The basic swinging problem**: There are a variety of ways in which an object might swing in a circle. Let us choose the following situation as our prototype swinging problem: An object connected to a cord, cable, rope, or similar material is swinging in a horizontal circle in the air (but, of course, we will neglect the air unless directed otherwise). For example, a monkey may tie a banana to a string, grab the other end of the string, and whirl the banana in a horizontal circle by making small (assumed to be negligible) motions with his hand. As another example, a toy airplane with a motor and propeller may be tied to a string, the other end of the string may be fixed to the ceiling, and the airplane flies in a horizontal circle with constant speed (in the case of the airplane, however, we better not neglect the air). The important thing to realize is that in both examples the object travels in a horizontal circle, but the string is not horizontal. Rather, the string sweeps out a cone, as illustrated on the following page.

---

[207] It's important not to confuse the tangential direction, ($T$), with the magnitude of the tension force, $T = \|\vec{\mathbf{T}}\|$. The difference should be clear from the context: The tangential direction will appear as a subscript, as in $a_T$, and as the label of the tangential direction in a FBD.

## 4 Uniform Circular Motion

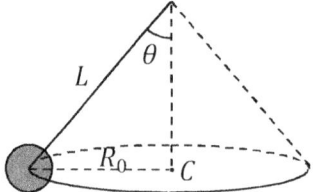

**FBD for the basic swinging problem**: There are two forces acting on the object in the basic swinging problem illustrated above – tension, $\vec{T}$, along the cable (or rope, cord, etc.) and weight, $m\vec{g}$, pulling downward. To setup our coordinate system, we first identify the center ($C$) of the circle. The inward ($in$) axis is then directed from the instantaneous position of the object to the center of the circle. In the side view of the diagram below, corresponding to the instantaneous position in the diagram above, the inward direction is instantaneously horizontal and to the right. The tangential ($T$) direction (not to be confused with the magnitude of the tension force) is tangent to the circle – it is instantaneously coming out of the page in the side view below. The third independent direction, $z$, is perpendicular to the plane of rotation, and is therefore vertical.

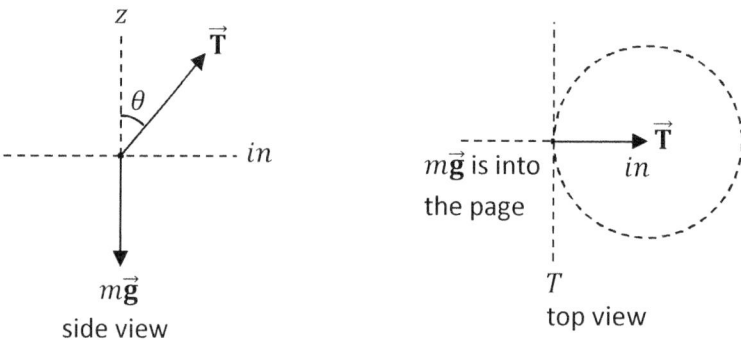

**Newton's second law applied to the basic swinging problem**: Set the sum of the inward components of the forces equal to $ma_c$ and the sums of the other two independent components equal to zero for UCM:

$$\sum_{i=1}^{2} F_{i,in} = ma_c \quad , \quad \sum_{i=1}^{2} F_{i,T} = 0 \quad , \quad \sum_{i=1}^{2} F_{i,z} = 0 \quad \text{(UCM)}$$

$T\sin\theta = ma_c \quad , \quad 0 = 0 \quad , \quad T\cos\theta - mg = 0 \quad \text{(basic swinging problem)}$

$T\sin\theta = ma_c \quad , \quad T\cos\theta = mg \quad \text{(basic swinging problem)}$

**Eliminating the tension**: The most efficient way to eliminate the tension in the above equations is to divide the two equations:[208]

$$\tan\theta = \frac{a_c}{g} \quad \text{(basic swinging problem)}$$

---

[208] If $a = b$ and $c = d$, it follows that $\frac{a}{c} = \frac{b}{d}$, since the numerators and denominators are both equal.

Since $a_c = v^2/R_0$, the equation for the angle that the cord (or rope, cable, etc.) makes with the vertical can be expressed as

$$\tan\theta = \frac{v^2}{gR_0} \quad \text{(basic swinging problem)}$$

This result should make sense conceptually: The faster the object spins, the greater the angle that the cord will make with the vertical.

> **Note**. If you memorize the above equation, but don't learn the strategy, you won't be able to solve similar problems that involve different FBD's. Therefore, it's much better to learn the problem-solving strategy than to memorize derived equations. Some instructors will only give full credit if you begin your solutions from fundamental starting equations.

**Eliminating the angle**: It's also possible to solve for the magnitude of the tension force, $T$, without knowing $\theta$ and without finding $\theta$ first. The most direct way to do this is to square the two equations that resulted from the sums of the components of the forces and add them together:

$$T^2 \sin^2\theta = m^2 a_c^2 \quad , \quad T^2 \cos^2\theta = m^2 g^2 \quad \text{(basic swinging problem)}$$
$$T^2 \sin^2\theta + T^2 \cos^2\theta = m^2(a_c^2 + g^2) \quad \text{(basic swinging problem)}$$
$$T^2 = m^2(a_c^2 + g^2) \quad \text{(basic swinging problem)}$$

Recall from trig that $\sin^2\theta + \cos^2\theta = 1$.

**Variations of the basic swinging problem**: There are several different ways that the basic swinging problem can be varied. These variations have different FBD's, but the strategy for drawing the FBD and applying Newton's second law is similar. Here are some common variations of the swinging problem:
- An object that is on horizontal frictionless ground is connected to a cord and swung in a horizontal circle as the object travels in a circle on the ground. The cord may make an angle with the ground, but in this situation it's also possible to have a horizontal cord.[209] The FBD for this problem includes a normal force.
- A toy airplane that is connected to the ceiling by a string can travel in a horizontal circle. This is very much like the basic swinging problem, except for the addition of two tangential forces – a force of propulsion, $\vec{F}_p$, and a force of air resistance, $\vec{F}_a$, which cancel out. Thus, the mathematics is identical to that of the basic swinging problem.
- A rigid material – such as a metal rod – may be used instead of a cord, cable, rope, etc. In this case, it is inappropriate to call the force a tension force, so use a different symbol, such as $\vec{F}_R$, to represent the force. Also, if the material is rigid, $\vec{F}_R$ may not have the same direction that a tension force would have along a cord, cable, rope, etc. We will see this in following examples.

---

[209] In this case, the ground provides a vertical support force to compensate for gravity.

- If a rigid material – such as a metal rod – is used, the object need not travel in a horizontal circle. If the material is like a cord, cable, or rope, the object will not travel with constant speed – i.e. with UCM – unless the object connected to it travels in a horizontal circle. However, if the connecting material is rigid, the person or motor can force the object to travel with UCM regardless of the orientation of the circle. We will also consider this in following examples.
- A Ferris Wheel is an amusement park ride built with rigid materials where passengers travel in vertical circles. A significant portion of the ride typically occurs with constant speed – hence with UCM. Passengers sit on seats on a Ferris Wheel ride, and experience a net force, $\vec{F}_s$, from the seat cushions, safety belt, and floor of the passenger car. This is not merely a normal force from the seat back or the seat bottom, but a combined effect from multiple surfaces. Therefore, the direction of $\vec{F}_s$ cannot be deduced by inspecting the geometry of the seats. However, the approximate direction of $\vec{F}_s$ can be determined, conceptually, by thinking about what direction it must point toward in order for there to be a net inward force and also to make the sum of the tangential and the sum of the $z$-components of the forces equal zero; the precise direction can be found from the math. We will consider a Ferris Wheel in a coming example.
- There is an amusement park ride where passengers travel in neither horizontal nor vertical circles, but in a circle for which the plane of rotation makes an angle $\theta$ with the vertical. Similar to the Ferris Wheel problem, the approximate direction of $\vec{F}_s$ can be determined, conceptually, by thinking about what direction it must point toward in order for there to be a net inward force and also to make the sum of the tangential and the sum of the $z$-components of the forces equal zero. This variation is considered in one of the problems at the end of the chapter.
- A centrifuge is a "gravity-defying" amusement park ride where people stand against the wall of a cylindrical room, the room begins to spin, and the floor is pulled away. The wall exerts a normal force on the people, supplying the centripetal force, which results in an upward force of static friction to balance gravity's downward pull. As long as the angular speed and coefficient of static friction are sufficient,[210] the passengers will feel like they are defying gravity. A similar idea can be applied to the design of a space station – such that the centrifugal force that inhabitants of the space station would perceive would closely mimic gravity. The size and angular speed of the space station determine the centripetal acceleration at the outer walls. This is useful for creating an "artificial gravitational field" in a region of space where gravity is weak; or in a free-fall orbit where gravity may be significant, but passengers would otherwise feel weightless. Centrifuge problems will be included with the end-of-chapter exercises.
- A person can grab the handle of a bucket full of water and whirl the bucket in a circle. This is similar to the amusement park ride except that the force exerted by the bucket on the water, $\vec{F}_b$, may be of interest – especially, if the circle is vertical and you are interested in whether the water will fall out. Similar to $\vec{F}_s$, the net force exerted on the water, $\vec{F}_b$, by the bucket includes a combination of normal forces – from the sides as well as the bottom (i.e. the side that's usually considered to be the bottom) – and therefore $\vec{F}_b$ may not be perpendicular to the bottom of the pail. As with $\vec{F}_s$, the approximate direction of $\vec{F}_b$ can be reasoned conceptually and computed mathematically. We will consider a whirling buck in the next example.

---

[210] Sometimes people get drenched from a water ride and then board the centrifuge, and find themselves slipping due to reduced friction.

**Conceptual Example.** A monkey grabs a bucket of water and whirls the bucket with constant speed in a vertical circle. How is it possible that the water might not fall out when the bucket is up-side-down?

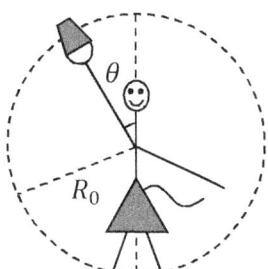

It is instructive to consider the FBD for the water when the bucket is in its topmost position. Two forces act on the water – its weight, $m\vec{g}$, directed downwards, and the force that the bucket exerts on the water, $\vec{F}_b$. In the topmost position, $\vec{F}_b$ must be directed downward – it doesn't have a tangential component, for example, since the sum of the tangential components of the forces must be zero and no other forces have a tangential component that needs to be balanced (in this position only).

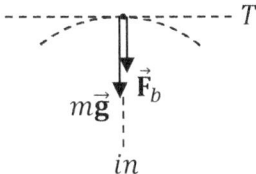

The sum of the inward components of the forces is

$$\sum_{i=1}^{2} F_{i,in} = ma_c$$

$$F_b + mg = \frac{mv^2}{R_0}$$

What determines if the water falls out of the bucket? The answer is that the water remains in the bucket as long as $F_b > 0$ – i.e. if the bucket is not exerting a force on the water, the water will fall out. This may seem ironic at first: The water doesn't fall out of the bucket as long as the bucket pushes the water downward! Really, this means that the bucket is moving fast enough tangentially that the water experiences a 'centrifugal force.' From an inertial perspective, the bucket is exerting a downward force on the water to prevent it from rising upward (due to the inertia it had from the upward part of the swing) – this is the real centripetal force in the problem. If the bucket is moving too slowly, the water won't reach the surface of the bucket's base when the bucket reaches its topmost position, and in this case the monkey will get wet.[211] The critical speed, $v_c > R_0 g$ is found by setting $F_b < 0$.

---

[211] It's kind of like projectile motion – enough vertical speed is needed to reach the maximum height of the base.

**Example.** An 80-kg chimpanzee sits 30 m from the center of a Ferris wheel that has a period of 20 s. What are the magnitude and direction of the force exerted on the chimpanzee by the Ferris wheel when the chimpanzee is halfway up?

Follow the UCM problem-solving strategy. First, identify the knowns and desired unknown:

$$T = 20 \text{ s} \quad , \quad R_0 = 30 \text{ m} \quad , \quad m = 80 \text{ kg}$$

The forces acting on the chimpanzee include the chimpanzee's weight, $m\vec{g}$, and the force exerted by the Ferris wheel, $\vec{F}_s$. The support force $\vec{F}_s$ is a combined effect exerted by the seat back, seat bottom, safety restraints, and related contact forces. From the question, we are interested in the position where the chimpanzee is halfway up. Since $m\vec{g}$ points downward, $\vec{F}_s$ must have both an upward and an inward component, as illustrated below: The inward component of $\vec{F}_s$ supplies the needed centripetal force, while the upward component of $\vec{F}_s$ balances $m\vec{g}$ so that the sum of the tangential components of the forces will be zero (so that the Ferris wheel travels with UCM).

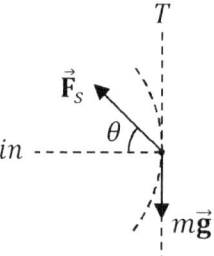

Applying Newton's second law, the inward component of acceleration is $a_c$ and the tangential component of acceleration is zero, $a_T = 0$:

$$\sum_{i=1}^{2} F_{i,in} = ma_c \quad , \quad \sum_{i=1}^{2} F_{i,T} = 0$$
$$F_s \cos\theta = ma_c \quad , \quad F_s \sin\theta = mg$$

From the period, $T$, we can find $\omega_0$, from which we can find $a_c$:

$$\omega_0 = \frac{2\pi}{T} = \frac{2\pi}{20} = 0.31 \text{ rad/s}$$
$$a_c = R_0 \omega_0^2 = (30)\left(\frac{\pi}{10}\right)^2 = 3.0 \text{ m/s}^2$$

The angle that $\vec{F}_s$ makes with the horizontal can be obtained efficiently through division:

$$\tan\theta = \frac{g}{a_c}$$

$$\theta = \tan^{-1}\left(\frac{g}{a_c}\right) = \tan^{-1}\left(\frac{9.81}{3.0}\right) = 73°$$

The magnitude of $\vec{F}_s$ is then found to be:

$$F_s = \frac{mg}{\sin\theta} = \frac{(80)(9.81)}{\sin(73°)} = 820 \text{ N}$$

**Note**. The support force, $\vec{F}_s$, that the Ferris wheel exerts on a passenger is a combined effect from multiple forces, including the seat bottom, seat back, safety belt, and similar contact forces. Since it is not exerted by a single surface, it does not have an obvious direction like a single normal force would. However, the approximate direction of $\vec{F}_s$ can be reasoned conceptually in UCM by considering Newton's second law – i.e. the sum of the force vectors must have an inward direction, and the components of the tangential and $z$-directions must each add up to zero. The precise direction of $\vec{F}_s$ can be determined mathematically by applying the problem-solving strategy for UCM.

## 4.5 Rounding a Turn

**Driving a car**: We all have experience with circular motion when we round a turn. In this section, we will explore the physics involved in rounding a turn, and see how it impacts icy road conditions. The physics of driving a car is somewhat more complicated than many of the UCM problems that we consider in first-year physics. For example, a car has wheels that roll rather than slide. For another, we know that resistive forces are not really negligible, since we must frequently pump fuel into our cars (or, if you have an electric car, you must frequently charge its battery). Let us consider a few of these details briefly before we attack the problem of rounding a turn.

**Rolling wheels**: Wheels introduce two complications: Wheels have rotational inertia in addition to the usual (linear) inertia, and friction doesn't have the same effect on rolling objects as it does on sliding objects. When a rigid object (i.e. one that won't fly apart or change shape during rotation) is rotated, it has a natural tendency to maintain constant angular momentum – analogous to an object's inertia, which is a natural tendency to maintain constant momentum. This additional type of inertia is important when we are concerned with problems that involve angular momentum changes (in part, if not in whole), and so won't impact our discussions of UCM. We will return to the concepts of rotational inertia and angular momentum in Chapter 7, where they will be very much relevant.

Unlike rotational inertia, friction plays a key role in rounding a turn. Let us first consider the effect that friction has on a wheel that rolls in a straight line, and later we will return to the issue of turning. If you place a quarter on a smooth, level concrete floor so that it faces heads or tails up, and give the quarter a flick with your finger, it won't slide too far because friction has a significant effect on its sliding. However, if you stand the quarter on its edge with one hand, and let go as you flick it with the other, it will roll a very long distance (if reasonably balanced) in comparison.

In the case of sliding, friction solely acts to overcome the quarter's inertia. Friction supplies the net external force to decelerate the quarter, quickly bringing it to a halt. In the case of rolling, however, when friction acts against the direction of motion, friction actually results in a torque in the forward direction. Thus, friction does not actually act to decelerate the quarter (nor does it act to accelerate it). Under the most idealized conditions, a quarter could roll without slipping on a horizontal surface indefinitely. We will explore this much more fully, including a mathematical treatment, in Chapter 7, and resolve some common questions that you may have about this at that time. In the meantime, it's important to realize that friction does not have the same effect on rolling objects as it has on sliding objects, which is important for a car that has wheels.[212]

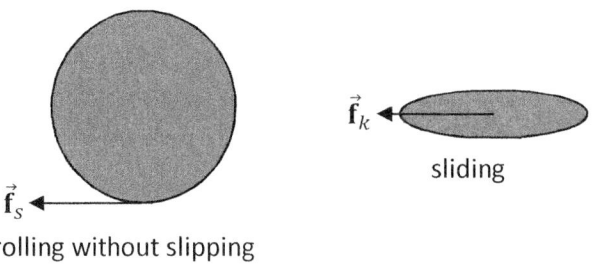

rolling without slipping

sliding

It's also worth noting that the force of kinetic friction, $\vec{f}_k$, acts on an object while it slides, but that the force of static friction, $\vec{f}_s$, acts on an object when it rolls without slipping. When an object slides across the ground, the same surface is always touching the ground, but when an object rolls along the ground, a different point is always touching the ground. In terms of energy, which we will explore in the next chapter, the force of kinetic friction subtracts mechanical energy from the object and converts it into heat, whereas the force of static friction does not. The force of static friction, and not kinetic friction, acts in the most idealized rolling.

**Resistive forces**: A real quarter, wheel, or other rolling object does not experience the most theoretically ideal rolling, but eventually comes to a stop on horizontal ground due to resistive forces. The two most obvious resistive forces include friction and air resistance. As we have seen, friction does not have the same resistive effect on a rolling object as it has on a sliding object. For our discussion of rounding a turn, it will not matter whether the resistive force is caused by friction, air resistance, or a combination of these, so we will speak of a general resistive force, $\vec{f}_R$, which opposes the velocity of the center of mass of the object.[213]

---

[212] The case of a quarter rolling without slipping along a horizontal surface, where no other forces act on the quarter besides its weight, normal force, and static friction, is a bit counterintuitive. On the horizontal, friction acts in a direction of a decelerating force, but also in the direction an accelerating torque – yet the quarter is neither decelerating nor accelerating along the horizontal with idealized rolling. Recall that static friction involves an inequality – i.e. its magnitude can be less than $\mu N$ (as little as zero!). Zero friction forces resolves this seeming paradox, until you consider that friction is needed to make the quarter roll rather than slide: If you remove friction by greasing up the cement, the quarter won't have enough traction to get a grip and roll without slipping. In perfect horizontal rolling, friction is needed just to initiate the rolling, then friction equals zero.
[213] Observe that different parts of a rolling object have different speeds (but the same angular speed).

**Drive force**: A car obviously experiences resistive forces: If you're driving along a level road and suddenly put the car in neutral, it will definitely lose speed. Thus, in order to be able to drive with constant speed, as in the case of UCM, a car must have a drive force. Usually, the engine burns fuel, converting thermal energy to do useful work (with what is termed a 'heat engine' in thermodynamics), but you could eject steam like rocket or sail with the wind like a ship, so we will speak on general terms of some vague drive force, $\vec{F}_D$. For our present discussions, the drive force serves to allow us to consider UCM, for without it the car would lose speed.

It is interesting to note that when the drive force is applied to gain speed, the force of static friction between the tires and the ground supplies the torque that accelerates the car. This is a clear conceptual example where static friction does not play the usual resistive role that kinetic friction does.

**Driving in a circle on horizontal ground with constant speed**: Like anything else, a car has a natural tendency to travel in a straight line with constant momentum. So if you are driving a car and wish to make a turn, a net outside force must be exerted on the car in order cause the change in direction. If you are driving in a horizontal circle with constant speed, the net external force must be directed toward the center of the circle.[214] Ponder the following question for a moment before you read on.

> **Conceptual Question**. When you round a turn, what force causes the car to change direction?

The answer is related to the physics of walking: You walk by pushing the ground (or floor) one direction, receiving a net external force through Newton's third law. If you pour a can of oil on cement and then try to walk on it, you will find that walking is much more difficult when friction is removed. That is, you need to be able to exert a force on the ground in order for the ground to push you forward.

A car similarly changes direction when the driver turns the steering wheel. When you rotate the steering wheel to the left, the tires rotate left, and as they do so the tires exert a force on the ground, pushing the ground to the right. According to Newton's third law, the ground exerts a force equal in magnitude, but pushing the car to the left. The car thus turns left. As the car rounds the turn to the left, the tires continually push the ground to the right (outward), and the ground continually pushes the car to the left (inward). The ground supplies the centripetal force that changes the car's direction.

The mutual force exerted between the tires and the ground is a force of static friction. The tires are not sliding inward or outward, but are rolling tangentially. The ground pushes the tires inward, causing the car to change direction, and the tires roll tangentially along a circular path (in the case of UCM). This inward force of static friction does not subtract mechanical energy from the system and convert it into heat (as would be the case with kinetic friction acting on a sliding object).

The force of static friction has a magnitude that is less than or equal to $\mu N$, and gives rise to a centripetal acceleration, where $a_c = v^2/R_0$. Conceptually, we can see that if the car travels too fast or if there is not enough friction, the static friction force will not be large enough to round the turn safely.

---

[214] The net external force must also have a centripetal component in order to cause the car to change direction even if the speed is changing. However, although the net external force will have an inward component in the case of rounding a turn while changing speed, it will also have a net tangential component (since tangential acceleration is required to change an object's speed), and so the net external force will not point straight to the center of the circle. In UCM, though, which is the subject of this chapter, objects travel with constant speed, and so only have centripetal acceleration.

**Example.** A monkey drives a car in a circle on horizontal ground with constant speed. Derive an inequality that relates the coefficient of static friction to the radius of the circle and the speed of the car.

Let us begin by drawing a FBD. Weight, $m\vec{g}$, pulls the car downward, while the ground pushes upward with a normal force, $\vec{N}$; the drive force, $\vec{F}_D$, pushes tangentially forward, while the resistive force, $\vec{F}_R$, pulls tangentially backward; and the force of static friction, $\vec{f}_s$, pushes the car inward.[215]

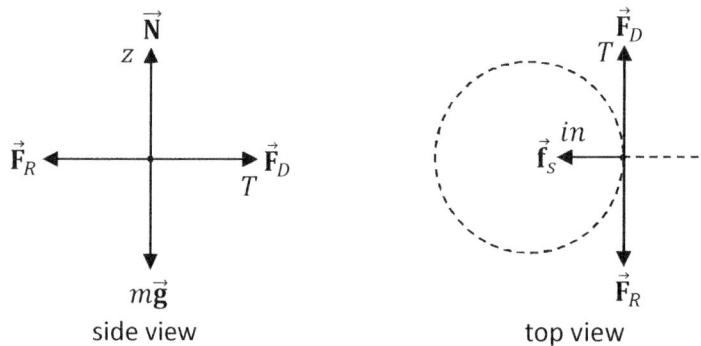

side view        top view

Applying Newton's second law to this problem, we find that:

$$\sum_{i=1}^{5} F_{i,in} = ma_c \quad , \quad \sum_{i=1}^{5} F_{i,T} = 0 \quad , \quad \sum_{i=1}^{5} F_{i,z} = 0$$
$$f_s = ma_c \quad , \quad F_D - F_R = 0 \quad , \quad N - mg = 0$$

The first and last equations can be combined according to the inequality, $f_s \leq \mu N$:

$$ma_c \leq \mu m g$$
$$v^2 \leq \mu R_0 g$$

where we applied the equation for centripetal acceleration in the final step. This result should make sense conceptually: If you drive too fast, if the curve is too sharp (i.e. $R_0$ is too small), or if there is not enough friction between the tires and ground (as on an icy or oily road), the force of static friction won't be able to supply as much force as the car needs to round the turn. As with many physics problems, the mass of the car (and driver and other contents) cancels out: A heavier car does have a greater normal force and thus a proportionately larger friction force, but more mass also means more inertia to overcome to change the car's direction, and these two effects compensate one another perfectly.

---

[215] Technically, there are two static friction forces – one directed inward that changes the direction of the car, and one directed backward associated with the rolling of the tires. In order to try to avoid confusion, we have lumped this second static friction force together with the resistive forces in $\vec{F}_R$, even though, as we have discussed, static friction does not have the role of a decelerating resistive force in the case of idealized rolling. This will not be an issue, mathematically or conceptually, in the current chapter; it's the inward force that's significant here.

**A banked curve**: Racetracks often feature banked curves as the normal force contributes to the centripetal force – i.e. to the sum of the inward components of the forces – in this case. In this way, a banked curve thus allows cars to round the turn at a greater speed. A banked curve also allows a car to round a turn on nearly frictionless road conditions. The following example considers a banked curve without friction, while a banked curve with friction is included in the end-of-chapter exercises.

**Banking angle**: The angle that a banked curve makes with the horizontal is called the banking angle.

**Example.** A circular racetrack features a constant banking angle, $\theta$ – i.e. the entire racetrack is a banked curve such that the racetrack looks like a slice of a cone. A monkey drives around the racetrack with constant speed, traveling in a horizontal circle (even though the road itself is sloped). The car is instantaneously heading out of the page in the illustration below. Assuming that the road is perfectly frictionless, derive an equation that relates the speed of the car to the banking angle and radius, $R_0$.

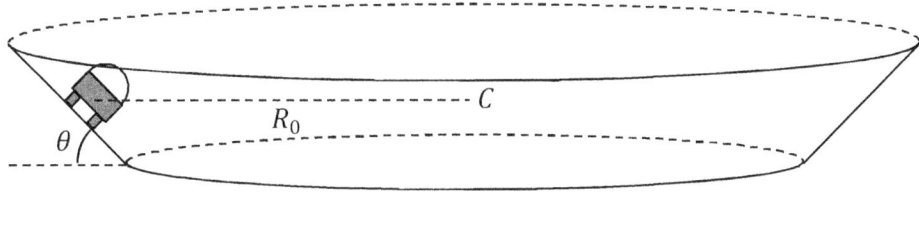

side view

First, draw and label a FBD for the car. The only forces acting on the car are its weight, $m\vec{g}$, pulling downward, and normal force, $\vec{N}$, perpendicular to the slope of the road.[216]

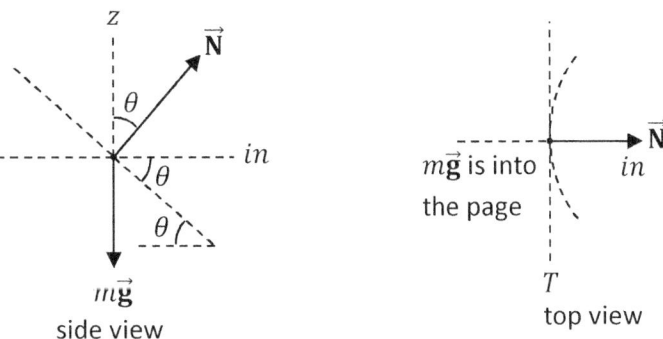

side view　　　　　　　　　　top view

Applying Newton's second law to the car,

---

[216] Since the road is assumed to be perfectly frictionless, the engine won't be able to accelerate the wheels (and therefore the car, too) through traction. As long as we're ignoring friction, we may as well ignore drive force and air resistance, too. In this case, the tangential forces – which would cancel one another anyway – won't appear in the FBD or sums of the components of the forces.

$$\sum_{i=1}^{2} F_{i,in} = ma_c \quad , \quad \sum_{i=1}^{2} F_{i,T} = 0 \quad , \quad \sum_{i=1}^{2} F_{i,z} = 0$$

$$N \sin\theta = ma_c \quad , \quad N\cos\theta - mg = 0$$

$$N \sin\theta = \frac{mv^2}{R_0} \quad , \quad N\cos\theta = mg$$

$$\tan\theta = \frac{v^2}{gR_0}$$

## 4.6 Satellite Motion

**Satellite**: An object that orbits another object is termed a satellite. For example, moons are satellites of planets, and planets, comets, and asteroids are satellites of stars. A satellite can also be a man-made object, such as numerous satellites that orbit the earth in order to aid communication.

**Satellites vs. projectiles**: Recall that a projectile is an object that travels through the air or space. A projectile launched from earth becomes a satellite if it has enough speed to go into orbit. It is easy to visualize the similarity between satellites and projectiles – as Galileo originally did – by imagining launching projectiles horizontally from the top of a very tall mountain. If the projectile is given enough initial speed, it could travel all the way around in a circle, making a complete orbit – and hence become a satellite.

**Satellite in a circular orbit**: Since we are considering UCM in this chapter, let us consider a satellite that travels in a circular orbit (and hence has constant speed). Let us also neglect the gravitational attraction that the satellite has to other celestial objects, except, of course, for the celestial body that it orbits (otherwise, the orbit would deviate from a perfect circle). In this case, the only force acting on the satellite is its gravitational attraction to the celestial body that it orbits.

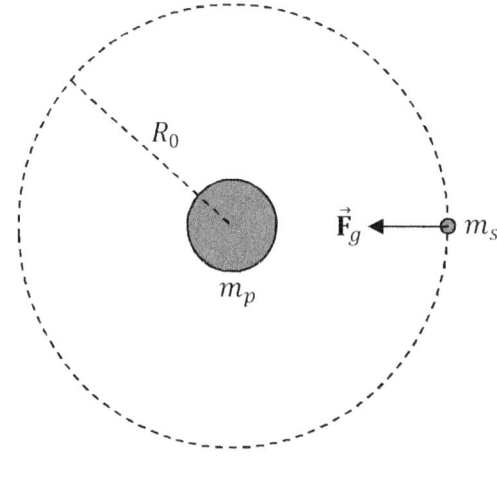

Applying Newton's second law to a satellite in UCM,[217]

$$\sum_{i=1}^{1} F_{i,in} = m_s a_c$$

$$\frac{Gm_p m_s}{R_0^2} = m_s a_c \quad \text{(satellite in UCM)}$$

$$\frac{Gm_p}{R_0^2} = \frac{v^2}{R_0} \quad \text{(satellite in UCM)}$$

$$v = \sqrt{\frac{Gm_p}{R_0}} \quad \text{(satellite in UCM)}$$

where we applied the formula for centripetal acceleration, $a_c = v^2/R_0$, in the third line.

**Important Distinction.** The mass of the satellite, $m_s$, cancels out in the two-body problem; only the mass of the celestial body that the satellite orbits, $m_p$, affects its motion.

**Important Distinction.** Note that $R_0$ is the radius of the satellite's orbit, not to be confused with the radius of the planet, $R_p$.[218]

**Important Distinction.** The centripetal acceleration, $a_c$, of the satellite equals the gravitational acceleration, $g$, of the satellite in UCM. However, it's important to realize that this is the gravitational acceleration at the location of the satellite, and not at the surface of the planet.

**Note.** It's better to learn and understand the strategy for solving satellite problems than to memorize the derived equations. Some instructors will not permit the derived satellite equations, like the final equation above, to be used as starting equations, but will instead expect you to be able to repeat the derivation. The satellite strategy for UCM is the same as the strategy for UCM in general, where Newton's law of universal gravitation supplies the centripetal force.

**Conceptual Example.** Since the moon and earth both attract one another, why doesn't the moon crash into the earth? Similarly, why doesn't the earth crash into the sun?

---

[217] The satellite's mass, $m_s$, appears on the right-hand side of Newton's second law because we are summing the forces acting on the satellite.

[218] Technically, $R_0$ extends from the center of mass (called the barycenter in this context) of the two-body system to the satellite, and so is shorter than the separation between the centers of the two bodies. However, this difference is negligible if the planet is much more massive than the satellite (i.e. if $m_p \gg m_s$). Otherwise, if $m_s$ is significant compared to $m_p$, it is necessary to first determine the location of the center of mass of the two objects (see Chapter 6) in order to determine $R_0$.

## 4 Uniform Circular Motion

The moon, planets, asteroids, and other celestial bodies have inertia – a natural tendency to maintain constant momentum. As such, the natural path of any such body is a straight line. If the sun's gravitational field suddenly vanished, its planets would go off on tangents.[219] For a planet (or moon) that travels in a nearly circular orbit, the gravitational pull is mostly centripetal – i.e. it has the effect of changing the planet's (or moon's) direction; the speed only changes if the orbital distance varies, as with an elliptical orbit. Recall the similarity between a projectile and a satellite: The object will complete an orbit around the central body as long as it has enough tangential speed. If the moon suddenly had less tangential speed, it would travel along a much more elliptical orbit, and if it suddenly lost enough tangential speed, it would crash into the earth (but that's one thing you don't have to worry about – since the moon has inertia, it's not going to run out of tangential speed). Similarly, the earth is not in danger of crashing into the sun because it has plenty of tangential speed for its average orbital radius.

**Example**. Two satellites are each traveling with UCM around the same planet. One satellite's orbital radius is twice that of the other: $R_{20} = 2R_{10}$. What is the ratio of their speeds?

Recall that we had previously derived the formula,

$$v = \sqrt{\frac{Gm_p}{R_0}}$$

The ratio of the satellites' speeds can be found by writing the above equation for each satellite using subscripts for any symbols that are different (namely, $v$ and $R_0$), and dividing the two equations:

$$\frac{v_2}{v_1} = \sqrt{\frac{R_{10}}{R_{20}}} = \frac{1}{\sqrt{2}}$$

That is, the satellite with the smaller orbit (corresponding to $R_{10}$) has the greater speed ($v_1$) by a factor of $\sqrt{2}$.

**Kepler's third law**: According to Kepler's third law of planetary motion, the square of the orbital period of each planet is directly proportional to the cube of the semi-major axis of its elliptical orbit:

$$T^2 \propto a^3 \quad \text{(Kepler's third law applied to UCM)}$$

The semi-major axis, $a$, is one-half the length of the major axis of the ellipse (see the figure below).

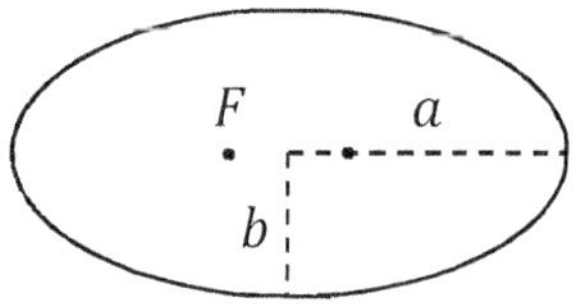

$F$ = focus
$a$ = semi-major axis
$b$ = semi-minor axis

---

[219] They wouldn't all travel in perfectly straight lines, as they would interact with one another gravitationally.

Kepler's third law applies to satellites, in general, which travel in closed (namely, elliptical or circular) orbits.[220] The constant of proportionality depends upon the mass of the central body, so you may only use Kepler's third law to compare orbits of two satellites orbiting the same central body. That is, you can't use it to compare the period of one of Jupiter's moons to the period of one of Saturn's moons, for example.

> **Notation**. The proportionality symbol ($\propto$) is used to mathematically state that the quantities on each side of it are proportional to one another.[221]

**Kepler's third law for a satellite in UCM**: In the limit that an ellipse becomes more and more circular, both the semi-major ($a$) and semi-minor ($b$) axes of the ellipse become equal, approaching the radius ($R_0$) of the resulting circle. Therefore, for a satellite in UCM,

$$T^2 \propto R_0^3 \quad \text{(Kepler's third law)}$$

This equation can be derived by applying the UCM strategy to the satellite, using the fact that the speed of a satellite in UCM equals its circumference divided by its period, $v = 2\pi R_0/T$:

$$\sum_{i=1}^{1} F_{i,in} = m_s a_c$$

$$\frac{G m_p m_s}{R_0^2} = m_s a_c \quad \text{(satellite in UCM)}$$

$$\frac{G m_p}{R_0^2} = \frac{v^2}{R_0} = \frac{4\pi^2 R_0}{T^2} \quad \text{(satellite in UCM)}$$

$$T^2 = \frac{4\pi^2 R_0^3}{G m_p} \quad \text{(satellite in UCM)}$$

In doing so, we discovered that the proportionality constant equals $4\pi^2/(Gm_p)$.

**Geosynchronous satellite**: A particularly practical satellite is one that is geosynchronous – i.e. synchronized with the earth's rotation. A geosynchronous satellite therefore has a period of 24 hours. A geosynchronous satellite traveling above earth's equator would appear stationary in the sky relative to observers on earth. If satellite TV companies didn't use geosynchronous satellites, you'd have to constantly go up to the roof to reposition your receiver.

> **Hint**. If you see the word 'geosynchronous' in a problem, you know that the period of the satellite is 24 hours.

---

[220] The orbits can only be perfect ellipses or circles if the only gravitational force acting on the satellite is exerted by the central body. Thus, real orbits may be approximate ellipses or circles, and feature wobbles to some degree.
[221] Although the proportionality symbol ($\propto$) looks like a curly alpha ($\alpha$), those who call the proportionality symbol alpha do so incorrectly. Like it or not, the official name for it is the proportionality symbol.

**Weightlessness**: A satellite is in free-fall in its natural circular or elliptical orbit. Therefore, a passenger inside such a satellite experiences a feeling of weightlessness, and any objects not fastened to the satellite or held by a passenger would float relative to the satellite. Of course, the satellite, passengers, and contents do have weight. Indeed, in our derivation of the formulas for a satellite with UCM we began by using the satellite's weight for the centripetal force. Rather, the feeling of weightlessness has to do with the satellite, passengers, and contents all accelerating at the rate of gravitational acceleration (since the centripetal acceleration of the satellite equals gravitational acceleration at the location of the satellite). Thus, no contents in the satellite fall relative to the satellite when released from rest relative to the satellite (but the contents and satellite are all falling relative to the central body); and the walls of the satellite do not exert a normal force to supply a feeling of weight. It is very much like riding in an elevator car with a cut cable. See Sec. 3.3 where the notion of apparent weight was first introduced and Sec. 3.9 on gravitational forces for more information on weightlessness.

> **Note**. The case of a satellite with an elliptical orbit will be discussed in the next chapter, in the context of gravitational potential energy where the change in altitude is significant (and not the $mgh$ formula for gravitational potential energy that applies only to uniform gravitational fields).

## Conceptual Questions

1. In contrast to the role that the angle $\theta$ plays in 2D polar coordinates, in circular motion and rotation we allow $\theta$ to be negative or exceed $2\pi$, and you cannot add or subtract multiples of $2\pi$ to $\theta$. For example, in UCM, the angles $\pi$ rad and $5\pi$ rad are distinctly different, whereas they are the same in ordinary 2D polar coordinates. Explain, conceptually, why we must make such a distinction in UCM. For example, what is the physical distinction between $-\pi$ rad and $\pi$ rad in UCM? Also, what impact does the difference in the range of $\theta$ in UCM compared to 2D polar coordinates have on the net angular displacement, $\Delta\theta$?
2. Regarding the earth's daily spin about its axis, which parts of the earth have the greatest and least (A) angular speeds and (B) tangential speeds?
3. A monkey twirls a baton in a circle about an axis through the center of the baton and perpendicular to the baton, such that the ends of the baton travel in circles with a constant tangential speed of 10 m/s. (A) Determine the tangential speed of each of the following points: the center of the baton, the two points one-third of the baton's length from either end, the two points one-quarter of the baton's length from either end. (B) What can you say about the angular speed of each of these points?
4. A monkey twirls a baton in a circle about an axis through the center of the baton and perpendicular to the baton, such that the ends of the baton travel in circles with a constant tangential speed. (A) Which points on the baton have the greatest centripetal acceleration? Where, if anywhere, is the centripetal acceleration zero? How does the centripetal acceleration of the endpoints compare to the centripetal acceleration of the two points one-quarter of the baton's length from either end? (Give a numerical factor.) (B) What can you say about the angular acceleration of each of these points? What can you say about the tangential acceleration of each of these points?
5. How is it possible for a monkey to run with constant speed, yet still be accelerating? What type(s) of acceleration does the monkey have in this case? Which type(s) of acceleration are zero in this case?

6. A stationary duck in a still pond kicks its feet at regular intervals, creating a series of outward-spreading concentric ripples. (A) If you count how many times a ripple passes under a leaf in the pond per minute, which physical quantity are you measuring directly? (B) If you measure the time it takes for one ripple to reach the leaf just after another ripple passed through the leaf, which physical quantity are you measuring directly?

7. A necklace dangles from the rearview mirror of a car. (A) As the car rounds a turn to the right, indicate the direction of each of the following (relative to the forward-facing driver – which is a monkey, of course): the direction the necklace leans, the direction of the car's acceleration, the direction of the necklace's acceleration, the direction of the net force exerted on the car, the direction of the net force exerted on the necklace, the direction that the driver feels pushed, the direction that the driver is pushed relative to an inertial reference frame. (B) From an inertial perspective, explain the direction that the monkey feels pushed. (C) What type of force does the monkey perceive that he is pushed with? Note that this answer differs from the answer to part (B). (D) From an inertial perspective, explain the direction that the monkey is actually pushed. Note that this answer also differs from the answer to part (B). (E) What type of force actually pushes on the monkey from an inertial perspective. Note that this answer corresponds to your answer from part (D).

8. A monkey travels in a counterclockwise circle centered at the origin with UCM, beginning at $\theta = \pi/2$ rad. Make three separate plots, sketching $x$, $v_x$, and $a_x$ (not to be confused with $s_T$, $v_T$, and $a_T$ or $a_c$) as functions of time. Note that you are only sketching the $x$-component of the motion.

9. A monkey travels with constant velocity while riding a unicycle. A mark on the tire travels in a circle with UCM relative to the center of the tire. What type of path does the mark travel along relative to the ground?

10. A monkey travels with constant velocity while riding a unicycle. (A) How does the speed of the center of the tire relate to the angular speed of the tire? (B) How does the net displacement of the unicycle relate to the net angular displacement of the tire?

11. A monkey drives in a circle with constant speed. If the monkey were to double his speed and still travel in a circle of the same radius, indicate by what numerical factor each of the following physical quantities would change: his angular speed, his period, his frequency, his acceleration, his net force.

12. A monkey drives in a circle with constant speed. If the monkey were to travel in a circle with twice the diameter while driving with the same constant speed, indicate by what numerical factor each of the following physical quantities would change: his angular speed, his period, his frequency, his acceleration, his net force.

13. (A) If you take the total distance that an object travels with UCM and divide by the circumference, what do you get? (B) If you take the total time that an object travels with UCM and divide by the period, what do you get?

14. A monkey ties one end of a thread to a banana and whirls the banana in a horizontal circle. If the monkey whirls the banana too fast, the thread will snap. Why?

15. A monkey ties one end of a thread to a banana and whirls the banana in a horizontal circle. Of course, the thread is not horizontal, but traces out a cone. What happens to the angle that the thread makes with the horizontal as the monkey whirls the banana faster and faster?[222]

---

[222] Note that it makes a huge difference if we ask about the angle that the thread makes with the horizontal or the vertical.

16. A monkey rides a Ferris wheel with constant angular speed. (A) At which points is the magnitude of the support force (from the seat, belt, cushions, etc.) exerted on the monkey greatest and least? (B) Is it possible for the support force to be zero at any of these positions? Explain. Physically, what would happen to the monkey in this case if he were not wearing his seatbelt?

17. When a racecar rounds a turn along a banked curve with friction that has a constant banking angle and constant banking radius, a range of speeds is actually possible for which the car can travel without sliding up or down the bank. Explain. Also, draw FBD's for the two extreme cases (they are different). Which FBD applies if the car rounds the turn with the greatest speed without sliding up the bank?

18. Is it possible for a satellite to orbit the earth with UCM with a period of one month and have an orbital radius smaller than the moon's average orbital radius? Explain.

19. Planet Chimp has four times as much mass as earth. If a satellite of Planet Chimp and a satellite of earth both travel with UCM at the same orbital radius, indicate by what numerical factor each of the following physical quantities would be different for the satellite orbiting Planet Chimp: the speed of the satellite, the angular speed of the satellite, the acceleration of the satellite, the period of the satellite, the frequency of the satellite.

20. Planet Chimp has four times as much mass as earth. If a satellite of Planet Chimp and a satellite of earth both travel with UCM at the same orbital speed, indicate by what numerical factor each of the following physical quantities would be different for the satellite orbiting Planet Chimp: the radius of the satellite's orbit, the angular speed of the satellite, the acceleration of the satellite, the period of the satellite, the frequency of the satellite.

## Practice Problems

### Basic rotational quantities

1. (A) What are the angular speeds of the second, minute, and hour hands of a clock in SI units? (B) How many radians does the hour-hand sweep out in 7.0 hours?

2. A baton is thrown straight upward. It spins with a frequency of 6.0 Hz about an axis perpendicular to the baton. The baton returns to its starting position in 3.0 s. What is the length of the baton?

3. (A) What is the angular speed of the earth in its rotation about its axis? (B) What tangential speed would an object have if it were located near the equator? (C) What tangential speed would an object have if it was located at a latitude of 30° (this is the angle above the equator from the center of the earth). (D) What centripetal acceleration would an object have if it were located near the equator? Compare your answer to gravitational acceleration. (E) What would the period of earth's rotation need to be in order for an object near the equator to have a centripetal acceleration equal to gravitational acceleration? Would the object be effectively twice as heavy, weightless, or something else? Explain.

4. Assume earth's orbit to be circular. (A) What is the angular speed of the earth in its orbit around the sun? (B) What is the tangential speed of the earth in its orbit around the sun? (C) What are the magnitude and direction of the centripetal acceleration of the earth? (D) What are the magnitude and direction of the centripetal force acting on the earth? (E) Compute the force that the earth exerts on the sun according to Newton's law of universal gravitation, and compare the result with your answer to (D) – computed by other means. Comment on this.

## Centripetal acceleration and centripetal forces

5. A 20-kg monkey who discovers buoyancy while taking a bath suddenly runs around the streets of Sillinews shouting *Eek! Eek!* The monkey runs with constant speed in a circle with a diameter of 400 m, completing ¾ of a revolution in 5.0 minutes. (A) Find the period. (B) Find the frequency. (C) Find the monkey's angular speed. (D) Find the monkey's speed. (E) Find the total distance the monkey travels. (F) Find the magnitude of the monkey's net displacement. (G) Find the acceleration of the monkey. (H) Find the net force acting on the monkey. (I) What is the minimum coefficient of friction needed between the monkey's shoes and the ground for the monkey to run as described without slipping?

6. A monkey ties a 200-g banana to its tail and spins it in a horizontal circle with 3.00-m diameter at a constant rate of 50.0 rev/min. (A) What is the angular speed of the banana? (B) What is the tangential speed of the banana? (C) What angle does the monkey's tail make with the horizontal? (D) How long is the monkey's tail? (E) What is the acceleration of the banana? (F) What kind of acceleration is this? (G) What is the direction of the acceleration? (H) What is the tension in the monkey's tail?

7. A monkey spins a 500-g alarm clock by its power cord in a horizontal circle with 4.0-m diameter. Each minute the alarm clock completes 15 revolutions. (A) Find the period. (B) Find the frequency. (C) Find the angular speed. (D) Find the tangential speed. (E) Find the acceleration. (F) Find the tension. (G) Find the angle that the cord makes with the vertical.

8. A monkey takes the 500-g glass eye out of his socket and connects it to a string. The monkey then spins the glass eye in a horizontal circle with an angular speed of $\sqrt{3}/2$ rad/sec. The string makes an angle of $60°$ with the vertical. (A) Find the period. (B) Find the tension in the string. (C) Find the acceleration of the eye. (D) How long is the string? (E) By what factor would the angular speed need to increase/decrease in order for the string to make an angle of $30°$ with the vertical?

9. You can 'defy' gravity on an amusement park centrifuge, which consists of a large cylindrical room with horizontal sides. You stand against the wall. First, the centrifuge rotates, then the floor disappears! Yet, you cling to the wall, rather than slide down the wall (unless you get soaking wet on a water ride just before entering the centrifuge). Derive an equation for the period, $T$, of rotation of the centrifuge needed to prevent you from sliding down the wall in terms of the radius, $R_0$, of the cylinder, the coefficient of static friction, $\mu_s$, between you and the wall, and/or $g$.

10. An amusement park ride consists of a 20-m diameter horizontal disk high above the ground. Several swings are suspended from the disk near its edge with 15-m long chains. As the disk spins, the swings extend outward at an angle of $50°$ from the vertical. How fast are the passengers moving?

11. A circular racetrack has a diameter of 200 m (from the top view, of course) and constant banking angle of $\theta = 20°$ (as viewed below, with the racecar headed out of the page). (A) What is the maximum constant speed a racecar can have if there is no friction? (B) What is the maximum constant speed a racecar can have if the coefficient of static friction between the road and tires is 0.40?

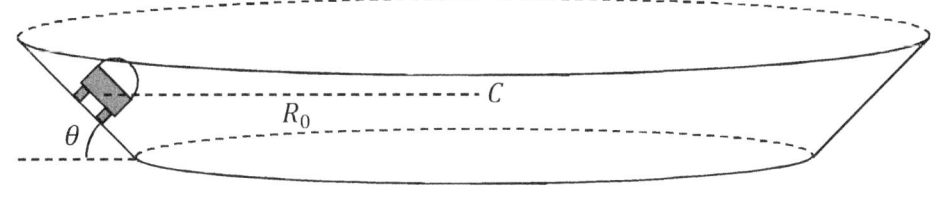

side view

12. A 50-kg monkey drives a 200-kg truck with a constant speed of 20 m/s along a semicircular hill with a radius of 80 m, as shown below. (A) What is the angular speed of the truck? (B) What is the normal force acting on the truck in the position shown? (C) What is the fastest speed that the truck could have in this problem and not lose contact with the ground at the top of the hill?

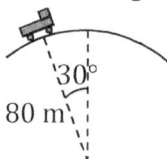

13. Little Isaac Newton the $12^{th}$ buys a monster truck in the local physics store. The 500-g toy monster truck drives along a vertical loop with 10-m diameter with constant speed – without falling away from the track! The period is $\pi$ seconds. Neglect the height of the truck compared to the diameter of the loop. (A) What is the speed of the monster truck? (B) Solve for the normal force for each of the positions shown in the diagram below. (C) What is the least speed the monster truck can have if the normal force is to be positive at every point on the loop?

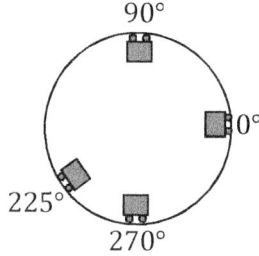

14. A monkey whirls a bucket of water with combined mass $m$ in a vertical circle of radius $R_0$ with constant angular speed $\omega_0$, as shown below. Express your answers using $m$, $\omega_0$, $R_0$, $\theta$, and/or $g$ only. (A) What is the speed of the bucket? (B) What is the period of the bucket? (C) What is the frequency of the bucket? (D) What is the acceleration of the bucket? (E) Draw a free-body diagram for the bucket in the position shown. (F) Sum the components of the forces. (G) What is the force of the monkey's pull in the position shown? (H) What is the slowest speed that the bucket could have without spilling any water at the top of the arc?

**Satellite motion**

15. What must be (A) the speed and (B) altitude of a geosynchronous satellite in a circular orbit around the earth (above the equator)?

16. Supermonkey flies to the top of a mountain, where the altitude is 100 km above the earth's sea level. He throws a banana horizontally. He throws it so fast that he catches it! (A) How fast did he throw the banana? (B) For how much time did Supermonkey have to wait for the banana to return?

17. Lemur and Chimp live on Banana planet, which orbits Banana star in a circular orbit. Lemur weighs 200 N on Banana planet, while Chimp weighs 350 N on Banana planet. Below are astronomical data for Banana planet:

Radius of Banana Planet: $1.45 \times 10^5$ m
Mass of Banana Planet: $9.24 \times 10^{23}$ kg
Distance between Banana Planet and Banana Star: $6.59 \times 10^9$ m
Mass of Banana Star: $3.72 \times 10^{26}$ kg
One 'Day' on Banana Planet: 8.23 hours
One 'Year' on Banana Planet: 720 days

(A) What is the value of gravitational acceleration near the surface of Banana planet? (B) What is Chimp's mass? (C) How much does Lemur weigh on the earth? (D) What force does Banana star exert on Banana planet? (E) What is the angular speed of Banana planet in its orbit around Banana star? (F) What is the tangential speed of Banana planet in its orbit around Banana star? (G) What is the acceleration of Banana planet? (H) What is the net force exerted on Banana planet? (I) What would be the period of a banana satellite orbiting Banana planet at a constant altitude of 200 km above the surface of Banana planet?

18. Below are astronomical data for Planet $\Phi_6$ and its moon:

Mass of $\Phi_6$: $6.0 \times 10^{23}$ kg  Mass of $\Phi_6$'s moon: $9.0 \times 10^{21}$ kg
Radius of $\Phi_6$: $4.0 \times 10^6$ m  Radius of $\Phi_6$'s moon: $2.0 \times 10^5$ m
Orbital radius of $\Phi_6$: $8.0 \times 10^{10}$ m  Orbital radius of $\Phi_6$'s moon: $4.0 \times 10^7$ m

(A) If you stand on the surface of $\Phi_6$ and throw a rock into the air, what will be its acceleration? (B) What force does $\Phi_6$ exert on its moon? (C) What is the speed of $\Phi_6$'s moon in its circular orbit around $\Phi_6$? (D) What is the period of $\Phi_6$'s moon in its circular orbit around $\Phi_6$? (E) A common myth is that $\Phi_6$'s moon is made of green banana. What can you calculate to resolve this myth? Calculate this quantity.

19. The little green monkeys are thinking of changing the laws of physics. In particular, they wish to modify Newton's law of universal gravitation to have the form:

$$F = H \frac{m_1^2 m_2^3}{r^4}$$

without modifying Newton's second law, where $m_2$ orbits $m_1$. (A) What are the SI units of $H$? (B) Derive an equation for the speed of a satellite of mass $m_2$ orbiting a planet of mass $m_1$ in a circular orbit of radius $r$ in terms of $m_1, m_2, r$, and/or $H$ only. (C) Derive an equation for the period of a satellite of mass $m_2$ orbiting a planet of mass $m_1$ in a circular orbit of radius $r$ in terms of $m_1, m_2, r$, and/or $H$ only.

20. Derive an equation for how far a satellite deviates from its tangential inertia as a function of time, and find the speed and acceleration as functions of time that correspond to this distance. Compare this acceleration with the centripetal acceleration of the satellite. Comment on this comparison.

# 5 Work and Energy

**Equations of motion**: According to Newton's second law, an object accelerates when a net force acts upon it. The equations of motion for the object can be derived by applying Newton's second law, which is a second-order differential equation. It also involves vector addition, so Newton's second law really represents three equations of motion – one for each component of motion in three-dimensional space. In principle, these differential equations can be solved, but real-world problems are often so complex that they must be solved numerically rather than algebraically. Algebraically or numerically, given the initial position and velocity of the object, Newton's second law provides a means of predicting the position, velocity, and acceleration at some later time. However, Newton's second law is not the only way of obtaining equations of motions for an object; conservation of energy provides an equivalent formulation – equivalent, yet quite different in some significant conceptual and strategic ways.

**Work and energy**: Everybody's heard of energy – it's an important concept in everyday life. We use electrical energy to see in the dark and to power electronics and appliances. We convert heat energy to do mechanical work when we drive a car. We eat meals and snacks every day, so that our bodies can make use of this stored biological energy. The energy that we utilize every day ultimately derives from the sun. We are all familiar with these forms of energy, but unless you have specifically studied the concept of energy, you would probably find energy to be a rather challenging physics concept to define. Energy in itself is a fairly abstract idea, but the conversion of energy from one form into another and using energy to do useful work are probably more tangible. In this chapter, we will first discuss work, which is more directly tangible than energy, and then we will learn that all forms of energy relate to work. We will discuss work in terms of force and displacement, which are more tangible yet – a force is a push or pull that you can feel directly, and a displacement is something that you can see. So if you find the important concept of energy to be abstract, you may find it easier to understand energy by thinking of it in terms of work, and even further breaking work down into force and displacement. When we define work mathematically, it will be in terms of an integral that will probably seem to have a strange form at first; but despite the way that work looks mathematically, you can still understand it conceptually in terms of force and displacement. Remember to pay special attention to the conceptual meaning of work, power, and the various forms of energy in this chapter, and not to get caught up just in the mathematics.

**Conservation of energy**: The total energy of the universe (or any isolated system) is observed to be conserved. Energy comes in a variety of forms – from mechanical to electrical to heat to many others – and it may be converted from one form to another, but energy cannot be created or destroyed. This is arguably the most useful concept in physics (and is also useful in all other branches of science), and applies to every branch of physics – even such widely diverse fields as cosmology and nanotechnology. Like Newton's second law, conservation of energy allows us to derive the equations of motion for an object; but compared to Newton's second law, conservation of energy has some distinct advantages.

Energy is a scalar quantity. Ultimately, energy derives from force and displacement, which are both vectors, but in such a way as to make a scalar quantity from the force and displacement vectors. So energy doesn't have a direction. When we apply Newton's second law, we work directly with force vectors, and draw FBD's and apply trig to resolve the force vectors into components. Newton's second law then gives three equations of motion – one for each component – for each object. Conservation of energy gives a single scalar equation. Also, Newton's second law is a second-order differential equation, while conservation of energy is a first-order differential equation – which is simpler, in principle, to solve. Most advanced formulations of physics make use of conservation of energy (such as the Hamiltonian formulation of classical and quantum mechanics) rather than Newton's second law. However, there are times when Newton's second law is more convenient than conservation of energy. For example, if you want to relate forces to acceleration, Newton's second law is far more direct. If you want to relate speed to position, however, conservation of energy will be much more convenient.

## 5.1 The Scalar Product

**Scalar product**: The scalar product is one of a few useful methods of multiplying vectors together.[223] The scalar product is also called the dot product and the inner product.[224] Given two vectors, $\vec{A}$ and $\vec{B}$, the scalar product is written with a dot,[225] as $\vec{A} \cdot \vec{B}$, and can be expressed in one of two different forms:

$$\vec{A} \cdot \vec{B} = AB \cos \theta = A_x B_x + A_y B_y + A_z B_z$$

where $\theta$ is the angle between $\vec{A}$ and $\vec{B}$. Note that $0 \leq \theta \leq 180°$. The form $AB \cos \theta$ is convenient when you know the magnitude of each vector and the angle between them, while the form $A_x B_x + A_y B_y + A_z B_z$ is convenient when you know the components of each vector. It is very handy to know both forms of the scalar product, as one form is usually much more convenient than the other. Shortly, we shall see that both forms of the scalar product are equivalent to one another.

Geometrically, the scalar product can be interpreted as the product of the magnitude of one vector and the projection of the other vector onto the first (bearing in mind that the projection can be positive or negative). The projection of one vector onto another is illustrated in the following diagram.

---

[223] Two others include the vector product, which we will learn in Chapter 7, and the outer product of linear algebra, which is useful for expressing the completeness relation in group theory.

[224] It's only appropriate to call it the inner product in the context of linear algebra. If you're not working with row and column vectors, either call it the scalar product or the dot product.

[225] You can write the product of the numbers 3 and 4 as $3 \times 4$ or $3 \cdot 4$, but the symbols $\times$ and $\cdot$ are absolutely not interchangeable in the context of vectors. The quantity $\vec{A} \cdot \vec{B}$ represents the scalar product, which results in a scalar, whereas the quantity $\vec{A} \times \vec{B}$ represents the vector product, which results in a vector. It's not just that one is a scalar the other is a vector – the scalar result and vector result, and the method of computing each, are considerably different. If you're wondering whether the scalar product is the magnitude of the vector product, it's not – the distinction is not so simple. In Chapter 7, you should come to see how the magnitude of the vector product is related to the scalar product.

5 Work and Energy

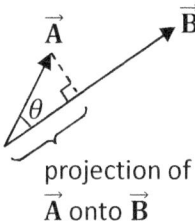

projection of
$\vec{A}$ onto $\vec{B}$

The scalar product between two vectors depends not only on the magnitudes of the vectors, but also on the angle between them. When two vectors are parallel, $\vec{A} \cdot \vec{B}$ equals $AB$; when two vectors are perpendicular, $\vec{A} \cdot \vec{B}$ is zero; and when two vectors are anti-parallel, $\vec{A} \cdot \vec{B}$ equals $-AB$. In general, we can say that $-AB \leq \vec{A} \cdot \vec{B} \leq AB$.

> **Note**. It's important to realize that the scalar product results in a scalar. This should be obvious from the name if you get into the habit of calling it the scalar product, but after students learn the vector product, they too often confuse the forms of the results, and write vectors for their answers when using the scalar product. Remember, the given vectors have magnitudes and directions, but the result of applying the scalar product does not have a direction.

**Magnitude in terms of the scalar product**: The scalar product between any vector and itself equals the magnitude of the vector squared:

$$\vec{A} \cdot \vec{A} = A^2 = A_x^2 + A_y^2 + A_z^2$$
$$A = \sqrt{\vec{A} \cdot \vec{A}}$$

**Algebra of the scalar product**: From the two forms of the scalar product, it's easy to see that the scalar product is both commutative and distributive:[226]

$$\vec{A} \cdot \vec{B} = \vec{B} \cdot \vec{A}$$
$$\vec{A} \cdot (\vec{B} + \vec{C}) = \vec{A} \cdot \vec{B} + \vec{A} \cdot \vec{C}$$

However, the associative property – which is $a(bc) = (ab)c$ for multiplying numbers, for example – does not apply to the scalar product, since $\vec{A}(\vec{B} \cdot \vec{C})$ is generally not equal to $(\vec{A} \cdot \vec{B})\vec{C}$. You might at first wonder if $\vec{A}(\vec{B} \cdot \vec{C})$ is ambiguous, but if you consider that $\vec{B} \cdot \vec{C}$ is a scalar, you should see that it can only mean to first evaluate the scalar product $\vec{B} \cdot \vec{C}$ and then multiply this scalar times the vector; the vector $\vec{A}$ is not itself involved in any scalar product in $\vec{A}(\vec{B} \cdot \vec{C})$.

---

[226] You might wonder if it's trivial to check such things, but as you learn more and more new kinds of quantities to add, subtract, multiply, and divide, you'll see that very often the algebra of such quantities is not always commutative, associative, or distributive – and the rules for the basic arithmetic becomes very important to know. For example, you'll see in Chapter 7 that the vector product is not commutative.

**Scalar product of unit vectors**: Consider the scalar product of the unit vector $\hat{\imath}$ with itself – i.e. $\hat{\imath} \cdot \hat{\imath}$. Since the magnitude of any unit vector is one, this equals one times one times the cosine of zero degrees (also one): $\hat{\imath} \cdot \hat{\imath} = 1$. In contrast, $\hat{\imath} \cdot \hat{\jmath}$ is zero because $\hat{\imath}$ and $\hat{\jmath}$ are perpendicular (so we get $\cos 90° = 0$). It is handy to know the scalar product between various pairs of Cartesian unit vectors:

$$\hat{\imath} \cdot \hat{\imath} = 1 \quad , \quad \hat{\jmath} \cdot \hat{\jmath} = 1 \quad , \quad \hat{k} \cdot \hat{k} = 1$$
$$\hat{\imath} \cdot \hat{\jmath} = 0 \quad , \quad \hat{\imath} \cdot \hat{k} = 0 \quad , \quad \hat{\jmath} \cdot \hat{k} = 0$$

**Law of Cosines**: Consider the resultant of $\vec{A}$ and $\vec{B}$, which is $\vec{R} = \vec{A} + \vec{B}$. Let us explore what happens when we compute $\vec{R} \cdot \vec{R}$:

$$R^2 = \vec{R} \cdot \vec{R} = (\vec{A} + \vec{B}) \cdot (\vec{A} + \vec{B}) = \vec{A} \cdot \vec{A} + 2\vec{A} \cdot \vec{B} + \vec{B} \cdot \vec{B}$$
$$R^2 = A^2 + 2AB \cos \theta + B^2$$

This result is the law of cosines.[227] Since $\vec{R}$ is the resultant of $\vec{A}$ and $\vec{B}$, $\vec{A}$ and $\vec{B}$ join tip-to-tail to form $\vec{R}$, as illustrated below. The law of cosines requires that $R^2 = A^2 + 2AB \cos \theta + B^2$.

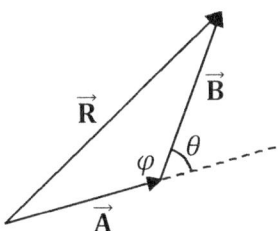

**Equivalence of the two forms of the scalar product**: In order to see that the two forms of the scalar product – i.e. $\vec{A} \cdot \vec{B} = AB \cos \theta$ and $\vec{A} \cdot \vec{B} = A_x B_x + A_y B_y + A_z B_z$ – are equivalent, multiply $\vec{A} \cdot \vec{B}$ out explicitly after expressing $\vec{A}$ and $\vec{B}$ as $\vec{A} = A_x \hat{\imath} + A_y \hat{\jmath} + A_z \hat{k}$ and $\vec{B} = B_x \hat{\imath} + B_y \hat{\jmath} + B_z \hat{k}$:

$$\vec{A} \cdot \vec{B} = AB \cos \theta = (A_x \hat{\imath} + A_y \hat{\jmath} + A_z \hat{k}) \cdot (B_x \hat{\imath} + B_y \hat{\jmath} + B_z \hat{k})$$
$$AB \cos \theta = A_x B_x + A_y B_y + A_z B_z$$

where we used the fact that $\hat{\imath} \cdot \hat{\imath} = 1, \hat{\imath} \cdot \hat{\jmath} = 0$, and so on.

---

[227] If you are used to seeing the law of cosines written with a minus sign, as in $R^2 = A^2 - 2AB \cos \varphi + B^2$, it's because we are working with a different angle. In math, when you learn about the law of cosines, you work with the angle $\varphi$, which is the angle opposite to the vector $\vec{R}$. In the context of the scalar product, we are working with the angle $\theta$, which is the angle between the vectors $\vec{A}$ and $\vec{B}$. Observe that $\theta$ is the supplement to $\varphi$ (see the illustration above). Therefore, $\cos \theta = \cos(180° - \varphi) = \cos 180° \cos \varphi - \sin 180° \sin \varphi = -\cos \varphi$, which resolves the mystery of the difference in sign: $A^2 + 2AB \cos \theta + B^2 = A^2 - 2AB \cos \varphi + B^2$.

**Example.** Determine the scalar product between $\vec{A} = 2\hat{i} + 3\hat{j} - \hat{k}$ and $\vec{B} = 3\hat{i} + 2\hat{k}$.
Since we know the components of $\vec{A}$ and $\vec{B}$, use the component form of the scalar product:

$$\vec{A} \cdot \vec{B} = A_x B_x + A_y B_y + A_z B_z = 2 \cdot 3 + 3 \cdot 0 - 1 \cdot 2 = 6 - 2 = 4$$

**Example.** Determine the angle between $\vec{A} = -4\sqrt{3}\hat{i} + 4\hat{j}$ and $\vec{B} = \hat{i} - \sqrt{3}\hat{j}$.
We can solve for the angle between two vectors by combining both forms of the scalar product:

$$AB \cos\theta = A_x B_x + A_y B_y + A_z B_z$$

$$\cos\theta \sqrt{A_x^2 + A_y^2 + A_z^2} \sqrt{B_x^2 + B_y^2 + B_z^2} = A_x B_x + A_y B_y + A_z B_z$$

$$\cos\theta \sqrt{\left(-4\sqrt{3}\right)^2 + (4)^2} \sqrt{(1)^2 + \left(-\sqrt{3}\right)^2} = -4\sqrt{3} \cdot 1 + 4 \cdot \left(-\sqrt{3}\right)$$

$$(8)(2)\cos\theta = -8\sqrt{3}$$

$$\theta = \cos^{-1}\left(-\frac{\sqrt{3}}{2}\right) = 150°$$

## 5.2 Work and Power

**Work:** Conceptually, work is done by a force when the force contributes toward (or against) the displacement of an object. Mathematically, we can express work at the differential scale as $dW = \vec{F} \cdot d\vec{s}$; this is an infinitesimal amount of work, $dW$, done by a finite force, $\vec{F}$, acting over an infinitesimal displacement, $d\vec{s}$. Looking at this mathematical definition, we see that in order for work to be done, a force must act on an object, the object must be displaced, and the force in question must have a component along (or against) the displacement (because the scalar product is zero if the two vectors are perpendicular) – all three of these conditions must be met in order for work to be done. The finite amount of work done over a finite displacement can be found through integration:

$$W = \int_i^f \vec{F} \cdot d\vec{s}$$

The SI unit of work is the Joule (J). One Joule equates to a Newton times a meter: $1\,\text{J} = 1\,\text{Nm} = \text{kg}\,\text{m}^2/\text{s}^2$.

**Conceptual Example.** If an object is displaced, is work necessarily done?
No. An object can be displaced according to its own inertia. Being displaced does not imply that a force is acting on the object. Three conditions must be met in order for work to be done – there must be a force, a displacement, and the force in question must have a component along the displacement.

**Work done by a specified force**: In general, there are multiple forces acting on an object. We can compute the work done by any – or some, or all – of these forces. It's ambiguous to ask, "How much work is done?" because the force hasn't been specified. It's important to identify the force in question. For example, if a monkey pulls a crate of bananas, and the question asks how much work is done by gravity, it would be totally incorrect to use the force of the monkey's pull. Every (well-written) question about work will specify the force in question. However, you must read the problem carefully and think through work questions because sometimes the wording may be subtle. For example, if a problem asks you to find the net work done, the 'force' to use is implied – 'net' work implies use the 'net' force (i.e. the vector sum of the forces acting on the object). Also, make sure that you use a force. If a question asks you to find the work done by gravity, the force is not $\vec{g}$, but $m\vec{g}$.

> **Note**. Read each problem carefully to determine which force(s) to use when calculating how much work is done when an object is displaced.

> **Conceptual Example**. A waiter balances a tray of 800-g of bananas on his shoulder while traveling 5 m horizontally. How much work is done by gravity?
> None! The force in question – weight, $m\vec{g}$ – is vertical, while the displacement is horizontal. The scalar product $\vec{F} \cdot d\vec{s}$ – which can be written as $F \cos \theta \, ds$ – is zero over the entire displacement because $\vec{F}$ and $d\vec{s}$ are perpendicular, so no work is done. A force must have a component along the direction of the displacement (over some section of the path) in order for work to be done.

**Sign of work**: Work can be positive, negative, or zero. Work is said to be done by the object (or system) when it is positive, and work is done against the object (or system) when it is negative.

**Differential displacement vector**: Recall the differential displacement vector from Sec. 2.3:

$$d\vec{s} = \hat{\imath} dx + \hat{\jmath} dy + \hat{k} dz$$

**Differential arc length**: Recall the differential arc length from Sec. 2.3, which can be expressed in general terms in Cartesian, cylindrical, or spherical coordinates as follows:

$$ds = \sqrt{dx^2 + dy^2 + dz^2}$$
$$ds = \sqrt{dr_c^2 + (r_c d\theta)^2 + dz^2} \quad \text{(cylindrical)}$$
$$ds = \sqrt{dr^2 + (rd\theta)^2 + (r \sin \theta \, d\varphi)^2} \quad \text{(spherical)}$$

**Differential displacement vector and differential arc length**: Recall that the differential displacement vector, $d\vec{s}$, and the differential arc length, $ds$, are related by $d\vec{s} = \hat{t} ds$, where $\hat{t}$ is a unit tangent; and that the differential arc length equals the magnitude of the differential displacement vector: $ds = \|d\vec{s}\|$. These two relations – $d\vec{s} = \hat{t} ds$ and $ds = \|d\vec{s}\|$ – only hold in general at the differential scale; the finite net displacement, $\Delta \vec{r}$, and total distance traveled, $\Delta s \geq 0$, cannot be related so simply in general terms because $\Delta \vec{r}$ is path-independent, whereas $\Delta s$ depends upon the path of integration.

5 Work and Energy

**Important Distinction**. It's important to distinguish between the differential displacement vector, $d\vec{s}$, and the differential arc length, $ds$; and also to distinguish between the net displacement, $\Delta\vec{r} = \int_i^f d\vec{s}$, and the total distance traveled, $\Delta s = \int_i^f ds$. The distinction between $\Delta\vec{r}$ and $\Delta s$ is more profound than the fact that one is a vector while the other is a scalar; in general, $\Delta s$ is not the magnitude of $\Delta\vec{r}$ because $\Delta\vec{r}$ is path-independent (it depends only upon the endpoints), whereas $\Delta s$ is path-dependent.

**Note**. Although the net displacement integral $\Delta\vec{r} = \int_i^f d\vec{s}$ is path-independent – i.e. for a given initial and final position, the same result is obtained regardless of the path taken – the work integral $W = \int_i^f \vec{F} \cdot d\vec{s}$ is in general path-dependent. We will return to this distinction in Sec. 5.6.

**The work integral**: The work integral can be performed two different ways, based on the two different ways of computing the scalar product. The integrand, $\vec{F} \cdot d\vec{s}$, can be expressed as $F \cos\theta\, ds$ or as $F_x dx + F_y dy + F_z dz$:

$$W = \int_{s=s_0}^{s} F \cos\theta\, ds$$

$$W = \int_{x=x_0}^{x} F_x dx + \int_{y=y_0}^{y} F_y dy + \int_{z=z_0}^{z} F_z dz$$

In general, the form $F \cos\theta\, ds$ results in a more complicated integral, as it can be challenging to express the magnitude of the force, $F$, as a function of $s$, and the angle between the force and instantaneous displacement, $\theta$, is generally a variable that can also be challenging to express in terms of $s$. It is usually much simpler to determine the Cartesian components of the force – $F_x$, $F_y$, and $F_z$ – as functions of $x$, $y$, and $z$. However, if you know that the magnitude of the force and the angle between the force and displacement remain constant over the path of integration, then in this special case the form $F \cos\theta\, ds$ becomes relatively simple. When either the magnitude of the force changes or the direction between the force and displacement is a variable, then the other form of integration is to be preferred. Forthfollowing, we will work out several symbolic examples of how to determine the work done by a variety of common forces (like weight, normal force, friction, and Hooke's law) to illustrate how to perform the work integral. Sometimes, it may also be useful to work in cylindrical or spherical coordinates (we will see an example of this in the case of work done by gravity).

**Special cases of the work integral**: The simplest case (other than zero force or zero displacement) of the work integral corresponds to a constant force – i.e. constant in both magnitude and direction. If the force is constant, we don't even need to use either of the two forms discussed in the previous note. If the force is constant in both magnitude and direction, we can pull it out of the integral:

$$W = \vec{F} \cdot \int_i^f d\vec{s} = \vec{F} \cdot \Delta\vec{r} = F\|\Delta\vec{r}\| \cos \theta = F_x \Delta x + F_y \Delta y + F_z \Delta z \quad \text{(if } \vec{F} \text{ is const.)}$$

It's important to realize that the above expression involves the net displacement, $\Delta\vec{r}$, and not the total distance traveled, $\Delta s$. In particular, the magnitude of the net displacement, $\|\Delta\vec{r}\|$, does not equal the total distance traveled, $\Delta s$, if the path involves any change in direction (i.e. if the path is curved, or if the path is straight and the object reverses direction). In the above expression, $\theta$ is the angle between the force and the net displacement. The above expression is true as long as both the magnitude and direction of the force are constant, and is independent of the path taken (it depends only on the endpoints of the path).

Another special case of interest is when the force has constant magnitude, but does not have constant direction, and yet the angle between $\vec{F}$ and $d\vec{s}$ is constant throughout the path. For example, a monkey may be pulling a box with a constant magnitude of force, dragging the box in a circle so that the direction of the force is changing, but always pulling 30° above the horizontal so that $\theta$ is constantly 30°; here, $\vec{F}$ is changing direction because the monkey is walking around in a circle, but $\theta$ is constant. Note the important distinction between this and the previous case: Here, $\vec{F}$ is not constant in direction, but $\theta$ is constant, whereas previously $\vec{F}$ was constant in direction, but $\theta$ was not necessarily constant throughout the path. In this case, $\vec{F}$ cannot come out of the integral because it's direction is not constant, but $F = \|\vec{F}\|$ can come out of the integral because its magnitude is constant. Furthermore, $\theta$ can come out of the integral because – although the direction of the force is changing in this case – the differential displacement vector is changing with the force such that $\theta$ is constant. Thus, in this case the work integral simplifies to

$$W = \int_{s=s_0}^{s} F \cos \theta \, ds = F \cos \theta \int_{s=s_0}^{s} ds = F \Delta s \cos \theta \quad \text{(if } F \text{ and } \theta \text{ are const.)}$$

Notice that this result is different from the previous result: $F \Delta s \cos \theta$ involves the total distance traveled, whereas $F\|\Delta\vec{r}\| \cos \theta$ involves the magnitude of the net displacement.

---

**Note.** The formulas $W = F \Delta s \cos \theta$ and $W = F\|\Delta\vec{r}\| \cos \theta$ can only be used when the specified force has constant magnitude, and also when either the direction of the force (in the equation with $\Delta s$) or the angle between the force and displacement (in the equation with $\|\Delta\vec{r}\|$) is constant over the path. Otherwise, you must integrate from scratch to derive an equation for the work done by a force.

---

**Work done by gravity:** We will derive two different formulas for the work done by gravity: One formula will apply only in the presence of a uniform gravitational field, which will hold to very good approximation for an object that makes a small change in altitude compared to the radius of a planet; and the second formula will apply more generally, allowing for significant changes in altitude, but will still be limited to work done in a gravitational field due to a single isolated spherically symmetric mass (like a planet). We will also discuss how to generalize the procedure to multiple sources.

First, consider an object outside of a large massive sphere, such as a planet or moon, that is making only a small change in altitude compared to the radius of the sphere. This would be true of most human motion near the surface of the earth, except climbing a very tall mountain; but this would not be true for launching a rocket into space. For such small changes in altitude, $\vec{g}$ is approximately uniform, such that the gravitational force exerted on the object is $m\vec{g}$. Treating $\vec{g}$ as a constant, we can pull it out of the integral. Setting up a coordinate system with $+y$ directed away from the planet's center (i.e. upward), $m\vec{g} = -mg\hat{\jmath}$. Note that $m\vec{g}$ has only a $y$-component. The work done by gravity for an object that is displaced in a uniform gravitational field can be found from the component form of the integral:

$$W_g = \int_{y=y_0}^{y} F_y \, dy = -\int_{y=y_0}^{y} mg \, dy = -mg(y - y_0) \quad (\text{if } \vec{g} \text{ is const.})$$

The work done by gravity is positive if the object is displaced downward ($y < y_0$) and negative if the object is displaced upward ($y > y_0$). The work done by gravity depends only on altitude (for a single source). Notice that we did not need to specify the path along which the object was displaced in order to derive the result that $W_g = -mg(y - y_0)$. This is because gravity is a conservative force, such that the work done by gravity is path-independent (a point which we will reconsider in Sec. 5.5).

If the object makes a significant change in altitude outside of a single spherically symmetric massive object, we must instead use the equation $\vec{g} = -G\frac{m_p}{R^2}\hat{R}$, so that $m\vec{g} = -G\frac{m_p m}{R^2}\hat{R}$ according to Newton's law of universal gravitation (see Sec. 3.9). In this case, it is useful to use spherical coordinates with the origin at the center of the planet, such that $\hat{r} = \hat{R}$ and $r = R$. In this case, $d\vec{s} = \hat{r} \, dr + \hat{\theta} \, r d\theta + \hat{\varphi} \, r \sin\theta \, d\varphi$. The work integral becomes

$$W_g = \int_i^f \vec{F} \cdot d\vec{s} = -\int_i^f G\frac{m_p m}{R^2}\hat{R} \cdot \left(\hat{r} \, dr + \hat{\theta} \, r d\theta + \hat{\varphi} \, r \sin\theta \, d\varphi\right) \quad \text{(single point-mass)}$$

$$W_g = -Gm_p m \int_{R=R_0}^{R} \frac{dR}{R^2} = Gm_p m \left(\frac{1}{R} - \frac{1}{R_0}\right) \quad \text{(single point-mass)}$$

Note that the signs are consistent with the previous case. For example, $W_g$ is positive if $R < R_0$.

If there are multiple spherically symmetric sources for which the gravitational field is significant in the problem (such as is the case for a rocket traveling between the earth and the moon), you can find the work done by gravity with respect to each source separately and then combine them together, bearing in mind that $R$ and $R_0$ are measured differently for each source – from the final and initial positions of the object, respectively, to the center of mass of each source.

---

**Note.** The formula $W_g = -mg(y - y_0)$ is only valid for a uniform gravitational field. For a significant change in altitude, you must instead use the equation involving $R$ and $R_0$; and if there is more than one source with a significant gravitational field in the region of space involved in the problem, you must use this equation for each source and add the results together.

---

**Work done by normal force**: Recall that the normal force, $\vec{N}$, is perpendicular to the surface. If an object is displaced along a surface, $d\vec{s}$ is tangential to the surface. In this case, $\vec{N}$ and $d\vec{s}$ are perpendicular, such that $\vec{N} \cdot d\vec{s} = 0$, and the work done by the normal force is zero: $W_N = 0$.

If a monkey lifts a box off of a surface, $\vec{N}$ no longer acts on the box as soon as it leaves the surface, so that the work done by the normal force is still zero.

In order for normal force to do work, an object has to be in contact with the surface while being displaced toward or away from the surface – as during a collision. If you hit a tennis ball against a wall, normal force does work compressing the tennis ball during the collision, as the center of mass of the tennis ball is displaced a couple of centimeters over a very short period of time.

**Work done by friction**: The force of friction acts against the direction of the displacement. Therefore, $\vec{f} \cdot d\vec{s} = f \cos 180° \, ds = -f ds$. Since the force of friction equals the coefficient of friction times the normal force, $f_k = \mu_k N$,[228] the work done by friction can generally be expressed as[229]

$$W_f = -\mu_k \int_{s=s_0}^{s} N ds$$

This can be a complicated integral for work done along a curved surface. For an object traveling along a curved hill, for example, $N$ varies as the slope of the hill varies; also, if the hill is curved, there will be centripetal acceleration, and $N$ will depend upon the centripetal acceleration, which itself depends upon the speed and radius of curvature.

The work done by friction is much easier to compute for an object sliding along an inclined plane. In this case, there is no centripetal acceleration, and $N$ will be constant unless there is some variable push or pull in the problem with a normal component. For a constant normal force, the work done by friction simplifies to

$$W_f = -\mu_k N \Delta s \quad (\text{if } N \text{ is const.})$$

where $\Delta s$ is the total distance traveled (not the net displacement). For work done by friction along an inclined plane (or along a horizontal), first draw a FBD, then apply Newton's second law to derive an equation for the normal force, and then use this expression to derive an equation for the work done.

The work done by friction is negative because friction removes mechanical energy from the system – converting it into heat, just like rubbing your hands together makes you feel slightly warmer on a very cold day. If you drag a large box across the ground, just after you might observe that the ground (and bottom of the box) are a little warmer. Friction is an example of a nonconservative force, which is path-dependent. The greater the distance traveled, the more mechanical energy friction removes from the system. We will explore the concepts of conservative and nonconservative work more in Sec.'s 5.5 and 5.6.

---

[228] Only kinetic friction is relevant for work, since the object is not being displaced when static friction is acting.

[229] Note that $\int_{s=s_0}^{s} N ds$ is generally nonzero, whereas $\int_i^f \vec{N} \cdot d\vec{s}$ is generally zero.

**Example.** A 20-kg box of bananas slides 5.0 m down a 30° incline. The coefficient of friction between the box of bananas and the incline is 0.30. How much work is done by friction?

We first draw a FBD so that we can determine the normal force.

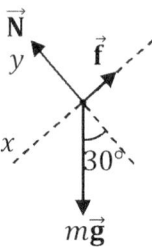

We only need to sum the $y$-components of the forces in order to solve for the normal force:

$$\sum_{i=1}^{3} F_{iy} = ma_y$$
$$N - mg \cos 30° = 0$$
$$N = \frac{mg\sqrt{3}}{2} = \frac{(20)(9.81)\sqrt{3}}{2} = 98\sqrt{3} \text{ N}$$

Since $N$ is a constant in this problem, the work done by friction equals

$$W_f = -\mu_k N \Delta s = -(0.30)(98\sqrt{3})(5.0) = -0.25 \text{ kJ}$$

**Work done stretching or compressing a spring:** Consider a spring that is being stretched from equilibrium, for which Hooke's law applies (see Sec. 3.6). We will setup our coordinate system with the $+x$-axis in the direction of the stretch, and designate the equilibrium position by $x_e$. When the spring is stretched from equilibrium, the restoring force equals $\vec{F}_r = -k(x - x_e)\hat{i}$. The differential displacement vector for this 1D problem is $d\vec{s} = \hat{i}\, dx$. The work done by the restoring force when the spring stretches from equilibrium is

$$W_r = \int_i^f \vec{F}_r \cdot d\vec{s} = -k \int_{x=x_e}^{x} (x - x_e) dx = -\frac{k(x - x_e)^2}{2} \quad \text{(stretching from equil.)}$$

If the spring is instead compressed from equilibrium, $\vec{F}_r = +k(x - x_e)\hat{i}$, but $d\vec{s} = -\hat{i}\, dx$ because the motion is directed along the negative $x$-axis:

$$W_r = -\frac{k(x - x_e)^2}{2} \quad \text{(compressing from equil.)}$$

If the spring is already stretched, but returning to equilibrium, then $\vec{F}_r \cdot d\vec{s}$ is positive and

$$W_r = +\frac{k(x-x_e)^2}{2} \quad \text{(compressing to equil.)}$$

Similarly, if the spring is already compressed, but returning to equilibrium, then

$$W_r = +\frac{k(x-x_e)^2}{2} \quad \text{(stretching to equil.)}$$

Finally, if the spring is stretching or compressing and neither the initial nor the final position is at equilibrium, you will have to integrate from the initial position, $x_0$, to the final position, $x$, noting that $x_0 \neq x_e$. Furthermore, if the spring is heading toward equilibrium during part of the trip and away from equilibrium during part of the trip, you will have to split the integral into two pieces, noting that $\vec{F}_r \cdot d\vec{s}$ is positive when the spring is stretching or compressing toward equilibrium and $\vec{F}_r \cdot d\vec{s}$ is negative when it is stretching or compressing away from equilibrium.

> **Hint.** If you setup your coordinate system with the origin at equilibrium, then $x_e = 0$, which simplifies the notation somewhat.

**Example.** A monkey pulls a 40-kg box of bananas 8.0 m along a horizontal surface with a constant force of 100 N at an angle of 30° above the horizontal. Determine the work done by the monkey's pull.

We choose to setup our coordinate system with $+x$ along the direction of the displacement and $+y$ vertically upward. The differential displacement vector is then $d\vec{s} = \hat{i}\, dx$, and the force of the monkey's pull can be expressed as $\vec{F}_p = \hat{i}\, F_p \cos 30° + \hat{j}\, F_p \sin 30°$.[230] The work done by the monkey's pull is then given by[231]

$$W_p = \int_i^f \vec{F}_p \cdot d\vec{s} = \int_{x=x_0}^{x} F_p \cos 30°\, dx = \frac{F_p \Delta x \sqrt{3}}{2} = \frac{(100)(8.0)\sqrt{3}}{2} = 400\sqrt{3}\text{ J}$$

**Example.** Determine the work done by the force $\vec{F} = xy\hat{i} - x^2\hat{j}$ (where SI units have been suppressed) acting on a banana that is displaced along the parabola $y = 2x^2$ over the interval $1 \leq x \leq 3$.

---

[230] Since we will be using the symbol $P$ for power, in the context of work and power it's a good idea to avoid using the symbol $\vec{P}$ for a push or a pull, so we will use $\vec{F}_p$ for the monkey's pull.

[231] Notice that we didn't need to know the mass of the box of bananas. It's good experience for an occasional problem to give some extraneous information. In the laboratory, you can measure anything you please (provided that you have the proper equipment), such as the temperature of the box, but not everything you measure will be relevant for what you wish to calculate. So from a practical perspective, it's important to be able to identify which quantities are relevant and which are irrelevant. This also encourages you to think – instead of just trying to figure out what to do with all of the given numbers to get the right answer.

It is convenient to use the component form of the work integral:

$$W = \int_{x=x_0}^{x} F_x dx + \int_{y=y_0}^{y} F_y dy = \int_{x=1}^{3} xy\,dx - \int_{y=2}^{18} x^2 dy$$

where the limits for the $y$-integration were found using the equation for the parabolic path, $y = 2x^2$, over the specified interval. When we integrate over $x$, we must use the equation $y = 2x^2$ to express $y$ in terms of $x$ (since $y$ is not a constant for this path of integration). Similarly, when we integrate over $y$, we must use the equation $y = 2x^2$ to express $x$ in terms of $y$. The work done by the force is

$$W = \int_{x=1}^{3} x \cdot 2x^2\,dx - \int_{y=2}^{18} \frac{y}{2} dy = \left[\frac{x^4}{2}\right]_{x=1}^{3} - \left[\frac{y^2}{4}\right]_{y=2}^{18} = \frac{81-1}{2} - \frac{324-4}{4} = 40 - 80 = -40\,\text{J}$$

**Net work**: If a problem asks about the net work done, it's not asking about the work done by a particular force, but is instead asking about the work done by the net force:

$$W_{net} = \int_{i}^{f} \left(\sum \vec{F}_{ext}\right) \cdot d\vec{s}$$

For a single object with constant mass, if you happen to know the object's acceleration, it will probably be more convenient to substitute $m\vec{a}$ for $\sum \vec{F}_{ext}$ (according to Newton's second law) than to sum the components of the forces directly. An alternative way to think of the net work is the sum of the work done by the various forces acting on the object:

$$W_{net} = \sum_{i=1}^{N} W_i$$

This is particularly handy if you have already found the work done by one or more forces in the problem.

**Work done by conservative forces**: If the work done displacing an object from some initial point, $(x_0, y_0, z_0)$, to some final point, $(x, y, z)$, is the same regardless of the path taken, the force is said to be conservative. We saw this in the case of work done by gravity, for example, which only depends upon the initial and final altitude, but not how the object gets from one altitude to the other. We will explore conservative forces further in Sec. 5.5.

**Work done by nonconservative forces**: If the work done by a force is path-dependent, the force is said to be nonconservative. In the case of friction, we found that the work done was $-\mu_k N \Delta s$ if $N$ was constant. It's easy to see that a path from $(x_0, y_0, z_0)$ to $(x, y, z)$ that covers more distance results in more $|W_f|$, and so $W_f$ is path-dependent. We will explore nonconservative forces further in Sec. 5.6.

**Power**: The instantaneous rate at which work is done is called power:

$$P = \frac{dW}{dt}$$

This is a derivative of work with respect to time. The SI unit of power is the Watt (W),[232] which is related to other SI units by $1 \text{ W} = 1 \text{ J/s} = 1 \text{ Nm/s} = 1 \text{ kg m}^2/\text{s}^3$.

Like work, problems will ask for the power delivered by a specific force – e.g. the power delivered by a motor means to use whichever force in the problem is exerted by the motor. Also, net power means to use the net work.

**Example**. Derive an equation for the power delivered by a motor for which the work is $W = 3t^2$, where $t_0 = 0$ and where SI units have been suppressed.
Simply take a derivative of work with respect to time:[233]

$$P = \frac{dW}{dt} = 6t$$

**Example**. A car accelerates from rest according to $\vec{a} = \beta t \hat{\imath}$, where $\beta$ is a constant. Derive an equation for the net power delivered to the car.
First, we must derive an expression for the net work, which is the work done by the net force. From Newton's second law, the net force equals $m\vec{a} = \beta m t \hat{\imath}$. Since the car starts from rest, its displacement will be along $\hat{\imath}$. The net work done by the car is then

$$W_{net} = \int_i^f \left(\sum \vec{F}_{ext}\right) \cdot d\vec{s} = \int_i^f m\vec{a} \cdot d\vec{s} = \beta m \int_{x=x_0}^x t\, dx$$

Time is definitely not constant, but increases as the car accelerates, so we must first express $t$ in terms of $x$ before evaluating this integral. We can do this by integrating:

$$a_x = \frac{dv_x}{dt} = \beta m t$$

---

[232] You should be able to distinguish between the symbol work, $W$, and the unit for power, the Watt (W), from the context. A lightbulb may be rated 150 W, for example, and it wouldn't make any sense to read it as 150 works. Also, if by now you're in the habit of writing $m\vec{g}$ for weight, you won't confuse weight with work or Watts.

[233] This simple example illustrates the relationship between work and power, but in practice problems tend to be more involved. Usually, you first express force as a function of position so that you can integrate force over displacement to find work. But then you know work as a function of position rather than time. So then you must use the equations of motion to write work as a function of time before you take the derivative. You must also be careful to write a general expression for work, and not to make the mistake of evaluating the work at a particular position (because, of course, you must apply a derivative before you evaluate the function at a specific point). We illustrate this strategy in the next example.

# 5 Work and Energy

$$\int_{v_x=0}^{v_x} dv_x = \beta m \int_{t=0}^{t} t\, dt$$

$$v_x = \frac{\beta m \Delta t^2}{2}$$

Since the car accelerated from rest, $v_{x0} = 0$. The position can be found after expressing $v_x$ as $dx/dt$:

$$v_x = \frac{dx}{dt} = \frac{\beta m t^2}{2}$$

$$\int_{x=x_0}^{x} dx = \int_{t=0}^{t} \frac{\beta m t^2}{2}\, dt$$

$$x = x_0 + \frac{\beta m t^3}{6}$$

Inverting this expression,

$$t = \left[\frac{6(x-x_0)}{\beta m}\right]^{1/3}$$

Now we can perform the work integral:

$$W_{net} = \beta m \int_{x=x_0}^{x} \left[\frac{6(x-x_0)}{\beta m}\right]^{1/3} dx = \frac{3 \cdot 6^{1/3} \beta^{2/3} m^{2/3} (x-x_0)^{4/3}}{4} = \frac{3^{4/3} \beta^{2/3} m^{2/3} (x-x_0)^{4/3}}{2^{5/3}}$$

where we wrote $6^{1/3}$ as $2^{1/3} 3^{1/3}$ and 4 as $2^2$. Now we need to express the work done as a function of time before we can take the derivative. To do this, first solve for $x - x_0$ in terms of $t$:

$$x - x_0 = \frac{\beta m t^3}{6}$$

Now substitute this expression into the work equation:

$$W_{net} = \frac{3^{4/3} \beta^{2/3} m^{2/3}}{2^{5/3}} \left(\frac{\beta m t^3}{2 \cdot 3}\right)^{4/3} = \frac{\beta^2 m^2 t^4}{8}$$

Finally, the net power can be found by differentiating the net work with respect to time:

$$P_{net} = \frac{dW_{net}}{dt} = \frac{\beta^2 m^2 t^3}{2}$$

**Special cases of instantaneous power**: If the force is constant in both magnitude and direction, the work done by the force is $W = \vec{F} \cdot \Delta \vec{r} = F\|\Delta \vec{r}\| \cos\theta = F_x \Delta x + F_y \Delta y + F_z \Delta z$, in which case the power delivered by the force is

$$P = \vec{F} \cdot \vec{v} = Fv \cos\theta = F_x v_x + F_y v_y + F_z v_z \quad (\text{if } \vec{F} \text{ is const.})$$

where $\vec{v}$ is the velocity at the moment at which you would like to know the instantaneous power.

If the force has constant magnitude, but does not have constant direction, and yet the angle between $\vec{F}$ and $d\vec{s}$ is constant throughout the path, then the work done by the force is $W = F\Delta s \cos\theta$, for which the power delivered by the force is

$$P = Fv \cos\theta \quad (\text{if } F \text{ and } \theta \text{ are const.})$$

where $v$ is the speed at the moment at which you would like to evaluate the instantaneous power.

Unlike work, note that these two special cases yield the same result – a consequence of the fact that the speed is the magnitude of the velocity, $v = \|\vec{v}\|$, whereas the total distance traveled is in general not equal to the magnitude of the net displacement, $\Delta s \geq \|\Delta \vec{r}\|$.

> **Note.** The equation $P = Fv \cos\theta$ is a special case, which only applies when the force is constant in both magnitude and direction or when the force has constant magnitude and the angle between the force and the displacement is also constant. Otherwise, you must first express work as a function of time and then take a derivative with respect to time in order to find the instantaneous power delivered.

**Average power**: According to the mean-value theorem, the average power equals the work done divided by the corresponding time interval:

$$\bar{P} = \frac{\Delta W}{\Delta t} \quad (\text{average power})$$

> **Note.** Only divide the work done by the elapsed time if you are finding average power. If instead you want (instantaneous) power, you need to take a derivative of work with respect to time instead.

**Special cases of average power**: If the force is constant in both magnitude and direction, the average power equals

$$\bar{P} = \vec{F} \cdot \bar{\vec{v}} = F\|\bar{\vec{v}}\| \cos\theta \quad (\text{average power if } \vec{F} \text{ is const.})$$

where $\bar{\vec{v}}$ is the average velocity. If the force has constant magnitude, but does not have constant direction, and yet the angle between $\vec{F}$ and $d\vec{s}$ is constant throughout the path, the average power is

$$\bar{P} = F\bar{v} \cos\theta \quad (\text{average power if } F \text{ and } \theta \text{ are const.})$$

where $\bar{v}$ is the average speed. In contrast to instantaneous power, the expressions for average power are different in these two cases because average speed is generally not equal to the magnitude of the average velocity, $\bar{v} \geq \|\bar{\vec{v}}\|$.

**Horsepower**: A common unit of power in the context of motors is the horsepower (hp). The conversion between horsepower and Watts is 1 hp = 746 W.

## 5.3 Potential and Kinetic Energy

**Energy**: When we say that an object has energy, we mean that it has the ability to do work. Energy comes in many forms: Some examples include mechanical energy, potential energy, kinetic energy, electromagnetic energy, heat energy, and nuclear energy. All forms of energy refer to work that can be done. Recall that work is done when a force contributes toward the displacement of an object. When an object has energy – which is the ability to do work – this energy refers to one or more ways that a force can contribute toward the displacement of the object.

If you're already moving and you wish to keep moving, nothing has to happen – your wish is automatically granted by virtue of your inertia. But if you're at rest and you wish to move (i.e. be displaced), or if you're moving and you wish to change your speed, then you need a net external force to overcome your inertia. Another way of looking at this is that you need energy if you would like to change your speed. Since energy is the ability to do work, you can use energy to cause a force to contribute toward your displacement – this force can cause you to change speed (i.e. accelerate), and in the process you will be displaced.

As you strive to understand what energy is, try to see how each form of energy relates to work that can be done, and specifically how it relates to a force that can contribute toward a displacement.

Energy has the same SI unit as work – the Joule (J).

**Mechanical energy**: In this volume of this text, we are focused on fundamental concepts of classical mechanics – i.e. the study of the motion of classical objects. Mechanics excludes quantum mechanics – where the microscopic interactions of particles produce quantum effects like electrons tunneling through a barrier – and Einstein's theory of relativity – which describes what happens when particles travel nearly as fast as light, how gravity causes spacetime to be curved, and how rest-mass can be converted into energy. Mechanics also excludes other branches of physics, such as electromagnetism and thermodynamics. So mechanical energy relates to energy associated with motion, gravity, and springs, for example, and excludes electromagnetic energy, heat, and nuclear energy, for example. In this volume, we will mainly need to distinguish between mechanical energy and heat.

**Heat energy**: Heat is a concept from thermodynamics, which is relevant for mechanics because sometimes mechanical energy is lost or gained due to exchanges of heat. When resistive forces like friction and air resistance do work, mechanical energy is lost, but the total energy of the universe will be conserved as the mechanical energy lost by the system is gained through the surroundings in the form of heat. Heat will be one form of nonconservative work that mechanical energy can be converted into.

A macroscopic object is made up of a very large number (one mole of the substance consists of Avogadro's number – $6.02 \times 10^{23}$) of molecules (or pairs of ions bonded together, or pure elements, as the case may be). When the object gains or loses heat, this thermodynamic form of energy relates to work that these molecules can do. One effect is that the temperature of the object may change – meaning that the average speed of the molecules changes. Another effect is that the object as a whole can do work by expanding (as when a metal rod becomes longer when heated, or as when a gas pushes against a piston to increase its volume).

**Internal energy**: Another form of energy that is relevant for mechanics is internal energy. A macroscopic object inherently has internal energy – which you can visualize if you think about what's going on inside the object. Even if the object is at rest, the large number of molecules that it is composed of are moving around, and so they have energy associated with their motion (called kinetic energy). The molecules also have energy associated with molecular bonds. The electrons have electromagnetic energy associated with their attraction to the protons and repulsion from other electrons. The protons and neutrons have nuclear energy that binds them together in the nucleus. All of these forms of energy make up the internal energy of the object. Mechanical energy can be converted into internal energy, and vice-versa. We will see this, for example, when we study collisions in the next chapter.[234]

> **Note**. We're going to focus almost exclusively on mechanical energy. We mention heat and internal energy only to have some idea where the energy goes when mechanical energy is lost or gained through nonconservative work. We will occasionally refer to heat or internal energy directly, in which case these definitions can help you get a sense of what this means.

**Two classes of energy**: All forms of energy can be classified as potential energy or kinetic energy (or heat). We will now define these two terms, and then consider these two forms of energy in detail.

**Potential energy**: Work that an object can do by changing position is called potential energy. Two examples of mechanical potential energy include gravitational potential energy and potential energy stored in a spring. Gravitational potential energy is work that an object can do by changing altitude, and spring potential energy is work that a spring can do by compressing or stretching.

It may be tempting to think of potential energy as "the potential to do work," but this would be a conceptual mistake because all forms of energy refer to the ability to do work. Potential energy can be thought of as stored energy. For example, a banana sitting atop a table has stored gravitational energy – simply push the banana off the edge of the table and gravity will instantly do work on it. Similarly, a spring that is compressed from equilibrium has stored energy – simply let go and the spring will stretch back to equilibrium. Mechanical wind-up toys have energy that is stored in them after they are wound up, and as soon as you release the screw, they do work.

---

[234] The first law of thermodynamics represents conservation of energy as it relates to heat, internal energy, and work that a macroscopic system can do by expanding. The change in internal energy of the system equals the heat added to the system (or subtracted from the system) minus the work that the system does by expanding (or contracting) plus chemical work that can be done by gaining (or losing) molecules.

For reasons that we will investigate in Sec. 5.5, it is only physically meaningful to associate potential energy with conservative forces. Mathematically, we will define the change in potential energy associated with a particular conservative force as the negative of the work done by that force:

$$\Delta U = U - U_0 = -W_c = -\int_i^f \vec{F}_c \cdot d\vec{s}$$

We will apply this mathematical definition to derive equations for potential energy for gravity and springs later in this section.

**Kinetic energy**: Work that can be done by changing speed is termed kinetic energy. Objects can actually have two kinds of kinetic energy: Translational kinetic energy is work that an object can do by changing its tangential speed, and rotational kinetic energy is work that an object can do by changing its angular speed. We will speak of kinetic energy and mean only translational kinetic energy until we reach Chapter 7, where we will study rotations.

Kinetic energy is sometimes referred to as the energy of motion. An object that is moving can do work by changing its speed. Keep in mind that an object in motion is not necessarily doing work – since it has inertia, it has a natural tendency to maintain its motion. Rather, an object in motion has kinetic energy in the sense that it can do work by changing speed. If the net external force acting on the object is zero, it won't change speed. If the net force is nonzero, it will change speed – in this case, net work is done by (or on) the object. When we apply this concept mathematically, we will see that the equation for kinetic energy is naturally interpreted as work that an object can do by coming to rest (i.e. by losing all of its speed). Since the net external force causes an object to change momentum, according to Newton's second law, we will mathematically define the change in kinetic energy as the work done by the net force:

$$\Delta K = K - K_0 = W_{net} = \int_i^f \left(\sum \vec{F}_{ext}\right) \cdot d\vec{s}$$

We will apply this mathematical definition to derive an equation for kinetic energy later in this section.

**Gravitational potential energy**: A massive object in the presence of a gravitational field has gravitational potential energy: Gravity is pulling downward on an object that is on or in the vicinity of a planet, and will do work on the object if it is not restrained by other forces (such as a normal force). If the object begins from rest and its weight is the only force acting on it, gravity will do work on the object, causing it to be displaced downward in a state of free fall. If the object is initially heading upward in free fall, gravity does work to reduce its speed. If you wish to climb a mountain, you must do work against gravity; and if you are unfortunate enough to lose your step and grip, gravity will naturally do work to bring you back down the mountain. Gravitational potential energy represents work that can be done by changing altitude. We will derive two useful equations for gravitational potential energy – one for a uniform gravitational field and one for a single spherically symmetric massive object (such as a planet) that allows for significant variations in altitude.

Let us first consider the case of a uniform gravitational field. Taking the $+y$-direction to be upward, the gravitational force in this case is $\vec{F}_g = -mg\hat{j}$, where $g$ is a constant. We can derive an equation for the gravitational potential energy of a massive object in a uniform gravitational field by starting with the mathematical definition of potential energy:

$$U_g - U_{g0} = -W_g = -\int_i^f \vec{F}_g \cdot d\vec{s} = \int_{y=y_0}^{y} mg \, dy = mg(y - y_0) \quad (\text{if } \vec{g} \text{ is const.})$$

In this context, it is conventional to work with $h$ for height, instead of using the symbol $y$, such that

$$U_g - U_{g0} = mgh - mgh_0 \quad (\text{if } \vec{g} \text{ is const.})$$

Separating initial from final, we obtain an expression for the gravitational potential energy of a massive object at a height, $h$, in a uniform gravitational field:

$$U_g = mgh \quad (\text{if } \vec{g} \text{ is const.})$$

Now an interesting issue arises: Where do we measure height, $h$, from? You might want to measure it relative to the ground. Someone else might want to measure it relative to the center of the earth. If you throw a ball up and it lands on a roof, you might want to measure $h$ from where the ball left your hand. It turns out not to matter: You can measure $h$ from any reference height of your choosing, so long as you measure all of the heights in the problem from the same reference height. This seeming arbitrariness is associated with a constant of integration: If you want to find the potential energy at specific position, there is an arbitrary constant of integration involved; but if you want to find the change in potential energy between two positions, this corresponds to a definite (rather than indefinite) integral, and the constant of integration cancels out. If you throw a ball upward, we will all agree on how much the height changes, regardless of where we choose our reference heights to be. Only the change in potential energy is measureable, and everyone will agree on this.

Note that $mgh$ may be negative. If you calculate the gravitational potential energy of an object that lies below your reference height, $h$ will be negative. You will still obtain the correct change in potential energy for an object that changes height.

Now let us consider a massive object in the presence of a (non-uniform) gravitational field created by a single spherically symmetric object, such as a planet. In this case, the force is given by Newton's law of universal gravitation (see Sec. 3.9): $\vec{F}_g = -G\frac{m_p m}{R^2}\hat{R}$. Again, we start with the definition of potential energy, with the origin (but not the reference point!) at the center of the planet:

$$U_g - U_{g0} = -W_g = -\int_i^f \vec{F}_g \cdot d\vec{s} = \int_i^f G\frac{m_p m}{R^2}\hat{R} \cdot \left(\hat{r} \, dr + \hat{\theta} \, r d\theta + \hat{\varphi} \, r \sin\theta \, d\varphi\right) \quad (\text{single point-mass})$$

$$U_g - U_{g0} = Gm_p m \int_{R=R_0}^{R} \frac{dR}{R^2} = -Gm_p m \left(\frac{1}{R} - \frac{1}{R_0}\right) \quad (\text{single point-mass})$$

Separating initial from final, we obtain an expression for the gravitational potential energy of a massive object in the presence of a single spherically symmetric massive object (which is the same as that of a point-mass):

$$U_g = -\frac{Gm_p m}{R} \quad \text{(single point-mass)}$$

Here, $R$ is measured from the center of the planet (or moon or other astronomical body). However, the reference "height" is actually infinitely far away! When we used the formula $mgh$, the reference height marked the position where $mgh$ was zero. When we use the formula $-Gm_p m/R$, the reference height marks where $-Gm_p m/R$ is zero (and not where $R$ is zero). If you recognize that the formula $-Gm_p m/R$ does not apply when $R$ is less than the radius of the planet (a point that was made in Sec. 3.9), you will see that it doesn't make sense to consider $R$ equal to zero. Putting the reference height infinitely far away is conceptually significant: It corresponds to how much work must be done to escape the gravitational pull of the planet. We will apply this concept in the next section to derive an expression for the escape speed of a rocket.

In the formula $U_g = -Gm_p m/R$, gravitational potential energy is always negative, and approaches its maximum value of zero only in the limit that $R$ becomes infinite. It is typical for binding potential energies to be defined so that they are always non-positive. Atomic, molecular, and nuclear binding energies are similarly negative for attractive forces, but are positive for repulsive forces. An electron's energy in the electron cloud of an atom is negative, for example, where zero energy corresponds to the ionization energy that must be added to free that particular electron. However, not all attractive potential energies are defined so that potential energy is negative; for example, we will see that a mass connected to a spring is attracted to equilibrium, but in this case the potential energy is nonnegative. Once again, you can always add an arbitrary constant of integration to potential energy; it's the change in potential energy that really matters.

If there are multiple spherically symmetric sources (e.g. if you have a rocket in between a planet and a moon), you must use $U_g = -Gm_p m/R$ for each source (measuring each $R$ from the object to the planet's — or moon's, or star's — center) and add all of these up to find the total gravitational potential energy.

> **Note**. When calculating the gravitational potential energy of an object in a uniform gravitational field, you must choose a reference height from which to measure $mgh$. This is analogous to setting up a coordinate system and choosing where to put the origin. Measurable physics will depend only on how much the potential energy changes, so your choice of reference height will cancel out in the end. When using $-Gm_p m/R$ (for problems where the object may make a significant change in altitude), the reference point is actually infinitely far away, but we still measure $R$ from the object to the center of the planet.

**Spring potential energy**: Energy is stored in a spring when the spring is not at equilibrium. When a spring is compressed or stretched from equilibrium, the spring exerts a restoring force in an effort to restore equilibrium. This is work that the spring can do by returning to equilibrium.

For a spring, the natural reference point from which to measure spring potential energy is the equilibrium position. Therefore, the spring will have no potential energy when it is at the equilibrium position (since it is happy to remain at equilibrium, the spring does not want to do work when it is at this position). We will setup our coordinate system with the $+x$-axis in the direction of the stretch, and designate the equilibrium position by $x_e$. Assuming that Hooke's law applies to the spring, when the spring is stretched from equilibrium, the restoring force equals $\vec{F}_r = -k(x - x_e)\hat{i}$; and when the spring is compressed from equilibrium, the restoring force is $\vec{F}_r = k(x - x_e)\hat{i}$ (see Sec. 3.6). Either way, if the spring is approaching equilibrium, $\vec{F}_r$ is parallel to $d\vec{s}$, such that $\vec{F}_r \cdot d\vec{s}$ is positive. Let us calculate the change in potential energy of the spring as it returns to equilibrium (where the final potential energy is zero).[235] Since the final potential energy is zero, let us write 0 for it and use $U_s$ for the initial potential energy (rather than $U_{s0}$):[236]

$$0 - U_s = -W_s = -\int_i^f \vec{F}_r \cdot d\vec{s} = \int_{x=x}^{x_e} k(x - x_e)dx = \left[\frac{k(x-x_e)^2}{2}\right]_{x=x}^{x_e}$$

$$0 - U_s = -\frac{k(x-x_e)^2}{2}$$

$$U_s = \frac{k(x-x_e)^2}{2}$$

Observe that $U_s$ equals zero at $x_e$ (which is our reference position), and that $U_s$ maximum when the spring is furthest from equilibrium. Also, $U_s$ is nonnegative; it's positive regardless of whether the spring is stretched or compressed from equilibrium.

**Note.** In general, there may be more than one kind of potential energy in a problem. Add up all of the potential energies to find the total potential energy of an object (or system of objects).

**Translational kinetic energy**: We will derive an equation for the translational kinetic energy of an object that has constant mass from the mathematical definition of kinetic energy:

$$\Delta K = K - K_0 = W_{net} = \int_i^f \left(\sum \vec{F}_{ext}\right) \cdot d\vec{s}$$

First, we will separate the net external force into components:

---

[235] The subscript $r$ designates the restoring force associated with Hooke's law, and the subscript $s$ refers to spring.

[236] The signs are tricky here. We know that $\vec{F}_r \cdot d\vec{s}$ is positive when the spring heads toward equilibrium. If the spring is stretching back to equilibrium, $\vec{F}_r$ is along $\hat{i}$ and $d\vec{s}$ is along $\hat{i}$ (and $x - x_e$ is negative the whole way); and if the spring is compressing, $\vec{F}_r$ is along $-\hat{i}$ and $d\vec{s}$ is along $-\hat{i}$ (and $x - x_e$ is positive the whole way). Either way, the result of the integral must be positive, and the overall minus sign that precedes the integral will persist. We temporarily absorbed the minus sign into the integral, since it appears again in the end when we evaluate the integral over the limits. At the very least, you should agree conceptually with the sign of the final result.

## 5 Work and Energy

$$\sum \vec{F}_{ext} = \hat{t}\sum F_T + \hat{n}_i \sum F_{in}$$

where $\hat{t}$ is a unit tangent (parallel to the instantaneous velocity, $\vec{v}$) and $\hat{n}_i$ is a unit normal that points toward the instantaneous center of curvature (i.e. along the centripetal acceleration, $\vec{a}_c$). Since $\hat{t}$ parallel to $d\vec{s}$ and $\hat{n}$ is perpendicular to $d\vec{s}$, only the first sum will contribute to the scalar product:

$$K - K_0 = \int_i^f \left(\sum F_T\right) ds$$

For a single object with constant mass, Newton's second law in component form states that the net tangential force equals mass times the tangential component of acceleration:

$$K - K_0 = \int_i^f ma_T ds$$

The tangential component of acceleration equals a derivative of tangential velocity with respect to time:

$$K - K_0 = m\int_i^f \frac{dv_T}{dt} ds$$

Bear in mind that $dv_T/dt$ is finite, while $ds$ is infinitesimal. Using the chain rule, we can rewrite $\frac{dv_T}{dt}ds$ as $dv_T \frac{ds}{dt} = v_T dv_T$:

$$K - K_0 = m\int_i^f v_T dv_T = \frac{mv_T^2}{2} - \frac{mv_{T0}^2}{2}$$

Since the speed equals the absolute value of the tangential velocity, and the tangential velocity is squared in the equation for kinetic energy, we may discard the subscript $T$ and just work with the speed, $v$. Separating the initial and final quantities, we obtain an equation for the kinetic energy of an object:

$$K = \frac{mv^2}{2}$$

The kinetic energy is quadratic in the speed – i.e. proportional to the square of the speed. For example, if the speed of an object doubles, its kinetic energy quadruples: $K(2v) = \frac{m(2v)^2}{2} = 2mv^2 = 4K(v)$.

**Work-energy theorem**: The work energy theorem states that the net work done on an object equals its change in kinetic energy, which is really what we used as our definition of kinetic energy:

$$W_{net} = \Delta K = K - K_0 = \frac{mv^2}{2} - \frac{mv_0^2}{2}$$

> **Note.** Many students who attempt to apply the work-energy theorem to solve a problem make mistakes by not considering all of the work – both conservative and nonconservative by all of the forces acting in the problem – or by not accounting for the signs correctly. If you want to use the work-energy theorem, you could just as well apply the strategy of conservation of energy (which we consider in detail, with examples, in the next section). Conservation of energy will naturally account for all kinds of work, separating conservative and nonconservative terms and also separating initial from final, and will be a conceptually friendly way (compared to the work-energy theorem) to solve problems that relate speed to position.

## 5.4 Conservation of Energy

**Conservation of energy from the work-energy theorem**: According to the work-energy theorem, the net work done on an object (or system of objects) equals its change in kinetic energy:

$$W_{net} = K - K_0$$

The net work includes work done by conservative and nonconservative forces. We can write this explicitly as

$$W_c + W_{nc} = K - K_0$$

The work done by the conservative forces equals the negative of the change in the total potential energy:

$$-(U - U_0) + W_{nc} = K - K_0$$

Now we reorganize the terms, putting initial on the left and final on the right:

$$U_0 + K_0 + W_{nc} = U + K$$

Note that $U_0 + K_0 = E_0$ is the total initial mechanical energy and $U + K = E$ is the total final mechanical energy:

$$E_0 + W_{nc} = E$$

This equation expresses conservation of energy. The total mechanical energy, $E$, is conserved for a system unless the system exchanges energy with its surroundings through nonconservative work, $W_{nc}$. The total mechanical energy of a system may not be conserved, but the total energy of the system plus its surroundings is always conserved.

**Conservation of energy**: The equation

$$U_0 + K_0 + W_{nc} = U + K$$

represents conservation of energy for a system, where $U_0$ is the total initial potential energy, $K_0$ is the total initial kinetic energy, $W_{nc}$ is work done by any nonconservative forces acting on the system, $U$ is the total final potential energy, and $K$ is the total final kinetic energy. We will use the above equation as the starting point for any solution where we apply the principle of conservation of energy.

**Potential energies**: The total potential energy of an object will include one or more kinds of potential energy. Here is a review of the types of potential energy that we considered in the previous section:
- The gravitational potential energy of an object in a uniform gravitational field is $U_g = mgh$. You must define a reference height from which to measure $h_0$ and $h$ for each object in the system. Note that $U_g$ is zero at the reference height.
- The gravitational potential energy of an object in the gravitational field created by a spherically symmetric distribution of mass – where the altitude changes significantly in the problem such that $g$ is not approximately uniform – is $U_g = -Gm_pm/R$, where $R$ is measured from the center of mass of the source to the center of mass of the object. In this case, $U_g$ is only zero infinitely far away from the source.
- The potential energy of a (1D) system that obeys Hooke's law is $U_s = \frac{1}{2}k(x - x_e)^2$, where $x_e$ represents the equilibrium position. Note that $U_s$ is zero when the system is at equilibrium. If you setup your coordinate system with the origin at equilibrium, then $x_e$ will be zero.

> **Note.** It's possible for $U_0$ and/or $U$ to include more than one term – i.e. there may be more than one type of potential energy initially or finally (such as gravitational potential energy plus spring potential energy). If there is more than one object in the system, both $U_0$ and $U$ will include terms for the gravitational potential energy of each object in the system (which will have different positions, in general). Be sure to use different symbols for the initial and final positions (i.e. $h_0$ and $h$, $R_0$ and $R$, and $x_0$ and $x$). When using $mgh$, you must first define a reference height, and use the same reference mark to measure both $h_0$ and $h$. Similarly, you must choose an origin before you can measure $x_0$ and $x$ (and also $x_e$).

**Kinetic energies**: Each object in the system will have translational kinetic energy equal to $mv_0^2/2$ initially and $mv^2/2$ finally if its center of mass is moving. If an object is at rest, its kinetic energy will be zero. If an object is rotating, you will also need to include rotational kinetic energy (see Chapter 7). However, if it is revolving (traveling in a circular path, as opposed to spinning about an axis), you may use translational kinetic energy instead (since its center of mass is moving in a circle).

**Work done by nonconservative forces**: If there are any nonconservative forces acting on the system, you can calculate the work done the same way that you can calculate the work done by any force – as described in Sec. 5.2:

$$W_{nc} = \int_i^f \vec{\mathbf{F}}_{nc} \cdot d\vec{\mathbf{s}}$$

There are two main types of nonconservative forces to consider: (1) Resistive forces like friction and air resistance, for which $W_{nc}$ is negative because resistive forces oppose the displacement; and (2) external tension forces (and other external forces with a similar effect), for which $W_{nc}$ may be positive or negative. In either case, the work done by the nonconservative force can be found by applying the integral above, following the examples of Sec. 5.2 (the work done by friction was one of the examples).

When there is friction or air resistance in a problem (remember that we neglect air resistance unless stated otherwise), it will be obvious that you must account for $W_{nc}$.[237] So the trickier case is when a problem has an external tension force (or other external force – but not an external force like weight, which we know to be conservative – with a similar effect). For example, recall Atwood's machine (see Sec. 3.4), which consists of two masses connected to the ends of a cord that passes over a pulley (which we will presently take to be frictionless to illustrate the main point). If you define your system as just one of the masses, you will need to treat tension as a nonconservative force – since this tension force affects how the mass changes speed. If instead you define your system as the pair of masses, the work done by the two tension forces will cancel out (since the tension forces are equal according to Newton's third law, provided that the pulley is frictionless).

**Applying conservation of energy**: Since potential energy depends on position, nonconservative work depends on displacement (change in position), and kinetic energy depends on speed, it is particularly useful to apply the principle of conservation of energy to relate speed to position. Conservation of energy is a very general strategy – like Newton's second law – that applies to a wide variety of problems from all branches of physics (in contrast to the strategy for solving uniform acceleration problems, for example, which applies only to those few problems where acceleration is constant).

To apply conservation of energy, first you have to define what your system consists of (i.e. which objects are part of the system); then anything else is part of the surroundings. You also need to define which positions to label as initial and final: One position should correspond to a point in the motion where you know information about the various types of energy, and the other position should correspond to the unknown that you are solving for. You may need to redefine which points you label as initial and final in multi-part problems (and if so, you better be careful not to use the old initial values as the new initial values if the initial position has changed, and the same for the final). Next, choose reference positions from which to measure height and/or the position of a spring (when relevant). Then write symbolic expressions for the total initial and final potential and kinetic energies, and also any nonconservative work that may be done. It's very important to understand when each kind of potential energy and kinetic energy is or isn't zero. After expressing conservation of energy, you are ready to proceed to solve for the unknown(s). We will illustrate this strategy in several examples.

---

[237] Exception: When we study rolling without slipping in Chapter 7, there will be friction, yet $W_{nc}$ will be zero!

**Note.** Newton's second law is particularly useful for relating forces to acceleration. Conservation of energy is very convenient for relating speed to position. If you just need to relate speed to position, and you don't know and aren't solving for acceleration, it will probably be much more efficient to use conservation of energy to solve a problem than Newton's second law. However, if you know acceleration or are looking for acceleration, Newton's second law will probably be more efficient.

## Problem-Solving Strategy for Conservation of Energy

0. It is useful to apply the principle of conservation to relate speed to position. It may also be a useful prelude to deriving equations of motion for a system.
1. Define the system. The system may consist of a single object, or it may consist of multiple objects. Only choose objects for which you can account for mechanical potential or kinetic energy – a motor, for example, involves other kinds of energies, too. Whatever is not part of the system is part of the surroundings. You will write down potential and kinetic energy terms for the system only, and the nonconservative work term will account for energy exchanges between the system and surroundings.[238]
2. Choose the reference position(s) from which to measure the relevant forms of potential energy. This includes defining a reference height for a uniform gravitational field and an origin from which to measure the position of a spring.
3. Identify which points in the motion to label as initial, $i$, and final, $f$. One point should correspond to your unknown, and another point should correspond to where you know information. Sketch the motion and label these points. (In a multi-part problem, you may need $i_1$ and $f_1$, $i_2$ and $f_2$, etc., and you must be wary that what is initial in one part may not be the same initial in another part – and similarly for final).
4. Determine what types of potential and kinetic energy there are in the initial and final position, and also what type of nonconservative work is done, if any. Gravitational potential energy is zero at the reference height (or where $R$ is infinite if the field is not uniform), spring potential energy is zero when the spring is at equilibrium, and kinetic energy is zero where an object is at rest. Use different subscripts for initial and final quantities, such as $v_0$ and $v$ for initial and final speed.
5. Write symbolic equations for each type of energy, and also for any nonconservative work (by performing the work integral). Plug all of these equations into the equation for conservation of energy, $U_0 + K_0 + W_{nc} = U + K$. Note that there may be multiple terms for any one of the symbols in this equation (i.e. there may be multiple kinds of potential energy – e.g. there could be both a spring and gravity in the problem).
6. Count your knowns and unknowns. In algebra courses, you generally need $N$ equations in $N$ unknowns, but beware that in physics courses, you can often solve a system of $N$ equations with fewer than $N$ unknowns – the reason is that some variables, like mass, often cancel out after substitutions.
7. If you are trying to derive the equations of motion for a system, write the speed as $v = ds/dt$ and proceed to solve this first-order differential equation by separating variables (if possible). We will consider this last step in more detail (with an example) later in this section.

---

[238] All heat exchanges will be considered as an exchange of energy between the system and surroundings, even if the heat is transformed into the internal energy of an object within the system. We're writing mechanical energy only for the system, and any other form of energy we will associate with an exchange with the surroundings.

**Conceptual Example.** A monkey who is holding a banana stands on the roof of a building. The monkey wants to throw the banana and watch it 'splat' when it strikes the horizontal ground below. Assuming that the initial speed of the banana will be the same regardless of which way the banana is thrown, and assuming that whichever way the monkey throws the banana that the initial height of the banana will be the same when it leaves his hand, at what angle relative to the horizontal should the monkey throw the banana so that it has the greatest speed when it lands?

Think about this and see if you can reason out the answer before you read on. No cheating now! It would be tedious to apply the projectile motion strategy and solve for the final speed in terms of the initial angle – especially, since the equations of projectile separately involve $v_x$ and $v_y$ (the final components of velocity), from which the final speed equals $v = \sqrt{v_x^2 + v_y^2}$. You would need to apply trig to resolve the initial velocity vector into components, combine multiple equations together algebraically, and then proceed to determine what launch angle maximizes the final speed.

In comparison, it is far simple to use conservation of energy as a guide. Note that conservation of energy will efficiently relate the initial height of the banana to the final speed, and since it is a scalar equation, we don't need to use trig to resolve vectors into components. We choose the initial position to be just after the banana leaves the monkey's hand and the final position to be just before impact.[239] Unless the monkey uses a rocket launcher, the banana won't make a significant change in altitude, so we can use $mgh_0$ for the initial gravitational potential energy. We choose to measure $h_0$ relative to the ground, so that the final height is zero.

Note that the initial potential energy is the same value, $mgh_0$, and the final potential energy is zero regardless of which way the banana is thrown. The initial kinetic energy, $\frac{1}{2}mv_0^2$, is similarly the same regardless of the direction of the throw. We always neglect air resistance unless stated otherwise, so $W_{nc}$ is zero. Now look at the conservation of energy equation, $U_0 + K_0 + W_{nc} = U + K$. Since $U$ and $W_{nc}$ are zero in this example, $U_0 + K_0 = K$. We have already reasoned that $U_0$ and $K_0$ are the same value regardless of which way the banana is thrown, which implies that $K$ is also independent of the launch angle. Therefore, no matter which way the monkey throws the banana, its final speed will be the same.

---

**Example.** A monkey throws a banana upward with an initial speed of 20 m/s at some angle from a height of 12 m above the ground. The banana rises to a maximum height of 22 m above the ground. What is the speed of the banana at the top of its trajectory?

We could solve this as a projectile motion problem, but since we are relating speed to position, we can also use conservation of energy. We choose the banana to be the system, the initial point to be just after release, the final position to be at the top of the trajectory, and measure the heights relative to the ground. There are no nonconservative forces acting. Conservation of energy leads to

$$U_0 + K_0 + W_{nc} = U + K$$

---

[239] The force of the throw and the collision with the ground involve nonconservative work, which we don't have enough information to account for. It's a major conceptual mistake to think that the 'initial' speed is zero because the monkey hasn't thrown it yet (as you should instead be thinking of what the speed is just after release) or that the 'final' speed is zero because it's lying on the ground (instead, think how fast it's moving just before impact).

$$mgh_0 + \frac{mv_0^2}{2} = mgh + \frac{mv^2}{2}$$

$$v = \sqrt{v_0^2 + 2g(h_0 - h)} = \sqrt{(20)^2 + 2(9.81)(12-22)} = 14 \text{ m/s}$$

**Example.** A monkey slides from rest down a frictionless hill in the shape of a quarter-circle with a 10-m radius, as illustrated below. Determine the monkey's speed at the bottom.

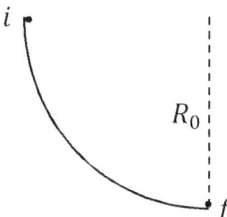

First, note that the equations of uniform acceleration do not apply because the acceleration is not constant (neither in direction nor magnitude). Conservation of energy does apply, however, and is convenient since we are solving for the final speed and we know information about the change in position. We define the system to be just the monkey. We labeled our choice of initial and final positions on the diagram above. We choose the final position as our reference height. The initial kinetic energy is zero because the monkey begins at rest, and the final potential energy is zero because the final position lies at our reference height. Note that the initial height equals the radius of the quarter-circle. There are only conservative forces acting on the monkey. Conservation of energy results in:

$$U_0 + K_0 + W_{nc} = U + K$$
$$mgR_0 + 0 + 0 = 0 + \frac{mv^2}{2}$$
$$v = \sqrt{2gR_0} = \sqrt{(2)(9.81)(10)} = 14 \text{ m/s}$$

**Example.** A box of bananas slides from rest down a 30° incline. The coefficient of friction between the box and the incline is 0.25. What is the speed of the box after sliding 8.0 m down the incline?

Conservation of energy provides an efficient calculation of the final speed (compared to applying Newton's second law combined with 1DUA). Mechanical energy is not conserved for the box of bananas; the mechanical energy lost in the form of heat is accounted for through $W_{nc}$, which is the work done by the friction force. The natural choice for the system is just the box of bananas, and the natural choice for initial and final are rest and after sliding 8.0 m. We choose the reference height to be at the final position:

$$U_0 + K_0 + W_{nc} = U + K$$
$$mgh_0 + 0 + \int_i^f \vec{\mathbf{f}} \cdot d\vec{\mathbf{s}} = \frac{mv^2}{2}$$

$$mgh_0 + \int_{x=0}^{8} \mu N \cos 180° \, dx = \frac{mv^2}{2}$$

The normal force can be found readily by drawing a FBD and summing the components of the forces perpendicular to the incline:

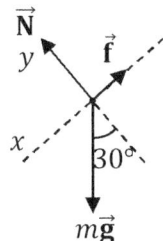

$$\sum_{i=1}^{3} F_{iy} = ma_y$$
$$N - mg \cos 30° = 0$$
$$N = \frac{mg\sqrt{3}}{2}$$

Substituting this back into the equation for conservation of energy, and relating the height to the displacement through trig – i.e. $h = \Delta x \sin 30°$ – we find that:

$$mg\Delta x \sin 30° - \mu \frac{mg\sqrt{3}}{2} \Delta x = \frac{mv^2}{2}$$

$$v = \sqrt{g\Delta x - \mu g \Delta x \sqrt{3}} = \sqrt{g\Delta x (1 - \mu\sqrt{3})} = \sqrt{(9.8)(8.0)[1 - (0.25)\sqrt{3}]} = 6.7 \text{ m/s}$$

---

**Example.** An Atwood's machine (see Sec. 3.4) consists of two bananas suspended in midair[240] attached to a string that passes over a pulley. Neglect friction between the cord and pulley. One banana of mass $m_1$ is initially at a height $h_{10}$ above the (horizontal) floor, and the other banana of mass $m_2 > m_1$ has initial height $h_{20}$ relative to the floor. The system is released from rest. Derive an equation for the final speed of the system just before the banana of mass $m_2$ strikes the floor (assuming that the other banana has not yet reached the pulley).

We proceed to conserve energy for the system of two bananas, choosing our reference height to be the floor, the initial position to be the point of release and the final position to be just before the heavier banana strikes the floor. There are two terms for the initial potential energy and the final kinetic energy – one for each banana – and there is one term for the final potential energy (since one banana – the one that rises – is above the floor in the final position):

---

[240] Or, more precisely, mid-vacuum, since we will neglect air resistance.

# 5 Work and Energy

$$U_0 + K_0 + W_{nc} = U + K$$

$$m_1 g h_{10} + m_2 g h_{20} + 0 + 0 = m_1 g(h_{10} + h_{20}) + \frac{m_1 v^2}{2} + \frac{m_2 v^2}{2}$$

$$(m_2 - m_1) g h_{20} = \frac{(m_1 + m_2) v^2}{2}$$

$$v = \sqrt{2 g h_{20} \frac{m_2 - m_1}{m_1 + m_2}}$$

**Conservation of energy for oscillatory systems**: Consider an oscillatory system in the absence of any resistive forces (like friction and air resistance). One characteristic of such oscillations is a conservative restoring force or restoring torque. For example, Hooke's law features a restoring force, and a pendulum features a restoring torque. Associated with the restoring force or torque is a potential energy. In the case of a spring, there is spring potential energy, and in the case of a pendulum, the restoring torque is exerted by gravity, for which the potential energy is gravitational.

If you make the natural choice of putting the reference position at equilibrium, the spring potential energy of a spring – or the gravitational potential energy of a pendulum – is zero when the system is at equilibrium and maximum when the system is furthest from equilibrium. Conversely, the kinetic energy is maximum at the instant when the system is passing through the equilibrium position and zero at the turning points. At any other position, the system has both potential and kinetic energy.

**Note**. The kinetic energy of a spring or pendulum is greatest at equilibrium. If you put the system at equilibrium at rest it will just stay at equilibrium and not oscillate, but if you displace the system from equilibrium and release it from rest, the system will oscillate back and forth and be moving fastest when it passes through the equilibrium position. So for problems where there is oscillation, the system is not at rest at the equilibrium position.

**Example**. A simple pendulum[241] of length $L$ and mass $m$ is released from rest with the string making an angle $\theta$ with the vertical. Derive an equation for the speed of the pendulum as it passes through the equilibrium position.

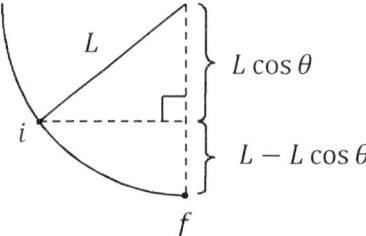

---

[241] A simple pendulum consists of a pointlike bob connected to an inextensible string of negligible mass, which is suspended from a fixed point of support and oscillates back and forth in a plane in the presence of a uniform (vertical) gravitational field in the absence of air resistance. If the pendulum bob is not approximately pointlike or if the string has significant mass, for example, then the pendulum is not classified as 'simple.'

We define the system to be the pendulum, choose the reference height to be at the bottom of the swing, the initial position to be the point of release, and the final position to be at the bottom:

$$U_0 + K_0 + W_{nc} = U + K$$
$$mgh_0 + 0 + 0 = 0 + \frac{mv^2}{2}$$
$$g(L - L\cos\theta) = \frac{v^2}{2}$$
$$v = \sqrt{2gL(1 - \cos\theta)}$$

**Hint.** It's worth studying the 'sailboat'-looking diagram in the preceding problem, as it comes up in many standard physics problems. You'll want to be able to reproduce this and apply this trigonometry.

**Example.** One end of a horizontal 32 N/m spring is fixed to a vertical wall, while a 2.0-kg box of banana-shaped chocolates is connected to its free end. There is no friction between the box and the horizontal. The box is displaced 5.0-cm from the equilibrium position and released from rest. What is the speed of the box when the spring is compressed 4.0 cm from equilibrium?

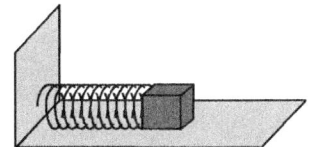

We measure the spring potential energy relative to equilibrium. Gravitational potential energy is not relevant for this problem since the motion is horizontal (the height doesn't change – so we can choose the reference height so that the initial and final heights are both zero). We choose the system to be the spring and box of banana-shaped chocolates, the initial position to correspond to the point of release, and the final position to be when the spring is compressed 4.0 cm from equilibrium. For this choice, the spring potential energy is maximum initially and there is no kinetic energy initially, and there are both potential and kinetic energy finally. We use $x_m$ to represent the maximum displacement from equilibrium (5.0 cm) and $x$ to represent a lesser displacement (in this case, 4.0 cm):

$$U_0 + K_0 + W_{nc} = U + K$$
$$\frac{kx_m^2}{2} + 0 + 0 = \frac{kx^2}{2} + \frac{mv^2}{2}$$
$$v = \sqrt{\frac{k}{m}(x_m^2 - x^2)} = \sqrt{\frac{32}{2.0}(0.050^2 - 0.040^2)} = 12 \text{ cm/s}$$

**Roller coaster loop:** Consider a roller coaster track that features a loop-the-loop, such as the one illustrated on the following page. A basic question to ask is whether any unrestrained items in the coaster may fall out when it passes through the loop – a concern passengers would have if any of the safety harnesses and/or seat belts were to malfunction during the ride.

5 Work and Energy

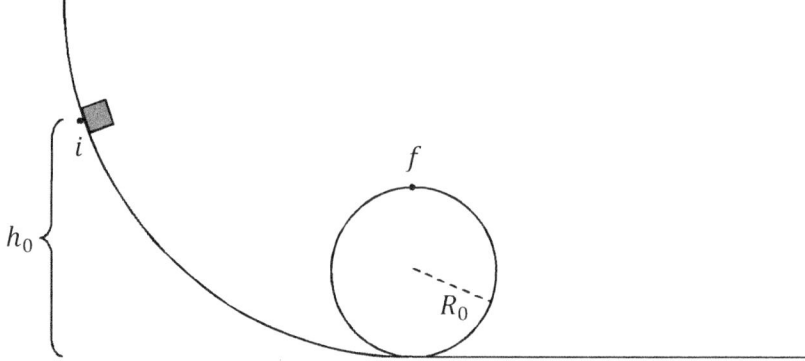

In order to illustrate how the principle of conservation of energy can be applied to answer this question, let us model the situation with a block that slides along a frictionless surface that features a circular loop of radius $R_0$. For specificity, we will assume that the block begins from rest from a height $h_0$ relative to the bottom of the circular loop, and we will derive an equation for the minimum value of $h_0$ needed in order for the block not to fall away from the surface of the loop.

The main underlying concept involved is to solve for the normal force, $N$, exerted by the surface on the block. If $N$ is nonnegative, this means that the block is still in contact with the track. The critical point is obviously at the top of the loop. We can solve for $N$ at the top of the loop by drawing a FBD and applying Newton's second law. Since the block is traveling in a circle in the loop, it is convenient to work with inward and tangential directions. At the top of the loop, the tangential component of acceleration is instantaneously zero ($a_T = 0$) because no forces have tangential components at this position, but the centripetal acceleration, $a_c$, is nonzero because the block is changing direction (see Sec.'s 4.2-4.5).

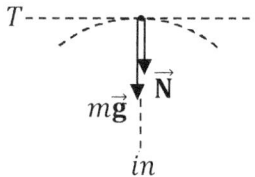

$$\sum_{i=1}^{2} F_{i,in} = ma_c$$

$$N + mg = \frac{mv^2}{R_0}$$

$$N = m\left(\frac{v^2}{R_0} - g\right)$$

We want the normal force to be nonnegative in order for the block not to lose contact with the track:

$$N > 0 \quad \Rightarrow \quad \frac{v^2}{R_0} - g > 0 \quad \Rightarrow \quad v^2 > R_0 g$$

Now we apply the principle of conservation of energy to relate the initial height of the block to its speed at the top of the loop. We choose the system to be just the block, the reference height to be at the bottom of the loop, the initial position to be the point of release, and the final position to be at the top of the loop:

$$U_0 + K_0 + W_{nc} = U + K$$
$$mgh_0 + 0 + 0 = mg2R_0 + \frac{mv^2}{2}$$
$$v^2 = 2gh_0 - 4gR_0$$

Substituting this expression into the inequality from the normal force, we find that

$$2gh_0 - 4gR_0 > R_0 g$$
$$h_0 > \frac{5R_0}{2}$$

That is, if the initial height of the block (relative to the bottom of the loop) is at least 1.25 times the diameter of the loop, the block will make it through the loop without falling away from the surface (if the block begins from rest). In practice though, the initial height should be somewhat higher to account for resistive forces such as friction and air resistance.

> **Hint.** If you want to know whether or not an object is in contact with a surface, solve for the normal force exerted by the surface on the object. The object is in contact with the surface if the normal force is nonnegative.

**Escape speed:** The escape speed, $v_E$, of a projectile is the minimum speed necessary to escape the pull of gravity. For a projectile launched from the surface of a spherically symmetric planet of radius $R_p$, the escape speed can be computed by conserving energy for the projectile. We define the system to be just the projectile, choose the initial position to be at the surface of the planet, and the final position to be infinitely far away. The minimum speed needed to escape the gravitational pull of the planet is that which is just enough to get infinitely far away (so the final kinetic energy will be zero). We must use the expression $-Gm_p m/R$ for gravitational potential energy for this problem (since $g$ is not constant). The final gravitational potential energy is zero – and maximum since the initial gravitational potential energy is negative. Conservation of energy yields

$$U_0 + K_0 + W_{nc} = U + K$$
$$-\frac{Gm_p m}{R_p} + \frac{mv_E^2}{2} + 0 = 0 + 0$$
$$v_E = \sqrt{\frac{2Gm_p}{R_p}}$$

## 5 Work and Energy

**Equations of motion**: The equations that relate position to time, velocity to time, and acceleration to time for an object or system are called the equations of motion. One way to arrive at the equations of motion is to apply Newton's second law and integrate the resulting differential equation. For a single object of constant mass, Newton's second law can be expressed as

$$\sum_{i=1}^{N} F_{i,x} = m\frac{d^2x}{dt^2} \quad , \quad \sum_{i=1}^{N} F_{i,y} = m\frac{d^2y}{dt^2} \quad , \quad \sum_{i=1}^{N} F_{i,z} = m\frac{d^2z}{dt^2}$$

where, for example, $a_x = d^2x/dt^2$ is the $x$-component of acceleration. We applied this approach in Sec. 3.8, for example, in the context of air resistance.

An alternative method of obtaining the equations of motion is to apply the principle of conservation of energy, write the final kinetic energy as[242]

$$K = \frac{m}{2}\left(\frac{dx}{dt}\right)^2$$

and then solve the resulting differential equation; the initial kinetic energy, $K_0 = mv_0^2/2$, is treated as a constant. There are a couple of advantages of using the conservation of energy method to derive the equations of motion: For one, it is not necessary to work with components, since energy is a scalar; and secondly, conservation of energy leads to a first-order differential equation, whereas Newton's second law is a second-order differential equation (the difference being whether the equation involves second-order derivatives with respect to time).

---

**Example.** One end of a horizontal spring is fixed to a vertical wall, while a box of bananas is connected to its free end. There is no friction between the box and the horizontal. The box is displaced from the equilibrium position and released from rest at $x = x_m$, where $x = 0$ at equilibrium. Derive an equation for the position of the box as a function of time.

We choose the initial position to be the point of release – where the kinetic energy is zero – and the final position to be a variable, $x$. Conservation of energy results in

$$U_0 + K_0 + W_{nc} = U + K$$
$$\frac{kx_m^2}{2} + 0 + 0 = \frac{kx^2}{2} + \frac{mv^2}{2}$$

Now we solve for the final speed and set it equal to $dx/dt$:

---

[242] Some notes: If there is more than one object in the system, there will be multiple terms of this form; if there is rotation involved in the problem, there will also be rotational kinetic energy; and depending on how you choose initial and final, it may be convenient to treat the initial kinetic energy as the variable and the final kinetic energy as the constant (especially if the final kinetic energy is zero).

$$v = \frac{dx}{dt} = \sqrt{\frac{k}{m}(x_m^2 - x^2)}$$

Next, we separate the two variables – i.e. $x$ and $t$ – on opposite sides of the equation, and then integrate. It is convenient to perform definite integrals from $x_m$ at $t_0$ to $x$ at $t$ (rather than indefinite integrals and later determine the significance of the constant of integration):

$$\int_{x=x_m}^{x} \frac{dx}{\sqrt{\frac{k}{m}(x_m^2 - x^2)}} = \int_{t=t_0}^{t} dt = t - t_0$$

The integral over $x$ can be performed via trig sub: $x \equiv x_m \sin\theta$, such that $dx = x_m \cos\theta\, d\theta$:

$$t - t_0 = \sqrt{\frac{m}{k}} \int_{x=x_m}^{x} \frac{x_m \cos\theta\, d\theta}{\sqrt{(x_m^2 - x_m^2 \sin^2\theta)}} = \sqrt{\frac{m}{k}} \int_{\theta=\theta_0}^{\theta} d\theta = \sqrt{\frac{m}{k}} \left[\sin^{-1}\left(\frac{x}{x_m}\right)\right]_{x=x_m}^{x}$$

$$\sqrt{\frac{k}{m}}(t - t_0) = \sin^{-1}\left(\frac{x}{x_m}\right) - \sin^{-1} 1 = \sin^{-1}\left(\frac{x}{x_m}\right) - \frac{\pi}{2}$$

$$x(t) = x_m \sin\left[\sqrt{\frac{k}{m}}(t - t_0) + \frac{\pi}{2}\right]$$

Thus, the spring oscillates back and forth sinusoidally.

## 5.5 Conservative Force Fields

**Path integral**: An integral of the form $\int_i^f \vec{A} \cdot d\vec{s}$ is called a path integral because, in general, it depends not only on the initial and final positions, but also on the path taken. In the case where such an integral is path-dependent, it is necessary to specify both the path and the endpoints in order to avoid ambiguity. The work integral, $\int_i^f \vec{F} \cdot d\vec{s}$, is an example of the path integral.

**Closed integral**: An integral over a closed path is said to be a closed integral. A path is closed if the initial and final positions are coincident. The symbol $\oint$ is used to indicate that an integral is over a closed path; the ordinary integration symbol $\int$ is over an open path. A closed integral features a symbol to designate the path beneath the integration symbol – as in $\oint_C$ – in lieu of initial and final positions (which are the same). Here, the symbol $C$ may more generally specify a particular closed path such as an ellipse or a rectangle. It is also conventional to traverse the path in a counterclockwise sense (and associate a minus sign with a clockwise sense).

From the integrals encountered in first-year calculus, it's a natural question to wonder *how* a closed integral might be nonzero. We'll illustrate this with examples in the following section. Presently, we shall focus on the cases where such an integral is zero.

**Line integral**: A closed integral of the form $\oint_C \vec{A} \cdot d\vec{s}$ is called a line integral (though the closed path may consist of a curve – or a combination of curves – rather than lines). The net work done over a closed path, $\oint_C \vec{F} \cdot d\vec{s}$, is an example of a line integral.

**Conservative force fields**: One way to define a conservative force field is to state that a force field is conservative if its line integral is zero for any closed path:

$$\oint_C \vec{F}_c \cdot d\vec{s} = 0 \quad \text{for any closed path } C \text{ (for a conservative force)}$$

Such is the case when the work integral is path-independent.

---

**Example.** Show that $\oint_C \vec{F} \cdot d\vec{s} = 0$ for the conservative force $\vec{F} = y\hat{\imath} + x\hat{\jmath}$ (where SI units have been suppressed) by computing the line integral along the following closed path: from $(-1,0)$ to $(2,0)$, from $(2,0)$ to $(-1,1)$, and from $(-1,1)$ to $(-1,0)$.

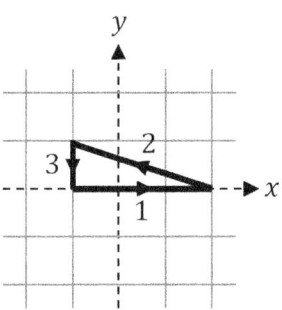

This closed integral can be expressed as the sum of three open integrals:

$$\oint_C \vec{F} \cdot d\vec{s} = \int_1 \vec{F} \cdot d\vec{s} + \int_2 \vec{F} \cdot d\vec{s} + \int_3 \vec{F} \cdot d\vec{s}$$

Since $F_x = y$ and $F_y = x$, each of these open integrals can be written as

$$\int_i \vec{F} \cdot d\vec{s} = \int_{x=x_0}^{x} F_x dx + \int_{y=y_0}^{y} F_y dy = \int_{x=x_0}^{x} y\, dx + \int_{y=y_0}^{y} x\, dy$$

Along the first path, $y = 0$; along the second path, $y = -\frac{x}{3} + \frac{2}{3}$; and for the third path, $x = -1$. These are the equations for the line segments that form the closed path. When integrating over $x$, if $y$ appears in the integrand, use the equation of the path to express it in terms of $x$; and, similarly, when integrating over $y$, if $x$ appears in the integrand, use the equation of the path to express it in terms of $y$:

$$\int_1 \vec{F} \cdot d\vec{s} = \int_{x=-1}^{2} y dx + \int_{y=0}^{0} x dy = \int_{x=-1}^{2} 0 dx + 0 = 0 + 0 = 0$$

$$\int_2 \vec{F} \cdot d\vec{s} = \int_{x=2}^{-1} y dx + \int_{y=0}^{1} x dy = \int_{x=2}^{-1} \left(-\frac{x}{3} + \frac{2}{3}\right) dx + \int_{y=0}^{1} (2 - 3y) dy$$

$$= \left[-\frac{x^2}{6} + \frac{2x}{3}\right]_{x=2}^{-1} + \left[2y - \frac{3y^2}{2}\right]_{y=0}^{1}$$

$$= \left[-\frac{(-1)^2}{6} + \frac{2(-1)}{3} + \frac{(2)^2}{6} - \frac{2(2)}{3}\right] + \left[2(1) - \frac{3(1)^2}{2} - 2(0) + \frac{3(0)^2}{2}\right]$$

$$= -\frac{1}{6} - \frac{2}{3} + \frac{2}{3} - \frac{4}{3} + 2 - \frac{3}{2} = -1$$

$$\int_3 \vec{F} \cdot d\vec{s} = \int_{x=-1}^{-1} y dx + \int_{y=1}^{0} x dy = 0 + \int_{y=1}^{0} (-1) dy = [-y]_{y=1}^{0} = -0 + 1 = 1$$

Therefore, $\oint_C \vec{F} \cdot d\vec{s} = 0 + (-1) + 1 = -1 + 1 = 0$.[243]

**Path-independence of gravitational work**: Recall the work done by gravity due to a single spherically symmetric massive object (such as a planet): For small changes in altitude, the work done by gravity is $W_g = -mg(y - y_0)$, and otherwise it is $W_g = Gm_p m \left(\frac{1}{R} - \frac{1}{R_0}\right)$. In either case, the work done by gravity depends only upon the change in altitude. Therefore, if you want to compute how much work is done by gravity when an object is displaced, the actual path does not matter – you need only look at the initial and final positions and determine how much the height has changed.

For example, consider a monkey climbing a jungle gym. If the monkey climbs 4 m upward, then 2 m across, and then 1 m downward, the work done by the monkey only depends upon the net change in altitude of 3 m; if the monkey had simply climbed straight upward for 3 m and stopped, the same amount of work would have been done by gravity.[244]

**Analogy with net displacement and total distance traveled**: The distinction between conservative and nonconservative forces is analogous to the distinction between the net displacement and the total distance traveled. Like net displacement, the work done by a conservative force is path-independent, and like the total distance traveled, the work done by a nonconservative force is path-dependent.

---

[243] This doesn't prove that $\vec{F}$ is conservative, since we only considered a single path, and not all possible paths.
[244] Really, if the monkey climbs upward, the monkey is doing work 'against' gravity, so the work done 'by' gravity is negative. Work is very much "preposition-dependent" (a little tidbit of verbal physics).

5 Work and Energy

**Partial derivative**: Consider a function of two or more independent variables, such as $f(x,y)$. It is sometimes useful to take a partial derivative of such a multivariable function. If you take a partial derivative with respect to one variable, hold the other independent variables constant. A partial derivative of $f$ with respect to $x$ is expressed as $\partial f/\partial x$, and it is computed just like an ordinary derivative except that the independent variable $y$ is held constant. Similarly, a partial derivative of $f$ with respect to $y$ is expressed as $\partial f/\partial y$, and it is taken with the independent variable $x$ held constant. The symbol $\partial$ is the partial differentiation operator.[245]

It's important that the variables be independent. For example, if $f = 2xu$ and $u = 3x$, the variables $x$ and $u$ are not independent; rather, $u$ is dependent upon $x$ as it always equals 3 times $x$. In this case, you could express $f$ as strictly a function of $x$ by substitution: $f = 2x(3x) = 6x^2$. So you could take an ordinary derivative with respect to $x$, and $f$ is actually a single-variable function.

The variables of orthogonal coordinate systems are independent variables. For example, the Cartesian coordinates $x$, $y$, and $z$ are all independent; similarly, the spherical coordinates $r$, $\theta$, and $\varphi$ are all independent. However, $x$ is not independent of $r$, $\theta$, and $\varphi$, since $x = r\cos\varphi\sin\theta$. If you mix the variables of different coordinate systems together, they are generally not independent.

---

**Example**. Find the partial derivatives of $f = 3xy^2 - 2x^3$ with respect to $x$ and $y$.
When taking a partial derivative of $f$ with respect to $x$, hold $y$ constant:

$$\frac{\partial f}{\partial x} = \frac{\partial}{\partial x}(3xy^2 - 2x^3) = 3y^2 - 6x^2$$

Similarly, when taking a partial derivative of $f$ with respect to $y$, hold $x$ constant:

$$\frac{\partial f}{\partial y} = \frac{\partial}{\partial y}(3xy^2 - 2x^3) = 6xy$$

---

**Gradient**: The gradient of a function is a vector function obtained from a scalar function, for which the components are found by taking partial derivatives with respect to the Cartesian coordinates as follows:

$$\vec{\nabla}f = \hat{\mathbf{i}}\frac{\partial f}{\partial x} + \hat{\mathbf{j}}\frac{\partial f}{\partial y} + \hat{\mathbf{k}}\frac{\partial f}{\partial z}$$

The symbol for the gradient, $\nabla$, is called del. The vector function $\vec{\nabla}f$ reads as "the gradient of $f$." The gradient of $f$ evaluated at the point $(x, y, z)$ has a direction that is perpendicular to the surface defined by setting $f(x, y, z) = $ const. and its components describe the slope of the surface.

---

[245] Since the partial derivative symbol $\partial$ derives as a variation of the symbol $\delta$ (lowercase delta), and since it is "partial," it would make sense to call it "del," which is my personal preference. The symbol for the gradient, $\nabla$, is universally called del, but this introduces no more confusion than having lowercase and uppercase alphabetic symbols of the same name – e.g. $\gamma$ and $\Gamma$ are both called gamma. The symbol $\partial$ is sometimes pronounced "der," but to me "der" just doesn't sound like the sort of intelligent syllable to be uttered frequently in an advanced math or physics course....

**Example**. Find the gradient of $f(x,y) = 4x^2/y^2$.

Take partial derivatives of the function with respect to each of the independent variables and use these for the Cartesian components of the gradient:

$$\vec{\nabla} f = \hat{\mathbf{i}} \frac{\partial f}{\partial x} + \hat{\mathbf{j}} \frac{\partial f}{\partial y} + \hat{\mathbf{k}} \frac{\partial f}{\partial z} = \frac{8x}{y^2}\hat{\mathbf{i}} - \frac{8x^2}{y^3}\hat{\mathbf{j}}$$

**Relationship between potential energy and force**: In Sec. 5.3, we defined the potential energy associated with a conservative force as follows: The change in potential energy equals the negative of the work done by its associated conservative force:

$$\Delta U = U - U_0 = -W_c = -\int_i^f \vec{\mathbf{F}}_c \cdot d\vec{\mathbf{s}}$$

This is the integral relation between potential energy and the corresponding conservative force. It can also be expressed in differential form as $dU = -dW_c = \vec{\mathbf{F}}_c \cdot d\vec{\mathbf{s}}$.

First, let us explore a simpler analogy. The relationship between net displacement, $\Delta \vec{\mathbf{r}}$, and velocity, $\vec{\mathbf{v}}$, can be expressed in integral, $\Delta \vec{\mathbf{r}} = \int_{t=t_0}^{t} \vec{\mathbf{v}} dt$, or differential form, $\vec{\mathbf{v}} = d\vec{\mathbf{r}}/dt$.

The potential energy integral is a little more complicated than the net displacement integral. In the net displacement integral, velocity is expressed as a function of one variable, $t$; but in the potential energy's path integral, force is most generally a function of three coordinates, $x$, $y$, and $z$. The path integral is inherently an integral over a specified path in 3D space. The velocity is 3D, but each component of velocity can be expressed in terms of one parameter, $t$, and so the integral is effectively 1D. In contrast, a force of the form $\vec{\mathbf{F}} = x^2 y^3 \hat{\mathbf{i}} + x^3 y^2 \hat{\mathbf{j}}$ leads to the integrals $\int_{x=x_0}^{x} x^2 y^3 dx + \int_{y=y_0}^{y} x^3 y^2 dy$, which have 2 (or more, in general, as there may also be a $z$ involved) variables in the integrand (and then the equation of the path must be applied in order to perform the integration).

The differential form of the potential energy integral does not involve an ordinary derivative with respect to one variable, but since $\vec{\mathbf{F}}_c$ generally depends on 3 coordinates $(x,y,z)$, it involves the gradient (which involves partial derivatives with respect to each of the 3 coordinates):

$$\vec{\mathbf{F}}_c = -\vec{\nabla} U$$

**Example**. Find the force associated with the potential energy $U = cx/y$, where $c$ is a constant. Also, where is the natural reference point for this potential energy?

The force is easily found by taking the negative of the gradient of the potential energy:

$$\vec{\mathbf{F}} = -\vec{\nabla} U = -\hat{\mathbf{i}} \frac{\partial U}{\partial x} - \hat{\mathbf{j}} \frac{\partial U}{\partial y} - \hat{\mathbf{k}} \frac{\partial U}{\partial x} = -\frac{c}{y}\hat{\mathbf{i}} + \frac{cx}{y^2}\hat{\mathbf{j}}$$

The natural reference point is the point $(x, y)$ that makes $U$ zero; in this case, $U$ is zero at $(0, \infty)$.

**Existence of potential energy for conservative forces**: We can conceptually understand why there is a distinction between two kinds of forces, which we classify as conservative and nonconservative, by considering the potential energy function in 2D, $U(x, y)$. In 2D, the associated force is given by

$$\vec{F} = -\vec{\nabla}U = -\hat{i}\frac{\partial U}{\partial x} - \hat{j}\frac{\partial U}{\partial y} = F_x\hat{i} + F_y\hat{j}$$

Therefore, the components of the force are related to the potential energy by

$$F_x = -\frac{\partial U}{\partial x} \quad , \quad F_y = -\frac{\partial U}{\partial y}$$

In mathematics, a multivariable function $U$ is a 'well-behaved' function if its mixed second derivatives are equal (i.e. if the order in which the partial derivatives are taken does not matter) according to:

$$\frac{\partial^2 U}{\partial x \partial y} = \frac{\partial^2 U}{\partial y \partial x} \quad \text{(conservative force in 2D)}$$

This equation can also be expressed in terms of the components of the forces as

$$\frac{\partial F_x}{\partial y} = \frac{\partial F_y}{\partial x} \quad \text{(conservative force in 2D)}$$

In 2D, the above equation is true for conservative forces, but is not true for nonconservative forces. In 3D, it is also necessary to check the analogous equations involving $z$.[246]

The distinction between conservative and nonconservative forces has to do with whether or not the mixed second derivatives of the associated potential energy would be equal. A well-behaved potential energy function only exists for conservative forces. The analogous quantity is not meaningful for nonconservative forces.

---

**Example.** Is the force $\vec{F} = xy^2\hat{i} + x^2y\hat{j}$ conservative or nonconservative?

The simplest way to check if a force is conservative is to see if it satisfies the above relation:

$$\frac{\partial F_x}{\partial y} = \frac{\partial}{\partial y}(xy^2) = 2xy$$
$$\frac{\partial F_y}{\partial x} = \frac{\partial}{\partial x}(x^2y) = 2xy$$

Since $\frac{\partial F_x}{\partial y} = \frac{\partial F_y}{\partial x}$, this 2D force is conservative.

---

[246] Namely, $\frac{\partial F_x}{\partial z} = \frac{\partial F_z}{\partial x}$ and $\frac{\partial F_z}{\partial y} = \frac{\partial F_y}{\partial z}$.

**Equilibrium conditions**: Suppose that the potential energy of an object can be expressed as a function of a single distance coordinate, which we call $R$: $U(R)$. This could be the $r$ of spherical coordinates, for example (as in a problem with a gravitational field, where $U$ depends on $r$, but not $\theta$ or $\varphi$), or it could be $x - x_e$ in a Hooke's law problem, or it could be the distance between two molecules interacting with one another, for example. Importantly, we are assuming that $U$ is not a function of two (or more) independent coordinates, and that the one coordinate is a distance (not an angle).

The conservative force associated with this potential energy is given by the gradient, which reduces to a derivative with respect to $R$ for this single-variable problem:

$$\vec{F}_c = -\vec{\nabla}U = -\hat{R}\frac{dU}{dR} \quad \text{(single-variable potential)}$$

The $R$-component of the force is

$$F_c = -\frac{dU}{dR} \quad \text{(single-variable potential)}$$

Now let us further assume that $\vec{F}_c$ is the net external force acting on the object. In this case, Newton's second law states that $\vec{F}_c$ is equal to $m\vec{a}$. If the object is in an equilibrium position, its acceleration is instantaneously zero (but its velocity may not be, if it is in dynamic rather than static equilibrium; in dynamic equilibrium, as when a spring is oscillating, equilibrium does not persist, but is just momentary). This means that both $\vec{F}_c$ and $dU/dR$ equal zero at equilibrium:

$$\frac{dU}{dR} = 0 \quad \text{(single-variable potential at equilibrium)}$$

Therefore, if you know the equation for the potential energy of an object, and if this equation depends only on a single coordinate (which is a distance), and if the net external force is given by $\vec{F}_c$, then the conditions for equilibrium (i.e. the values of $R$ for which the object would be at equilibrium) can be obtained by setting $dU/dR$ equal to zero.

If the potential energy depends upon two or more coordinates, instead set the gradient, $\vec{\nabla}U$, equal to zero and demand that each component equal zero (i.e. $\partial U/\partial x$, $\partial U/\partial y$, and $\partial U/\partial z$ in Cartesian coordinates). If $\vec{F}_c$ is not the net external force, but the only forces other than $\vec{F}_c$ are resistive forces like friction and air resistance, $dU/dR = 0$ (or more generally $\vec{\nabla}U = 0$) still provides the equilibrium conditions; the difference is that the resistive forces will eventually bring the object to rest (not necessarily at an equilibrium position).

Since the derivative, $dU/dR$, represents the slope of a graph of $U$ as a function of $R$, if you make such a graph, the equilibrium positions correspond to horizontal tangents.

**Stability conditions**: Equilibrium can be stable, neutral, or unstable. When an object (or system) receives a small displacement from equilibrium, it tends to oscillate about equilibrium if it is stable and tends to go away from equilibrium if it is unstable.

Let us continue to consider an object for which $U$ is a function of $R$ only, where the associated $\vec{F}_c$ is the net external force acting on the object such that $dU/dR = 0$ at the equilibrium positions. In this case, the second derivative determines whether equilibrium is stable, neutral, or unstable:

$$\frac{dU}{dR} = 0 \text{ and } \frac{d^2U}{dR^2} > 0 \quad \text{(stable equilibrium; single-variable potential)}$$
$$\frac{dU}{dR} = 0 \text{ and } \frac{d^2U}{dR^2} = 0 \quad \text{(neutral equilibrium; single-variable potential)}$$
$$\frac{dU}{dR} = 0 \text{ and } \frac{d^2U}{dR^2} < 0 \quad \text{(unstable equilibrium; single-variable potential)}$$

Looking at a graph of $U$ as a function of $R$, stable equilibrium corresponds to a point where the slope is zero and the curve is concave up,[247] neutral equilibrium corresponds to a point of inflection where the slope is instantaneously zero, and unstable equilibrium corresponds to a point where the slope is zero and the curve is concave down. Examples of each are illustrated in the graph below.

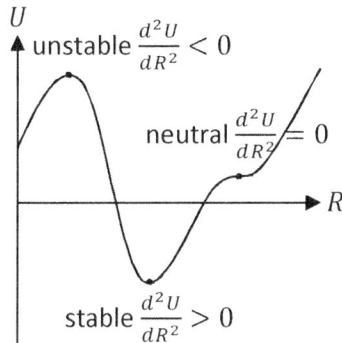

Consider an object at a given equilibrium position, where $F_c = -dU/dR = 0$. Suppose that the object is given a small displacement from equilibrium, so that $F_c = -dU/dR$ is no longer equal to zero. If $d^2U/dR^2 > 0$ in the vicinity of equilibrium, the function $U$ has a local minimum at $R$: If the object is displaced in the $+R$-direction, the force $F_c = -dU/dR$ pushes the object in the $-R$-direction (because the curve has positive slope there), and if the object is displaced in the $-R$-direction, the force $F_c = -dU/dR$ pushes the object in the $+R$-direction (because the curve has negative slope there). In this case, the object tends to oscillate about equilibrium. Thus, $dU/dR = 0$ and $d^2U/dR^2 > 0$ at a point of stable equilibrium, which correspond to a local minimum for $U(R)$.

If instead $d^2U/dR^2 < 0$ in the vicinity of equilibrium, the function $U$ has a local maximum at $R$: If the object is displaced in the $+R$-direction, the force $F_c = -dU/dR$ pushes the object in the $+R$-direction (because the curve has negative slope there), and if the object is displaced in the $-R$-direction, the force $F_c = -dU/dR$ pushes the object in the $-R$-direction (because the curve has positive slope there). In this case, the object tends to go away from equilibrium. Thus, $dU/dR = 0$ and $d^2U/dR^2 < 0$ at a point of unstable equilibrium, which correspond to a local maximum for $U(R)$.

---

[247] 'Concave up' and 'concave down' are sometimes instead called 'concave' and 'convex,' respectively.

## 5.6 Nonconservative Work

**How a closed integral can be nonzero**: When you first see a closed integral, it's natural wonder how an integral can be zero when the initial and final positions are the same. For example, for the 'well-behaved' functions of calculus, an integral of the form $\int_{x=x_0}^{x_0} f(x)dx$ is generally zero. The line integral, $\oint_C \vec{A} \cdot d\vec{s}$, has a different form than does $\int_{x=x_0}^{x_0} f(x)dx$. However, the line integral may be expressed in a similar form by expressing the scalar products in terms of the components of $\vec{A}$ and $d\vec{s}$: $\oint_C \vec{A} \cdot d\vec{s} = \int_{x=x_0}^{x_0} A_x dx + \int_{y=y_0}^{y_0} A_y dy + \int_{z=z_0}^{z_0} A_z dz$. There is a key difference between the integral $\int_{x=x_0}^{x_0} f(x)dx$ and the integrals like $\int_{x=x_0}^{x_0} A_x dx$: The components of $\vec{A}$ may be functions of all three coordinates, whereas the integral $\int_{x=x_0}^{x_0} f(x)dx$ involves an integrand that depends only on $x$. The integral $\oint_C \vec{A} \cdot d\vec{s}$ depends on the path, in general, because the components are generally functions of $x$, $y$, and $z$ – e.g. $A_x(x,y,z)$.

Let us illustrate how $\oint_C \vec{A} \cdot d\vec{s}$ can be nonzero with a simple example. Suppose that the path is a circle lying in the $xy$ plane, centered about the origin and traversed counterclockwise from the $+z$-axis. Then the differential displacement vector is $d\vec{s} = R_0 d\theta \hat{\boldsymbol{\theta}}$; it has a magnitude of $R_0 d\theta$ (where $R_0$ is the radius of the circular path) and a direction of $\hat{\boldsymbol{\theta}}$, which is tangent to the circle in a counterclockwise sense. It is easy to see that the line integral is nonzero if $\vec{A} = \hat{\boldsymbol{\theta}}$. Since $\vec{A}$ is parallel to $d\vec{s}$, the scalar product is simply $\vec{A} \cdot d\vec{s} = \hat{\boldsymbol{\theta}} \cdot R_0 d\theta \hat{\boldsymbol{\theta}} = R_0 d\theta$. The line integral is therefore $\oint_C \vec{A} \cdot d\vec{s} = \int_{\theta=0}^{2\pi} R_0 d\theta = 2\pi R_0$, which is a nonzero result over a closed path. We will see another example involving Cartesian coordinates when we discuss work done by conservative forces shortly.

**How a closed integral can be nonzero in 1D**: The line integral $\oint_C \vec{A} \cdot d\vec{s}$ can be zero even in 1D, where $\oint_C \vec{A} \cdot d\vec{s} = \int_{x=x_0}^{x_0} A_x(x)dx$. Here, $A_x$ is a function of $x$ only, but it won't be quite the 'well-behaved' function considered in calculus courses. Consider the path $(1,0) \to (3,0) \to (1,0)$, which is a closed path in 1D. To see an example of how the line integral can be zero in 1D, consider the function $A_x(x)$ be defined as follows:

$$A_x(x) = \begin{cases} x & \text{if } dx > 0 \\ -x & \text{if } dx < 0 \end{cases}$$

Now the integral $\int_{x=x_0}^{x_0} A_x(x)dx$ must be split into two integrals – one along the path $(1,0) \to (3,0)$, where $x$ is increasing and so $dx > 0$, and another along the path $(3,0) \to (1,0)$, where $x$ is decreasing and so $dx < 0$: $\int_{x=1}^{1} A_x(x)dx = \int_{x=1}^{3} A_x(x)dx + \int_{x=3}^{1} A_x(x)dx = \int_{x=1}^{3} x dx - \int_{x=3}^{1} x dx = \left[\frac{x^2}{2}\right]_{x=1}^{3} - \left[\frac{x^2}{2}\right]_{x=3}^{1} = \left(\frac{9}{2} - \frac{1}{2}\right) - \left(\frac{1}{2} - \frac{9}{2}\right) = 4 - (-4) = 8$.

This particular function, $A_x(x)$, that we used may seem like a strange function, but actually its main property is actually very similar to a very physical function – friction. If you push a box of bananas to the north, and then push it south, returning the box to its starting position, friction acts south during the first part of the trip and then acts north during the second part of the trip. The line integral over the friction force is nonzero over any nonzero path – even in 1D; the friction force does nonconservative work, and so always has a nonzero line integral. (Of course, the function $A_x(x)$ that we used was not the friction force, since it is proportional to $x$, but direction-wise it behaves similar to friction in 1D, always opposing the direction of motion.)

**Work done by nonconservative forces**: If the line integral for the work done by a force is nonzero for some closed path, then the force is said to be nonconservative:[248]

$$\oint_C \vec{F}_{nc} \cdot d\vec{s} \neq 0 \quad \text{for some closed path } C \text{ (for a non-conservative force)}$$

Such is the case when the work integral is path-dependent. That is, going from $(x_0, y_0, z_0)$ to $(x, y, z)$, the work done is generally different for different paths.

**Example.** Show that the force $\vec{F} = y\hat{\imath} - x\hat{\jmath}$ (where SI units have been suppressed) is nonconservative by computing the line integral $\oint_C \vec{F} \cdot d\vec{s}$ along the following closed path: from $(-1,0)$ to $(2,0)$, from $(2,0)$ to $(-1,1)$, and from $(-1,1)$ to $(-1,0)$.

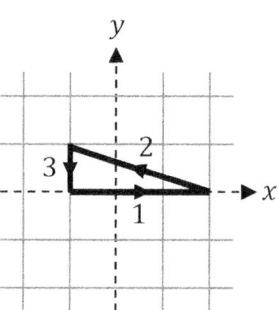

This closed integral can be expressed as the sum of three open integrals:

$$\oint_C \vec{F} \cdot d\vec{s} = \int_1 \vec{F} \cdot d\vec{s} + \int_2 \vec{F} \cdot d\vec{s} + \int_3 \vec{F} \cdot d\vec{s}$$

---

[248] There just needs to exist at least one closed path for which the line integral is nonzero, and then the force is determined to be nonconservative. It's possible for the line integral to be zero for some closed paths, but nonzero for others – if there exists at least one closed path for which the line integral is nonzero, the force is nonconservative. Contrast this with a conservative force, where the line integral must be zero for every possible closed path. Note the roles of the words 'any' and 'some' in the two definitions: 'Some' implies the existence of at least one, while 'any' implies that it must be true for all.

Since $F_x = y$ and $F_y = -x$, each of these open integrals can be written as

$$\int_i \vec{F} \cdot d\vec{s} = \int_{x=x_0}^{x} F_x dx + \int_{y=y_0}^{y} F_y dy = \int_{x=x_0}^{x} y dx - \int_{y=y_0}^{y} x dy$$

Along the first path, $y = 0$; along the second path, $y = -\frac{x}{3} + \frac{2}{3}$; and for the third path, $x = -1$. These are the equations for the line segments that form the closed path. When integrating over $x$, if $y$ appears in the integrand, use the equation of the path to express it in terms of $x$; and, similarly, when integrating over $y$, if $x$ appears in the integrand, use the equation of the path to express it in terms of $y$:

$$\int_1 \vec{F} \cdot d\vec{s} = \int_{x=-1}^{2} y dx - \int_{y=0}^{0} x dy = \int_{x=-1}^{2} 0 dx - 0 = 0 - 0 = 0$$

$$\int_2 \vec{F} \cdot d\vec{s} = \int_{x=2}^{-1} y dx - \int_{y=0}^{1} x dy = \int_{x=2}^{-1} \left(-\frac{x}{3} + \frac{2}{3}\right) dx - \int_{y=0}^{1} (2 - 3y) dy$$

$$= \left[-\frac{x^2}{6} + \frac{2x}{3}\right]_{x=2}^{-1} - \left[2y - \frac{3y^2}{2}\right]_{y=0}^{1}$$

$$= \left[-\frac{(-1)^2}{6} + \frac{2(-1)}{3} + \frac{(2)^2}{6} - \frac{2(2)}{3}\right] - \left[2(1) - \frac{3(1)^2}{2} - 2(0) + \frac{3(0)^2}{2}\right]$$

$$= -\frac{1}{6} - \frac{2}{3} + \frac{2}{3} - \frac{4}{3} - 2 + \frac{3}{2} = -2$$

$$\int_3 \vec{F} \cdot d\vec{s} = \int_{x=-1}^{-1} y dx - \int_{y=1}^{0} x dy = 0 - \int_{y=1}^{0} (-1) dy = -[-y]_{y=1}^{0} = 0 - 1 = -1$$

Therefore, $\oint_C \vec{F} \cdot d\vec{s} = 0 - 2 - 1 = -3$ and the force is nonconservative. It is instructive to compare the subtle, but significant, distinctions between this example and the similar example from the previous section.

**Path-dependence of friction**: The magnitude of the force of kinetic friction equals the coefficient of kinetic friction times the normal force: $f_k = \mu_k N$. The direction of the friction force opposes the displacement of the object: $\vec{f}_k \updownarrow d\vec{s}$.[249] The work done by the friction force is

$$W_f = -\mu_k \int_{s=s_0}^{S} N ds$$

If the normal force is constant, this reduces to

---
[249] The symbol $\updownarrow$ reads, "…is anti-parallel to…"

$$W_f = -\mu_k N \Delta s \quad \text{(if } N \text{ is const.)}$$

where $\Delta s$ is the total distance traveled. In this case, it's easy to see that $W_f$ will be negative (i.e. nonzero) for any path (open or closed) since the total distance traveled must be positive. The work done by friction is also negative even if normal force is not constant (the case along a curved incline, for example).

**Line integral for resistive forces**: Resistive forces, like friction and air resistance, always do negative work because they oppose the motion – and so $\vec{F}_{nc} \updownarrow d\vec{s}$. The line integral for resistive forces is therefore negative for any closed path (and open integrals for the work done over an open path is also negative for resistive forces). Resistive forces subtract mechanical energy from the system.

Non-resistive forces that do nonconservative work may add mechanical energy to a system. A motor, for example, that is external to a system may supply mechanical energy so that the system can do work. So it's possible for $W_{nc}$ to be positive. For resistive forces, however, $W_{nc}$ is strictly negative.

**Non-existence of potential energy for nonconservative forces**: The partial derivative test determines whether or not a force is conservative or nonconservative. In 2D,

$$\frac{\partial F_x}{\partial y} = \frac{\partial F_y}{\partial x} \quad \text{(conservative force in 2D)}$$
$$\frac{\partial F_x}{\partial y} \neq \frac{\partial F_y}{\partial x} \quad \text{(nonconservative force in 2D)}$$

In 3D, one must also check the analogous equations among $y$ and $z$ and also among $x$ and $z$; all three pairs of partial derivatives must be equal in order for the force to be conservative. The mixed partial derivatives like $\frac{\partial^2 U}{\partial x \partial y}$ and $\frac{\partial^2 U}{\partial y \partial x}$ will only be equal if the force is conservative, and so a 'well-behaved' function $U$ for which $\vec{F} = -\vec{\nabla}U$ only exists for conservative forces. Hence, we only associate the concept of potential energy with conservative forces, and when we apply the principle of conservation of energy, we treat the nonconservative forces separately.

**Work done by friction along a curved surface**: Let us consider a block sliding down a circular arc, as drawn below, with friction to illustrate the calculation of nonconservative work done along a curved surface. The radius of the circle is $R_0$, the block has mass $m$, and the coefficient of (kinetic) friction between the block and arc is $\mu$.

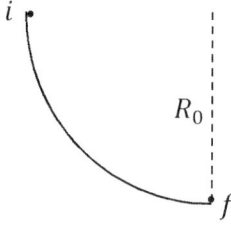

The work done by friction is

$$W_f = -\mu_k \int_{s=s_0}^{s} N\,ds$$

For this curved incline, the magnitude of the normal force is not constant, but varies with the speed of the block. We can derive an equation for the normal force by drawing a FBD (at an arbitrary point down the curved incline – specified by the angle $\theta$ with the vertical in the diagram below) and applying Newton's second law. For circular motion, we work with the inward ($in$) and tangential ($T$) directions. The sum of the inward components of the forces equals $ma_c$, where the centripetal acceleration, $a_c$, is not zero because the block is changing direction. The sum of the tangential components equals $ma_T$, where the tangential component of the acceleration, $a_T$, is also nonzero because the block is changing speed. However, we only need the inward components to solve for the normal force. Observe that the normal force, $\vec{N}$, is directly inward, while the weight, $m\vec{g}$, has an outward component (which will come with a minus sign in the sum of the inward components).

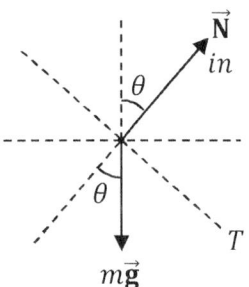

$$\sum_{i=1}^{2} F_{i,in} = ma_c$$
$$N - mg\cos\theta = \frac{mv^2}{R_0}$$

In principle, we could substitute this expression for $N$ into the work integral and proceed to integrate. Neither $v$ nor $\theta$ are constant, so some substitutions are needed to express $v$ and $\theta$ in terms of $s$ before the expression can be integrated. However, let's look ahead to a more practical calculation. Suppose that we wish to apply the principle of conservation of energy to find the final speed of the block. In this case, you can see that we will run into a problem. In the equation for $W_{nc}$, we need to know $v$ so that we can integrate, but $v$ is what we wish to solve for. The way around this is to express conservation of energy in terms of differentials as follows:

$$dW_{nc} = dU + dK$$

To see that this equation expresses conservation of energy, consider that $dU$ is an infinitesimal change in potential energy – i.e. a very small $(U - U_0)$ – and $dK$ is an infinitesimal change in kinetic energy – i.e. a very small $(K - K_0)$. Starting with $U_0 + K_0 + W_{nc} = U + K$, we can regroup this as $W_{nc} = (U - U_0) + (K - K_0)$, and then infinitesimal changes lead to $dW_{nc} = dU + dK$.

For definiteness, let us release the block from rest at the top of the quarter-circle, and take the initial position to be the point of release and the final position to be at the bottom of the arc. In terms of differentials, $dU = mgdy$, where $y$ is the height, which we choose to measure from the final position at the bottom. Note that $dy$ and therefore $dU$ are negative since $y$ is decreasing. We can obtain a similar expression for $dK$ by differentiating $K = mv^2/2$ with respect to $v$: $dK = mvdv$. Both $dv$ and $dK$ are positive since $v$ is increasing. The differential nonconservative work is $dW_{nc} = \vec{F}_{nc} \cdot d\vec{s}$. Since the nonconservative force is the (kinetic) friction force, and since the magnitude of the friction force is $f_k = \mu_k N$ and its direction opposes $d\vec{s}$, the differential nonconservative work can be expressed as $dW_{nc} = -\mu_k N ds$. Substituting these differentials into the differential form of conservation of energy,

$$-\mu_k N ds = mgdy + mvdv$$

Next, we plug in the equation for normal force, which we had derived earlier:

$$-\mu_k \left( mg \cos\theta + \frac{mv^2}{R_0} \right) ds = mgdy + mvdv$$

Since the arc is circular, $ds = R_0 d\theta$. Based on how we defined $\theta$, $y = R_0(1 - \cos\theta)$, as in the pendulum example of Sec. 5.4. Implicit differentiation of this equation leads to $dy = R_0 \sin\theta \, d\theta$. Substituting these expressions into our conservation of energy equation,

$$-\mu_k \left( mg \cos\theta + \frac{mv^2}{R_0} \right) R_0 d\theta = mgR_0 \sin\theta \, d\theta + mvdv$$

Now we have just two variables in the equation – $v$ and $\theta$. In principle, we can integrate; as $\theta$ varies from 90° to 0°, $v$ varies from 0 to $v$ (these are the limits of integration). This differential equation is somewhat complicated since one cannot algebraically separate $v$ and $\theta$ on opposite sides of the equation (such that neither side contains both a $v$ and $\theta$, whether differential or finite). However, a numerical technique for solving this first-order differential equation could be applied to solve for the final speed of the block. Most real-world problems involve numerical techniques.

## Conceptual Questions

1. Comment on the relationship between the direction of two vectors if the scalar product between them: equals the product of their magnitudes, is zero, equals the negative of the product of their magnitudes, is negative, is positive. Draw two vectors for which the scalar product is negative.
2. Vector $\vec{A}$ doesn't have a $z$-component. Vector $\vec{B}$ only has a $z$-component. What does $\vec{A} \cdot \vec{B}$ equal?
3. How can you use the scalar product to determine the angle between two given vectors?
4. A monkey connects one end of a string to a banana and holds the free end. When the monkey is initially at rest, the string hangs vertically. Indicate whether or not work is done by the tension force in each of the following cases, and justify your answers: the monkey gains speed traveling in a straight line, the monkey travels with constant velocity, the monkey decelerates in a straight line. In each case, the monkey travels along horizontal ground.
5. A monkey travels with UCM. Does the centripetal force (which may not be a specific force, but equals the sum of the inward components of the forces acting on the monkey) do any work? Explain.
6. In each of the following cases, indicate whether or not the work by the specified force is positive, negative, or zero: the work done by gravity as a monkey climbs a ladder, the work done by friction as a monkey slides down a hill, the work done by the normal force as a monkey drives up a hill, the work done by gravity as a barrel of monkeys rolls down a hill, the work done by friction as a box of bananas slides up an incline, the work done by gravity as a monkey drives along a level road.
7. In each of the following cases, indicate whether or not the work done by gravity is positive, negative, or zero: a rocket blasts off from earth, a satellite travels with uniform circular motion, a planet travels in an elliptical orbit from perihelion (closest approach to the sun) to aphelion (furthest approach), a meteor falls toward earth's surface.
8. In each of the following cases, indicate whether or not the work done by the restoring force is positive, negative, or zero: a spring stretches away from equilibrium, a spring compresses away from equilibrium, a compressed spring stretches toward equilibrium, a stretched spring compresses toward equilibrium.
9. Two monkeys with the same mass enter a building through the same door. One monkey walks 100 m east, walks upstairs for four stories, and then walks 50 m north. The other monkey walks 20 m north, rides an elevator four stories up, and then walks 30 m north. Which monkey did more work against gravity? Explain.
10. One monkey travels in a straight line with constant speed. Another monkey with the same mass travels along a parabolic path with constant speed. (Neither path is necessarily horizontal.) In which case is more net work done on the monkey? Explain.
11. Two identical charged particles travel in a region where there is a uniform electric field, such that the force exerted on either particle is always 10 N to the north. Both charged particles start at point A and finish at point B, which is due north of (and level with) point A. One charged particle travels along a straight line, while the other travels along a semicircle. In which case is more work done by the electric field? Explain.

12. One monkey pulls a box of bananas horizontally along a straight line from point A to point B along a level path with a constant force of 100 N towards point B. Another monkey pulls a box of bananas of identical mass from point A to point B with a horizontal force that has a constant magnitude of 100 N tangential to a (horizontal) semicircular path. In which case is more work done by the monkey's push? Explain. Also, explain why the answers to Questions 11 and 12 differ.

13. How is it possible that $ds$ equals the magnitude of $d\vec{s}$, but $\int_i^f ds$ is generally greater than the magnitude of $\int_i^f d\vec{s}$?

14. Is the integral $\int_i^f \vec{v}\, dt$ path-dependent or path-independent? Explain.

15. Is the integral $\int_i^f v\, dt$ path-dependent or path-independent? Explain.

16. A flying saucer travels from Mars to Neptune to Saturn. Another flying saucer travels from Mars to Venus to Saturn. Both flying saucers have a method of propulsion which does not change the mass of the flying saucer (unlike a rocket that ejects steam or burns fuel, for example). Both flying saucers take off from and land at (virtually) the same places at the same time. Which flying saucer does more work against gravity? Explain.

17. A monkey drags a box of bananas from point A to point B along a straight line. Another monkey drags a box of bananas of the same mass from point A to point B along a semicircular path. In both cases the ground is horizontal, both monkeys pull the boxes with horizontal forces, the coefficient of friction between the ground and box is the same in both cases. In which case is the work done by friction more negative? Explain.

18. Electric companies' bills are based on kilowatt-hours (kWh). (A) What physical quantities could have units of kWh? (B) Express a kWh in terms of the SI unit of energy (Joules), meters, and/or seconds only, including any numerical factor needed to express the equality.

19. A motor pulls a box of bananas with a constant horizontal force of 200 N to the east, displacing the box in a straight line from point A to point B, where point B is due east of (and level with) point A. A second motor pulls a box of bananas of the same mass with a constant horizontal force of 200 N to the east, displacing the box along a semicircular path from point A to point B (yes, this is possible, as there may be other forces acting on the box, and the direction of the box's initial velocity also affects its path). Both boxes travel with constant speed. Which motor delivers more average power?

20. A motor pulls a box of bananas with a constant horizontal force of 200 N to the east, displacing the box in a straight line from point A to point B, where point B is due east of (and level with) point A. A second motor pulls a box of bananas of the same mass with a constant horizontal force of 200 N that is always tangential to the path of the box, while the box is displaced along a semicircular path from point A to point B. Both boxes travel with constant speed. Which motor delivers more average power? Also, explain why the answers to Questions 19 and 20 differ.

21. For each of the following sources of energy used on earth, indicate if the sun is ultimately responsible for the origin of that energy source, and if so, explain how: wind energy, nuclear energy, energy of tides, energy from coal, energy from natural gas.

22. (A) Two monkeys with the same mass are running; one monkey is running twice as fast as the other. What is the ratio of their kinetic energies? (B) Two monkeys are running with the same speed; one monkey has twice the mass of the other. What is the ratio of their kinetic energies? (C) Two monkeys are running; one monkey has both twice the speed and twice the mass of the other. What is the ratio of their kinetic energies?

An Advanced Introduction to Calculus-Based Physics (Mechanics)

23. If you take a partial derivative of kinetic energy with respect to speed for an object of constant mass, what do you get? (Give a one-word answer.)

24. Express the kinetic energy of an object in terms of its momentum and mass only.

25. During any problems, once you define the system and surroundings, there are a myriad of forces that could potentially affect the system. Following are a couple of examples. (A) A monkey in the surroundings could come over at any time and exert some force on the system. If such an external monkey does exert a force on the system, where does the equation for conservation of energy account for it? (B) Some astronomical event could suddenly disrupt the system. If this happens, where does it figure into the equation for conservation of energy? (C) Does a potential energy term exist for every type of force that could potentially affect the system? If not, for which types of forces do potential energy terms exist, and where does the equation for conservation of energy account for other types of forces for which potential energy terms do not exist?

26. A monkey throws a banana straight upward. As the banana travels upward, describe happens to each of the following physical quantities: its potential energy, its kinetic energy, its total mechanical energy.

27. A monkey releases a banana from rest. As the banana falls downward, describe happens to each of the following physical quantities: its potential energy, its kinetic energy, its total mechanical energy.

28. A monkey ties a cord to a banana and uses it as a pendulum. As the banana oscillates back and forth, indicate where each of the following physical quantities are minimum and also where they are maximum: potential energy, kinetic energy, total mechanical energy.

29. A monkey connects a banana to a spring. As the banana oscillates back and forth, indicate where each of the following physical quantities are minimum and also where they are maximum: potential energy, kinetic energy, total mechanical energy.

30. For each of the following processes, describe the transformation of energy that occurs: a banana rises upward, a stretched spring compresses toward equilibrium, a pendulum swings downward, a box of bananas sliding along a horizontal comes to rest, a fan rotates after being plugged into an AC outlet, a car burns fuel to gain speed along a level road, a monkey at rest suddenly starts running on level ground.

31. When a rocket blasts off from earth, is it gaining or losing gravitational potential energy? In light of this, does it make sense for the rocket to approach zero potential energy as it becomes infinitely far away from the earth? Explain.

32. Where the system is defined as the box of bananas, indicate whether each of the following forces are conservative or nonconservative: a stretched spring pushes a box of bananas back to equilibrium, the earth pulls a box of bananas down an incline, friction decelerates a sliding box of bananas, a monkey pushes a box of bananas up a ramp, a truck's motor accelerates a box of bananas.

33. A topological contour map consists of a set of closed curves, called equipotentials, where gravitational potential energy is constant. (A) What information does a gradient provide regarding a contour map? (B) How can you visually determine where the magnitude of the gradient is smaller or larger on a contour map? (C) How can you visually determine the direction of the gradient at any point on a contour map? (D) How can you visually determine equilibrium positions on a contour map? (E) How can you visually determine whether equilibrium positions on a contour map are stable, neutral, or unstable? Geographically, what do stable, neutral, and unstable equilibrium positions look like?

# 5 Work and Energy

## Practice Problems

### The scalar product, work, and power

1. A monkey pulls a wagon full of bananas with a constant force $\vec{F} = 5\hat{i} + 4\hat{j}$ along a displacement $\vec{s} = -32\hat{i} + 16\hat{j}$, where SI units have been suppressed for convenience. (A) Compute the scalar product of the force and displacement vectors [using the dot product between unit vectors – do not use any angles or cosines in part (A)] to find the work done by the monkey in pulling the wagon. (B) Find the magnitude and direction of the force exerted by the monkey. (C) Find the magnitude and direction of the net displacement of the wagon. (D) Find the angle between the force and the net displacement from the directions found in parts (B) and (C). (E) Verify that $Fs \cos \theta$ agrees with the scalar product computed in part (A).

2. A monkey paints an X on each face of a cube by painting lines along all of the face diagonals. Use the scalar product to determine the angle between any two of these painted lines that meet at a corner.

3. Evaluate the following scalar products, where all of these unit vectors correspond either to Cartesian coordinates or 2D polar coordinates: $\hat{r} \cdot \hat{i}$, $\hat{j} \cdot \hat{\theta}$, $\hat{r} \cdot \hat{r}$, $\hat{r} \cdot \hat{\theta}$, and $\hat{\theta} \cdot \hat{\theta}$.

4. Evaluate the following scalar products, where all of these unit vectors correspond either to Cartesian coordinates or spherical coordinates (except $\hat{r}_c$, which belongs to cylindrical coordinates): $\hat{r} \cdot \hat{i}$, $\hat{\varphi} \cdot \hat{j}$, $\hat{k} \cdot \hat{\theta}$, $\hat{r} \cdot \hat{r}$, $\hat{r} \cdot \hat{\varphi}$, $\hat{\varphi} \cdot \hat{\varphi}$, and $\hat{r} \cdot \hat{r}_c$.

5. Monkeys work hard to earn their bananas at Monko, Inc. (A) A monkey walks up a 60° incline. (i) When computing the work done by gravity, what is the angle between $\vec{F}$ and $\vec{s}$? (ii) When computing the work done by the normal force, what is the angle between $\vec{F}$ and $\vec{s}$? (iii) When computing the work done by friction, what is the angle between $\vec{F}$ and $\vec{s}$? (iv) When computing the work done by gravity, what is the force? (v) When computing the work done on the monkey, what is the force? (B) A monkey exerts a force $\vec{F} = 2\hat{i} - 2\sqrt{3}\hat{j}$ on a box of bananas, displacing the box of bananas along $\vec{s} = 24\hat{i} - 8\sqrt{3}\hat{j}$ with an average speed of 4.0 m/s, where SI units have been suppressed in the vector expressions. (i) How much work does the monkey do? (ii) What is the magnitude of $\vec{F}$? (iii) What is the direction of $\vec{s}$? (iv) What average power does the monkey deliver to the box of bananas?

6. A monkey with a white collar sits at a desk and contemplates the work done and power delivered by various hypothetical forces, where SI units have been suppressed. (A) How much work is done by a force $\vec{F} = 3\hat{i} - 2\hat{j}$ that displaces an object according to $\vec{s} = 2\hat{i} - \hat{j}$? (B) How much work is done by a force $F = 2 \sin\left(\frac{\pi x}{12}\right)$ that displaces an object from $x = 2.0$ m to $x = 10.0$ m? (C) How much power is delivered at $t = 4.0$ s if $W = 4t^{5/2}$.

7. A 400-g box of bananas rests on the floor. The coefficient of friction between the box of bananas and the floor is 0.60. (A) A monkey pushes the box of bananas horizontally with a force of 5.0 N for 7.0 seconds. How much work did the monkey do? (B) A monkey pulls the box of bananas at an angle of 40° above the horizontal with a constant force of 5.0 N until the speed of the box is 10 m/s. How much work did the friction force do? (C) A monkey lifts the box of bananas 1.5 m in the air. How much work did the monkey do against gravity? (D) A monkey pulls the box of bananas at an angle of 80° above the horizontal with a constant force of 5.0 N for 7.0 seconds. How much work did the friction force do? (E) A monkey pulls the box of bananas at an angle of 20° above the horizontal with a constant force of 5.0 N. The monkey delivers 2.4 hp to the box of bananas. What is the average speed of the bananas?

8. Fun, being the antithesis of work, is naturally defined as the negative of the work:
$$Fun = -W$$
A 50-kg chimpanzee swings from a chandelier, slides down a 35° banister, lands on a skateboard, and crashes into a cabinet of fine china. Now that's fun! The chimpanzee slides down the 20-m long banister with an initial rate of 10 m/s. The coefficient of kinetic friction between the chimpanzee and the banister is 0.30. (A) How much fun is provided by friction as the chimpanzee slides down the banister? (B) How much fun is provided by the normal force as the chimpanzee slides down the banister? (C) How much fun does the chimpanzee have while sliding down the banister? (D) What is the average power delivered to the chimpanzee by gravity?

9. Fun, being the antithesis of work, is naturally defined as the negative of the work:
$$Fun = -W$$
SI units are suppressed. (A) A puppy playfully tugs on the tail of a kitten with a force $\vec{F} = 3\hat{i} - 4\hat{j}$, while the displacement of the kitten is $\vec{s} = 5\hat{i} + 2\hat{j}$. How much fun does the kitten have playing with the puppy? (B) A kitten nudges a mouse with a force $\vec{F} = -3(x-4)^2\hat{i}$ from $x = 0$ to $x = 6.0$ m. How much fun does the kitten provide to the mouse? (C) The amount of fun a kitten has playing with a ball of yarn is $Fun = -3(t-2)^2$. How much power does the kitten deliver to the ball of yarn at $t = 6.0$ seconds? What is the average power delivered to the ball of yarn during the first 6.0 seconds?

10. The more you work, the less you play. The more you play, the less you work. Thus, a monkey hypothesizes that
$$Play + Work = 1000 \text{ J}$$
Knowledge is power. Therefore, a monkey predicts that
$$Ignorance + Power = 1000 \text{ W}$$
A monkey exerts a force of 200 N to pull a 28-kg box of bananas with an initial velocity of $\sqrt{3}$ m/s a distance of $\sqrt{12}$ m along the ground, as illustrated below. The coefficient of friction between the box and ground is $1/\sqrt{3}$. (A) How much does the normal force play with the box of bananas? (B) How much does friction play with the box of bananas? (C) How much does gravity play with the box of bananas? (D) How much does the box of bananas play? (This is analogous to, "How much work is done on the box of bananas?") (E) How much ignorance does the monkey deliver to the box of bananas, on average? **Hint**: Which force is associated with the monkey?

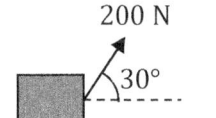

11. A monkey supposes that leisure is the inverse of work:
$$Leisure = \frac{1}{W}$$
(A) A puppy playfully tugs on the tail of a kitten with a force $\vec{F} = 2\hat{i} - \hat{j} + 3\hat{k}$, while the displacement of the kitten is $\vec{s} = 3\hat{i} + 2\hat{j} - 4\hat{k}$, where SI units have been suppressed. How much leisure does the puppy have? (B) A 15-kg puppy drags a 5.0-kg wedding dress 40 m along a 30° muddy incline. How much leisure does gravity have? (C) A 20-kg puppy runs along horizontal ground with an acceleration of 5.0 m/s² for 8.0 seconds. How much does leisure does the normal force have? (D) A 10-kg puppy rides on a 20-kg sled on horizontal ground for 50 m. The coefficient of friction between the ground and the sled is 1/2. How much leisure does friction have?

12. A monkey with a blue collar pulls an 8.0-kg box of bananas a distance of 6.0 m along a 60° incline with a force of 120 N, as illustrated below. The coefficient of friction between the box and incline is 0.50. (A) How much work does the monkey do? (B) How much work does gravity do? (C) How much nonconservative work is done? (D) Starting from rest, what is the average power delivered by the monkey's pull?

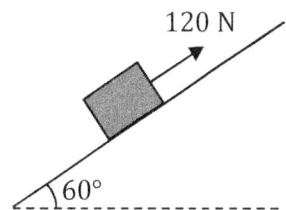

13. A monkey pushes a 4.0-kg box of precious banana gems with a force $\vec{F} = (☺x^2 + ☹x)\hat{i}$, where ☺ = 3.0 and ☹ = −5.0 in their appropriate SI units. The banana travels from (−1 m, 4 m) to (2 m, 4 m) along a straight line. The initial speed of the banana is 5.0 m/s. (A) What are the SI units of ☺ and ☹? (B) How much work did the monkey do? (C) What is the final speed of the banana? (D) Is the work path-independent?

14. Monk − James Monk, aka secret agent $00\pi$ − is pursuing his nemesis, Osamonk Been Rotten, in a high-speed car chase. (A) $00\pi$'s sporty car is powered by a force $\vec{F} = 5x^2y\hat{i} − 10x^2y^3\hat{j}$ as his car travels along the curve $y = 4x^2$ over the interval $0 \le x \le 2.0$, where the force is expressed in kN and position is expressed in km. How much work is done by the car's engine? (B) Osamonk Been Rotten's monster truck delivers a power $P = 6t^2 − 4t + 2$ over the interval $0 \le t \le 8.0$, where the power delivered is expressed in kW and time is expressed in SI units. How much work is done by the truck's engine (assuming the initial work to be zero)?

15. A motor does work $W = \beta t^3$ to accelerate a 200-kg bananamobile from rest, where beta ($\beta$) is 500 in SI units. (A) What instantaneous horsepower does the motor deliver to the bananamobile at $t = 5.0$ sec? (B) The bananamobile, presently moving horizontally 50 m/s, exerts a braking force given by $F(x) = −\mu x^2$, where mu ($\mu$) is a constant. If the car comes to rest in 200 m, what is $\mu$?

**Conservation of energy**

16. A 50-kg bananamobile is traveling 30 m/s down the highway. How much work is required to bring the bananamobile to a stop?

17. A miniature 200-g monkey skis 150 m down a 25° frictionless incline on banana peels. The initial speed of the monkey is 20 m/s. (A) How much does the potential energy of the monkey change? (B) What is initial kinetic energy of the monkey? (C) What is the final kinetic energy of the monkey? (D) What is the final speed of the monkey? (E) What is the work done by gravity? (F) What is the work done by the normal force? (G) How much power does gravity deliver to the monkey, on average?

18. A 600-g banana is thrown upward at some angle at a rate of 30 m/s off the top of a 100-m high hill carved in the shape of a monkey. The maximum height of the banana is 140 m above the ground. (A) Use energy methods to find the speed of the banana at the top of its trajectory. (B) At what angle is the banana thrown?

19. A miniature 200-g monkey skis 150 m down a 25° incline on a 400-g sled. The initial speed of the monkey is 20 m/s. The coefficient of friction between the sled and the incline is 0.30. (A) How much energy is lost to friction? (B) Use energy methods to find the final speed of the monkey.

20. A miniature 80-g monkey slides 60 m down a 35° incline with an initial speed of 20 m/s. The monkey loses 10% of its total initial energy (relative to the bottom) to friction. What is the final speed?

21. A monkey stands on the roof of a 10-m tall house and throws an adorable kitty cat upward (but not straight upward) with an initial speed of 5.0 m/s as part of his science fair project to determine if cats always do land on their feet. The cat does, in fact, land on its feet after losing 36% of its energy (relative to the ground) to air resistance. What is the cat's speed just before landing on the ground?

22. While the monkeys are biting their nails at Point $A$ in the illustration below, the car suddenly drops. The combined mass of the car and monkeys is 400 kg. The ride is frictionless between Points $A$ and $B$, but there is friction between Points $B$ and $C$. The ride comes to rest at Point $C$. (A) What is the speed of the monkeys at Point $B$? (B) How much nonconservative work is done between Points $B$ and $C$? (C) What is the coefficient of friction between the car and surface between Points $B$ and $C$?

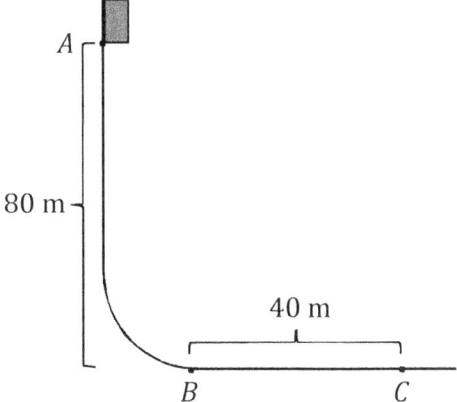

23. Monkeys wait in line for hours to ride on Free Fall at Monkey Mountain. While the monkeys are biting their nails at Point $A$, the car suddenly drops (see illustration above). The ride applies a nonconservative braking force between Points $B$ and $C$ given by $\vec{F}(x) = -\beta x^2 \hat{\imath}$. The combined mass of the car and monkeys is 400 kg. (A) What is the maximum speed of the monkeys? (B) What value of $\beta$ is required in order to bring the car to a stop at Point $C$?

24. An eskimonkey of mass $m$ slides from rest from the top of a frictionless hemispherical igloo of radius $R_0$ as illustrated below. (A) What is the initial potential energy of the eskimonkey relative to the ground? (B) What is the initial kinetic energy of the eskimonkey? (C) What is the potential energy of the eskimonkey relative to the ground when the eskimonkey is at angle $\theta$? (D) What is the speed of the eskimonkey when the eskimonkey is at angle $\theta$? (E) What force does the igloo exert on the eskimonkey when the eskimonkey is at angle $\theta$? (F) What is the speed of the eskimonkey when the normal force equals zero? (G) At what critical angle, $\theta_c$, does the monkey lose contact with the igloo?

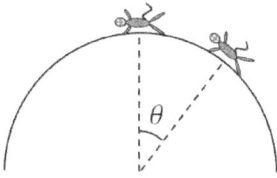

25. A lemur of mass $m$ swings back-and-forth on a vine of length $L$ suspended from the branch of a banana tree. The vine makes a maximum angle of $\theta$ with the vertical. Take the height of the lemur to be zero when the vine is vertical. Also, take initial to be maximum angle, and final to be vertical. (A) Derive equations for the initial and final height of the lemur. (B) Derive equations for the initial and final velocity of the lemur. (C) Derive equations for the initial and final acceleration of the lemur. (D) Derive equations for the initial and final tension in the vine. (E) If $\theta = 30°$, what is the angle of the vine when the lemur's potential energy is half its maximum value? (It's not 15°! Why not?)

26. A 400-g banana oscillates with amplitude 2.5 cm at the end of a spring whose constant is 1.0 N/cm. Find the speed of the banana when its displacement from the equilibrium position is (A) 2.5 cm, (B) 0 cm, and (C) 0.50 cm.

27. In the frictionless roller coaster illustrated below, a 20-kg box of bananas passes through the 35-m diameter loop once and then compresses a spring 8.0 m from its equilibrium position (after which it repeats its journey in reverse). In the position shown, the box of bananas is 60 m above the ground and moving 20 m/s. (A) What is the speed of the box of bananas at point $A$? (B) What is the normal force at point $A$? (C) What is the spring constant?

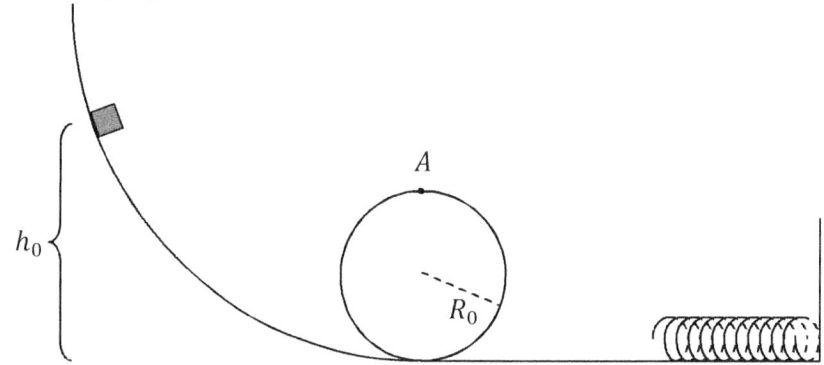

28. In the roller coaster illustrated below, a 6.0-kg box of bananas is launched to the right by a spring with constant 200 N/m that is compressed 3.0 m from equilibrium. The diameter of the loop is 10-m. Neglect friction. (A) How much energy is stored in the compressed spring? (B) How high up the hill does the box of bananas get (before falling back down)? (C) What is the normal force at point $A$? (D) What percentage of the box of bananas' initial energy can the box of bananas afford to lose (to any nonconservative forces) and still complete the loop without falling away from the track?

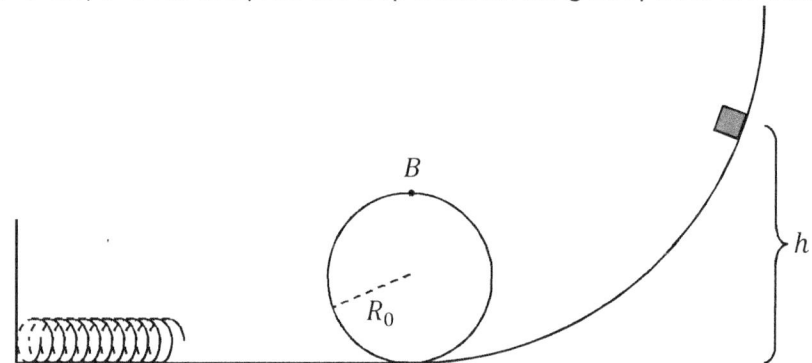

29. A 2.0-kg box of miniature bananas slides down a 30° frictionless incline with an initial speed of 5.0 m/s. The box of miniature bananas slides 20 m before reaching a spring with a spring constant of 22 N/m. (A) How fast is the box of miniature bananas sliding just before reaching the spring? (B) What is the maximum compression of the spring from equilibrium?

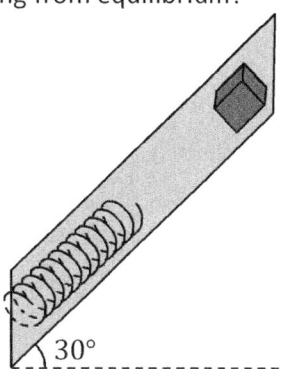

30. It turns out that the kitty cat in Problem 21 has a godfather who is a ferocious lion. The lion places the 15-kg monkey atop a vertical spring with constant 8.0 N/cm, compresses the spring 2.0 m from equilibrium, and releases the spring from rest. How high does the monkey travel (above equilibrium) if he loses 25% of his spring potential energy to air resistance?

31. As illustrated below, a $10\pi$-kg monkey sleeps on a $30\pi$-kg bed. The spring constant is $30\pi$ N/m. The spring is presently stretched $20\pi$-m from equilibrium. This elaborate alarm clock cuts the cord at 3:14 a.m., thereby sending the bed and monkey into simple harmonic oscillation. What a way to start the day! The coefficient of friction between the floor and the bed is 0.20. (A) What will be the speed of the bed when it returns to equilibrium? (B) What is the speed of the bed after traveling $10\pi$ m to the right of its initial position? (C) What must be the coefficient of friction between the bed and the monkey in order to prevent the monkey from sliding on the bed during oscillation?

32. A spring with a constant of 60 N/m and natural length of 12.0 m is attached to a wall on the left, while another spring with a constant of 40 N/m and natural length of 18.0 m is attached to a wall on the right. As illustrated on the following page, there is a 20-kg box of bananas connected to both two springs. The two walls are 41.0 m apart, and the box is 1.0 m wide. Neglect friction. (A) When the system is in equilibrium, how far is the box of bananas from the left wall? (B) If the box of bananas is released from rest from a position 12.0 m from the left wall, what will its speed be when it is 16.0 m from the left wall?

# 5 Work and Energy

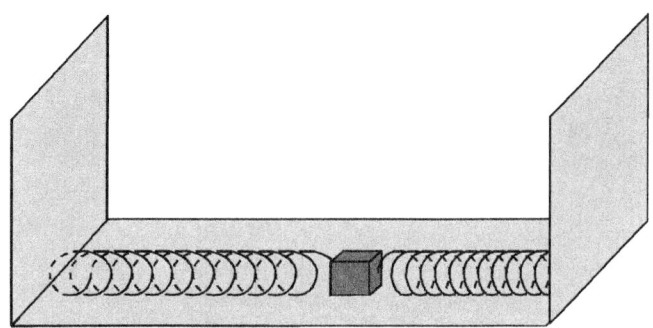

33. Below are astronomical data for Banana planet:
   - Radius of Banana Planet: $1.45 \times 10^5$ m
   - Mass of Banana Planet: $9.24 \times 10^{23}$ kg
   - Distance between Banana Planet and Banana Star: $6.59 \times 10^9$ m
   - Mass of Banana Star: $3.72 \times 10^{26}$ kg
   - One 'Day' on Banana Planet: 8.23 hours
   - One 'Year' on Banana Planet: 720 days

(A) A 3000-kg Banana-synchronous satellite (i.e. synchronized with the planet's rotation about its axis, above the equator) is to be moved such that it is twice as far away from the center of the planet. Apply the satellite strategy of Sec. 4.6 to determine the radius and speed of the initial and final (circular) orbits. (B) How much work must be done in order to make this change in (circular) orbits? (C) Use conservation of energy to determine the escape speed of an object on the surface of Banana planet.

34. Neglecting air resistance, if a projectile (so it will be in free fall, unlike a rocket that ejects steam) is launched from earth with escape speed with a path that takes it to the moon, how fast will it be moving when it reaches the moon's surface?

## Conservative and nonconservative forces

35. All of the monkeys chase you with physics notes. Ahhhh! They catch you and torment you with physics problems. Consider the following equations, where SI units have been suppressed:

$$\vec{F}_1 = 4\hat{\imath} + 7\hat{\jmath}$$
$$\vec{F}_2 = (2x^2 + 3x)\hat{\imath}$$
$$W_3(t) = 2t^2 - 5t + 4$$
$$U_4(r) = \frac{9}{r^4} - \frac{4}{r^2}$$

(A) How much work is done by $\vec{F}_1$ in moving an object along the displacement $\vec{s}_1 = 12\hat{\imath} - 5\hat{\jmath}$? (B) What is the potential energy associated with $\vec{F}_2$? (C) Evaluate the power for $W_3$ at 3.0 seconds. (D) Locate the positions of stable equilibrium for a particle with potential energy $U_4$. (E) For the first two equations, indicate whether or not the force is conservative.

36. When a monkey is asked whether or not he is conservative, he responds that he is as conservative as the force $\vec{F}_1 = 2x^2 y\hat{\imath} - xy^2\hat{\jmath}$, where SI units have been suppressed. (A) Find the work done by the force from (0,0) to (3,2) along the straight line that connects the two endpoints. (B) Compare with the work done by the force from (0,0) to (3,2) along the path $(0,0) \to (3,0) \to (3,2)$. (C) Is the force conservative?

37. Monkeymatics: 2 Monkeys + 1 Banana = 1 Fight. SI units have been suppressed. (A) Find the partial derivative of $3x^2y + 7x^3\sqrt{y} - 4(2-y)^2$ with respect to $x$. (B) Find the partial derivative of $3x^2y + 7x^3\sqrt{y} - 4(2-y)^2$ with respect to $y$. (C) Find the gradient of $\frac{x}{y} - 3xy^2 + \frac{y^2}{x^2}$. (D) Is the force $\vec{F} = 4xy^2\hat{i} + 4x^2y\hat{j}$ conservative? (E) Is the force $\vec{F} = 4x^2y\hat{i} + 4xy^2\hat{j}$ conservative? (F) Given $U(x,y) = \gamma x - \lambda y^2$, find the SI units of $\gamma$ and $\lambda$. (G) Given $U(x,y) = 3x - 2y^2$, find the magnitude and direction of $\vec{F}$. (H) Given $(r) = r^3 + 3r^2 - 16r + 9$, find the equilibrium positions and characterize each as stable, neutral, or unstable. (I) Given $\vec{F}(t) = 8t^2 + 16t - 32$, find the impulse from $t = 0$ to $t = 30$ ms. (**Hint:** Impulse is defined in Sec. 6.2.) (J) For the force $\vec{F} = 2\hat{i} + 4x^2\hat{j}$, compute the work done from $(0,0)$ to $(2,2)$ along the path $(0,0) \to (0,2) \to (2,2)$ and compare it to the work done from $(0,0)$ to $(2,2)$ along the path $(0,0) \to (2,0) \to (2,2)$. Is this force conservative? (K) For the force $\vec{F} = 4x^2\hat{i} + 2\hat{j}$, compute the work done from $(0,0)$ to $(2,2)$ along the path $(0,0) \to (0,2) \to (2,2)$ and compare it to the work done from $(0,0)$ to $(2,2)$ along the path $(0,0) \to (2,0) \to (2,2)$. Is this force conservative?

38. A monkey wearing a pink collar is doing research on the physics of tennis serves. She hypothesizes that the force delivered by the tennis racket is $\vec{F} = kr^2 \sin(2\theta)\hat{\theta}$, where $k$ is a constant and $r$ and $\theta$ are the 2D polar coordinates that describe the position of the tennis racket. Either prove that this force is conservative or show that it is not.

39. Compute the work done by the force $\vec{F} = y^2\hat{i} + x^2\hat{j}$ along the following closed path: $(1,0) \to (2,2) \to (1,2) \to (1,0)$. SI units have been suppressed. Is this force conservative? If your answer seems inconsistent with your work integral, explain.

40. Compute the work done by the force $\vec{F} = x^2\hat{i} + y^2\hat{j}$ along the following closed path: $(1,0) \to (2,2) \to (1,2) \to (1,0)$. SI units have been suppressed. Is this force conservative? If your answer seems inconsistent with your work integral, explain.

41. Compute the work done by the force $\vec{F} = r\hat{r}$ along the following closed path: a circle centered about the origin, starting and ending at $(4,0)$ in a counterclockwise sense. SI units have been suppressed. Is this force conservative? If your answer seems inconsistent with your work integral, explain.

42. Compute the work done by the force $\vec{F} = r\hat{\theta}$ along the following closed path: a circle centered about the origin, starting and ending at $(4,0)$ in a counterclockwise sense. SI units have been suppressed. Is this force conservative? If your answer seems inconsistent with your work integral, explain.

# 6 Systems of Objects

**Systems of objects**: Many problems involve multiple objects interacting with one another. For example, atoms consist of electrons in orbit around a positively charged nucleus, for which the electrons are attracted to the nucleus and repelled by other electrons; the solar system consists of a star, planets, and moons that all attract one another gravitationally; and two cars exert forces on one another when they collide.

**Center of the system**: When working with a system of objects, it is often useful to know where its center lies. There is actually more than one way to define what is meant by the 'center' of the system – e.g. there is a geometric center, a center of mass, a center of weight, etc. In Sec. 6.4, we will see that the center of mass has particular significance in the context of collisions. It will also have significance in Chapter 7, where we consider rotations – an object naturally tends to rotate about an axis through its center of mass (though it can also be forced to rotate about a different axis).

**Interactions within the system**: When treating a system of objects as a whole, all of the forces that are internal to the system cancel out according to Newton's third law. For this reason, we will see that only external forces affect the motion of the center of mass. However, the internal forces do affect the motion of the individual objects that make up the system. One result of these internal forces is that objects within the system tend to change velocity. We will see that it is particularly convenient to work with momentum – mass times velocity – in order to relate the initial and final velocities of objects in a system. Specifically, we will see that momentum is often conserved during collisions, and that the principle of conservation of momentum allows us to predict the final velocities of the objects in the system from the initial velocities and the masses of the objects.

**Rocket propulsion**: We will finally treat the problem, in Sec. 6.9, where the mass of an object is variable, and apply the more general form of Newton's second law – i.e. working with $d\vec{p}/dt$ instead of $m\vec{a}$. Rocket propulsion is the prototype for the variable-mass problem: A rocket ejects a substance such as steam in order to change its velocity. Thinking of the rocket plus the steam as the system, one may apply the principle of conservation of momentum to determine the final velocity of the rocket.

## 6.1 Center of Mass

**Arithmetic mean**: If you make $N$ measurements of a quantity, $x$, the arithmetic mean provides the simplest measure of the average value of $x$ – denoted by $\bar{x}$:[250]

---

[250] This notation is introduced in Sec. 1.2 and also appears in Chapters 6-7. Here, $\sum_{i=1}^{N} x_i = x_1 + x_2 + \cdots + x_N$.

$$\bar{x} = \frac{1}{N}\sum_{i=1}^{N} x_i \quad \text{(arithmetic mean)}$$

where $x_i$ designates the $i^{th}$ measurement. This is the first type of average that all students first encounter in math courses, and is the most intuitive way to average values; it's what first comes to mind when we think of the word 'average.' However, the arithmetic mean is not always the most appropriate method of averaging values. For example, sometimes the measurements must be weighted differently according to weighting factors – such an average is called a weighted average, which we consider next.

**Weighted average**: If you make $N$ measurements of a quantity, $x$, a weighted average can be found by including weighting factors, $w_i$, as follows:

$$\bar{x} = \frac{\sum_{i=1}^{N} w_i x_i}{\sum_{i=1}^{N} w_i} \quad \text{(weighted average)}$$

Note that if all factors are weighted equally, the weighted average simplifies to the arithmetic mean. In general, however, the weighting factors are not all equal. We will see an example of the weighted average when we consider how to determine the center of mass of a system.

**Center of mass for a discrete system**: Imagine a 50-kg monkey sitting at $x = 0$ and a 500-kg gorilla sitting at $x = 10$ m. If you proceed to average the values of $x$ using the arithmetic mean $\bar{x} = (x_1 + x_2)/2 = 5.0$ m, what you are finding is really a center of position and not a center of mass. Conceptually, it should seem reasonable that the center of mass should be closer to the gorilla than to the monkey, since the gorilla has more mass than the monkey. Therefore, instead of using the arithmetic mean, we adopt the weighted average – using mass as the weighting factor – as our definition of center of mass.

Given a system of $N$ objects with masses $m_1, m_2, \cdots, m_N$, the center of mass of the system is specified by the following position vector:

$$\vec{r}_{cm} = \frac{\sum_{i=1}^{N} m_i \vec{r}_i}{\sum_{i=1}^{N} m_i} = \frac{1}{M}\sum_{i=1}^{N} m_i \vec{r}_i$$

The position vector $\vec{r}_{cm}$ extends from the origin to the center of mass of the system, while each position vector $\vec{r}_i$ extends from the origin to the center of mass of the $i^{th}$ object. The total mass of the system is

$$M = \sum_{i=1}^{N} m_i$$

**Components of the center of mass vector**: The center of mass position vector can be expressed in terms of components as

## 6 Systems of Objects

$$\vec{r}_{cm} = x_{cm}\hat{i} + y_{cm}\hat{j} + z_{cm}\hat{k}$$

The components of the center of mass position vector are the coordinates of the center of mass. These components can be found directly as follows:

$$x_{cm} = \frac{1}{M}\sum_{i=1}^{N} m_i x_i \quad , \quad y_{cm} = \frac{1}{M}\sum_{i=1}^{N} m_i y_i \quad , \quad z_{cm} = \frac{1}{M}\sum_{i=1}^{N} m_i z_i$$

---

**Example.** Ten 200-g bananas are positioned as follows: 3 lie at $(4.0\text{ m}, -1.0\text{ m})$, 2 lie at $(-2.0\text{ m}, 6.0\text{ m})$, and 5 lie at $(0, -3.0\text{ m})$. Where is the center of mass of the system of 10 bananas?

It is convenient to think of this system as being divided into 3 objects with masses of 600 g, 400 g, and 1000 g.

$$x_{cm} = \frac{m_1 x_1 + m_2 x_2 + m_3 x_3}{M} = \frac{(600)(4.0) + (400)(-2.0) + (1000)(0)}{(10)(200)} = 0.80 \text{ m}$$

$$y_{cm} = \frac{m_1 y_1 + m_2 y_2 + m_3 y_3}{M} = \frac{(600)(-1.0) + (400)(6.0) + (1000)(-3.0)}{(10)(200)} = -0.60 \text{ m}$$

The center of mass of the system lies at $(0.80\text{ m}, -0.60\text{ m})$.

---

**Center of gravity**: Similar to center of mass, there is another quantity known as center of gravity. The distinction is that the magnitude of the weight is used for the weighting factor instead of mass. For a discrete system of objects,

$$\vec{r}_{cg} = \frac{\sum_{i=1}^{N} W_i \vec{r}_i}{\sum_{i=1}^{N} W_i} = \frac{\sum_{i=1}^{N} m_i g_i \vec{r}_i}{\sum_{i=1}^{N} m_i g_i}$$

Observe that the center of gravity is the same as the center of mass in the presence of a uniform gravitational field (since in that case all of the $g_i$'s will be the same and cancel out). The distinction between center of gravity and center of mass is only important when the system is large enough that there is a significant variation in gravity throughout the system.

For example, there is a small, yet significant, difference between the moon's center of mass and its center of gravity because earth's gravitational field is a little stronger on the near side of the moon compared to the far side of the moon. The moon naturally rotates about an axis through its center of mass, but the earth exerts a gravitational force that acts on average at the moon's center of gravity. This distinction results in a small, but observable, gravitational torque exerted on the moon.

**Discrete system versus a continuous object**: The summation formulas for center of mass are appropriate for a system of multiple objects (a discrete system), where you already know the center of mass of each individual object in the system. Sometimes you need to find the center of mass of a single continuous object via integration (described in the remainder of this section). Both methods may be needed in combination: You might first need to integrate for each object, and then sum for the system.

**Center of mass for a continuous object**: Given a single continuous object like a cone or a banana, in principle one could use the summation formulas to find the center of mass, but in practice this is not practical for an object that consists of Avogadro's number of particles to sum over. However, such a sum is a Riemann sum, which means that we can instead determine the center of mass via integration. The center of mass position vector for a continuous object is

$$\vec{r}_{cm} = \frac{1}{M} \int \vec{r} \, dm$$

and its components can be found directly from

$$x_{cm} = \frac{1}{M} \int x \, dm \quad , \quad y_{cm} = \frac{1}{M} \int y \, dm \quad , \quad z_{cm} = \frac{1}{M} \int z \, dm$$

In the remainder of this section, we will build up and discuss a technique for performing these integrals. An important concept underlying this technique is that, in general, neither $\vec{r}$, $x$, $y$, nor $z$ are constant, and therefore cannot be pulled out of the integration. Our strategy will involve rewriting $dm$ in terms of the coordinates and then integrating.

**Density**: You probably know the formula for the density of a uniform object without any inaccessible hollow regions – mass over volume. However, not all objects have a uniform distribution of mass – e.g. astronomical objects tend to be much more dense in their core than at their surface. In order to accommodate non-uniform mass distributions, density is defined in terms of calculus. We will also work with three different kinds of density, depending upon whether the mass distribution is effectively 1D, 2D, or 3D:

$$dm = \begin{cases} \lambda \, ds & \text{in 1D} \\ \sigma \, dA & \text{in 2D} \\ \rho \, dV & \text{in 3D} \end{cases}$$

where lambda ($\lambda$), sigma ($\sigma$), and rho ($\rho$) are the densities in 1D, 2D, and 3D; and where $ds$ is the differential arc length, $dA$ is the differential area element, and $dV$ is the differential volume element. The densities have SI units of $[\lambda]_u = \text{kg/m}$, $[\sigma]_u = \text{kg/m}^2$, and $[\rho]_u = \text{kg/m}^3$.

For a uniform distribution of mass, $\lambda$ simplifies to mass per unit length, $\sigma$ becomes mass per unit area, and $\rho$ reduces to the usual mass per unit volume. For a non-uniform distribution of mass, you must express the densities using differentials, and not simply think of the densities as ratios.

> **Note.** It's important to realize that the 3D density, $\rho$, is not always equal to the mass of the object divided by its volume, but for a non-uniform distribution of mass must be expressed using $dm = \rho \, dV$; and similarly for the 1D and 2D densities. For non-uniform distributions, the densities are not constants, and so may not be pulled out of the center of mass integral; instead, a problem will give you an expression to use for the non-uniform density, which you must substitute into the integral.

# 6 Systems of Objects

**Differential arc length**: The differential arc length, $ds$, has a different form in each coordinate system.

The simplest case is a linear object oriented along a single axis, for which two of the Cartesian coordinates will be constant – e.g. for a linear object that lies on (or parallel to) the $x$-axis, $ds = dx$.

For a curve lying in the $xy$ plane, the differential arc length can be expressed as $ds = \sqrt{dx^2 + dy^2}$ or $ds = \sqrt{dr^2 + r^2 d\theta^2}$. The latter equation is well-suited for a circle, since then $ds = r d\theta$, and the former is well-suited for a line, since then $dy/dx$ is constant and equal to the slope. For a noncircular curve, factor the differential arc length as $ds = \sqrt{1 + \left(\frac{dy}{dx}\right)^2}\, dx$ or $ds = \sqrt{\left(\frac{dr}{d\theta}\right)^2 + r^2}\, d\theta$ and then use the equation of the curve to take the derivative. For example, for the parabola $y = 3x^2 - 1$, $\frac{dy}{dx} = 6x$ and $ds = \sqrt{1 + 36x^2}\, dx$.

Most generally, for a curve such as a helix that winds through 3D space:

$$ds = \sqrt{dx^2 + dy^2 + dz^2} \quad \text{(Cartesian)}$$

$$ds = \sqrt{dr_c^2 + r_c^2 d\theta^2 + dz^2} \quad \text{(cylindrical)}$$

$$ds = \sqrt{dr^2 + r^2 \sin^2\theta\, d\varphi^2 + r^2 d\theta^2} \quad \text{(spherical)}$$

| | |
|---|---|
| linear object parallel to $x$-axis | $ds = dx$ |
| linear object parallel to $y$-axis | $ds = dy$ |
| linear object parallel to $z$-axis | $ds = dz$ |
| linear object lying parallel to $xy$ plane | $ds = \sqrt{1 + (\text{slope})^2}$ |
| circular arc about origin in $xy$ plane | $ds = r d\theta$ |
| curve lying in $xy$ plane in Cartesian | $ds = \sqrt{1 + \left(\frac{dy}{dx}\right)^2}\, dx$ |
| curve lying in $xy$ plane in 2D polar | $ds = \sqrt{r^2 + \left(\frac{dr}{d\theta}\right)^2}\, d\theta$ |
| general curve in Cartesian | $ds = \sqrt{1 + \left(\frac{dy}{dx}\right)^2 + \left(\frac{dz}{dx}\right)^2}\, dx$ |
| general curve in cylindrical | $ds = \sqrt{r_c^2 + \left(\frac{dr_c}{d\theta}\right)^2 + \left(\frac{dz}{d\theta}\right)^2}\, d\theta$ |
| general curve in spherical | $ds = \sqrt{r^2 + \left(\frac{dr}{d\theta}\right)^2 + r^2 \sin^2\theta \left(\frac{d\varphi}{d\theta}\right)^2}\, d\theta$ |

**Differential area elements**: The differential area element, $dA$, is constructed from pairs of independent differential arc lengths (i.e. the differential arc lengths correspond to small displacements in mutually perpendicular directions), and has a different form in each coordinate system.

The simplest differential area element is for a plane surface lying in (or parallel to) the $xy$ plane. In this case, $dA = dxdy$ in Cartesian coordinates and $dA = rdrd\theta$ in 2D polar coordinates. The former is more natural for a plane surface bounded by straight lines, such as a rectangle or triangle, and the latter is more convenient if the surface is bounded (in part or in whole) by a circular arc. Otherwise, decide whether it is more convenient to write the equations of the bounding curves in Cartesian or 2D polar coordinates to decide which coordinate system to use.

To write $dA$ for an infinitesimal section of the surface area of a sphere, it is most convenient to work with spherical coordinates – in this case $dA = r^2 \sin\theta\, d\theta\, d\varphi$. For the body of a right-circular cylinder, $dA$ is expressed most concisely in cylindrical coordinates: $dA = r_c d\theta dz$.

Observe that the simplest differential area elements are constructed from pairs of differential arc lengths. In Cartesian coordinates, the elementary differential arc lengths are $dx$, $dy$, and $dz$ (whereas the general arc length is $\sqrt{dx^2 + dy^2 + dz^2}$). For a planar surface lying in the $yz$ plane, for example, we construct $dA$ by multiplying $dy$ and $dz$ together: $dA = dydz$.

In 2D polar coordinates, the elementary differential arc lengths are $dr$ and $rd\theta$, from which $dA = rdrd\theta$. In cylindrical coordinates, the elementary differential arc lengths are $dr_c$, $d\theta$, and $dz$. For the body of a right-circular cylinder, for example, $r_c$ is constant, and we multiply the other two differential arc lengths together to form $dA = r_c d\theta dz$. In spherical coordinates, the elementary differential arc lengths are $dr$ (outward), $rd\theta$ (a circular longitude), and $r\sin\theta\, d\varphi$ (a circular latitude). For the surface of a sphere, $r$ is constant and we obtain $dA = r^2 \sin\theta\, d\theta\, d\varphi$.

Most generally, none of the coordinates will be constant, and we obtain $dA$ using the 3D generalization of the Pythagorean theorem for the three independent pairs of differential arc lengths:

$$dA = \sqrt{(dxdy)^2 + (dydz)^2 + (dxdz)^2} \quad \text{(Cartesian)}$$
$$dA = \sqrt{(rdr_c d\theta)^2 + (dr_c dz)^2 + (r_c d\theta dz)^2} \quad \text{(cylindrical)}$$
$$dA = \sqrt{(r\sin\theta\, drd\varphi)^2 + (rdrd\theta)^2 + (r^2 \sin\theta\, d\theta d\varphi)^2} \quad \text{(spherical)}$$

| | |
|---|---|
| planar surface parallel to $xy$ plane | $dA = dxdy$ |
| planar surface parallel to $yz$ plane | $dA = dydz$ |
| planar surface parallel to $xz$ plane | $dA = dxdz$ |
| planar surface in 2D polar | $dA = rdrd\theta$ |
| body of a right-circular cylinder | $dA = r_c d\theta dz$ |
| surface of a sphere | $dA = r^2 \sin\theta\, d\theta\, d\varphi$ |
| general surface in Cartesian | $dA = \sqrt{(dxdy)^2 + (dydz)^2 + (dxdz)^2}$ |
| general surface in cylindrical | $dA = \sqrt{(rdr_c d\theta)^2 + (dr_c dz)^2 + (r_c d\theta dz)^2}$ |
| general surface in spherical | $dA = \sqrt{(r\sin\theta\, drd\varphi)^2 + (rdrd\theta)^2 + (r^2 \sin\theta\, d\theta d\varphi)^2}$ |

**Differential volume elements**: The differential volume element, $dV$, is constructed by multiplying all three independent differential arc lengths together:

$$dV = dxdydz \quad \text{(Cartesian)}$$
$$dV = r_c dr_c d\theta dz \quad \text{(cylindrical)}$$
$$dV = r^2 \sin\theta\, drd\theta d\varphi \quad \text{(spherical)}$$

**Multiple integrals**: The center of mass integral is either a single, double, or triple integral. An integral of the form $\int_C f(x,y,z)ds$ is an ordinary single integral over a path $C$; an integral of the form $\int_S f(x,y,z)dA$ is a double integral over a surface $S$; and an integral of the form $\int_V f(x,y,z)dV$ is a triple integral over a volume $V$. Therefore, it is sometimes necessary to perform double or triple integrals in order to determine the center of mass of a continuous object.

Let us consider an integral of the form $\int_S f(x,y)dA$. The function $f(x,y)$ is the integrand; it could be of the form $f(x,y) = x$ or $f(x,y) = 3xy^2$, for example. The differential area element can be expressed as $dxdy$ or $rdrd\theta$ — in the latter case, we would think of the function in terms of 2D polar coordinates as $f(r,\theta)$. So we have in mind a double integral of the form $\int_{x=x_0}^{x} \int_{y=y_0}^{y} f(x,y)dxdy$ or $\int_{r=r_0}^{r} \int_{\theta=\theta_0}^{\theta} f(r,\theta)rdrd\theta$.

Such a double integral is very much like a single integral except for a couple of important differences. For one, the upper and/or lower limit(s) of one integral are usually functions of the other independent variable. That is, when you imagine dividing the surface $S$ over which you are integrating up into infinitesimal chunks, $dA$, in Cartesian coordinates you visualize the Riemann sum of all of the $dA$'s as a checkerboard pattern, while in 2D polar coordinates you visualize this as pie slices, as illustrated below. Let one coordinate — either $x$ or $y$ in Cartesian, or $r$ or $\theta$ in 2D polar — have its full freedom and range between two constants — i.e. its lower and upper limits. The other coordinate will, in general, have one or two variable limits. For example, if you give $x$ its full freedom, then for each possible value of $x$, $y$ will range over a different set of values. Each value of $x$ defines a strip over which $y$ ranges. In this case, the lower and upper limits of $y$ will be functions of $x$.

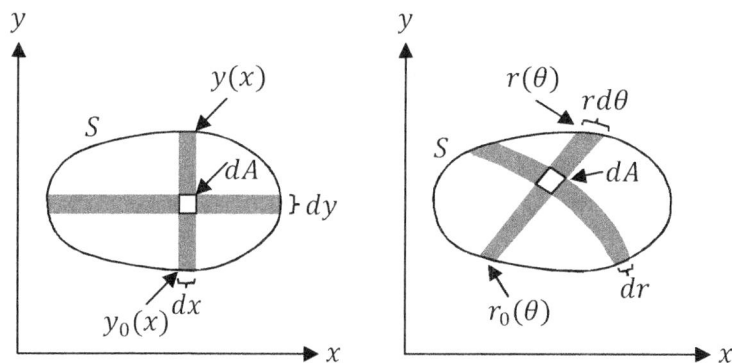

If either variable has variable limits, you must perform that integral first. There are two simple cases where neither variable will have variable limits in 2D: If the surface is a rectangle and you use Cartesian coordinates, or if the surface is a pie slice or circle (centered about the origin) and you use 2D polar coordinates. Otherwise, one coordinate will have at least one variable limit (the lower limit, upper limit, or both). This integral must be performed first.

When integrating over one coordinate, treat its independent coordinate as if it is a constant (until you complete the first definite integral; afterward, you must treat it as a variable, of course).

A triple integral works much the same way. We illustrate the techniques of double and triple integration in the examples that follow.

**Note.** If you integrate over a rectangular surface in Cartesian coordinates (with sides parallel to the Cartesian coordinates) in 2D, a pie slice or circle (centered about the origin) in 2D polar coordinates, a rectangular volume in Cartesian coordinates (with sides parallel to the elementary Cartesian planes) in 3D, a cylindrical surface or volume in cylindrical coordinates, or a spherical surface or volume in spherical coordinates, then all of the limits will be constant. In any other situation, at least one integration variable will have variable limits. It's very important not to use all constant limits in these other situations, which is a common mistake among students who obtain incorrect answers. If any of the integration variables has variable limits, you must integrate over that variable first. When integrating over one variable, treat the other independent coordinates as constants.

**Example.** Use the technique of double integration to determine the area of the triangle bounded by the $x$- and $y$-axes and the line that connects $(0, h)$ to $(b, 0)$.

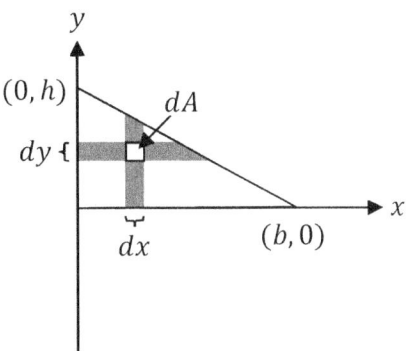

It is convenient to work with Cartesian coordinates. We choose to let $x$ have its full range: $0 \leq x \leq b$. Then for each value of $x$, $y$ begins at zero, but the upper limit of $y$ depends on the value of $x$. Specifically, the upper limit of $y$ is given by the equation of the line that connects $(0, h)$ to $(b, 0)$. The slope of this line is $m = (0 - h)/(b - 0) = -h/b$ and the $y$-intercept is $h$. Therefore, the equation of this line is $y = h - hx/b$; this is the upper limit for $y$ in the double integral:

$$A = \int_S dA = \int_{x=0}^{x=b} \int_{y=0}^{h-hx/b} dxdy$$

We can't integrate over $x$ yet because there is an $x$ in the limit of the $y$ integral. Hence, we must first integrate over $y$. We treat $x$ as a constant when integrating over $y$:

$$A = \int_{x=0}^{x=b} \left( \int_{y=0}^{h-\frac{hx}{b}} dy \right) dx = \int_{x=0}^{b} [y]_{y=0}^{h-\frac{hx}{b}} dx = \int_{x=0}^{b} \left( h - \frac{hx}{b} - 0 \right) dx = \left[ hx - \frac{hx^2}{2b} \right]_{x=0}^{b} = \frac{bh}{2}$$

> **Example**. Use triple and double integrals to derive formulas for the volume and surface area of a solid sphere of radius $R_0$.
>
> The volume of a sphere is most easily found using a triple integral in spherical coordinates.[251] When integrating over the volume of a sphere in spherical coordinates, none of the limits are variable. For every combination of $\varphi$ and $\theta$, $r$ varies from 0 to $R_0$. To cover every point in the solid sphere, $\varphi$ varies from 0 to $2\pi$ and $\theta$ varies from 0 to $\pi$:[252]
>
> $$V = \int_V dV = \int_{r=0}^{R_0} \int_{\varphi=0}^{2\pi} \int_{\theta=0}^{\pi} r^2 \sin\theta \, dr d\varphi d\theta$$
>
> The order is irrelevant here since all of the limits are constant. Observe that the integrand is separable:
>
> $$V = \int_{r=0}^{R_0} r^2 dr \int_{\varphi=0}^{2\pi} d\varphi \int_{\theta=0}^{\pi} \sin\theta \, d\theta = \left[\frac{r^3}{3}\right]_{r=0}^{R_0} [\varphi]_{\varphi=0}^{2\pi} [-\cos\theta]_{\theta=0}^{\pi} = \frac{4\pi R_0^3}{3}$$
>
> The surface area of a sphere can be found by a similar double integral. In this case, $r = R_0$ is a constant:
>
> $$A = \int_S dA = \int_{\varphi=0}^{2\pi} \int_{\theta=0}^{\pi} R_0^2 \sin\theta \, d\varphi d\theta = 4\pi R_0^2$$

**Total mass of a continuous object**: The total mass, $M$, of a continuous object can be found by integrating $M = \int dm$ and writing $dm$ as $\lambda ds$, $\sigma dA$, or $\rho dV$, depending upon the dimensionality. This same technique allows you to express the proportionality constant in terms of the total mass when the distribution of mass is non-uniform, as illustrated in the next example.

---

[251] It's possible to perform single integrals using techniques from first-semester calculus (e.g. volume of revolution) to derive these volume and surface area formulas. However, these single integral methods do not generalize to physics integrals like center of mass and moment of inertia, which include a function in the integrand. That is, integrals of the form $\int_S f(x,y,z) dA$ and $\int_V f(x,y,z) dV$ must be performed as double or triple integrals, in general (and not through revolution techniques). Center of mass integrals sometimes take a different form in vector calculus, but we set them up conceptually in physics, and the physics integrals very naturally accommodate non-uniform densities.

[252] For fixed $\varphi$ and $\theta$, varying $r$ from 0 to $R_0$ produces a line segment extending from the origin to the surface of the sphere. Still holding $\theta$ fixed, varying $\varphi$ from 0 to $2\pi$ (and still varying $r$ from 0 to $R_0$) sweeps the line segment into a right-circular cone of half-angle $\theta$ (which is symmetric about the z-axis). Sweeping $\theta$ from 0 to $\pi$ (and still varying both $r$ and $\varphi$) produces cones that point both up and down, and cover every point on the sphere. It is intuitive to many students to want both $\varphi$ and $\theta$ to vary from 0 to $2\pi$, but that would cover every point on the sphere twice. (If you're wondering if you could instead vary $\varphi$ from 0 to $\pi$ and $\theta$ from 0 to $2\pi$, the answer is negative, which you should be able to see if you try to visualize what this means geometrically. However, keep in mind that in math courses, the use of $\varphi$ and $\theta$ is generally backwards compared to physics courses.)

**Example**. A solid disc of radius $R_0$ has non-uniform density $\sigma = \beta r$, where beta ($\beta$) is a positive constant. Derive an equation that relates the total mass of the disc, $M$, to the constant $\beta$.

We integrate over $dm$, writing $dm = \sigma dA$ since the solid disc is a surface (2D). A disc is circular, for which 2D polar coordinates are well-suited – so $dA = r dr d\theta$. The limits of integration are $0 \leq r \leq R_0$ and $0 \leq \theta \leq 2\pi$:

$$M = \int dm = \int_S \sigma dA = \int_{r=0}^{R_0} \int_{\theta=0}^{2\pi} (\beta r) r dr d\theta = \frac{2\pi \beta R_0^3}{3}$$

## Problem-Solving Strategy for Performing Center of Mass Integrals

0. Use this technique to determine the center of mass of an object that has a continuous distribution of mass (as opposed to a discrete system of pointlike masses).
1. Determine which coordinate system is best-suited to the geometry of the object.
2. Write $dm = \lambda ds$ if the object looks like a line or a curve, write $dm = \sigma dA$ if the object looks like a surface, and write $dm = \rho dV$ if the object looks like a 3D solid.
3. Write the differential element from Step 2 – i.e. $ds$, $dA$, or $dV$ – as appropriate for the geometry of the object in the coordinate system that you are using. See the tables on the previous pages.
4. Determine the integration limits. Choose one variable to have its full range and write down its lower and upper limits. Make a sketch and determine whether or not the limits of the second/third coordinates depend on the value of first coordinate. For rectangular objects in Cartesian coordinates, circular objects or pie slices in 2D polar coordinates (centered at the origin), right-circular cylinders in cylindrical coordinates coaxial with $z$, or spheres in spherical coordinates (centered at the origin), all of the limits will be constants. Otherwise, you should generally expect to have at least one variable limit.
5. Perform the integral $M = \int dm$ using the substitutions from Steps 2-3 and the integration limits established in Step 4. If the density is uniform (i.e. constant), this will provide a relation between the total mass and the density of the object; if the object is non-uniform (i.e. it is specified as a function of the coordinates in the problem), this will provide a relation between the total mass and the proportionality constant from the density equation.
6. Choose the vector integral for the center of mass position vector or the component integrals for the Cartesian coordinates of the center of mass. For the vector integral, write $\vec{r} = x\hat{i} + y\hat{j} + z\hat{k}$ in Cartesian coordinates (in 3D), $\vec{r} = r\cos\theta\,\hat{i} + r\sin\theta\,\hat{j}$ in 2D polar coordinates, $\vec{r} = r_c\cos\theta\,\hat{i} + r_c\sin\theta\,\hat{j} + z\hat{k}$ in cylindrical coordinates, or $\vec{r} = r\cos\varphi\sin\theta\,\hat{i} + r\sin\varphi\sin\theta\,\hat{j} + r\cos\theta\,\hat{k}$ for spherical coordinates. If $r$ is constant – as for a circular arc or a very thin spherical shell – use $R_0$ for $r$ (but not $\vec{r}$!). (Since $\hat{r}$ and $\hat{\theta}$ are not constants, but point in different directions at different points in space, it is better to work exclusively with Cartesian unit vectors, which are always constant, when performing integrals, regardless of which coordinate system you are working with.)
7. Write down the starting equation for the center of mass integral in vector or component form. Make all of the substitutions from Steps 2-3 and 5-6 (being careful not to omit any factors) and use the integration limits from Step 4. If doing a double or triple integral, first integrate over any coordinate(s) that have variable limits. When integrating over one variable, hold its independent variables constant.

**Note.** This integration technique is particularly important in physics. We will apply a similar technique for moment of inertia in Chapter 7, and the technique will grow in subsequent volumes on electricity and magnetism. Many first-year textbooks only touch on the center of mass and moment of inertia integrals lightly, expecting many students to be just learning calculus when taking first-semester physics. However, it is the norm to consider electric and magnetic field integrals in detail in second-semester physics. This is a sudden change to a more advanced mathematical treatment in the typical first-year physics curriculum, which makes second-semester physics very challenging. Much more calculus has been incorporated in this text, which can make the transformation to the mathematics of electricity and magnetism much smoother. The more effort you put toward attempting to understand the center of mass and moment of inertia integrals as well as you can now, the more it will pay dividends when you study the integration techniques of electricity and magnetism.

**Note.** The Cartesian unit vectors – i.e. $\hat{\imath}$, $\hat{\jmath}$, and $\hat{k}$ – are constant and therefore can be pulled out of the integrals. Other unit vectors – such as $\hat{r}$ and $\hat{\theta}$ – vary in direction over the course of the integration, and therefore cannot be pulled out of the integrals. Therefore, it is convenient to express all vectors in terms of Cartesian unit vectors, even if you are not using Cartesian coordinates.

**Example.** A non-uniform rod has one end at the origin, the other end at $(L, 0)$, and non-uniform density $\lambda = \beta x$, where $\beta$ is a constant. Where is the center of mass of the rod?

The rod is linear, so we write $dm = \lambda ds$, where $ds = dx$ since the rod extends only along the $x$-axis. The limits of integration are $0 \leq x \leq L$. We first integrate to express the total mass of the rod in terms of the proportionality constant $\beta$:

$$M = \int dm = \int_C \lambda ds = \int_{x=0}^{L} (\beta x) dx = \frac{\beta L^2}{2}$$

The $x$-component of the center of mass of the rod is

$$x_{CM} = \frac{1}{M} \int x\, dm = \frac{1}{M} \int_C x \lambda ds = \frac{1}{M} \int_{x=0}^{L} x(\beta x) dx = \frac{\beta L^3}{3M}$$

Now we eliminate $\beta$ using our previous expression for the total mass of the rod:

$$x_{CM} = \frac{2L}{3}$$

Conceptually, the density of the rod increases with increasing $x$, so it should make sense that the center of mass lies beyond the geometric center (at $x = L/2$). It's always a good idea to check that your mathematical result for the center of mass agrees with your conceptual expectations.

**Example.** A uniform solid equilateral triangle has corners at $(L/2,0)$, $(0, L\sqrt{3}/2)$, and $(-L/2, 0)$. Determine the coordinates of its center of mass.

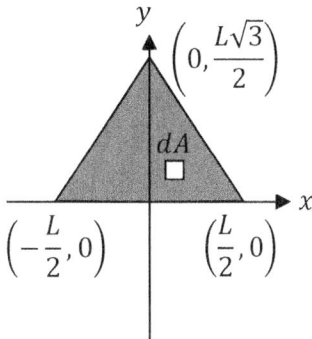

This solid triangle is a surface, so $dm = \sigma dA$. We choose to work with Cartesian coordinates, so $dA = dxdy$. We will need to split the integrals into two separate integrals – one for the left half and one for the right half of the triangle (since the equations for the right and left sides are different). For the right half, we may allow $x$ to vary from 0 to $L/2$, but then $y$ varies from 0 to $\frac{L\sqrt{3}}{2} - x\sqrt{3}$ (that's the equation for the line that you obtain after finding the slope and $y$-intercept); for the left half, $-\frac{L}{2} \leq x \leq 0$ and $0 \leq y \leq \frac{L\sqrt{3}}{2} + x\sqrt{3}$. The total mass of this uniform equilateral triangle is

$$M = \int dm = \int_S \sigma dA = \sigma \int_S dA = \sigma A = \sigma \frac{1}{2} L \frac{L\sqrt{3}}{2} = \frac{\sigma L^2 \sqrt{3}}{4}$$

Since the triangle is uniform, the density $\sigma$ is constant, so we pulled it out of the integral. Then $\int_S dA$ trivially equals the area of the triangle, which is one-half the base times the height. (However, when the density is non-uniform, you must do the actual integration.) The $x$-component of the center of mass is zero by symmetry:

$$x_{CM} = 0$$

The $y$-component of the center of mass is:

$$y_{CM} = \frac{1}{M} \int y\, dm = \frac{1}{M} \int_S y\sigma dA = \frac{4}{\sigma L^2 \sqrt{3}} \left( \int_{x=-\frac{L}{2}}^{0} \int_{y=0}^{\frac{L\sqrt{3}}{2}+x\sqrt{3}} \sigma y\, dxdy + \int_{x=0}^{\frac{L}{2}} \int_{y=0}^{\frac{L\sqrt{3}}{2}-x\sqrt{3}} \sigma y\, dxdy \right)$$

$$y_{CM} = \frac{8}{L^2\sqrt{3}} \int_{x=0}^{\frac{L}{2}} \int_{y=0}^{\frac{L\sqrt{3}}{2}-x\sqrt{3}} y\, dxdy$$

Here, we again invoked symmetry, realizing that the double integrals for the two halves of the triangle would give the same contribution toward $y_{CM}$.[253] We must perform the integration over $y$ first since its upper limit depends on $x$:

$$y_{CM} = \frac{8}{L^2\sqrt{3}} \int_{x=0}^{\frac{L}{2}} \left[\frac{y^2}{2}\right]_{y=0}^{\frac{L\sqrt{3}}{2}-x\sqrt{3}} dx = \frac{4}{L^2\sqrt{3}} \int_{x=0}^{\frac{L}{2}} \left(\frac{L\sqrt{3}}{2} - x\sqrt{3}\right)^2 dx$$

$$y_{CM} = \frac{4}{L^2\sqrt{3}} \left[\frac{-1}{3\sqrt{3}}\left(\frac{L\sqrt{3}}{2} - x\sqrt{3}\right)^3\right]_{x=0}^{\frac{L}{2}} = \frac{4}{9L^2}\left(\frac{L\sqrt{3}}{2}\right)^3 = \frac{L\sqrt{3}}{6}$$

Since the height of the triangle is $\frac{L\sqrt{3}}{2}$, we see that the center of mass lies on the bisector, a distance that is one-third along the height from the base – i.e. $y_{CM} = H/3$, where $H$ is the height of the triangle. Geometrically, this is the point where all three bisectors meet, known as the centroid, and it should come as no surprise that the center of mass lies at the centroid.

**Example.** A solid uniform hemisphere (i.e. one-half of a sphere) of radius $R_0$ has its planar side lying in the $xy$ plane centered about the origin and extends in the positive $z$-direction. Determine the location of its center of mass.

We write $dm = \rho dV$ for this 3D solid, and express $dV = r^2 \sin\theta \, dr d\varphi d\theta$ in spherical coordinates. The integration limits are $0 \leq r \leq R_0$, $0 \leq \varphi \leq 2\pi$, and $0 \leq \theta \leq \pi/2$ ($\theta$ only receives one-half its full range for this hemisphere). The total mass of this uniform hemisphere is

$$M = \int dm = \int_V \rho dV = \rho \int_V dV = \rho V = \frac{2\pi\rho R_0^3}{3}$$

since the volume of the hemisphere is one-half that of a full sphere ($4\pi R_0^3/3$). The center of mass vector is

$$\vec{r}_{cm} = \frac{1}{M}\int \vec{r} dm = \frac{3}{2\pi\rho R_0^3}\int_V (r\cos\varphi\sin\theta \,\hat{\mathbf{i}} + r\sin\varphi\sin\theta\,\hat{\mathbf{j}} + r\cos\theta\,\hat{\mathbf{k}})\rho dV$$

---

[253] You must be careful when making symmetry arguments. For example, if there had been a non-uniform density that was a function of $x$, or if the triangle had not been symmetric about the $y$-axis, then $x_{CM}$ would not have been zero. If you argue that one component of the center of mass of an object is zero by symmetry, and turn out to be wrong – you won't have any work to show toward earning partial credit, so you had better be sure. In this case, if you do the math, you'll see that it agrees with our conceptual expectation that $x_{CM} = 0$. Regarding the $y_{CM}$ integrals, we used the fact that $\int_{x=-x}^{x_0} f(x)dx + \int_{x=x_0}^{x} f(x)dx = -\int_{x=x_0}^{-x} f(x)dx + \int_{x=x_0}^{x} f(x)dx = \int_{x=x_0}^{x} f(x)dx + \int_{x=x_0}^{x} f(x)dx = 2\int_{x=x_0}^{x} f(x)dx$ (noting that in the upper limit of the $x$-integration, when the limits of $x$ are reversed, the sign of $x$ must also change here, too); where $-\int_{x=x_0}^{-x} f(x)dx = \int_{x=x_0}^{x} f(x)dx$ because $f(x)$ is odd – i.e. $f(-x) = -f(x)$. If you do the two separate integrals correctly, you will find that they are indeed equal.

$$\vec{r}_{cm} = \frac{3}{2\pi R_0^3} \int_{r=0}^{R_0} \int_{\varphi=0}^{2\pi} \int_{\theta=0}^{\pi/2} (r\cos\varphi \sin\theta\, \hat{\mathbf{i}} + r\sin\varphi \sin\theta\, \hat{\mathbf{j}} + r\cos\theta\, \hat{\mathbf{k}}) r^2 \sin\theta\, dr d\varphi d\theta$$

Since the hemisphere is symmetric about the $z$-axis, the $x$- and $y$-components of the center of mass must be zero:

$$\vec{r}_{cm} = \frac{3}{2\pi R_0^3} \int_{r=0}^{R_0} \int_{\varphi=0}^{2\pi} \int_{\theta=0}^{\pi/2} r\cos\theta\, \hat{\mathbf{k}}\, r^2 \sin\theta\, dr d\varphi d\theta$$

$$\vec{r}_{cm} = \frac{3\hat{\mathbf{k}}}{2\pi R_0^3} \int_{r=0}^{R_0} r^3 dr \int_{\varphi=0}^{2\pi} d\varphi \int_{\theta=0}^{\pi/2} \cos\theta \sin\theta\, d\theta$$

$$\vec{r}_{cm} = \frac{3\hat{\mathbf{k}}}{2\pi R_0^3} \frac{R_0^4}{4} 2\pi \int_{\theta=0}^{\pi/2} \frac{\sin 2\theta}{2} d\theta$$

$$\vec{r}_{cm} = \frac{3 R_0 \hat{\mathbf{k}}}{4} \left[ -\frac{\cos 2\theta}{4} \right]_{\theta=0}^{\pi/2} = \frac{3 R_0 \hat{\mathbf{k}}}{8}$$

where we used the trigonometric identity $\sin 2\theta = 2 \sin\theta \cos\theta$. Conceptually, it should make sense that the center of mass of the hemisphere lies a little lower than one-half its height, since there is clearly more mass below the half-height mark than above it.

**Locating the center of mass of a continuous linear object**: The center of mass is the object's balancing point: Place the object on a fulcrum (a point of support about which an object can rotate – e.g. you can balance a meterstick on your finger, where your finger is the fulcrum, and the point of support of a teeter totter, or seesaw, serves as a fulcrum) and determine experimentally where you need to place the fulcrum in order to balance the object. The center of mass is the point where the net external torque is zero (as described in Chapter 7).

**Locating the center of mass of a continuous planar object**: The center of mass of a continuous planar object can similarly be found by locating the balancing point: Hold the object so that the plane of the object is horizontal, and determine where a fulcrum can be placed underneath in order to balance the object.

Alternatively, the center of mass of a continuous planar object in a uniform gravitational field can be found experimentally as follows:
- First, suspend the object from a single point. Draw a vertical line passing through this point of suspension.
- Second, suspend the object from another point that does not lie on the first vertical line. Draw a new vertical line passing through this second point of suspension.

• The two lines intersect at the center of gravity of the object, which is the same as the center of mass provided that the object is in a uniform gravitational field.

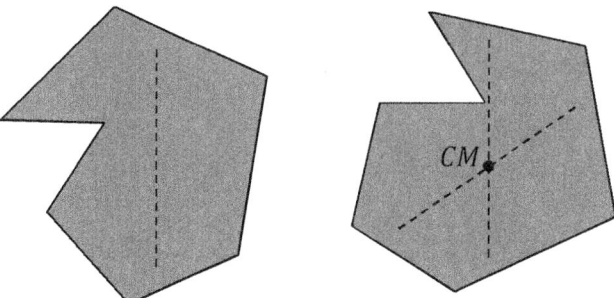

**Locating the center of mass of a continuous 3D object**: For a continuous 3D solid – that is neither approximately linear nor planar – its center of mass can also be determined using the two previous experimental methods – i.e. using a fulcrum to locate the balancing point or suspending the solid in a uniform gravitational field. If balancing the object on a fulcrum, the center of mass lies vertically above the balancing point, and if suspending the object from a single fixed point of support, in equilibrium the center of mass lies vertically below the point of support. Since the object is not planar, you can't draw two lines through the object, but you can visualize two different lines made by balancing or suspending the object from two different points – or you can setup a coordinate system and mathematically solve for the location of the center of mass.

## 6.2 Momentum and Impulse

**Momentum**: Recall the definition of momentum from Sec. 3.1: The momentum, $\vec{p}$, of an object is a vector quantity that equals the product of the object's mass, $m$, and its velocity, $\vec{v}$:

$$\vec{p} = m\vec{v}$$

The SI units for momentum may be expressed as kg·m/s or as N·s.

According to the most general form of Newton's second law – which is $\sum \vec{F}_{ext} = \frac{d\vec{p}}{dt}$, and not $\sum \vec{F}_{ext} = m\vec{a}$ (as $\sum \vec{F}_{ext} = m\vec{a}$ only applies when mass is constant, and must be interpreted carefully for a system of objects with constant mass) – the net external force acting on an object causes its momentum to change. Therefore, Newton's first law most generally states that an object has a natural tendency to maintain constant momentum. In this chapter, we will explore the more general form of Newton's second law. In the context of collisions, we will examine how to interpret the acceleration when $\sum \vec{F}_{ext} = m\vec{a}$ is applied to a system of objects where the individual objects do not all have the same acceleration, and we will see the convenience of working with momentum. In Sec. 6.9, we will treat the problem of varying mass (applied to rocket propulsion), where it is necessary to work with momentum rather than acceleration.

**Momentum for a system of objects**: The total momentum for a system of $N$ objects can be found from vector addition:

$$\vec{\mathbf{p}}_{tot} = \sum_{i=1}^{N} \vec{\mathbf{p}}_i = \sum_{i=1}^{N} m_i \vec{\mathbf{v}}_i$$

When Newton's second law is applied to a system of objects, the net external force acting on the system describes how the total momentum of the system changes:

$$\sum_{system} \vec{\mathbf{F}}_{ext} = \frac{d\vec{\mathbf{p}}_{tot}}{dt}$$

This follows by writing Newton's second law for each object individually, and then adding the equations together. Note that any objects internal to the system cancel out in the overall sum according to Newton's third law. The internal forces do affect individual objects; internal forces affect individual momenta, but internal forces do not affect the total momentum of the system. It's important to be careful about defining which objects are included in your system so that you properly account for which forces are internal and which are external to the system.

**Impulse**: An object receives an impulse, $\vec{\mathbf{J}}$, when its momentum changes. The impulse is simply defined as the change in an object's momentum:

$$\vec{\mathbf{J}} = \Delta\vec{\mathbf{p}} = m\Delta\vec{\mathbf{v}} = m(\vec{\mathbf{v}} - \vec{\mathbf{v}}_0)$$

Impulse has the same SI units as momentum – i.e. kg·m/s or N·s.[254]

**Momentum transfer**: When two objects collide, each object receives an impulse – i.e. each object changes momentum as a result of the collision force (along with any external forces that may be acting on the system). Momentum is transferred from one object to the other during the collision. If momentum is conserved for the system, the amount of momentum that one object gains compensates for the amount of momentum that the other loses. We will explore the conditions under which momentum is or is not conserved in the following section.

**Collision forces**: When two objects collide, they exert equal and opposite forces on one another according to Newton's third law:

---

[254] Impulse is a useful concept conceptually for analyzing collisions. It's not a numerical quantity that's given or asked for directly in many textbook problems, but standardized exams like to give impulse as a numerical value in occasional problems. This happens once in a while on the problem-solving component (i.e. compared to the multiple choice component) of standardized exams, in which case students who have forgotten what impulse is – and can't deduce its meaning from the units given – will be completely stuck on the problem. You should take note of this if you may be signing up to take a standardized physics exam.

$$\vec{F}_{12} = -\vec{F}_{21}$$

where $\vec{F}_{ij}$ represents the force that object $j$ exerts on object $i$. If you define the system to include both objects, these forces are internal and cancel out; but if you define each object to be its own separate system, then each of these forces becomes an external force that affects the momentum of the other object (objects don't exert forces on themselves[255]).

> **Important Distinction.** When we apply the principle of conservation of momentum in subsequent sections, we will be thinking of the system as containing all of the objects involved in collisions. Thus, the collision forces will be internal and will cancel out as far as the total momentum of the system is concerned (but these collision forces will obviously be responsible for individual objects changing momentum).
>
> When we work with impulse, we usually apply the concept to individual objects (i.e. thinking of the system as consisting of a single object). In this case, the collision force does not cancel out, and directly affects the change in momentum of the object. This distinction between working with a system of objects for conservation of momentum and working with an isolated object for impulse is important, and is the source of many lost points in students' solutions to questions and problems concerning impulse or average collision forces.

**Average collision force**: Suppose that two objects collide and that the collision force is the net external force acting on either individual object. There may be other forces acting on either object, but the collision force can still be the net force acting on either object if the other forces cancel out (e.g. if normal force balances the weight, then these two forces cancel out). In this case, according to Newton's second law,

$$\sum_{object\ 1} \vec{F}_{ext} = \vec{F}_{12} = \frac{d\vec{p}_1}{dt} \quad , \quad \sum_{object\ 2} \vec{F}_{ext} = \vec{F}_{21} = \frac{d\vec{p}_2}{dt} \quad \text{(2-object collision)}$$

and from Newton's third law,

$$\vec{F}_{21} = -\vec{F}_{12} \quad , \quad \frac{d\vec{p}_2}{dt} = -\frac{d\vec{p}_1}{dt} \quad \text{(2-object collision)}$$

Using the mean-value theorem from calculus, the average collision force can be expressed as

---

[255] Well, you might want to disagree with this statement if you punch yourself in the face – i.e. both your hand and face will feel that you can exert a force on yourself. However, if you define yourself to be the system, both your hand and face are part of the same system, and the net force that you exert on yourself when you punch yourself in the face is zero. The forces do indeed cancel out, and therefore do not contribute toward changing your overall momentum. They do, nonetheless, change the momentum of individual parts of your system – i.e. your hand and your face. Perhaps it would be better (it's more precise, but also more tedious) to state that an object can't exert a net force on itself, but it can exert pairs of internal forces on itself that do affect the momentum of individual parts, but do not affect the momentum of the whole.

$$\vec{\bar{F}}_{12} = \frac{\Delta \vec{p}_1}{\Delta t} = \frac{m_1 \Delta \vec{v}_1}{\Delta t} = \frac{m_1(\vec{v}_1 - \vec{v}_{10})}{\Delta t} = \frac{\vec{J}_1}{\Delta t} \quad \text{(2-object collision, } \vec{\bar{F}}_{12} = \vec{\bar{F}}_{ext,1})$$

$$\vec{\bar{F}}_{21} = \frac{\Delta \vec{p}_2}{\Delta t} = \frac{m_2 \Delta \vec{v}_2}{\Delta t} = \frac{m_2(\vec{v}_2 - \vec{v}_{20})}{\Delta t} = \frac{\vec{J}_2}{\Delta t} = -\vec{\bar{F}}_{12} \quad \text{(2-object collision, } \vec{\bar{F}}_{21} = \vec{\bar{F}}_{ext,2})$$

To see how this relates to the mean-value theorem, compare with the definition of average velocity (Sec. 2.3).

If the net external force acting on each object is not equal to the collision force, then you must consider all of the forces acting on the object – not just the collision force – when applying Newton's second law. Nonetheless, you can still use the change in momentum – as in the formulas above – to solve for the average net external force, $\overline{\sum \vec{F}_{ext}}$, acting on either object, as long as you realize that this does not also equal the average collision force, $\vec{\bar{F}}_{ij}$, if the average net external force is not equal to the average collision force.

---

**Important Distinction.** The equations for the average collision force involve the impulse (i.e. the change in momentum) of a single object, and not the total change in momentum of the system. When students get in the habit of conserving momentum for collisions, they also develop the habit of working with the total momentum of the system; and then when they suddenly see a question about the average collision force acting on an object, it's a common mistake to work with the change in the total momentum of the system instead of the change in momentum of a single object.

---

**Note.** Observe the importance of the overbar: The quantity $\vec{\bar{F}}_{21}$ is the average collision force exerted on object 2 by object 1, whereas $\vec{F}_{21}$ is the instantaneous collision force exerted on object 2 by object 1. The average collision force involves the impulse (change in momentum), whereas the instantaneous collision force involves a derivative of momentum with respect to time. The collision force is generally non-uniform, so the two terms (i.e. average and instantaneous) are not interchangeable.

---

**Conceptual Example.** A monkey jumps off of the roof of a single-story building. Which is better for the monkey when he lands on the ground – keeping his legs stretched stiff or allowing his knees to flex?

You can reason the answer to this question using the concept of impulse. The monkey's impulse during the collision with the ground equals the monkey's change in momentum. The monkey's initial momentum, $\vec{p}_0$, for his collision with the ground equals his momentum just before impact – this will be the same whether or not he flexes his knees during the collision. The monkey's final momentum for his collision with the ground equals zero: $\vec{p} = 0$. The monkey's impulse is therefore $\vec{J} = \Delta \vec{p} = \vec{p} - \vec{p}_0 = -\vec{p}_0$, regardless of whether or not the monkey flexes his knees or keeps them stiff.

The only force acting on the monkey during the collision (recalling that we neglect air resistance unless the problem mentions it) is the force that the earth exerts on the monkey (note that this force is not simply equal to his weight during impact). Therefore, the monkey's impulse equals the average collision force times the duration of the collision: $\vec{J} = \vec{\bar{F}} \Delta t$. By flexing his knees – rather than keeping them stiff – the monkey increases the time interval of the collision, resulting in less average force exerted on the monkey during the collision – which is much safer for the monkey.

> **Example.** A monkey drops a 200-g banana from rest from a height of 5.0 m above the floor. The banana is in contact with the floor for 250 ms before coming to rest. What is the magnitude of the average force exerted on the banana during impact?
>
> This is just like the previous conceptual example, but numerical. So let us apply the equations that we reasoned out for the previous example: $\vec{F} = \vec{J}/\Delta t = \Delta \vec{p}/\Delta t = (\vec{p} - \vec{p}_0)/\Delta t = -\vec{p}_0/\Delta t$, such that $\|\vec{F}\| = mv_0/\Delta t$. It's important to realize that $v_0$ refers here to the initial speed just before impact (and not the speed of the banana when it was released).
>
> We can apply the principle of conservation of energy to find the speed, $v_0$, of the banana just before impact. For this, we choose initial to be just after release and final to be just before impact (i.e. initial for the collision is final for conservation of energy on the way down),[256] and the reference point to be the position of the banana just before impact (we assume that the given 5.0 m corresponds to the banana's descent to this position). When we conserve energy for the banana, we will use subscripts $i$ and $f$, so as not to confuse the initial and final speeds – $v_i$ and $v_f$ – of conservation of energy with the initial and final speeds – $v_0$ and $v$ – of the collision. The banana has just gravitational potential energy at the initial position, and only kinetic energy at the final position:
>
> $$U_i + K_i + W_{nc} = U_f + K_f$$
> $$mgh_i + 0 + 0 = 0 + \frac{mv_f^2}{2}$$
> $$v_f = \sqrt{2gh_i}$$
>
> Since the final speed for conservation of energy is the initial speed for the collision (i.e. $v_f = v_0$),
>
> $$\|\vec{F}\| = \frac{mv_0}{\Delta t} = \frac{m\sqrt{2gh_i}}{\Delta t} = \frac{(0.20)\sqrt{(2)(9.81)(5.0)}}{0.25} = 7.9 \text{ N}$$
>
> remembering that mass needs to be in kilograms and time needs to be in seconds in order for force to be in Newtons. (Observe that the average collision force exceeds the banana's weight.)

## 6.3 Conservation of Momentum

**Newton's second law for a system of objects**: Consider a system of $N$ objects, which may be interacting (i.e. colliding) with one another as well as some external agent (such as gravity). Each object in the system may be changing momentum individually according to Newton's second law, which is expressed most generally in terms of momentum as

---

[256] As we will learn in the following sections, mechanical energy is not conserved for this completely inelastic collision, but is only conserved for the banana's downward descent up to the point of contact.

$$\sum_{\substack{j=1 \\ j \neq i}}^{N} \vec{\mathbf{F}}_{ij} + \sum_{e} \vec{\mathbf{F}}_{ie} = \frac{d\vec{\mathbf{p}}_i}{dt}$$

where $\vec{\mathbf{p}}_i$ is the momentum of object $i$ (where $1 \leq i \leq N$), $\vec{\mathbf{F}}_{ij}$ is the internal force that object $j$ (from within the system) exerts on object $i$, and $\vec{\mathbf{F}}_{ie}$ is one of the external forces exerted on object $i$ (from outside the system). The first sum is over the index $j$, which ranges from 1 to $N$, but skips $j = i$ (because an object does not exert a force on itself); the first sum is over the internal forces acting on object $i$. The second sum is over the index $e$, which represents forces exerted on object $i$ by external agents (such as the earth's gravitational field[257]). In the above equation, we think of object $i$ as being the only object in the system when we write down Newton's second law for a single object: Although each force $\vec{\mathbf{F}}_{ij}$ is internal to the system of $N$ objects, it is external to object $i$. Each force $\vec{\mathbf{F}}_{ij}$ influences the momentum of object $i$, but – as we shall see shortly – does not influence the total momentum of the system.

An equation of the form above applies to each individual object in the system. Let us add these equations together to obtain Newton's second law for the system of $N$ objects as a whole:

$$\sum_{i=1}^{N} \sum_{\substack{j=1 \\ j \neq i}}^{N} \vec{\mathbf{F}}_{ij} + \sum_{i=1}^{N} \sum_{e} \vec{\mathbf{F}}_{ie} = \sum_{i=1}^{N} \frac{d\vec{\mathbf{p}}_i}{dt}$$

The rightmost summation equals a derivative of the total momentum of the system with respect to time since $\vec{\mathbf{p}}_{tot} = \vec{\mathbf{p}}_1 + \vec{\mathbf{p}}_2 + \cdots + \vec{\mathbf{p}}_N$:

$$\sum_{i=1}^{N} \frac{d\vec{\mathbf{p}}_i}{dt} = \frac{d}{dt} \sum_{i=1}^{N} \vec{\mathbf{p}}_i = \frac{d\vec{\mathbf{p}}_{tot}}{dt}$$

The second sum on the left side of Newton's second law for the system as a whole is just the net external force acting on the whole system:

---

[257] Unless you want the earth to be part of your system. For example, if you drop a basketball, it has a collision with the earth – you may treat the earth as external and ask what happens to the basketball before, during, and after the collision, in which case the basketball's weight and also the force of the collision are external forces; or you may treat the system of the basketball plus the earth, in which case these forces are internal, in which case you will learn how the basketball and earth both behave relative to the center of mass of the system (see the next section). In this case, the basketball's mass is insignificant compared to the earth, and so the earth's motion is not significantly influenced by the basketball – in which case you could treat the earth as an immovable object to excellent approximation. However, if you work with an astronomical object (like the moon or the sun, or like a large asteroid headed toward earth), the distinction between whether the earth is within the system or without becomes much more important.

# 6 Systems of Objects

$$\sum_{i=1}^{N}\sum_{e} \vec{F}_{ie} = \sum_{system} \vec{F}_{ext}$$

The leftmost sum in the equation for Newton's second law acting on the system as a whole vanishes by Newton's third law (this sum represents the sum of the internal forces acting on the system):

$$\sum_{i=1}^{N}\sum_{\substack{j=1 \\ j \neq i}}^{N} \vec{F}_{ij} = \sum_{i=1}^{N}\sum_{\substack{j=1 \\ i<j}}^{N} (\vec{F}_{ij} + \vec{F}_{ji}) = \sum_{i=1}^{N}\sum_{\substack{j=1 \\ i<j}}^{N} (\vec{F}_{ij} - \vec{F}_{ij}) = 0$$

The first double sum includes all values of both $i$ and $j$ from 1 to $N$, excluding $i = j$; the second and third double sums include all values of $i$ from 1 to $N$ and all values of $j$ from $i + 1$ to $N$ (i.e. $j$ is not less than $i$). The distinction between these two cases is that the first sum includes both $i = 2, j = 3$ and $i = 3, j = 2$, for example, whereas the second case does not include $i = 3, j = 2$. Yet the two double sums are the same because the case $i = 3, j = 2$ is covered by including both sets of forces, $\vec{F}_{ij} + \vec{F}_{ji}$. According to Newton's third law, $\vec{F}_{ji} = -\vec{F}_{ij}$.

Putting all of this together, the net external force acting on the system of $N$ objects becomes

$$\sum_{system} \vec{F}_{ext} = \frac{d\vec{p}_{tot}}{dt}$$

**Conditions for conservation of momentum**: According to Newton's second law, the total momentum of a system of $N$ objects is conserved – i.e. constant – if the net external force acting on the system equals zero:[258]

$$\sum_{system} \vec{F}_{ext} = 0 \quad \Rightarrow \quad \vec{p}_{tot} = const.$$

Therefore, if you want to know whether the total momentum of a system is conserved, you should draw a FBD and apply Newton's second law for the system as a whole. If the net external force acting on the system is zero, then the total momentum of the system is conserved; conversely, if the total momentum of the system is conserved, this implies that the net external force acting on the system is zero.

---

[258] In the classroom, I derive Newton's second law for the system as a prelude to discussing the principle of conservation of momentum, as I have done here. I then ask students when the total momentum of the system would be conserved. There is always a student who bases his/her answer on the equation $\vec{p} = m\vec{v}$, answering that the total momentum of the system would be constant if all of the objects had constant mass and constant velocity. Well, the total momentum would definitely be constant in that case, but this case is too trivial to be practical. It's important to realize that every object in the system can change both speed and direction, and yet the total momentum of the system can still be constant – this is the case provided that the net external force acting on the system is zero. Indeed, when the net external force acting on the system is zero, the principle of conservation of momentum will allow us to *predict* the final velocities of the particles after interacting with one another.

**Conservation of momentum**: When the total momentum of a system of $N$ objects is conserved, this means that the total initial momentum of the system equals the total final momentum of the system:

$$\vec{\mathbf{p}}_{tot,0} = \vec{\mathbf{p}}_{tot} \quad \text{(if momentum is conserved)}$$

$$\sum_{i=1}^{N} \vec{\mathbf{p}}_{i0} = \sum_{i=1}^{N} \vec{\mathbf{p}}_i \quad \text{(if momentum is conserved)}$$

$$\vec{\mathbf{p}}_{10} + \vec{\mathbf{p}}_{20} + \cdots + \vec{\mathbf{p}}_{N0} = \vec{\mathbf{p}}_1 + \vec{\mathbf{p}}_2 + \cdots + \vec{\mathbf{p}}_N \quad \text{(if momentum is conserved)}$$

where $\vec{\mathbf{p}}_{i0}$ represents the initial momentum (both magnitude and direction) of object $i$, and $\vec{\mathbf{p}}_i$ is the final momentum (both magnitude and direction) of object $i$. In Sec.'s 6.5-6.6, we will see that the principle of conservation of momentum is particularly useful for solving problems pertaining to collisions; we will discuss the strategy for applying the principle of conservation of momentum in those sections. In Sec. 6.4, we will also see that the principle of conservation of momentum has special significance for the center of mass of the system.

> **Note**. Conservation of momentum is mathematically expressed as a vector addition problem. This means that you have to resolve each momentum vector into components, and separately conserve momentum for each component (as described in Sec. 6.6). It would be a major conceptual/strategic mistake to naïvely plug the magnitudes of the momenta in for the vectors in the equation for conservation of momentum.

> **Conceptual Example.** On horizontal frictionless ice, a box of bananas is sliding east toward a box of coconuts that is sliding west. The two boxes collide. Is momentum conserved for this system of two boxes?
> 
> The question equates to determining whether or not the net external force acting on the system (defined in the problem to be both boxes) is zero. We first draw a FBD for the system. Each box has weight acting downward and normal force acting upward. The boxes also exert horizontal forces on one another during the collision, but these are internal to the system. The sum of the vertical components of the forces must be zero because the boxes can't accelerate vertically in this situation. The only horizontal forces – the collision forces – are internal forces which cancel out according to Newton's third law, so the sum of the horizontal components of the forces is also zero. Therefore, the net external force acting on the system is zero, so the total momentum of this system is conserved.
>
>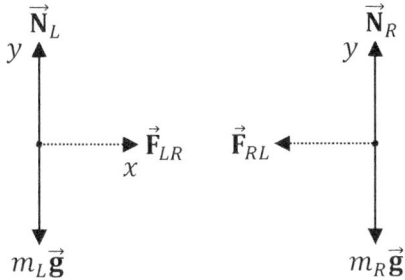

**Conceptual Example**. A monkey jumps off the edge of the roof of a building, falling straight downward, vertically. On the ground directly beneath the falling monkey, a gorilla leaps straight up into the air. The gorilla and monkey collide in midair. Is momentum conserved for this system of gorilla plus monkey?

We again answer this question by determining, conceptually, whether or not the net external force acting on the system is zero. The only external forces acting on the system – defined in the problem as the gorilla plus the monkey – are the weights of the gorilla and monkey, which both act downward; the collision forces are internal to the system and cancel out. Since all of the external forces act downward, the net external force is definitely nonzero. Therefore, the total momentum of this system is nonzero. Both objects in the system are being accelerated downward due to the gravitational pull of the earth, so the system is gaining momentum vertically downward.

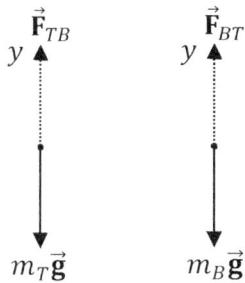

**Collisions**: The principle of conservation of momentum is particularly useful for analyzing collisions. For example, to the extent that resistive forces are negligible, the total momentum is conserved for collisions between billiard balls[259] on a level pool table. Even if the net external force acting on the system is nonzero – as is the case for two cars colliding on an inclined road – momentum is still often approximately conserved for collisions since the duration of most collisions is relatively short (so there is not much time for the total momentum of the system to change during the collision; but in this case, the total momentum of the system may be changing significantly before and after the collision).

**Net impulse**: If the total momentum of a system is conserved, the net impulse – i.e. the sum of the individual impulses – equals zero:

$$\vec{p}_{10} + \vec{p}_{20} + \cdots + \vec{p}_{N0} = \vec{p}_1 + \vec{p}_2 + \cdots + \vec{p}_N \quad \text{(if momentum is conserved)}$$
$$\vec{p}_1 - \vec{p}_{10} + \vec{p}_2 - \vec{p}_{20} + \cdots + \vec{p}_N - \vec{p}_{N0} = 0 \quad \text{(if momentum is conserved)}$$
$$\vec{J}_1 + \vec{J}_2 + \cdots + \vec{J}_N = 0 \quad \text{(if momentum is conserved)}$$
$$\sum_{i=1}^{N} \vec{J}_i = \vec{J}_{tot} = 0 \quad \text{(if momentum is conserved)}$$

---

[259] However, collisions between the billiard balls and the edges of the pool table are a separate matter. The resistive forces between the feet of the pool table are significant, as they prevent the pool table from moving during the game, and the (very massive, compared to the billiard balls) pool table provides a nearly perfect reflection of the billiard balls' momenta during collisions with its edges. Resistive forces between the surface of the pool table may also be significant external forces, affecting the spin and English of the billiard balls.

**On the conservation and non-conservation of momentum and mechanical energy**: It's important to realize that momentum and mechanical energy are not always conserved. The total energy of the universe (or any completely isolated system) is always conserved, but the total energy of a particular system may change by exchanging energy with its surroundings through $W_{nc}$. The mechanical energy of any system (isolated or not) may change, as it may be transformed into other forms of energy, such as heat or electrical energy. The total momentum of a system is not always conserved; that's the case when the net external force acting on a system is nonzero.

## 6.4 Motion of the Center of Mass

**Position of the center of mass**: Recall the formula for the center of mass vector for a discrete system of $N$ objects:

$$\vec{r}_{cm} = \frac{1}{M} \sum_{i=1}^{N} m_i \vec{r}_i$$

where $M = \sum_{i=1}^{N} m_i$ is the total mass of the system. The center of mass position vector, $\vec{r}_{cm}$, specifies the instantaneous location of the center of mass of the system relative to the origin of some specified coordinate system.

**Velocity of the center of mass**: The velocity of the center of mass can be found by differentiating the center of mass position vector with respect to time:

$$\vec{v}_{cm} = \frac{d\vec{r}_{cm}}{dt} = \frac{d}{dt}\left(\frac{1}{M}\sum_{i=1}^{N} m_i \vec{r}_i\right)$$

$$\vec{v}_{cm} = \frac{1}{M}\sum_{i=1}^{N} m_i \frac{d\vec{r}_i}{dt} = \frac{1}{M}\sum_{i=1}^{N} m_i \vec{v}_i \quad (\text{if } M = \text{const.})$$

assuming that the total mass of the system, $M$, is constant. The velocity of the center of mass, $\vec{v}_{cm}$, specifies both how fast the center of mass is instantaneously moving – namely, $v_{cm} = \|\vec{v}_{cm}\|$ – in addition to which way the center of mass is instantaneously heading.

**Total momentum and the center of mass**: Since the total momentum of the system equals $\vec{p}_{tot} = \sum_{i=1}^{N} m_i \vec{v}_i$, the velocity of the center of mass may be expressed as

$$\vec{v}_{cm} = \frac{1}{M}\sum_{i=1}^{N} m_i \vec{v}_i = \frac{\vec{p}_{tot}}{M} \quad (\text{if } M = \text{const.})$$

6 Systems of Objects

Therefore, the total momentum of the system equals the product of the total mass of the system and the velocity of the center of mass:

$$\vec{\mathbf{p}}_{tot} = M\vec{\mathbf{v}}_{cm} \quad (\text{if } M = \text{const.})$$

**Newton's second law and the center of mass**: Recall Newton's second law for a system of objects:

$$\sum_{system} \vec{\mathbf{F}}_{ext} = \frac{d\vec{\mathbf{p}}_{tot}}{dt}$$

This can be expressed in terms of the velocity of the center of mass as

$$\sum_{system} \vec{\mathbf{F}}_{ext} = M\frac{d\vec{\mathbf{v}}_{cm}}{dt} \quad (\text{if } M = \text{const.})$$

**Acceleration of the center of mass**: Differentiating the velocity of the center of mass with respect to time yields the acceleration of the center of mass:

$$\vec{\mathbf{a}}_{cm} = \frac{d\vec{\mathbf{v}}_{cm}}{dt}$$

Combining the two equations above,

$$\sum_{system} \vec{\mathbf{F}}_{ext} = M\vec{\mathbf{a}}_{cm} \quad (\text{if } M = \text{const.})$$

We interpret this result conceptually as follows: The system as a whole behaves as a single object with mass $M$, located at the center of mass by $\vec{\mathbf{r}}_{cm}$, which has instantaneous velocity $\vec{\mathbf{v}}_{cm}$ and instantaneous acceleration $\vec{\mathbf{a}}_{cm}$ (provided that $M = \text{const.}$).

**Conservation of momentum and the center of mass**: We have already observed that the total momentum of the system is conserved if the net external force is zero. The above equation shows that if the total momentum of the system is conserved, the center of mass of the system does not accelerate – i.e. the velocity of the center of mass is constant. It is sometimes useful, both mathematically and conceptually, to know that the center of mass travels with constant speed in a straight line when the total momentum of the system is conserved.

**Example**. A 20-kg monkey stands on one end of a 12-m long, 60-kg plank. The plank lies on top of horizontal frictionless ice. There is friction, however, between the plank and the monkey. The monkey is initially at rest relative to the plank, which is initially moving 3.0 m/s to the east relative to the ice. Then the monkey begins walking across the plank at a speed of 5.0 m/s to the east relative to the ice. What is the velocity of the plank relative to the ice as the monkey walks across it?

First, defining the system as the monkey plus the plank, note that the net external force acting on the system is zero. There is weight pulling downward on both the monkey and plank, and the ice exerts an upward normal force on the plank equal to the weight of the plank plus the weight of the monkey in order to prevent the system from accelerating vertically. There are also normal and friction force pairs between the monkey and the plank, which are internal to the system and cancel out according to Newton's third law.

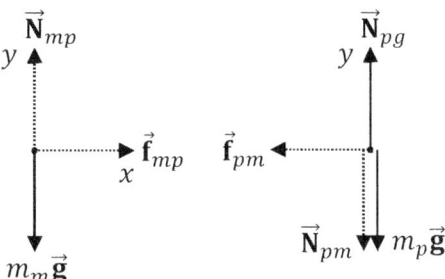

Since the net external force is zero, we could apply the principal of conservation of momentum — a strategy that we will explore in the following two sections. We can also solve this problem by working with the center of mass, which we shall do presently. Recall that if the net external force is zero, the center of mass of the system travels with constant velocity. In this problem, both the monkey and plank had the same initial velocity — 3.0 m/s to the east relative to the ice. Therefore, the center of mass of the system was initially moving 3.0 m/s to the east. According to Newton's second law for this system of objects, the center of mass must continue moving 3.0 m/s to the east even as the monkey walks across the plank (since the center of mass must maintain constant velocity when the net external force acting on the system is zero).

Recalling that $\vec{v}_{cm} = \frac{1}{M}\sum_{i=1}^{N} m_i \vec{v}_i$ if $M$ is constant, we find that:

$$\vec{v}_{cm} = \frac{1}{M}\sum_{i=1}^{N} m_i \vec{v}_i = \frac{m_m \vec{v}_m + m_p \vec{v}_p}{M}$$
$$m_m \vec{v}_m + m_p \vec{v}_p = M\vec{v}_{cm}$$
$$\vec{v}_p = \frac{M\vec{v}_{cm} - m_m \vec{v}_m}{m_p}$$

Setting up a coordinate system such that the $+x$-axis heads east,

$$\vec{v}_p = \frac{(80)(3.0\hat{\imath}) - (20)(5.0\hat{\imath})}{(60)} = 2.3 \text{ m/s } \hat{\imath}$$

relative to the ice. Note that we must be consistent and measure all of the velocities relative to the same reference frame in this calculation. We have chosen the ice as the reference frame from which to measure the velocities.

> Observe that we didn't need to find the center of mass of the system to solve this problem; we just needed to know the velocity of the center of mass. We also didn't need to use the given length of the plank. (In lab, you can measure many more quantities than you actually need to know, and must learn to identify those quantities that are relevant for your calculations.)

**Reference frames**: It is important to bear in mind that momentum, like velocity, is measured relative to a particular reference frame (i.e. a coordinate system). Different observers may choose to measure velocity relative to different reference frames, which may be in relative motion. The laws of physics do not depend upon your choice of reference frame, and the physical laws will have the same mathematical form in any inertial reference frame – i.e. a reference frame that is moving with constant velocity. However, the values of the velocities are relative, and depend on the reference frame. All of the velocities in the problem are shifted in one reference frame compared to another reference frame.

If one observer measures the velocity of an object with respect to one reference frame, and another observer measures the velocity of the same object with respect to a different reference frame, the velocities are related through the concept of relative velocity (first discussed in Sec. 2.6). Specifically, suppose that observer $A$ is moving with instantaneous velocity $\vec{v}_{AB}$ relative to observer $B$. If $\vec{v}_{oA}$ is the velocity of an object relative to observer $A$ and $\vec{v}_{oB}$ is the velocity of the same object relative to observer $B$, then

$$\vec{v}_{oB} = \vec{v}_{oA} + \vec{v}_{AB}$$

If $A$ and $B$ are both inertial observers – i.e. they both travel with constant velocity – then $\vec{v}_{AB}$ will be a constant velocity vector.

When applying the principle of conservation of momentum – or when relating the center of mass momentum to the individual momenta of the system – it is important to choose one reference frame and measure all of the velocities relative to that same reference frame. It's also important to choose an inertial reference frame.

> **Notation.** In the relative velocity equation, the notation $\vec{v}_{ij}$ represents the velocity of $i$ relative to $j$.

> **Note.** To be sure that you apply the relative velocity equation correctly, first examine the velocity of one observer with respect to the other observer, and note with respect to whom the relative velocity, $\vec{v}_{AB}$, is measured; in the equation above, $\vec{v}_{AB}$ is measured relative to observer $B$. Note that $\vec{v}_{AB}$ is added to the velocity measured by the *other* observer; in the above equation, $\vec{v}_{AB}$ is added to the velocity, $\vec{v}_{oA}$, measured by observer $A$. Also, observe that $\vec{v}_{oA} = \vec{v}_{oB} + \vec{v}_{BA}$.

> **Note.** When applying the principle of conservation of momentum or relating the center of mass momentum to the individual momenta of the system, be sure to measure all of the velocities relative to the same inertial reference frame.

**The order of indices in the relative velocity**: Swapping the indices of the relative velocity comes at the cost of a minus sign. For example, if $\vec{v}_{ij}$ is the velocity of $i$ relative to $j$, then $\vec{v}_{ji}$ is the velocity of $j$ relative to $i$, which is the negative of the first:

$$\vec{v}_{ji} = -\vec{v}_{ij}$$

For example, suppose that a monkey standing on the ground throws a banana with a velocity of 15 m/s to the west relative to the ground. Let us choose the $+x$-axis to point east. The velocity of the banana relative to the monkey[260] is $\vec{v}_{bm} = -15$ m/s $\hat{\imath}$. The velocity of the monkey relative to the banana is $\vec{v}_{bm} = +15$ m/s $\hat{\imath}$. The monkey sees the banana flying west with a speed of 15 m/s, while a bug on the banana would see the monkey moving east with the same speed (relative to the bug and banana).

---

**Conceptual Example.** An orangutan standing on the ground watches a train pass by with a velocity of 30 m/s to the north. A monkey is running on the roof of the train, traveling 10 m/s to the north relative to the train. What are (A) the velocity of the monkey relative to the orangutan, (B) the velocity of the orangutan relative to the train, (C) the velocity of the train relative to the monkey, and (D) the velocity of the orangutan relative to the monkey?

Apply the relative velocity equation, being careful to put the relative velocity on the correct side of the equation. We choose to setup our coordinate system with the $+y$-axis oriented to the north. (A) The monkey's velocity relative to the orangutan is $\vec{v}_{mo} = \vec{v}_{mt} + \vec{v}_{to} = 10$ m/s $\hat{\jmath} + 30$ m/s $\hat{\jmath} = 40$ m/s $\hat{\jmath}$. (B) The orangutan's velocity relative to the train is simply $\vec{v}_{ot} = -\vec{v}_{to} = -30$ m/s $\hat{\jmath}$. (C) The train's velocity relative to the monkey is simply $\vec{v}_{tm} = -\vec{v}_{mt} = -10$ m/s $\hat{\jmath}$. (D) The orangutan's velocity relative to the monkey is $\vec{v}_{om} = \vec{v}_{ot} + \vec{v}_{tm} = -30$ m/s $\hat{\jmath} - 10$ m/s $\hat{\jmath} = -40$ m/s $\hat{\jmath}$, which agrees with $\vec{v}_{om} = -\vec{v}_{mo}$.

---

**Center of mass frame**: If you setup a coordinate system such that the origin lies at the center of mass, where the coordinate system moves with the same velocity as the center of mass (so that the origin always lies at the center of mass), this is called the center of mass frame (CM frame). The CM frame is the simplest reference frame to work with mathematically and conceptually, but it is not a practical reference frame from which to measure the velocities.

If the total momentum of the system is conserved, the center of mass travels with constant velocity, in which case the CM frame is an inertial reference frame. Recall the equation for the velocity of the center of mass:

$$\vec{v}_{cm} = \frac{1}{M} \sum_{i=1}^{N} m_i \vec{v}_i \quad (\text{if } M = \text{const.})$$

Of course, the center of mass is stationary relative to the CM frame, so

---

[260] To illustrate the main idea here, let us assume that the monkey's recoil is negligible such that the monkey remains stationary relative to the ground. We will generally *not* neglect recoil, but will include the recoil velocity when applying the principle of conservation of momentum.

6 Systems of Objects

$$\sum_{i=1}^{N} m_i \vec{v}_i = m_1\vec{v}_1 + m_2\vec{v}_2 + \cdots + m_N\vec{v}_N = 0 \quad (\text{C. M. frame}, M = \text{const.}, \sum_{system} \vec{F}_{ext} = 0)$$

The above equation must hold true for both the initial and final velocities, so we can also write

$$\sum_{i=1}^{N} m_i \vec{v}_{i0} = m_1\vec{v}_{10} + m_2\vec{v}_{20} + \cdots + m_N\vec{v}_{N0} = 0 \quad (\text{C. M. frame}, M = \text{const.}, \sum_{system} \vec{F}_{ext} = 0)$$

For a two-particle collision where the net external force equals zero (so that the total momentum of the system is conserved), these equations become

$$\begin{aligned} m_1\vec{v}_1 &= -m_2\vec{v}_2 \\ m_1\vec{v}_{10} &= -m_2\vec{v}_{20} \end{aligned} \quad (\text{C. M. frame}, N = 2, M = \text{const.}, \sum_{system} \vec{F}_{ext} = 0)$$

**Lab frame**: If one of the objects in the system is at rest initially, the reference frame is called the lab frame. The lab frame is particularly useful experimentally. For example, if you place a thin gold foil at rest and bombard it with a beam of charged particles, the initial and final measurements of all of the velocities are most readily made in the lab frame (where the gold foil is at rest). The CM frame is simpler mathematically, but first you must make several measurements and perform a calculation just to find the center of mass, and then you need to use the relative velocity equation to determine all of the velocities relative to the center of mass if you wish to use the CM frame.

---

**Example**. In the previous example involving the monkey that was running along a plank, identify the initial and final velocities of both the monkey and plank in the CM frame.

For convenience, below we tabulate all of the velocities that we had found in that example.

Initial velocity of plank relative to ice: $\vec{v}_{pi0} = 3.0 \text{ m/s}\,\hat{\imath}$
Initial velocity of monkey relative to ice: $\vec{v}_{mi0} = 3.0 \text{ m/s}\,\hat{\imath}$
Velocity of center of mass relative to ice: $\vec{v}_{cm,i} = 3.0 \text{ m/s}\,\hat{\imath}$
Final velocity of monkey relative to ice: $\vec{v}_{mi} = 5.0 \text{ m/s}\,\hat{\imath}$
Final velocity of plank relative to ice: $\vec{v}_{pi} = 2.3 \text{ m/s}\,\hat{\imath}$

The initial velocities of the monkey and plank relative to the CM are both zero:

$$\vec{v}_{p0,cm} = \vec{v}_{m0,cm} = 0$$

Their final velocities relative to the CM frame are:

$$\begin{aligned} \vec{v}_{p,cm} &= \vec{v}_{pi} + \vec{v}_{i,cm} = 2.3 \text{ m/s}\,\hat{\imath} + (-3.0 \text{ m/s}\,\hat{\imath}) = -0.7 \text{ m/s}\,\hat{\imath} \\ \vec{v}_{m,cm} &= \vec{v}_{mi} + \vec{v}_{i,cm} = 5.0 \text{ m/s}\,\hat{\imath} + (-3.0 \text{ m/s}\,\hat{\imath}) = 2.0 \text{ m/s}\,\hat{\imath} \end{aligned}$$

## 6.5 Collisions in 1D

**Types of collisions**: Collisions are either elastic or inelastic. Of the inelastic variety, a special case is the completely inelastic collision. These types of collisions are described below.

**Elastic collisions**: A collision between two (or more) objects is elastic if none of the objects experiences a change in its internal energy (review Sec. 5.3). In this case, there will also be no heat exchanges, $W_{nc}$ will be zero for the complete system (i.e. all of the objects involved in the elastic collision), and the total mechanical energy of the system will be conserved.

Since $W_{nc} = 0$ for an elastic collision, only conservative forces do work, and the net external force is related to the total potential energy by $\sum \vec{F}_{ext} = -\vec{\nabla} U$ (see Sec. 5.5). If the total momentum of the system is conserved for the collision, the net external force acting on the system must be zero. In this case, $\sum \vec{F}_{ext} = -\vec{\nabla} U = 0$, which will hold in general if the total potential energy is constant. Since $W_{nc} = 0$ and $U = U_0$ for an elastic collision that conserves momentum, the total kinetic energy of the system is conserved (in addition to the total momentum):

$$\vec{p}_{10} + \vec{p}_{20} + \cdots + \vec{p}_{N0} = \vec{p}_1 + \vec{p}_2 + \cdots + \vec{p}_N \quad \text{(if } \sum_{system} \vec{F}_{ext} = 0\text{)}$$

$$K_{10} + K_{20} + \cdots + K_{N0} = K_1 + K_2 + \cdots + K_N \quad \text{(elastic collision, } \sum_{system} \vec{F}_{ext} = 0\text{)}$$

Notice that the definition of elastic only involves conservation of mechanical energy – the definition says nothing about whether or not the objects "bounce off" one another. If the objects stick together, the collision is definitely not elastic (it is instead completely inelastic); but if the objects bounce off one another, you need more information to determine whether or not the collision is elastic or inelastic (all you know is that the collision is not completely inelastic).

**Inelastic collisions**: A collision is inelastic if any of the objects experiences a change in its internal energy. This change in internal energy may or may not be used to exchange heat between objects in the system, or between the system and the surroundings. The change in internal energy will result in nonconservative work being done, so mechanical energy is not conserved for inelastic collisions. Only the total momentum of the system may be conserved for inelastic collisions.

$$\vec{p}_{10} + \vec{p}_{20} + \cdots + \vec{p}_{N0} = \vec{p}_1 + \vec{p}_2 + \cdots + \vec{p}_N \quad \text{(if } \sum_{system} \vec{F}_{ext} = 0\text{)}$$

$K$ is not conserved for inelastic collisions

Note that the objects may or may not stick together in an inelastic collision. The collision is completely inelastic if they stick together, but if they bounce off, the collision may be elastic or inelastic. The determining factor is whether or not mechanical energy is conserved. Only collisions between particles can be perfectly elastic. However, macroscopic collisions may be approximately elastic.

**Completely inelastic collisions**: A collision is completely inelastic if all of the colliding objects stick together after the collision. Since the collision is inelastic, mechanical energy is not conserved. In addition, since the objects stick together, in a completely inelastic collision all of the final velocities are equal:

$$\vec{p}_{10} + \vec{p}_{20} + \cdots + \vec{p}_{N0} = M\vec{v} \quad \text{(completely inelastic, } \sum_{system} \vec{F}_{ext} = 0\text{)}$$

$K$ is not conserved for completely inelastic collisions

where $M$ is the total mass of the system and $\vec{v}_1 = \vec{v}_2 = \cdots = \vec{v}_N \equiv \vec{v}$ is the final velocity of each object.

**Inverse completely inelastic collision**: If all of the objects in the system are stuck together initially, and then separate after the collision, such a collision is just like a completely inelastic collision run in reverse. In this case,

$$M\vec{v}_0 = \vec{p}_1 + \vec{p}_2 + \cdots + \vec{p}_N \quad \text{(completely inelastic, } \sum_{system} \vec{F}_{ext} = 0\text{)}$$

$K$ is not conserved for inverse completely inelastic collisions

where $\vec{v}_{10} = \vec{v}_{20} = \cdots = \vec{v}_{N0} \equiv \vec{v}_0$ is the initial velocity of each object.

---

**Common Mistake.** It's intuitive for most students to associate the word 'elastic' with bounciness, and so students who do not pay proper attention to the precise definitions of the terminology for collisions often mistake an elastic collision to mean that the objects bounce off, and inelastic to mean that they stick together. But you should see the problem with this intuition – what would the distinction be between inelastic and completely inelastic?

The correct distinction between elastic and inelastic is whether or not mechanical energy is conserved, not whether or not the objects stick together. Whether or not the objects stick together distinguishes an inelastic collision from a completely inelastic collision. Remember, completely inelastic means that the objects stick together, elastic means that mechanical energy is conserved and the objects don't stick together, and inelastic means that mechanical energy is not conserved and the objects may or may not stick together. So if the objects don't stick together, you need more information in order to determine whether or not the collision is elastic or inelastic.

---

**Which quantities are conserved for collisions**: The total momentum and/or the total kinetic energy of the system may be conserved during collisions. If the net external force is zero, the total momentum of the system is conserved – regardless of what kind of collision occurs. Even if the net external force is not zero, if the collision is short-lived (i.e. occurs over a short duration), the total momentum of the system may be approximately conserved from just before to just after the collision, since the net external force may not be acting over a long enough time interval to make significant changes to the overall momentum. Therefore, in practice, we generally apply the principal of conservation of momentum to collisions under most situations to good approximation. However, if the net external force is nonzero, it's important to restrict yourself just to the short time interval of the collision.

If the net external force is zero and the collision is elastic, then the total kinetic energy of the system is also conserved. Again, if the collision is short-lived, even if the net external force is nonzero, both the total momentum and the total kinetic energy of the elastic collision may be conserved to good approximation for the short time interval of the collision.

If the collision is inelastic (whether or not it is completely inelastic), then the total kinetic energy of the system is not conserved. In this case, the total momentum of the system is still conserved if the net external force is zero (and may be approximately conserved during the collision interval even if the net external force is nonzero).

| Type of collision | Is momentum conserved? | Is kinetic energy conserved? |
|---|---|---|
| elastic | yes | yes |
| inelastic | yes | no |
| completely inelastic | yes | no |

**Conservation of momentum in 1D**: A collision is 1D if all of the initial and final velocities of the objects are collinear. A head-on collision between two objects is 1D; if the collision is instead off-center, the final velocities will not be collinear with the initial velocities, and the resulting collision is 2D (as described in the next section).

We can describe the position of an object in 1D motion with a single coordinate, which may call $x$. Conservation of momentum in 1D can then be expressed as conservation of the $x$-components of momentum:

$$m_1 v_{1x0} + m_2 v_{2x0} + \cdots + m_N v_{Nx0} = m_1 v_{1x} + m_2 v_{2x} + \cdots + m_N v_{Nx} \quad (1D, \sum F_x = 0)$$

where we have assumed that the mass of each object has remained constant. It's important to realize that this equation involves $x$-components of velocity, which may be positive or negative, and not speeds (which can only be nonnegative).

---

**Common Mistake**. It's a common mistake for students to neglect minus signs in the above equation. You must setup a coordinate system – i.e. define which way is $+x$ for a 1D collision – and then write a negative $x$-component of velocity for an object that is headed in the $-x$-direction (initially or finally). For example, if you define east to be $+$, any object headed west will have a negative $x$-component of velocity.

However, only include explicit minus signs when plugging in numerical values for negative components of velocities. Don't try to anticipate the signs of the unknowns that you are solving for – let the algebra determine the signs (and hence directions) of the unknowns.

---

**Conservation of mechanical energy for a 1D elastic collision**: Recall that the equation for the kinetic energy for an object is $K = mv^2/2$. In 1D, we may write this as $K = mv_x^2/2$. As we have discussed previously, the total kinetic energy of a system of objects is conserved for an elastic collision for which the net external force is zero. For such a 1D elastic collision, we may write this as

$$\frac{m_1 v_{1x0}^2}{2} + \frac{m_2 v_{2x0}^2}{2} + \cdots + \frac{m_N v_{Nx0}^2}{2} = \frac{m_1 v_{1x}^2}{2} + \frac{m_2 v_{2x}^2}{2} + \cdots + \frac{m_N v_{Nx}^2}{2} \quad (1D, \text{elastic}, \sum F_x = 0)\,[261]$$

where we have again assumed that the mass of each object remains constant. We will continue to make this assumption until we reach Sec. 6.9.

**The velocity equation**: For a 1D elastic collision between two objects (where the net external force is zero, or where the duration of the collision is short enough that the net external force does not have a significant effect over the duration of the collision), two equations apply – one for conservation of momentum, another for conservation of kinetic energy:

$$m_1 v_{1x0} + m_2 v_{2x0} = m_1 v_{1x} + m_2 v_{2x}$$
$$\frac{m_1 v_{1x0}^2}{2} + \frac{m_2 v_{2x0}^2}{2} = \frac{m_1 v_{1x}^2}{2} + \frac{m_2 v_{2x}^2}{2} \quad (1D, \text{elastic}, \sum F_x = 0, 2 \text{ objects})$$

If both of the masses are known and if both initial velocities are known, these equations allow us to predict the two final velocities. However, the algebra can be somewhat cumbersome since one equation is quadratic in the $x$-component of velocity.[262] It turns out that we can derive a third equation – called the velocity equation – from the two equations above. The velocity equation (as we shall see in a moment) is linear in the $x$-component of velocity, unlike the kinetic energy conservation equation, which is quadratic in the $x$-component of velocity. The purpose of the velocity equation is that it may be used in lieu of the kinetic energy conservation equation (but only for a 1D elastic collision between two objects – if the collision is 2D, or if it is inelastic, or if it involves three or more objects colliding at once, you may not use the velocity equation). This simplifies the algebra for a large class of problems.

The simplest way to derive the velocity equation is to exploit the obvious symmetry in the two equations above – i.e. the equations have the same structure (the only difference is that the $x$-component of velocity is squared in the second equation). First, we will regroup terms to separate the masses on opposite sides of the equations:

$$m_1 v_{1x0} - m_1 v_{1x} = m_2 v_{2x} - m_2 v_{2x0}$$
$$m_1 v_{1x0}^2 - m_1 v_{1x}^2 = m_2 v_{2x}^2 - m_2 v_{2x0}^2 \quad (1D, \text{elastic}, \sum F_x = 0, 2 \text{ objects})$$

Next, we will factor out the masses on each side of the equation:

$$m_1(v_{1x0} - v_{1x}) = m_2(v_{2x} - v_{2x0})$$
$$m_1(v_{1x0}^2 - v_{1x}^2) = m_2(v_{2x}^2 - v_{2x0}^2) \quad (1D, \text{elastic}, \sum F_x = 0, 2 \text{ objects})$$

Now we will solve for the ratio of the masses from both equations:

---

[261] Yes, you may cancel the 2's. Including them when you first write the equation demonstrates conceptually that your starting point is to conserve kinetic energy (in addition to momentum) for an elastic collision.
[262] Of course, there is no trig involved in 1D, so in 1D 'the $x$-component of velocity' only differs from the speed by a possible minus sign.

$$\frac{m_1}{m_2} = \frac{v_{2x} - v_{2x0}}{v_{1x0} - v_{1x}} = \frac{v_{2x}^2 - v_{2x0}^2}{v_{1x0}^2 - v_{1x}^2} \quad (1\text{D, elastic}, \sum F_x = 0, 2 \text{ objects})$$

Recall from algebra that $x^2 - y^2 = (x+y)(x-y)$. Applying this algebraic identity to the above equation,[263]

$$\frac{v_{2x} - v_{2x0}}{v_{1x0} - v_{1x}} = \left(\frac{v_{2x} + v_{2x0}}{v_{1x0} + v_{1x}}\right)\left(\frac{v_{2x} - v_{2x0}}{v_{1x0} - v_{1x}}\right) \quad (1\text{D, elastic}, \sum F_x = 0, 2 \text{ objects})$$

Examine the structure of the above equation: It has the form $a = ba$, which only holds true if $b = 1$. Therefore,

$$\frac{v_{2x} + v_{2x0}}{v_{1x0} + v_{1x}} = 1 \quad (1\text{D, elastic}, \sum F_x = 0, 2 \text{ objects})$$

which we may rewrite as

$$v_{1x0} + v_{1x} = v_{2x0} + v_{2x} \quad (1\text{D, elastic}, \sum F_x = 0, 2 \text{ objects})$$

The above equation is known as the velocity equation. The convenience of the velocity equation is that you may combine conservation of momentum with the velocity equation (instead of conserving kinetic energy) to solve a problem with a 1D elastic collision between two objects.

**Note.** You may only use the velocity equation to solve problems where there is a 1D elastic collision between two objects – and then you must also combine this equation with conservation of momentum in order to solve the problem. If the collision is 2D (or 3D), or if the collision is inelastic (or completely inelastic), or if three or more objects collide together at once, you may not use the velocity equation.

**Note.** The velocity equation is not a conservation law; it does not express conservation of velocity. Observe that the structure of the velocity equation is different from conservation laws: The two sides of the velocity equation separate object 1 from object 2, whereas the two sides of an equation corresponding to a conservation law instead separate initial from final. This is actually the source of many mistakes. Once you develop the habit of writing down conservation of momentum, you will be accustomed to having initial on one side and final on the other. As a result, many students inadvertently write down the velocity equation incorrectly, putting initial on one side and final on the other – whereas the velocity equation is correctly written with different objects on the two sides of the equation.

Remember, the $x$-components of velocity may be negative, even when applying the velocity equation.

---

[263] There is a problem, algebraically, if $v_{1x0} = v_{1x}$, since in that case we would be dividing by zero on both sides of the equation. However, if $v_{1x0} = v_{1x}$, all of the velocities will be unchanged, which is not really much of a collision.

# Problem-Solving Strategy for 1D Collisions

0. Apply this strategy to solve problems involving 1D collisions (for 2D collisions, see the following section) between two or more objects (and other problems where the net external force acting on the system is zero and the objects change velocity, but which may not look like 'collisions'). In order for the total momentum of the system to be conserved, the net external force acting on the system must be zero. If the net external force is nonzero, the duration of the collision may still be short enough that the total momentum of the system is approximately conserved over the time interval of the collision. (Otherwise, instead of applying this strategy, you should begin by applying Newton's second law to the system as a whole, setting the net external force equal to a derivative of the total momentum of the system with respect to time.) **Note**: If the masses of the objects change during the problem,[264] instead of following this strategy, see Sec. 6.9.

1. Setup a coordinate system – i.e. choose which way to call $+x$. Choose an inertial reference frame and measure all of the velocities relative to this same reference frame.

2. First, express conservation of momentum for the 1D collision, regardless of what type of collision occurs: $m_1 v_{1x0} + m_2 v_{2x0} + \cdots + m_N v_{Nx0} = m_1 v_{1x} + m_2 v_{2x} + \cdots + m_N v_{Nx}$.

3. If you are told that the 1D collision is elastic (or, equivalently, if you are told that the total mechanical energy of the system is conserved for the collision), and if the 1D collision involves only two objects colliding at once, in addition to conserving momentum, also use the velocity equation: $v_{1x0} + v_{1x} = v_{2x0} + v_{2x}$. If the 1D collision is elastic, but two or more objects collide together at once, write down conservation of kinetic energy, $\frac{m_1 v_{1x0}^2}{2} + \frac{m_2 v_{2x0}^2}{2} + \cdots + \frac{m_N v_{Nx0}^2}{2} = \frac{m_1 v_{1x}^2}{2} + \frac{m_2 v_{2x}^2}{2} + \cdots + \frac{m_N v_{Nx}^2}{2}$, instead of using the velocity equation. (Either way, you still need to use conservation of momentum.)

4. If you are told that the 1D collision is completely inelastic (or if you know that all of the objects stick together), set the final $x$-components of velocity equal to one another in the conservation of momentum equation: $m_1 v_{1x0} + m_2 v_{2x0} + \cdots + m_N v_{Nx0} = (m_1 + m_2 + \cdots + m_N) v_x$.

5. If you have an inverse completely inelastic collision – i.e. the objects are stuck together initially, but not finally – then set the initial $x$-components of velocity equal to one another in the conservation of momentum equation: $(m_1 + m_2 + \cdots + m_N) v_{x0} = m_1 v_{1x} + m_2 v_{2x} + \cdots + m_N v_{Nx}$.

6. If you are not told whether or not the collision is elastic, inelastic, or completely inelastic (and are not given other information to figure this out), you may only use conservation of momentum. (That is, you can't use the velocity equation or kinetic energy conservation unless you know that the collision is elastic, and you can't set the final velocities or initial velocities equal unless you know that the collision is completely inelastic or its inverse.) Instead, you will be given additional information in the problem – either an additional known, or information that you may express in an equation. For example, if the problem tells you that a percentage of kinetic energy is lost, you may write an equation of the form $\left(\frac{m_1 v_{1x0}^2}{2} + \frac{m_2 v_{2x0}^2}{2} + \cdots + \frac{m_N v_{Nx0}^2}{2}\right) \gamma = \frac{m_1 v_{1x}^2}{2} + \frac{m_2 v_{2x}^2}{2} + \cdots + \frac{m_N v_{Nx}^2}{2}$, where $\gamma$ (the Greek letter gamma) represents the fraction of kinetic energy that remains (1 minus the fraction that is lost).

7. As usual, look for $N$ equations in $N$ unknowns ($N$ being the number of unknowns, not objects).

---

[264] This means: If the mass of any object is continually varying over time, as in the case of a rocket ejecting steam. If instead there is a simple (instantaneous) regrouping of the objects – like a banana splits into two pieces – as occurs during a completely inelastic collision (or its inverse) you may still apply this strategy.

**Note.** Be sure to use the correct equations for the type of collision that you have. Don't make assumptions about the collision – look for language that identifies the collision, such as the word 'elastic' or the phrase 'stick together.' If the problem doesn't make it clear what type of collision occurs, only use conservation of momentum (unless later in the problem you are finally able to determine whether or not mechanical energy is conserved and whether or not the initial or final velocities of the objects are equal).

Conserve momentum for all collisions. Only use the velocity equation for a 1D elastic collision involving only two objects colliding at once. Only use kinetic energy conservation for a more complicated elastic collision, where the velocity equation does not apply.

| Type of 1D collision | Equations to use |
|---|---|
| Elastic 1D collision, between 2 objects | $m_1 v_{1x0} + m_2 v_{2x0} = m_1 v_{1x} + m_2 v_{2x}$ <br> $v_{1x0} + v_{1x} = v_{2x0} + v_{2x}$ |
| Elastic 1D collision, 3 or more objects collide at once | $m_1 v_{1x0} + m_2 v_{2x0} + \cdots + m_N v_{Nx0} = m_1 v_{1x} + m_2 v_{2x} + \cdots + m_N v_{Nx}$ <br> $\frac{m_1 v_{1x0}^2}{2} + \frac{m_2 v_{2x0}^2}{2} + \cdots + \frac{m_N v_{Nx0}^2}{2} = \frac{m_1 v_{1x}^2}{2} + \frac{m_2 v_{2x}^2}{2} + \cdots + \frac{m_N v_{Nx}^2}{2}$ |
| Completely inelastic 1D collision | $m_1 v_{1x0} + m_2 v_{2x0} + \cdots + m_N v_{Nx0} = (m_1 + m_2 + \cdots + m_N) v_x$ |
| Inverse completely inelastic 1D collision | $(m_1 + m_2 + \cdots + m_N) v_{x0} = m_1 v_{1x} + m_2 v_{2x} + \cdots + m_N v_{Nx}$ |
| Inelastic 1D collision | $m_1 v_{1x0} + m_2 v_{2x0} + \cdots + m_N v_{Nx0} = m_1 v_{1x} + m_2 v_{2x} + \cdots + m_N v_{Nx}$ |

**Example.** A 20-kg monkey stands on one end of a 12-m long, 60-kg plank. The plank lies on top of horizontal frictionless ice. There is friction, however, between the plank and the monkey. The monkey is initially at rest relative to the plank, which is initially moving 3.0 m/s to the east relative to the ice. Then the monkey begins walking across the plank at a speed of 5.0 m/s to the east relative to the ice. What is the velocity of the plank relative to the ice as the monkey walks across it?

This is identical to one of the examples of the preceding section. Previously, we solved this problem by applying the principle of center of mass momentum. We will now solve the same problem by applying the principle of conservation of momentum. We have already shown in the preceding section that the net external force acting on this system is zero. Therefore, the total momentum of the system (defined as the monkey plus the plank) is conserved.

Observe that this is an inverse completely inelastic collision, since the initial velocities of the plank and monkey are the same. We choose the $+x$-direction to be east, and measure all of the velocities relative to the ice.

$$(m_m + m_p) v_{x0} = m_m v_{mx} + m_p v_{px}$$
$$v_{px} = \frac{(m_m + m_p) v_{x0} - m_m v_{mx}}{m_p}$$
$$v_{px} = \frac{(20 + 60)(3.0) - (20)(5.0)}{60} = 2.3 \text{ m/s}$$

**Example**. A 30.0-kg box of bananas traveling 12 m/s to the north on horizontal frictionless ice collides head-on with a 60.0-kg box of coconuts traveling 3.0 m/s to the south. The collision is elastic. Determine the final velocities of the two boxes after the collision. All velocities are relative to the ice.

We choose the $+x$-axis to point north. Observe that the net external force acting on the system of two boxes is zero: The weights of the two boxes cancel the normal forces, since neither box will accelerate vertically, and the collision force is internal to the system of two boxes. We use conservation of momentum in conjunction with the velocity equation for this 1D elastic collision between two boxes:

$$m_1 v_{1x0} + m_2 v_{2x0} = m_1 v_{1x} + m_2 v_{2x}$$
$$(30.0)(12.0) + (60.0)(-3.0) = 30.0 v_{1x} + 60.0 v_{2x}$$
$$360 - 180 = 180 = 30.0 v_{1x} + 60.0 v_{2x} \ (*)$$
$$v_{1x0} + v_{1x} = v_{2x0} + v_{2x}$$
$$12.0 + v_{1x} = -3.0 + v_{2x} \ (*)$$

Note that $v_{2x0} = -3.0$ m/s since south is along the $-x$-direction. There are two equations in two unknowns, which are marked with asterisks (*). This system of equations is easily solved through substitution:[265]

$$v_{2x} = 12.0 + 3.0 + v_{1x} = 15.0 + v_{1x}$$
$$180 = 30.0 v_{1x} + 60.0 v_{2x} = 30.0 v_{1x} + 60.0(15.0 + v_{1x})$$
$$180 = 30.0 v_{1x} + 900 + 60.0 v_{1x}$$
$$90 v_{1x} = -720$$
$$v_{1x} = -8.0 \text{ m/s} \quad \Rightarrow \quad \vec{v}_1 = -8.0 \text{ m/s} \, \hat{\imath}$$
$$v_{2x} = 15.0 + v_{1x} = 15.0 + (-8.0) = 7.0 \text{ m/s} \quad \Rightarrow \quad \vec{v}_2 = 7.0 \text{ m/s} \, \hat{\imath}$$

The answers can easily be checked by substituting them into the original equations:

$$180 = 30.0 v_{1x} + 60.0 v_{2x} = (30.0)(-8.0) + (60.0)(7.0) = -240 + 420 = 180 \ \checkmark$$
$$12.0 + v_{1x} = 12.0 + (-8.0) = 4.0 \quad , \quad -3.0 + v_{2x} = -3.0 + 7.0 = 4.0 \ \checkmark$$

**Example**. A 50-kg monkey traveling 4.0 m/s to the east on horizontal frictionless ice collides head-on with a 25-kg box of bananas traveling 6.0 m/s to the west. Both velocities are relative to the ice. The monkey grabs onto the box during the collision and doesn't let go. Determine the final velocity of the box and monkey after the collision.

---

[265] Some textbooks derive 'master formulas' for the 1D elastic collision between two objects. In many ways, it's better to understand the strategy and work with a minimum set of fundamental equations. Also, practice with algebra helps to build and maintain algebra fluency. The algebra is actually pretty simple here, especially since we are using the velocity equation (you would get more algebra practice if you used kinetic energy conservation instead of the velocity equation). It's not just algebra, though: When you begin with two fundamental equations, you can better understand the strategy behind the math in conceptual terms. When you first write down conservation of momentum, you see exactly what you're doing; if instead you begin with a 'master formula,' where the algebra has already been done for you, and all that's left is to plug-and-chug, you really miss out on the concepts. Finally, if you get used to using the master formulas, you'll really be stuck if you need to solve a problem with a more complicated elastic collision – such as one in 2D.

We setup our coordinate system with the $+x$-axis directed east. The net external force acting on this system is zero, for the same reasons as in the previous example. We apply the principle of conservation of momentum, and since this 1D collision is completely inelastic, we set the final velocities equal to each other:

$$m_1 v_{1x0} + m_2 v_{2x0} = (m_1 + m_2) v_x$$
$$(50)(4.0) + (25)(-6.0) = 200 - 150 = 50 = (50+25) v_x$$
$$v_x = \frac{50}{75} = \frac{2}{3} \text{ m/s} \quad \Rightarrow \quad \vec{v} = 0.67 \text{ m/s } \hat{\imath}$$

**1D elastic collision between objects of equal mass**: An interesting result occurs when two objects of equal mass collide elastically and head-on (so that the collision is 1D). In this case, the masses cancel out of the conservation of momentum equation:

$$m v_{1x0} + m v_{2x0} = m v_{1x} + m v_{2x} \quad (1\text{D}, m_1 = m_2, \sum F_x = 0, 2 \text{ objects})$$
$$v_{1x0} + v_{2x0} = v_{1x} + v_{2x} \quad (1\text{D}, m_1 = m_2, \sum F_x = 0, 2 \text{ objects})$$

The velocity equation also applies to this 1D elastic collision between two objects:

$$v_{1x0} + v_{1x} = v_{2x0} + v_{2x} \quad (1\text{D, elastic}, \sum F_x = 0, 2 \text{ objects})$$

Let us subtract the velocity equation from the equation above it:

$$v_{2x0} - v_{1x} = v_{1x} - v_{2x0} \quad (1\text{D, elastic}, m_1 = m_2, \sum F_x = 0, 2 \text{ objects})$$
$$v_{1x} = v_{2x0} \quad (1\text{D, elastic}, m_1 = m_2, \sum F_x = 0, 2 \text{ objects})$$

That is, the final velocity of the first object equals the initial velocity of the second object. Substituting this result into the velocity equation, we also find that

$$v_{2x} = v_{1x0} \quad (1\text{D, elastic}, m_1 = m_2, \sum F_x = 0, 2 \text{ objects})$$

Similarly, the final velocity of the second object equals the initial velocity of the first object. That is, the two objects swap velocities as a result of the collision. As a special case, if one object is at rest prior to the collision, the moving objects stops and the object at rest begins moving with the same velocity as the other object had to begin with (again, only if the masses are equal and the collision is head-on).

Note that velocity-swapping results only for a 1D elastic collision between two objects of equal mass. For example, if two billiard balls of identical mass collide head-on (disregarding friction between the balls and the pool table, such that spin does not affect the motion), the two balls should swap velocities. However, if the collision is not head-on, it will be 2D and the above result does not apply. Similarly, the velocities do not swap if the masses are unequal or if the collision is inelastic.

# 6 Systems of Objects

## 6.6 Collisions in 2D

**Conservation of momentum in 2D**: Conservation of momentum involves vector addition:

$$\vec{\mathbf{p}}_{10} + \vec{\mathbf{p}}_{20} + \cdots + \vec{\mathbf{p}}_{N0} = \vec{\mathbf{p}}_1 + \vec{\mathbf{p}}_2 + \cdots + \vec{\mathbf{p}}_N \quad \text{(if } \sum_{system} \vec{\mathbf{F}}_{ext} = 0\text{)}$$

In 2D, each momentum vector has two components: $\vec{\mathbf{p}}_i = p_{ix}\hat{\mathbf{i}} + p_{iy}\hat{\mathbf{j}}$. Therefore, conservation of momentum in 2D can be expressed with a pair of equations – one for each component of momentum:

$$\begin{aligned} m_1 v_{1x0} + m_2 v_{2x0} + \cdots + m_N v_{Nx0} &= m_1 v_{1x} + m_2 v_{2x} + \cdots + m_N v_{Nx} \\ m_1 v_{1y0} + m_2 v_{2y0} + \cdots + m_N v_{Ny0} &= m_1 v_{1y} + m_2 v_{2y} + \cdots + m_N v_{Ny} \end{aligned} \quad \text{(2D, } \sum_{system} \vec{\mathbf{F}}_{ext} = 0\text{)}$$

As usual, the components of the initial and final velocities are related to the speeds via the Pythagorean theorem, $v_i = \sqrt{v_{ix}^2 + v_{iy}^2}$, and the direction of each vector is specified by an angle measured counterclockwise from the +x-axis, $\theta_i = \tan^{-1}(v_{iy}/v_{ix})$.

**Elastic collisions in 2D**: For an elastic collision, KE energy is conserved in addition to momentum:

$$\begin{aligned} m_1 v_{1x0} + m_2 v_{2x0} + \cdots + m_N v_{Nx0} &= m_1 v_{1x} + m_2 v_{2x} + \cdots + m_N v_{Nx} \\ m_1 v_{1y0} + m_2 v_{2y0} + \cdots + m_N v_{Ny0} &= m_1 v_{1y} + m_2 v_{2y} + \cdots + m_N v_{Ny} \\ \frac{m_1 v_{10}^2}{2} + \frac{m_2 v_{20}^2}{2} + \cdots + \frac{m_N v_{N0}^2}{2} &= \frac{m_1 v_1^2}{2} + \frac{m_2 v_2^2}{2} + \cdots + \frac{m_N v_N^2}{2} \end{aligned} \quad \text{(2D, elastic, } \sum_{system} \vec{\mathbf{F}}_{ext} = 0\text{)}$$

**Completely inelastic collisions in 2D**: For a completely inelastic collision, only momentum is conserved, but the final velocities are equal:

$$\begin{aligned} m_1 v_{1x0} + m_2 v_{2x0} + \cdots + m_N v_{Nx0} &= (m_1 + m_2 + \cdots + m_N) v_x \\ m_1 v_{1y0} + m_2 v_{2y0} + \cdots + m_N v_{Ny0} &= (m_1 + m_2 + \cdots + m_N) v_y \end{aligned} \quad \text{(2D, completely inelastic, } \sum_{system} \vec{\mathbf{F}}_{ext} = 0\text{)}$$

**Inverse completely inelastic collisions in 2D**: The initial velocities are equal in an inverse completely inelastic collision:

$$\begin{aligned} (m_1 + m_2 + \cdots + m_N) v_{x0} &= m_1 v_{1x} + m_2 v_{2x} + \cdots + m_N v_{Nx} \\ (m_1 + m_2 + \cdots + m_N) v_{y0} &= m_1 v_{1y} + m_2 v_{2y} + \cdots + m_N v_{Ny} \end{aligned} \quad \text{(2D, completely inelastic, } \sum_{system} \vec{\mathbf{F}}_{ext} = 0\text{)}$$

# Problem-Solving Strategy for 2D Collisions

0. Apply this strategy to solve problems involving 2D collisions between two or more objects (and other problems where the net external force acting on the system is zero and the objects change velocity, but which may not look like 'collisions'). In order for the total momentum of the system to be conserved, the net external force acting on the system must be zero – or the duration of the collision must be short enough that the total momentum of the system is approximately conserved over the time interval of the collision. (Otherwise, instead of applying this strategy, you should begin by applying Newton's second law to the system as a whole, setting the net external force equal to a derivative of the total momentum of the system with respect to time.) **Note**: If the masses of the objects change during the problem, instead of following this strategy, see Sec. 6.9.

1. Setup an $xy$ coordinate system. Choose an inertial reference frame (e.g. the lab frame or the CM frame) and measure all of the velocities relative to this same reference frame.

2. First, express conservation of momentum for the 2D collision, regardless of what type of collision occurs: $m_1 v_{1x0} + m_2 v_{2x0} + \cdots + m_N v_{Nx0} = m_1 v_{1x} + m_2 v_{2x} + \cdots + m_N v_{Nx}$ and $m_1 v_{1y0} + m_2 v_{2y0} + \cdots + m_N v_{Ny0} = m_1 v_{1y} + m_2 v_{2y} + \cdots + m_N v_{Ny}$.

3. If you are told that the 2D collision is elastic (or, equivalently, if you are told that the total mechanical energy of the system is conserved for the collision), write down conservation of kinetic energy, $\frac{m_1 v_{10}^2}{2} + \frac{m_2 v_{20}^2}{2} + \cdots + \frac{m_N v_{N0}^2}{2} = \frac{m_1 v_1^2}{2} + \frac{m_2 v_2^2}{2} + \cdots + \frac{m_N v_N^2}{2}$, in addition to momentum conservation.

4. If you are told that the 1D collision is completely inelastic (or if you know that all of the objects stick together), set the final velocities equal to one another in the conservation of momentum equations: $m_1 v_{1x0} + m_2 v_{2x0} + \cdots + m_N v_{Nx0} = (m_1 + m_2 + \cdots + m_N) v_x$ and $m_1 v_{1y0} + m_2 v_{2y0} + \cdots + m_N v_{Ny0} = (m_1 + m_2 + \cdots + m_N) v_y$.

5. If you have an inverse completely inelastic collision – i.e. the objects are stuck together initially, but not finally – then set the initial velocities equal to one another in the conservation of momentum equations: $(m_1 + m_2 + \cdots + m_N) v_{x0} = m_1 v_{1x} + m_2 v_{2x} + \cdots + m_N v_{Nx}$ and $(m_1 + m_2 + \cdots + m_N) v_{y0} = m_1 v_{1y} + m_2 v_{2y} + \cdots + m_N v_{Ny}$.

6. Use trig to relate the components of the velocities to the speeds and directions: $v_{ix} = v_i \cos \theta_i$ and $v_{iy} = v_i \sin \theta_i$, where each $\theta_i$ is measured counterclockwise relative to the $+x$-axis.

7. As usual, look for $N$ equations in $N$ unknowns ($N$ being the number of unknowns, not objects).

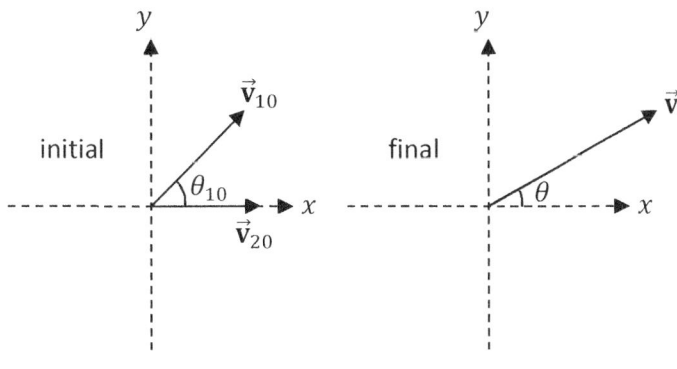

# 6 Systems of Objects

**Example.** As illustrated on the previous page, a monkey drives a car with a combined mass of 200 kg with a velocity 40 m/s northeast, and collides with another monkey driving a car with a combined mass of 150 kg with a velocity 30 m/s east. The cars stick together after the collision. Neglecting friction between the tires and road, determine the final velocity of the cars.

Conserve the $x$- and $y$-components of momentum and set the final velocities equal to one another for this 2D completely inelastic collision:

$$m_1 v_{1x0} + m_2 v_{2x0} = (m_1 + m_2) v_x$$
$$m_1 v_{1y0} + m_2 v_{2y0} = (m_1 + m_2) v_y$$

We must use trig to express the given speeds and directions in terms of the components of the initial velocities:[266]

$$v_{1x0} = v_{10} \cos \theta_{10} = 40 \cos 45° = 20\sqrt{2} \text{ m/s}$$
$$v_{1y0} = v_{10} \sin \theta_{10} = 40 \sin 45° = 20\sqrt{2} \text{ m/s}$$
$$v_{2x0} = v_{20} \cos \theta_{20} = 30 \cos 0° = 30 \text{ m/s}$$
$$v_{2y0} = v_{20} \sin \theta_{20} = 30 \sin 0° = 0$$

Substituting these into the equations for conservation of momentum,

$$(200)(20\sqrt{2}) + (150)(30) = 350 v_x$$
$$(200)(20\sqrt{2}) = 350 v_y$$
$$v_x = 29 \text{ m/s} \quad , \quad v_y = 16 \text{ m/s}$$

Apply trig once again to determine the magnitude and direction of the final velocity:[267]

$$v = \sqrt{v_x^2 + v_y^2} = \sqrt{29^2 + 16^2} = 33 \text{ m/s}$$
$$\theta = \tan^{-1}\left(\frac{v_y}{v_x}\right) = \tan^{-1}\left(\frac{16}{29}\right) = 29°$$

---

**Note.** It would be a mistake in the previous example to simply add the given velocities using vector addition, as that would not account for the effect of the masses. Instead, conservation of momentum (which is vector addition of momentum, rather than velocity) correctly accounts for both the masses and velocities. Symbolically, the distinction is that $\vec{p}_{10} + \vec{p}_{20} + \cdots + \vec{p}_{N0} = \vec{p}_1 + \vec{p}_2 + \cdots + \vec{p}_N$ for a collision where the net external force is zero, but $\vec{v}_{10} + \vec{v}_{20} + \cdots + \vec{v}_{N0} \neq \vec{v}_1 + \vec{v}_2 + \cdots + \vec{v}_N$ (unless all of the masses are equal).

---

[266] In many problems, the initial angles may lie outside of the first quadrant, and so some of the components of the initial velocities may be negative. Don't ignore the signs!

[267] When using the inverse tangent on your calculator, remember to add 180° if $v_x < 0$.

**2D collisions between two objects in the lab frame**: Consider a collision between two objects in the lab frame. One object is initially at rest in the lab frame; we choose to label this as object 2; it is called the target. Therefore, $\vec{v}_{20,L} = 0$, where the subscript $L$ designates that the velocity is measured relative to the lab (and hence the target). If there is to be a collision, object 1 must be directed toward object 2. If the collision is off-center (rather than head-on), the final velocities will not be collinear with the initial velocities, but all of the velocities – initial and final – will be coplanar (as you should be able to see conceptually). Hence, such a collision will be 2D.

It is convenient to setup a coordinate system such that the $+x$-axis is oriented along the initial velocity of object 1, such that $\theta_{10} = 0°$. In this case, the directions of the final velocities can be interpreted as deflection angles, as illustrated below. Note that one of the deflection angles will lie in the first quadrant, and the other will lie in the second quadrant (not necessarily as drawn below). Therefore, one of the two final velocities – either $\vec{v}_{1L}$ or $\vec{v}_{2L}$ – will have a negative $y$-component (not necessarily $\vec{v}_{1L}$, as depicted below).

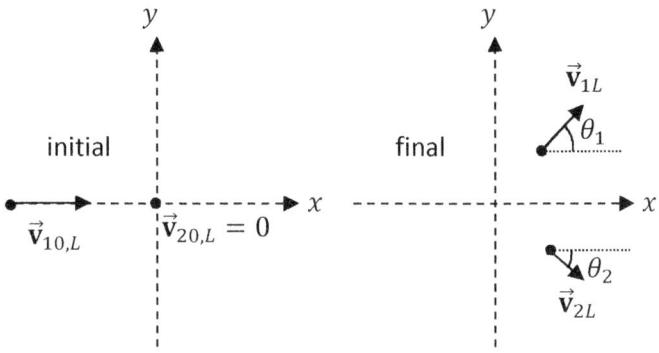

Applying conservation of momentum to this 2D collision, we find that

$$m_1 v_{10,L} = m_1 v_{1L} \cos\theta_1 + m_2 v_{2L} \cos\theta_2$$
$$0 = m_1 v_{1L} \sin\theta_1 + m_2 v_{2L} \sin\theta_2$$
(lab frame, 2D, 2 objects, $\sum_{system} \vec{F}_{ext} = 0$)

If the masses and initial velocities are all known, there are 4 final parameters to predict: $v_{1L}$, $v_{2L}$, $\theta_1$, and $\theta_2$. There are two equations above. If the collision is elastic, kinetic energy conservation will provide a third equation. In order to obtain $N$ equations in $N$ unknowns, in this case, we must also examine the direction of the impulse. We will return to this last part when we discuss the impact parameter at the end of this section. If instead the collision is completely inelastic, there will only be 2 final parameters (since $v_{2L} = v_{1L}$ and $\theta_{1L} = \theta_{2L}$ will be equal, as in the last example).

**2D collisions between two objects in the CM frame**: We now consider a collision between two objects in the CM frame. The center of mass is at rest in the CM frame (assuming, as we generally do for collisions, that the net external force acting on the system is zero, or at least does not have a significant effect over the time interval of the collision). Recalling that the total momentum of the system equals the total mass of the system times the velocity of the center of mass, $\vec{p}_{tot} = M\vec{v}_{cm}$, the total momentum of the system will be zero in the CM frame if momentum is conserved:

$$\vec{p}_{10} + \vec{p}_{20} = \vec{p}_1 + \vec{p}_2 = 0 \quad \text{(CM frame, 2 objects, } \sum_{system} \vec{F}_{ext} = 0\text{)}$$

Therefore, $\vec{p}_{10}$ and $\vec{p}_{20}$ are anti-parallel, and $\vec{p}_1$ and $\vec{p}_2$ are also anti-parallel, in the CM frame (but the initial and final momenta will not be collinear, in general):

$$\vec{p}_{20} = -\vec{p}_{10} \quad , \quad \vec{p}_2 = -\vec{p}_1 \quad \text{(CM frame, 2 objects, } \sum_{system} \vec{F}_{ext} = 0\text{)}$$

Resolving these vectors into components, and applying $\theta_{20} = \theta_{10} + \pi$ and $\theta_2 = \theta_1 + \pi$,[268] we find[269]

$$\begin{aligned} m_2 v_{20,cm} \cos\theta_{10} &= m_1 v_{10,cm} \cos\theta_{10} \\ m_2 v_{20,cm} \sin\theta_{10} &= m_1 v_{10,cm} \sin\theta_{10} \\ m_2 v_{2,cm} \cos\theta_1 &= m_1 v_{1,cm} \cos\theta_1 \\ m_2 v_{2,cm} \sin\theta_1 &= m_1 v_{1,cm} \sin\theta_1 \end{aligned} \quad \text{(CM frame, 2D, 2 objects, } \sum_{system} \vec{F}_{ext} = 0\text{)}$$

where the subscript $cm$ designates that the speed is measured relative to the center of mass, and where we have utilized that fact that the sine and cosine functions both reflect upon a shift of 180° – e.g. $\cos(\theta + \pi) = -\cos\theta$ and $\sin(\theta + \pi) = -\sin\theta$.

It is convenient to setup our coordinate system such that the $+x$-axis points along $\vec{p}_{10}$, such that $\theta_{10}$ and $\theta_{20}$ are 0° and 180°, respectively. In this case, a single scattering angle specifies the axis of the final velocities, as illustrated below. In general, $\vec{v}_{1,cm}$ may lie in a different quadrant than the case shown below, which will affect the individual signs of the components of the velocities.

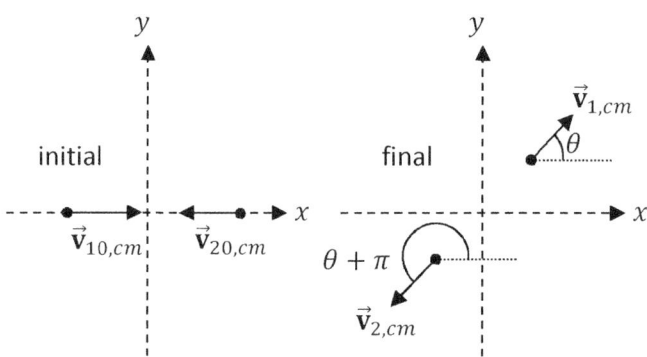

In component form, with these coordinates, conservation of momentum for this collision simplifies to

$$\begin{aligned} m_2 v_{20,cm} &= m_1 v_{10,cm} \\ m_2 v_{2,cm} &= -m_1 v_{1,cm} \end{aligned} \quad \text{(CM frame, 2D, 2 objects, } \sum_{system} \vec{F}_{ext} = 0\text{)}$$

---

[268] This is what equal and opposite momenta (both initially and finally), stated in the previous equations, means.
[269] Of course you can cancel the trig functions, but first you should verify the simplifications that we made.

**2D Elastic collision between equal masses in the lab frame**: Consider a 2D elastic collision between two objects that have equal masses ($m_2 = m_1$). In this case, conservation of momentum and conservation of kinetic energy simplify to

$$v_{10,L} = v_{1L} \cos\theta_1 + v_{2L} \cos\theta_2$$
$$v_{2L} \sin\theta_2 = -v_{1L} \sin\theta_1 \quad \text{(lab frame, 2D elastic, 2 objects, } m_2 = m_1, \sum_{system} \vec{F}_{ext} = 0\text{)}$$
$$v_{10,L}^2 = v_{1L}^2 + v_{2L}^2$$

We can eliminate $\theta_2$ from the top two equations by solving for $v_{2L} \cos\theta_2$ and $v_{2L} \sin\theta_2$ in each equation, then using the trig identity $\cos^2\theta_2 + \sin^2\theta_2 = 1$:

$$v_{2L} \cos\theta_2 = v_{10,L} - v_{1L} \cos\theta_1 \quad \text{lab frame}$$
$$v_{2L} \sin\theta_2 = -v_{1L} \sin\theta_1 \quad \text{2D elastic}$$
$$v_{2L}^2 \cos^2\theta_2 + v_{2L}^2 \sin^2\theta_2 = \left(v_{10,L} - v_{1L} \cos\theta_1\right)^2 + (-v_{1L} \sin\theta_1)^2 \quad \begin{array}{l}\text{2 objects}\\ m_2 = m_1\end{array}$$
$$v_{2L}^2 = v_{10,L}^2 - 2v_{10,L}v_{1L} \cos\theta_1 + v_{1L}^2 \cos^2\theta_1 + v_{1L}^2 \sin^2\theta_1$$
$$v_{2L}^2 = v_{10,L}^2 - 2v_{10,L}v_{1L} \cos\theta_1 + v_{1L}^2 \qquad \sum_{system} \vec{F}_{ext} = 0$$

We now combine this equation with the kinetic energy conservation equation:

$$v_{2L}^2 = v_{10,L}^2 - 2v_{10,L}v_{1L} \cos\theta_1 + v_{1L}^2 = v_{10,L}^2 - v_{1L}^2 \quad \text{lab frame, 2D elastic}$$
$$-2v_{10,L}v_{1L} \cos\theta_1 + v_{1L}^2 = -v_{1L}^2 \qquad \text{2 objects, } m_2 = m_1$$
$$v_{1L} = v_{10,L} \cos\theta_1 \qquad \sum_{system} \vec{F}_{ext} = 0$$

Substituting this back into the conservation of kinetic energy equation,

$$v_{10,L}^2 = v_{1L}^2 + v_{2L}^2 = v_{10,L}^2 \cos^2\theta_1 + v_{2L}^2 \quad \text{lab frame, 2D elasic}$$
$$v_{10,L}^2 - v_{10,L}^2 \cos^2\theta_1 = v_{10,L}^2 \sin^2\theta_1 = v_{2L}^2 \quad \text{2 objects, } m_2 = m_1$$
$$v_{2L} = |v_{10,L} \sin\theta_1| \qquad \sum_{system} \vec{F}_{ext} = 0$$

where the absolute values are needed to ensure a nonnegative speed $v_{2L}$ since $\theta_1$ may lie in Quadrant IV, in general. The final speeds can then be predicted from the following pair of equations if the deflection angle is known; and the deflection angle itself can be predicted by measuring the impact parameter (to be discussed shortly).

$$v_{1L} = v_{10,L} \cos\theta_1$$
$$v_{2L} = \pm v_{10,L} \sin\theta_1 \quad \text{(lab frame, 2D elastic, 2 objects, } m_2 = m_1, \sum_{system} \vec{F}_{ext} = 0\text{)}$$

We obtain an interesting constraint on the deflection angles, $\theta_1$ and $\theta_2$, by substituting this pair of equations into the equation for conservation of the $y$-component of momentum:

$$v_{2L}\sin\theta_2 = -v_{1L}\sin\theta_1$$
$$\pm v_{10,L}\sin\theta_1\sin\theta_2 = v_{10,L}\cos\theta_1\sin\theta_1$$
$$\pm\sin\theta_2 = \cos\theta_1 = \sin(90° - \theta_1)$$
$$|\theta_1| + |\theta_2| = 90°$$

(lab frame, 2D elastic, 2 objects, $m_2 = m_1$, $\sum_{system}\vec{F}_{ext} = 0$)

where we used the trig identity $\cos x = \sin(90° - x)$ and where we wrote our equations such that a Quadrant IV angle is written as a negative Quadrant I angle. (The previous trig identity is easily verified by drawing a right triangle: You will readily see that the cosine of one angle equals the sine of its complement, since the adjacent of one angle is the opposite of the other.)

We have thus proven that the deflection angles of the lab frame are perpendicular to one another for a 2D elastic collision between two objects (think about this). Therefore, neglecting spin and English and any other complications introduced by friction, we predict that if one billiard ball strikes another with identical mass,[270] if the collision is not head-on, the two balls should travel at right angles to one another after the collision – i.e. their final velocities should be perpendicular. This information is useful to predict which way the cue ball will travel after the collision – so that you can avoid scratching on the shot, and also to help you setup the next shot.

**Impact parameter**: The impact parameter, $b$, measures how much off-center a two-body collision is. In the lab frame, the impact parameter can be measured as follows, and as illustrated below:
- Dray a ray along $\vec{v}_{10,L}$ – i.e. the velocity of the object that is moving toward the target initially.
- Draw a second ray parallel to the first ray, but which passes through the center of the target.
- The impact parameter is the distance between these two lines.

stationary target

For a hard collision – i.e. when the objects make direct contact, as in the case of two billiard balls colliding – the impact parameter tells you where object 1 will strike the target (object 2). The impact parameter also has significance for collisions where the two objects do not actually 'touch' – as in the case of a proton scattering off of a positively charged nucleus, or an asteroid scattering as it approaches the earth (we will see an example of scattering in Sec. 6.8). In the case of scattering, the impact parameter is interpreted as what the distance of closest approach would be if the two objects did not interact with one another. Note that objects can have a significant collision in the case of scattering, even though geometrically object 1 may appear to be lined up to 'miss' object 2; objects can interact at a distance, as in the case of gravitational or electromagnetic forces, for example.

---

[270] However, the cue ball is slightly more massive than the numbered billiard balls for a regulation billiard set.

**Relating the impact parameter to the deflection angle for a hard collision between hard spheres**: If the two objects engage in a hard collision, the initial line of sight of object 1 must be directed within the shape of the target (object 2). Consider a hard collision between two spheres (or they could be hockey pucks, with identical results, ignoring frictional and related spin effects) in the lab frame, as illustrated below.

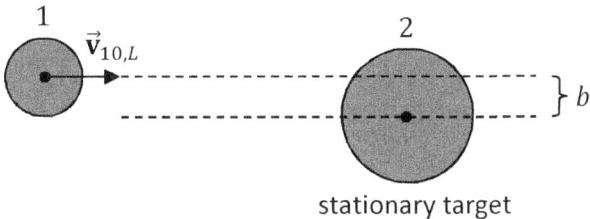

stationary target

The two spheres exert mutual impulses on one another during the collision. The collision forces are equal and opposite according to Newton's third law. For a hard collision, the collision force acts only during the brief time interval over which the collision occurs. The collision forces are internal to the system of two spheres, but external to each individual sphere; these collision forces cause the change in momentum of each sphere during the collision according to Newton's second law, while conserving the total momentum of the system (provided that the net external force equals zero), also according to Newton's second law. This collision force causes the deflection. As a result, the deflection of the target (object 2), characterized by $\theta_2$, is directly along the collision force in the lab frame. Therefore, we can apply trig to relate the impact parameter, $b$, to the deflection angle, $\theta_1$, (object 1) for a hard collision:

$$\sin \theta_2 = \frac{b}{R_1 + R_2} \quad \text{(hard collision, elastic, 2 spheres)}$$

where $R_1$ and $R_2$ are the radii of the respective spheres.

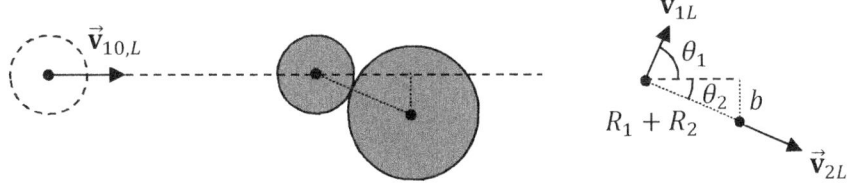

Measurement of the impact parameter, masses, sizes of the spheres, and initial velocities then allows us to predict the two final velocities for a 2D elastic collision between two hard spheres.

**2D Elastic collision between equal masses in the CM frame**: We now consider a 2D elastic collision between two objects that have equal masses ($m_2 = m_1$) in the CM frame. Let us first cancel the masses in the equations that we had previously obtained for conservation of momentum for such a collision in the CM frame:

$$v_{20,cm} = v_{10,cm}$$
$$v_{2,cm} = v_{1,cm}$$
(CM frame, 2D collision, 2 objects, $m_2 = m_1$, $\sum_{system} \vec{F}_{ext} = 0$)

That is, since the momenta are equal and opposite both initially and finally in the CM frame, if the masses of the two objects are equal, it follows that the velocities must also be equal and opposite initially and finally in the CM frame:

$$\vec{v}_{20,cm} = -\vec{v}_{10,cm}$$
$$\vec{v}_{2,cm} = -\vec{v}_{1,cm}$$
(CM frame, 2D collision, 2 objects, $m_2 = m_1$, $\sum_{system} \vec{F}_{ext} = 0$)

Additionally, conservation of kinetic energy for this elastic collision yields

$$v_{10,cm}^2 + v_{20,cm}^2 = v_{1,cm}^2 + v_{2,cm}^2 \quad \text{(CM frame, 2D elastic, 2 objects, } m_2 = m_1, \sum_{system} \vec{F}_{ext} = 0)$$

which reduces to $v_{10,cm}^2 = v_{1,cm}^2$, and therefore

$$v_{10,cm} = v_{20,cm} = v_{1,cm} = v_{2,cm} \quad \text{(CM frame, 2D elastic, 2 objects, } m_2 = m_1, \sum_{system} \vec{F}_{ext} = 0)$$

Note that although the speeds are equal, the velocities differ. These mathematical results are indeed much simpler in the CM frame compared to the lab frame, though the measurements themselves are not as practical for the CM velocities as they are for the lab velocities.

## 6.7 The Ballistic Pendulum

**Ballistic pendulum**: A ballistic pendulum consists of a block of wood that is suspended vertically by multiple strings (or chains, or a single rigid rod that is free to rotate through a vertical plane),[271] as illustrated on the following page. When a bullet is fired horizontally into the block, the block rises upward to some maximum angle, $\theta_m$, before swinging back down.

---

[271] But not a single thread. We don't want the block to rotate in a horizontal plane – multiple threads connected to the block in multiple positions (say, near the top four corners) prevent this unwanted horizontal rotation.

**Conservation laws for the ballistic pendulum**: During the collision, the bullet and block exert equal and opposite collision forces on one another, and the upward support forces in the support strings balance the weight of the bullet and block. Defining the system as the bullet plus the block, the net external force acting on the system is zero during the collision. The total momentum of the system is therefore conserved during the collision. Mechanical energy is lost during the collision for this completely inelastic collision. During the swing upward, the strings are at an angle, and so no longer balance the weight of the system. Thus, momentum is not conserved for the swing upward. However, mechanical energy is conserved for the swing upward, neglecting resistive forces that may act on the system.

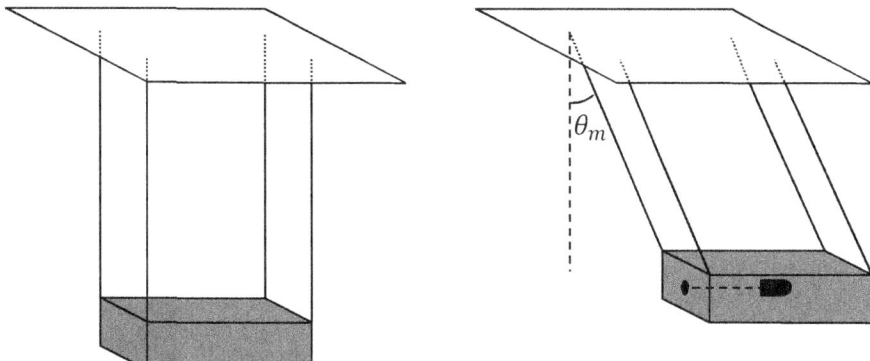

**Conservation of momentum for the collision in the ballistic pendulum**: Assuming that the duration of the collision is short-lived compared to the time it takes for the ballistic pendulum to swing upward – quite reasonable considering typical bullet speeds – the collision itself is essentially 1D (purely horizontal). We setup our coordinate system with $x$ being horizontal, along the motion of the bullet, and $y$ being vertically upward. Conservation of momentum for this completely inelastic collision yields

$$m_p v_p = (m_p + m_b) v_b \quad \text{(ballistic pendulum)}$$

where we have made the following definitions:[272]
- $m_p$ is the mass of the bullet (projectile).
- $v_p$ is the initial speed of the bullet.
- $m_b$ is the mass of the block (before being struck by the bullet).
- $v_b$ is the speed of the block just after the collision.

**Conservation of energy for the swing upward in the ballistic pendulum**: We choose our reference height to be the center of mass of the system immediately after the collision, while the strings are still vertical. We choose the initial position for conservation of energy to be immediately after the collision, and the final position to be at the maximum height, just before returning back down. Conservation of energy for the swing upward gives

---

[272] We designated the $p$ for projectile and the $b$ for block. Unfortunately, 'bullet' also starts with a 'b,' and pendulum also starts with a 'p,' so you need to be careful not to mix these up.

$$U_0 + K_0 + W_{nc} = U + K$$

$$\frac{(m_p + m_b)v_b^2}{2} = (m_p + m_b)gh_m \quad \text{(ballistic pendulum)}$$

$$v_b^2 = 2gh_m \quad \text{(ballistic pendulum)}$$

This can be expressed in terms of $\theta_m$ by applying trig (see the diagram below):

$$v_b = \sqrt{2gL(1 - \cos\theta_m)} \quad \text{(ballistic pendulum)}$$

where $L$ is the length of each string (assumed all to be of the same length).

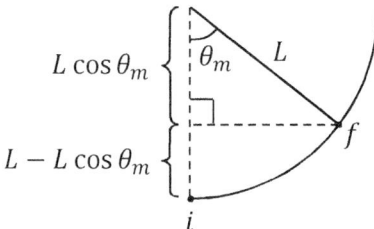

**Combined results for conservation of momentum and energy for the ballistic pendulum**: We obtain the following equation when we combine our two previous results for the ballistic pendulum:

$$v_p = \frac{(m_p + m_b)\sqrt{2gL(1 - \cos\theta_m)}}{m_p} \quad \text{(ballistic pendulum)}$$

**Loss of overall mechanical energy for the ballistic pendulum**: If you proceed to conserve energy from the point just prior to impact to the maximum height of the block, you will find that some mechanical energy has been lost (meaning that $W_{nc} \neq 0$, and hence that there are some internal energy changes, which in this case result in a transfer of energy in the form of heat – you can safely bet that the bullet and block will be hot just after the collision). This should come as no surprise, though, since mechanical energy is not conserved for the completely inelastic collision:

$$U_0 + K_0 + W_{nc} = U + K$$

$$\frac{m_p v_p^2}{2} + W_{nc} = (m_p + m_b)gh_m \quad \text{(ballistic pendulum)}$$

$$W_{nc} = (m_p + m_b)gh_m - \frac{m_p v_p^2}{2} \quad \text{(ballistic pendulum)}$$

$$W_{nc} = (m_p + m_b)gL(1 - \cos\theta_m) - \frac{(m_p + m_b)^2 gL(1 - \cos\theta_m)}{m_p} \quad \text{(ballistic pendulum)}$$

$$W_{nc} = [m_p - (m_p + m_b)]\frac{(m_p + m_b)gL(1 - \cos\theta_m)}{m_p} \quad \text{(ballistic pendulum)}$$

$$W_{nc} = -\frac{m_b}{m_p}(m_p + m_b)gL(1 - \cos\theta_m) < 0 \quad \text{(ballistic pendulum)}$$

> **Note**. Don't memorize or directly apply the results for the ballistic pendulum. Instead, learn the strategy for how to separately apply conservation of momentum and energy to derive the results. This is much more useful. For one, it will allow you to solve problems that look similar, but are somewhat different. Since the results for the ballistic pendulum don't apply to problems that are a little different, they are of very limited use in comparison to learning the technique.

## 6.8 Scattering

**Scattering in the lab frame**: Consider a pointlike particle (particle 1) heading toward the neighborhood of a pointlike target (particle 2) that is initially at rest, which corresponds to two-body scattering in the lab frame. Particle 1 is not heading directly toward particle 2, but is aimed slightly off-center, such that the collision is 2D. The impact parameter, $b$, represents the nearest distance that particle 1 would approach particle 2 if the two particles were completely non-interacting. However, due to a force of attraction or repulsion – which may be gravitational, as in the case of an asteroid, or electromagnetic, as in the case of a proton in a beam that is passing near a nucleus – the actual distance of closest approach is different because particle 1 will take a curved path, and particle 2 will also move as a result of the collision.

Contrast scattering with a hard collision. When two billiard balls collide in a hard collision, for example, the two billiard balls only interact with one another during the brief moment when physical contact is made. In scattering however, the force acts over a distance (often in the form of an inverse-square law, like gravity), and the particles might not physically 'touch' one another. In the case of scattering, the deflection occurs over the course of a smooth curve, as illustrated below for a repulsive interaction between the particles.

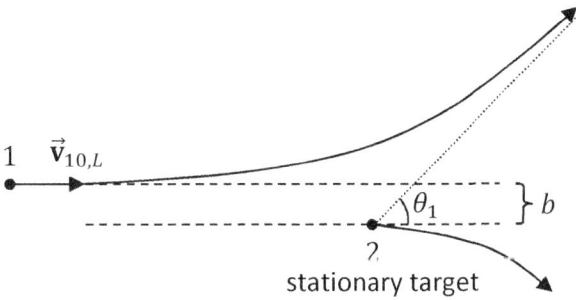

**Scattering angle**: The deflection of particle 1 is measured from the initial position of the target to the asymptotic inertial path that particle 1 approaches in its final state in the lab frame, as illustrated above. This deflection angle, $\theta_1$, characterizes the scattering in the lab frame.

6 Systems of Objects

**Scattering cross section**: The cross section, $\sigma$,[273] is an abstract measurement of area that quantifies the probability that the two particles will scatter significantly. If the interaction depends on an inverse-square law of gravity or electricity, for example, then if the impact parameter, $b$, is too large, the force between the two particles will be quite small; in this case, particle 1 will be virtually undeflected, and particle 2 will remain virtually stationary. A greater value of the cross section means that significant scattering can occur for a larger impact parameter, while a smaller value of the cross section means that the incident particle needs to be aimed more closely toward the target in order to achieve a significant deflection.

This notion of area originates from analogy with a hard collision between two spheres. Hard scattering between spheres is a hit-or-miss collision, unlike the scattering between charged particles or between astronomical objects. If the incident sphere is directed so that it will make physical contact with the second sphere, then there will be a collision – i.e. if $b < R_1 + R_2$ (see the last diagram at the end of Sec. 6.6). For hard scattering between two spheres, the area $\pi(R_1 + R_2)^2$ represents a physical target at which the incident sphere must be aimed in order to achieve a collision.

For the 'soft' scattering that we are discussing in this section, the cross section provides an abstract area in analogy to the more tangible collision area of hard collisions. If the incident particle is aimed within the area of the cross section (centered about the target), significant deflection will occur.

**The impact area**: The impact parameter defines a circle with an area of $\pi b^2$ centered about the line of approach. Scattering will be significant if $\pi b^2$ does not exceed the scattering cross section.

**The impact ring**: A differential variation in the impact parameter, $db$, defines an infinitesimally thin ring of inner radius $b$ and outer radius $b + db$, as illustrated below. The differential area of the thin ring is $dA = 2\pi b\, db$.

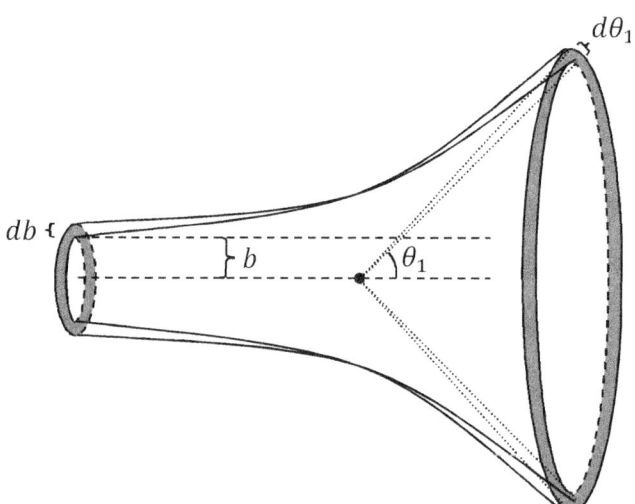

---

[273] This is the Greek symbol sigma ($\sigma$), which we have already seen in the context of densities in the center of mass integrals. It is intuitive to try to draw sigma right to left, and then finish with a circle, but if you attempt this, you will probably find it difficult to close the circle properly. A better technique is to begin by drawing a clockwise circle starting with a downward motion, and finish with the left-to-right line segment when you reach the top.

**Scattering cone**: The scattering angle, $\theta_1$, defines a scattering cone with a vertex at the scattering center – i.e. at the initial position of the target. The half-angle of the cone is $\theta_1$.

**Scattering ring**: The differential variation in the impact parameter, $db$, corresponds to a differential variation in the scattering angle, $d\theta_1$. This defines a scattering ring corresponding to the infinitesimal window between $\theta_1$ and $\theta_1 + d\theta_1$. The impact ring corresponds to the scattering ring – i.e. the variation in $b$ produces a variation in $\theta_1$. Observe that an increase in $b$ decreases $\theta_1$.

**Solid angle**: Analogous to how $ds = rd\theta$ relates a differential arc length to a differential angle, we can write $dA = r^2 d\Omega$ to relate the differential area element to a differential solid angle. The uppercase Greek letter omega ($\Omega$) is called the solid angle, and the solid angle obtained from $\Omega = \int d\Omega$ is a sort of two-dimensional angle. The SI unit of the solid angle is the steradian, and, similar to the definition of the radian, it expresses a fractional area, and gets swept under the rug in the context of square meters.

Have you ever noticed that the sun and moon appear about the same 'size' when you look at them in the sky? (But don't look go out and look at the sun if you value your retinas!) Of course, the actual size of the sun is much greater than that of the moon. When you look at the sun and moon, your eye perceives a measure of 'area.' This is not the actual cross-sectional area of the sun or moon – because, again, the sun has much greater cross-sectional area than the moon. What your eye is perceiving is solid angle. When you view an object, a vertical angle specifies the height of the object, and a horizontal angle specifies the width of the object; the solid angle combines these two independent angular directions together.

Just as $2\pi$ radians corresponds to a full circle, $4\pi$ steradians corresponds to a full sphere, as calculated in the next example. Expressing $dA$ in spherical coordinates for a fixed value of $r$, $dA = r^2 \sin\theta \, dr d\theta d\varphi$, we obtain a formula for the differential solid angle:

$$d\Omega = \sin\theta \, d\theta d\varphi$$

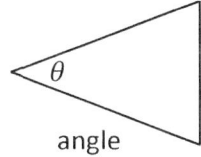
angle

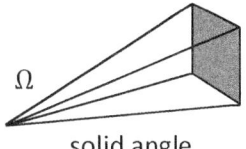
solid angle

---

**Example**. Measuring solid angle from the center of a sphere, show that the solid angle for the complete sphere is $4\pi$ steradians.

Performing the double integral over the full sphere, we obtain

$$\Omega = \int d\Omega = \int_{\varphi=0}^{2\pi} \int_{\theta=0}^{\pi} \sin\theta \, d\theta d\varphi = 2\pi[-\cos\theta]_{\theta=0}^{\pi} = 4\pi \text{ steradians}$$

**Differential and total cross sections**: The total cross section, $\sigma$, can be expressed in terms of what is termed a differential cross section, $d\sigma/d\Omega$. It is conventional, though somewhat confusing, to use the symbol $\sigma(\theta_1)$ to represent the differential cross section: $\sigma(\theta_1) \equiv d\sigma/d\Omega \neq \sigma$. Observe a distinction between the differential cross section, $\sigma(\theta_1)$, and the total cross section, $\sigma$; these are not the same thing.[274] In addition to the notation, the terminology is also a little confusing. What we frequently call a differential cross section is actually finite: $d\sigma$ and $d\Omega$ are each infinitesimal, but $\sigma(\theta_1) = d\sigma/d\Omega$ is finite (just like the infinitesimal $d\vec{r}$ displacement and infinitesimal time $dt$ conspire to make a finite velocity $\vec{v} = d\vec{r}/dt$). The 'differential' cross section is really $\sigma(\theta_1)d\Omega$, which equals $d\sigma$. Nonetheless, it's customary to refer to $\sigma(\theta_1)$, which is $d\sigma/d\Omega$, as the differential cross section.

The total cross section, $\sigma$, is related to the differential cross section, $\sigma(\theta_1)$, through

$$\sigma = \int \sigma(\theta_1) d\Omega$$

since $\int \sigma(\theta_1) d\Omega = \int \frac{d\sigma}{d\Omega} d\Omega = \int d\sigma = \sigma$, where the integration is over the full range of $\varphi$ from 0 to $2\pi$ and the physically possible range of $\theta_1$ (which is 0 to $\pi$ if the target is more massive than the incident particle, but a smaller range, 0 to some $\theta_1^{max}$, if the incident particle is more massive than the target).

The total cross section, $\sigma$, is the abstract area for which there will be significant scattering. The differential cross section, $\sigma(\theta_1)$, is how large the impact area needs to be in order to produce scattering at a specified scattering angle, $\theta_1$. More precisely, the area of the impact ring, $2\pi b db$, is related to the area of the scattering ring through

$$2\pi b db = -\sigma(\theta_1) 2\pi \sin\theta_1 \, d\theta_1$$

where the $2\pi \sin\theta_1 \, d\theta_1$ comes from integrating $d\Omega$ of the scattering ring over $\varphi$. Observe that the range of $\varphi$ makes a ring for a fixed $\theta_1$. The minus sign reflects that $\theta_1$ decreases as $b$ increases, which should make sense conceptually: There is less scattering if the collision is more off-center. Solving for the differential cross section, we obtain

$$\sigma(\theta_1) = -\frac{b}{\sin\theta_1} \frac{db}{d\theta_1}$$

Once the relationship between the impact parameter and the scattering angle in the lab frame is established, both the differential cross section and total cross section can be obtained. In practice, one does not fire a single incident particle upon the target, but a beam of such particles. Experimentally, one measures the number of particles that scatter as a function of the scattering angle, which can be related to the differential cross section. Namely, $I\sigma(\theta_1) = dN/d\Omega$, where $I$ is the beam intensity (the number of particles per unit time per unit area) and $dN/d\Omega$ is the number of particles scattered through the ring defined by $\theta_1$ and $\theta_1 + d\theta_1$ per unit time.

---

[274] This confusing notation – when you first see it, but once you understand it, you should get used to it – is typical of distribution functions. There is a similar distinction, for example, between the radiancy, $R_T$, and spectral radiancy, $R_T(\nu)$, in the context of blackbody radiation.

**Example**. Ernest Rutherford directed a beam of alpha particles (positively charged helium nuclei, $He^{2+}$ or $2p2n$) at a thin metal foil. Most of the alpha particles passed straight through, virtually undeflected; but some scattered at large scattering angles (i.e. up to $\theta_1 = 180°$). In this way, Rutherford discovered that matter consists of mostly empty space, with most of the mass and all of the positive charge concentrated into extremely tiny regions of space – the nuclei. In the case of Rutherford scattering between a positively charged incident particle and a positively charged nucleus, the impact parameter is related to the scattering angle by $b = a\csc^2(\theta_1/2)$, where $a$ is a positive constant with dimensions of length. Determine the differential cross section for Rutherford scattering, and interpret the results conceptually.

Since we know $b$ as a function of $\theta_1$, we can simply take a derivative and substitute the result into the formula for the differential cross section:

$$\sigma(\theta_1) = -\frac{b}{\sin\theta_1}\frac{db}{d\theta_1} = -\frac{b}{\sin\theta_1}\frac{d}{d\theta_1}\left[a\csc^2\left(\frac{\theta_1}{2}\right)\right] = -\frac{ab}{\sin\theta_1}\frac{\cos\left(\frac{\theta_1}{2}\right)}{\sin^3\left(\frac{\theta_1}{2}\right)}$$

$$\sigma(\theta_1) = -\frac{ab}{2\sin\left(\frac{\theta_1}{2}\right)\cos\left(\frac{\theta_1}{2}\right)}\frac{\cos\left(\frac{\theta_1}{2}\right)}{\sin^3\left(\frac{\theta_1}{2}\right)} = -\frac{ab}{2\sin^4\left(\frac{\theta_1}{2}\right)} = -\frac{a^2}{2\sin^6\left(\frac{\theta_1}{2}\right)}$$

where we used the trig identity $\sin 2x = 2\sin x\cos x$.

Let us first examine the given relationship, $b = a\csc^2(\theta_1/2)$, which we can write as

$$b = \frac{a}{\sin^2(\theta_1/2)}$$

The sine function, $\sin(\theta_1/2)$, has a maximum value of 1 at $\theta_1/2 = 90°$ (meaning $\theta_1 = 180°$), and is zero at $\theta_1 = 0°$. Therefore, this formula agrees with our expectation that the scattering angle will be greater for a small impact parameter and about $0°$ for a large impact parameter.

Now consider the differential cross section that we obtained,

$$\sigma(\theta_1) = -\frac{a^2}{2\sin^6\left(\frac{\theta_1}{2}\right)} = -\frac{b^3}{2a}$$

This shows that there is smaller effective target area to achieve greater scattering angles, but a larger effective target area for which there is less scattering. We also see expected the proportionality between $\sigma(\theta_1)$ and $b$: A greater impact parameter corresponds to a greater effective target area.

It is interesting to note that the total cross section, $\sigma$, is infinite for Rutherford scattering. This indicates that there is always some degree of scattering, even for very large impact parameters (though the scattering angle is very small for a very large impact parameter), since the electric force is an inverse-square law like gravity with infinite range. In practice, however, a screening factor (for the effect of electrons surrounding the nucleus) needs to be included for large impact parameters, which yields a finite total cross section, such that there is a finite target area over which scattering occurs.

## 6.9 Rocket Motion

**Rocket propulsion**: Let us consider the concept of rocket propulsion as our prototype for variable-mass problems. Out in space, a rocket can travel along its natural free-fall orbit for 'free' – i.e. it is not necessary to burn fuel to do this. However, if the pilot wishes for the rocket to travel along a trajectory that does not correspond to a free-fall orbit, the rocket must burn fuel and eject the gases in order for the rocket to change its course.

The main idea behind rocket propulsion is Newton's third law: The rocket exerts a force on the gases, ejecting them in one direction, and therefore the gases exert an equal force on the rocket in the opposite direction. Mathematically, however, Newton's third law is too simple to be useful to solve any problem by itself. The most general form of Newton's second law – i.e. allowing for variable mass – allows us to compute the instantaneous velocity and position of the rocket at some specified time. If the net external force acting on the rocket is zero – meaning that it's far enough away from external gravitational fields that their influence is negligible – then the instantaneous velocities of the rocket and gases (i.e. the complete system, so that the forces between the rocket and gases are internal and therefore cancel out) can be related through conservation of momentum.

**Continual change in mass**: It's important to realize that a rocket ejects gases over a finite period of time, and not all at once. If the pilot simply launches a missile instead of continually ejecting gases, then the rocket problem would be extremely simple: We would simply treat the missile launch as an inverse completely inelastic collision. However, if the pilot ejects gases over an extended period of time, eventually ejecting just as much mass as the missile, and at the same speed relative to the rocket, the outcome is actually different in this case. It's important to realize that we can't solve the rocket propulsion problem by treating the gases as if it were all ejected at once, but must work with differentials and allow for a continual transfer of mass. We illustrate this numerically in the following example. Conceptually, the distinction is this: If you eject all of the gases with the same speed relative to the rocket, the rocket is constantly changing speed as a result, so different parts of the gases travel different speeds relative to some inertial reference frame; whereas if you eject all of the gases at once, all of the gases will have the same speed relative to an inertial reference frame.

---

**Example**. A 40-kg monkey is sitting on a 60-kg plank. The plank is on top of horizontal frictionless ice, while the monkey is secured to the plank (so he does not slip or fall off, but remains fastened to the plank at all times). The monkey has two 20-kg bowling balls. The system is initially at rest. The monkey throws each bowling ball 10 m/s south relative to the plank. Determine the final velocity of the monkey and plank if he throws (A) both bowling balls at once and (B) the bowling balls one at a time.

If we define the system to include the monkey, plank, and both bowling balls, then the net external force acting on the system is zero:[275] The weights are balanced by normal forces, and the collision forces are all internal. We choose the $+x$-axis to point north, and measure all velocities relative to the ice. Conservation of momentum for the inverse completely inelastic collision in part (A) yields

---

[275] Technically, the bowling balls will gain downward momentum as they fall downward. Since this is vertical momentum, the total horizontal momentum of the system is still conserved, so the height of the monkey is irrelevant to the problem.

$$(m_1 + m_2 + m_m + m_p)v_{x0} = (m_1 + m_2)v_{bi,x} + (m_m + m_p)v_{pi,x}$$
$$0 = (20 + 20)(-10) + (40 + 60)v_{pi,x}$$
$$v_{pi,x} = 4.0 \text{ m/s} \quad \Rightarrow \quad \vec{v}_{pi} = 4.0 \text{ m/s } \hat{\imath}$$

For part (B), we must conserve momentum separately for the throw of each bowling ball. For the first bowling ball,

$$(m_1 + m_2 + m_m + m_p)v_{x0} = (m_1)v_{bi,x} + (m_2 + m_m + m_p)v_{pi,x}$$
$$0 = (20)(-10) + (20 + 40 + 60)v_{pi,x}$$
$$v_{pi,x} = 1.7 \text{ m/s} \quad \Rightarrow \quad \vec{v}_{pi} = 1.7 \text{ m/s } \hat{\imath}$$

Note that the second bowling ball is thrown 10 m/s south relative to the plank, which is 8.3 m/s south relative to the ice. For the second bowling ball, we take the system to be exclude the first bowling ball (which will simply cancel out on both sides of the equation if you choose to include it instead). Also, the final $x$-component of the velocity of the plank, $v_{pi,x}$, for the first throw is now the initial $x$-component of the velocity of the plank, $v_{pi,x0}$, for the second throw:

$$(m_2 + m_m + m_p)v_{pi,x0} = (m_2)v_{bi,x} + (m_m + m_p)v_{pi,x}$$
$$(20 + 40 + 60)(1.7) = (20)(-8.3) + (40 + 60)v_{pi,x}$$
$$v_{pi,x} = 3.7 \text{ m/s} \quad \Rightarrow \quad \vec{v}_{pi} = 3.7 \text{ m/s } \hat{\imath}$$

The final velocity of the plank and monkey is a little less, 3.7 m/s $\hat{\imath}$ compared to 4.0 m/s $\hat{\imath}$, when the two bowling balls are thrown one at a time rather than all at once.

**The rocket propulsion problem**: We have in mind a rocket that continually ejects gases. We make the following definitions:
- $m$ equals the mass of the rocket plus the unburned fuel.
- $\vec{v}$ is the instantaneous velocity of the rocket relative to an inertial reference frame.
- $\vec{u}_{gr}$ is the instantaneous velocity of the ejected gases (not all of the gases, but just that which is ejected at that instant) relative to the rocket.
- $\vec{v} - \vec{u}_{gr}$ is the instantaneous velocity of the ejected gases relative to the inertial reference frame (of course, the same inertial reference frame from which $\vec{v}$ is measured).

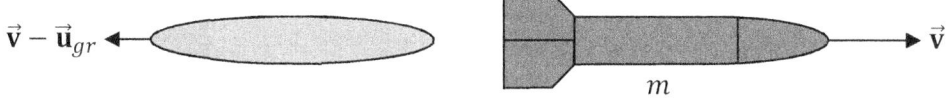

**Newton's second law for rocket propulsion**: Let us define the system to be the rocket plus the unburned fuel plus the ejected gases. In this case, Newton's second law is

## 6 Systems of Objects

$$\sum_{\substack{rocket \\ +unburned\ fuel \\ +ejected\ gases}} \vec{F}_{ext} = \frac{d\vec{p}_{tot}}{dt} = \lim_{\Delta t \to 0} \frac{\Delta \vec{p}_{tot}}{\Delta t} = \lim_{\Delta t \to 0} \frac{\vec{p}_{tot}(t + \Delta t) - \vec{p}_{tot}(t)}{\Delta t}$$

Note that only the gases instantaneously being ejected matter: Gas that has already been ejected has inertia, and so is no longer changing momentum; whereas gas that is just now being ejected is presently changing momentum.

The total momentum of the system is only conserved if the net external force acting on the system is zero. In general, we expect rockets to be in the presence of a significant gravitational field, so we do not expect the total momentum of the system to be conserved.

We write the initial total momentum as

$$\vec{p}_{tot}(t) = m(t)\vec{v}(t)$$

where $m(t)$ is the instantaneous mass of the rocket at time $t$ and $\vec{v}(t)$ is the instantaneous velocity of the rocket at time $t$ (relative to an inertial reference frame; we may choose any inertial reference frame, but we must use the same inertial reference frame throughout the problem).

At a slightly later time, $t + \Delta t$, the mass of the rocket is reduced to $m(t + \Delta t) = m(t) - \Delta m$, where $\Delta m$ is the mass of gas that is ejected over the short time interval $\Delta t$. The velocity of the rocket increases to $\vec{v}(t + \Delta t) = \vec{v}(t) + \Delta \vec{v}$ and the gas ejected over this short duration has momentum $\Delta m [\vec{v}(t) - \vec{u}_{gr}]$ relative to the inertial reference frame. The final total momentum is

$$\vec{p}_{tot}(t + \Delta t) = [m(t) - \Delta m][\vec{v}(t) + \Delta \vec{v}] + \Delta m [\vec{v}(t) - \vec{u}_{gr}]$$
$$\vec{p}_{tot}(t + \Delta t) = m(t)\vec{v}(t) + m(t)\Delta \vec{v} - \Delta m \vec{v}(t) - \Delta m \Delta \vec{v} + \Delta m \vec{v}(t) - \Delta m \vec{u}_{gr}$$
$$\vec{p}_{tot}(t + \Delta t) = m(t)\vec{v}(t) + m(t)\Delta \vec{v} - \Delta m \Delta \vec{v} - \Delta m \vec{u}_{gr}$$

Substituting these momentum expressions into Newton's second law,

$$\sum_{\substack{rocket \\ +unburned\ fuel \\ +ejected\ gases}} \vec{F}_{ext} = \lim_{\Delta t \to 0} \frac{m(t)\vec{v}(t) + m(t)\Delta \vec{v} - \Delta m \Delta \vec{v} - \Delta m \vec{u}_{gr} - m(t)\vec{v}(t)}{\Delta t}$$

$$\sum_{\substack{rocket \\ +unburned\ fuel \\ +ejected\ gases}} \vec{F}_{ext} = \lim_{\Delta t \to 0} \frac{m(t)\Delta \vec{v} - \Delta m \Delta \vec{v} - \Delta m \vec{u}_{gr}}{\Delta t}$$

$$\sum_{\substack{rocket \\ +unburned\ fuel \\ +ejected\ gases}} \vec{F}_{ext} = m\frac{d\vec{v}}{dt} + \vec{u}_{gr}\frac{dm}{dt}$$

to first order (i.e. neglecting terms of order $\Delta^2$), where the 'missing' minus sign was absorbed by the fact that $dm/dt$ is negative, whereas $\Delta m$ was defined as a positive mass – so $\lim_{\Delta t \to 0} \Delta m/\Delta t = -dm/dt$.

The net external force does not equal $m\vec{a}$, but includes a correction term because the mass of the rocket is changing. The preceding equation is the most general form of Newton's second law.

**A note on the variable-mass form of Newton's second law**: Suppose that we defined the system to include only the rocket and unburned fuel – i.e. we excluded the ejected gases, treating it as external to the system. Naïvely, we would want to state that the net external force acting on the rocket and unburned gases equals $d\vec{p}_{tot}/dt$, where $\vec{p}_{tot}$ would include the momentum of the rocket and unburned fuel only. Since $\vec{p}_{tot} = m\vec{v}$, we would be tempted to write $d\vec{p}_{tot} = md\vec{v}/dt + \vec{v}dm/dt$. It's important to realize that this intuitive strategy does not work: It does not agree with the equation that we just derived (there is a $\vec{v}$ in place of $\vec{u}_{gr}$).

The correct procedure is to apply Newton's second law to a system defined in such a way that the total mass of the system is constant. This is what we did previously to derive the correct result. Once you have done this, you can – if you want – separate the resulting expression among the subsystems to obtain Newton's second law for a subsystem for which the mass is not constant, but varying.

**Thrust**: The rocket's thrust is defined as

$$\vec{F}_{thrust} = -\vec{u}_{gr}\frac{dm}{dt}$$

In terms of the thrust, Newton's second law can be expressed as

$$\vec{F}_{thrust} + \sum_{\substack{rocket \\ +unburned\ fuel \\ +ejected\ gases}} \vec{F}_{ext} = m\frac{d\vec{v}}{dt}$$

The thrust is the rocket's reaction to ejecting the gases. It is important to remember that $dm/dt$ is negative since the rocket is losing mass. The net external force acting on the system is usually the net gravitational force acting on the system (including only the gases that are instantaneously ejected, and not those that have already been ejected). Thus, we see that the effect of the thrust is to cause the rocket to deviate from its natural free-fall orbit.

**1D rocket propulsion through a uniform gravitational field with a constant burn rate**: Consider a rocket that is launched directly away from the center of mass of a planet. In this case, the net external force acting on the system equals the weight of the rocket, $m\vec{g}$:

$$\vec{F}_{thrust} + m\vec{g} = m\frac{d\vec{v}}{dt} \quad \text{(1D free fall)}$$

Choosing $+y$ to be vertically upward, the thrust is upward, $\vec{F}_{thrust} = F_{thrust}\hat{j}$, and the weight is downward, $m\vec{g} = -mg\hat{j}$:

$$F_{thrust} - mg = m\frac{dv_y}{dt} \quad \text{(1D free fall)}$$

Recalling the definition of thrust, we can write this as

$$-u_{gr}\frac{dm}{dt} - mg = m\frac{dv_y}{dt} \quad \text{(1D free fall)}$$

Note that $-u_{gr}dm/dt > 0$ since $dm/dt < 0$. We denote the rate at which the rocket burns fuel by $R_b \equiv |dm/dt|$. This allows us to rewrite $dt$ in terms of $dm$:

$$u_{gr}R_b - mg = -R_b m\frac{dv_y}{dm} \quad \text{(1D free fall)}$$

If the change in altitude is not too large compared to the radius of the planet, the gravitational field will be approximately uniform. If fuel is also burned at a constant rate, then both $g$ and $R_b$ will both be constants, allowing us to separate variables and perform the integration:

$$\int_{m=m_0}^{m}\left(\frac{u_{gr}R_b}{m} - g\right)dm = -\int_{v_y=v_{y0}}^{v_y} R_b dv_y \quad \text{(1D free fall)}$$

$$u_{gr}R_b \ln\left(\frac{m}{m_0}\right) - g(m - m_0) = -R_b(v_y - v_{y0}) \quad \text{(1D free fall, } \vec{g} = \text{const.}, R_b = \text{const.)}$$

$$v_y = v_{y0} + \frac{g(m - m_0)}{R_b} + u_{gr}\ln\left(\frac{m_0}{m}\right) \quad \text{(1D free fall, } \vec{g} = \text{const.}, R_b = \text{const.)}$$

where $m_0$ is the initial mass of the rocket plus fuel and where we used the fact that $\ln(x) = -\ln(1/x)$. Since the burn rate was assumed to be constant, $dm = -R_b dt$ implies that $m - m_0 = -R_b t_b$, where $t_b$ is the burn time. In terms of the burn time,

$$v_y = v_{y0} - gt_b + u_{gr}\ln\left(\frac{m_0}{m}\right) \quad \text{(1D free fall, } \vec{g} = \text{const.}, R_b = \text{const.)}$$

The first two terms are the usual result for free-fall in a uniform gravitational field, and the third term is the contribution from burning fuel to eject gases.

**1D rocket propulsion through zero-gravity:** If the rocket is moving through a region of space where the net gravitational field is insignificant compared to the thrust, we may set $g \approx 0$ to obtain the following result:

$$-u_{gr}\frac{dm}{dt} = m\frac{dv_y}{dt} \quad (1D, \Sigma\vec{F}_{ext} = 0)$$

An Advanced Introduction to Calculus-Based Physics (Mechanics)

$$-\int_{m=m_0}^{m} u_{gr} \frac{dm}{m} = \int_{v_y=v_{y0}}^{v_y} dv_y \quad (1D, \Sigma\vec{F}_{ext} = 0)$$

$$-u_{gr} \ln\left(\frac{m}{m_0}\right) = (v_y - v_{y0}) \quad (1D, \Sigma\vec{F}_{ext} = 0)$$

$$v_y = v_{y0} + u_{gr} \ln\left(\frac{m_0}{m}\right) \quad (1D, \Sigma\vec{F}_{ext} = 0)$$

Note that in this case we have made no assumptions about the burn rate.

**Payload**: The mass of the rocket without any fuel, $m_p$, is called the payload. The payload is often stated as a percentage of the initial mass (including the fuel), $r_p$:

$$r_p = \frac{m_p}{m_0}$$

The initial mass can be expressed in terms of the payload and the total fuel, $m_{f0}$, as

$$m_0 = m_p + m_{f0}$$

The final mass equals the payload plus the remaining fuel, $m_f$:

$$m = m_p + m_f$$

The mass ratio in the equations of motion can then be expressed as

$$\frac{m_0}{m} = \frac{m_0}{m_p + m_f} = \frac{m_p + m_{f0}}{m_p + m_f}$$

When all of the fuel is spent, this becomes

$$\frac{m_0}{m} = \frac{m_0}{m_p} = \frac{1}{r_p} \quad (100\% \text{ of fuel spent})$$

In this way, knowing the payload as a percentage allows us to directly compute the final velocity of the rocket in the equations of motion.

**Example**. A monkey launches a homemade rocket straight upward from rest. The initial mass of the rocket is 80% water, which is ejected in the form of steam with an exhaust speed of 200 m/s relative to the rocket with a constant burn rate of 50 kg/s. Burnout occurs 10 s after launch. Determine the velocity and acceleration of the rocket when burnout occurs, assuming that $g \approx$ const.

Choosing $+y$ to point upward, the $y$-component of the velocity when burnout occurs is

$$v_y(t_{burnout}) = v_{y0} - gt_b + u_{gr} \ln\left(\frac{m_0}{m}\right) = 0 - (9.81)(10) + (200)\ln\left(\frac{1}{r_p}\right)$$

The payload is 20%, so $r_p = 1/5$. At burnout,

$$v_y(t_{burnout}) = -98 + (200)\ln(5) = 0.22\,\frac{\text{km}}{\text{s}} \quad \Rightarrow \quad \vec{v}(t_{burnout}) = 0.22\,\frac{\text{km}}{\text{s}}\hat{\jmath}$$

The thrust of the rocket is

$$F_{thrust} = -u_{gr}\frac{dm}{dt} = -(200)(-50) = 10 \text{ kN}$$

The acceleration of the rocket can be found from Newton's second law:

$$F_{thrust} - mg = ma_y$$
$$a_y = \frac{F_{thrust}}{m} - g$$

At burnout, the mass of the rocket equals the payload (as a mass, not as a percentage, of course):

$$a_y(t_{burnout}) = \frac{F_{thrust}}{m_p} - g$$

We must determine the mass of the payload before we can find the acceleration. Since the burn rate is constant, $m - m_0 = -R_b t_b$ and $m_p - m_0 = -R_b t_{burnout}$. The initial mass of the fuel was therefore $m_{f0} = m_0 - m_p = R_b t_{burnout} = 50(10) = 500$ kg. Since the payload is 20%,

$$r_p = \frac{m_p}{m_0} = \frac{m_0 - m_{f0}}{m_0} = \frac{1}{5}$$
$$5m_0 - 5m_{f0} = m_0$$
$$4m_0 = 5m_{f0}$$
$$m_0 = \frac{5m_{f0}}{4} = \frac{5(500)}{4} = 625 \text{ kg}$$

The payload is

$$m_p = m_0 - m_{f0} = 625 - 500 = 125 \text{ kg}$$

The acceleration of the rocket at burnout is thus found to be

$$a_y(t_{burnout}) = \frac{F_{thrust}}{m_p} - g = \frac{10{,}000}{125} - 9.81 = 70\,\frac{\text{m}}{\text{s}^2} \quad \Rightarrow \quad \vec{a}(t_{burnout}) = 70\,\frac{\text{m}}{\text{s}^2}\hat{\jmath}$$

> Just as the water is completely exhausted, the rocket is accelerating upward at approximately seven gravities.

## Conceptual Questions

1. The formulas for average velocity and average speed in physics are weighted averages. (A) What physical quantity serves as the weighting factor? (B) Give a conceptual example where an object travels with two or more different constant speeds during the problem, where the average speed does not equal the arithmetic mean of the speeds. Explain how the weighting factor identified in part (A) affects the average speed in this case.

2. How is it possible for the center of mass of an object to lie outside of the object? Think of a handful of simple shapes where the center of mass would like outside of the object. Draw each shape and label the position of approximate center of mass.

3. By analogy with center of mass and center of gravity, how would you define the center of electric charge and center of electric field for a system of $N$ point-charges, where an electric charge $q$ experiences a force $q\vec{\mathbf{E}}$ in the presence of an electric field $\vec{\mathbf{E}}$?

4. Which is more dense – an ice cube that you pull out of the freezer or a cup of water sitting on a table? As a result, what will happen when you place the ice cube in a cup of water? Is the relationship between the density of ice and the density of liquid water typical of most materials? Evidently, what happens to water as it freezes? Why is this a major problem when the temperature is below the freezing point of water on a cold winter night?

5. A monkey draws a circle in the $xy$ plane, centered about the origin, and then erases half of it – the half with negative $y$-values. The monkey wants to determine the area of the remaining semicircle by performing a double integral. For each case below, indicate the limits of each variable that the monkey needs to use: (A) The monkey wants to integrate over $x$ first and then $y$; (B) the monkey wants to integrate over $y$ first and then $x$; and (C) the monkey wants to integrate in 2D polar coordinates.

6. A monkey bakes a square cookie with a uniform distribution of mass. The monkey then cuts a circular cookie out of the square cookie. Given the coordinates of the circular hole relative to the center of the square, and the size of the hole, how can you determine the location of the center of mass of the remainder of the square cookie?

7. A monkey bakes a square cookie with a uniform distribution of mass. The monkey then cuts a large circular cookie out of the square cookie – so large that the diameter of the circular hole equals the edge length of the square. The circular hole is concentric with the square. (A) Determine the ratio of the mass of the circular cut out cookie to the mass of the original square cookie. (B) Determine the ratio of the mass of the circular cut-out cookie to the mass of the remainder of the square cookie.

8. Two monkeys riding golf cars are playing chicken: They are driving toward one another with the same speed and mass, but opposite directions. At the last moment, each monkey veers off the side of the cart path. One monkey crashes into a large tree. The other monkey crashes through several small trees before coming to rest. Use the concept of impulse to explain which collision is more likely to have more severe damage.

9. One box of bananas slides down a frictionless incline as another box of bananas slides up the incline. The two boxes of bananas collide on the incline. Is the total momentum of the system (defined as the two boxes of bananas) conserved for the collision? If not: If you resolve the momentum of each box of bananas into components, can one component of momentum be conserved for the collision? If so, in what direction must this component be oriented?

10. As a monkey slides down the frictionless wedge illustrated below, the wedge itself moves. Is the total momentum of the system (defined as the monkey plus the wedge) conserved for this collision? If not: If you resolve the momentum vector into components, can one component of momentum be conserved for the collision? If so, in what direction must this component be oriented? Which way does the wedge move as the monkey falls? Use the principle of conservation of momentum to explain this.

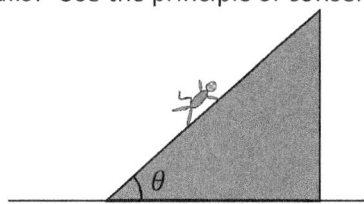

11. A banana falls toward the earth. As the banana falls, is the total momentum of the system conserved if we define the system as the banana? How about if we define the system as the banana plus the earth? Therefore, what happens to the earth as the banana falls? Explain first in terms of conservation of momentum, and then again in terms of one of Newton's laws. Why do we notice the motion of the banana, but not the motion of the earth?

12. An astronaut is repairing a spaceship in a free-fall orbit when the astronaut loses his tie rope. Fortunately, the astronaut is still holding his wrench. Explain how the principle of conservation of momentum can help the astronaut return to the spaceship. According to this principle, what should the astronaut do in order to return to the spaceship?

13. A monkey stands at the north end of a plank on horizontal frictionless ice. There is friction between the monkey and the plank. Use the principle of conservation of momentum to explain what happens when (A) the monkey walks south along the plank, (B) the monkey stops as he reaches the south end of the plank, and (C) the monkey leaps east off the south end of the plank. Also, describe what happens to the center of mass of the system (defined as the monkey plus the plank) in each case.

14. A monkey slides along horizontal frictionless ice. The monkey collides with a watermelon head-on. Explain whether momentum and/or mechanical energy are conserved for the system (defined as the monkey plus the watermelon) if (A) the monkey uses his hands to hold onto the watermelon and (B) if the monkey doesn't use his hands, but his head collides with the watermelon so that the watermelon bounces off of his head. Also, what type of collision occurs in each case? Finally, describe what happens to the center of mass of the system (defined as the monkey plus the watermelon) in each case.

15. A monkey, holding a watermelon, rides a plank with a velocity to the east on horizontal frictionless ice. Use the principle of conservation of momentum in 1D to explain what will happen to the monkey and plank if the monkey throws the watermelon directly to the (A) east and (B) west.

16. A monkey, holding a watermelon, rides a plank with a velocity to the east on horizontal frictionless ice. Use the principle of conservation of momentum in 2D to explain what will happen to the monkey and plank, approximately, if the monkey throws the watermelon directly to the north relative to the ice.

17. A monkey, holding a watermelon, rides a plank with a velocity to the east on horizontal frictionless ice. Use the principle of conservation of momentum in 2D to explain which way the monkey should throw the watermelon, approximately, if the monkey wants to change direction so that he will be heading due north.

18. A monkey's uncle heads north relative to the monkey. The monkey's aunt heads east relative to the monkey. Which way does the monkey's uncle head relative to the monkey's aunt? Which way does the monkey's aunt head relative to the monkey's uncle?

19. Imagine playing billiards without any effects of English, friction, or air resistance. Assume that the cue ball has the same mass as the other billiard balls.[276] Each collision described below is elastic. (A) A cue ball collides-head on with a billiard ball that was at rest. What happens to each ball? (B) Two billiard balls are at rest, touching one another. A cue ball collinear with these strikes one ball. What happens to each ball?

20. Is the total mechanical energy of the system (defined as bullet plus pendulum) conserved for the ballistic pendulum (from just before impact to maximum height)? If not, identify the energy transformation(s) by which mechanical energy is exchanged, and indicate whether the system gains or loses mechanical energy. Is the total momentum of the system conserved for the ballistic pendulum (from just before impact to maximum height)? If not, how could you redefine the system such that the total momentum of the system is conserved? Evidently, what must this other object do during the pendulum's upward swing? Why don't we notice this?

21. Consider Rutherford's scattering experiment. Rutherford observed that most of the alpha particles passed straight through the metal foil, virtually undeflected, while a very small percentage of alpha particles deflected, sometimes with very large scattering angles (even deflecting backward). What conclusions was Rutherford able to draw from this experiment?

22. When sunlight scatters off of particles in earth's atmosphere, the intensity of sunlight scattered at a scattering angle, $\theta_1$, depends upon the wavelength of light, $\lambda$, according to $I \propto 1/\lambda^4$, where $\propto$ means that the two sides are proportional to one another (rather than equal). This is known as Rayleigh scattering. Red light has longer wavelength, while violet light has shorter wavelength, and all of the colors are in order according to the acronym ROY G. BIV (red, orange, yellow, green, blue, indigo,[277] and violet). How does this formula explain why we see blue when we look in the sky? Also, use the formula to explain why the sun appears red at sunrise and sunset,[278] but yellow at noon.

23. A monkey puts his boat on a trailer and hitches it to his truck. While the monkey is heading down an incline, he puts the truck in neutral. The monkey's brother unhitches the trailer. Describe the motion of both the trailer (with boat) and truck precisely.

24. A monkey drives a dump truck filled with dirt. While the monkey is heading down an incline, he puts the truck in neutral. Describe the motion of the truck precisely as (A) dirt simply trickles out of the top of dump truck and (B) another monkey shovels dirt out of the dump truck, throwing the dirt backward.

---

[276] That's not quite the case with regulation billiard balls, where the cue ball is slightly heavier.

[277] Evidently, indigo is just too complicated and/or abstract, so the acronym is being shortened to ROY G. BV in K12 textbooks. Poor indigo – it's enough to make a monkey teary-eyed.

[278] Of course, if you value your retinas, you should not look directly at the sun (or even strongly reflected sunlight).

25. An astronaut is aboard a spaceship. When the astronaut faces the direction of the spaceship's velocity, he is looking fore, which is toward the bow. Directly opposite is aft, which is toward the stern. While facing fore, port is toward the astronaut's left, while starboard is toward the astronaut's right. In terms of the directions fore, aft, port, and starboard, use the principle of conservation of momentum to explain what the spaceship needs to do in order to (A) speed up (B) slow down (C) turn port ninety degrees (D) turn starboard ninety degrees.

26. If the payload is 25%, what fraction of the rocket's initial mass is fuel? What symbolic ratio expresses the percentage of the rocket's initial mass that is fuel?

27. For a rocket traveling upward in a uniform gravitational field that is downward, what must be the magnitude and direction of a rocket's thrust in order to (A) have an acceleration of $g$ upward, have an acceleration of $5g$ upward, (C) have a constant velocity, (D) have a deceleration with a magnitude of $2g$, and (E) have a deceleration with a magnitude of $g/2$. Express your answer in terms of the instantaneous weight of the rocket.

28. A rocket travels upward in a uniform gravitational field that is downward. The payload is 33%. The rocket has a constant thrust that is six times the initial weight of the rocket. Find the ratio of the initial acceleration of the rocket to the acceleration at burnout.

## Practice Problems

### Center of mass

1. A 30-kg monkey is 20-m away from a 90-kg monkey. A bunch of bananas lies at their center of mass. Where, exactly, are the bananas?

2. Find the location of the center of mass of the following system of three monkeys: a 100-kg monkey located at $(10\,\text{m}, 0)$; a 150-kg monkey located at $(-4.0\,\text{m}, 6.0\,\text{m})$; and a 200-kg monkey located at $(-2.0\,\text{m}, -8.0\,\text{m})$.

3. An 80-kg chimpanzee is at $(-4.0\,\text{m}, 2.0\,\text{m})$ and a 20-kg lemur is at $(12.0\,\text{m}, -3.0\,\text{m})$. In addition, there is an invisible 40-kg monkey somewhere. The center of mass of the whole system (i.e. the chimpanzee, lemur, and invisible monkey) is at $(-2.0\,\text{m}, 1.0\,\text{m})$. Where is the invisible monkey? That is, find his $x$- and $y$-coordinates.

4. The following freshly baked banana cookies have uniform density: A rectangular cookie of mass $4m$ has corners at $(0,0)$, $(2L, 0)$, $(0, L)$, and $(2L, L)$; a circular cookie of mass $6m$ has diameter $2L$ and center located at $(4L, 2L)$; and a long, thin cookie of mass $5m$ has endpoints $(-2L, 2L)$ and $(0, -3L)$. Find the center of mass of the system.

5. A chimpanzee and an orangutan have a center of mass that is twice as far from the chimpanzee as it is from the orangutan. Their combined mass is 150-kg. Determine the mass of each.

6. A 50-kg monkey stands at one end of a 10-m long plank. A fulcrum is placed underneath the plank, 3.0 m away from the monkey. The system is in static equilibrium — that is, the fulcrum is positioned beneath the center of mass of the system. What is the mass of the plank?

7. Hippy Longtalking wishes to lift a 20-m long, 200-kg platform with a 150-kg man standing 4.0 m from one end. If she lifts the platform in the center, it will tip over. Where should she lift the platform such that it won't tip over?

8. The top view of the single-story building below is filled with monkeys with approximately uniform density. Find the location of the center of mass of the monkeys.

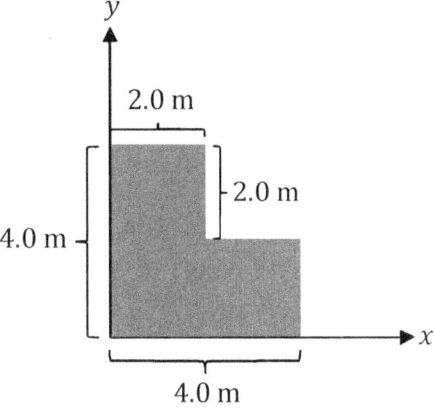

9. The banana-cream pie illustrated below has uniform density. Find the location of its center of mass.

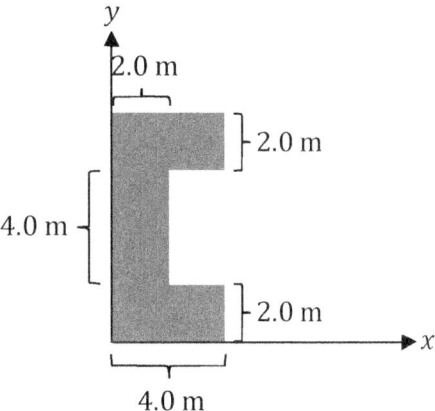

10. Consider the map of Monkey Island drawn below. A treasure chest of golden bananas is buried beneath the center of mass of Monkey Island. Monkey Island has uniform density. Monkey Island is the entire shaded region (rectangle and circles). Where is the treasure?

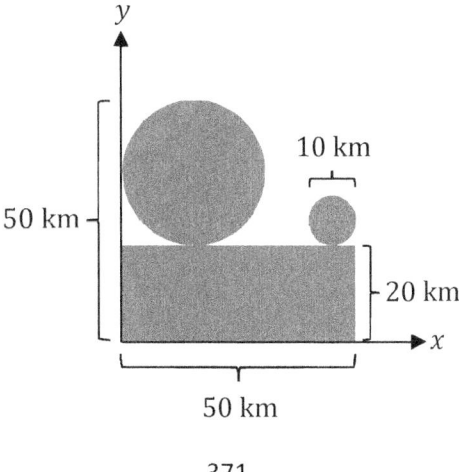

11. A monkey balances a 40.0-kg wooden mallet with uniform density on a fulcrum as illustrated below. The monkey then saws the mallet into two pieces at its center of gravity. How much does each end weigh? **Notes**: The handle and the head are not the two ends. The two ends are the two pieces of the mallet that remain after sawing – 18.0 cm of the handle being one end, and 2.0 cm of handle connected to the head being the other end. Also, this mallet is rectangular – it does not have a cylindrical handle or head, for example. (The answer is not 192 N. Why not?)

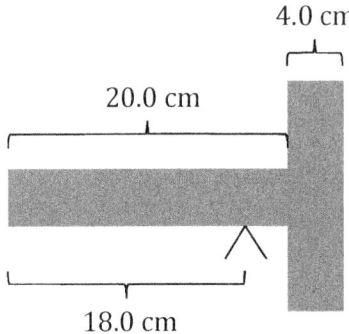

12. A banana-cream pie of radius $R_0$ and uniform mass density, $\sigma$, is centered about the origin. There is a circular hole of radius $R_0/2$ centered about the point $(R_0/2, 0)$. Find the center of mass of the banana-cream pie.

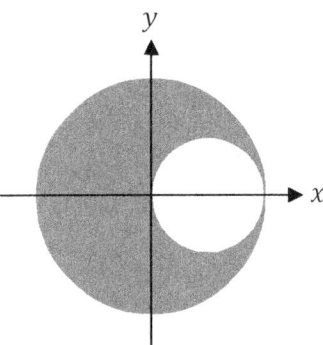

13. The shaded object illustrated below has uniform mass density, $\sigma$. If you stand in the center of mass of this object, one thousand monkeys will howl at you. Where is the center of mass?

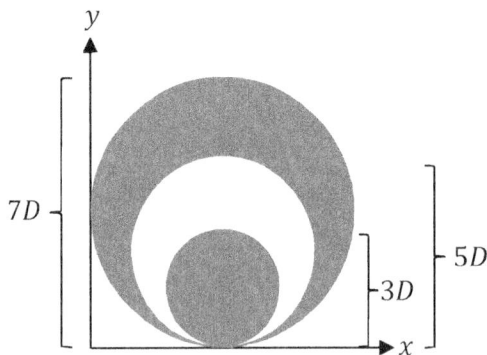

14. A monkeystick with endpoints $(0,0)$ and $(L,0)$ has non-uniform density $\lambda = ax^b$ where $a$ and $b$ are positive constants. Show that its center of mass lies at $x_{CM} = \frac{b+1}{b+2}L$.

15. Monkey Isle is the region bounded by the $y = x^2$ and the line $y = 4$. Monkey Isle has uniform density $\sigma = 2$. SI units have been suppressed. A monkey pirate buried a treasure in the center of mass of Monkey Isle. (A) Which of the three forms is appropriate for $dm$? (B) What are suitable integration limits? (C) Which of the integrals must be performed first? (D) Write a symbolic equation to find the mass of Monkey Isle. Make a series of clear substitutions to express your integrand in terms of $x$ and $y$ (with no other variables). Include the limits of your integrals. (E) Integrate over the first variable – determined by your answer to part (C). (F) Now integrate over the second variable. (G) Write a symbolic equation for $x_{CM}$. (H) Carry out steps (D) thru (F) to find $x_{CM}$. (I) Carry out steps (D) thru (F) to find $y_{CM}$.

16. Osamonk Been Rotten's evil lair is about to be destroyed in an explosion. Special agent $00\pi$ knows that a vault is located in the lair's center of mass – a good place to take cover. The lair is shaped like a solid semicircle with 50-m radius; the straight section runs along the $x$-axis and the rest extends in the positive $y$-direction, bisected by the $y$-axis. The lair has non-uniform mass density $\sigma(r) = \beta\sqrt{r}$, where $\beta$ is a constant. At what coordinates will special agent $00\pi$ find the vault?

17. The shaded semicircular ring illustrated below has inner radius $R_0$, outer radius $2R_0$, and non-uniform mass density $\sigma(r) = \beta r$, where $\beta$ is a constant. A monkey must balance this object on his fingertip in order to score well in this event. (A) Solve for $\beta$ in terms of $m$ and $R_0$. (B) Express the center of mass vector $\vec{r}_{CM}$ in terms of $R_0$ and Cartesian unit vectors.

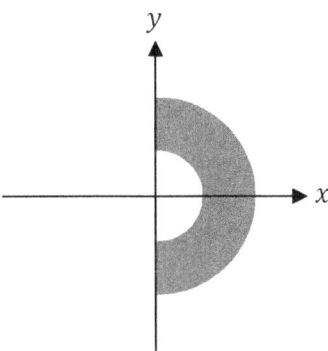

18. Find the center of mass of an infinitesimally thin banana of uniform mass per unit length, $\lambda$, in the shape of ¼ of a circle has endpoints $(R_0, 0)$ and $(0, R_0)$.

19. A monkey locates the center of mass of an equilateral triangle with edge length $L$, and then cuts the triangle into two pieces, as illustrated below. To the monkey's surprise, the two pieces have different mass. (A) Determine the coordinates of the triangle's center of mass, setting up a coordinate system with the origin at the bottom left corner with the $x$-axis horizontal. (B) Determine the ratio of the mass of the bottom piece to the mass of the top piece.

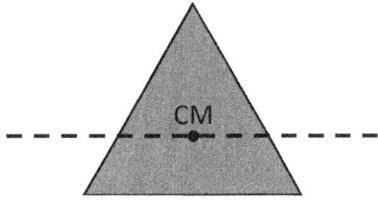

# 6 Systems of Objects

## Impulse, momentum, conservation of momentum, and collisions

20. A monkey drops a 50-g rubber-coated coconut from a height of 10.0 m. The coconut is in contact with the ground for 30 ms. The maximum height of the rebound is 8.0 m. (A) What is the momentum of the coconut just before striking the ground? (B) What is the momentum of the coconut just after striking the ground? (C) What is the average force exerted on the coconut during impact? (D) How much energy does the coconut lose during the collision?

21. A 50-kg monkey stands at the south end of a 150-kg plank. The 20-m long plank is on ice. There is friction between the monkey and plank, but not between the plank and ice. The monkey and plank are initially at rest. The monkey walks to the north at a constant rate of 2.0 m/s relative to the ice, continuing with this velocity as he walks off of the plank. (A) As the monkey walks along the plank, list the forces acting on the system. Which are internal, and which are external? What is the net external force? Is momentum conserved for the system? (B) What is the final velocity (include direction) of the plank? (C) How much kinetic energy has the monkey-plank system lost or gained (compared to the initial kinetic energy of the system)?

22. It's a jungle out there when monkeys drive bananamobiles on horizontal frictionless ice. (A) A 50-kg bananamobile traveling 8.0 m/s to the east collides head-on with a 100-kg bananamobile traveling 12.0 m/s to the west. The collision is completely inelastic. What is the final velocity of the bananamobiles? (B) A 600-kg bananamobile traveling 3.0 m/s to the east collides head-on with a 150-kg bananamobile traveling 9.0 m/s to the north. The collision is completely inelastic. What is the final velocity of the bananamobiles?

23. A 600-kg bananamobile traveling 90 m/s to the north collides head-on with a 400-kg bananamobile traveling 50 m/s to the south. The collision is elastic. Find the final velocities of the bananamobiles.

24. This sumo wrestling match is held on horizontal frictionless ice. A 20-kg sumo wrestling orangutan traveling 3.0 m/s to the north collides head-on with a 10-kg sumo wrestling orangutan traveling 8.0 m/s to the south. The orangutans bounce off each other after the collision, and the 20-kg sumo wrestling orangutan travels 4.0 m/s to the south after the collision. (A) Find the final speed of the 10-kg sumo wrestling orangutan. (B) Which way is the 10-kg orangutan moving after the collision? (C) Is the collision elastic? (D) If the sumo wrestling orangutans are in contact with each other for 200 ms, what is the average force they exert on one another during the collision?

25. This sumo wrestling match is held on horizontal frictionless ice. A 20-kg sumo wrestling orangutan traveling 3.0 m/s to the north collides head-on in an elastic collision with a 10-kg sumo wrestling orangutan traveling 9.0 m/s to the south. (A) Find the final velocity (including direction) of each sumo wrestling orangutan. (B) What percentage of the kinetic energy does the system lose if the collision is instead completely inelastic?

26. This sumo wrestling match is held on horizontal, frictionless ice. Consider the illustration on the next page. The orangutan on the left with mass $2m$ is moving toward the other orangutans with speed $15v_0$. The other orangutans are initially at rest. Assume all collisions to be head-on, elastic collisions. (A) Express the final velocities of all three orangutans in terms of $v_0$. Include direction. (B) How many collisions are there? Explain. (C) Suppose instead that the collisions are all completely inelastic. What percentage of the initial kinetic energy is lost after all of the collisions?

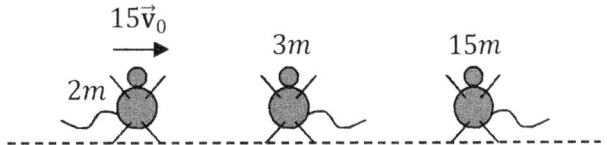

27. This sumo wrestling match is held on horizontal, frictionless ice. One orangutan with mass $4m$ is moving with speed $3v_0$. A second orangutan with mass $3m$ and is moving with speed $2v_0$. The orangutans are moving in the directions shown prior to the collision. They collide in a completely inelastic collision. Determine the magnitude and direction of the final velocity of each orangutan.

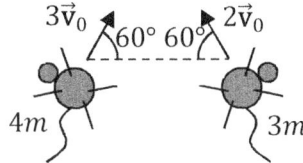

28. You find some monkeys playing marbles with coconuts. Upon closer inspection, you realize that the coconuts are painted like eyeballs. A coconut of mass $4m$ traveling $3v_0$ to the north collides head-on with a coconut of mass $8m$ traveling $2v_0$ to the south. The collision is elastic. Find the final velocities (include direction).

29. A 48.0-kg monkey is at rest on horizontal frictionless ice. Fortunately, he is holding a 12.0-kg watermelon. (A) If he throws the watermelon 8.0 m/s to the south, what will be the velocity (i.e. speed and direction) of the monkey? (B) Now that the monkey is moving, he decides that he would prefer to be at rest. The monkey has two 2.0-kg shoes that he can part with. How fast, and in what direction, must the monkey throw his shoes, relative to the ice, in order to come to a complete stop? (C) If the watermelon was in contact with the monkey's hand for 400 ms as he was throwing it in part (A), what was the average force of the monkey's throw?

30. A 20-kg monkey is at rest on horizontal frictionless ice. Fortunately, he is holding a 15-kg watermelon. (A) If he throws the watermelon 4.0 m/s to the south, what will be the velocity (i.e. speed and direction) of the monkey? (B) The monkey continues to move with the final velocity computed in (A) when he collides head-on with a 10-kg monkey traveling 9.0 m/s to the south. If the collision is elastic, what are the final velocities of each monkey?

31. A 40-kg monkey holding a 10-kg watermelon is moving 4.0 m/s to the east on horizontal frictionless ice. The monkey throws the watermelon 8.0 m/s to the north relative to the ice. Find the magnitude and direction of the monkey's final velocity.

32. A 30-kg monkey holding a 15-kg watermelon is moving 2.0 m/s to the northeast on horizontal frictionless ice. The monkey throws the watermelon 10.0 m/s relative to the ice. Which way should the monkey throw the watermelon so that he will be traveling directly to the east after throwing it?

33. Three monkeys are running toward each other. Viewed from above, they have the following velocities in the horizontal $xy$ plane, where the angles are measured counterclockwise from the $+x$-axis: A 40-kg monkey is running with velocity 5.0 m/s at an angle of $0°$, a 60-kg monkey is running with velocity 3.0 m/s at an angle of $40°$, and an 80-kg monkey is running with velocity 2.0 m/s at an angle of $200°$. The three monkeys collide at the same moment in a completely inelastic collision. (A) Find the final velocity of the trio. (B) How much energy is lost/gained during the collision? (C) What impulse does the 40-kg monkey experience during the collision?

34. A male ape of mass $m$ in distress stands on an island in a crocodile-infested river. A female ape of mass $2m$ swings down on a vine, catches the male ape of mass $m$ at the bottom of the arc, and just barely reaches the other side with the male ape in her arms. The initial angle is $\theta_0$, and the final angle is $\theta$. (A) Derive an equation for $\theta$. (B) What is $\theta$ if $\theta_0 = \frac{\pi}{2}$ rad? (C) What is $\theta_0$ if $\theta = \frac{\pi}{6}$ rad?

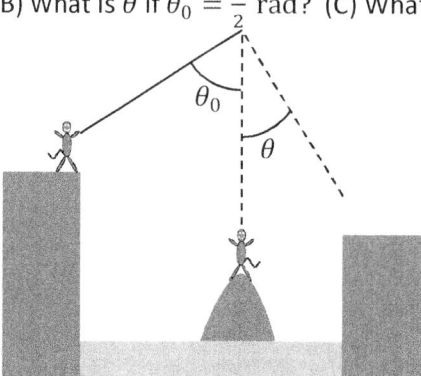

35. A 30-kg monkey, on the left, and 20-kg monkey, on the right, swing from 10-m long ropes as illustrated below. Each monkey begins from rest from the position shown. (The 30-kg monkey starts a little earlier in order to time their collision to occur at the bottom of their arcs.) The collision is completely inelastic. How high do the monkeys rise after the collision, and which way do they swing?

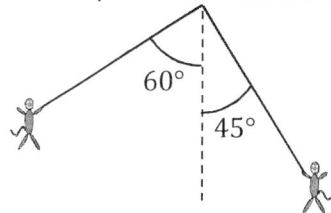

36. A 3.0-kg monkeyball heads 12.0 m/s along the $+x$-axis toward a 5.0-kg monkeyball that lies at rest at the origin. The collision is head-on and completely inelastic. (A) Express the initial velocities of each monkeyball in the lab frame. (B) Express the initial velocities of each monkeyball in the CM frame. (C) Determine the final velocities of each monkeyball in the lab frame. (D) Express the final velocities of each monkeyball in the CM frame.

37. A 4.0-kg monkeyball heads 8.0 m/s along the $+x$-axis toward a 6.0-kg monkeyball that lies at rest at the origin. Each monkeyball has a radius of 20 cm. The impact parameter is 10 cm. The collision is elastic. (A) Express the initial velocities of each monkeyball in the lab frame. (B) Express the initial velocities of each monkeyball in the CM frame. (C) Determine the final velocities of each monkeyball in the lab frame. (D) Express the final velocities of each monkeyball in the CM frame.

38. The Compton effect is observed when an x-ray photon is incident upon an approximately stationary electron. A photon (a particle of light) has zero rest-mass, yet carries momentum $p_\gamma = h/\lambda$ and energy $E_\gamma = hf$, where $h = 6.626 \times 10^{-34}$ J·s is Planck's constant, $\lambda$ is the wavelength of the photon, and $f$ is the frequency of the photon. The wavelength and frequency are related through $\lambda f = c$, where $c = 2.9979 \times 10^8$ m/s is the speed of light in vacuum. In a relativistic collision, the energy and momentum of an electron are related through $E_e^2 = p_e^2 c^2 + m_e^2 c^4$. Apply conservation of momentum and energy for this elastic collision to derive the equation for the Compton shift – i.e. the change in wavelength of the x-ray photon due to its scattering off of the electron target – in the lab frame:

$$\lambda - \lambda_0 = \frac{h}{m_e c}(1 - \cos\theta)$$

where $\theta$ is the scattering angle for the x-ray photon.

**Rocket propulsion**

39. A 300-kg gorilla standing at rest on horizontal frictionless ice holds a 100-kg banana-launcher (mass when empty) filled with 50 bananas, where each banana has a mass of 2.0 kg. The banana-launcher launches each banana horizontally with a speed of 60 m/s relative to the banana-launcher. Determine the final velocity of the gorilla after launching all 50 bananas, and compare with what the gorilla's final velocity would have been had all 50 bananas been launched simultaneously with a speed of 60 m/s relative to the banana-launcher.

40. A 2,000-kg rocket (not including the fuel) in deep space – i.e. where gravity is negligible compared to the thrust of the rocket – has a payload of 25% and an initial speed of 200 m/s relative to the nearest (yet quite distant) star. Fuel is ejected with an exhaust speed of 1500 m/s at a constant burn rate of 50 kg/s in order to accelerate the rocket to a final speed of 600 m/s. (A) What percentage of the mass of the fuel is spent to accomplish this? (B) How much time does this take?

41. A 3,000-kg rocket (not including the fuel) has 6,000 kg of fuel and an initial speed of 100 m/s as it accelerates upward away from earth. With a constant burn rate, the rocket accelerates to 500 m/s in 12 s. The exhaust speed is 2000 m/s. Assume that the rocket's initial altitude is low enough that any variation in gravitational acceleration from 9.81 m/s$^2$ may be neglected during this time interval. (A) What percentage of the mass of the fuel is spent to accomplish this? (B) What is the burn rate? (C) What is the thrust? (D) What is the initial acceleration? (E) What is the final acceleration?

# 7 Rotation

**General circular motion**: We considered a special case of circular motion in Chapter 4 – uniform circular motion, where an object traveled in a circle with constant speed. In this chapter, we will treat the more general case, where an object may gain or lose speed while traveling in a circle. Much of the content from Chapter 4 will be relevant to Chapter 7. Therefore, the reader may find it beneficial to review the material from Chapter 4 before proceeding. Some of the equations will still be applicable in the same form – e.g. the equation for centripetal acceleration. However, some equations and strategies from Chapter 4 do not apply to more general circular motion as they are, but need to be generalized. For example, the relationship between arc length and time must be generalized to account for angular acceleration. Of course, you won't have to guess: In physics, everything can be developed and explained in a logical fashion (but, of course, in physics, the student must also be willing to devote some time toward contemplating the concepts and working through the ideas for him- or herself before the content appears fully complete and logical).

**Rotation**: In addition to treating the general motion of objects traveling in circles, we will also allow objects to spin about some axis. When an object spins about an axis, we call this rotation, which is not quite the same as an object traveling in a circle. For example, when we speak of the earth's 'rotation,' we specifically mean how it spins about its axis, but when we want to describe the earth's orbit about the sun, we instead use the word 'revolution.' When an object rotates about an axis, different parts of the object travel with different speeds. For example, passengers on a merry-go-round travel faster if they are seated further away from the axis of rotation. Although their speeds differ, they all have the same angular speed. Rotation is somewhat more involved than an object traveling in a circular path. When we describe the rotation of an object, we will concern ourselves specifically with a rigid body – i.e. one that doesn't change shape or fly apart as it rotates.

**Analogies with linear motion**: Almost every concept and equation in rotation has an analogy with linear motion. If you understand the material from Chapters 1 thru 6 well, you should find it easy to master many of the concepts and equations from Chapter 7 by analogy.[279] At the end of Chapter 7, you should basically know two sets of equations – one set for linear motion and a very similar set for rotation. Instead of memorizing two sets of equations, though, you should just know one set of equations and understand the analogies between linear motion and rotation. It is well worth studying these analogies, since mastering the analogies will make it much easier to master the subject of rotation.

**Basic circular motion quantities**: In the following table, we summarize several basic circular motion quantities from Chapter 4, along with equations that apply without modification to general circular motion in Chapter 7. You may find it helpful to review this table before proceeding.

---

[279] However, moment of inertia (Sec. 7.2) and the vector product (Sec. 7.3) will seem very different.

| Basic Circular Motion Quantities ($R_0 = const.$) | |
|---|---|
| angular displacement, $\Delta\theta$ | $\Delta\theta = \theta - \theta_0$ |
| angular velocity, $\vec{\boldsymbol{\omega}}$ | $\|\vec{\boldsymbol{\omega}}\| = |d\theta/dt|$ |
| angular speed, $\omega$ | $\omega = \|\vec{\boldsymbol{\omega}}\|$ |
| angular acceleration, $\vec{\boldsymbol{\alpha}}$ | $\|\vec{\boldsymbol{\alpha}}\| = |d\omega/dt| = |d^2\theta/dt^2|$ |
| net tangential displacement, $\Delta s_T$ | $\Delta s_T = s_T - s_{T0} = R_0 \Delta\theta$ |
| arc length, $s$ | $s = |s_T|$ |
| tangential velocity, $v_T$ | $v_T = ds_T/dt$ , $|v_T| = R_0 \omega$ |
| tangential speed, $v$ | $v = |v_T|$ |
| tangential acceleration, $a_T$ | $a_T = dv_T/dt$ , $|a_T| = R_0 \alpha$ |
| centripetal acceleration, $a_c$ | $a_c = v_T^2/R_0$ |
| position vector, $\vec{\mathbf{r}}$ | $\vec{\mathbf{r}} = R_0 \hat{\mathbf{r}}$ |
| velocity, $\vec{\mathbf{v}}$ | $\vec{\mathbf{v}} = d\vec{\mathbf{r}}/dt = v_T \hat{\boldsymbol{\theta}}$ |
| acceleration, $\vec{\mathbf{a}}$ | $\vec{\mathbf{a}} = d\vec{\mathbf{v}}/dt = a_T \hat{\boldsymbol{\theta}} - a_c \hat{\mathbf{r}}$ <br> $a = \|\vec{\mathbf{a}}\| = \sqrt{a_T^2 + a_c^2}$ |

- The angular motion variables are $\Delta\theta$, $\vec{\boldsymbol{\omega}}$, and $\vec{\boldsymbol{\alpha}}$.
- The tangential motion variables are $\Delta s_T$, $v_T$, and $a_T$.
- The angular variables are related to the corresponding tangential variables by a factor of $R_0$:
  $$\Delta s_T = R_0 \Delta\theta \quad , \quad v_T = R_0 \omega \quad , \quad a_T = R_0 \alpha \quad \text{(circular motion)}$$
- The vector motion variables are $\vec{\mathbf{r}}$, $\vec{\mathbf{v}}$, and $\vec{\mathbf{a}}$.
- In circular motion, $\vec{\mathbf{r}}$ points outward, $\vec{\mathbf{v}}$ is tangential, and $\vec{\mathbf{a}}$ has both inward (centripetal) and tangential components, in general.
- There are four types of acceleration relevant to circular motion:
  - The angular acceleration, $\vec{\boldsymbol{\alpha}}$, describes how the angular velocity, $\vec{\boldsymbol{\omega}}$, changes in time.
  - The tangential acceleration, $a_T$, describes how the tangential velocity, $v_T$, changes in time. Up to a possible sign difference, the tangential acceleration describes how the speed, $v$, changes in time.
  - The centripetal acceleration, $a_c$, describes how the direction of the velocity changes in time.
  - The acceleration, $\vec{\mathbf{a}}$, describes how the velocity, $\vec{\mathbf{v}}$, changes in time, including how both the speed and direction change. In general, the acceleration, $\vec{\mathbf{a}}$, has both tangential and centripetal components, $a_T$ and $a_c$. (If a circular motion problem asks for acceleration, and does not specify angular, tangential, or centripetal, then the problem is referring to the acceleration vector.)
- The tangential variables $\Delta s_T$, $v_T$, and $a_T$ can be positive or negative, where the positive direction is counterclockwise (when the circle lies in the $xy$ plane and is viewed from the $+z$-axis). The arc length, $s$, and speed, $v$, are strictly nonnegative: $s = |s_T|$ and $v = |v_T|$.

7 Rotation

## 7.1 Angular Acceleration

**Circular motion coordinates**: For circular motion, it is convenient to setup a coordinate system with the circle lying in the $xy$ plane, centered about the origin. The positive sense of direction is based on the convention of measuring $\theta$ counterclockwise from the $+x$-axis, as viewed from the $+z$-axis. We adhere to this convention for circular motion throughout this text.

**Signs of the tangential motion variables**: The tangential variables $\Delta s_T$, $v_T$, and $a_T$ can be positive or negative (or zero), as described below:
- The net tangential displacement, $\Delta s_T$, has the same sign convention as $\Delta\theta$: It is positive if the object has a net counterclockwise displacement, and is negative if it has a net clockwise displacement.
- The tangential velocity, $v_T$, is positive if the object is instantaneously heading counterclockwise, and is negative if it is heading clockwise.
- The tangential acceleration, $a_T$, is positive if $v_T$ is positive and the object is speeding up or if $v_T$ is negative and the object is slowing down, and $a_T$ is negative if $v_T$ is positive and the object is slowing down or if $v_T$ is negative and the object is speeding up.

**Direction of the angular velocity**: The angular velocity, $\vec{\omega}$, has a direction that is perpendicular to the plane of rotation. With our usual choice of coordinates, $\vec{\omega}$ only has a z-component: $\vec{\omega}$ is along $+\hat{\mathbf{k}}$ if the object is instantaneously heading counterclockwise, and is along $-\hat{\mathbf{k}}$ if it is heading clockwise. That is,

$$\vec{\omega} = \omega_z \hat{\mathbf{k}} \quad \text{(circular motion)}$$

where $\omega_z$ has the same sign as $v_T$. The tangential velocity is related to $\omega_z$ by

$$v_T = R_0 \omega_z \quad \text{(circular motion)}$$

**Direction of the angular acceleration**: The angular acceleration, $\vec{\alpha}$, equals a derivative of the angular velocity, $\vec{\omega}$, with respect to time:

$$\vec{\alpha} = \frac{d\vec{\omega}}{dt}$$

In circular motion, with our usual choice of coordinates,

$$\vec{\alpha} = \alpha_z \hat{\mathbf{k}} \quad \text{(circular motion)}$$

where $\alpha_z$ has the same sign as $a_T$:

$$a_T = R_0 \alpha_z \quad \text{(circular motion)}$$

The angular acceleration, $\vec{\alpha}$, is along $+\hat{k}$ if $\vec{\omega}$ is along $+\hat{k}$ and the object is speeding up or if $\vec{\omega}$ is along $-\hat{k}$ and the object is slowing down, and $\vec{\alpha}$ is along $-\hat{k}$ if $\vec{\omega}$ is along $+\hat{k}$ and the object is slowing down or if $\vec{\omega}$ is along $-\hat{k}$ and the object is speeding up.

**Circular motion**: Since an object traveling along a circle has a single degree of freedom – i.e. it can only move clockwise or counterclockwise around the circle – circular motion is 1D motion. All of the content from Chapter 1, on 1D motion, applies to an object traveling in a circle. So if you want to find the average speed for an object traveling in a circle, if you have a graph of angular velocity as a function of time and want to find the angular acceleration, or if you have one object pursuing another around a circle, for example, you can directly apply the techniques from Chapter 1.

However, there is a difference between general circular motion and the 1D motion of Chapter 1: In general circular motion, there are more motion quantities to work with than we had considered in Chapter 1. For example, we have both angular and tangential variables. Also, we have vector motion quantities in the context of circular motion, which we did not consider until Chapter 2.

---

**Note**. You may find it useful to apply techniques from Chapter 1, on 1D motion, to solve problems relating to general circular motion.

---

**Net displacement in circular motion**: If you want to find the net displacement, $\Delta \vec{r}$, for object traveling in a circle, first determine the $(x, y)$ coordinates of the initial and final positions. The Cartesian coordinates are related to the polar coordinates by

$$x_0 = R_0 \cos \theta_0 \quad , \quad y_0 = R_0 \sin \theta_0 \quad , \quad x = R_0 \cos \theta \quad , \quad y_0 = R_0 \sin \theta \quad \text{(circular motion)}$$

The net displacement can then be found from

$$\Delta \vec{r} = (x - x_0)\hat{i} + (y - y_0)\hat{j} \quad \text{(2D motion)}$$

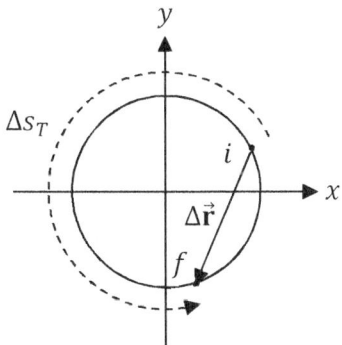

**Total distance traveled in circular motion**: The total distance traveled is $\Delta s = \sum_{i=1}^{N} |\Delta s_{T,i}|$, as described in Sec. 1.2. Note that $\Delta s \geq |\Delta s_T|$, even though $s = |s_T|$, since reversing direction always increases the arc length $s$, but not $s_T$. Also, $\|\Delta \vec{r}\| \leq |\Delta s_T|$ (see the above figure, e.g.).

7 Rotation

**Average angular velocity**: The average angular velocity, $\vec{\bar{\omega}}$, is analogous to the definition of average velocity. Recall that the average velocity, $\vec{\bar{v}}$, equals the net displacement divided by the elapsed time. Therefore, the average angular velocity equals the net angular displacement divided by the elapsed time times a unit vector that also provides its direction correctly:

$$\vec{\bar{\omega}} = \frac{\Delta\theta}{\Delta t}\hat{\mathbf{k}} \quad \text{(circular motion)}$$

**Average angular speed**: The average angular speed, $\bar{\omega}$, is analogous to the definition of average speed. Recall that the average speed, $\bar{v}$, equals the total distance traveled divided by the elapsed time, and differs from the magnitude of the average velocity if the object changes direction. The average angular speed can be found from

$$\bar{\omega} = \frac{\sum_{i=1}^{N}|\Delta\theta_i|}{\Delta t} = \frac{\Delta s}{R_0 \Delta t} = \frac{\bar{v}}{R_0} \quad \text{(circular motion)}$$

where $\Delta s$ is the total distance traveled (not to be confused with the net tangential displacement): $\Delta s \neq |\Delta s_T|$ if the object reverses direction one or more times.

**Average angular acceleration**: The average angular acceleration, $\vec{\bar{\alpha}}$, is analogous to the definition of average acceleration, $\vec{\bar{a}}$:

$$\vec{\bar{\alpha}} = \frac{\Delta\vec{\omega}}{\Delta t} = \frac{\Delta\omega_z}{\Delta t}\hat{\mathbf{k}} \quad \text{(circular motion)}$$

where $\Delta\vec{\omega} = \vec{\omega} - \vec{\omega}_0$ is the change in angular velocity of the object.

**Uniform angular velocity**: Notice that an object traveling in a circle with uniform angular velocity is the same as uniform circular motion, which we treated extensively in Chapter 4.

**Motion graphs for circular motion**: Motion graphs for circular motion can be analyzed using the techniques of Sec. 1.5. However, there are more motion quantities to work with in the case of circular motion – e.g. there are both tangential and angular variables, which can be related through the radius of the circle: $\Delta s_T = R_0 \Delta\theta$, $v_T = R_0 \omega_z$, and $a_T = R_0 \alpha_z$.

**Uniform angular acceleration**: An object that travels in a circle with constant angular acceleration travels with uniform angular acceleration, which can also be considered uniform tangential acceleration, since $a_{T0} = R_0 \alpha_{z0}$. [However, if a rigid extended object (or a rigid system of connected objects) is rotating, it would only make sense to describe the motion of the entire system as uniform angular acceleration – since the angular acceleration is the same for each part of the object (or system); but uniform tangential acceleration varies with the distance from the axis of rotation, and so is not the same throughout.] The strategy for solving uniform angular acceleration problems is just like that of 1DUA (it might help to review Sec. 1.6 before proceeding), but there are more equations to work with:

$$\omega_z - \omega_{z0} = \alpha_{z0}\Delta t \qquad v_T - v_{T0} = a_{T0}\Delta t$$

$$\Delta\theta = \omega_{z0}\Delta t + \frac{\alpha_{z0}\Delta t^2}{2} \qquad \Delta s_T = v_{T0}\Delta t + \frac{a_{T0}\Delta t^2}{2} \quad \text{(uniform angular acceleration)}$$

$$\omega_z^2 - \omega_{z0}^2 = 2\alpha_{z0}\Delta\theta \qquad v_T^2 - v_{T0}^2 = 2a_{T0}\Delta s_T$$

$$\Delta s_T = R_0\Delta\theta \;,\quad v_{T0} = R_0\omega_{z0} \;,\quad v_T = R_0\omega_z \;,\quad a_{T0} = R_0\alpha_{z0} \quad \text{(circular motion)}$$

When applying the equations of uniform angular acceleration to a problem where an object travels in a circle with constant angular acceleration, it's important to make sure that you work with $N$ independent equations in $N$ unknowns. Only two of the three purely angular equations are independent, and similarly for the purely tangential variables. If you're working exclusively with angular or tangential variables, you should therefore expect to know 3 of the 5 symbols (i.e. $\Delta\theta$, $\omega_{z0}$, $\omega_z$, $\alpha_{z0}$, and $\Delta t$ or $\Delta s_T$, $v_{T0}$, $v_T$, $a_{T0}$, and $\Delta t$). However, if you're mixing and matching (not necessarily by choice – the problem might give you the initial angular speed in conjunction with the net tangential displacement, for example), you should expect to know a fourth independent symbol (note, for example, that knowing that $\omega_{z0}$ equals zero is not independent of knowing that $v_{T0}$ is zero, but if they are both nonzero, knowing both allows you to find $R_0$).

### Problem-Solving Strategy for Uniform Angular Acceleration

0. This strategy only applies to problems where the angular acceleration is constant. If the angular acceleration is specified as a function of time, or if an object rolls down a curved hill, for example, then the angular acceleration is not constant and this strategy is not applicable.
1. Draw and label a diagram. Include the initial ($i$) and final ($f$) positions on your diagram. Also, label the positive sense of direction – clockwise or counterclockwise. The convention is to choose counterclockwise to be the positive direction: The rules for determining the signs are based on this convention, so if you decide to be a rebel and choose clockwise to be the positive direction, you need to think about how this affects the sign convention.
2. List your knowns and identify the desired unknown(s). If you can solve the problem with purely angular or purely tangential variables, you should expect to know 3 of the 5 symbols (i.e. $\Delta\theta$, $\omega_{z0}$, $\omega_z$, $\alpha_{z0}$, and $\Delta t$ or $\Delta s_T$, $v_{T0}$, $v_T$, $a_{T0}$, and $\Delta t$). If you need to mix-'n'-match to solve the problem, you should generally expect to know 4 independent symbols.
3. Look at the equations of uniform angular acceleration, uniform tangential acceleration, and the connecting equations (i.e. $\Delta s_T = R_0\Delta\theta$, $v_{T0} = R_0\omega_{z0}$, $v_T = R_0\omega_z$, and $a_{T0} = R_0\alpha_{z0}$). Sometimes you can solve for the unknown with a single equation, but other times it will be necessary to setup a system of $N$ independent equations in $N$ independent unknowns.
4. Some problems give you information that you can express mathematically – like "when the final angular speed equals twice the initial angular speed." In such a case, you will need to use this equation to establish $N$ equations in $N$ unknowns.

**Hint**. If you want to determine the net number of revolutions completed, solve for $\Delta\theta$ and convert it from radians to revolutions. (Alternatively, you can solve for $\Delta s_T$ and divide by the circumference.)

**Hint**. Look at the units to help decipher which quantities are given in the problems. For example, 4.5 rev/s can only be an angular speed (also note that the units are not SI) because none of the other symbols have these same units. Similarly, 3.2 m/s² can only be acceleration (it could be the tangential acceleration, but centripetal acceleration and the magnitude of the acceleration also share these same units – it can't be angular acceleration, though). If you're not sure which symbol a given number represents, at least make sure that you choose a symbol that has the same units.

**Math Note**. You must use radians when using the equations $\Delta s_T = R_0 \Delta\theta$, $v_{T0} = R_0 \omega_{z0}$, $v_T = R_0 \omega_z$, or $a_{T0} = R_0 \alpha_{z0}$. These equations do not work in revolutions or degrees. However, if you can solve a problem exclusively by using the angular variables, you may work with revolutions instead of radians.

**Important Distinction**. Note that $\omega_{z0}$ and $\omega_z$ are the z-components of the initial and final angular velocity, $\vec{\omega}_0$ and $\vec{\omega}$, respectively, and may be positive or negative because the object may travel clockwise or counterclockwise. The similar symbols, $\omega_0$ and $\omega$, represent the initial and final angular speed, respectively – i.e. $\omega_0 = \|\vec{\omega}_0\|$ and $\omega = \|\vec{\omega}\|$ – and are strictly nonnegative.

**Example**. A monkey runs around a 50-m diameter circle with an initial angular speed of 0.50 rev/min. The monkey completes two laps in 2.0 minutes. Determine the monkey's angular acceleration, tangential acceleration, initial centripetal acceleration, and the magnitude of the initial acceleration.

Begin with a labeled diagram, including the positive sense of direction.

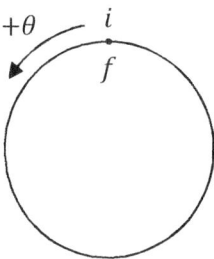

Tabulate the knowns and identify the desired unknown:

$$D = 50 \text{ m} \quad , \quad \omega_{z0} = 0.50 \text{ rev/min.} \quad , \quad \Delta t = 120 \text{ sec.} \quad , \quad \alpha = ?$$

We first convert the z-component of the initial angular velocity to SI units:

$$\omega_{z0} = 0.50 \, \frac{\text{rev}}{\text{min.}} \times \frac{1 \text{ min.}}{60 \text{ sec}} \times \frac{2\pi \text{ rad}}{1 \text{ rev}} = \frac{\pi}{60} \text{rad/s}$$

The radius of the circle is $R_0 = D/2 = 25$ m. Since the monkey completed two laps, $\Delta\theta = 2$ rev $= 4\pi$ rad. We now know 4 symbols from the set of equations for uniform angular acceleration: $R_0$, $\omega_{z0}$, $\Delta t$, and $\Delta\theta$. We can now solve for the angular acceleration:

$$\Delta\theta = \omega_{z0}\Delta t + \frac{\alpha_{z0}\Delta t^2}{2}$$

$$\alpha_{z0} = \frac{2(\Delta\theta - \omega_{z0}\Delta t)}{\Delta t^2} = \frac{2\left[4\pi - \left(\frac{\pi}{60}\right)(120)\right]}{(120)^2} = \frac{\pi}{3600}\,\text{rad/s}^2 \quad \Rightarrow \quad \vec{\alpha} = \frac{\pi}{3600}\,\text{rad/s}^2\,\hat{\mathbf{k}}$$

The tangential acceleration can be found from the angular acceleration by multiplying by the radius:

$$a_{T0} = R_0\alpha_{z0} = (25)\left(\frac{\pi}{3600}\right) = \frac{\pi}{144}\,\text{m/s}^2$$

The monkey's initial centripetal acceleration is

$$a_{c0} = \frac{v_{T0}^2}{R_0} = R_0\omega_{z0}^2 = (25)\left(\frac{\pi}{60}\right)^2 = \frac{\pi^2}{144}\,\text{m/s}^2$$

The magnitude of the monkey's initial acceleration is

$$a_0 = \sqrt{a_{T0}^2 + a_{c0}^2} = \sqrt{\left(\frac{\pi}{144}\right)^2 + \left(\frac{\pi^2}{144}\right)^2} = \frac{\pi\sqrt{1+\pi^2}}{144}\,\text{m/s}^2$$

**Example.** A monkey on ice skates is spinning in a circle. Her angular speed changes from 2.0 rev/s to 20.0 rev/s in 3.0 seconds as she brings her arms inward. Assuming her angular acceleration to be uniform,[280] how many revolutions does she complete during these 3.0 seconds?

We don't yet know if the monkey will complete an integral number of revolutions, so we didn't place the initial and final positions at the same point in the diagram below.

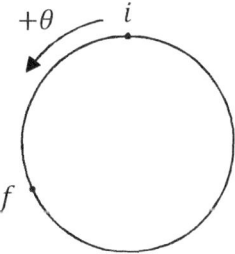

We are given 3 symbols from the angular set of motion variables, and we can determine the number of revolutions after first finding $\Delta\theta$:

---

[280] A real ice skater begins with one angular speed, held constant for some time, then angularly accelerates, reaching a new angular speed, and holds this constant for some time. Hence, uniform angular acceleration can only be a rough approximation at best, as the ice skater's angular acceleration must first increase from zero and then decrease back to zero to explain the constant angular speeds before and after the period of angular acceleration.

# 7 Rotation

$$\omega_{z0} = 2.0 \text{ rev/s} \quad , \quad \omega_z = 20.0 \text{ rev/s} \quad , \quad \Delta t = 3.0 \text{ sec.} \quad , \quad \Delta\theta = ?$$

There is no reason to convert revolutions to radians, since we don't need to work with the tangential variables to solve this problem. (Especially, since we're solving for the number of revolutions, such a conversion would be rather silly.) We can't solve for $\Delta\theta$ using a single equation, but we can solve for $\Delta\theta$ after first finding $\alpha$:[281]

$$\omega_z - \omega_{z0} = \alpha_{z0} \Delta t$$
$$\alpha_{z0} = \frac{\omega_z - \omega_{z0}}{\Delta t} = \frac{20.0 - 2.0}{3.0} = 6.0 \text{ rev/s}^2$$
$$\Delta\theta = \omega_{z0} \Delta t + \frac{\alpha_{z0} \Delta t^2}{2} = (2.0)(3.0) + \frac{(6.0)(3.0)^2}{2} = 33 \text{ rev}$$

(Now we know that the final position does lie at the initial position, but it's totally incorrect to guess this – there will be many problems where that's not the case.)

**Non-uniform angular acceleration**: Circular motion with non-uniform angular acceleration is just like the 1D non-uniform acceleration described in Sec. 1.8, except that – like the case of uniform angular acceleration – there are two sets of equations of motion, one set for the angular variables and another set for the tangential variables:

$$\omega_z = \frac{d\theta}{dt} \quad , \quad \alpha_z = \frac{d\omega_z}{dt} = \frac{d^2\theta}{dt^2} \quad ; \quad v_T = \frac{ds_T}{dt} \quad , \quad a_T = \frac{dv_T}{dt} = \frac{d^2 s_T}{dt^2}$$

$$\omega_z - \omega_{z0} = \int_{t=t_0}^{t} \alpha_z dt \quad , \quad \Delta\theta = \int_{t=t_0}^{t} \omega_z dt \quad ; \quad v_T - v_{T0} = \int_{t=t_0}^{t} a_T dt \quad , \quad \Delta s_T = \int_{t=t_0}^{t} v_T dt$$

$$\Delta s_T = R_0 \Delta\theta \quad , \quad v_{T0} = R_0 \omega_{z0} \quad , \quad v_T = R_0 \omega_z \quad , \quad a_T = R_0 \alpha_z \quad \text{(circular motion)}$$

**Example.** A monkey runs around a circle with a radius of 15 m with an angular velocity of $\vec{\omega}(t) = 0.25t^2 \hat{k}$, where SI units have been suppressed. How far does the monkey run over the interval $0 \leq t \leq 2.0$ s?

We can find $\Delta\theta$ by integrating $\omega_z$ over time:

$$\Delta\theta = \int_{t=t_0}^{t} \omega_z dt = \int_{t=0}^{2} \frac{t^2}{4} dt = \left[\frac{t^3}{12}\right]_{t=0}^{2} = \frac{2}{3} \text{ rad}$$

The net tangential displacement is then $\Delta s_T = R_0 \Delta\theta = (15)(2/3) = 10$ m. Since $\omega_z = .25t^2$ is nonnegative, the monkey does not change direction, and so the total distance traveled equals the absolute value of the net tangential displacement: $\Delta s = \Delta s_T = 10$ m.

---

[281] You can't always solve for something else first, using a single equation. Sometimes you just have to setup a system of equations, and solve them simultaneously.

**Multiple moving objects in circular motion**: If there are multiple objects moving in a circle, the strategy from Sec. 1.9 can be applied, where there are two underlying sets of equations – one set for the angular quantities and another set for the tangential quantities. Note that if one object moves clockwise and another moves counterclockwise, though they travel in opposite directions, they will still meet up; in this case, the sum of the absolute values of the individual displacements to the point where they meet can be related[282] to the circumference.

> **Example**. Two monkeys start at the same point on a circular jogging track. One monkey runs from rest with a uniform angular acceleration of $\vec{\alpha}_1 = 0.06 \text{ rad/s}^2 \, \hat{\mathbf{k}}$, while the other runs with a constant angular velocity of $\vec{\omega}_2 = -0.4 \text{ rad/s} \, \hat{\mathbf{k}}$. When do the two monkeys meet?
>
>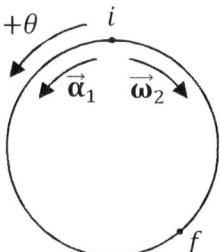
>
> Strategically, we will write an equation for the angular displacement of each monkey, then write an equation that relates the two angular displacements, and combine these equations to solve for the time. The first monkey travels with uniform angular acceleration:
>
> $$\Delta\theta_1 = \omega_{z10}\Delta t + \frac{\alpha_{z10}\Delta t^2}{2} = 0.03\Delta t^2$$
>
> The second monkey travels with uniform angular velocity (so $\vec{\alpha}_2 = 0$):
>
> $$\Delta\theta_2 = \omega_{z20}\Delta t = -0.4\Delta t$$
>
> The sum of the absolute values of the angular displacements must be $2\pi$ rad when the monkeys meet, since together they will have traveled full-circle. Since $\Delta\theta_2$ is negative, we write this as
>
> $$\Delta\theta_1 - \Delta\theta_2 = 2\pi$$
> $$0.03\Delta t^2 - (-0.4\Delta t) = 2\pi$$
> $$0.03\Delta t^2 + 0.4\Delta t - 2\pi = 0$$
> $$\Delta t = \frac{-0.4 \pm \sqrt{(0.4)^2 - (4)(0.03)(-2\pi)}}{0.06} = 9.3 \text{ s}$$

---

[282] Note that the sum of the absolute values of their $\Delta s_T$'s might not be equal the circumference because the two objects might not start in the same position.

## 7.2 Moment of Inertia

**Kinetic energy of a pointlike object traveling in a circle**: Recall that the (translational) kinetic energy of an object is $K_t = mv^2/2$. If the object travels in a circle and the object is pointlike – i.e. small in size compared to the radius of the circle[283] – then we can substitute $v = R_0\omega$ into the formula for kinetic energy to obtain $K_r = mR_0^2\omega^2/2$. If we define the moment of inertia of the pointlike object to be $I \equiv mR_0^2$, we can express the kinetic energy as rotational kinetic energy:

$$K_r = \frac{I\omega^2}{2} \quad \text{(rotational KE)}$$

In general, an object can have two kinds of kinetic energy – translational kinetic energy and rotational kinetic energy. If you grab a banana by its stem and throw it in the air, the banana's center of mass will travel along a parabola, while the banana rotates about its center of mass, as illustrated below. The banana has translational kinetic energy due to the motion of its center of mass along the parabola, and rotational kinetic energy due to its rotation about an axis passing through its center of mass.

**Moment of inertia of a rigid body**: The above formula for rotational kinetic energy applies to any rigid body[284] that is rotating, and serves mathematically to define the moment of inertia of an object. However, the formula for the moment of inertia of an object (or system of objects) is generally not $mR_0^2$ – that formula only applies to pointlike objects. The SI units of moment of inertia are kg·m$^2$.

---

[283] If the object is not small compared to the radius of the circle, then the part of the object nearest to the center of the circle will travel significantly faster than the part of the object furthest from the center of the circle.
[284] If the body is not rigid, it may deform or fall apart during rotation, which is a much more complicated situation. In the context of rotation, we shall restrict ourselves to rigid bodies.

Comparison of the formulas for translational kinetic energy, $K_t = mv^2/2$, and rotational kinetic energy, $K_r = I\omega^2/2$, reveals that the moment of inertia of an object, $I$, is the rotational analog for mass. Recall that mass is a measure of an object's inertia. Analogously, the moment of inertia provides a measure of the object's rotational inertia. We will elaborate on this point following the next example.

**Moment of inertia for a pointlike object**: The moment of inertia of a pointlike object is

$$I = mr_\perp^2 \quad \text{(pointlike)}$$

where $r_\perp$ is the shortest distance from the pointlike object to the axis of rotation – i.e. radius of the circle in which the pointlike object is traveling (or would travel[285]). An object is pointlike if it is small in size compared to $r_\perp$.

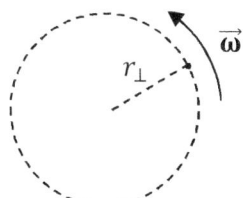

**Kinetic energy of a rotating system of pointlike objects (with rigid massless connectors)**: Imagine a system of $N$ pointlike masses connected together by rigid rods that have negligible mass compared to the pointlike masses.[286] Such a system is illustrated on the following page. The total kinetic energy of such a system is

$$K_r = \sum_{i=1}^{N} \frac{m_i v_i^2}{2} = \frac{1}{2}\sum_{i=1}^{N} m_i r_{\perp i}^2 \omega^2 = \frac{\omega^2}{2}\sum_{i=1}^{N} m_i r_{\perp i}^2 = \frac{\omega^2}{2}\sum_{i=1}^{N} I_i = \frac{I\omega^2}{2} \quad \text{(rigid system of point-masses)}$$

since each pointlike object has speed $v_i = r_{\perp i}\omega$, where $r_{\perp i}$ is the shortest distance from object $i$ to the axis of rotation.[287] Observe that each object in the system rotates with the same angular velocity, $\vec{\omega}$ (in both magnitude and direction), but not the same velocity, $\vec{v}_i$ (which varies, in general, from object to object both in magnitude and direction). Each pointlike object has moment of inertia $I_i = m_i r_{\perp i}^2$, giving rise to a total moment of inertia, $I = \sum_{i=1}^{N} I_i = \sum_{i=1}^{N} m_i r_{\perp i}^2$, for the system.

---

[285] The object does not actually have to be moving in order to speak of its moment of inertia. However, the quantitative value of the moment of inertia does depend upon the axis of rotation – an object has a different moment of inertia about every different possible axis of rotation. Therefore, if the object is not actually rotating, if you wish to speak of its moment of inertia, you must specify which axis of rotation it is computed relative to.

[286] We will learn how to account for the moment of inertia of the rods when we discuss the moment of inertia of an extended rigid body (and later the moment of inertia for a system of extended rigid bodies). In the meantime, let us assume that the rods have negligible mass such that the system effectively consists only of pointlike objects; the purpose of the rods is only to keep the system rigid so that every object rotates with the same angular velocity.

[287] Note that each $r_{\perp i}$ is perpendicular to the axis of rotation.

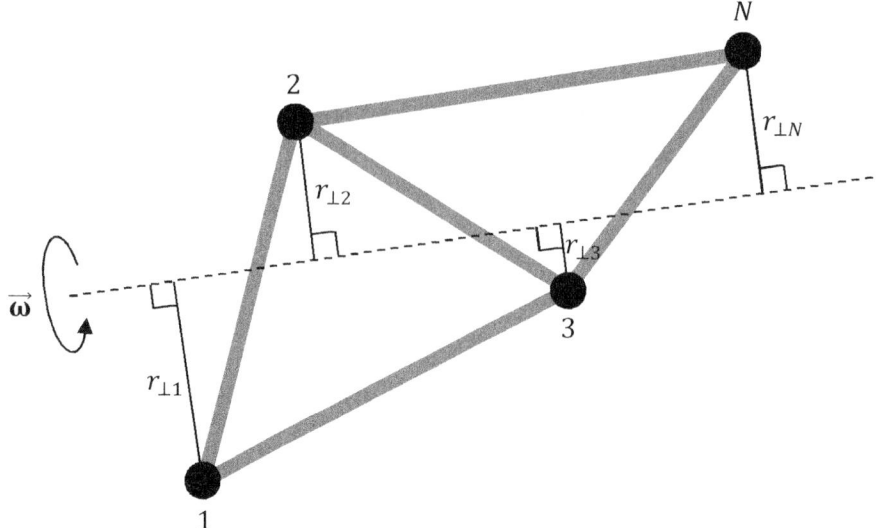

**Moment of inertia for a system of pointlike objects (with rigid massless connectors):** We can write the total kinetic energy of a system of $N$ pointlike objects connected with relatively massless rigid rods as $K = I\omega^2/2$ by identifying the total moment of inertia of the system about the specified axis of rotation as the sum of the moments of inertia of the pointlike objects about the same axis of rotation:

$$I = \sum_{i=1}^{N} I_i = \sum_{i=1}^{N} m_i r_{\perp i}^2$$

where $r_{\perp i}$ is the shortest distance from object $i$ to the axis of rotation.

**Axis of rotation:** When a rigid body rotates, every point of the rigid body travels in a circle about the axis of rotation. The axis of rotation is an abstract line; every point of the rigid body travels in a circle centered about this line.

If the object is hinged – i.e. fixed at one point – the axis of rotation will pass through the hinge. For example, consider the rod illustrated on the next page. The rod lies on a fulcrum, which serves as a hinge. In this case, each part of the rod travels in a clockwise circle in the plane of the page; the axis of rotation is perpendicular to the page, passing through the rod.

If the object rotates freely – i.e. it is not hinged – the axis of rotation will pass through the center of mass of the object.[288] The banana illustrated on the following page is shown rotating about a vertical axis that passes through its center of mass. Every part of the banana travels in a horizontal circle centered about this vertical axis. Compare with the rotating banana illustrated a couple of pages ago, for which the axis of rotation is perpendicular to the page and passing through its center of mass.

---

[288] Of course, there is an infinite set of possible axes of rotation that pass through the object's center of mass. Therefore, phrases of the sort, "...the moment of inertia about the center of mass...," would be quite ambiguous.

An Advanced Introduction to Calculus-Based Physics (Mechanics)

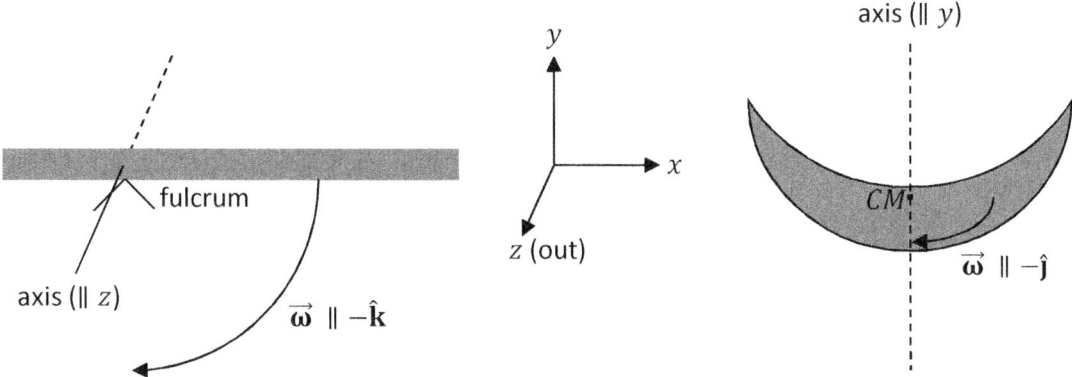

**Example.** The system of point-masses illustrated below is connected by lightweight (i.e. compared to the point-masses) rigid rods. All three objects lie in the $xy$ plane. Determine the moment of inertia of the system about the $x$-axis, the $y$-axis, and the $z$-axis.

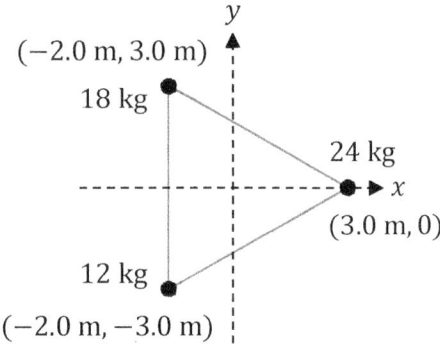

We apply the formula for the moment of inertia of a system of point-masses, assuming massless rigid connecting rods, where $r_{\perp i}$ is the shortest distance from object $i$ to the axis of rotation:

$$I = m_1 r_{\perp 1}^2 + m_2 r_{\perp 2}^2 + m_3 r_{\perp 3}^2$$
$$I_x = (18)(3.0)^2 + (12)(3.0)^2 + (24)(0)^2 = 270 \text{ kg·m}^2$$
$$I_y = (18)(2.0)^2 + (12)(2.0)^2 + (24)(3.0)^2 = 336 \text{ kg·m}^2$$
$$I_z = (18)(3.0^2 + 2.0^2) + (12)(3.0^2 + 2.0^2) + (24)(3.0)^2 = 606 \text{ kg·m}^2$$

where we used the Pythagorean to determine the distance of the 12- and 18-kg masses from the $z$-axis (which is the same as the distance to the origin in this case, since the objects lie in the $xy$ plane).

**Dependence of moment of inertia on the axis of rotation**: As the previous example illustrates, the moment of inertia of an object or system depends upon the axis of rotation. Therefore, it's ambiguous to speak of "the moment of inertia of an object (or system)" without specifying a particular axis of rotation. The moment of inertia depends upon how the mass is distributed about the axis of rotation.

**Dependence of moment of inertia on the distribution of mass**: All of the formulas for moment of inertia – whether for a system of pointlike objects or a single extended rigid body, or a system of extended rigid bodies – involve $r_\perp^2$, where the $r_\perp$ (or $r_{\perp i}$'s) is the distance to the axis of rotation – and also involve the mass(es) of the object(s). Conceptually, this means that the moment of inertia of the object or system is greater if more of the mass is far away from the axis of rotation and smaller if more of the mass is closer to the axis of rotation.

---

**Hint**. If you're trying to compare the rotational properties (like angular acceleration) of two different rigid bodies or systems, first compare their moments of inertia about the specified axes of rotation. Moment of inertia is greater when more of the mass is far from the axis of rotation, and smaller if more of the mass is closer to the axis of rotation.

---

**Rotational inertia**: Comparing the formulas for translational kinetic energy, $K_t = mv^2/2$, and rotational kinetic energy, $K_r = I\omega^2/2$, we see that moment of inertia, $I$, is the rotational analog for mass, $m$. Recall that mass is a measure of an object's inertia – a natural tendency to maintain constant momentum. Analogously, moment of inertia is a measure of a rigid body's (or system's) rotational inertia – a natural tendency of a rigid body (or system) to maintain constant angular momentum.

For a single object of constant mass, inertia is a natural tendency of the object to maintain constant velocity, $\vec{v}$. Similarly, for a single rigid body of constant moment of inertia, rotational inertia is a natural tendency of the rigid body to maintain constant angular velocity, $\vec{\omega}$.

The mass of an object tells you how easy or difficult it is to change the momentum of the object: For an object of constant mass, a greater net external force is needed to achieve a given acceleration for an object of greater mass. Analogously, the moment of inertia of a rigid body tells you how easy or difficult it is to change the angular momentum (defined more precisely in Sec. 7.8) of the rigid body: For a rigid body of constant moment of inertia, a greater net external torque (see Sec.'s 7.4-7.5) is needed to achieve a given angular acceleration for a rigid body of greater moment of inertia.

It is easier to change the angular momentum (i.e. less net external torque is needed) of a rigid body when more of its mass is closer to the axis of rotation, since in that case it has a smaller moment of inertia.

---

**Conceptual Example**. Does a coin have a greater moment of inertia when it is flipping through the air or when it is rolling along a table?

When a coin rolls along a table, the axis of rotation is perpendicular to the coin and passing through its center. When a coin flips through the air, the axis of rotation bisects the coin along a diameter. The two cases are illustrated on the following page.

In the case of rolling, only the center of the coin lies on the axis of rotation, and every point on the circumference is a distance of one radius from the axis of rotation. In the case of flipping, there is an entire diameter of points coincident with the axis of rotation, and every point on the circumference except for the two extreme points is closer than one radius to the axis of rotation. Overall, the moment of inertia of the coin is smaller when it is flipping than when it is rolling.

This means that it's easier to change the angular momentum of the coin by flipping it than by rolling it. For example, starting from rest, it's easier to give the coin a greater angular speed just after release by flipping it than by rolling it.

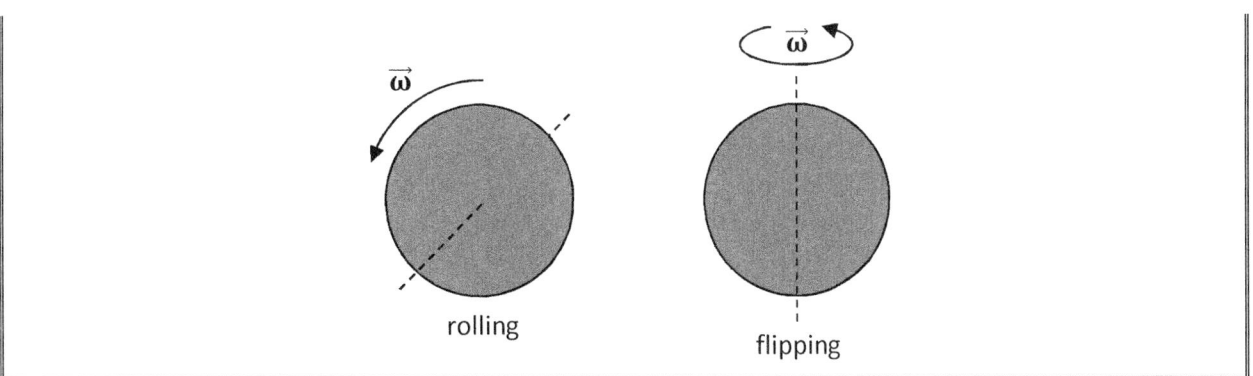

rolling    flipping

**Moment of inertia for an extended rigid body**: If we imagine dividing an extended rigid body – such as a rod or a disc – into Avogadro's number (or so) of molecules, and apply the summation formula, $I = \sum_{i=1}^{N} m_i r_{\perp i}^2$, to find the moment of inertia of the system, in this limit this Riemann sum becomes an integral:

$$I = \int r_\perp^2 dm$$

where $r_\perp$ is the shortest distance from each differential mass element, $dm$, to the axis of rotation. The strategy for carrying out the moment of inertia integral for an extended rigid body is similar to the strategy for performing the center of mass integral for a continuous object. It may be helpful to review Sec. 6.1. We use the same choice of substitutions to rewrite $dm$, we determine the limits of integration the same way, and we perform multiple integrals with the same technique. The differences are that the integrand includes $r_\perp^2$ instead of $\vec{r}$ (so moment of inertia is a scalar integral rather than a vector integral), and moment of inertia does not include a coefficient of $1/M$.

In order to perform the moment of inertia integration, you need to write an algebraic expression for $r_\perp$ in terms of the integration coordinates (i.e. $x$, $y$, and $z$; $r_c$, $\theta$, and $z$; or $r$, $\varphi$, and $\theta$). Draw and label a diagram to help visualize $r_\perp$. First, draw the object and the axis of rotation. Setup a coordinate system. Draw a representative differential mass element, $dm$. Don't draw $dm$ at a special point – like at the origin, on an axis (unless every point of the object lies on the axis), on the circumference (unless the object is an infinitesimally thin hoop), etc. That is, try to draw $dm$ out in the middle of the object. By placing $dm$ out in the open, you are more likely to write a general algebraic expression for $r_\perp$ that applies to every point in the object; if instead you draw $dm$ at a special point, you increase the chance that you will write an equation for $r_\perp$ that applies only to special points, and does not apply to every point in the object. Draw a line connecting $dm$ to the axis of rotation, which is perpendicular to the axis of rotation; label this as $r_\perp$. Look at your diagram and express $r_\perp$ in terms of the integration coordinates, noting that the coordinates of $dm$ are $(x, y, z)$, $(r_c, \theta, z)$, or $(r, \varphi, \theta)$. Such a diagram is illustrated on the following page.

When you complete the integration, express your answer for the moment of inertia of the extended rigid body in terms of the total mass of the object. Do this by performing the integral $M = \int dm$, as we did for the center of mass integrals (see Sec. 6.1), and use the result of this mass integral to substitute the total mass of the object into the result for the moment of inertia integration. This technique is illustrated in the following examples.

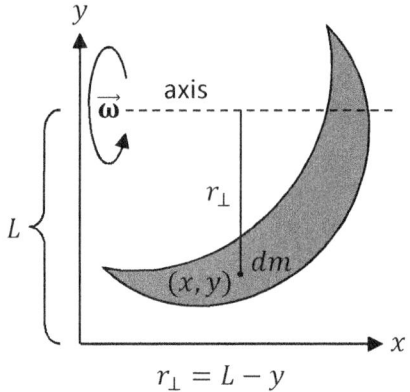

$$r_\perp = L - y$$

## Problem-Solving Strategy for Performing Moment of Inertia Integrals

0. Use this technique to derive an equation for the moment of inertia of an extended rigid body about a specified axis of rotation.
1. Determine which coordinate system is best-suited to the geometry of the object.
2. Write $dm = \lambda ds$ if the object looks like a line or a curve, write $dm = \sigma dA$ if the object looks like a surface, and write $dm = \rho dV$ if the object looks like a 3D solid.
3. Write the differential element from Step 2 – i.e. $ds$, $dA$, or $dV$ – as appropriate for the geometry of the object in the coordinate system that you are using. See the tables in Sec. 6.1.
4. Determine the integration limits. Choose one variable to have its full range and write down its lower and upper limits. Make a sketch and determine whether or not the limits of the second/third coordinates depend on the value of first coordinate. For rectangular objects in Cartesian coordinates, circular objects or pie slices in 2D polar coordinates or right-circular cylinders in cylindrical coordinates (coaxial with the natural $z$-axis), or spheres in spherical coordinates (centered at the origin), all of the limits will be constants. Otherwise, you should generally expect to have at least one variable limit.
5. Perform the integral $M = \int dm$ using the substitutions from Steps 2-3 and the integration limits established in Step 4. If the density is uniform (i.e. constant), this will provide a relation between the total mass and the density of the object; if the object is non-uniform (i.e. it is specified as a function of the coordinates in the problem), this will provide a relation between the total mass and the proportionality constant from the density equation.
6. Draw and label a diagram of the object, the coordinate system, the axis of rotation, a representative $dm$ out in the middle of the object (not at special points like the origin, an axis, or the boundary of the object), and $r_\perp$ from $dm$ to the axis of rotation (perpendicular to the axis of rotation). Note that the coordinates of $dm$ are $(x, y, z)$, $(r_c, \theta, z)$, or $(r, \varphi, \theta)$. Express $r_\perp$ algebraically in terms of the integration variables – i.e. $x$, $y$, and $z$; $r_c$, $\theta$, and $z$; or $r$, $\varphi$, and $\theta$.
7. Write down the starting equation for the moment of inertia integral. Make all of the substitutions from Steps 2-3 and 5-6 (being careful not to omit any factors) and use the integration limits from Step 4. Note that $r_\perp$ is squared in the integrand. If doing a double or triple integral, first integrate over any coordinate(s) that have variable limits. When integrating over one variable, hold its independent variables constant.

**Note**. The quantity $r_\perp$ extends from the axis of rotation to each $dm$. The integration over $dm$ will cover every $dm$ in the object. Draw $r_\perp$ perpendicular to the axis of rotation to $dm$. Express $r_\perp$ in terms of the integration variables, noting that the coordinates of $dm$ are $(x, y, z)$, $(r_c, \theta, z)$, or $(r, \varphi, \theta)$. Remember to square $r_\perp$ because $I = \int r_\perp^2 dm$.

**Example**. The rod illustrated below has non-uniform linear mass density $\lambda(x) = \beta x$, where $\beta$ is a positive constant. Derive an equation for the moment of inertia of the rod about the line $x = L/3$ in terms of the total mass of the rod, $M$, and the length of the rod, $L$.

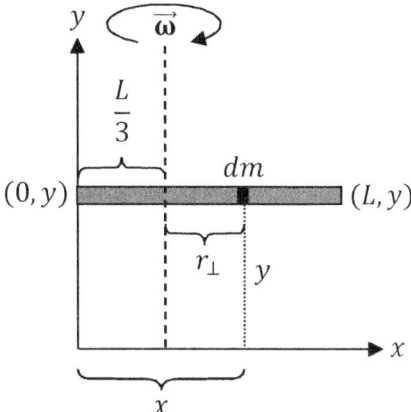

We choose to work with Cartesian coordinates since the rod is linear. The rod is effectively 1D, so we write $dm = \lambda ds$. Since we are given the linear mass density, $dm = \beta x ds$. The rod only extends along the $x$-direction, so $ds = dx$ and $dm = \beta x dx$. The limits of integration are over the possible positions of $dm$ in the rod: $0 \leq x \leq L$. We first integrate over $dm$ to relate $M$ to $\beta$:

$$M = \int dM = \int_{x=0}^{L} \beta x dx = \frac{\beta L^2}{2}$$

In the diagram above, we drew and labeled a representative $dm$, being careful not to place it on the boundary of the rod or on any axis.[289] We labeled $x$ and $y$ – the Cartesian coordinates of $dm$ – on the diagram, and drew $r_\perp$ from the axis of rotation to $dm$. Looking at the diagram, it should be clear that $r_\perp = x - L/3$ for the representative $dm$. Note that if we had drawn $dm$ to the left of the axis of rotation, we would have found that $r_\perp = L/3 - x$; this distinction is not important since $r_\perp$ is squared in the integrand. We are now prepared to perform the moment of inertia integral:

---

[289] If the rod were placed on the $x$-axis, for example, then it wouldn't be possible to draw a $dm$ that didn't lie on an axis. Avoid drawing $dm$ on any axis (coordinate axis or axis of rotation) whenever this is possible.

$$I = \int r_\perp^2 dm = \int_{x=0}^{L} \left(x - \frac{L}{3}\right)^2 \beta x\, dx = \beta \int_{x=0}^{L} \left(x^3 - \frac{2x^2 L}{3} + \frac{xL^2}{9}\right) dx$$

$$I = \beta \left(\frac{L^4}{4} - \frac{2L^4}{9} + \frac{L^4}{18}\right) = \frac{\beta L^4}{12} = \frac{2M}{L^2}\frac{L^4}{12} = \frac{ML^2}{6}$$

**Example.** The isosceles triangle illustrated below has uniform mass distribution. Derive an equation for its moment of inertia about the $y$-axis in terms of $b$, $h$, and/or $M$.

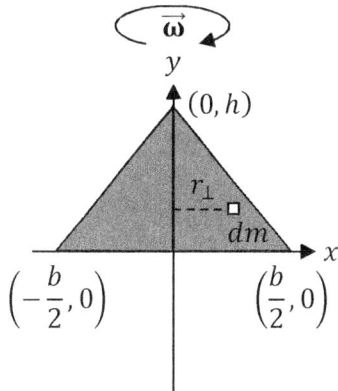

Working with Cartesian coordinates for this 2D triangle, $dm = \sigma dA = \sigma dx dy$. We allow $x$ to vary from 0 to $b/2$; since the triangle is symmetric about the axis of rotation and since the triangle is uniform, we choose to integrate the right-half only and double the result. For each $x$, $y$ has a minimum of 0, but the maximum depends on the value of $x$. The equation for the line, $y = h - 2hx/b$, represents the upper limit of $y$ for positive $x$. The total mass of the triangle is related to its density via

$$M = \int dm = \int_S \sigma dA = \sigma \int_S dA = \sigma A = \sigma \frac{1}{2} bh$$

Since the density is uniform, the total mass is proportional to the area, and we don't need to integrate to find the area of the triangle; but we will need to integrate to find moment of inertia.

As illustrated above, every $dm$ lies a distance $r_\perp = x$ from the axis of rotation:

$$I = \int r_\perp^2 dm = 2 \int_{x=0}^{\frac{b}{2}} \int_{y=0}^{h - \frac{2hx}{b}} x^2 \sigma dx dy = 2\sigma \int_{x=0}^{\frac{b}{2}} x^2 \left(h - \frac{2hx}{b}\right) dx = 2\sigma \int_{x=0}^{\frac{b}{2}} \left(hx^2 - \frac{2hx^3}{b}\right) dx$$

$$I = 2\sigma \left(\frac{hb^3}{24} - \frac{hb^3}{32}\right) = 2 \frac{2M}{bh} \frac{hb^3}{8} \left(\frac{1}{3} - \frac{1}{4}\right) = \frac{Mb^2}{24}$$

**Example**. Derive an equation for the moment of inertia of a solid uniform sphere about an axis that passes through its center in terms of its radius, $R_0$, and total mass, $M$.

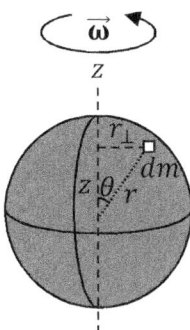

Working in spherical coordinates, $dm = \rho dV = \rho r^2 \sin\theta \, dr d\varphi d\theta$ for this 3D solid, and the limits of integration are $0 \leq r \leq R_0$, $0 \leq \varphi \leq 2\pi$, and $0 \leq \theta \leq \pi$. The total mass of the sphere is related to the density via

$$M = \int dm = \int_V \rho dV = \rho V = \frac{4\pi R_0^3 \rho}{3}$$

since $\rho$ is constant for this uniform sphere.

By symmetry, the moment of inertia of the sphere is the same for any axis of rotation (since it's uniform), so for convenience we choose the $z$-axis for the axis of rotation. Each $dm$ is a distance $\sqrt{x^2 + y^2}$ from the axis of rotation, which can also be expressed as $r_\perp = r\sin\theta$ (as illustrated above). The moment of inertia of the sphere about the $z$-axis is

$$I = \int r_\perp^2 dm = \int_{r=0}^{R_0} \int_{\varphi=0}^{2\pi} \int_{\theta=0}^{\pi} (r^2 \sin^2\theta)\rho r^2 \sin\theta \, dr d\varphi d\theta = \rho \frac{R_0^5}{5} 2\pi \int_{\theta=0}^{\pi} \sin^3\theta \, d\theta$$

$$I = \rho \frac{R_0^5}{5} 2\pi \left( \int_{\theta=0}^{\pi} \sin\theta \, d\theta - \int_{\theta=0}^{\pi} \sin\theta \cos^2\theta \, d\theta \right)$$

since $\sin^3\theta = \sin\theta \sin^2\theta = \sin\theta(1 - \cos^2\theta)$. The second integral can be performed with the substitution $u \equiv \cos\theta$, $du = -\sin\theta \, d\theta$:

$$I = \frac{3M}{4\pi R_0^3} \frac{2\pi R_0^5}{5} \left( [-\cos\theta]_{\theta=0}^{\pi} + \int_{u=1}^{-1} u^2 du \right) = \frac{3MR_0^2}{10} \left( 2 + \left[\frac{u^3}{3}\right]_{u=1}^{-1} \right)$$

$$I = \frac{3MR_0^2}{10} \left( 2 - \frac{2}{3} \right) = \frac{2MR_0^2}{5}$$

**Moments of inertia for common geometries about common axes of rotation**: The following table provides the moment of inertia of several common geometries of rigid bodies about common axes of rotation, assuming uniform density. This table can be used for a few purposes:

- Most conveniently, if you need to use the formula for the moment of inertia of a uniform extended rigid body with a common geometry about a common axis of rotation (and the problem does not ask you to derive this formula), you can look it up in the following table.
- If a problem asks you to derive an equation for the moment of inertia of a uniform extended rigid body with a common geometry – like a rod, disc, or sphere – about a common axis of rotation, you can use this table to check your answer.
- If the axis of rotation differs from those listed in the table for a given rigid body, as long as the axis of rotation is parallel to an axis of rotation listed in the table, you can use the formula from the table in conjunction with the parallel-axis theorem (described later in this section).
- If the rigid body has non-uniform density or if the axis of rotation is not parallel to any axes listed in the table, it may still be helpful, conceptually, to compare your answer to one or more formulas in the table to see if the formula that you derive via integration seems reasonable.

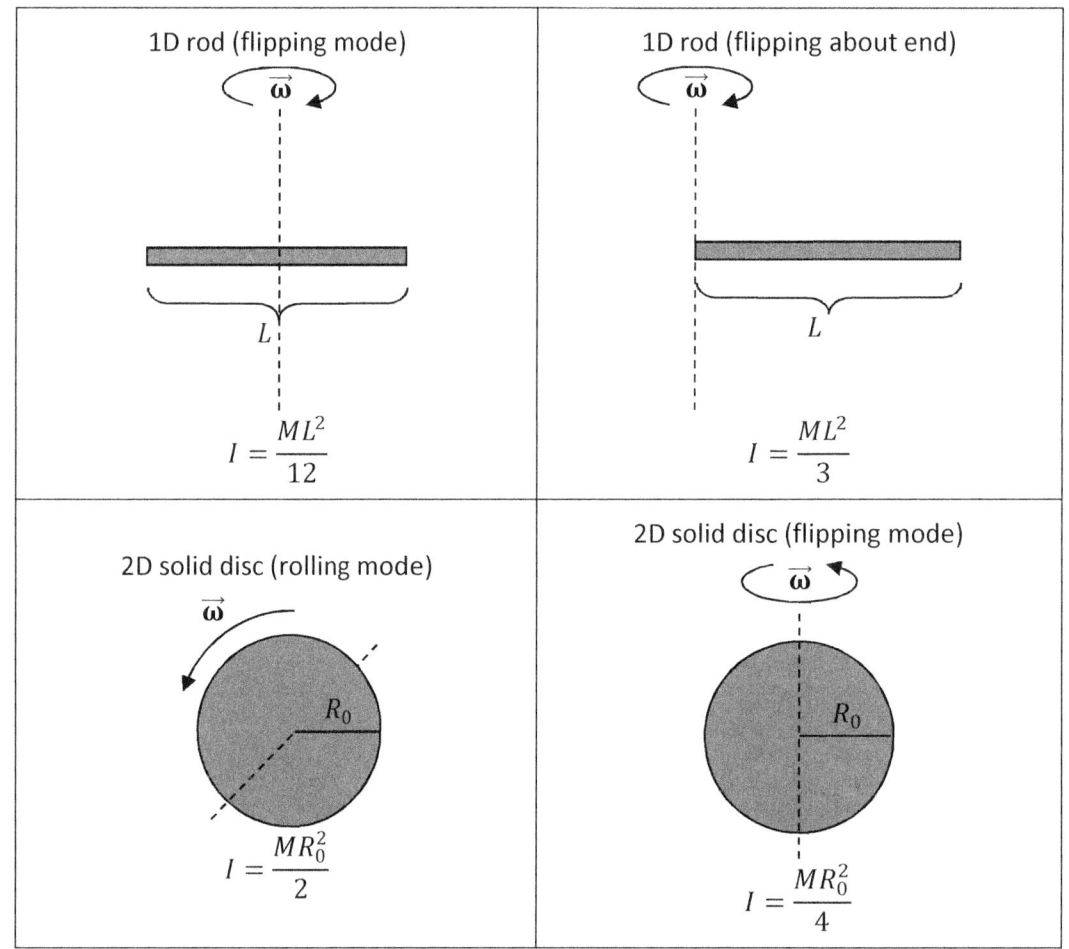

| 1D ring (rolling mode) 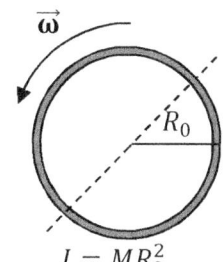 $I = MR_0^2$ | 1D ring (flipping mode) 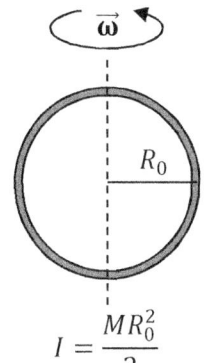 $I = \dfrac{MR_0^2}{2}$ |
| --- | --- |
| 2D ring (rolling mode) 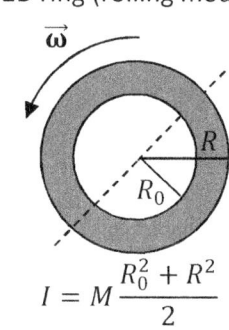 $I = M\dfrac{R_0^2 + R^2}{2}$ | 2D ring (flipping mode) 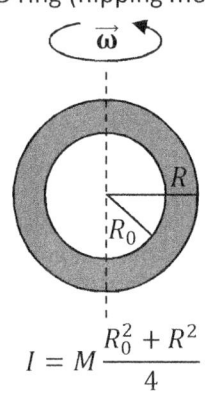 $I = M\dfrac{R_0^2 + R^2}{4}$ |
| 3D solid cylinder (rolling mode) 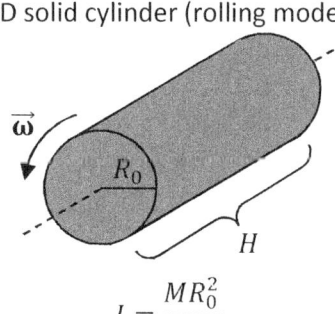 $I = \dfrac{MR_0^2}{2}$ | 2D hollow cylinder (rolling mode) 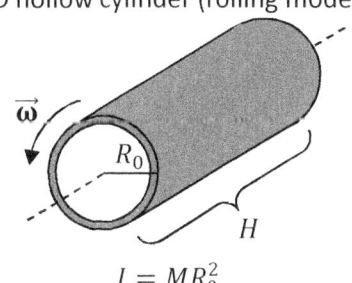 $I = MR_0^2$ |

| | |
|---|---|
| 3D hollow cylinder (rolling mode)<br>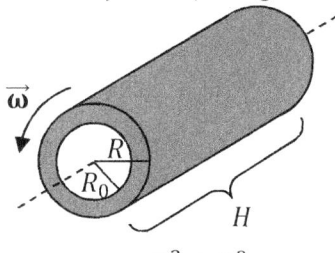<br>$$I = M\frac{R_0^2 + R^2}{2}$$ | 3D solid sphere (rolling mode)<br>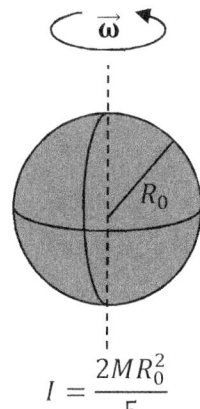<br>$$I = \frac{2MR_0^2}{5}$$ |
| 2D hollow sphere (rolling mode)<br>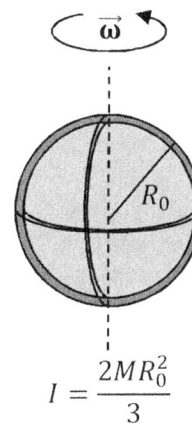<br>$$I = \frac{2MR_0^2}{3}$$ | 3D hollow sphere (rolling mode)<br>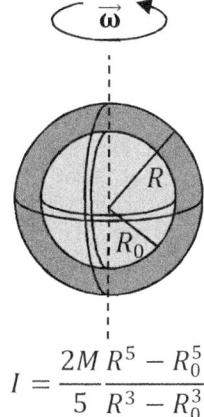<br>$$I = \frac{2M}{5}\frac{R^5 - R_0^5}{R^3 - R_0^3}$$ |
| 3D solid cube about an axis through its center and perpendicular to a face<br>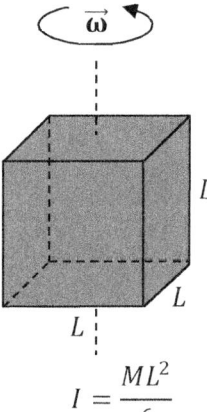<br>$$I = \frac{ML^2}{6}$$ | 2D rectangle about an axis through its center and perpendicular to the rectangle<br>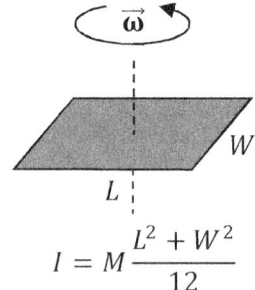<br>$$I = M\frac{L^2 + W^2}{12}$$ |

| 2D rectangle (flipping mode) | 3D solid cone about its symmetry axis |
|---|---|
| 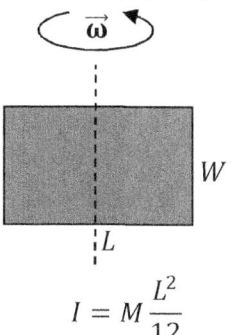 $$I = M\frac{L^2}{12}$$ | 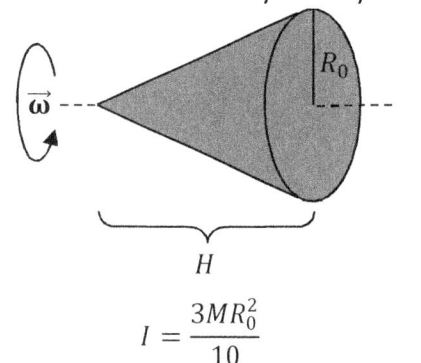 $$I = \frac{3MR_0^2}{10}$$ |
| 3D solid single-holed ring torus (rolling mode) | 2D ellipse about a symmetric bisector |
| 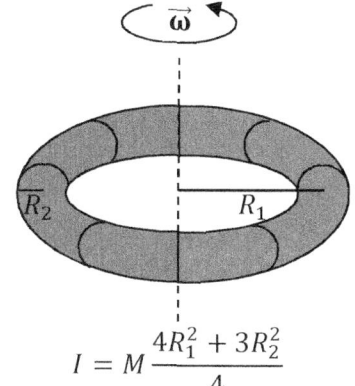 $$I = M\frac{4R_1^2 + 3R_2^2}{4}$$ | 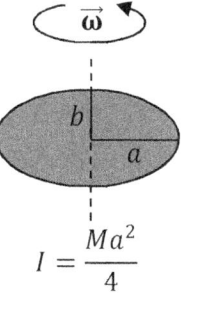 $$I = \frac{Ma^2}{4}$$ |

- Rolling mode refers to a round object that rolls or otherwise rotates such that the axis of rotation passes through its center and is perpendicular to its circular cross section (as drawn).
- Flipping mode refers to a planar object that rotes about an axis within the plane of the object that passes through its center (as drawn).
- A linear rod with finite thickness that is small compared to its length will have approximately the same moment of inertia as a 1D rod.
- A 1D hollow ring is infinitesimally thin, whereas a 2D hollow ring has finite thickness. If the thickness of a 2D hollow ring is small compared to its radius, its moment of inertia will be approximately the same as that of a 1D hollow ring. Observe that the moment of inertia of the rolling 2D hollow ring agrees with the formulas for the rolling 1D hollow ring and rolling 2D solid disc in appropriate limits (and similarly for flipping mode):

$$\lim_{R_0 \to 0} I_{\text{rolling 2D ring}} = M\frac{R_0^2 + R^2}{2} = \frac{MR^2}{2} = I_{\text{rolling 2D disc}}$$

$$\lim_{R \to R_0} I_{\text{rolling 2D ring}} = M\frac{R_0^2 + R^2}{2} = MR^2 = I_{\text{rolling 1D ring}}$$

7 Rotation

- A 2D hollow sphere is infinitesimally thin, whereas a 3D hollow sphere has finite thickness. If the thickness of a 3D hollow sphere is small compared to its radius, its moment of inertia will be approximately the same as that of a 2D hollow sphere. Observe that the moment of inertia of the 3D hollow sphere agrees with the formulas for the 2D hollow sphere and 3D solid sphere in appropriate limits:[290]

$$\lim_{R_0 \to 0} I_{\text{rolling 3D hollow sphere}} = \frac{2M}{5} \frac{R^5 - R_0^5}{R^3 - R_0^3} = \frac{2MR^2}{5} = I_{\text{rolling 3D solid sphere}}$$

$$\lim_{R \to R_0} I_{\text{rolling 3D hollow sphere}} = \frac{2M}{5} \frac{R^5 - R_0^5}{R^3 - R_0^3} = \frac{2M}{5} \frac{\frac{d}{dR}\left(R^5 - R_0^5\right)}{\frac{d}{dR}\left(R^3 - R_0^3\right)} = \frac{2M}{5} \frac{5R^4}{3R^2}$$

$$= \frac{2MR^2}{3} = I_{\text{rolling 2D hollow sphere}}$$

where we applied l'Hôpital's rule[291] since both the numerator, $R^5 - R_0^5$, and denominator, $R^3 - R_0^3$, approached zero in the limit.

**Radius of gyration**: The formula for the moment of inertia of a rigid body (or rigid system) about a given axis can be cast in the form $I = Mk^2$, where $k$ is the radius of gyration about the specified axis. Conceptually, we interpret the radius of gyration, $k$, as the radius of the equivalent thin ring (or hoop) – i.e. the infinitesimally thin ring that would have the same moment of inertia as the rigid body about the specified axis of rotation, assuming that the thin ring and rigid body have the same mass, $M$ (since the moment of inertia of a thin ring is $MR_0^2$ in rolling mode). That is, if you were to reshape the rigid body in a thin ring of radius $k$, so that all of its mass was a distance $k$ from the axis of rotation, the radius of gyration would be the radius needed to make this thin ring have the same moment of inertia (in rolling mode) as the original rigid body about the specified axis of rotation.

**Example**. How does the radius of gyration of a uniform solid sphere rotating about an axis coincident with its diameter compare to the radius of the sphere?

Observe that the specified axis corresponds to rolling mode (even if the sphere is simply spinning rather than rolling). The formula for the moment of inertia of a uniform solid sphere in rolling mode is $I = 2MR_0^2/5$. The corresponding radius of gyration is $k = \sqrt{I/M} = \sqrt{2R_0^2/5} = R_0\sqrt{2/5} = 0.63\,R_0$. The radius of gyration is 63% as large as the radius of the solid sphere in rolling mode.

---

[290] Unfortunately, a few websites that provide tables of moments of inertia list an incorrect formula for the 3D hollow sphere of finite thickness. Apparently, they tried to intuitively generalize the formula for the 2D hollow ring of finite thickness to a sphere, instead of applying calculus directly to the 3D hollow sphere of finite thickness. You can easily check that a formula of the form $2M(R_0^2 + R^2)/5$ would not agree with both the 3D solid sphere and the 2D hollow sphere in appropriate limits, and therefore cannot possibly be the moment of inertia of the 3D hollow sphere of finite thickness.

[291] Recall l'Hôpital's rule from calculus: If you're taking the limit of a ratio of the form $\lim_{x \to a} \frac{f(x)}{g(x)}$ and both the numerator and denominator approach zero in this limit, this limit is equal to $\lim_{x \to a} \frac{df/dx}{dg/dx}$.

**Parallel-axis theorem**: Suppose that you know the moment of inertia of a rigid body (or system) about an axis, which we call the $c$-axis, which passes through the center of mass of the rigid body (or system). Furthermore, suppose that you want to find the moment of inertia of the rigid body about a different axis, which we call the $p$-axis, which is parallel to the first axis ($p \parallel c$). If this is the case, the two moments of inertia are related by the parallel-axis theorem:

$$I_p = I_c + Mh^2$$

where $I_p$ and $I_c$ are the moments of inertia of the rigid body about the $p$- and $c$-axes, respectively, $M$ is the mass of the rigid body, and $h$ is the separation between the two axes. A case where the parallel-axis theorem could be applied is illustrated in the following example. In practice, the parallel-axis theorem is especially useful when a common geometry is rotating about an axis that's not so common, but which is parallel to an axis listed in a table of moments of inertia. We will prove the parallel-axis theorem in Sec. 7.7, where we discuss rotational and total kinetic energy in more detail.

**Note**. The parallel-axis theorem does not apply if neither axis passes through the center of mass of the rigid body (or system) or if the two axes are not parallel.

**Example**. Show that the two moments of inertia for flipping a uniform rod listed in the table of moments of inertia for common geometries – one about the center of the rod and the other about its end, both perpendicular to the rod – agree with the parallel-axis theorem.

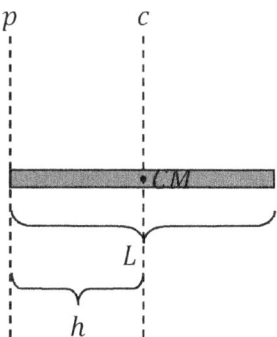

The two axes illustrated above are separated by a distance $h = L/2$. Using the formula for the center of mass of the uniform rod about an axis through its center and perpendicular to the rod, $I_c = ML^2/12$, we find

$$I_p = I_c + Mh^2 = \frac{ML^2}{12} + M\left(\frac{L}{2}\right)^2 = ML^2\left(\frac{1}{12} + \frac{1}{4}\right) = \frac{ML^2}{3}$$

which agrees with the moment of inertia formula from the table for a uniform rod about an axis through its end and perpendicular to the rod.

## 7 Rotation

**Moment of inertia for a rigid system of extended rigid bodies**: The total moment of inertia of a rigid system of $N$ extended rigid bodies and/or pointlike masses about a specified axis of rotation can be found by adding together all of the moments of inertia about that axis of rotation:

$$I = \sum_{i=1}^{N} I_i$$

It's important to make sure that you determine all of the moments of inertia about the same axis of rotation before adding them together.

**Example.** A monkey glues the end of a 20-g, 12-cm long uniform wooden rod to a 5.0-g uniform solid wooden sphere with an 8.0-cm diameter. Determine the moment of inertia of the combined object about an axis perpendicular to the rod that passes through the free end of the rod, as illustrated below.

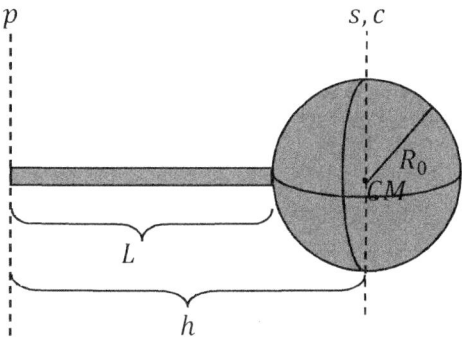

The moment of inertia of a uniform rod rotating about an axis through its end and perpendicular to the rod can be found from the table:

$$I_{r,p} = \frac{M_r L^2}{3} = \frac{(20)(12)^2}{3} = 960 \text{ g·cm}^2$$

The moment of inertia of a solid sphere is not listed in the table for the specified axis, but we can first find its moment of inertia about an axis through its center and then use the parallel-axis theorem, where $h$ is the distance from the free end of the rod to the center of the sphere:

$$I_{s,c} = \frac{2M_s R_0^2}{5} = \frac{(2)(5.0)(4.0)^2}{5} = 32 \text{ g·cm}^2$$
$$I_{s,p} = I_{s,c} + M_s h^2 = 32 + (5.0)(16)^2 = 1312 \text{ g·cm}^2$$

Since $1.0 \text{ m}^2 = (100 \text{ cm})^2 = 10^4 \text{ cm}^2$ and $1 \text{ kg} = 10^3 \text{ g}$, the total moment of inertia is

$$I_{r,p} + I_{s,p} = 2.2 \times 10^{-4} \text{ kg·m}^2$$

## 7.3 The Vector Product

**Determinants**: A matrix is an array of numbers. To find the determinant of a 2 × 2 matrix, multiply the numbers on the main diagonal (top left times bottom right) and subtract the product of the numbers on the cross diagonal (bottom left times top right[292]) as follows:

$$\begin{vmatrix} a & b \\ c & d \end{vmatrix} = ad - cb$$

One way to generalize the determinant of a 2 × 2 matrix to the determinant of a 3 × 3 matrix is to use the method of cofactors. The elements of the top row of the matrix serve as cofactors. Blocking out the row and column of a cofactor yields its submatrix. For example, we have illustrated the submatrix of the cofactor $a$ in the 3 × 3 matrix below:

$$\begin{pmatrix} a & b & c \\ d & e & f \\ g & h & i \end{pmatrix}$$

Using the method of cofactors, the determinant of a 3 × 3 matrix can be found by treating each element of the top row as a cofactor, multiplying each cofactor by the determinant of its corresponding 2 × 2 submatrix, and adding these factors together with alternating signs as follows:

$$\begin{vmatrix} a & b & c \\ d & e & f \\ g & h & i \end{vmatrix} = a \begin{vmatrix} e & f \\ h & i \end{vmatrix} - b \begin{vmatrix} d & f \\ g & i \end{vmatrix} + c \begin{vmatrix} d & e \\ g & h \end{vmatrix} = aei + bfg + cdh - afh - bdi - ceg$$

The cofactor method naturally generalizes to $N \times N$ matrices, and serves as a particularly convenient form for the vector product (to be discussed shortly), as the unit vectors serve as the cofactors.

Many students prefer a shortcut method for computing the determinant of a 3 × 3 matrix. However, if you ever take an advanced math course and need to take the determinant of a 4 × 4 or higher-dimensional square matrix, the cofactor method will generalize naturally, but the following shortcut method will not. The shortcut method for the determinant of a 3 × 3 matrix involves adding two columns to the right of the matrix by repeating the two leftmost columns, multiplying down the three diagonals and then up the three diagonals, adding the terms together with negative signs for the upward multiplications as follows:

$$\begin{vmatrix} a & b & c & a & b \\ d & e & f & d & e \\ g & h & i & g & h \end{vmatrix} = aei + bfg + cdh - ceg - bdi - afh$$

---

[292] Technically, if there are any differential operators in the matrix elements, you should act with the top right element before the bottom left element, as in the case of the curl operator from vector calculus.

**Notation**. When parentheses enclose an array of numbers – as in $\begin{pmatrix} a & b \\ c & d \end{pmatrix}$ – we are referring to the matrix itself, but when straight lines (not to be confused with absolute values) enclose the matrix – as in $\begin{vmatrix} a & b \\ c & d \end{vmatrix}$ – this means to take the determinant of the matrix. The determinant is sometimes also expressed using the abbreviation $det$, as in $det\begin{pmatrix} a & b \\ c & d \end{pmatrix}$, instead of using straight lines.

**Example**. Find the determinant of the matrix $\begin{pmatrix} 2 & -1 \\ 3 & 4 \end{pmatrix}$.

Following the technique described previously,

$$\begin{vmatrix} 2 & -1 \\ 3 & 4 \end{vmatrix} = (2)(4) - (3)(-1) = 8 + 3 = 11$$

**Example**. Find the determinant of the matrix $\begin{pmatrix} 2 & 1 & -1 \\ 1 & 3 & 0 \\ 2 & -1 & 4 \end{pmatrix}$.

Applying the cofactor method,

$$\begin{vmatrix} 2 & 1 & -1 \\ 1 & 3 & 0 \\ 2 & -1 & 4 \end{vmatrix} = 2\begin{vmatrix} 3 & 0 \\ -1 & 4 \end{vmatrix} - 1\begin{vmatrix} 1 & 0 \\ 2 & 4 \end{vmatrix} + (-1)\begin{vmatrix} 1 & 3 \\ 2 & -1 \end{vmatrix}$$
$$= 2[(3)(4) - (-1)(0)] - [(1)(4) - (2)(0)] - [(1)(-1) - (2)(3)] = 24 - 4 + 1 + 6 = 27$$

**The vector product**: A method of multiplying two vectors, $\vec{A}$ and $\vec{B}$, together and obtaining a vector as a result is known as the vector product. The vector product $\vec{A} \times \vec{B}$ is defined according to a $3 \times 3$ determinant[293] formed by placing the Cartesian unit vectors along the top row, the Cartesian components of the first vector along the middle row, and the Cartesian components of the second vector along the bottom row:

$$\vec{A} \times \vec{B} \equiv \begin{vmatrix} \hat{\imath} & \hat{\jmath} & \hat{k} \\ A_x & A_y & A_z \\ B_x & B_y & B_z \end{vmatrix} = \hat{\imath}\begin{vmatrix} A_y & A_z \\ B_y & B_z \end{vmatrix} - \hat{\jmath}\begin{vmatrix} A_x & A_z \\ B_x & B_z \end{vmatrix} + \hat{k}\begin{vmatrix} A_x & A_y \\ B_x & B_y \end{vmatrix}$$
$$\vec{A} \times \vec{B} = (A_y B_z - A_z B_y)\hat{\imath} - (A_x B_z - A_z B_x)\hat{\jmath} + (A_x B_y - A_y B_x)\hat{k}$$
$$\vec{A} \times \vec{B} = A_y B_z \hat{\imath} - A_z B_y \hat{\imath} + A_z B_x \hat{\jmath} - A_x B_z \hat{\jmath} + A_x B_y \hat{k} - A_y B_x \hat{k}$$

---

[293] Although the determinant is a valid operation for any $N \times N$ matrix, the algebra of the vector product is only mathematically valid in 3D and 8D space, and only physically meaningful in 3D space. However, if you take a course in mathematical physics, you will learn about higher-rank tensors, and learn that the vector product can also be expressed as a tensor multiplication, replacing the pseudovector with a higher-rank tensor. Whereas the vector product does not generalize to higher-dimensional space, the tensor product form (which is said to be the dual of the pseudovector form in 3D) does generalize to higher dimensions. So if you're wondering how physics formulas that involve the vector product could be valid in a higher-dimensional spacetime predicted by superstring theory, for example, the answer is that it must first be cast in its dual tensor form and then generalized.

The vector product is also called the cross product. It's important to distinguish between the vector product, $\vec{A} \times \vec{B}$, and the scalar product, $\vec{A} \cdot \vec{B}$ (see Sec. 5.1). The vector product, $\vec{A} \times \vec{B}$, involves a determinant and results in a vector, whereas the scalar product, $\vec{A} \cdot \vec{B}$, results in a scalar.

**Notation.** The cross ($\times$) is used to designate the vector (or cross) product, as in $\vec{A} \times \vec{B}$, which results in a vector, whereas the dot ($\cdot$) is used to designate the scalar (or dot) product, as in $\vec{A} \cdot \vec{B}$. When multiplying ordinary numbers, we use the cross and dot interchangeably – i.e. $3 \times 2 = 3 \cdot 2 = (3)(2) = 6$. However, you may not swap the cross and dot in the context of vector multiplication – i.e. $\vec{A} \times \vec{B}$ is much different from $\vec{A} \cdot \vec{B}$.

**Example.** Given $\vec{A} = 2\hat{\imath} + \hat{\jmath} - 3\hat{k}$ and $\vec{B} = 4\hat{\imath} + 2\hat{k}$, determine $\vec{C}$ where $\vec{C} = \vec{A} \times \vec{B}$.

First form a matrix by placing the Cartesian unit vectors – in the order $\hat{\imath}, \hat{\jmath}, \hat{k}$ – in the top row, the components of $\vec{A}$ in the middle row, and the components of $\vec{B}$ in the bottom row. Then take the determinant of this matrix:

$$\vec{C} = \vec{A} \times \vec{B} = \begin{vmatrix} \hat{\imath} & \hat{\jmath} & \hat{k} \\ A_x & A_y & A_z \\ B_x & B_y & B_z \end{vmatrix} = \begin{vmatrix} \hat{\imath} & \hat{\jmath} & \hat{k} \\ 2 & 1 & -3 \\ 4 & 0 & 2 \end{vmatrix}$$

$$\vec{C} = \hat{\imath}\begin{vmatrix} 1 & -3 \\ 0 & 2 \end{vmatrix} - \hat{\jmath}\begin{vmatrix} 2 & -3 \\ 4 & 2 \end{vmatrix} + \hat{k}\begin{vmatrix} 2 & 1 \\ 4 & 0 \end{vmatrix}$$

$$\vec{C} = [(1)(2) - (0)(-3)]\hat{\imath} - [(2)(2) - (4)(-3)]\hat{\jmath} + [(2)(0) - (4)(1)]\hat{k}$$

$$\vec{C} = 2\hat{\imath} - 16\hat{\jmath} - 4\hat{k}$$

**Magnitude of the vector product**: The magnitude of the vector product, $\|\vec{A} \times \vec{B}\|$, turns out to have a form that is similar to the scalar product,[294] except involving a sine instead of a cosine:

$$\|\vec{A} \times \vec{B}\| = AB \sin \theta$$

where, like the scalar product, $\theta$ is the angle between $\vec{A}$ and $\vec{B}$. Recall that $0 \leq \theta \leq 180°$. When the two vectors are parallel or anti-parallel, $\|\vec{A} \times \vec{B}\|$ is zero; and when the two vectors are perpendicular, $\|\vec{A} \times \vec{B}\|$ equals $AB$.[295] We prove the above relation toward the end of this section.

Geometrically, the magnitude of the vector product is interpreted as the area[296] of the parallelogram formed by joining $\vec{A}$ and $\vec{B}$ tip-to-tail, as illustrated in the diagram on the following page.

---

[294] Recall that the scalar product can be expressed as $\vec{A} \cdot \vec{B} = AB \cos \theta$.

[295] Compare with the scalar product: When two vectors are parallel, $\vec{A} \cdot \vec{B}$ equals $AB$; when two vectors are perpendicular, $\vec{A} \cdot \vec{B}$ is zero; and when two vectors are anti-parallel $\vec{A} \cdot \vec{B}$, equals $-AB$.

[296] This 'area' will only have SI units of square meters like the ordinary geometric area if the two vectors are position or displacement vectors – i.e. vectors that are individually made from distances. If either vector is a different type of vector, such as a velocity vector or a force vector, the 'area' of the parallelogram will generally have different SI units (than square meters).

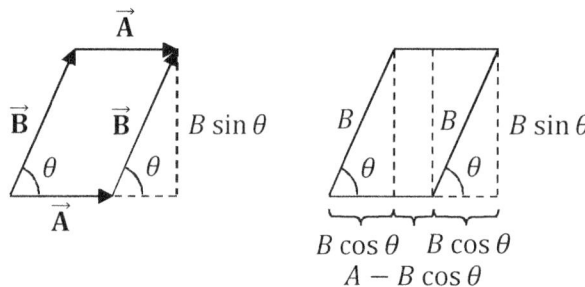

The parallelogram illustrated above by joining $\vec{A}$ and $\vec{B}$ together tip-to-tail can be divided up into two right triangles and one rectangle in between. The area of this parallelogram is

$$A_p = 2A_t + A_r = 2\frac{1}{2}(B\cos\theta)(B\sin\theta) + (A - B\cos\theta)(B\sin\theta) = AB\sin\theta \quad \text{(parallelogram)}$$

**Hint.** If you only need the magnitude of the vector product, and don't need to express the result in terms of unit vectors, it is more efficient to use the formula $\|\vec{A} \times \vec{B}\| = AB\sin\theta$ instead of using the determinant method.

**Example.** Two vectors, $\vec{A}$ and $\vec{B}$, have equal magnitude. A third vector, $\vec{C}$, has a magnitude equal to $A^2/2$. If the three vectors are related by $\vec{C} = \vec{A} \times \vec{B}$, what is the angle between $\vec{A}$ and $\vec{B}$?

Apply the formula for the magnitude of the vector product:

$$C = \|\vec{A} \times \vec{B}\| = AB\sin\theta$$
$$\frac{A^2}{2} = A^2 \sin\theta$$
$$\theta = \sin^{-1}\left(\frac{1}{2}\right) = 30° \text{ or } 150°$$

**Algebra of the vector product**: Looking at the determinant definition of the vector product, it's easy to see (by swapping the order of $\vec{A}$ and $\vec{B}$) that the vector product is anti-commutative:[297]

$$\vec{B} \times \vec{A} = -\vec{A} \times \vec{B}$$

The vector product is distributive (left as an exercise at the end of the chapter):

$$\vec{A} \times (\vec{B} + \vec{C}) = \vec{A} \times \vec{B} + \vec{A} \times \vec{C}$$

However, the vector product is not associative, since $\vec{A} \times (\vec{B} \times \vec{C})$ is not generally equal to $(\vec{A} \times \vec{B}) \times \vec{C}$.

---

[297] The scalar product, in contrast, is commutative: $\vec{B} \cdot \vec{A} = \vec{A} \cdot \vec{B}$.

**Vector product of unit vectors**: Consider the vector product $\hat{\imath} \times \hat{\jmath}$. The best way to think of this is to define $\vec{A} = \hat{\imath}$ and $\vec{B} = \hat{\jmath}$. We can always write any vector in terms of its components as $\vec{A} = A_x\hat{\imath} + A_y\hat{\jmath} + A_z\hat{k}$. Therefore, for $\vec{A} = \hat{\imath}$, $A_x = 1$ and $A_y = A_z = 0$. Similarly, for $\vec{B} = \hat{\jmath}$, $B_x = B_z = 0$ and $B_y = 1$. We can evaluate $\hat{\imath} \times \hat{\jmath}$ in determinant form using these components:[298]

$$\hat{\imath} \times \hat{\jmath} = \begin{vmatrix} \hat{\imath} & \hat{\jmath} & \hat{k} \\ 1 & 0 & 0 \\ 0 & 1 & 0 \end{vmatrix} = \hat{k}$$

You can similarly work out the vector product between other pairs of Cartesian unit vectors:[299]

$$\hat{\imath} \times \hat{\imath} = 0 \quad , \quad \hat{\jmath} \times \hat{\jmath} = 0 \quad , \quad \hat{k} \times \hat{k} = 0$$
$$\hat{\imath} \times \hat{\jmath} = \hat{k} \quad , \quad \hat{\imath} \times \hat{k} = -\hat{\jmath} \quad , \quad \hat{\jmath} \times \hat{k} = \hat{\imath}$$
$$\hat{\jmath} \times \hat{\imath} = -\hat{k} \quad , \quad \hat{k} \times \hat{\imath} = \hat{\jmath} \quad , \quad \hat{k} \times \hat{\jmath} = -\hat{\imath}$$

Here is one way to remember which of these are positive and which are negative. The vector product is said to be cyclic. That is, the unit vectors follow a circular order, according to the following sequence, $\hat{\imath}, \hat{\jmath}, \hat{k}, \hat{\imath}, \hat{\jmath}$ ... What this means is that $\hat{\imath} \times \hat{\jmath}$ equals $+\hat{k}$, $\hat{\jmath} \times \hat{k}$ equals $+\hat{\imath}$, and $\hat{k} \times \hat{\imath}$ equals $+\hat{\jmath}$ because they follow this cyclic order (where $\hat{\imath}$ comes after $\hat{k}$). If the unit vectors proceed in the reverse order, the result is negative, as in $\hat{k} \times \hat{\jmath} = -\hat{\imath}$.

If you want to find the vector product between unit vectors of other coordinate systems, the most straightforward way to proceed is to express all of the unit vectors in terms of Cartesian unit vectors. This is illustrated in the following example.

**Example**. Evaluate $\hat{r}_c \times \hat{\imath}$, where $\hat{r}_c$ is the radial unit vector of cylindrical coordinates.

From Sec. 2.4, we know that $\hat{r}_c = \hat{\imath} \cos\theta + \hat{\jmath} \sin\theta$. Therefore, $\hat{r}_c$ has an $x$-component of $\cos\theta$ and a $y$-component of $\sin\theta$, such that

$$\hat{r}_c \times \hat{\imath} = \begin{vmatrix} \hat{\imath} & \hat{\jmath} & \hat{k} \\ \cos\theta & \sin\theta & 0 \\ 1 & 0 & 0 \end{vmatrix} = -\hat{k} \sin\theta$$

**Direction of the vector product**: The direction of $\vec{A} \times \vec{B}$ is perpendicular to $\vec{A}$ and also to $\vec{B}$. If $\vec{C} = \vec{A} \times \vec{B}$, then $\vec{C}$ is perpendicular to $\vec{A}$ (i.e. $\vec{C} \perp \vec{A}$) and $\vec{C}$ is perpendicular to $\vec{B}$ (i.e. $\vec{C} \perp \vec{B}$). Put another way, $\vec{C}$ is perpendicular to the $\vec{A}$-$\vec{B}$ plane (i.e. the plane that contains the two vectors $\vec{A}$ and $\vec{B}$). However, it is important to realize that while $\vec{A}$ and $\vec{B}$ are each perpendicular to $\vec{C}$, $\vec{A}$ and $\vec{B}$ may not be perpendicular to one another.

---

[298] Alternatively, you can use the right-hand rule, described later in this section. The right-hand rule correctly predicts the result of the vector product between two unit vectors of the same coordinate system. When mixing unit vectors of different coordinate systems, the right-hand rule gives only the direction (not the magnitude).

[299] Compare with the scalar product between Cartesian unit vectors: $\hat{\imath} \cdot \hat{\imath} = 1, \hat{\jmath} \cdot \hat{\jmath} = 1, \hat{k} \cdot \hat{k} = 1, \hat{\imath} \cdot \hat{\jmath} = 0, \hat{\imath} \cdot \hat{k} = 0$, and $\hat{\jmath} \cdot \hat{k} = 0$.

If $\vec{C} = \vec{A} \times \vec{B}$, its direction is perpendicular to the plane formed by the vectors $\vec{A}$ and $\vec{B}$. However, this information alone does not fully specify the direction of $\vec{C}$: There are two possible directions perpendicular to a plane. For example, up and down are two directions that are both perpendicular to a level floor. If you know the directions of the vectors $\vec{A}$ and $\vec{B}$, you can determine the direction of $\vec{C}$ qualitatively using the right-hand rule (which we shall describe in a moment).

The proof that the vector product, $\vec{A} \times \vec{B}$, is perpendicular to the plane formed by the vectors $\vec{A}$ and $\vec{B}$ is left as a directed exercise at the end of the chapter.

**Arrows in 3D space**: Drawing the vectors associated with the vector product is inherently 3D, since the vector product is perpendicular to the plane formed by the two vectors from which it is calculated. As a consequence, when drawing diagrams to illustrate how the directions of the three vectors – i.e. $\vec{A}$, $\vec{B}$, and $\vec{A} \times \vec{B}$ – are related, it is often necessary to draw arrows going into or out of the page. We will use the symbol $\otimes$ to represent an arrow going into the page, and $\odot$ to represent an arrow coming out of the page. The symbol $\otimes$ looks like the tail feathers that you would see if a physical arrow really were stuck into the page, and $\odot$ looks like the tip of the arrow that you would see if a physical arrow really were coming out of the page toward you.

---

**Notation.** The symbol $\otimes$ is an arrow directed into the page and the symbol $\odot$ is an arrow coming out of the page.

---

**Right-hand rule**: The direction of the vector product can be found by using your right hand. Here is one way to use the right-hand rule:[300]
- Extend the fingers of your right hand[301] such that they are coplanar with your palm. Also extend your thumb so that it is coplanar with your palm, but perpendicular to your extended fingers.
- Point your extended fingers along $\vec{A}$, leaving your fingers and thumb extended, as previously noted. There is more than one way to point your fingers toward $\vec{A}$: If you literally twist your arm, your fingers will still be directed toward $\vec{A}$.

---

[300] This is not the only way to form a right-hand rule. For one, it's not really necessary to curl your fingers: You could just position two fingers and a thumb so that they are mutually perpendicular. Then your forefinger (also called index finger; the one nearest your thumb) could extend along $\vec{A}$, your middle finger along could extend along $\vec{B}$, and your thumb would point along $\vec{A} \times \vec{B}$. Since the vector product is cyclic, we could instead point the middle finger toward $\vec{A}$, the thumb along $\vec{B}$, and then the forefinger would point along $\vec{A} \times \vec{B}$; or we could point the thumb along $\vec{A}$, the forefinger along $\vec{B}$, and then the middle finger would point along $\vec{A} \times \vec{B}$. The problem with this three-finger right-hand rule method is that the forefinger must be extended outward, while the middle finger must be bent perpendicular to the palm: Having seen students inadvertently extend the middle finger and bend the forefinger instead, I realized that the finger-curling right-hand rule is slightly more foolproof. If you want to be tedious, you can also devise cyclic variations of the finger-curling method; and then we can work out some left-hand rules, too.

[301] It's amazing how many times I've given a quiz on the right-hand rule and observed students using their left hands! Even students who are right-handed: If you're right-handed, you have a pencil in your right hand during the quiz, and so you instinctively proceed to use your free hand for the right-hand rule... Oops!

- Now twist your arm so that your extended fingers still point along $\vec{A}$, but so that your palm faces[302] $\vec{B}$, still leaving your fingers and thumb extended, as previously noted.
- Check that you have done this correctly: If you curl your fingers naturally,[303] you should be able to point your fingers along $\vec{B}$; if you uncurl your fingers back to their original extended position, they should now point along $\vec{A}$. Both of these must be true or you are not ready to proceed to the next step.
- Your thumb should still remain in its original extended position at all times. Your thumb should be perpendicular to your fingers whether they are extended or curled.
- With your fingers pointing along $\vec{A}$ when extended and along $\vec{B}$ when curled, your thumb will point along $\vec{A} \times \vec{B}$, assuming that it's still in its original extended position.[304]
- Check that $\vec{A} \times \vec{B}$ is perpendicular to $\vec{A}$ and that $\vec{A} \times \vec{B}$ is also perpendicular to $\vec{B}$. If not, you have a problem.

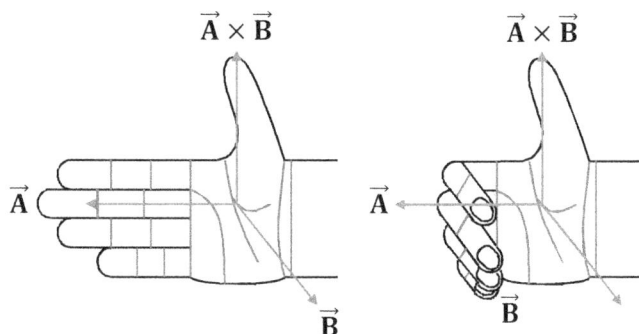

**Conceptual Example.** Given that $\vec{A}$ is directed upward (↑) and that $\vec{B}$ is directed to the left (←), determine the direction of $\vec{A} \times \vec{B}$.

Point the fingers of your right hand along $\vec{A}$, which in this exercise is upward (↑). Twist your arm until you can curl your fingers toward $\vec{B}$, which in this exercise is left (←). If done correctly, your extended thumb will point out of the page (⊙), which is the direction of $\vec{A} \times \vec{B}$.

---

[302] If you visualize an arrow coming out of your palm, perpendicular to your palm, it will point along $\vec{B}$.

[303] I've seen some very 'gifted' double-jointed right hands – i.e. students who can curl their fingers in ways that were 'natural' for them, but quite 'unnatural' to most. If you're similarly gifted with your right hand, make sure that you curl your fingers in a way that most humans naturally curl their fingers.

[304] Every year, a student can't refrain from asking, "What happens if I lose my right hand before the test?" The obvious answer is that you either have to use your left hand and negate the result, or if you don't have any hands at the time of the test, you must simply visualize your right hand in your mind. However, so that other students don't obtain an unfair advantage, we'll tie their hands behind their backs. Most likely, the state superintendent of education will happen to be taking a tour of the school and walk into class during the exam, and we'll all have a blast. Ask a stupid question, get a creative answer. My second favorite right-hand rule question is, "Can I tattoo the words 'right' and 'left' on my arms before the exam?" Indubitably, a few clever (!) students have made some subtle indication on their arms or sleeves, or worn a watch on one hand, so as to help remember which is which.

## 7 Rotation

> **Conceptual Example.** Given that $\vec{v}$ is directed into the page ($\otimes$) and that $\vec{B}$ is directed downward ($\downarrow$), determine the direction of $\vec{F}$, where $\vec{F} = q\vec{v} \times \vec{B}$, assuming that $q$ is positive.
>
> Point the fingers of your right hand along the first vector, $\vec{v}$, which in this exercise is into the page ($\otimes$). Twist your arm until you can curl your fingers toward $\vec{B}$, which in this exercise is downward ($\downarrow$). If done correctly, your thumb will point to the left ($\leftarrow$), which is the direction of $\vec{F}$. (If the constant $q$ had been negative, then the direction of $\vec{F}$ would be negated – i.e. $\vec{F}$ would instead point to the right.)

**Inverting the vector product calculation**: Given two vectors, $\vec{A}$ and $\vec{B}$, which are not collinear, the vector product, $\vec{A} \times \vec{B}$, is uniquely determined: Enter the Cartesian components of $\vec{A}$ and $\vec{B}$ into the determinant form of the vector product, and you will obtain a unique result for $\vec{A} \times \vec{B}$.

However, suppose that instead you are given $\vec{A}$ and $\vec{C}$, where $\vec{C} = \vec{A} \times \vec{B}$, but wish to determine $\vec{B}$. In this case, the vector $\vec{B}$ could one of an infinite set of vectors that satisfies the relation $\vec{C} = \vec{A} \times \vec{B}$. Therefore, if you know $\vec{A}$ and $\vec{A} \times \vec{B}$, you must have additional information in order to uniquely determine $\vec{B}$.

One way to understand this distinction is to examine the formula for the magnitude of the vector product, $\|\vec{A} \times \vec{B}\| = AB \sin \theta$. Given $\vec{A}$ and $\vec{B}$, the angle between them, $\theta$, can be determined, from which $\|\vec{A} \times \vec{B}\|$ can be determined. The direction of $\vec{A} \times \vec{B}$ is uniquely determined through the right-hand rule. (Alternatively, you can examine the determinant form of the vector product, which gives a vector from which both the magnitude and direction can be determined.)

In the other case, we are given $\vec{A}$ and $\vec{C}$, where $\vec{C} = \vec{A} \times \vec{B}$, and wish to determine the magnitude and direction of $\vec{B}$. The magnitudes of these vectors are related by $C = AB \sin \theta$. Knowing $C$ and $A$ leaves two unknowns, $B \sin \theta$. The angle between $\vec{A}$ and $\vec{C}$ is 90° (just like the angle between $\vec{B}$ and $\vec{C}$), which does not help to determine the angle between $\vec{A}$ and $\vec{B}$, which is $\theta$. There is a whole plane of possible directions of $\vec{B}$; for each possible value of $\theta$, $\vec{B}$ has a different magnitude, in order to satisfy $C = AB \sin \theta$.

**Jacobi identity**: The cyclic property of the vector product is embodied by the Jacobi identity (derived in the last example of this section):

$$\vec{A} \times (\vec{B} \times \vec{C}) + \vec{B} \times (\vec{C} \times \vec{A}) + \vec{C} \times (\vec{A} \times \vec{B}) = 0$$

**The triple vector product**: It is often convenient to express the triple vector product, $\vec{A} \times (\vec{B} \times \vec{C})$, in terms of scalar products using the BAC-CAB rule (the name of the rule being helpful for remembering the equation):

$$\vec{A} \times (\vec{B} \times \vec{C}) = \vec{B}(\vec{A} \cdot \vec{C}) - \vec{C}(\vec{A} \cdot \vec{B})$$

Observe that there is no ambiguity in an expression such as $\vec{B}(\vec{A} \cdot \vec{C})$: Since $\vec{A} \cdot \vec{C}$ is a scalar, $\vec{B}(\vec{A} \cdot \vec{C})$ can only mean to multiply the scalar $\vec{A} \cdot \vec{C}$ times the vector $\vec{B}$. We leave the proof of the BAC-CAB rule as an exercise at the end of the chapter.

**Kronecker delta**: Many of the equations involving scalar and vector products can be derived most concisely by introducing two multi-index symbols, called the Kronecker delta[305] and the Levi-Civita symbol. The Kronecker delta, $\delta_{ij}$, equals 1 if $i = j$ and 0 otherwise:

$$\delta_{ij} = \begin{cases} 1 & \text{if } i = j \\ 0 & \text{if } i \neq j \end{cases}$$

**Double summation**: Two indices, $i$ and $j$, vary in the double summation below. This double summation consists of 9 terms – one for each pair of values for $i$ and $j$:

$$\sum_{i=1}^{3}\sum_{j=1}^{3} A_i B_j = A_1 B_1 + A_1 B_2 + A_1 B_3 + A_2 B_1 + A_2 B_2 + A_2 B_3 + A_3 B_1 + A_3 B_2 + A_3 B_3$$

where $A_1$, $A_2$, and $A_3$ represent the Cartesian components – i.e. $A_x$, $A_y$, and $A_z$, respectively – of the vector $\vec{A}$, and similarly for $\vec{B}$. When both indices vary over the same range, we often express the double summation more concisely as

$$\sum_{i,j=1}^{3} A_i B_j \equiv \sum_{i=1}^{3}\sum_{j=1}^{3} A_i B_j$$

---

**Notation.** In summation notation, it is convenient to refer to the Cartesian components – i.e. $A_x$, $A_y$, and $A_z$ – of a vector with numerical subscripts as $A_1$, $A_2$, and $A_3$, respectively, where the subscripts 1, 2, and 3 correspond to the Cartesian coordinates $x$, $y$, and $z$.

---

**Contracting indices using the Kronecker delta**: The Kronecker delta has the effect of 'contracting' the two indices that appear in its subscripts. For example, the following double summation featuring a Kronecker delta simplifies to a single summation:

$$\sum_{i,j=1}^{3} A_i B_j \, \delta_{ij} = \sum_{i=1}^{3} A_i B_i$$

All of the terms where the two indices, $i$ and $j$, are not equal are zero because $\delta_{ij}$ when $i \neq j$. In this way, the Kronecker delta 'contracts' two indices. When you see a Kronecker delta multiplying other factors that share one or both of its indices, set the two indices equal in the other factors multiplying the Kronecker delta and remove the summation over the index that you eliminate (as illustrated in the above equation and coming examples).

---

[305] It's important to include the name Kronecker when referring to the Kronecker delta symbol because there are two important delta's ($\delta$) in mathematical physics – the Kronecker delta that we are discussing here, and a delta function which means something much different.

# 7 Rotation

**Scalar product in terms of the Kronecker delta**: The scalar product can be written in terms of the Kronecker delta as follows:

$$\vec{\mathbf{A}} \cdot \vec{\mathbf{B}} = A_x B_x + A_y B_y + A_z B_z = \sum_{i=1}^{3} A_i B_i = \sum_{i,j=1}^{3} A_i B_j \delta_{ij}$$

**Example.** Show that $\sum_{j=1}^{3} \delta_{ij}\delta_{jk} = \delta_{ik}$, where the subscripts $i$ and $k$ are elements of the set $\{1,2,3\}$ – i.e. $i,k \in \{1,2,3\}$.

The Kronecker delta has the effect of contracting indices. That is, we can remove the $\delta_{jk}$ from the sum, set $j = k$, and remove the sum over $j$, which yields $\delta_{ik}$. Conceptually, $\delta_{ij}\delta_{jk}$ can only be nonzero if $i = j$ and $j = k$, and since we are summing over all possible values of $j$, this equates to saying that $\delta_{ij}\delta_{jk}$ can only be nonzero if $i = k$ – which is exactly what $\delta_{ik}$ says.

The equation $\sum_{j=1}^{3} \delta_{ij}\delta_{jk} = \delta_{ik}$ really represents 9 different equations – one for each possible pair of $i$ and $k$. For example, for $i = 1$ and $k = 2$, the equation reads $\sum_{j=1}^{3} \delta_{1j}\delta_{j2} = \delta_{12}$. For each of the 9 equations, there are 3 terms on the left-hand side. For example, for $i = 1$ and $k = 2$, the equation reads $\delta_{11}\delta_{12} + \delta_{12}\delta_{22} + \delta_{13}\delta_{32} = \delta_{12}$ in expanded form. You can easily verify that $\delta_{11}\delta_{12} + \delta_{12}\delta_{22} + \delta_{13}\delta_{32}$ and $\delta_{12}$ are both zero. As a second example, when $i = 3$ and $k = 3$, the equation reads $\delta_{31}\delta_{13} + \delta_{32}\delta_{23} + \delta_{33}\delta_{33} = \delta_{33}$, which simplifies to $0 + 0 + 1 = 1$.

**Notation.** The braces, $\{\}$, are utilized to denote the elements of a set. For example $i \in \{1,2,\cdots,N\}$, denotes that $i$ is a member of the set $\{1,2,\cdots N\}$, where $\in$ is the set membership symbol.[306]

**Free versus dummy indices**: An index that is being summed over is referred to as a 'dummy index,' whereas an index that is not being summed over is referred to as a 'free index.' For example, in the equation $\sum_{i=1}^{3} A_i \delta_{ij} = A_j$, the subscript $i$ is a dummy index and the subscript $j$ is a free index. The dummy indices are repeated within the sum – e.g. the index $i$ appears twice in $\sum_{i=1}^{3} A_i \delta_{ij} = A_j$. A free index appears just once within the sum. In this notation,[307] there is a conservation of indices – i.e. freedom is the same in each term of the equation. For example, in $\sum_{i=1}^{3} A_i \delta_{ij} = A_j$, the free index $j$ appears in every term throughout the equation (and appears exactly once in each term).

---

[306] The set membership symbol ($\in$) is derived from the lowercase Greek symbol epsilon ($\varepsilon$), which comes in a variety of forms. The lunate variation of the epsilon symbol ($\epsilon$) looks very similar, yet has a distinctly different usage, meaning something entirely different from set membership, as we will see momentarily.

[307] This notation is called tensor notation. It turns out that scalars and vectors are special cases of a more general multi-component object called a tensor. A scalar is a single-component tensor of rank zero, and a vector is a three-component tensor of rank one (in 3D space). All advanced work in vector and tensor analysis is done using tensor notation, and everybody gets lazy and drops the summation symbols – instead, there is an implied summation over repeated (dummy) indices in any term. If you're interested in tensors, look for a text on tensor calculus or mathematical physics.

**Levi-Civita symbol**: Whereas the Kronecker delta is a symmetric tensor ($\delta_{ji} = \delta_{ij}$), the Levi-Civita symbol, $\epsilon_{ijk}$,[308] is an anti-symmetric tensor:

$$\epsilon_{ijk} = \begin{cases} 1 & \text{for } \epsilon_{123}, \epsilon_{231}, \text{ and } \epsilon_{312}\text{[309]} \\ -1 & \text{for } \epsilon_{213}, \epsilon_{321}, \text{ and } \epsilon_{132} \\ 0 & \text{for repeated indices} \end{cases}$$

So $\epsilon_{112}$ and $\epsilon_{232}$, for example, both equal zero because indices have been repeated.

**Vector product in terms of the Levi-Civita symbol**: The vector product can be expressed concisely as a summation using the Levi-Civita symbol:

$$\vec{A} \times \vec{B} = \sum_{i,j,k=1}^{3} \epsilon_{ijk} A_i B_j \hat{x}_k$$

where $\hat{x}_1$, $\hat{x}_2$, and $\hat{x}_3$ represent the Cartesian unit vectors – i.e. $\hat{i}$, $\hat{j}$, and $\hat{k}$, respectively. If you write out this triple summation long-hand, there are 27 terms. However, 21 of these terms are zero because $\epsilon_{ijk}$ will have repeated indices. The 6 nonzero terms are

$$\sum_{i,j,k=1}^{3} \epsilon_{ijk} A_i B_j \hat{x}_k$$
$$= \epsilon_{123} A_1 B_2 \hat{x}_3 + \epsilon_{213} A_2 B_1 \hat{x}_3 + \epsilon_{231} A_2 B_3 \hat{x}_1 + \epsilon_{321} A_3 B_2 \hat{x}_1 + \epsilon_{312} A_3 B_1 \hat{x}_2 + \epsilon_{132} A_1 B_3 \hat{x}_2$$
$$= (A_2 B_3 - A_3 B_2) \hat{x}_1 + (A_3 B_1 - A_1 B_3) \hat{x}_2 + (A_1 B_2 - A_2 B_1) \hat{x}_3$$
$$= A_y B_z \hat{i} - A_z B_y \hat{i} + A_z B_x \hat{j} - A_x B_z \hat{j} + A_x B_y \hat{k} - A_y B_x \hat{k} = \vec{A} \times \vec{B}$$

If $\vec{C} = \vec{A} \times \vec{B}$, we can write a similar expression for the $k^{th}$-component of $\vec{C}$:

$$C_k = \sum_{i,j=1}^{3} \epsilon_{ijk} A_i B_j$$

since

$$\vec{C} = \sum_{k=1}^{3} C_k \hat{x}_k$$

---

[308] The Levi-Civita symbol is represented by the lunate variation of the epsilon symbol ($\epsilon$), which appears very similar to the set membership symbol ($\in$). More significant is the difference in their meaning and usage.
[309] $\epsilon_{231}$ and $\epsilon_{312}$ are said to be cyclic permutations of $\epsilon_{123}$, while $\epsilon_{321}$ and $\epsilon_{132}$ are cyclic permutations of $\epsilon_{213}$.

> **Notation.** In tensor notation, the Cartesian unit vectors – i.e. $\hat{\mathbf{i}}$, $\hat{\mathbf{j}}$, and $\hat{\mathbf{k}}$ – are expressed more concisely by using numbers for the subscripts – i.e. $\hat{\mathbf{x}}_1$, $\hat{\mathbf{x}}_2$, and $\hat{\mathbf{x}}_3$, respectively.

**Useful identities among Levi-Civita symbols**: It is often useful to realize that two adjacent indices can be swapped at the cost of a minus sign. For example,

$$\epsilon_{ikj} = -\epsilon_{ijk} \quad , \quad \epsilon_{jik} = -\epsilon_{ijk}$$

When a pair of Levi-Civita symbols appear in a summation, the following identities may prove useful:

$$\sum_{i,j=1}^{3} \epsilon_{ijk}\epsilon_{ijm} = 2\delta_{km}$$

$$\sum_{i=1}^{3} \epsilon_{ijk}\epsilon_{imn} = \delta_{jm}\delta_{kn} - \delta_{jn}\delta_{km}$$

The first summation above is easier to understand. If $k = m$, the only nonzero terms will come when $i \neq j$ and $i \neq k$, and there are two such terms – one where $\epsilon_{ijk}$ is $+1$ and one where $\epsilon_{ijk}$ is $-1$. In this case, the summation equals $(+1)^2 + (-1)^2 = 2$, which agrees with the right-hand side of the equation for $k = m$. For example, if $k = m = 2$, the only nonzero terms come when $i = 1$ and $j = 3$ and when $i = 3$ and $j = 1$: $\sum_{i,j=1}^{3} \epsilon_{ij2}\epsilon_{ij2} = \epsilon_{132}\epsilon_{132} + \epsilon_{312}\epsilon_{312} = (+1)^2 + (-1)^2 = 2 = 2\delta_{22}$. On the other hand, if $k \neq m$, every combination of $i$ and $j$ gives zero, and the equation simply states that $0 = 0$. For example, if $k = 1$ and $m = 2$, the only way that $\epsilon_{ijk}$ can be nonzero is if $i = 2$ and $j = 3$ or if $i = 3$ and $j = 2$, which means that $\epsilon_{ijm}$ has a repeated index, and therefore $\sum_{i,j=1}^{3} \epsilon_{ij1}\epsilon_{ij2} = 0$, which agrees with $2\delta_{12} = 0$.

The second summation is a little more involved, but if you invest the time to understand it conceptually, it will be much easier to remember the formula and you will also have better success applying it. The second summation arises frequently in vector product proofs, and so it is well worth the effort to understand it.

In the second summation, the only way that $\epsilon_{ijk}\epsilon_{imn}$ can be nonzero is if there are no repeated indices in both $\epsilon_{ijk}$ and $\epsilon_{imn}$ – i.e. if $j \neq k$ and that $m \neq n$ and if $j$, $k$, $m$, and $n$ do not include a 1, a 2, and a 3 (as opposed to only two of these digits, rather than all three). Some thought should reveal that this will only occur under two conditions: If $j = m$ and $k = n$ or if $j = n$ and $k = m$. In the former case, $\epsilon_{ijk}\epsilon_{imn}$ equals $+1$, and in the latter case, $\epsilon_{ijk}\epsilon_{imn}$ equals $-1$, when we sum over the dummy index $i$. The delta functions mathematically state these conceptual conditions: $\delta_{jm}\delta_{kn}$ states that $\epsilon_{ijk}\epsilon_{imn}$ will equal $+1$ if $j = m$ and $k = n$, and $\delta_{jn}\delta_{km}$ states that $\epsilon_{ijk}\epsilon_{imn}$ will equal $-1$ if $j = n$ and $k = m$ (for some value of $i$, which is guaranteed to happen since we are summing over all possible values of $i$). Now we address the two cases where $\epsilon_{ijk}\epsilon_{imn}$ would not have any nonzero terms: If $j = k$ or $m = n$, the summation will be zero; and if $j$, $k$, $m$, and $n$ include a 1, a 2, and a 3, it won't be possible for any $i$ to avoid repeating an index. If you consider $\delta_{jm}\delta_{kn} - \delta_{jn}\delta_{km}$ carefully, you will see that it's zero if $j = k$ or if $m = n$; and it's also zero if $j$, $k$, $m$, and $n$ include a 1, a 2, and a 3. Try to think this through, work out a few representative numerical examples, and try thinking it through again.

Here are a few representative examples to help illustrate the logic of the previous paragraph:

$$\sum_{i=1}^{3} \epsilon_{ijk}\epsilon_{imn} = \delta_{jm}\delta_{kn} - \delta_{jn}\delta_{km}$$

$j = 2, k = 3, m = 2, n = 3:$ $\quad \sum_{i=1}^{3} \epsilon_{i23}\epsilon_{i23} = 1 \quad, \quad \delta_{22}\delta_{33} - \delta_{23}\delta_{32} = 1$

$j = 2, k = 3, m = 3, n = 2:$ $\quad \sum_{i=1}^{3} \epsilon_{i23}\epsilon_{i32} = -1 \quad, \quad \delta_{23}\delta_{32} - \delta_{22}\delta_{33} = -1$

$j = 2, k = 3, m = 2, n = 2:$ $\quad \sum_{i=1}^{3} \epsilon_{i23}\epsilon_{i22} = 0 \quad, \quad \delta_{22}\delta_{32} - \delta_{22}\delta_{32} = 0$

$j = 2, k = 3, m = 1, n = 2:$ $\quad \sum_{i=1}^{3} \epsilon_{i23}\epsilon_{i12} = 0 \quad, \quad \delta_{21}\delta_{32} - \delta_{22}\delta_{31} = 0$

When it's possible to do so, it's very useful to express a pair of Levi-Civita symbols in terms of Kronecker deltas so that you may then use the Kronecker deltas to contract indices. We illustrate this in the following example and the derivation that follows it. This strategy is particularly useful in proving a variety of relations that involve vector products.

---

**Example**. Simplify the expression $\sum_{i,j,m=1}^{3} \epsilon_{ijk}\epsilon_{ijm} A_m$.

Using the identity $\sum_{i,j=1}^{3} \epsilon_{ijk}\epsilon_{ijm} = 2\delta_{km}$, this expression becomes $\sum_{m=1}^{3} 2\delta_{km} A_m$. The Kronecker delta contracts the indices $k$ and $m$, resulting in $2A_k$. All together,

$$\sum_{i,j,m=1}^{3} \epsilon_{ijk}\epsilon_{ijm} A_m = \sum_{m=1}^{3} 2\delta_{km} A_m = 2A_k$$

---

**Derivation of the formula for the magnitude of the vector product**: Let us begin with the expression $\vec{C} = \vec{A} \times \vec{B}$ and proceed to derive an equation for $C = \|\vec{C}\|$ in terms of $A = \|\vec{A}\|$, $B = \|\vec{B}\|$, and the angle between them, $\theta$. In tensor notation, the vector product can be expressed as

$$\vec{C} = \sum_{k=1}^{3} C_k \hat{x}_k = \sum_{i,j,k=1}^{3} \epsilon_{ijk} A_i B_j \hat{x}_k = \vec{A} \times \vec{B}$$

We find the scalar product of $\vec{C}$ with itself, which equals its magnitude squared:

$$\vec{C} \cdot \vec{C} = \sum_{k=1}^{3} C_k \hat{x}_k \cdot \sum_{n=1}^{3} C_n \hat{x}_n = C^2 = \sum_{i,j,k=1}^{3} \epsilon_{ijk} A_i B_j \hat{x}_k \cdot \sum_{\ell,m,n=1}^{3} \epsilon_{\ell mn} A_\ell B_m \hat{x}_n$$

We can collect and regroup this as

$$C^2 = \sum_{i,j,k,\ell,m,n=1}^{3} \epsilon_{ijk}\epsilon_{\ell mn} A_i A_\ell B_j B_m \hat{\mathbf{x}}_k \cdot \hat{\mathbf{x}}_n$$

The scalar product between unit vectors, $\hat{\mathbf{x}}_k \cdot \hat{\mathbf{x}}_n$, equals 1 if $k = n$ and 0 if $k \neq n$, which we can express using a Kronecker delta as $\hat{\mathbf{x}}_k \cdot \hat{\mathbf{x}}_n = \delta_{kn}$:

$$C^2 = \sum_{i,j,k,\ell,m,n=1}^{3} \epsilon_{ijk}\epsilon_{\ell mn} A_i A_\ell B_j B_m \delta_{kn}$$

The Kronecker delta allows us to contract indices, removing the sum over $n$ and changing the $n$ to a $k$:

$$C^2 = \sum_{i,j,k,\ell,m=1}^{3} \epsilon_{ijk}\epsilon_{\ell mk} A_i A_\ell B_j B_m$$

Now we apply one of the identities for pairs of Levi-Civita symbols (noting that $\epsilon_{kij} = \epsilon_{ijk}$):

$$C^2 = \sum_{i,j,\ell,m=1}^{3} (\delta_{i\ell}\delta_{jm} - \delta_{im}\delta_{j\ell}) A_i A_\ell B_j B_m$$

Let us separate this into two separate multi-summations:

$$C^2 = \sum_{i,j,\ell,m=1}^{3} \delta_{i\ell}\delta_{jm} A_i A_\ell B_j B_m - \sum_{i,j,\ell,m=1}^{3} \delta_{im}\delta_{j\ell} A_i A_\ell B_j B_m$$

Each Kronecker delta allows us to contract indices:

$$C^2 = \sum_{i,j=1}^{3} A_i A_i B_j B_j - \sum_{i,j=1}^{3} A_i A_j B_j B_i = \sum_{i=1}^{3} A_i A_i \sum_{j=1}^{3} B_j B_j - \sum_{i=1}^{3} A_i B_i \sum_{j=1}^{3} A_j B_j$$

Noting that $A^2 = \sum_{i=1}^{3} A_i A_i$, $B^2 = \sum_{j=1}^{3} B_j B_j$, and $\vec{\mathbf{A}} \cdot \vec{\mathbf{B}} = \sum_{i=1}^{3} A_i B_i$, this simplifies to

$$C^2 = A^2 B^2 - \left(\vec{\mathbf{A}} \cdot \vec{\mathbf{B}}\right)^2$$

Recalling that $\vec{\mathbf{A}} \cdot \vec{\mathbf{B}} = AB \cos\theta$,

$$C^2 = A^2B^2 - A^2B^2\cos^2\theta = A^2B^2\sin^2\theta$$
$$C = \|\vec{A}\times\vec{B}\| = AB\sin\theta$$

where we used the trigonometric identity $\sin^2\theta + \cos^2\theta = 1$.

---

**Example**. Derive the Jacobi identity using tensor notation.

We begin by writing the triple vector product with a Levi-Civita symbol:

$$\vec{A}\times(\vec{B}\times\vec{C}) = \vec{A}\times\sum_{i,j,k=1}^{3}\epsilon_{ijk}B_iC_j\hat{x}_k = \sum_{i,j,k,\ell,m=1}^{3}\epsilon_{\ell km}\epsilon_{ijk}A_\ell B_i C_j\hat{x}_m$$

Examine this last step carefully: It has the form $\sum_{k,\ell,m=1}^{3}\epsilon_{\ell km}A_\ell(\vec{B}\times\vec{C})_k\hat{x}_m$, where $(\vec{B}\times\vec{C})_k = \sum_{i,j=1}^{3}\epsilon_{ijk}B_iC_j$. We now apply this result to express the sum of three cyclic triple products:

$$\vec{A}\times(\vec{B}\times\vec{C}) + \vec{B}\times(\vec{C}\times\vec{A}) + \vec{C}\times(\vec{A}\times\vec{B})$$
$$= \vec{A}\times\sum_{i,j,k=1}^{3}\epsilon_{ijk}B_iC_j\hat{x}_k + \vec{B}\times\sum_{i,j,k=1}^{3}\epsilon_{ijk}C_iA_j\hat{x}_k + \vec{C}\times\sum_{i,j,k=1}^{3}\epsilon_{ijk}A_iB_j\hat{x}_k$$
$$= \sum_{i,j,k,\ell,m=1}^{3}\epsilon_{\ell km}\epsilon_{ijk}A_\ell B_i C_j\hat{x}_m + \sum_{i,j,k,\ell,m=1}^{3}\epsilon_{\ell km}\epsilon_{ijk}B_\ell C_i A_j\hat{x}_m + \sum_{i,j,k,\ell,m=1}^{3}\epsilon_{\ell km}\epsilon_{ijk}C_\ell A_i B_j\hat{x}_m$$
$$= -\sum_{i,j,k,\ell,m=1}^{3}\epsilon_{\ell mk}\epsilon_{ijk}\bigl(A_\ell B_i C_j + B_\ell C_i A_j + C_\ell A_i B_j\bigr)\hat{x}_m$$

where swapping adjacent indices in $\epsilon_{\ell km}$ to form $\epsilon_{\ell mk}$ (so that the common index would match the position of $\epsilon_{ijk}$) costs a minus sign. We now exchange this pair of Levi-Civita symbols for Kronecker deltas, and then contract indices:

$$= -\sum_{i,j,\ell,m=1}^{3}\bigl(\delta_{\ell i}\delta_{mj} - \delta_{\ell j}\delta_{mi}\bigr)\bigl(A_\ell B_i C_j + B_\ell C_i A_j + C_\ell A_i B_j\bigr)\hat{x}_m$$
$$= -\sum_{i,j=1}^{3}\bigl(A_iB_iC_j\hat{x}_j - A_jB_iC_j\hat{x}_i + B_iC_iA_j\hat{x}_j - B_jC_iA_j\hat{x}_i + C_iA_iB_j\hat{x}_j - C_jA_iB_j\hat{x}_i\bigr)$$
$$= -(\vec{A}\cdot\vec{B})\vec{C} + (\vec{A}\cdot\vec{C})\vec{B} - (\vec{B}\cdot\vec{C})\vec{A} + (\vec{B}\cdot\vec{A})\vec{C} - (\vec{C}\cdot\vec{A})\vec{B} + (\vec{C}\cdot\vec{B})\vec{A} = 0$$

since $\sum_{i,j=1}^{3}A_iB_i = \vec{A}\cdot\vec{B}$ and $\sum_{i=1}^{3}A_i\hat{x}_i = \vec{A}$. The final result is zero because the scalar product is commutative – e.g. $\vec{A}\cdot\vec{B} = \vec{B}\cdot\vec{A}$ (unlike the vector product).

## 7.4 Torque

**Torque**: Recall that a force, $\vec{F}$, is a push or a pull, and a net external force causes an object to change momentum (or accelerate). In rotation, there is an analogous quantity called torque, $\vec{\tau}$.[310] Torque involves both a force and a position vector that describes how the force is applied. A net external torque causes a rigid body (or system) to change angular momentum (or have angular acceleration). We define torque mathematically as a vector product,

$$\vec{\tau} = \vec{r}_\perp \times \vec{F}$$

where $\vec{r}_\perp$ is a vector distance that extends from the axis of rotation (perpendicularly) to the point where the force is applied, as illustrated below.

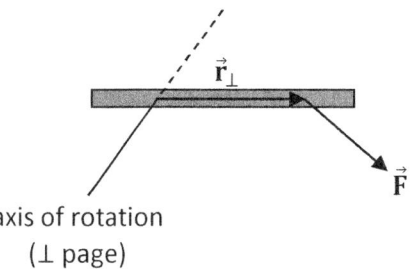

axis of rotation
($\perp$ page)

The SI units of torque are N·m.[311] However, although a Newton times a meter equates to a Joule (1 J = 1 N·m), torque is not expressed in Joules. The reason is conceptual: The Joule is reserved exclusively for physical quantities that are a measure of work or energy; torque is neither a measure of work nor energy.

The fact that torque, work, and energy all have SI units of N·m, but only work and energy are expressed in Joules, is actually related to the concept of the radian. Recall the definition of arc length for an object traveling in a circle of radius $R_0$: $s = R_0 \theta$. This formula states that a meter must equal a meter times a radian. There is a similar relationship between rotational work and torque: $W_r = \int_{\theta=\theta_0}^{\theta} \tau d\theta$, which is analogous to the equation for translational work, $W_t = \int_i^f \vec{F} \cdot d\vec{s}$. From the equation $W_r = \int_{\theta=\theta_0}^{\theta} \tau d\theta$, we can see that a Joule equals a Newton-meter times a radian:

$$1 \text{ J} = 1 \text{ N·m·rad} \neq 1 \text{ N·m}$$

---

**Important Distinction.** A net external force, $\sum \vec{F}_{ext}$, causes an object to change momentum (i.e. accelerate), whereas a net external torque, $\sum \vec{\tau}_{ext}$, causes a rigid body to change angular momentum (i.e. have angular acceleration). Torque is the rotational analog for force, but is distinctly different.

---

[310] The lowercase Greek symbol tau ($\tau$) is used to represent torque.
[311] Order is important: If you write mN instead of Nm, it will be confused with milliNewtons.

**Common Mistake**. Students often think of torque as something that is needed to cause rotation, but this statement is imprecise: Rigid bodies have a natural tendency – called rotational inertia – to maintain constant angular momentum. Once a rigid body is rotating, you don't need anything to keep it rotating with constant angular momentum – that's what it would do in the absence of a net external torque. If you think that torque is something you need to cause rotation, that's only true if the rigid body is initially at rest. If the rigid body is rotating to begin with, a torque is what you actually need to *stop* its rotation! It is most precise to think of a net external torque as something that is needed to cause the angular momentum of a rigid body to change.

It is similarly incorrect to think that force is something needed to cause motion – since an object in motion has a natural tendency to maintain constant momentum. A net external force is something that is needed to cause the (translational or linear) momentum to change.

**Conceptual Example**. Suppose that a door is at rest and partly ajar (so that the handle doesn't need to be turned in order to open it further). A monkey wishes to open the door so that he may walk through the doorway. Since the door is at rest, the monkey must change its angular momentum in order to open the door, which means that the monkey must apply a net external torque on the door.

If the monkey applies a force, $\vec{F}$, to the door – i.e. pushes or pulls on the door – the door may or may not change its angular momentum. The monkey must apply a torque, $\vec{\tau}$, in order to open the door. Since $\vec{\tau} = \vec{r}_\perp \times \vec{F}$, the monkey must exert a force, $\vec{F}$, in order to apply a torque, $\vec{\tau}$, to the door. Note that the vector $\vec{r}_\perp$ plays a significant role in whether or not the door will open; $\vec{r}_\perp$ describes *how* the force is applied. Not any force, $\vec{F}$, will result in a torque, $\vec{\tau}$: The monkey must push or pull the door in an acceptable way in order to change the door's angular momentum. A torque will definitely open the door further, but a force may not. Let us consider a few specific cases.

First, the monkey grabs the door by the hinges and pulls as hard as he can. Why doesn't the door open? If you push or pull on the axis of rotation, $\vec{r}_\perp = 0$, so there is no torque.

Next, the monkey grabs both handles and pulls directly away from the hinges. Why doesn't the door open now? This time, $\vec{r}_\perp \neq 0$, but $\vec{r}_\perp$ is parallel to $\vec{F}$ – i.e. $\vec{r}_\perp \parallel \vec{F}$: The monkey is pulling on the door in a direction that is away from the hinges, and $\vec{r}_\perp$ extends from the hinges to the end of the door where the monkey applies the force.

The monkey will get the most torque out of his force if he pushes perpendicularly to the door, and if he pushes on the door near the edge that is furthest from the hinges (that's where the doorknob is, of course). Since the magnitude of the vector product is $\|\vec{\tau}\| = \|\vec{r}_\perp \times \vec{F}\| = r_\perp F \sin\theta$, pushing perpendicularly to the door makes $\theta = 90°$, and pushes the door near the handle makes $r_\perp = \|\vec{r}_\perp\|$ large.

**Hint**. Any force that acts on the axis of rotation results in zero torque, and any force that points directly toward or away from the axis of rotation results in zero torque. If the force is applied at the axis of rotation, $\vec{r}_\perp = 0$; this is like pushing or pulling the door on the hinges. If the force is applied elsewhere, but points directly toward or away from the axis of rotation, $\theta = 0°$ or $180°$, such that $\vec{r}_\perp \times \vec{F} = 0$; this is like pulling the handles directly away from the hinges.

7 Rotation

**Direction of the torque:** Since the formula for torque, $\vec{\tau} = \vec{r}_\perp \times \vec{F}$, is a vector product, the direction of the torque can be determined qualitatively using the right-hand rule, as illustrated in the next example.

**Conceptual Example.** A rod was initially balanced on a fulcrum when a banana was placed on the uniform rod, near its right end, causing the system to develop angular momentum. Determine the direction of the torque that the banana exerts on the rod.

The force, $\vec{F}$, in question is the banana's weight, which is directed downward ($\downarrow$). The vector $\vec{r}_\perp$ extends from the fulcrum to the banana. Thus, $\vec{r}_\perp$ is directed to the right ($\rightarrow$). Since formula for the torque exerted by the banana is $\vec{\tau} = \vec{r}_\perp \times \vec{F}$, according to the right-hand rule, we point our fingers along $\vec{r}_\perp$ ($\rightarrow$), curl them toward $\vec{F}$ ($\downarrow$), and our thumb points along $\vec{\tau}$, which is directed into the page ($\otimes$).

**Example.** A force $\vec{F} = 60\,\text{N}\,\hat{\jmath} - 30\,\text{N}\,\hat{k}$ is exerted on a rigid body at the point $(2.0\,\text{m}, 4.0\,\text{m}, 0)$. The rigid body is hinged in such a way that it rotates about the $y$-axis. Determine the torque vector.

The vector $\vec{r}_\perp$ extends perpendicularly from the axis of rotation – in this case, the $y$-axis – to the point where the force is applied. The distance from the $y$-axis to the point $(2.0\,\text{m}, 4.0\,\text{m}, 0)$ is $2.0\,\text{m}$. Therefore, $\vec{r}_\perp = 2.0\,\text{m}\,\hat{\imath}$. The torque can be found from the determinant method of the vector product:

$$\vec{\tau} = \vec{r}_\perp \times \vec{F} = \begin{vmatrix} \hat{\imath} & \hat{\jmath} & \hat{k} \\ r_{\perp x} & r_{\perp y} & r_{\perp z} \\ F_x & F_y & F_z \end{vmatrix} = \begin{vmatrix} \hat{\imath} & \hat{\jmath} & \hat{k} \\ 2.0 & 0 & 0 \\ 0 & 60 & -30 \end{vmatrix}$$

$$\vec{\tau} = \hat{\imath}\begin{vmatrix} 0 & 0 \\ 60 & -30 \end{vmatrix} - \hat{\jmath}\begin{vmatrix} 2.0 & 0 \\ 0 & -30 \end{vmatrix} + \hat{k}\begin{vmatrix} 2.0 & 0 \\ 0 & 60 \end{vmatrix} = 60\,\text{Nm}\,\hat{\jmath} + 120\,\text{Nm}\,\hat{k}$$

**Magnitude of the torque:** Since the formula for torque involves a vector product, the magnitude of the torque, $\tau = \|\vec{\tau}\|$, can be found from

$$\tau = r_\perp F \sin\theta$$

where $\theta$ is the angle between $\vec{r}_\perp$ and $\vec{F}$. Note that $0° \leq \theta \leq 180°$, which means to choose the smaller of the two angles between $\vec{r}_\perp$ and $\vec{F}$.

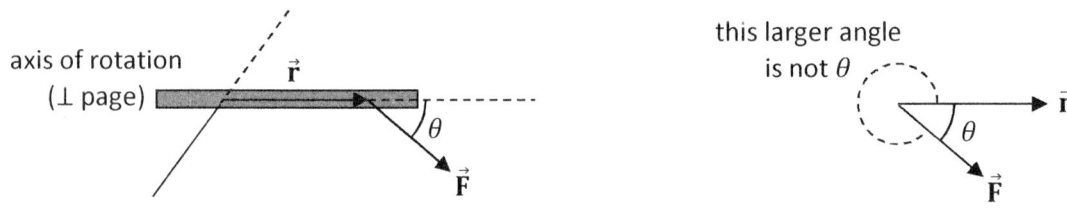

**Example.** In the diagram below, the 20.0-cm long uniform rod is hinged about its center so that it may rotate within the plane of the page – its axis of rotation will be perpendicular to the paper and pass through the hinge. The white block has a mass of 25 g and the black block has a mass of 10 g. Determine the magnitude and direction of the instantaneous net torque exerted on the rod.

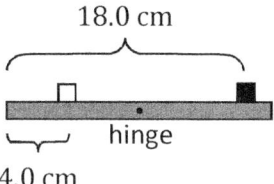

Choosing $+x$ to be horizontally to the right and $+y$ to be vertically upward, $+z$ comes out of the page for a right-handed coordinate system. The torque due to the white block's weight is along $\hat{\mathbf{k}}$ (out of the page), while the torque due to the black block's weight is along $-\hat{\mathbf{k}}$ (into the page):

$$\sum_{i=1}^{2} \vec{\boldsymbol{\tau}}_{ext,i} = \vec{\boldsymbol{\tau}}_w + \vec{\boldsymbol{\tau}}_b = r_{\perp w} m_w g \sin 90° \,\hat{\mathbf{k}} - r_{\perp b} m_b g \sin 90° \,\hat{\mathbf{k}}$$
$$= (0.060)(0.025)(9.81)\hat{\mathbf{k}} - (0.080)(0.010)(9.81)\hat{\mathbf{k}} = 0.0069 \text{ N·m }\hat{\mathbf{k}}$$

We utilized the fact that each torque has only a $z$-component, and therefore each torque can be written as its magnitude, $\tau = r_\perp F \sin\theta$, times $+\hat{\mathbf{k}}$ or $-\hat{\mathbf{k}}$. Also, we converted the distances to meters and the masses to kilograms in order to obtain torque in N·m.

**Torque due to gravity**: When an extended rigid body's own weight exerts a torque on itself, gravity is pulling on every molecule of the rigid body. However, on average the weight acts on the center of gravity of the object, which is approximately the same as the object's center of mass if the object is small compared to the size of the planet. Therefore, we can think of a torque exerted by gravity as if the entire weight of the rigid body were acting sole at its center of gravity.

**Note.** When you need to find the torque exerted on a rigid body by gravity, treat the problem as if the object's weight were exerted at the rigid body's center of gravity.

**Example.** The 12.0-m long, 60-kg uniform rod illustrated below is hinged at the left end. Determine the magnitude of the instantaneous torque exerted on the rod.

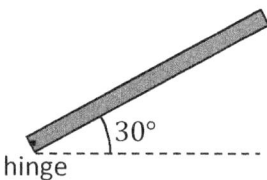

7 Rotation

> The weight of the rod effectively acts on the center of mass of the uniform[312] rod, which is 6.0 m from the axis of rotation. If you draw and label a diagram of $\vec{r}_\perp$ and $\vec{F}$, you will see that $\theta = 120°$: $\vec{r}_\perp$ is directed 30° above the horizontal, while $\vec{F}$ is directed downward. The magnitude of the instantaneous[313] torque exerted on the rod by gravity is
>
> $$\tau = r_\perp mg \sin\theta = (6.0)(60)(9.81)\sin 120° = 3.1 \text{ kN·m}$$

**Common Mistake.** Many students would incorrectly use $\theta = 30°$ in the previous problem, obtaining an incorrect answer for torque. Study the previous example to convince yourself that the correct angle to use is $\theta = 120°$. If you always take the time to draw and label a diagram of $\vec{r}_\perp$ and $\vec{F}$, you can avoid making similar conceptual mistakes.

**Lever arm**: The combination $r_\perp \sin\theta$ is referred to as the lever arm, $\ell$:

$$\ell = r_\perp \sin\theta$$

The magnitude of the torque is thus related to the lever arm via

$$\tau = F\ell$$

**Natural versus forced rotation**: A rigid body or system may rotate naturally, or it may be hinged. When a rigid body or system rotates naturally, it rotates about an axis that passes through its center of mass. For example, if you grab one end of a banana and throw it through the air, you will observe that the banana naturally rotates about its center of mass and that the center of mass travels through a parabolic arc.

A hinge or fulcrum can be used to force a rigid body or system to rotate about a different axis. (Of course, you can also place a hinge at the center of mass.) If you see that a rigid body is hinged or that it is supported by a fulcrum, the axis of rotation will pass through the hinge or fulcrum.

## 7.5 Summing the Torques

**Vector product relationship between velocity and angular velocity for a revolving point-mass**: The velocity, $\vec{v}$, and angular velocity, $\vec{\omega}$, of a revolving point-mass are related through a vector product:[314]

---

[312] If the rod's density were non-uniform, you would first need to integrate to determine the location of its center of mass (following the technique described in Sec. 6.1).

[313] In this and other examples, the torque may change in magnitude and/or direction as the rigid body rotates. Here, we are calculating the instantaneous torque only for the instant illustrated in the diagram.

[314] After learning uniform circular motion (Chapter 4), students are familiar with the relationship between speed and angular speed in circular motion, $v = R_0\omega$, and so are accustomed to writing the distance first and angular speed last. Perhaps this is the reason that most students intuitively want to write $\vec{r}_\perp \times \vec{\omega}$ instead of $\vec{\omega} \times \vec{r}_\perp$.

$$\vec{v} = \vec{\omega} \times \vec{r}_\perp \quad \text{(revolving point-mass)}$$

where $\vec{r}_\perp$ extends perpendicularly from the axis of rotation to the position of the point-mass. The speed, $v = \|\vec{v}\|$, and angular speed, $\omega = \|\vec{\omega}\|$, of a point-mass are therefore related by

$$v = r_\perp \omega \sin\theta \quad \text{(revolving point-mass)}$$

We see that $v = R_0 \omega$ for a point-mass traveling in a circle about a fixed[315] axis of rotation.

**Vector product form for the acceleration of a revolving point-mass**: The acceleration, $\vec{a}$, of a revolving point-mass can also be expressed in terms of vector products:

$$\vec{a} = \vec{\alpha} \times \vec{r}_\perp + \vec{\omega} \times \vec{v} \quad \text{(revolving point-mass)}$$

which includes the usual tangential and centripetal components of acceleration:

$$a_T \hat{\boldsymbol{\theta}} = \vec{\alpha} \times \vec{r}_\perp \quad , \quad -a_c \hat{\mathbf{r}} = \vec{\omega} \times \vec{v} \quad \text{(revolving point-mass)}$$

For a point-mass traveling in a circle about a fixed axis of rotation, these reduce to $|a_T| = R_0 \alpha$ and $a_c = v^2/R_0 = R_0 \omega^2$.[316]

**Net external torque acting on a rigid body**: Let us consider a rigid body that is purely rotating (so there is no translation). Let us further assume that the moment of inertia and axis of rotation all remain constant. According to the work-energy theorem (see the end of Sec. 5.3), the net work done on the rigid system equals its change in kinetic energy. In differential form, this can be expressed as[317]

$$dW_{net} = dK$$

In general, the total work includes both translational and rotational terms, $W_{net} = W_t + W_r = \int_i^f \vec{F} \cdot d\vec{s} + \int_{\theta=\theta_0}^{\theta} \tau d\theta$, but we are presently treating a system that is only rotating:

---

[315] The axis of a top, for example, is not fixed. The axis of the top sweeps out a cone, while the top spins about this moving axis of rotation. See Sec. 7.9.

[316] Recall that $\alpha$ is the magnitude of the angular acceleration, $\vec{\alpha}$, and so is inherently nonnegative. Components of angular acceleration, such as $\alpha_z$, however, may be negative. The tangential acceleration, $a_T$, is a component of the acceleration, $\vec{a} = a_T \hat{\boldsymbol{\theta}} - a_c \hat{\mathbf{r}}$, along $\hat{\boldsymbol{\theta}}$, and so may be negative. The formula for centripetal acceleration, $a_c$, provides a nonnegative result, and the centripetal acceleration is inherently inward – i.e. along $-\hat{\mathbf{r}}$. When we use the 2D polar unit vectors $\hat{\mathbf{r}}$ and $\hat{\boldsymbol{\theta}}$, it must be understood that they lie in the instantaneous plane of rotation and have an instantaneous origin lying at the center of rotation, such that $-\hat{\mathbf{r}}$ is inward.

[317] In practice, if you are solving a problem and wish to relate position to speed, it is conceptually and strategically better to apply the principle of conservation of energy than the equivalent work-energy theorem. Many students who try to use the work-energy theorem instead often make mistakes trying to correctly account for all of the work done by the conservative and nonconservative forces. Conservation of energy conveniently separates all of the types of different work/energy terms, and also separates initial from final.

## 7 Rotation

$$dW_r = dK \quad \text{(pure rotation)}$$
$$\left(\sum \tau_{ext}\right) d\theta = I\omega d\omega \quad (I = const.)$$

Since $K = I\omega^2/2$, implicit differentiation leads to $dK = I\omega d\omega$ provided that the moment of inertia is constant, which we used above. We now introduce time differentials to both sides of the equation:

$$\left(\sum \tau_{ext}\right)\frac{d\theta}{dt} = I\omega\frac{d\omega}{dt} \quad (I = const.)$$
$$\left(\sum \tau_{ext}\right)\omega = I\omega\alpha \quad (I = const.)$$
$$\sum \vec{\tau}_{ext} = I\vec{\alpha} \quad (I = const.)$$

We will discuss the more general case in Sec.'s 7.8-7.9, which involves angular momentum.

**Analog of Newton's second law for rigid bodies**: According to Newton's second law, the net external force acting on an object or system equals a derivative of the total momentum with respect to time, $\sum \vec{F}_{ext} = d\vec{p}/dt$. We often apply this to objects with constant mass, for which Newton's second law simplifies to $\sum \vec{F}_{ext} = m\vec{a}$.

The rotational equivalent of Newton's second law is most generally

$$\sum \vec{\tau}_{ext} = \frac{d\vec{L}}{dt}$$

where $\vec{L}$ is the angular momentum of the rigid body or system. We will return to this general form in Sec.'s 7.8-7.9, where we discuss angular momentum. There are many problems in the context of rotation where moment of inertia changes, for which the above equation must be used. When moment of inertia is constant, the above equation simplifies to

$$\sum \vec{\tau}_{ext} = I\vec{\alpha} \quad (I = const.)$$

The above equations are easy to remember if you understand that torque is the rotational analog for force, angular momentum is the analog for linear momentum, angular acceleration is the angular analog for acceleration, and moment of inertia is the analog for mass. If you recall that mass is a measure of inertia, this will help you to remember that moment of inertia is its angular equivalent. Most of the equations in rotation are very similar to equations from linear motion, where the only difference is substituting the angular analogs for the linear quantities.

**Analog of Newton's first law for rigid bodies**: According to Newton's first law, an object has a natural tendency to maintain constant (linear) momentum, $\vec{p}$. If the object has constant mass, this implies that the object also has a natural tendency to maintain constant velocity, $\vec{v}$. A net external force, $\sum \vec{F}_{ext}$, is needed to change an object's momentum.

Analogously, a rigid body has a natural tendency to maintain constant angular momentum, $\vec{L}$. If the rigid body has constant moment of inertia, this implies that the rigid body also has a natural tendency to maintain constant angular velocity, $\vec{\omega}$ – or to have zero angular acceleration, $\vec{\alpha}$. A net external torque, $\sum \vec{\tau}_{ext}$, is needed to change a rigid body's angular momentum.

## Problem-Solving Strategy for Summing the Torques

0. This strategy is useful for relating torques to moment of inertia and angular acceleration when moment of inertia is constant, and also useful for relating torques to one another when there is no angular acceleration. When moment of inertia is not constant, the net external torque equals a derivative of angular momentum with respect to time (see Sec.'s 7.8-7.9).
1. Draw and label an extended FBD for any rigid bodies[318] for which rotation is relevant, using appropriate subscripts. Draw and label the forces at the points where they act on the rigid body. In the case of a force like gravity that acts throughout the rigid body, draw where it effectively acts (namely, the center of mass). Draw and label the $\vec{r}_\perp$'s and $\theta$'s.
2. Also draw and label ordinary FBD's for any objects for which rotation is not relevant, using appropriate subscripts. Draw and label the forces acting on these objects.
3. Setup a coordinate system (not necessarily Cartesian). It is wise to choose one coordinate along the direction of acceleration; the remaining coordinates must be chosen such that all of the independent coordinates are mutually perpendicular to one another. This choice ensures that (at least) two components of acceleration will be zero. The coordinate for the rotation of the rigid body must be consistent with the coordinates for the translation of other objects in the problem (as illustrated in coming examples). If there is a pulley in the problem, it is useful to think of the pulley as bending a coordinate axis – see Sec. 3.4.
4. Sum the torques for each rigid body for which rotation is relevant. Set the sum of the torques equal to $I\vec{\alpha}$ if the moment of inertia is constant; otherwise, set it equal to $d\vec{L}/dt$. If the rigid body rotates about an axis through its center of mass, you must first determine where its center of mass lies.
5. Sum the components of forces separately for each object for each coordinate axis. You will need to apply trig to resolve each force into components. If the rigid body is translating in addition to rotating, you will need to sum the components of the forces for it, too. (As we will see in the next section, this is sometimes useful even if the rigid body is not translating.)
6. Write each torque in terms of the corresponding force.
7. If there is friction in the problem, you will need to apply the equation for friction (see Sec. 3.5). However, when friction is causing a rigid body to roll or rotate without slipping, the inequality for static friction will generally apply. In the case of rolling or rotation without slipping, it may be possible to relate the friction force to other forces in the problem, as we will see in coming examples.
8. If there is a spring in the problem, you will need to apply Hooke's law (see Sec. 3.6).
9. Count your knowns and unknowns. In algebra courses, you generally need $N$ equations in $N$ unknowns, but beware that in physics courses, you can often solve a system of $N$ equations with fewer than $N$ unknowns – the reason is that some variables, like mass, often cancel out after substitutions.

---

[318] A rigid body may be an extended rigid object, a rigid system of pointlike objects, or a rigid system of extended rigid bodies; the distinction only affects *how* you calculate the moment of inertia – i.e. with a sum or integral.

# 7 Rotation

**Example.** The uniform rod illustrated below is hinged at its left end. Derive an equation for the instantaneous angular acceleration of the rod in terms of gravitational acceleration, the length of the rod, and the angle that the rod makes with the vertical, $\theta$.

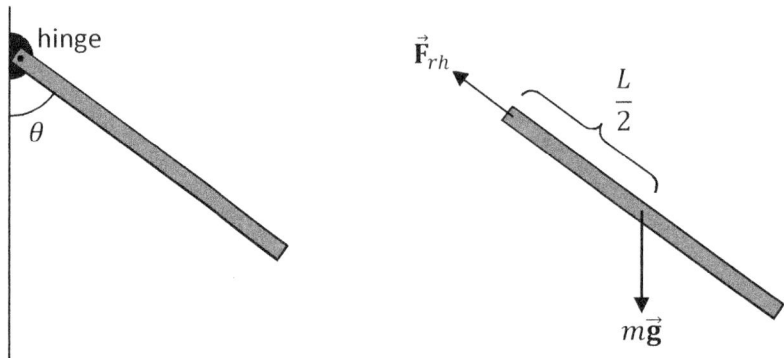

We may treat this problem as a pure rotation[319] about an axis through the hinge and perpendicular to the page. In this case, we only need to sum the torques. We drew an extended FBD for the rod to help visualize the $\vec{r}_\perp$'s and $\theta$'s needed to calculate the torques. The force exerted on the rod by the hingepin, $\vec{F}_{rh}$, does not result in a torque since $\vec{r}_{\perp h} = 0$. The only torque is exerted by the weight of the rod, which acts at the center of mass of the rod on average. For this uniform rod, the center of mass lies at the rod's center, such that $\vec{r}_{\perp g} = L/2$. It turns out that the angle needed for $r_{\perp g} mg \sin \theta_g$ is the same as the angle that the rod makes with the vertical.[320] We setup our coordinate system with $+x$ horizontally to the right, $+y$ vertically upward, and $+z$ out of the page. The net torque and angular acceleration have only $z$-components; the net torque has a negative $z$-component according to the right-hand rule. The moment of inertia of the rod is constant during this problem, so the net external torque equals $I\vec{\alpha}$. The moment of inertia of the uniform rod about its end is $mL^2/3$:

$$\sum_{i=1}^{2} \tau_{iz} = I\alpha_z$$
$$0 - r_{\perp g} mg \sin \theta_g = I\alpha_z$$
$$-\frac{L}{2} mg \sin \theta = \frac{mL^2}{3} \alpha_z$$
$$\alpha_z = -\frac{3g}{2L} \sin \theta$$

---

[319] Equivalently, we may treat the rod as a combination of rotation about an axis through its center of mass and perpendicular to the page and translation of the center of mass in a circular arc. In this case, the moment of inertia will be different because the axis of rotation is different. If you carry out the physics correctly, you will obtain the same result for the angular acceleration applying either perspective. We will return to this choice of perspectives in Sec. 7.7.

[320] The angle that you need for $r_\perp F \sin \theta$ will not always coincidentally be an angle that happens to be labeled in the problem. Always take the time to draw the direction of $\vec{r}_\perp$ and $\vec{F}$; $\theta$ is the angle between them.

**Pulleys**: In Chapter 3, we assumed that the pulleys were frictionless. We will now consider a more realistic case, where the cord rotates with the pulley without slipping. If the cord is accelerating, the pulley must also experience a corresponding angular acceleration in order for the cord not to slip. The cord thus exerts a net torque on the pulley. Therefore, the tensions in the two sections of the cord divided by the pulley must not be equal in magnitude; this difference in tensions results in a net torque on the pulley, causing its angular acceleration.

Consider the diagram below. In Chapter 3, we would have assumed the pulley to be frictionless. In that case, the two tensions would have been equal in magnitude. Now in Chapter 7, we will treat the case where the cord rotates with the pulley without slipping. In this case, the tensions in each section of the cord have different magnitudes. The greater of the two tensions determines the direction of the acceleration. The magnitude of the force of static friction between the cord and pulley equals the difference in the magnitudes of the tension forces: $f_s = |T_{Rp} - T_{Lp}|$. Recall that $f_s \leq \mu_s N$ for static friction; the force of static friction has an upper limit, but will only equal what is necessary to prevent slipping. Since the difference in the tension magnitudes equals the magnitude of the friction force, it's possible to solve many rotating-without-slipping problems without using the expression $f_s \leq \mu_s N$, as illustrated in the next example. The problem with using $f_s \leq \mu_s N$ is that the inequality may very well apply, allowing for an indefinite friction force; the expression $f_s = |T_{Rp} - T_{Lp}|$, on the other hand, leads to a definite value for the magnitude of the friction force.

Observe that Newton's third law is not violated by having $T_{Lp} \neq T_{Rp}$. Rather, there are equal and opposite tension forces in the left cord, and equal and opposite tension forces in the right cord: The force that the left box exerts on the pulley, $\vec{T}_{pL}$, is equal and opposite to the force that the pulley exerts on the left box, $\vec{T}_{Lp} = -\vec{T}_{pL}$, and the force that the right box exerts on the pulley, $\vec{T}_{pR}$, is equal and opposite to the force that the pulley exerts on the right box, $\vec{T}_{Rp} = -\vec{T}_{pR}$.

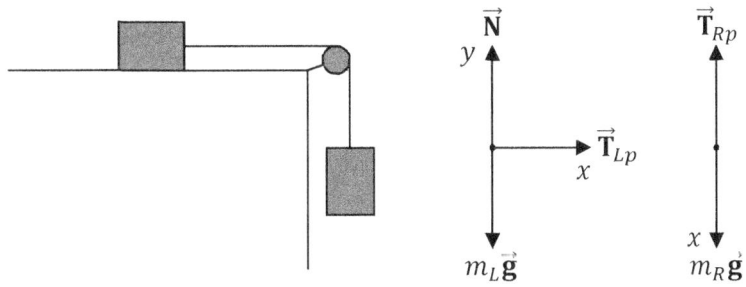

**Common Mistake**. The tensions in the two sections of a cord divided by a pulley do not have equal magnitude if there is friction between the cord and pulley – i.e. if the cord rotates with the pulley. It's a common mistake for students to set these tensions equal in problems where the cord rotates with the pulley without slipping (since they are in the habit of doing this from Chapter 3, where all of the pulleys were assumed to be frictionless). In the chapter on rotation, the pulleys are generally not assumed to be frictionless, and so the tensions in the two sections of the cord will not have equal magnitudes.

**Example.** A monkey makes Atwood's machine by suspending bananas of mass $m_L$ and $m_R > m_L$ from the two ends of a cord and passing the cord over a pulley of mass $m_p$. The cord rotates with the pulley without slipping. The pulley is a uniform solid disc. Derive an equation for the acceleration of the masses in terms of the suspended masses and the value gravitational acceleration.

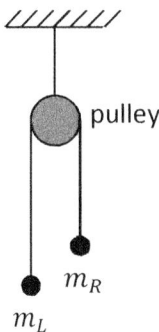

We begin by drawing FBD's for each suspended mass along with an extended FBD for the pulley. We define our coordinate system in the diagram below. As usual with a pulley problem, we choose to have one axis (in this case, the $+y$-axis) bend around the pulley. With this choice, $a_{Ly} = a_{Ry} = a_{Tp} = R_p \alpha_z \equiv a_y$, where $a_{Tp}$ and $\alpha_z$ are the tangential and angular, respectively, accelerations of the (edge of the) pulley and $R_p$ is the radius of the pulley. Note that the positive sense of $\alpha_z$ must match the positive sense of $a_y$. We don't need to relate the friction force between the cord and pulley to normal force (exerted between the cord and pulley), since we can instead relate the magnitude of the friction force to the difference in the tension magnitudes. We sum the $z$-components of the torques for the rotating pulley and the $y$-components of the forces for the translating suspended masses. Note that the weight and support force of the pulley do not exert torques on the pulley since they are coincident with the center of mass of the pulley (and the axis of rotation passes through its center, perpendicular to the page). The moment of inertia of the pulley is that of a solid disc in rolling mode, $m_p R_p^2 / 2$.

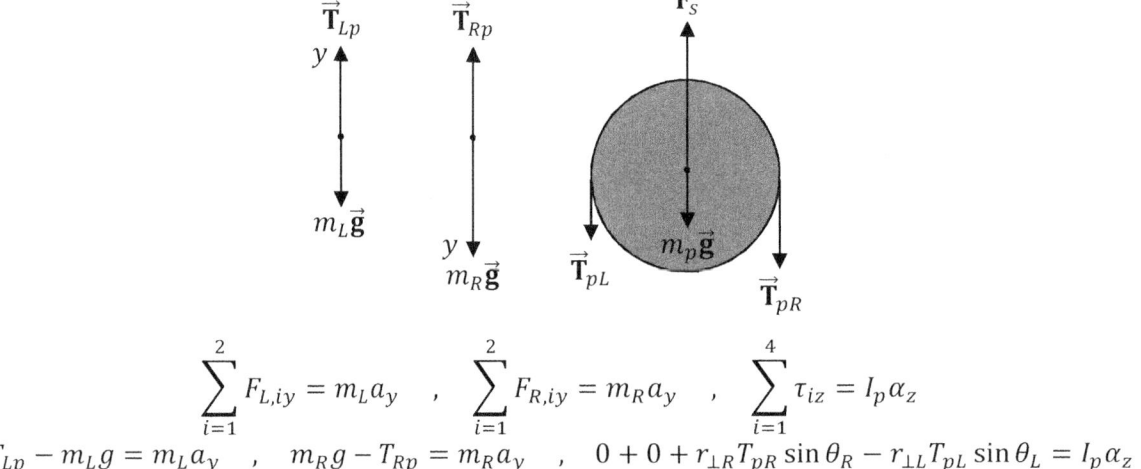

$$\sum_{i=1}^{2} F_{L,iy} = m_L a_y \quad , \quad \sum_{i=1}^{2} F_{R,iy} = m_R a_y \quad , \quad \sum_{i=1}^{4} \tau_{iz} = I_p \alpha_z$$

$$T_{Lp} - m_L g = m_L a_y \quad , \quad m_R g - T_{Rp} = m_R a_y \quad , \quad 0 + 0 + r_{\perp R} T_{pR} \sin \theta_R - r_{\perp L} T_{pL} \sin \theta_L = I_p \alpha_z$$

It is convenient to first simplify the sum of the torques equation. The two tensions pull tangentially on the pulley, such that the tension forces are perpendicular to the corresponding $\vec{r}_\perp$'s – i.e. $\theta_R = \theta_L = 90°$. Also, $r_{\perp R} = r_{\perp L} = R_p$. We plug in the equation for the moment of inertia of the pulley, cancel one of the $R_p$'s on both sides, and combine the remaining $R_p$ with $\alpha_z$ to make $a_{Tp} = R_p \alpha_z \equiv a_y$:

$$R_p T_{pR} \sin 90° - R_p T_{pL} \sin 90° = \frac{m_p R_p^2 \alpha_z}{2}$$

$$T_{pR} - T_{pL} = \frac{m_p a_y}{2}$$

According to Newton's third law, the two tension force pairs in each section of the cord have equal magnitude: $T_{pR} = T_{Rp}$ and $T_{pL} = T_{Lp}$ (but $T_{Rp} \neq T_{Lp}$). It is convenient to add the three equations together – i.e. two from summing components of forces and one from summing the torques – as this algebraic strategy cancels the unknown tensions:

$$m_R g - m_L g = m_R a_y + m_L a_y + \frac{m_p a_y}{2}$$

$$a_y = \frac{m_R - m_L}{m_R + m_L + \frac{m_p}{2}} g$$

Observe that this expression agrees with the frictionless Atwood machine (Sec. 3.4) in the limit that the pulley has negligible mass (i.e. $m_p \ll m_{L,R}$).

---

**Hint.** When summing components of forces as well as torques in a pulley problem, write the angular acceleration of the pulley in terms of the tangential acceleration of the cord using $a_T = R_p \alpha_z$, and set the tangential acceleration of the cord equal to the acceleration of the masses that are attached to the cord; the masses and cord must all have the same acceleration since they are connected (assuming that the cord does not stretch and remains taut).

---

**Hint.** Tensions act tangentially on a pulley so that $\theta = 90°$ and $r_\perp = R_p$ in $r_\perp T \sin \theta$, regardless of whether the tension force is horizontal, vertical, or directed at some other angle.

---

**Example.** Derive an equation for the vertical acceleration of a descending yo-yo that is made by connecting two solid discs, each of mass $m_D$ and radius $R_D$, to a thin hollow rod of mass $m_R$ and radius $R_R < R_D$, neglecting the thickness of the cord.

An extended FBD for the yo-yo is illustrated on the following page. We choose $+y$ to be vertically upward and $+z$ to be directed to the left (along the axis of the yo-yo). The yo-yo is not in free fall because a tension force is pulling it upward. Thus, we expect its acceleration to be less than $g$. We sum both the torques and vertical components of the forces for the yo-yo, since the center of mass of the yo-yo is itself accelerating. Since the two discs and hollow rod share a common axis of rotation – they all rotate in rolling mode – the total moment of inertia of the yo-yo is

$$I_y = 2I_D + I_R = 2\frac{m_D R_D^2}{2} + m_R R_R^2 = m_D R_D^2 + m_R R_R^2$$

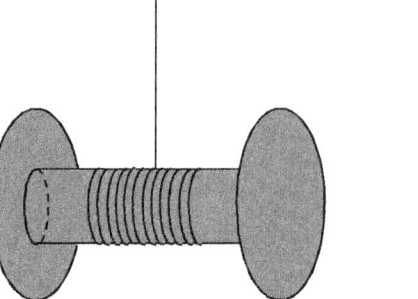

Note that only the tension exerts a torque on the yo-yo, and it acts a distance $R_R$ from the axis of rotation. Summing the vertical components of the forces and the components of the torques,

$$\sum_{i=1}^{2} F_{iy} = m_y a_y \quad , \quad \sum_{i=1}^{2} \tau_{iz} = I_y \alpha_z$$

$$T - m_y g = m_y a_y \quad , \quad 0 - r_\perp T \sin\theta = (m_D R_D^2 + m_R R_R^2)\alpha_z$$

$$T - (2m_D + m_R)g = (2m_D + m_R)a_y \quad , \quad -R_R T \sin 90° = (m_D R_D^2 + m_R R_R^2)\frac{a_y}{R_R}$$

$$T - (2m_D + m_R)g = (2m_D + m_R)a_y \quad , \quad -T = (m_D R_D^2 + m_R R_R^2)\frac{a_y}{R_R^2}$$

$$-(2m_D + m_R)g = (2m_D + m_R)a_y + (m_D R_D^2 + m_R R_R^2)\frac{a_y}{R_R^2}$$

$$a_y = -\frac{2m_D + m_R}{2m_D + 2m_R + m_D \frac{R_D^2}{R_R^2}} g$$

**Rolling without slipping**: Up until now, we have allowed objects to slide, but not to roll. We will now allow objects to roll without slipping. In the case of sliding, the object is translating, and the net external force (if there is one) overcomes the object's inertia (natural tendency to maintain constant linear momentum). When an object slides, friction is a resistive force that has the effect (if acting alone) of reducing the object's momentum.

When a rigid body rolls without slipping, the net external force still overcomes the rigid body's inertia; in addition, the net external torque overcomes the rigid body's moment of inertia (natural tendency to maintain constant angular momentum). When a rigid body rolls without slipping, the friction force acts against the direction of motion, affecting the sum of the forces in a way that would reduce the object's momentum; but at the same time, the friction force exerts a torque on the rigid body that acts in the direction of motion. The torque exerted by friction on a rolling rigid body is not a resistive torque. Unlike sliding, in the case of rolling without slipping, the friction force does not result in nonconservative work; the friction force does not subtract mechanical energy from the system (converting it into internal energy or heat) when an object rolls without slipping.

Another thing that may seem odd about the friction force in the case of rolling without slipping is that it is static friction, not kinetic friction. When a box slides, the same part of the box is always in contact with the ground, and mechanical energy is converted to internal energy and/or heat. When a ball rolls without slipping, no part of the ball is in contact with the ground for more than one instant, and no mechanical energy is lost.

In Sec. 7.7, we will see that it is easy to make conceptual mistakes trying to reason what will happen when an object rolls without slipping if your reasoning is based on the usual effect of the friction force. Instead, you are more likely to reason the concepts out correctly if you think in terms of the combination of (translational) inertia and rotational inertia. See the examples in Sec. 7.7.

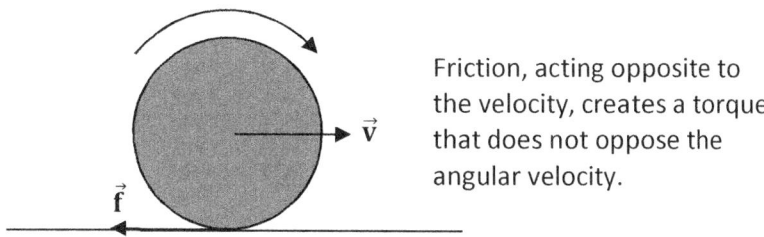

Friction, acting opposite to the velocity, creates a torque that does not oppose the angular velocity.

**Example.** A uniform solid sphere rolls without slipping down an incline that makes an angle $\theta$ with the horizontal. Derive an expression for the magnitude of the sphere's acceleration. Also, derive an inequality for the minimum coefficient of static friction needed to allow the sphere to roll without slipping.

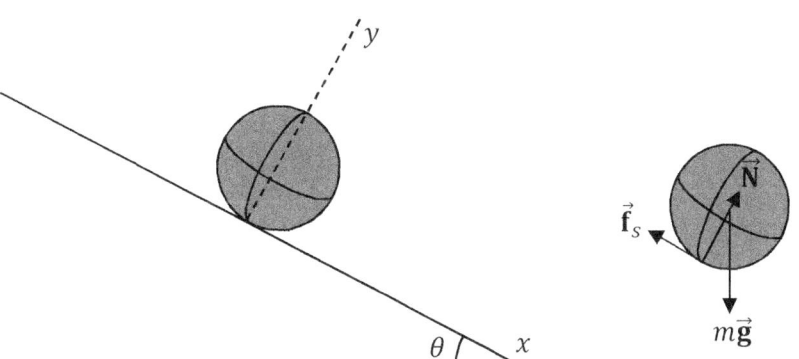

We drew an extended FBD for the rigid body in the diagram above, showing where friction and normal forces act on the sphere and where its weight acts on average. Normal force and weight are collinear with the center of mass of the sphere, and therefore exert no torques. We choose $+x$ to be down the incline, $+y$ to be along the normal force, and therefore $+z$ is directed out of the page for a right-handed coordinate system. We now sum the components of the forces and torques:

$$\sum_{i=1}^{3} F_{ix} = ma_x \quad , \quad \sum_{i=1}^{3} F_{iy} = ma_y \quad , \quad \sum_{i=1}^{3} \tau_{iz} = I_s \alpha_z$$

# 7 Rotation

$$mg \sin \theta - f_s = ma_x \quad , \quad N - mg \cos \theta = 0 \quad , \quad f_s R_0 \sin 90° = \frac{2}{5} m R_0^2 \alpha_z$$

The force of friction is tangential to the sphere, so $\theta_f = 90°$ for its torque. The moment of inertia of a rolling, uniform, solid sphere is $2mR_0^2/5$. The linear acceleration of the sphere's center of mass, $a_x$, down the incline equals the tangential acceleration of the sphere, $a_x = a_T = R_0 \alpha_z$:

$$f_s = \frac{2}{5} m a_x$$
$$mg \sin \theta - \frac{2}{5} m a_x = m a_x$$
$$a_x = \frac{5}{7} g \sin \theta$$

This is 5/7 what the acceleration of the sphere would be if it instead slid without friction.

The minimum coefficient of friction needed for the sphere to roll without slipping can be found by using the relation $f_s \leq \mu_s N$:

$$f_s \leq \mu_s N$$
$$\frac{2}{5} m a_x \leq \mu_s m g \cos \theta$$
$$\frac{2}{7} g \sin \theta \leq \mu_s g \cos \theta$$
$$\mu_{s,min} = \frac{2}{7} \tan \theta$$

**Rolling without slipping along a horizontal surface**: In the previous example, the surface becomes horizontal in the limit that $\theta$ approaches zero. In this limit, the acceleration of the sphere becomes zero and the minimum coefficient of static friction also becomes zero. The friction force only has a horizontal component in this limit, so the friction force must also be zero (otherwise the sphere would decelerate, contradicting the result that the acceleration approaches zero in this limit). So if a car is coasting in neutral along a horizontal, if the car were in a vacuum and if it could roll with perfect smoothness without slipping (and if the tires were not sticky), it wouldn't need any gas to coast along a level highway with constant speed. Note that there does need to be friction for the wheels to begin rolling on the road in the first place; friction gets the wheels rolling in the first place (for a car that begins from rest), and once they roll without slipping, then in this idealized horizontal rolling friction then becomes zero.

However, accelerating and decelerating would still require gas even in these ideal conditions. The force of static friction actually accelerates the car, propelling the tires forward through torque. Not only does friction not play its usual resistive role in rolling without slipping, but friction can actually serve to increase the speed of the car. You should now be able to return to Chapter 3, where we first discussed the effect of friction in the case of driving, and better understand how the force of friction relates to the acceleration and deceleration of a car. Another interesting feature of the car is that only the wheels and axles rotate: The full mass of the car serves as the car's inertia, but only the wheels and axles serve as the car's rotational inertia. As we will see in Sec. 7.7, the ratio of the rotating mass to the translating mass has a very significant effect on the car's acceleration.

## 7.6 Static Equilibrium

**Static equilibrium**: A rigid body in static equilibrium is stationary – it is neither translating nor rotating.

**Dynamic equilibrium**: A rigid body is in dynamic equilibrium if it is not changing linear or angular momentum. If the mass and moment of inertia of the rigid body are constant, the velocity of the rigid body's center of mass and the angular velocity of the rigid body are both constant (one, but not both, may also be zero) if the rigid body is in dynamic equilibrium.

**Equilibrium**: In both types of equilibrium – i.e. static and dynamic – the net torque and net force acting on the rigid body (or system) are both zero; in static equilibrium, the speed of its center of mass and its angular speed are also zero. The sum of the forces and the sum of the torques acting on a rigid body must both be zero for a rigid body (or system) in equilibrium.

---

### Problem-Solving Strategy for Equilibrium Problems

0. Apply this strategy when a rigid body is in static or dynamic equilibrium (and the moment of inertia of the rigid body is constant).
1. Draw and label an extended FBD for each rigid body, using appropriate subscripts. Draw and label the forces where they act on the rigid body. In the case of a force like gravity that acts throughout the rigid body, draw where it effectively acts (namely, the center of mass). Draw and label the $\vec{r}_\perp$'s and $\theta$'s.
2. Setup a coordinate system (not necessarily Cartesian). The coordinate for the rotation of the rigid body must be consistent with the coordinates for the translation of the rigid body and any other objects in the problem (as illustrated in previous and coming examples). If there is a pulley in the problem, it is useful to think of the pulley as bending a coordinate axis – see Sec. 3.4.
4. Sum the torques for each rigid body for which rotation is relevant. Set the sum of the torques equal to 0. If the rigid body is in static equilibrium, you may choose any convenient axis for the axis of rotation because it is not rotating about any axis.
5. Sum the components of forces separately for each object for each coordinate axis. You will need to apply trig to resolve each force into components. Set the sum of the components of the forces acting on each object equal to zero.
6. Write each torque in terms of the corresponding force.
7. Count your knowns and unknowns. In algebra courses, you generally need $N$ equations in $N$ unknowns, but beware that in physics courses, you can often solve a system of $N$ equations with fewer than $N$ unknowns – the reason is that some variables, like mass, often cancel out after substitutions.

---

**Note**. If a rigid body is in static equilibrium, you have the freedom to pick any axis of rotation that may be convenient, since the rigid body is not rotating about any axis. It is usually most convenient to choose an axis of rotation that would make the greatest number of torques equal to zero.

---

**Note**. You may be able to solve some equilibrium problems without summing both torques and forces.

# 7 Rotation

**Conceptual Example.** The mallet illustrated below is cut into two pieces by sawing it at its center of mass. Which end of the mallet, if any, weighs more?

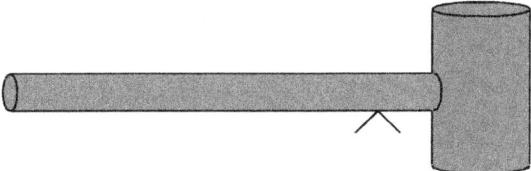

We assume that the handle and head are separately uniform, and that the head is not less dense than the handle. The center of mass is found by placing a fulcrum under the mallet at the position where the mallet is in static equilibrium. It is convenient to think of the mallet as two separate objects – the left and right pieces formed by sawing it into two pieces. Prior to sawing, there are two torques acting on the system – one due to the weight of the left end and one due to the weight of the right end. These two torques are equal and opposite when the mallet is in static equilibrium. In both cases, $\theta = 90°$. Therefore, $r_{\perp L} m_L g = r_{\perp R} m_R g$. Clearly, $r_{\perp L} > r_{\perp R}$, which implies that $m_R > m_L$. The head end weighs more than the handle end.[321] Conceptually, this is because the handle end is further away from the fulcrum, on average. One of the practice problems at the end of the chapter (with hints and an answer in the back of the book) is a quantitative version of this conceptual example.

**Common Mistake.** Many students intuitively expect the two ends to weigh the same in the preceding example. Even after learning, counterintuitively, during lecture – and agreeing with the reasoning in class – that the mallet end weighs more, they very often revert to their intuition on the final exam, answering the question incorrectly. Note that the torques – but not the forces! – are equal and opposite. The torques must be balanced, not the weights, to achieve rotational equilibrium.

**Example.** In the diagram below, the 30.0-cm long uniform rod is hinged about its center (such that it could rotate within the plane of the page). The white block has a mass of 15 kg and the black block has a mass of 10 kg. Where should the white block be placed in order to balance the system?

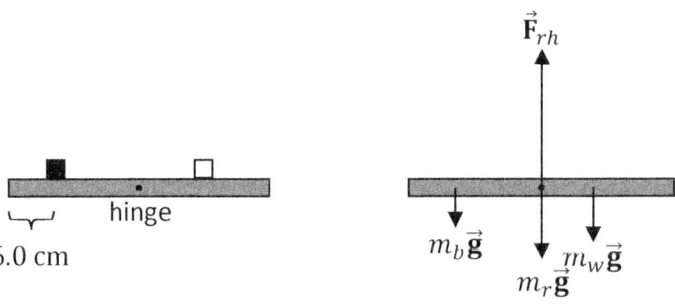

---

[321] Note that $m_R$ and $m_L$ do not refer to the masses of the head and handle, but to the masses of the right and left pieces of the mallet: $m_R$ includes the head plus the section of handle between the head and fulcrum, and $m_L$ includes only the portion of the handle to the left of the fulcrum.

# An Advanced Introduction to Calculus-Based Physics (Mechanics)

We choose $+x$ to point right, $+y$ to be up, and $+z$ to come out of the page, and the axis of rotation to be the natural axis through the hinge and perpendicular to the page. With this choice, neither the weight of the rod, $m_r\vec{g}$, nor the support force from the hinge, $\vec{F}_{rh}$, exerts a torque since they pull on the axis of rotation. Using the right-hand rule, the black block exerts a positive torque and the white block exerts a negative torque. The black block is $(30.0)/2 - 6.0 = 9.0$ cm from the axis:

$$\sum_{i=1}^{4} \tau_{iz} = I_r \alpha_z$$
$$0 + 0 + m_b g r_{\perp b} \sin 90° - m_w g r_{\perp w} \sin 90° = 0$$
$$m_b r_{\perp b} = m_w r_{\perp w}$$
$$r_{\perp w} = \frac{m_b r_{\perp b}}{m_w} = \frac{(10)(9.0)}{(15)} = 6.0 \text{ cm}$$

The white block must be placed 6.0 cm to the right of the hinge in order to balance the system.

**Example.** In the previous example, what is the magnitude of the force exerted by the hinge on the rod if the rod's mass is 100 kg?

We can solve for this force by summing the $y$-components of the forces acting on the rod:

$$\sum_{i=1}^{4} F_{iy} = m a_y$$
$$F_{rh} - m_r g - m_b g - m_w g = 0$$
$$F_{rh} = (m_r + m_b + m_w)g = (100 + 10 + 15)(9.81) = 1.2 \times 10^3 \text{ kg}$$

This is an important engineering consideration: If too much stress is exerted on the hinge, the system won't be able to sustain static equilibrium without becoming damaged.

**Example.** Solve for the position of the white block in the previous examples by using a different axis of rotation – one perpendicular to the page and passing through the left end of the rod.

Since the system will be in static equilibrium with the desired position of the white block, we are free to choose any axis of rotation – since the rod will not have angular momentum about any axis. With the specified axis, the force exerted by the hinge results in a positive torque, while the three weights exert negative torques. Note that $r_{\perp b}$ is now 6.0 cm (instead of 9.0 cm):

$$\sum_{i=1}^{4} \tau_{iz} = I_r \alpha_z$$
$$F_{rh} r_{\perp h} \sin 90° - m_r g r_{\perp r} \sin 90° - m_b g r_{\perp b} \sin 90° - m_w g r_{\perp w} \sin 90° = 0$$
$$F_{rh} r_{\perp h} - m_r g r_{\perp r} - m_b r_{\perp b} = m_w r_{\perp w}$$
$$r_{\perp w} = \frac{F_{rh} r_{\perp h} - m_r g r_{\perp r} - m_b g r_{\perp b}}{m_w g} = \frac{(1226)(15.0) - (100)(9.81)(15.0) - (10)(9.81)(6.0)}{(15)(9.81)}$$
$$r_{\perp w} = 21.0 \text{ cm}$$

# 7 Rotation

Observe that the answer is the same as in the previous examples: The white block must still be placed 6.0 cm to the right of the hinge in order to balance the system. Also note that the prior solution was more efficient, as the original axis of rotation made two of the torques equal to zero and did not require first finding the force exerted by the hinge.

**Example.** This time, we place the white block at the right end of the rod (i.e. the block's center of mass will be at the end of the rod); instead of asking where to move the white block, let us instead ask where to place a fulcrum to achieve static equilibrium.

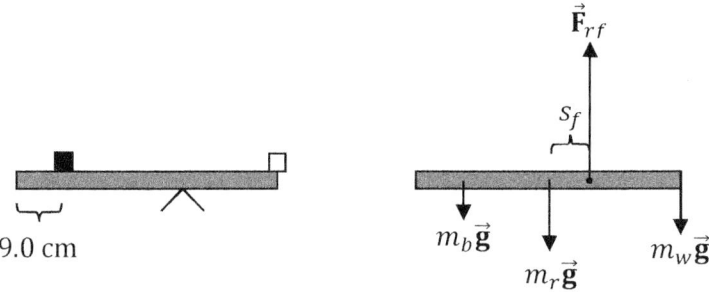

The weight of the rod will now also exert a torque, since the fulcrum must be off-center in order to achieve static equilibrium. Clearly, the fulcrum must be placed to the right of the rod's center of mass. Let us call the distance from the needed fulcrum position to the center of the rod $s_f$. All three torques depend on the position of the fulcrum, and so all three $r_\perp$'s will be functions of $s_f$. The weight of the black block and the rod itself exert positive $z$-components of torque (using the same directions for the coordinate axes again), while the weight of the white block exerts a negative $z$-component of torque:

$$\sum_{i=1}^{4} \tau_{iz} = I_r \alpha_z$$
$$F_{rf} r_{\perp f} \sin 90° + m_r g r_{\perp r} \sin 90° + m_b g r_{\perp b} \sin 90° - m_w g r_{\perp w} \sin 90° = 0$$
$$0 + m_r s_f + m_b (6.0 + s_f) = m_w (15.0 - s_f)$$
$$100 s_f + 10(6.0 + s_f) = 15(15.0 - s_f)$$
$$100 s_f + 60 + 10 s_f = 225 - 15 s_f$$
$$s_f = \frac{165}{125} = 1.3 \text{ cm}$$

The fulcrum needs to be positioned 1.3 cm to the right of the rod's center. Note that this problem could alternatively be solved by finding the center of mass of the two blocks and rod, since the force exerted by the fulcrum on the rod does not result in a torque.

**Example**. A box of bananas is suspended from the midpoint of a clothesline, as illustrated below. The system is in static equilibrium. Derive an equation that relates the magnitude of the tension in the clothesline to the angle that each section of the clothesline makes with the horizontal.

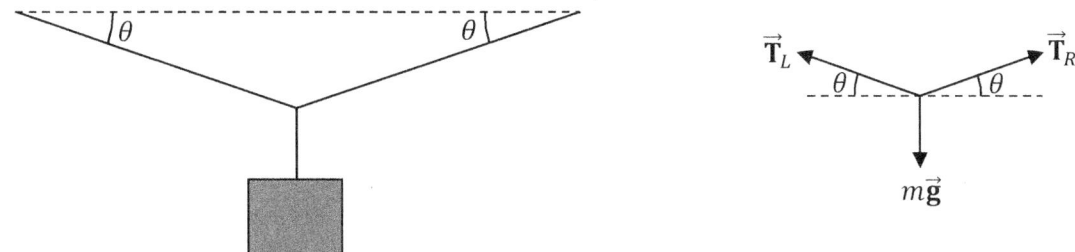

This problem can be solved by summing just the components of the forces. We choose $+x$ to point right and $+y$ to point up. It is convenient to sum the forces for the knot that connects the suspended box to the clothesline (which we can do because the acceleration of the knot equals zero). There are three tensions pulling on the knot. The downward tension equals the weight of the box of bananas because the box is not accelerating. The right and left tensions have equal magnitude – i.e. $T_L = T_R \equiv T$ – because the box is suspended at the midpoint of the clothesline; if the box is moved off-center, these tensions will no longer have equal magnitude, and their angles with the horizontal will no longer be equal (but the $x$-components of these tensions will still be equal and opposite):

$$\sum_{i=1}^{3} F_{ix} = ma_x \quad , \quad \sum_{i=1}^{3} F_{iy} = ma_y$$
$$T\cos\theta - T\cos\theta = 0 \quad , \quad T\sin\theta + T\sin\theta - mg = 0$$
$$T = \frac{mg}{2\sin\theta}$$

**Statics**: Setting the sums of the components of the net external forces and torques acting on a system yields important design considerations in the branch of mechanical engineering called statics. An example of such design criteria is the maximum load that a crane can support without snapping the suspension cable or causing more stress on the hingepin than it can handle.

**Boom**: A long pole, which may be horizontal or angled, that is hinged at one end (often, at a vertical wall or support post) and from which a load is suspended (often by a cable) is called a boom.

**Load**: A weight that is suspended from a boom (often by a support cable) is called a load. For example, a wrecking ball may be suspended from a boom to make a simple crane; in this case, the wrecking ball is the load.

**Hingepin force**: The hingepin exerts a force at one end of a boom to keep the system in static equilibrium. It's important to calculate the magnitude of this force to ensure that the hingepin is able to withstand it.

Both the magnitude and direction of the hingepin force depend upon the other forces acting on the boom. Since the direction is not known at the beginning of the problem, but is something that you can solve for, it's convenient to label the $x$- and $y$-components, $F_{bh,x}+F_{bh,y}$, of the hingepin force, $\vec{F}_{bh}$, in the extended FBD, from which the magnitude, $F_{bh}$, and direction, $\theta_{bh}$, may be computed. An example of this is provided in the following example.

**Example.** Derive equations for the tension in the (horizontal) tie rope and the magnitude of the force exerted on the uniform boom by the hingepin for the crane illustrated below in terms of the mass of the boom and the angle that the boom makes with the vertical.

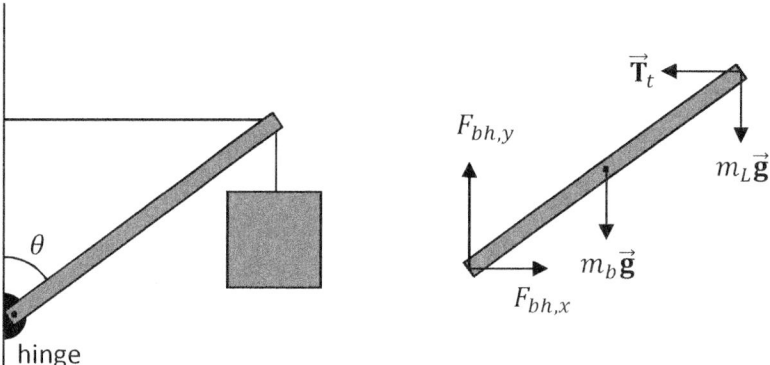

We choose $+x$ to be horizontal and $+y$ to be vertical, so $+z$ comes out of the page for a right-handed coordinate system.[322] We also choose the axis of rotation to be the $z$-axis, passing through the hinge. We resolved the force exerted on the boom by the hingepin into components in the extended FBD. Summing the components of the forces and torques exerted on the boom,

$$\sum_{i=1}^{4} F_{ix} = ma_x \quad , \quad \sum_{i=1}^{4} F_{iy} = ma_y \quad , \quad \sum_{i=1}^{4} \tau_{iz} = I_b \alpha_z$$

$$F_{bh,x} - T_t = 0 \quad , \quad F_{bh,y} - m_b g - m_L g = 0 \quad , \quad LT_t \cos\theta - Lm_L g \sin\theta - \frac{L}{2} m_b g \sin\theta = 0$$

$$F_{bh,x} = T_t \quad , \quad F_{bh,y} = m_b g + m_L g \quad , \quad T_t = m_L g \tan\theta + \frac{m_b g}{2} \tan\theta$$

$$F_{bh} = \sqrt{T_t^2 + (m_b + m_L)^2 g^2} \quad , \quad T_t = \left(m_L + \frac{m_b}{2}\right) g \tan\theta$$

where $L$ is the length of the boom. Recall that $\sin(180° - \theta) = \sin\theta$ and $\sin(90° + \theta) = \cos\theta$.

---

[322] The right-hand rule and corresponding signs for the vector product – such as the vector product between Cartesian unit vectors – apply if you use a right-handed coordinate system. To see if a coordinate system is right-handed, point the fingers of your right hand along $+x$, curl them toward $+y$, and check if your thumb is along $+z$.

**Stability**: Recall from Sec. 5.5 that a system with a single-variable potential energy is in equilibrium if $\frac{dU}{dR} = 0$. In addition to $\frac{dU}{dR} = 0$, the system is in stable equilibrium if $\frac{d^2U}{dR^2} > 0$, neutral equilibrium if $\frac{d^2U}{dR^2} = 0$, or unstable equilibrium if $\frac{d^2U}{dR^2} < 0$. Also recall that the conservative force associated with the potential energy is related to the potential energy by $\vec{F}_c = -\vec{\nabla}U$.

Consider a rigid body lying on a horizontal surface with no other forces acting on the system other than weight and normal force. Such a rigid body will be in equilibrium if its center of mass lies above a point of support – since in this case, the net external torque will be zero (the net external force is already zero because the normal force balances the weight, since the surface is horizontal). Three such rigid bodies in equilibrium are illustrated below.

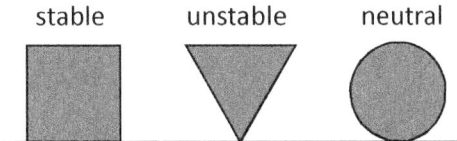

We can conceptually determine whether or not such a rigid body is in stable, neutral, or unstable equilibrium by considering whether or not the application of a horizontal force would have the effect of raising or lowering the center of mass of the rigid body. If a horizontal force would raise the center of mass, the rigid body is in stable equilibrium; if the horizontal force would lower the center of mass, the rigid body is in unstable equilibrium; and if the horizontal force would neither raise nor lower the center of mass, the rigid body is in neutral equilibrium.

**Note**. The center of mass of a rigid body may lie outside of the body itself, as in the examples below.

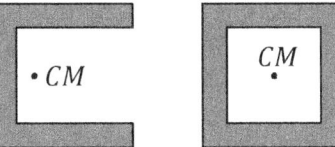

**Conceptual Example**. Three uniform bricks of width $L$ and equal mass are stacked off-center, as depicted below (where $L$ is measured horizontally, left/right, in the diagram). How far can each of the top two bricks overhang, relative to the edge of the brick below it, without causing the stack to fall over?

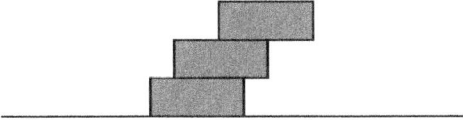

The top brick can overhang as much as $L/2$ over the brick beneath it; in this case, its center of mass will be above a point of support (i.e. its center of mass will lie above the edge of the middle brick). With the top brick overhanging its maximum amount, the middle brick can't overhang as much.

> We can't treat the middle brick by itself, as the brick above it is important. Instead, to determine how much the middle brick can overhang – with the brick above it overhanging by $L/2$ – we treat the top two bricks as a system. The center of mass of the top two bricks is (placing the origin at the center of the middle brick): $x_{CM,tm} = (m_t x_t + m_m x_m)/(m_t + m_m) = (0 + mL/2)(2m) = L/4$. The center of mass of the top two bricks is $L/4$ to the right of the center of the middle brick. This means that the middle brick can overhang as much as $L/4$ with the top brick overhanging $L/2$.

## 7.7 Rotational Kinetic Energy

**Translational kinetic energy**: An object has translational kinetic energy, $K_t$, if its center of mass is in motion:

$$K_t = \frac{mv^2}{2} \quad \text{(translational KE)}$$

where $v$ is the speed of the center of mass.[323]

**Rotational kinetic energy**: A rigid body has rotational kinetic energy, $K_r$, if it is rotating:

$$K_r = \frac{I\omega^2}{2} \quad \text{(rotational KE)}$$

where $I$ is the moment of inertia of the rigid body about the axis of rotation.

**Total kinetic energy**: If a rigid body's center of mass is moving and the rigid body is also rotating, the total kinetic energy of the object equals its translational kinetic energy plus its rotational kinetic energy:[324]

$$K = K_t + K_r = \frac{mv^2}{2} + \frac{I\omega^2}{2} \quad \text{(total KE)}$$

In the above formula, we treat the rigid body as rotating about an axis of rotation passing through its center of mass (not any axis, of course, but one corresponding to its sense of rotation).

---

[323] If the object is also rotating, different parts of the object will be traveling with different speeds. For example, if a door is shutting, although all parts of the door have the same angular speed, $\omega$, different parts of the door move with different speeds, $v$: Parts of the door furthest from the hinge travel faster (covering a greater distance in the same time) than parts of the door nearest to the hinges. Since different parts of the object may travel with different speeds, you must remember to use the speed of the center of mass in the formula for translational kinetic energy.

[324] Note that the subscript $t$ in $K_t$ means 'translational' and not 'total,' while the subscript-less $K$ means 'total.'

**Total kinetic energy from different perspectives**: Consider a rotating rigid body whose center of mass is traveling in a circle, where the axis of revolution of the center of mass is parallel to the axis of rotation of the rigid body. An example of such combined revolution and rotation is illustrated by the left diagram below. In this case, the total kinetic energy can be calculated from two different perspectives. As usual, you can compute the translational kinetic energy of the center of mass, using the total mass of the rigid body and the speed of the center of mass, and add this to the rotational kinetic energy of the rigid body about an axis through its center of mass. Alternatively, in this instance, you can think of the total kinetic energy as being purely rotational kinetic energy about the parallel axis passing through the center of revolution (i.e. the center of the circle that the center of mass is traveling in). These equivalent descriptions of the total kinetic energy serve as the basis for the parallel-axis theorem.

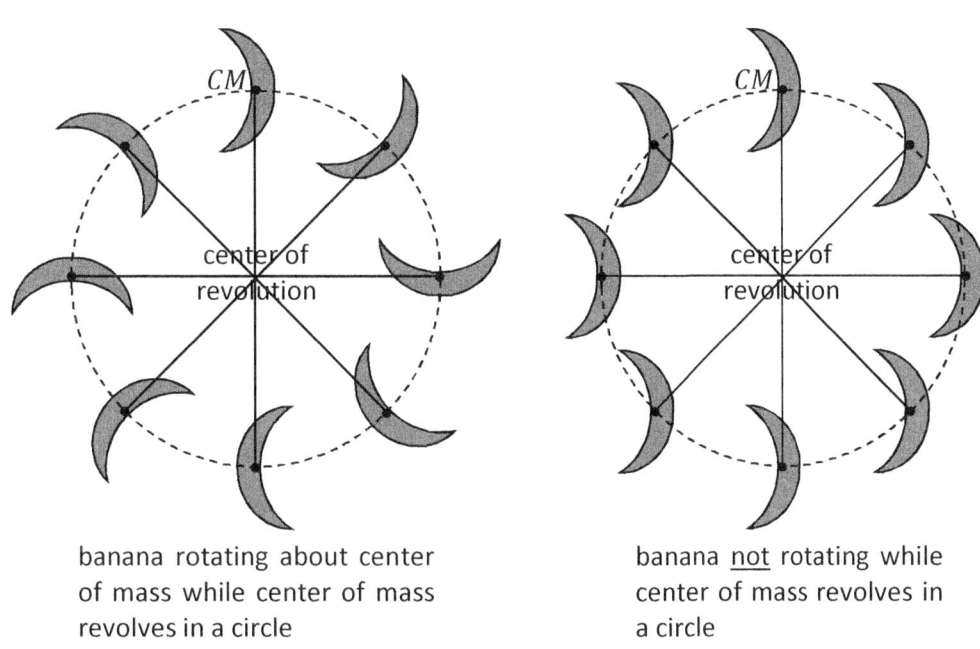

banana rotating about center of mass while center of mass revolves in a circle

banana not rotating while center of mass revolves in a circle

**Proof of the parallel-axis theorem**: Recall the parallel-axis theorem from Sec. 7.2. The parallel-axis theorem would apply to the banana rotating in the left diagram above, but not in the right diagram above: The angular velocity of the banana must be the same relative to both parallel axes, which is the case in the left diagram, but not the right diagram. The left diagram is what you would get if you were roasting a banana on a stick over a fire, and suddenly whirled the stick in a circle; the right diagram is what you get if you made a Ferris Wheel with banana-shaped seats. The moon orbits the earth like the diagram on the left – the moon completes one revolution about the earth and one rotation about its axis in approximately 27.3 days, so that the same face of the moon always points toward the earth. Thus, the parallel axis theorem applies to the moon's revolution about the earth. In contrast, the earth's rotation about its axis is not synchronized with its revolution about the sun – as its rotational period is 24 hours, while its period of revolution is 365 days. We will prove the parallel-axis theorem for a situation such as the left diagram, where the center of mass of the rigid body travels in a circle with an axis of revolution that is parallel to the axis of rotation through the center of mass, where the angular velocity of the rigid body is the same relative to each axis.

We simply equate the total kinetic energy of the rigid body from the two equivalent perspectives. In one case, the rigid body has translational kinetic energy associated with the revolution of its center of mass plus rotational kinetic energy about the rotational axis through its center of mass, and in the other case the rigid body has purely rotational kinetic energy about the axis of revolution through the center of revolution. Recall that the distance between the two parallel axes is $h$, as illustrated below. We label the axis of revolution as $p$, and the rotational axis through the center of mass as $c$.

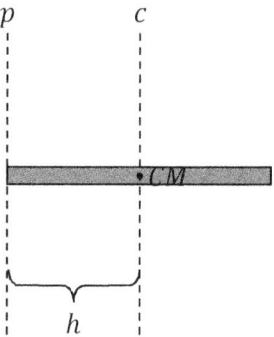

The total kinetic energy of the rigid body is:

$$K_c = K_p$$
$$K_{tc} + K_{rc} = K_{rp}$$
$$\frac{mv^2}{2} + \frac{I_c \omega^2}{2} = \frac{I_p \omega^2}{2}$$

Since the angular velocity is the same relative to each axis in the parallel-axis theorem, the speed of the center of mass relative to the center of revolution is $v = h\omega$:

$$\frac{mh^2\omega^2}{2} + \frac{I_c \omega^2}{2} = \frac{I_p \omega^2}{2}$$
$$I_p = I_c + mh^2$$

**Conservation of energy for rotating rigid bodies:** If a rigid body is rotating, when you conserve energy for the rotating rigid body, you must use the total kinetic energy of the rigid body – i.e. the translational kinetic energy associated with the motion of its center of mass plus the rotational kinetic energy of the rigid body about the corresponding axis through its center of mass. Except for allowing for two kinetic energy terms for each rigid body on each side of the equation, the strategy for applying the principle of conservation of energy is the same as for objects that are only translating. Recall that it is useful to conserve energy to relate initial and final speeds to initial and final positions. Also recall that you need to identify initial and final positions – one should be where you know information and the other should correspond to the unknown that you are solving for – and that you need to choose a reference height from which to measure gravitational potential energy.

## Problem-Solving Strategy for Conservation of Energy with Rotation

0. It is useful to apply the principle of conservation of energy to relate speed to position. It may also be a useful prelude to deriving equations of motion for a system. When the system includes a rotating rigid body, you must account for both the translational and rotational kinetic energy of the rigid body.
1. Define the system. The system may consist of a single object, or it may consist of multiple objects. Only choose objects for which you can account for mechanical potential or kinetic energy – a motor, for example, involves other kinds of energies, too. Whatever is not part of the system is part of the surroundings. You will write down potential and kinetic energy terms for the system only, and the nonconservative work term will account for energy exchanges between the system and surroundings.
2. Choose the reference position(s) from which to measure the relevant forms of potential energy. This includes defining a reference height for a uniform gravitational field and an origin from which to measure the position of a spring. Measure to the center of mass of a rigid body.
3. Identify which points in the motion to label as initial, $i$, and final, $f$. One point should correspond to your unknown, and another point should correspond to where you know information. Sketch the motion and label these points. (In a multi-part problem, you may need $i_1$ and $f_1$, $i_2$ and $f_2$, etc., and you must be wary that what is initial in one part may not be the same initial in another part, for example).
4. Determine what types of potential and kinetic energy there are in the initial and final position, and also what type of nonconservative work is done, if any. Gravitational potential energy is zero at the reference height (or where $R$ is infinite if the field is not uniform), spring potential energy is zero when the spring is at equilibrium, and kinetic energy is zero where an object is at rest. Use different subscripts for initial and final quantities, such as $v_0$ and $v$ for initial and final speed. Note that friction does not do work in the case of a rigid body rolling without slipping or a pulley rotating without its cord slipping.
5. Write symbolic equations for each type of energy, and also for any nonconservative work (by performing the work integral). Plug all of these equations into the equation for conservation of energy, $U_0 + K_0 + W_{nc} = U + K$. Note that there may be multiple terms for any one of the symbols in this equation (i.e. there may be multiple kinds of potential energy – e.g. there could be both a spring and gravity in the problem).
6. Count your knowns and unknowns. In algebra courses, you generally need $N$ equations in $N$ unknowns, but beware that in physics courses, you can often solve a system of $N$ equations with fewer than $N$ unknowns – the reason is that some variables, like mass, often cancel out after substitutions.
7. If you are trying to derive the equations of motion for a system, write the speed as $v = ds/dt$ and proceed to solve this first-order differential equation by separating variables (if possible). An example of this was considered at the end of Sec. 5.4.

**Note.** For a rotating rigid body whose center of mass is traveling in a circle, where the axis of revolution of the center of mass is parallel to the axis of rotation of the rigid body (and where the period of revolution of the center of mass equals the period of rotation about the axis through the center of mass), recall that there are two equivalent perspectives from which to measure the total kinetic energy. In this case, you may alternatively treat the total kinetic energy as pure rotational kinetic energy about the axis through the center of revolution (equal to the total kinetic energy obtained from the translation of the center of mass plus the rotational kinetic energy about an axis through the center of mass).

## 7 Rotation

**Friction is not nonconservative for rolling/rotating without slipping**: When a box slides with friction, the force of friction does nonconservative work on the box, subtracting mechanical energy from the box (converting it to the form of internal energy and/or heat energy). However, when a round rigid body rolls without slipping, friction does not do nonconservative work on the rigid body. Note that while the force of friction acts against the direction of motion, in the case of rolling, friction actually creates a torque in the forward direction. Contrast with the case of sliding, where friction exclusively acts against the motion. Also, note that the same surface of the box is always making contact during sliding, but each part of the round rigid body is only momentarily in contact with the surface when rolling without slipping.

Conceptually, it should be easy to convince yourself that friction does nonconservative work in the case of sliding, but not in the case of rolling without slipping, by considering a coin sliding versus rolling along a horizontal floor. If you slide the coin along the floor – i.e. you give it a temporary push and let go – the coin will quickly come to a halt because the force of friction converts all of its initial translational kinetic energy into internal energy and/or heat. If instead you set the coin rolling along the floor, it will continue much farther for the same initial speed because in this case friction is not doing nonconservative work. The coin eventually topples over and comes to rest because you don't have perfectly idealized rolling without slipping on a perfectly smooth horizontal floor with a perfectly round coin. Nonetheless, with this simple demonstration (try it) you can see a significant distinction between the effect of friction in sliding versus rolling without slipping.

---

**Note.** When conserving energy for an object sliding with friction, include a $W_{nc}$ term for the work done by friction; but when conserving energy for a round rigid body rolling without slipping (or a pulley rotating without slipping), $W_{nc} = 0$ even when there is friction in the problem to allow the rigid body to roll without slipping.

---

**Conceptual Example.** A ball starts from rest at the top of an incline. Would the ball travel faster at the bottom if the incline is frictionless or if there is enough friction that the ball rolls without slipping?

We can reason out the answer conceptually by thinking in terms of conservation of energy. Conservation of energy consists of the following forms of energy: $U_0 + K_0 + W_{nc} = U + K$. We choose the bottom of the incline as our reference height, so that $U = 0$. Since the ball starts from rest, $K_0 = 0$. In either case, $W_{nc} = 0$; whether it slides without friction[325] or rolls without slipping, no mechanical energy is lost through friction (in contrast to the case of sliding with friction, which is not one of the two cases considered in this problem). Therefore, in this problem $U_0 = K$. In either case, $U_0 = mgh_0$ is the same. Sliding without friction, the final kinetic energy is purely translational, $K_{slide} = K_t$, while rolling without slipping, the final kinetic energy includes both translational and rotational terms, $K_{roll} = K_t + K_r$. Sliding without friction, all of the initial gravitational potential goes toward increasing the tangential velocity; rolling without slipping, some also goes toward increasing the angular velocity. Therefore, the final tangential velocity is greater when the ball slides without friction.

---

[325] Friction is needed to create a torque to change the angular velocity of the ball. So if the ball begins from rest, it will slide rather than roll if the incline is perfectly frictionless. This is why it's difficult to drive on icy road conditions – reduced friction prevents the tires from getting a good grip on the road and changing the angular velocity of the wheels.

**Note**. Students often intuitively want to directly point to friction to determine in which case the ball wins the race in the previous example. However, if you try to use similar reasoning in the following example, it will backfire. It is better to make your argument in terms of conservation of energy or in terms of inertia and moment of inertia. In the previous example, we could alternatively explain that gravity has only to overcome the ball's (translational) inertia in the case of sliding without friction, but must overcome both the ball's (translational) inertia and its rotational inertia in the case of rolling without slipping, and therefore gives greater tangential acceleration in the case of sliding without friction. You can also explain the following example in terms of inertia and rotational inertia. Try it!

**Conceptual Example**. A monkey standing at the bottom of an incline bowls a bowling ball up an incline. Would the bowling ball travel farther up the incline if the incline is frictionless or if the incline has enough friction that the bowling ball rolls without slipping, assuming that in either case the bowling ball has the same initial tangential velocity?

We again examine conservation of energy, $U_0 + K_0 + W_{nc} = U + K$, conceptually.[326] Choosing our reference height to be at the bottom of the incline, this time $U_0 = 0$. Once again, $W_{nc} = 0$ in either case. When the ball stops traveling up the incline, $K = 0$. Therefore, in this example, $K_0 = U$. Sliding without friction, $K_{slide,0} = K_{t0}$, and rolling without slipping, $K_{roll,0} = K_{t0} + K_{r0}$. Rolling without slipping, the ball has more initial kinetic energy, and therefore travels further up the incline. Think about this carefully: The ball actually travels further up the incline if you add friction; if you remove friction, it travels a shorter distance up the incline (assuming the same initial tangential velocity in either case). It's a good exercise to see if you can reach the same conclusion by thinking in terms of (translational) inertia and rotational inertia. We will reconsider these same problems quantitatively in two examples, in case you wish to compare the algebraic results to these conceptual exercises.

**Example**. Starting from rest, a uniform solid sphere rolls without slipping down an inclined plane. Derive an equation for the speed of the solid sphere at the bottom of the incline as a function of its initial height above the bottom of the incline.

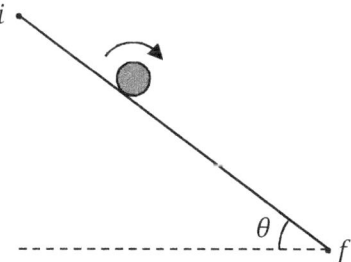

We choose the initial position to be the point of release and the final position to be at the bottom of the incline, and the reference height to correspond to the final position. With these choice, $K_0$ and $U$ are zero for this problem. Since the solid sphere rolls without slipping, $W_{nc} = 0$:

---

[326] You don't need to guess at physics concepts. Let the equations be your conceptual guide.

# 7 Rotation

$$U_0 + K_0 + W_{nc} = U + K$$
$$mgh_0 + 0 + 0 = 0 + K_t + K_r$$
$$mgh_0 = \frac{mv^2}{2} + \frac{I\omega^2}{2} = \frac{mv^2}{2} + \frac{2}{5}mR_0^2\frac{\omega^2}{2} = \frac{mv^2}{2} + \frac{mv^2}{5}$$
$$gh_0 = \frac{7v^2}{10}$$
$$v = \sqrt{\frac{10gh_0}{7}}$$

where $2mR_0^2/5$ is the moment of inertia of a solid sphere in rolling mode and where we used the relation $v = R_0\omega$. The solid sphere has both translational and rotational kinetic energy as it rolls down the incline. If the solid sphere had instead slid without friction, it should be easy to see that its final speed would have been $\sqrt{2gh_0}$, which is greater than $\sqrt{10gh_0/7}$.

**Example.** A uniform solid sphere rolls without slipping up an inclined plane. Derive an equation for the maximum height of the solid sphere from its initial position as a function of its initial speed.

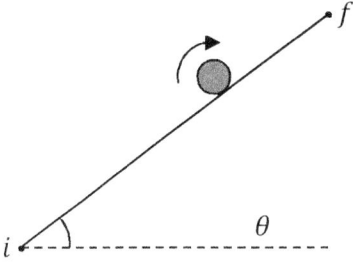

We choose the initial position to be the point of release and the final position to be at the top of the incline, and the reference height to correspond to the initial position. With these choice, $U_0$ and $K$ are zero for this problem. Since the solid sphere rolls without slipping, $W_{nc} = 0$:

$$U_0 + K_0 + W_{nc} = U + K$$
$$0 + K_{t0} + K_{r0} + 0 = mgh + 0$$
$$\frac{mv_0^2}{2} + \frac{I\omega_0^2}{2} = \frac{mv_0^2}{2} + \frac{2}{5}mR_0^2\frac{\omega_0^2}{2} = \frac{mv_0^2}{2} + \frac{mv_0^2}{5} = mgh$$
$$h = \frac{7v_0^2}{10g}$$

If the solid sphere instead slid without friction – with the same initial speed – its final height would have been $v_0^2/(2g)$, which is less than $7v_0^2/(10g)$. That is, the ball travels farther up the incline if there <u>is</u> friction!

> **Math Note**. When the moment of inertia formula is proportional to an $R_0^2$ and $v = R_0\omega$ – with the same $R_0$ as the moment of inertia formula – combine the $R_0^2$ from the moment of inertia formula with the $\omega^2$ in the expression for rotational kinetic energy, $K_r = I\omega^2/2$, to make $v^2$. By doing so, $R_0$ cancels out, and there is one less unknown. Study this in the previous examples.
>
> It is also important to realize that $R_0^2$ will not always cancel out. For one, the moment of inertia may not be proportional to $R_0^2$. For example, consider the yo-yo from Sec. 7.5, for which the moment of inertia in 'standard yo-yo mode' depends upon two different radii, and so is not proportional to a single radius. Secondly, it's possible to have $v = r\omega$, where $r \neq R_0$. This would be the case, for example, with a solid sphere rolling along a U-channel groove rather than an incline, making contact at the two edges of the U-channel, so that $r < R_0$; here, $R_0$ is the radius of the solid sphere, while $2r$ is the distance between the two parallel edges of the U-channel.

**Dependence on mass and size**: In the previous examples, the mass of the uniform solid sphere and the radius of the uniform solid sphere both cancelled out. The acceleration of a rigid body rolling without slipping often depends neither on the mass nor the size of the rigid body. The acceleration very much depends on how the mass is distributed, however – i.e. whether it is shaped like a solid sphere, solid cylinder, hollow sphere, hollow cylinder, etc.

As noted above, the size does not always cancel out, but there a few situations where size matters. Similarly, mass does not always cancel out in the formulas for the acceleration and/or final speed of a rigid body rolling or rotating without slipping. For example, when a car rolls without slipping, only a fraction of the mass (i.e. the wheels and axles) are rotating; the rest of the car is only translating. In mechanics lab, when a PasCar rolls down an incline without slipping, gravity pulls all of the PasCar's mass and all of the PasCar's mass has inertia, but only the wheels and axles have moment of inertia. Therefore, mass does not cancel out for a PasCar rolling without slipping down an incline.[327]

**Path-independence of rolling without slipping**: Recall that the work done by a conservative force is path-independent. Since friction does not do nonconservative work in the case of rolling without slipping, the final speed of the rolling rigid body does not depend upon the path taken – only the change in height (and the initial speed, the shape of the rigid body, etc.). Whether it rolls down an incline or a curved path is irrelevant[328] for its final speed (but not for its instantaneous acceleration).

---

[327] Interestingly, since the PasCar's wheels and axles are very light compared to the overall weight of the PasCar, its acceleration is approximately equal to what you would get if the PasCar slid without friction. So the effect of friction is negligible for a PasCar that rolls without slipping and has only a small fraction of its mass rotating. Therefore, the numerous students who cite friction as a source of error for the PasCar in their lab reports are not realizing that friction is much less significant than it would be if the PasCar were a block that slid down the incline. In fact, if the PasCar rolls without slipping up the incline and the goal of the lab is to measure how far it travels up the incline, as we found in the previous example, friction actually causes the PasCar to travel farther up the incline. Better sources of error to cite in a PasCar lab – since only a small fraction of the PasCar's mass is rotating – are air resistance, vibrations of the PasCar (i.e. not perfect rolling without slipping, but some vertical and/or side-to-side vibrations that affect its acceleration), and errors associated with the photogate and distance measurements. If a rigid body is rolling without slipping, it is usually easier to understand what's going on conceptually in terms of inertia and rotational inertia, and the effect of friction is usually more indirect and less intuitive.

[328] The shape of the path does matter if the rigid body slips or flies off of a convex path (becoming a projectile).

7 Rotation

**Conceptual Example**. A uniform solid ball and a uniform hollow ball both start from rest at the top of an incline and roll without slipping. Which ball reaches the bottom first?

First, note that it does not matter which ball is heavier, nor does it matter which ball is larger: Both the mass of the ball and the radius of the ball cancel out. What doesn't cancel out is the shape of the ball. The solid ball has a greater fraction of its mass nearer to the axis of rotation compared to the hollow ball. Therefore, if you express the moment of inertia of each ball as a coefficient times its mass times its radius squared, you will find that the coefficient is smaller for the solid ball: The coefficient is $\frac{2}{5}$ for a uniform solid sphere in rolling mode (since $I = \frac{2}{5}MR_0^2$), and will be greater than $\frac{2}{5}$ (as much as $\frac{2}{3}$) for a uniform hollow sphere in rolling mode (depending on the thickness of the hollow sphere).[329] Conservation of energy for either ball – with initial at rest, final at the bottom, and height measured from the bottom – yields $U_0 = K_t + K_r$. The smaller coefficient in $K_r$ leads to a greater final speed at the bottom[330] (to see this more clearly, study the math in the previous two examples).

**Conceptual Example**. A monkey has three cans of banana juice. He pulled one can from the freezer, so that the banana juice is frozen solid. He pulled a second can from the refrigerator, so that the banana juice is liquid. He pulled the last can from the garbage, which is empty. The monkey releases all three cans from rest on an incline. Assume that all three cans are perfectly round and roll without slipping.[331] Which can of banana juice will reach the bottom of the incline first?

You should be able to see that the empty bottle reaches last, since all of its mass is far from its axis of rotation. The trick to this problem is realizing that liquid banana juice does not have rotational kinetic energy, whereas the frozen banana juice does. Only a fraction of the total mass rotates in the case of the liquid banana juice. Mass does not cancel out in $U_0 = K_t + K_r$, which makes the $K_r$ smaller compared to $K_t$, resulting in a greater final speed. Therefore, the refrigerated can of banana juice wins the race. We will work out the math for this can of banana juice in the next example.[332]

**Example**. A can of liquid banana juice rolls from rest without slipping down an inclined plane. Derive an equation for the speed of the can of liquid banana juice at the bottom of the incline as a function of its initial height above the bottom of the incline.

---

[329] Since the question does not state how the balls compare in mass or size, it's incorrect to argue that the hollow ball has a greater moment of inertia than the solid ball: A large, heavy solid ball can have a greater moment of inertia than a small, light hollow ball. We can only say that a hollow ball will have a larger moment of inertia than a solid ball of the same mass and size. Since both mass and size will cancel out when we solve for the final speed, we can compare the coefficients in the moment of inertia formulas (provided that we express each moment of inertia formula in terms of mass and the radius squared – for a hollow sphere of finite thickness, this will require factoring the outer radius out, since the formula also involves an inner radius).

[330] Beware of similar questions where either mass or size do not cancel out, as in the following examples. In that case, you can't simply compare coefficients of moment of inertia formulas.

[331] If you try this with water bottles, you may notice that the frozen bottle is not so round and so does not roll with constant angular acceleration; also, the frozen and liquid bottles may slip.

[332] The math is also a little tricky, conceptually, for the frozen can of banana juice: You can't treat the frozen can of banana juice as a solid cylinder. Instead, you must treat it as a hollow cylinder (the can itself) plus a solid cylinder (for the frozen juice) because the can and juice do not have the same density.

We choose the initial position to be the point of release and the final position to be at the bottom of the incline, and the reference height to correspond to the final position. With these choices, $K_0$ and $U$ are zero; and since the solid sphere rolls without slipping, $W_{nc} = 0$:

$$U_0 + K_0 + W_{nc} = U + K$$
$$mgh_0 + 0 + 0 = 0 + K_t + K_r$$
$$(m_c + m_j)gh_0 = \frac{(m_c + m_j)v^2}{2} + \frac{I_c \omega^2}{2}$$
$$(m_c + m_j)gh_0 = \frac{(m_c + m_j)v^2}{2} + m_c R_0^2 \frac{\omega^2}{2}$$

where $m_c$ is the mass of the can, $m_j$ is the mass of the juice, and we approximated the can as a hollow cylinder of negligible thickness (so $I_c = m_c R_0^2$). Using $v = R_0 \omega$,

$$(m_c + m_j)gh_0 = \frac{(m_c + m_j)v^2}{2} + m_c \frac{v^2}{2}$$
$$v = \sqrt{\frac{(m_c + m_j)gh_0}{m_c + \frac{m_j}{2}}}$$

It's useful, conceptually, to check this formula in a couple of limits. First, if the can is empty, $m_j = 0$ and $v = \sqrt{gh_0}$, which agrees with the final speed for a thin hollow cylinder. Second, in the limit that the can is light compared to the juice, i.e. $m_c \ll m_j$, $v = \sqrt{2gh_0}$, which is the same final speed as an object that slides without friction. This shows that the smaller the fraction of the mass of the rigid body that rolls without slipping, the greater the final speed of the object. Contrast this with a can of frozen banana juice, where all of the mass is rotating.

This is why friction is a negligible source of error for a PasCar rolling without slipping down an inclined track (since the wheels make up only a negligible fraction of its total mass). Friction is very significant in making the car roll, but has an insignificant effect on its final speed.

## 7.8 Angular Momentum

**Angular momentum**: Analogous to the equation for (linear) momentum, $\vec{p} = m\vec{v}$, the equation for angular momentum, $\vec{L}$, is

$$\vec{L} = I \vec{\omega}$$

where $I$ is moment of inertia and $\vec{\omega}$ is the angular velocity. The SI units of angular momentum are $kg \cdot m^2/s$ or $N \cdot m \cdot s$.

# 7 Rotation

> **Important Distinction.** It's important not to confuse angular velocity, $\vec{\omega}$, with angular momentum, $\vec{L}$. They are related through the equation $\vec{L} = I\vec{\omega}$.

**Rotational inertia:** A rigid body has a natural tendency to maintain constant angular momentum, $\vec{L}$. If the moment of inertia of the rigid body remains constant, this reduces to a natural tendency to maintain constant angular velocity, $\vec{\omega}$ – or to have zero angular acceleration, $\vec{\alpha}$. There are many situations where a rigid body changes its moment of inertia during the problem – e.g. an ice skater decreases her moment of inertia when she brings her limbs inward. Therefore, it's important to think most generally in terms of angular momentum, $\vec{L}$. The natural tendency corresponds to no net external torque.

**Net external torque:** In general, the rotational analog of Newton's second law for a rigid system of rigid bodies is

$$\sum \vec{\tau}_{ext} = \frac{d\vec{L}}{dt}$$

A net external torque causes the angular momentum of the system to change. For a single rigid body with constant moment of inertia, $I$, this reduces to

$$\sum \vec{\tau}_{ext} = \frac{d(I\vec{\omega})}{dt} = I\frac{d\vec{\omega}}{dt} = I\vec{\alpha} \quad (I = const.)$$

**Angular momentum for a pointlike object:** There is an alternative to $\vec{L} = I\vec{\omega}$ for a revolving point-mass. We may write the angular momentum of a revolving pointlike object as

$$\vec{L} = \vec{r}_\perp \times \vec{p} \quad \text{(revolving point-mass)}$$

To see this, consider the net external torque in the case of a revolving pointlike object:

$$\sum \vec{\tau}_{ext} = \frac{d\vec{L}}{dt} = \frac{d}{dt}(\vec{r}_\perp \times \vec{p}) = \vec{r}_\perp \times \frac{d\vec{p}}{dt} + \frac{d\vec{r}}{dt} \times \vec{p} = \vec{r}_\perp \times \frac{d\vec{p}}{dt} + \vec{v} \times \vec{p} \quad \text{(revolving point-mass)}$$

$$\sum \vec{\tau}_{ext} = \vec{r}_\perp \times \sum \vec{F}_{ext} \quad \text{(revolving point-mass)}$$

First, we used the product rule to write the derivative of the vector product with respect to time as a sum of two terms. Next, $\vec{v} \times \vec{p} = 0$ because $\vec{p} = m\vec{v}$ and $\vec{v} \times \vec{v} = 0$. As expected, the net external torque acting on the point-mass equals $\vec{r}_\perp$ crossed into the net external force.

The magnitude of the angular momentum of a revolving pointlike object is

$$L = \|\vec{L}\| = \|\vec{r}_\perp \times \vec{p}\| = mvr_\perp \quad \text{(revolving point-mass)}$$

We could also obtain this result from $\vec{L} = I\vec{\omega}$, since $I = mr_\perp^2$ for a point-mass and $v = r_\perp \omega$.

**Angular momentum is relative**: The equation $\vec{L} = \vec{r}_\perp \times \vec{p}$ for a revolving point-mass depends on the axis of rotation, since $\vec{r}_\perp$ is measured from the axis of rotation. Similarly, in the equation $\vec{L} = I\vec{\omega}$, the moment of inertia depends on the axis of rotation. For example, if you want to know the earth's angular momentum, you get a different value if you measure its spin angular momentum (about its spin axis) or its orbital angular momentum (about an axis through the sun).

**Vector addition of angular velocity**: Suppose that a rigid body is simultaneously subject to two different angular velocity vectors, $\vec{\omega}_1$ and $\vec{\omega}_2$. For example, the earth has one angular velocity about its (tilted) axis of rotation and another angular velocity for its revolution about the sun. The net angular velocity can be found from vector addition: $\vec{\omega}_{net} = \vec{\omega}_1 + \vec{\omega}_2$.

**Vector addition of angular momentum**: Similarly, net angular momentum can be found from vector addition: $\vec{L}_{net} = \vec{L}_1 + \vec{L}_2 = I_1\vec{\omega}_1 + I_2\vec{\omega}_2$. It's important to realize that $I_1$ and $I_2$ will generally be different, since these moments of inertia correspond to different axes of rotation.

---

**Example.** A monkey rides a unicycle in a horizontal circle with a constant angular speed of 0.50 rad/s relative to a vertical axis through the center of the circle. The wheel of the unicycle has a constant angular speed of 2.0 rad/s relative to a horizontal axis through the center of the wheel and perpendicular to the wheel – i.e. along the axle. Determine the magnitude and direction of the net angular velocity of the wheel of the unicycle.

The net angular velocity is found through vector addition. Since the given angular velocities are perpendicular – one is vertical and the other is horizontal – we can use the Pythagorean theorem to find the magnitude of the net angular velocity:

$$\omega_{net} = \sqrt{\omega_1^2 + \omega_2^2} = \sqrt{(0.50)^2 + (2.0)^2} = 2.1 \text{ rad/s}$$

The direction of the net angular velocity is not constant, but the axis corresponding to the net angular velocity rotates as the wheel rolls. It always points the same amount above the horizontal, given by $\theta_{net}$, and also points away from the vertical axis through the center of the circle:

$$\theta_{net} = \tan^{-1}\left(\frac{\omega_1}{\omega_2}\right) = \tan^{-1}\left(\frac{0.50}{2.0}\right) = 14°$$

---

**Moment of inertia tensor**: Moment of inertia is not actually a scalar, nor is it a vector; it is a second-rank tensor.[333] The moment of inertia of an extended rigid body about the $x$-axis is $I_x = \int(y^2 + z^2)dm$, since $\sqrt{y^2 + z^2}$ is the shortest distance from each $dm$ to the axis of rotation. Similarly, $I_y = \int(x^2 + z^2)dm$ and $I_z = \int(x^2 + y^2)dm$ are the moments of inertia about the $y$- and $z$-axes, respectively. These are not components of a vector, though, because there are also off-diagonal terms.

---

[333] A vector is a first-rank tensor and a scalar is a zero-rank tensor. In more advanced course work, scalars, vectors, and more general tensors are precisely defined by how they transform under rotations of the coordinate system. If interested, consult a textbook on mathematical physics.

In addition to $I_x$, $I_y$, and $I_z$, there are also products of inertia: $I_{xy} = \int xy\,dm$, $I_{yz} = \int yz\,dm$, and $I_{xz} = \int xz\,dm$. These are six independent components of moment of inertia, which can be represented as elements of a symmetric $3 \times 3$ matrix:

$$I = \begin{pmatrix} I_x & I_{xy} & I_{xz} \\ I_{xy} & I_y & I_{yz} \\ I_{xz} & I_{yz} & I_z \end{pmatrix}$$

The equation $\vec{L} = I\vec{\omega}$ actually represents a matrix multiplication: A vector equals a square matrix times another vector. An interesting feature of this matrix multiplication is that $\vec{L}$ is not necessarily parallel to $\vec{\omega}$ (which would be the case if moment of inertia were a scalar instead of a second-rank tensor).[334]

Similarly, in the equation $\sum \vec{\tau}_{ext} = I\vec{\alpha}$, which applies only when the moment of inertia is constant, the direction of the net external torque and angular acceleration may not be parallel.

**Angular and linear analogies**: The table below shows which angular quantities are analogous to which linear quantities. Understanding this table can greatly reduce the number of equations that you need to memorize in order to solve mechanics problems, as there is an angular equation corresponding to most linear equations.

| Angular and Linear Analogies | |
|---|---|
| Angular Quantity[335] | Linear Quantity |
| angular displacement, $\Delta\theta$ | tangential displacement, $\Delta s_T$ |
| angular velocity, $\vec{\omega}$ | velocity, $\vec{v}$ |
| z-component of $\vec{\omega}$, $\omega_z$ | tangential velocity, $v_T$ |
| angular speed, $\omega$ | tangential speed, $v$ |
| angular acceleration, $\vec{\alpha}$ | acceleration, $\vec{a}$ |
| z-component of $\vec{\alpha}$, $\alpha_z$ | tangential acceleration, $a_T$ |
| moment of inertia, $I$ | mass, $m$ |
| torque, $\vec{\tau}$ | force, $\vec{F}$ |
| rotational kinetic energy, $K_r = \frac{I\omega^2}{2}$ | translational kinetic energy, $K_t = \frac{mv^2}{2}$ |
| rotational work, $W_r = \int \tau\,d\theta$ | translational work, $W_t = \int \vec{F} \cdot d\vec{s}$ |
| angular momentum, $\vec{L} = I\vec{\omega}$ | linear momentum, $\vec{p} = m\vec{v}$ |

- If the torque is constant, the rotational power is $P_r = \tau\omega$. Similarly, when the force is constant, the linear power is $P_t = \vec{F} \cdot \vec{v}$. More generally, power can always be found from $P = dW/dt$.

---

[334] In matrix mechanics, a scalar is represented as a multiple of the identity matrix, which has 1's along the main diagonal and zeroes off-diagonal – e.g. in 3D, the identity matrix is $\begin{pmatrix} 1 & 0 & 0 \\ 0 & 1 & 0 \\ 0 & 0 & 1 \end{pmatrix}$. Also, there are two types of vectors – row vectors and column vectors.

[335] Here, the z-component assumes rotation in the $xy$ plane.

## 7.9 Conservation of Angular Momentum

**Conservation of angular momentum**: The total angular momentum of a system is conserved when the net external torque acting on the system equals zero:

$$\sum_{system} \vec{\tau}_{ext} = 0 \quad \Rightarrow \quad \vec{L}_{tot} = const.$$

When the total angular momentum of a system is conserved, the total initial angular momentum equals the total final angular momentum:

$$\vec{L}_{10} + \vec{L}_{20} + \cdots + \vec{L}_{N0} = \vec{L}_1 + \vec{L}_2 + \cdots + \vec{L}_N \quad \text{(if angular momentum is conserved)}$$

If the angular velocities and axes of rotation of all of the rigid bodies are parallel, this simplifies to

$$I_{10}\omega_{1z0} + I_{20}\omega_{2z0} + \cdots + I_{N0}\omega_{Nz0} = I_1\omega_{1z} + I_2\omega_{2z} + \cdots + I_N\omega_{Nz} \quad \left(\sum \tau_z = 0\right)$$

where the $z$-axis is parallel to the rotation axes.

Recall that conservation of (linear) momentum was particularly useful in the context of collisions. Conservation of angular momentum can also be applied to rotational collisions. Conservation of angular momentum is also useful for many problems that don't look like collisions, but where either the mass is redistributed or the axis of rotation is not constant – i.e. problems where a change in moment of inertia is compensated by a change in angular velocity. We didn't explore many problems where mass changed in the context of linear momentum because most objects don't change mass in the middle of the problem (though there are exceptions, like the rocket). However, there are many standard physics problems where the moment of inertia changes during the problem.

Conservation of angular momentum applies to any problem where the net external torque acting on the system equals zero. You can draw an extended FBD to visualize all of the forces acting on each object along with the positions where they act in order to determine whether or not the net external torque acting on the system is zero. Remember that you are interested in the sum of the torques (but not forces), where each torque is related to a force by $\vec{\tau} = \vec{r}_\perp \times \vec{F}$, if you want to know whether or not the total angular momentum is conserved (the net external force, on the other hand, tells you whether or not the total linear momentum is conserved).

**Note**. Look out for problems that don't look like rotational collisions, but where the distribution of mass changes, for which it is useful to conserve angular momentum. If you're not sure if angular momentum is conserved, the ultimate test is to see if the net external torque acting on the system is zero. From a practical perspective, studying a variety of examples and solving a variety of problems where angular momentum is conserved is the best way to learn for what types of problems it is useful to conserve angular momentum.

# Problem-Solving Strategy for Conservation of Angular Momentum

0. This strategy applies to problems where the net external torque acting on the system is zero. This includes rotational collisions as well as many other problems where a change in moment of inertia causes a change in angular velocity.

1. Setup a coordinate system. It is convenient to choose the $z$-axis to be along the axis of rotation for problems where there is a single, common axis of rotation for all rotating objects. Define which objects are part of the system (being sure that the net external torque acting on the 'system' is zero).

2. Write down conservation of angular momentum for the system. If there is a common, fixed axis of rotation, which is the $z$-axis, conservation of angular momentum will have the form $I_{10}\omega_{1z0} + I_{20}\omega_{2z0} + \cdots + I_{N0}\omega_{Nz0} = I_1\omega_{1z} + I_2\omega_{2z} + \cdots + I_N\omega_{Nz}$. (Otherwise, you need to also write down this equation for $x$- and $y$-components, too.)

3. If two objects collide in a rotational collision that is elastic, also write down conservation of kinetic energy – which will involve rotational kinetic energy, $K_r = I\omega^2/2$, and may also involve translational kinetic energy, $K_t = mv^2/2$. If two objects collide in a completely inelastic rotational collision, set the final angular speeds equal; for an inverse completely inelastic rotational collision, set the initial angular speeds equal.

4. If the problem is not a rotational collision, but the net external torque on the system is zero, read the problem to make any other deductions that may help you relate the initial and final moments of inertia and/or angular velocities (i.e. in addition to conserving angular momentum).

5. As usual, look for $N$ equations in $N$ unknowns ($N$ being the number of unknowns, not objects).

**Conceptual Example.** A door is ajar and a monkey wishes to close it. However, the monkey is very lazy, so instead of closing the door by hand, he throws a ball of silly putty. The silly putty sticks to the door, and the door swings shut. Is the angular momentum of the system – defined as the door plus the silly putty – conserved for this process (from just before impact to just before the door is completely shut)?

To see whether or not the angular momentum of the system is conserved, consider the torques acting on the system. Each torque has the form $\vec{\tau} = \vec{r}_\perp \times \vec{F}$, where $\vec{r}_\perp$ extends perpendicularly from the axis of rotation to the point where the force is applied. The axis of rotation is vertical and passes through the hinges. The weight of the silly putty does not exert a torque – it has the effect of causing vertical translation, and is compensated by the force that adheres the silly putty to the door. The weight of the door is similarly compensated by a vertical force from the hinges, neither of which exert torques. The collision forces between the silly putty and door are internal to the system, and the corresponding internal torques cancel out. Any friction at the hinges results in a negligible torque because $\vec{r}_\perp$ is approximately equal to zero. (Recall also that we always neglect air resistance unless stated otherwise.) Therefore, the net external torque acting on the system is zero, which means that the total angular momentum of the system (silly putty plus door) is conserved for the specified process.

Even though the silly putty is not moving in a circle just before impact, when conserving angular momentum for this problem, we measure its angular momentum as $\vec{L} = \vec{r}_\perp \times \vec{p}$, where $\vec{r}_\perp$ corresponds to the axis of rotation that it is about to have. Note that (linear) momentum is not conserved as a net external force accelerates the center of mass of the system, and mechanical energy is not conserved for this completely inelastic rotational collision.

**Conceptual Example**. A tetherball consists of a ball tied to the top of a vertical pole. A monkey holds the tetherball some distance away from the pole and punches the tetherball. The tetherball then travels around the pole, getting closer and closer to the pole as the cord wraps around the pole. Is the angular momentum of the tetherball conserved for this process (from just after the monkey punches the tetherball to just before the cord is completely wound around the pole)?

First, note that the monkey's punch is irrelevant since the process is defined to begin just after the punch. The monkey's punch is responsible for the ball's initial velocity, after which the tetherball has inertia and so naturally wants to maintain that velocity (while various forces in the problem act to create acceleration).

Two forces act on the tetherball – one is exerted by the cord, and the other is exerted by gravity. Gravity does exert a torque on the tetherball: As a result, the tetherball will descend vertically as it travels around the pole. This is a net external torque, and therefore the angular momentum of the tetherball is not conserved for this process. (It is instructive to compare this example to the previous one, and to understand how gravity exerts a net external torque on the tetherball, but not on the door. What is going on in the door problem that is not having the same effect on the tetherball?)

**Conceptual Example**. An ice skater is spinning in a circle with one leg vertical, while her torso and other leg are horizontal and her arms are extended horizontally. In this position, she has as much mass as possible far from the (vertical) axis of rotation. Her initial angular speed is $3.0$ rev/s. Then she raises her torso vertically, while bringing her free leg downward vertically (but raised slightly so that it is not touching the ice) and raising her arms vertically above her head. In this position, she has all of her mass very close to the axis of rotation. Her final angular speed is $18.0$ rev/s. By what factor did her moment of inertia decrease as she changed the position of her leg, torso, and arms?

As we know from Sec. 7.6, the ice skater must maintain excellent balance: Her center of mass must always lie directly above her ice skate in order for her to avoid falling over. Since she does not fall over during the problem, we conclude that gravity and normal force do not contribute toward a net external torque. The only other force acting is friction between the ice skate and the ice, and $\vec{r}_\perp$ is so small for the corresponding torque that it is negligible. Therefore, the net external torque is approximately zero and the ice skater's angular momentum is approximately conserved.

Since there is just one object in the system, conservation of angular momentum reduces to $I_0\omega_0 = I\omega$. We know that the angular speed increases by a factor of 6; therefore, the ice skater's moment of inertia must decrease by a factor of 6 in order to conserve the ice skater's angular momentum. (Of course, you can't set up a similar proportion in multi-object problems, but must instead write out conservation of angular momentum explicitly and work through the algebra.)

**Example**. At a local park, a 50-kg monkey stands at the edge of a 200-kg merry-go-round with a 10.0-m diameter. The monkey and merry-go-round are initially at rest. The monkey begins walking counterclockwise with an angular speed of $0.40$ rad/s relative to the merry-go-round. What is the angular speed of the merry-go-round relative to the ground while the monkey is walking?

The weight of the merry-go-round is compensated by a support force; it does not exert a torque. The weight of the monkey is balanced by a normal force from the merry-go-round; the corresponding internal torques cancel out. Friction between the monkey and merry-go-round also results in cancelling internal torques. The net external torque is zero, so angular momentum is conserved.

We choose our reference frame to correspond to the system at rest. We place the $xy$ plane in the plane of the merry-go-round with the $z$-axis vertical through the center of the merry-go-round, coming upward. We label the monkey as object 1 and the merry-go-round as object 2. From Newton's third law, we know that the merry-go-round will spin clockwise (i.e. opposite to the monkey). When we conserve angular momentum, we must use vector addition (see the end of Sec. 2.6) in order to find the angular velocity of the monkey relative to our rest frame: $\omega_{2z} = 0.40$ rad/s $+ \omega_{1z}$. That is, the angular velocity of the monkey relative to the ground equals the angular velocity of the monkey relative to the merry-go-round plus the angular velocity of the merry-go-round relative to the ground (the latter being the addition of a negative number in this case). We approximate the merry-go-round as a solid disc and the monkey as pointlike (since the width of the monkey is small compared to the radius of the disc):

$$I_{10}\omega_{1z0} + I_{20}\omega_{2z0} = I_1\omega_{1z} + I_2\omega_{2z}$$
$$0 + 0 = \frac{m_1 R_1^2}{2}\omega_{1z} + m_2 R_2^2 \omega_{2z}$$
$$0 = \frac{m_1 R_1^2}{2}\omega_{1z} + m_2 R_2^2(0.4 + \omega_{1z})$$
$$0 = \frac{(200)(5.0)^2}{2}\omega_{1z} + (50)(5.0)^2(0.4 + \omega_{1z})$$
$$0 = 2500\omega_{1z} + 1250\omega_{1z} + 500$$
$$-500 = 3750\omega_{1z}$$
$$\omega_{1z} = -0.13 \text{ rad/s} \to \omega_1 = 0.13 \text{ rad/s}$$

**Example.** A solid 4.0-kg disc with 20-cm radius is initially spinning with an angular speed of 3.0 rad/s on horizontal frictionless ice. A thin 2.0-kg ring with 20-cm radius, which is initially not spinning, is gently lowered on top of the spinning disc, such that the ring and disc are coaxial. What is the angular speed of the system once the ring has been added?

The net external torque is clearly zero for the combined system (disc plus ring), so the total angular momentum is conserved. Choosing the $z$-axis to be the axis of the disc and ring,

$$I_{d0}\omega_{dz0} + I_{r0}\omega_{rz0} = I_d\omega_{dz} + I_r\omega_{rz}$$
$$\frac{m_d R_d^2}{2}\omega_{dz0} + 0 = \frac{m_d R_d^2}{2}\omega_{dz} + m_r R_r^2 \omega_{rz}$$

Assuming sufficient friction between the disc and ring, this rotational collision is completely inelastic, so the disc and ring have the same final angular speed. They also have the same radius, which cancels:

$$\frac{m_d}{2}\omega_{dz0} = \left(\frac{m_d}{2} + m_r\right)\omega_z$$
$$\frac{4.0}{2}3.0 = \left(\frac{4.0}{2} + 2.0\right)\omega_z$$
$$6.0 = 4.0\,\omega$$
$$\omega = 1.5 \text{ rad/s}$$

**Conservation laws**: There are three fundamental conservation laws in mechanics:
- conservation of energy.
- conservation of (linear) momentum.
- conservation of angular momentum.

The total mechanical energy of a system is conserved when no nonconservative work is done (such as that done by resistive forces). The total (linear) momentum of a system is conserved when the net external force acting on the system is zero. The total angular momentum of a system is conserved when the net external torque acting on the system is zero.

**Noether's theorem**: According to Noether's theorem, each fundamental conservation law of physics is associated with a type of symmetry: Conservation of energy corresponds to homogeneity (uniform throughout) in time, conservation of (linear) momentum corresponds to homogeneity in space, and conservation of angular momentum corresponds to isotropy (the same in all directions) in space. The symmetry described by Noether's theorem refers to the laws of motion and not to the system itself. For example, the earth itself is not perfectly isotropic, but when you write down the equations for the motion of the earth, you find that the equations are isotropic and so the earth's total angular momentum is conserved.

**Precession**: A top or a gyroscope experiences precession when the rigid body spins about an axis and when the axis itself is rotating; the axis of rotation sweeps out a cone, as illustrated below. There are two distinct angular velocities in this case: The spin rate, $\vec{\omega}_s$, refers to the rate at which the rigid body spins about the rigid body's axis (the spin axis), and the precession rate, $\vec{\omega}_p$, refers to the rate at which the axis of rotation itself rotates. There is spin angular momentum, $\vec{L}_s$, along the rigid body's spin axis, in addition to precessional angular momentum, $\vec{L}_p$, from the center of mass traveling in a circle as the rigid body precesses. The precessional angular momentum is insignificant compared to the spin angular momentum when the spin rate is very high; when the spin rate is too low, this is not the case and the rigid body topples over (in the case of a top).

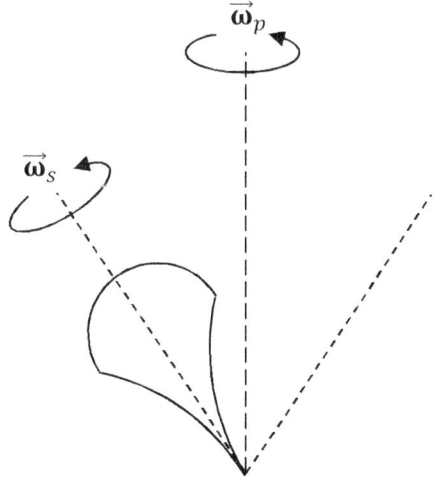

Consider the case of a top with a high enough spin rate that its total angular momentum is approximately along the precessing axis of rotation. There is a net external torque acting on the top, which is exerted by gravity. If the spin rate were low, the torque exerted by gravity would cause the top to topple over. However, we will see that this is not the case when the spin rate is high. The net external torque equals $\sum \vec{\tau}_{ext} = \frac{d\vec{L}}{dt}$. Any torque due to friction between the bottom of the top and the surface it spins on is negligible because $\vec{r}_\perp$ is very small. Therefore, the net external torque approximately equals the torque exerted by gravity. Given a snapshot of the top, the direction of the torque is perpendicular to both the spin axis and the weight of the top, while the direction of the angular momentum is along the spin axis (since we are assuming a high spin rate). The net external torque is parallel to the change in angular momentum, $d\vec{L}$. As a result, the axis of rotation precesses, in analogy with uniform circular motion. Recall that the centripetal force is perpendicular to the velocity, and causes the direction of the velocity to change without affecting the speed. Similarly, in the case of this precessing top, the net external torque causes the direction of the angular momentum to change, without changing its magnitude. The direction of the change in angular momentum is such that it causes the axis of rotation to sweep out a cone.

## Conceptual Questions

1. Two fleas are riding on a second-hand. One flea sits at the far end of the second hand (i.e. far from the center of rotation), while the other flea is halfway between the center of rotation and the first flea. Find the ratio of the fleas' (A) angular speeds, (B) tangential speeds, (C) centripetal accelerations, and (D) total distances traveled.
2. For each situation described, indicate whether each of the following types of acceleration are zero or nonzero – angular acceleration, tangential acceleration, and centripetal acceleration. (A) A banana falls straight downward. (B) A monkey runs in a horizontal circle with constant speed. (C) A monkey drives a car in a straight line along a level highway with the cruise control set. (D) A monkey drives up a steady incline with the cruise control set. (E) A pea slides down the inside of a circular bowl without friction.
3. A monkey has a baton. At first, the monkey rotates the baton about an axis that is perpendicular to the baton and passes through one end. Next, the monkey throws the baton up in the air and it rotates about an axis that is perpendicular to the baton and passes through its center. Finally, the monkey rolls the baton along the floor. Rank the moment of inertia of the baton in each case from smallest to greatest, and justify your order conceptually.
4. A monkey has a solid uniform equilateral triangle. At first, the monkey rotates the triangle about an axis that is perpendicular to the triangle and passes through its centroid (which is the center of mass of the triangle, where the three bisectors intersect). Next, the monkey rotates the triangle about one of its bisectors. Finally, the monkey rotates the triangle about an axis that is parallel to one edge and passes through its centroid. Rank the moment of inertia of the triangle in each case from smallest to greatest, and justify your order conceptually.
5. A monkey has two solid spheres of non-uniform density. One sphere's density increases with distance from the center, while the other sphere's density decreases with distance from the center. Which sphere has the greater moment of inertia? Explain.

6. A monkey rotates a thin ring and a solid disc of the same size and mass. The thin ring rotates about an axis that is perpendicular to the ring and passes through its center. The solid disc rotates about an axis that is perpendicular to the disc, but does not pass through its center. The moment of inertia is the same in each case. Precisely, where is the axis of rotation for the solid disc? What is its radius of gyration for this axis of rotation? Compare this to the distance that the furthest point on the disc is from the axis of rotation.

7. A parallelepiped is a 3D generalization of the parallelogram: It has three pairs of parallelograms on its faces. Show how both the scalar product and the vector product can be used together to determine the volume of the parallelepiped.

8. For each Cartesian coordinate system illustrated below, indicate if it is right- or left-handed.

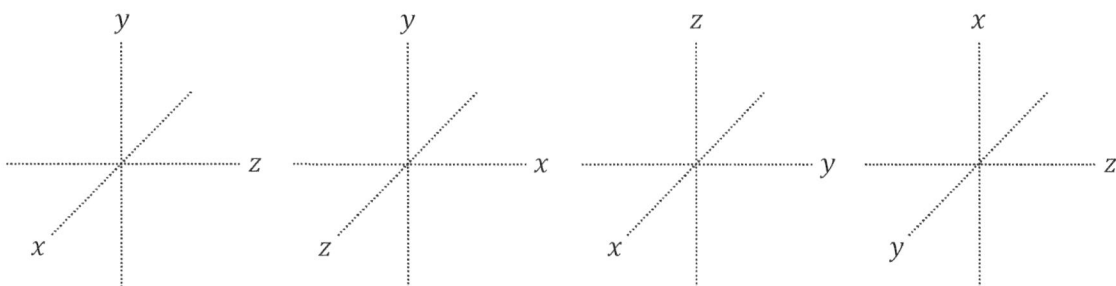

9. For each case below, use the right-hand rule to determine the direction of the indicated quantity.
   (A) $\vec{A}$ is →, $\vec{B}$ is ↓. Find the direction of $\vec{C} = \vec{A} \times \vec{B}$.
   (B) $\vec{A}$ is ↑, $\vec{B}$ is ←. Find the direction of $\vec{C} = \vec{B} \times \vec{A}$.
   (C) $\vec{A}$ is ⊙, $\vec{C}$ is ↑. Find the direction of $\vec{B} = \vec{A} \times \vec{C}$.
   (D) $\vec{r}_\perp$ is ←, $\vec{F}$ is ⊗. Find the direction of $\vec{\tau}$.
   (E) $\vec{r}_\perp$ is ↑, $\vec{F}$ is ↓. Find the direction of $\vec{\tau}$.
   (F) $\vec{r}_\perp$ is ↗, $\vec{F}$ is ↘. Find the direction of $\vec{\tau}$.
   (G) $\vec{r}_\perp$ is ↖, $\vec{p}$ is ⊙. Find the direction of $\vec{L}$ (point-mass).
   (H) $\vec{r}_\perp$ is ↑, $\vec{p}$ is ↙. Find the direction of $\vec{L}$ (point-mass).
   (I) $\vec{r}_\perp$ is ⊗, $\vec{p}$ is →. Find the direction of $\vec{L}$ (point-mass).
   (J) $\vec{r}_\perp$ is ↘, $\vec{\omega}$ is ↗. Find the direction of $\vec{v}$ (revolving point-mass).
   (K) $\vec{r}_\perp$ is ⊙, $\vec{\omega}$ is ←. Find the direction of $\vec{v}$ (revolving point-mass).
   (L) $\vec{r}_\perp$ is ↙, $\vec{\omega}$ is ⊗. Find the direction of $\vec{v}$ (revolving point-mass).
   (M) $\vec{r}_\perp$ is ↓, $\vec{\omega}$ is →, $\vec{\alpha} = 0$. Find the directions of $\vec{v}$ and $\vec{a}$ (revolving point-mass).

10. Use the right-hand rule (and conceptual understanding) to evaluate the following vector products in spherical coordinates: $\hat{r} \times \hat{r}$, $\hat{r} \times \hat{\theta}$, $\hat{\theta} \times \hat{r}$, $\hat{\theta} \times \hat{\theta}$, $\hat{r} \times \hat{\varphi}$, $\hat{\varphi} \times \hat{r}$, $\hat{\varphi} \times \hat{\varphi}$, $\hat{\theta} \times \hat{\varphi}$, and $\hat{\varphi} \times \hat{\theta}$.

11. For each equation below, indicate how many terms are on each side and how many of the terms are nonzero: (A) $\sum_{i=1}^{3} A_i^2 = B$, (B) $\sum_{i,j=1}^{3} A_i B_j \delta_{ij} = \sum_{i=1}^{3} C_i D_i$, (C) $C_k = \sum_{i,j=1}^{3} \epsilon_{ijk} A_i B_j$, (D) $\sum_{i=1}^{3} \epsilon_{ijk} \epsilon_{imn} = \delta_{jm}\delta_{kn} - \delta_{jn}\delta_{km}$, and (E) $C^2 = \sum_{i,j,k,\ell,m,n=1}^{3} \epsilon_{ijk} \epsilon_{\ell mn} A_i A_\ell B_j B_m \delta_{kn}$.

12. For each diagram on the following page, determine the values of $r_\perp$ and the angle $\theta$ involved in the equation $\tau = r_\perp F \sin\theta$ for the torque exerted by the indicated force. In addition, determine the direction of the torque that is exerted by the indicated force.

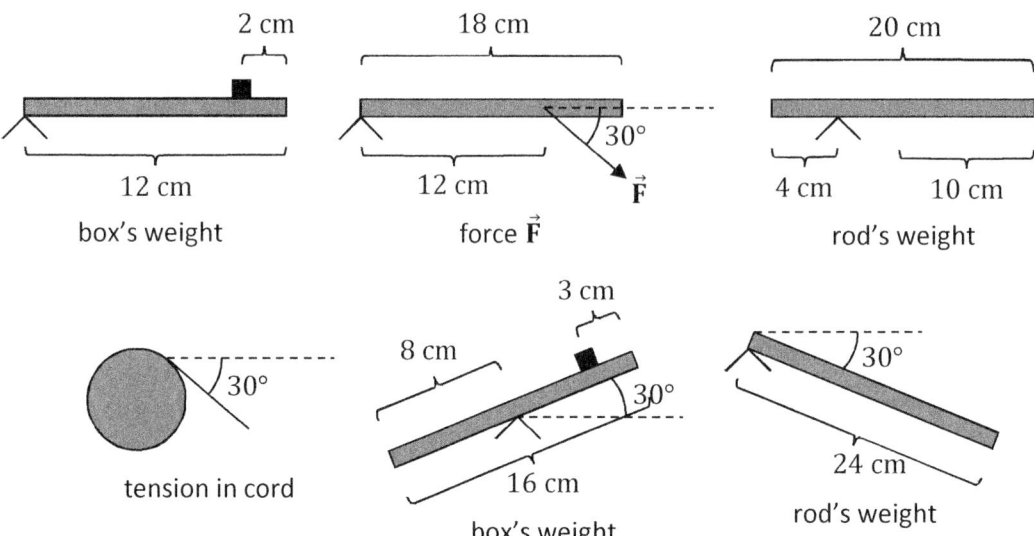

13. A monkey has threw screwdrivers: a screwdriver with a yellow handle, a screwdriver with a red handle that is twice as long as the one with the yellow handle (with an equally thick handle), and a screwdriver with a blue handle for which the handle is twice as thick as the red or yellow handles (but equally long as the red screwdriver. Rank the screwdrivers in their effectiveness at applying torque while tightening or loosening a screw. If there are any ties, indicate this.

14. A monkey nails two pieces of wood together, which have the same uniform density, to form an 'L.' The monkey balances the L on a fulcrum to locate its center of mass. The monkey then saws the L into two pieces at its balancing point, as illustrated below. Which end weighs more? Explain.

15. A monkey nails two pieces of wood together, which have the same uniform density, to form a 'L,' as in the previous problem. The monkey places a fulcrum beneath the point that bisects the mass of the L (i.e. the sides of the L to the left and right of the fulcrum have equal mass). Starting from rest, which way will the L rotate, if any?

16. If you are given $\vec{A}$ and $\vec{B}$, where $\vec{C} = \vec{A} \times \vec{B}$, it is straightforward to solve for $\vec{C}$. However, if you are given $\vec{A}$ and $\vec{C}$, when you proceed to solve for $\vec{B}$ using the determinant method, you will come across an algebra problem. Conceptually, explain why there is a problem, and identify what the algebraic issue is.

17. A solid disc rolls without slipping down an incline. The monkey is thinking that if he reduces the moment of inertia of the solid disc, it will reach the bottom of the incline sooner. So the monkey cuts the solid disc in such a way that it becomes a solid disc with half of its original diameter. Will the smaller solid disc reach the bottom of the incline sooner? If so, explain why, and determine by what factor it reaches the bottom sooner. If not, explain why the smaller moment of inertia doesn't increase its tangential acceleration.

18. Two solid discs of the same size roll without slipping down an incline. One is made out of aluminum, while the other is made out of lead (which is, of course, more dense). Both solid discs simultaneously roll down the incline from rest from the same height. Which solid disc reaches the bottom first? Explain.

19. A solid sphere and a solid cylinder roll without slipping up an incline with the same initial angular speed. Assuming each object to be uniformly dense, which rolls further up the incline? Explain.

20. A hollow sphere and a solid cylinder roll without slipping down an incline. They simultaneously start form rest from the same height and both are uniformly dense. Which object reaches the bottom of the incline first? Explain.

21. A monkey has three balls of the same size and construction: One is hollow, one is filled with liquid water, and the other is filled with frozen water. All three balls roll without slipping up an incline with the same initial angular speed. Which rolls further up the incline? Explain.

22. Write the angular analogy for each of the following linear expressions: $v_y = \sqrt{2a_y \Delta y}$, $p^2/2m$, $\int v\, dm$, and $dF/dt$.

23. Why is it incorrect to set $\vec{v}$ equal to $r_\perp \vec{\omega}$? Is this expression correct in either magnitude or direction? What is the correct expression?

24. A pendulum oscillates back and forth along a vertical circle. Neglect any resistive forces. Draw an extended FBD for the pendulum. Indicate whether or not each force individually exerts a torque on the pendulum. Is the next external torque zero or nonzero? Is the total angular momentum of the pendulum conserved? Explain.

25. A pendulum sweeps out a cone as its bob travels in a horizontal circle. Neglect any resistive forces. Draw an extended FBD for the pendulum. Indicate whether or not each force individually exerts a torque on the pendulum. Is the next external torque zero or nonzero? Is the total angular momentum of the pendulum conserved? Explain.

26. Neglecting any resistive forces, what happens to a spinning ice skater's moment of inertia, angular speed, and angular momentum as she extends her leg and arms outward?

27. A comet travels in an elliptical orbit with the center of mass of the comet-star system lying at one focus. Ignore the effects of any other astronomical bodies besides the comet and star. What force(s) act on the comet? For each force, indicate whether or not it exerts a torque on the comet. Is the total angular momentum of the comet conserved? Explain. What happens to the tangential speed of the comet as it approaches the sun?

28. A solid sphere and a hollow sphere each roll without slipping toward one another on a horizontal surface. The two balls collide elastically. There is friction in this problem between each ball and the horizontal and between the two balls when they collide. Is the net external force acting on the system (defined as the two spheres) zero or not zero? Is the net external torque acting on the system zero or not zero? Which of the following quantities are conserved for the collision: momentum, angular momentum, and/or mechanical energy? If any quantity is conserved only depending on the situation, explain.

7 Rotation

29. If everybody on the earth suddenly started walking to the east at the same time, what would happen to the angular speed of the earth? Explain. What will happen when everyone stops walking?

30. A large, heavily loaded, very long train is built to travel around the equator. The train is to run continuously without ever changing its speed. Which way should the train run to make the days last longer? Explain.

## Practice Problems

### Angular acceleration

1. A monkey puts his younger brother in a 70-cm diameter tire and gives it a big shove forward. It topples over 15 seconds later, after completing 60 revolutions. Assume that the tire uniformly decelerates. (A) How far does the tire travel? (B) What is the initial angular speed of the tire?

2. A monkey shows symptoms of being under the influence of too much physics. Specifically, his head begins spinning with an initial angular speed of $5\pi$ rad/s with an angular acceleration of $5\pi$ rad/s$^2$. His head completes 36 revolutions during this uniform angular acceleration. (A) What is the final angular speed of the monkey's head? (B) For how much time does the monkey's head rotate?

3. A monkey shows symptoms of physicsitis. Specifically, his stomach begins churning (from rest) with a uniform angular acceleration of $10\pi$ rad/s$^2$. His stomach completes 90 revolutions. (A) What is the final angular speed of the monkey's stomach? (B) For how much time does the monkey's stomach rotate?

4. Faster than a speeding photon,[336] more powerful than a black hole,[337] capable of leaping across the universe in a single bound,[338] Supermonk hears the plea of desperate physics students across the world: There just isn't enough time in the day to complete their homework assignments. Supermonk begins running (and, when necessary, swimming) to the east at a super-high speed in order to slow the earth's rotation. The earth uniformly decelerates until there are 36 hours in a 'day.' Supermonk runs continuously for 720 hours to achieve this amazing feat. (A) What is the earth's angular deceleration? (B) How many revolutions does the earth complete while Supermonk runs?

5. As you begin to read this problem, a monkey ties a rope to your leg, drags you outside, and sets you on a skateboard. The monkey spins you in a $60/\pi$-m diameter circle. Starting from rest, you rotate with uniform angular acceleration until your angular speed is $2\pi/3$ rad/s. You travel 300 m. (A) How many revolutions do you complete? (B) For how much time does the monkey spin you in a circle? (C) What is your tangential acceleration? (D) What is the magnitude of your acceleration at $t = 5.0$ sec?

6. A monkey puts her baby on the edge of a 4.0-m diameter merry-go-round, initially at rest. The baby holds on tight. The monkey grabs the edge of the merry-go-round and runs around in a circle, providing uniform angular acceleration to the merry-go-round. The monkey runs for 8.0 seconds, reaching a final speed of 4.0 m/s. (A) What is the baby's tangential acceleration? (B) What is the baby's angular acceleration? (C) What is the baby's final angular speed? (D) How far does the baby travel? (E) How many revolutions does the baby complete? (F) What is the baby's final centripetal acceleration?

---

[336] With an uncertainty of $\pm 300,000,000$ m/s.
[337] From a distance of 100,000 lightyears or more.
[338] Through a wormhole.

7. The minute hand of the Official Quiz Clock points to the 12 and is functioning normally until the quiz begins, at which time the minute hand develops a uniform angular acceleration of 6.0 rev/hr$^2$. The quiz is over when the minute hand points to the 3. How much quiz time do the students lose?

8. A monkey ties a rope to your leg, drags you outside, and sets you on a skateboard. The monkey spins you in an 8-m diameter circle. Starting from rest, you rotate:
  (i) with uniform angular acceleration from $t = 0$ to $t = 5.0$ sec.
  (ii) at a constant rate of 0.30 rev/sec from $t = 5.0$ sec sec to $t = 25.0$ sec sec
  (iii) with uniform deceleration for 15 rev from $t = 25.0$ sec sec until coming to rest

**Note:** All three parts of the trip fit together: For example, the angular speed at the end of trip (i) matches the angular speed at the beginning of trip (ii). **Hint:** You may need to look at the information for parts (ii) or (iii) when answering questions about parts (i) or (ii). (A) What is your angular acceleration during Part (i)? (B) What is the magnitude of your acceleration at $t = 4.0$ sec? (C) How many revolutions do you complete during Part (i)? (D) How far do you travel during Part (ii)? (E) What is the magnitude of your acceleration during Part (ii)? (F) What is your tangential acceleration during Part (iii)? (G) For how much time do you decelerate during Part (iii)? (H) What is your average angular speed for the entire trip?

9. A monkey throws a monkeyball up into the air with an initial angular speed of $\pi$ rad/s. The monkeyball completes 4.0 revolutions during its flight. (A) What is the angular speed of the monkeyball at the top of its trajectory? (B) How high does the monkeyball rise? (C) What is the diameter of the monkeyball?

10. The $z$-component of the angular velocity of a bumblebee (flying in a circle with a 0.50-m radius in the $xy$ plane) is plotted below as a function of time. (A) What is average angular acceleration of the bumblebee? (B) What is the initial angular acceleration of the bumblebee? (C) What is the angular acceleration of the bumblebee at $t = 6.0$ sec? (D) What is the maximum angular speed of the bumblebee? (E) What is the maximum speed of the bumblebee? (F) How far does the bumblebee travel? (G) What is the average $z$-component of the angular velocity of the bumblebee?

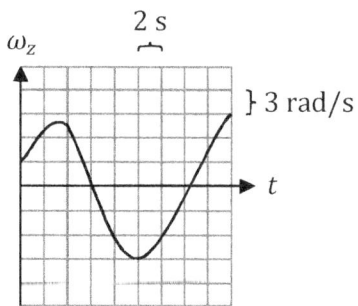

11. An identified flying monkey (IFM) flies in a circle with a radius of 20 m in the $xy$ plane, centered about the origin. The angular velocity of the IFM is given by $\vec{\omega} = \beta\sqrt{t}\,\hat{k}$, where $\beta = 4.0$ in SI units. The IFM's motion begins at $t = 0$. Determine the following at $t = 9.0$ sec: (A) the SI units of beta ($\beta$), (B) the speed, (B) the angular acceleration, (C) the centripetal acceleration, (D) the tangential acceleration, (E) the acceleration, and (F) the total distance traveled for the IFM.

12. Two monkeys jog around a circular track with a radius of one kilometer. One monkey jogs clockwise with a constant angular speed of 3.0 rev/hr, while his sister jogs counterclockwise with a constant angular acceleration of 5.0 rev/hr$^2$, starting from rest. They begin from the same position. (A) How much time passes before they meet? (B) How far has each monkey traveled during this time?

## 7 Rotation

### Moment of inertia

13. Find the moment of inertia of a uniform monkeystick about an axis perpendicular to the monkeystick and passing through the monkeystick a distance that is one-fourth of its length from one end.

14. A monkey uses recycled physics textbooks to manufacture the smartbell illustrated below. The smartbell consists of an infinitesimally thin, hollow sphere on the left and a solid sphere on the right connected by a uniform rod. The hollow sphere has a mass of 15.0 kg and radius of 10.0 m. The solid sphere has a mass of 20.0 kg and a radius of 5.0 m. The rod has a mass of 6.0 kg and length of 40.0 m. (Holes were not drilled into the spheres to connect the rod; instead, the surface of each sphere was welded to the rod.) Determine the moment of inertia of the smartbell about each of the following axes: (A) about the axis of the rod and (B) about the dashed axis illustrated below.

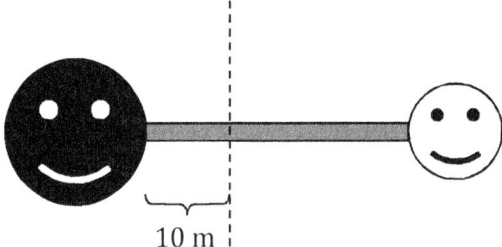

15. The object illustrated below consists of three point-like objects connected by rigid rods of negligible mass. Determine the moment of inertia of the system about (A) the $x$-axis, (B) the $y$-axis, (C) the $z$-axis, and (D) an axis parallel to the $z$-axis that passes through the center of mass of the system.

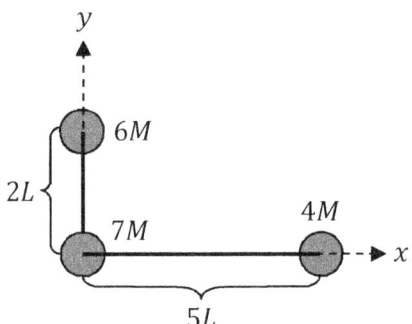

16. A monkey connects three small (point-like) 5.0-kg monkeyballs together with three 4.0-m long lightweight (relatively massless) rigid rods to form the equilateral triangle illustrated on the following page. Find the moment of inertia of this system about (A) the $x$-axis and (B) the $y$-axis.

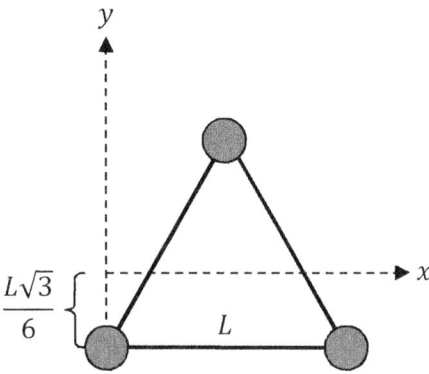

17. The two thin rings illustrated below are welded together at the point of contact. The large ring has mass $M_0$ and radius $R_0$, while the small ring has mass $2M_0/3$ and radius $2R_0/3$. Determine the moment of inertia of the system about an axis perpendicular to the rings and (A) through the center of the large ring and (B) through the center of mass of the system.

18. A uniform thin ring of radius $R_0$ lies in the $xy$ plane centered about the origin. Determine the moment of inertia of the ring about an axis that is parallel to the $y$-axis and passes through the point $(R_0/2,0)$.

19. A uniform solid disc of radius $R_0$ in the $xy$ plane centered about the origin has a circular hole cut out of it of radius $R_0/2$ centered about the point $(R_0/2,0)$. Determine the moment of inertia of this object about the $z$-axis.

20. Derive the perpendicular axis theorem for planar rigid bodies. That is, for a planar rigid body lying in the $xy$ plane, show that the moments of inertia about the $x$-, $y$-, and $z$-axes are related by $I_z = I_x + I_y$. After deriving this formula, verify that it works for the following objects illustrated in the table of moments of inertia for common geometries about common axes of rotation: a 2D solid disc and a 2D rectangle.

21. A uniform monkeystick of mass $M$ has endpoints $(0,0)$ and $(L,0)$. Integrate to derive an equation for its moment of inertia about the axis $x = L/4$ in terms of $M$ and $L$ only.

22. A non-uniform monkeystick of mass $M$ has endpoints $(0,0)$ and $(L,0)$ and linear mass density $\lambda = ax^b$ where $a$ and $b$ are positive constants. Derive an equation for its moment of inertia about the $y$-axis in terms of $M$, $L$, and/or $b$ only.

23. Perform the moment of inertia integration to derive an equation for the moments of inertia of a uniform solid disc in rolling mode and flipping mode. Check your answers with the table of moments of inertia for common geometries about common axes of rotation.

24. Perform the moment of inertia integration to derive an equation for the moment of inertia of a uniform thin ring in flipping mode. Check your answer with the table of moments of inertia for common geometries about common axes of rotation.

25. Perform the moment of inertia integration to derive an equation for the moments of inertia of a uniform solid rectangle about (A) an axis perpendicular to the rectangle and passing through its center

and (B) a symmetric bisector. Check your answers with the table of moments of inertia for common geometries about common axes of rotation.

26. Perform the moment of inertia integration to derive an equation for the moment of inertia of a uniform infinitesimally thin hollow sphere in rolling mode. Check your answer with the table of moments of inertia for common geometries about common axes of rotation.

27. Perform the moment of inertia integration to derive an equation for the moment of inertia of a uniform solid cone about its symmetry axis. Check your answer with the table of moments of inertia for common geometries about common axes of rotation.

28. Perform the moment of inertia integration to derive an equation for the moment of inertia of a uniform single-holed ring torus in rolling mode. Check your answer with the table of moments of inertia for common geometries about common axes of rotation.

29. Consider the uniform monkey tail of mass $M$ illustrated below; it has the shape of a semicircular arc of radius $R_0$. Express the moment of inertia of the monkey tail about the axis $y = R_0/4$ in terms of $M$ and $R_0$ only.

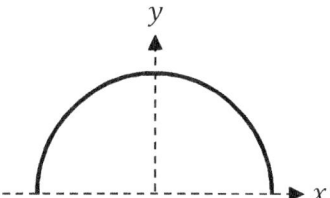

30. The slice of banana-cream pie illustrated below has radius $R_0$, mass $M$, and non-uniform density $\sigma = \beta r^2 \theta$, where $\beta$ is a positive constant. Derive an equation for the moment of inertia of the banana-cream pie about the $y$-axis in terms of $M$ and $R_0$ only.

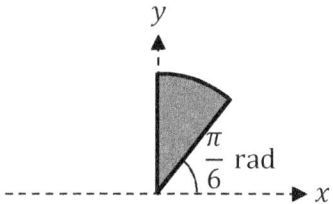

31. A spherical bananaball has mass $M$, radius $R_0$, and non-uniform density $\rho = \beta r$, where $\beta$ is a positive constant. Derive an equation for the moment of inertia of the bananaball in rolling mode in terms of $M$ and $R_0$ only.

32. The ghost of Isaac Newton reaches inside your ear and yanks out your brain! He then stretches your brain into a long straight line, and finally folds it into a thin rectangle. Isaac then rotates your rectangular brain as illustrated on the following page. Now your brain is an antenna capable of receiving physics signals with improved clarity.

Think of your brain as four separate rods (not a rectangular plate). The vertical rods have uniform linear mass density, while the horizontal rods have non-uniform linear mass densities $\lambda = \beta x$, where $\beta$ is a positive constant. The horizontal rods have twice as much mass as the vertical rods. The total mass of your brain is $M$.

(A) What is $\beta$? (B) Express the mass of each rod in terms of $M$. (C) Express the total moment of inertia of your brain about the indicated axis in terms of $M$ and $L$ only.

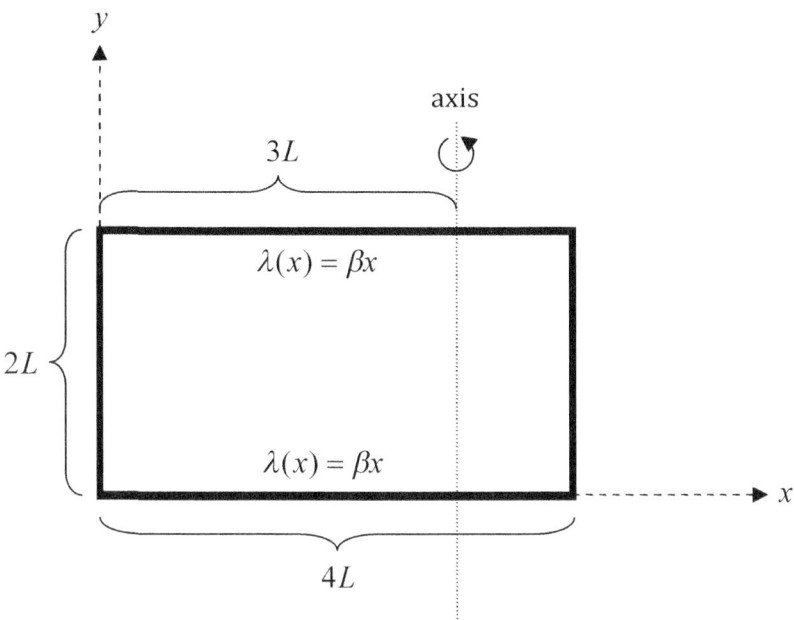

**Determinants, the vector product, and tensor notation**

33. Determine the determinant of the following matrix:
$$\begin{pmatrix} 5 & -3 \\ -2 & -4 \end{pmatrix}$$

34. Determine the determinant of the following matrix:
$$\begin{pmatrix} 3 & 1 & 2 \\ 0 & 4 & 6 \\ -1 & -2 & -1 \end{pmatrix}$$

35. Determine the determinant of the following matrix:
$$\begin{pmatrix} 2 & -2 & 2 \\ -1 & 2 & -3 \\ 4 & -3 & 2 \end{pmatrix}$$

36. Determine the determinant of the following matrix:
$$\begin{pmatrix} 1 & -1 & 2 & -2 \\ 3 & 1 & -1 & -3 \\ 2 & 4 & 4 & 2 \\ -2 & 2 & 2 & 2 \end{pmatrix}$$

37. Show that the determinant of an $N \times N$ matrix will be zero if any two rows (or columns) are the same. Show that it is also zero if one row (or column) is a simple multiple of another. Show that it is also zero if one row (or column) equals the sum of other rows (or columns). Show that it is zero, in general, if one row (or column) is a linear combination of other rows (or columns): That is, if the elements of one row can be obtained from the elements of another row with a formula of the form $R_i = a_1 R_1 + a_2 R_2 + \cdots + a_N R_N$, where the $\{a_i\}$ are constants.

38. What do you get when you cross monkeys with multiplication? (A) Find $\hat{\mathbf{k}} \times \hat{\mathbf{i}}$. (B) Find $\hat{\mathbf{i}} \times \hat{\mathbf{k}}$. (C) Find $\hat{\mathbf{j}} \times \hat{\mathbf{j}}$. (D) Find $\hat{\mathbf{i}} \times (\hat{\mathbf{i}} - \hat{\mathbf{j}})$. (E) Find $\vec{\mathbf{A}} \times \vec{\mathbf{B}}$, where $\vec{\mathbf{A}} = -2\hat{\mathbf{i}} - 3\hat{\mathbf{j}} + 4\hat{\mathbf{k}}$ and $\vec{\mathbf{B}} = -2\hat{\mathbf{j}} + \hat{\mathbf{k}}$. (F) Find $\vec{\mathbf{A}} \times \vec{\mathbf{B}}$, where $\vec{\mathbf{A}} = \hat{\mathbf{i}} - 2\hat{\mathbf{k}}$ and $\vec{\mathbf{B}} = -3\hat{\mathbf{i}} + 6\hat{\mathbf{k}}$. (G) Find $(\vec{\mathbf{A}} \times \vec{\mathbf{B}}) \times \vec{\mathbf{C}}$, where $\vec{\mathbf{A}} = 2\hat{\mathbf{i}} + \hat{\mathbf{j}}$, $\vec{\mathbf{B}} = \hat{\mathbf{j}} - 2\hat{\mathbf{k}}$, and $\vec{\mathbf{C}} = \hat{\mathbf{i}} - \hat{\mathbf{k}}$. (H) Find $\vec{\mathbf{A}} \times (\vec{\mathbf{B}} \times \vec{\mathbf{C}})$, where $\vec{\mathbf{A}} = 2\hat{\mathbf{i}} + \hat{\mathbf{j}}$, $\vec{\mathbf{B}} = \hat{\mathbf{j}} - 2\hat{\mathbf{k}}$, and $\vec{\mathbf{C}} = \hat{\mathbf{i}} - \hat{\mathbf{k}}$.

39. Given $\vec{\mathbf{A}} = 3\hat{\mathbf{i}} - \sqrt{3}\hat{\mathbf{j}} + 2\hat{\mathbf{k}}$ and $\vec{\mathbf{B}} = \sqrt{3}\hat{\mathbf{i}} - \hat{\mathbf{j}} + 2\sqrt{3}\hat{\mathbf{k}}$, find: (A) $\vec{\mathbf{A}} \cdot \vec{\mathbf{B}}$, (B) $\vec{\mathbf{A}} \times \vec{\mathbf{B}}$, and (C) the angle between $\vec{\mathbf{A}}$ and $\vec{\mathbf{B}}$.

40. Find the torques exerted on the following monkeys: (A) $\vec{\mathbf{r}}_\perp = \hat{\mathbf{k}}$, $\vec{\mathbf{F}} = \hat{\mathbf{j}}$; (B) $\vec{\mathbf{r}}_\perp = \hat{\mathbf{k}}$, $\vec{\mathbf{F}} = \hat{\mathbf{k}}$; (C) $\vec{\mathbf{r}}_\perp = \hat{\mathbf{k}}$, $\vec{\mathbf{F}} = \hat{\mathbf{i}} - \hat{\mathbf{j}}$; (D) $\vec{\mathbf{r}}_\perp = 2\hat{\mathbf{i}} + 3\hat{\mathbf{j}} - 4\hat{\mathbf{k}}$, $\vec{\mathbf{F}} = 2\hat{\mathbf{j}} - \hat{\mathbf{k}}$; (F) $\vec{\mathbf{r}}_\perp = -2\hat{\mathbf{i}} + 4\hat{\mathbf{k}}$, $\vec{\mathbf{F}} = \hat{\mathbf{i}} - 2\hat{\mathbf{k}}$.

41. The Powder Puff Pearls save the day by lassoing 40-kg No-Mo Yo-Yo and then flying around. The distance between No-Mo Yo-Yo and the axis of rotation is described by $\vec{\mathbf{r}}_\perp = 3\hat{\mathbf{i}} - 8\hat{\mathbf{j}} - 24\hat{\mathbf{k}}$. The force Possum exerts on No-Mo Yo-Yo is $\vec{\mathbf{F}} = 16\hat{\mathbf{i}} - 8\hat{\mathbf{j}} + 2\hat{\mathbf{k}}$. SI units have been suppressed for convenience. (A) Express the torque that the Possum exerts on the monkey in terms of Cartesian unit vectors. (B) What is the angle between $\vec{\mathbf{r}}_\perp$ and $\vec{\mathbf{F}}$?

42. Evaluate the following vector products, where all of these unit vectors correspond either to Cartesian coordinates or cylindrical coordinates: $\hat{\mathbf{r}}_c \times \hat{\mathbf{j}}$, $\hat{\mathbf{i}} \times \hat{\boldsymbol{\theta}}$, $\hat{\mathbf{r}}_c \times \hat{\mathbf{k}}$, $\hat{\mathbf{r}}_c \times \hat{\boldsymbol{\theta}}$, and $\hat{\mathbf{k}} \times \hat{\boldsymbol{\theta}}$.

43. Evaluate the following vector products, where all of these unit vectors correspond either to Cartesian coordinates or spherical coordinates (except $\hat{\mathbf{r}}_c$, which belongs to cylindrical coordinates): $\hat{\mathbf{r}} \times \hat{\mathbf{i}}$, $\hat{\boldsymbol{\varphi}} \times \hat{\mathbf{j}}$, $\hat{\mathbf{k}} \times \hat{\boldsymbol{\theta}}$, $\hat{\mathbf{r}} \times \hat{\mathbf{r}}$, $\hat{\mathbf{r}} \times \hat{\boldsymbol{\varphi}}$, $\hat{\boldsymbol{\varphi}} \times \hat{\boldsymbol{\varphi}}$, and $\hat{\mathbf{r}} \times \hat{\mathbf{r}}_c$.

44. Prove that $\vec{\mathbf{A}} \times \vec{\mathbf{B}}$ is perpendicular to both $\vec{\mathbf{A}}$ and $\vec{\mathbf{B}}$.

45. Evaluate the sum $\sum_{i=1}^{3} \delta_{ii}$.

46. Evaluate the double sum $\sum_{i,j=1}^{3} \delta_{ij} \delta_{ij}$.

47. Evaluate the double sum $\sum_{i,j=1}^{3} \delta_{ij} \epsilon_{ijk}$.

48. Evaluate the triple sum $\sum_{i,j,k=1}^{3} \epsilon_{ijk} \epsilon_{ijk}$.

49. Evaluate the triple sum $\sum_{i,j,k=1}^{3} \epsilon_{ijk} \epsilon_{ijm} A_k B_m$.

50. Evaluate the quadruple sum $\sum_{i,j,k,\ell=1}^{3} \delta_{ij} \delta_{k\ell} A_i A_j B_k B_\ell$.

51. Use tensor notation to prove that $\vec{\mathbf{B}} \times \vec{\mathbf{A}} = -\vec{\mathbf{A}} \times \vec{\mathbf{B}}$.

52. Use tensor notation to prove that the vector product is distributive – i.e. that $\vec{\mathbf{A}} \times (\vec{\mathbf{B}} + \vec{\mathbf{C}}) = \vec{\mathbf{A}} \times \vec{\mathbf{B}} + \vec{\mathbf{A}} \times \vec{\mathbf{C}}$.

53. Use tensor notation to derive the BAC-CAB rule for the triple vector product.

## Summing the torques

54. As illustrated on the following page, on Monksgiving, one 40-kg monkey hangs from one end of a 100-kg uniform rod that is 12.0 m long, another 60-kg monkey stands above the pivot (center of rotation), and a 20-kg turkey lies 1.0 m from the opposite end. Also, the standing monkey pulls on a rope as shown with a force of 200 N. (A) Find the torque exerted by the dangling monkey. (Relative signs are crucial.) (B) Find the torque exerted by the standing monkey's weight. (C) Find the torque exerted by the tension in the rope (pulled by the standing monkey). (D) Find the torque exerted by the turkey. (E) Which torque is missing? Find it. (F) What is the net torque exerted on the rod?

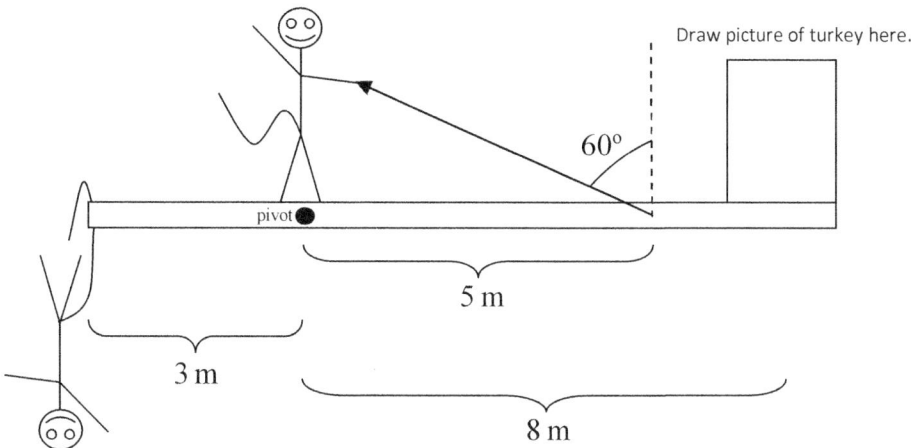

55. The uniform boom drawn below is 20.0-m in length and weighs 600 N. The pivot is located 5.0 m from the left end. Also, $\|\vec{F}_1\| = 2400$ N, $\|\vec{F}_2\| = 1800$ N, $\|\vec{F}_3\| = 3600$ N, and $\|\vec{F}_4\| = 1200$ N. (A) Find the torques exerted by $\vec{F}_1$, $\vec{F}_2$, $\vec{F}_3$, and $\vec{F}_4$. (B) What must be the magnitude of $\vec{P}$ for the system to be in rotational equilibrium?

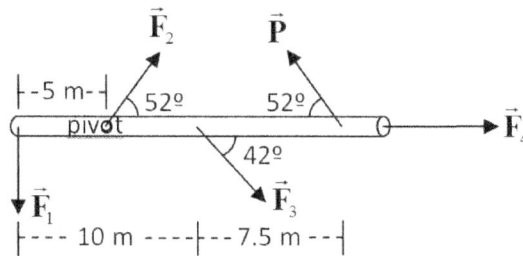

56. As illustrated below, a 60-kg monkey hangs from one end of a 12.0-m long rod, while an 80-kg monkey stands 4.0-m from the opposite end. A fulcrum rests beneath the center of the rod. (A) What torque is exerted by the hanging monkey? (B) What torque is exerted by the standing monkey? (C) Where, precisely, should a 40-kg box of bananas be placed in order to achieve static equilibrium?

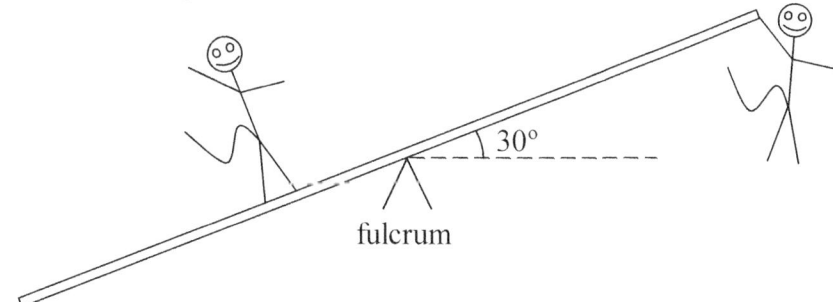

57. A monkey suspends 35 g from the 80-cm mark of a 100-g uniform meterstick. Where, precisely, must the fulcrum be placed in order to achieve static equilibrium?

58. A monkey places an apple at the 10-cm mark of a uniform meterstick, and an orange at the 80-cm mark. The fulcrum is underneath the 50-cm mark, and the system is in static equilibrium. The combined mass of the apple and orange is 300 g. What is the mass of the apple?

59. A monkey balances a 40-kg wooden mallet with uniform density on a fulcrum as illustrated below. The monkey then saws the mallet into two pieces at its center of gravity. Use the concepts of torque and static equilibrium to determine how much each end weighs. **Notes**: The handle and the head are not the two ends. The two ends are the two pieces of the mallet that remain after sawing – 18.0 cm of the handle being one end, and 2.0 cm of handle connected to the head being the other end. Also, this mallet is rectangular – it does not have a cylindrical handle or head, for example. (The answer is not 192 N. Why not?) Note that this same problem also appears with the Chapter 6 exercises, since it can also be solved from the concept of center of mass.

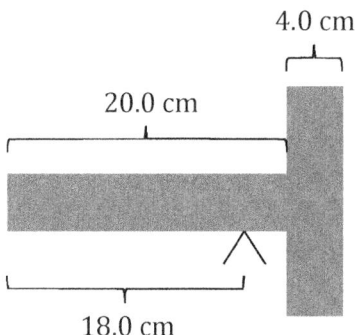

60. As illustrated below, a 75-kg monkey is suspended from a rope that passes over a 20-kg pulley and connects to a 50-kg box of bananas on horizontal ground. The coefficient of friction between the box of bananas and the ground is 0.40. The pulley is a solid disc with a diameter of 3.0 meters. The monkey is released from rest and falls 50 meters before landing in a barrel of physics notes below. Assume that there is no slipping between the cord and the pulley. (A) How many revolutions does the pulley complete as the monkey falls 50 meters? (B) What is the force of friction between the bananas and the ground? (C) Draw free-body diagrams for the monkey and box of bananas. Sum the forces for each object. Note that the two tensions are not equal due to friction between the cord and pulley. You will have two equations and three unknowns $(T_1, T_2, a_T)$. (D) What is the moment of inertia of the pulley? (E) Draw the pulley. Draw the forces that exert torques on the pulley. Sum the torques. This will give you a third equation with unknowns $(T_1, T_2, \alpha)$. Two of these four unknowns are related, which can be used to reduce the system to three equations in three unknowns. (F) Find the acceleration of the monkey. (G) Find the angular acceleration of the pulley. (H) Find the speed of the monkey just before landing in the barrel. (I) Find the final angular speed of the pulley. (J) Find the tension in each section of cord.

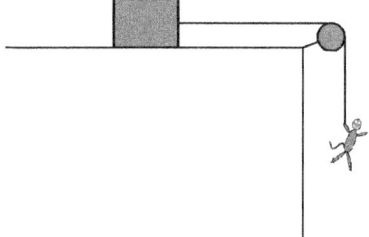

61. A 40-kg monkey is connected to a 20-kg box of bananas as illustrated below. The coefficient of friction between the box of bananas and the incline is $\sqrt{3}/3$. The cord rotates with the pulley without slipping. The 10-kg pulley is a solid disc. The system starts from rest. (A) Determine the acceleration of the system. (B) Determine the tension in each cord.

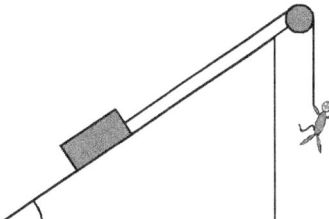

62. A 20-kg box of bananas is suspended from a clothesline, as illustrated below. The system is in static equilibrium. Determine the tension in each of the three sections of cord.

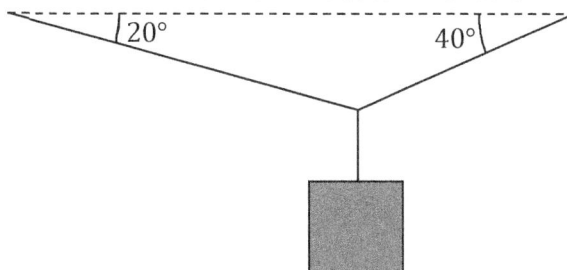

63. A monkey applies his knowledge of physics in order to store his precious bananas in static equilibrium. The uniform boom has a mass of 3.0 kg and is 12.0 m long; its lower end is connected to a hingepin. As illustrated below, the boom is supported by a tie rope that connects to a wall. The tie rope can sustain a maximum tension of 30 N. (A) What maximum load can the boom support without snapping the tie rope? (B) Find the maximum horizontal and vertical components of the force exerted on the hingepin. (C) Suppose that the bananas have a mass corresponding to the maximum load and the tie rope is cut. What will be the initial angular acceleration of the system?

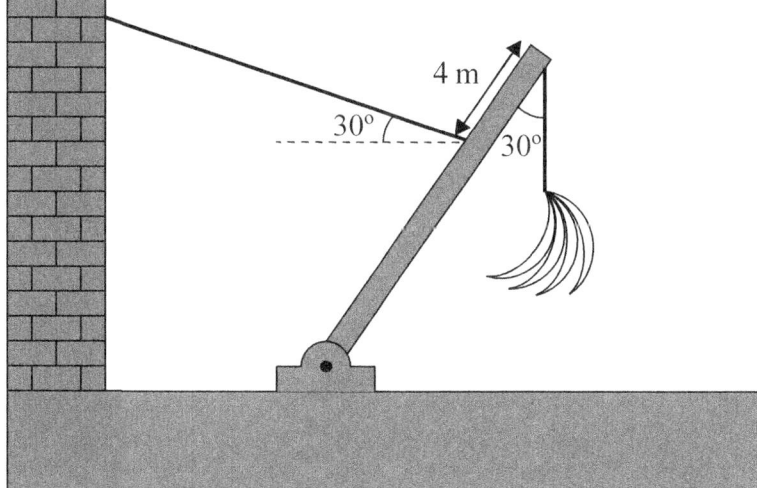

64. In the diagram below, the crane weighs 700 N and is 20.0 m long. The load weighs 6.0 kN and is connected 16.0 m from the hinge. The system is in static equilibrium. (A) What is the tension in the support cable? (B) Find the horizontal and vertical components of the force supplied by the hingepin.

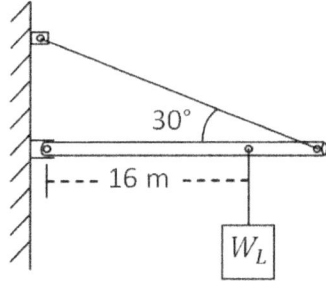

65. If $N$ uniform bricks of length $L$ are stacked together, but each brick is as far off-center as possible without the stack tipping over, derive a simple formula that predicts how far the $M^{th}$ brick will overhang (measured from the edge of the bottom brick).

66. Four uniform bricks of length $L$ are stacked together. Each brick overhangs the brick below it by the same amount (unlike the stack of the previous question). What maximum overhang allows the stack to be in static equilibrium?

67. A uniform hollow sphere rolls without slipping on an inclined plane. Derive an equation for the acceleration of the hollow sphere in terms of $g$ and the incline angle, $\theta$, only for the case that (A) the hollow sphere is rolling up the incline and (B) the hollow sphere is rolling down the incline. Derive an equation for the minimum coefficient of static friction needed for the hollow sphere to roll without slipping in terms of the mass, $m$, of the hollow sphere, gravitational acceleration, $g$, and the incline angle, $\theta$, only for the case that (C) the hollow sphere is rolling up the incline and (D) the hollow sphere is rolling down the incline.

## Rotational kinetic energy

68. A monkey standing at the top of a 30° incline pulls the glass eyes out of their sockets. One glass eye is a uniform solid sphere, while the other glass eye is a uniform hollow sphere. Both glass eyes start from rest from the same position on the incline. They roll without slipping a distance of 12 m down the incline. (A) Find the final speed of each sphere. Be sure to indicate which is which. (B) Use physics concepts to support your answer above for which glass eye wins the race.

69. Some monkeys are playing a friendly game of Boulder Dash, in which a monkey tries to outrun a uniform spherical boulder rolling from rest down a 56-m long, 30° incline. (A) What is the speed of the boulder at the bottom of the incline? (B) How much time does the monkey have to safely reach the bottom – assuming that the monkey runs with constant speed? (Unfortunately, he isn't given a head start.)

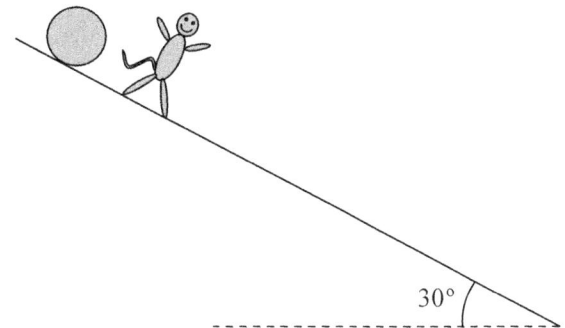

70. A monkey rolls a solid cylinder, hollow cylinder, solid sphere, and hollow sphere down a steady incline of 30°; these objects roll without slipping. He also slides a block of ice down the incline that is virtually frictionless. Each object starts from rest and travels a distance $L$ down the incline. Derive symbolic equations for the final speed of each of these objects in terms of $L$ and $g$ only. Use your answers to rank the objects from greatest to least acceleration.

71. Some coconuts arrange ten monkeys like bowling pens. The coconuts then climb a curved hill until they are 100 meters above the ground, from which point they release an object from rest, hoping to score a strike. (A) (i) Find the final speed of the object if it is a uniform solid sphere that rolls without slipping. (ii) How much faster would the sphere reach the bottom if it were to slide down the hill in the absence of friction? (B) Find the final speed of the object if it is a uniform cylindrical shell with inner radius equal to one-half the outer radius, which rolls without slipping. (C) If the object is a barrel of monkeys that rolls without slipping and the final speed is 40 m/s, find the moment of inertia of the barrel of monkeys as a decimal times $MR_0^2$. Does your answer seem plausible?

72. A monkey stands at the bottom of an incline of angle $\theta$ with a uniform hollow sphere. He bowls the hollow sphere up the incline with an initial speed $v_0$. Assume that the hollow sphere rolls without slipping for the entire trip. (A) Show that the hollow sphere travels a distance $L = \dfrac{5v_0^2}{6g\sin\theta}$ up the incline before running out of speed. (B) Show that if the hollow sphere instead slides without friction the hollow sphere travels a maximum distance $L = \dfrac{v_0^2}{2g\sin\theta}$ up the incline before running out of speed. (C) Explain why, in terms of physics concepts, the hollow sphere travels further up the incline when rolling without slipping than in the case where the incline is frictionless. (D) Which of the standard geometries (e.g. solid or hollow sphere or cylinder) would give the greatest value of $L$ in part (A)? Explain your reasoning in terms of physics concepts.

73. The 40-kg No-Mo Yo-Yo escapes on a 50-kg unicycle that has a 20-kg wheel (a uniform thin hoop) with 70-cm diameter. No-Mo Yo-Yo coasts 100-m down a 30° incline, starting from rest. The wheel rolls without slipping. (A) What is the speed of the unicycle at the bottom of the incline? (B) What is the acceleration of the unicycle? (C) What is the force of friction acting on the wheel? (D) What would be the final speed if the wheel were rolled by itself (without the unicycle and monkey)? (E) What would be the unicycle's speed if it were instead sliding without friction?

74. A monkey has three identical 100-g bottles: One bottle is empty, another bottle is filled with 300-g of liquid banana juice, and the third bottle has 300-g banana juice that is frozen. The three bottles roll without slipping up a 30° incline with the same initial speed of 60 m/s. (A) Find the maximum height of each bottle. Be sure to indicate which is which. (B) Explain the ordering of your answers in terms of physics concepts.

75. No-Mo Yo-Yo is an evil monkey with plans to destroy the world. Presently, he holds the free end of the 2.0-m cord of negligible mass wound around the Yo-Yo of Doom and releases it from rest. The Yo-Yo of Doom consists of two 30-g uniform solid disks of 6.0-cm diameter joined by a 10-g uniform hollow cylinder of 1.0-cm diameter. In order to save the world from destruction, the Powder Puff Pearls must stop the Yo-Yo of Doom before it completely unwinds. The cord rotates with the Yo-Yo of Doom without slipping. (A) What is the moment of inertia of the Yo-Yo of Doom about its axis of rotation? (B) Apply Newton's second law to find the acceleration of the Yo-Yo of Doom. (C) Conserve energy to find the speed of the Yo-Yo of Doom at the bottom. (D) What is the tension in the cord? (E) How much time do the Powder Puff Pearls have to save the world?

76. A 50-kg monkey drives a 1150-kg car up a 20° incline, uniformly accelerating from rest to 40 m/s in 8.0 seconds. The wheels (uniform solid disks) roll without slipping. Each wheel has a mass of 60 kg and radius of 15 cm. (A) What is the acceleration of the car up the incline? (B) How far does the car travel up the incline? (C) What is the angular acceleration of the wheels? (D) How many revolutions do the wheels complete? (E) What torque does the motor deliver to the wheels? (F) How much work is done against gravity?

77. A uniform solid sphere of radius $R_0$ rolls a distance $L$ without slipping down a V-channel inclined at an angle $\theta$ relative to the horizontal from rest. The spacing of the V-channel grooves is $d$, which is smaller than $2R_0$. Derive an equation for the final speed of the solid sphere in terms of $g$, $L$, $\theta$, $R_0$, and/or $d$ only. Also, interpret your result conceptually.

## Conservation of angular momentum

78. A monkey is performing her exercise in the ice skating competition of the Winter Monkolympics. In one part of her routine, she spins while standing tall, with her arms straight up in the air and her legs straight down, maintaining excellent balance by positioning her center of gravity above the point of contact between her skate and the ice. She completes 27 revolutions in 12.0 seconds. Then she quickly stretches her arms horizontally outward, extends one of her legs horizontally backward, and leans her torso and head horizontally forward, effectively increasing her moment of inertia by a factor of 3.0 – still maintaining her balance exquisitely. She spins like this for 8.0 seconds. How many revolutions does she complete in this extended position?

79. A monkey[339] is SO frustrated with his slow internet connection that he picks up his laptop, slams it against a brick wall, and then jumps high into the air and stomps on it. At that exact moment, the earth suddenly contracts until it has one-third of its initial radius. How long will a 'day' be now, and how many 'days' will there be in a year?

80. A 60-kg monkey is placed at the center of a merry-go-round – a large disk with 200-kg mass and 10-m diameter. A turkey spins the merry-go-round at a rate of 0.10 rev/sec and lets go. As the merry-go-round spins, the monkey walks outward until he reaches the edge. It takes the monkey 3.0 seconds to walk from the center to the edge. (A) Conserve angular momentum for the system (merry-go-round plus monkey) to find the angular speed of the merry-go-round when the monkey reaches the edge. (B) What is the average angular acceleration of the merry-go-round while the monkey is walking? (C) How many revolutions does the merry-go-round make while the monkey is walking?

---

[339] The very same monkey is rumored to have slapped his television set just before a famous New York blackout.

81. A hamster likes to run, but not in the traditional hamster wheel: She likes to run on an old-fashioned record player. A horizontal record is held in place by a small vertical axle. When the hamster climbs onto the record and runs in a clockwise circle, the record rotates counterclockwise. Neglect any friction between the record and the surface upon which it rests. The system is initially at rest. The 200-g hamster runs 50 cm/s (relative to the floor) in a circle with a radius of 6.0 cm, centered about the axle. The 400-g record has a diameter of 20 cm. What is the angular speed of the record?

82. A turkey throws a 20-kg monkey made out of silly putty at a 50-kg door with 60-cm width and 2.00-m height. The monkey travels 5.0 m/s horizontally when it strikes the door at a height of 1.00 m, 40 cm from the hinges. The monkey sticks to the door. The door is at rest prior to the collision. Take the system to include the monkey and the door. (A) Calculate the initial linear momentum of the system. (B) Calculate the initial energy of the system. (C) Calculate the initial angular momentum of the system relative to the hinges. (D) What is conserved for this process (from just before impact to after impact)? Justify your answer. (E) Calculate the angular speed of the door as it closes. (F) Calculate the percentage of energy lost or gained.

83. A solid sphere traveling 5.0 m/s to the right collides head-on with a hollow sphere of equal mass and radius traveling 3.0 m/s to the left in an elastic collision. The spheres have uniform densities. The spheres roll without slipping. (A) Explain why the velocity equation is not valid even though this is a one-dimensional elastic collision between two objects. (B) Find the final velocity of each sphere. (C) Is the total momentum of the system (defined as the two spheres) conserved? If not, explain.

84. As illustrated below, a 20-kg monkey on a frictionless table is connected to a turkey on the floor below by a light, inextensible cord that passes through a hole in the table. The turkey is initially pulling the cord such that the monkey initially slides at a rate of 5.0 m/s in a circle of 80-cm diameter centered about the hole. The turkey then increases the tension until the monkey slides in a circle 20-cm diameter. (A) Find the initial angular velocity of the monkey. (B) Find the initial moment of inertia of the monkey. (C) Find the initial acceleration of the monkey. (D) Find the initial tension in the cord. (E) Find the initial angular momentum of the monkey relative to the hole. (F) Find the initial net torque exerted on the monkey. (G) Which of the following are conserved as the tension is increased – linear momentum, energy, and/or angular momentum? Explain your answers. (H) Find the final velocity of the monkey. (I) Find the final angular velocity of the monkey. (J) Find the final moment of inertia of the monkey. (K) Find the final acceleration of the monkey. (L) Find the final tension in the cord.

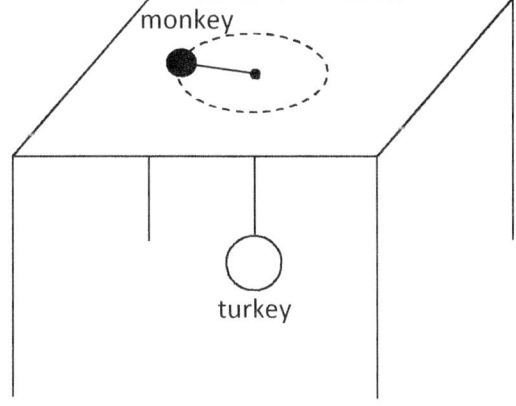

85. A monkey balances a spinning 3.0-kg coconut on his fingertip, as illustrated below. The coconut has the shape of a sphere with 12-cm diameter. It is initially spinning with an angular speed of 24 rev/s. While it is spinning, another monkey gently places a thin ring on top of the coconut. The ring has a radius of 4.0 cm and a mass of 2.0 kg. (A) What is the angular speed of the system after the ring is added? (B) What percentage of kinetic energy does the system lose? (C) Is mechanical energy lost, conserved, or gained in this process? If it is not conserved, where does the missing energy go or come from?

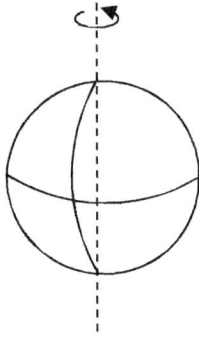 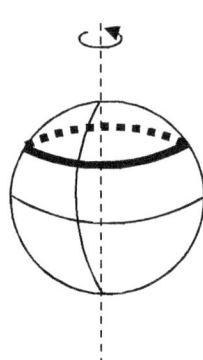

# Appendix A: Propagation of Errors

**Error**: There is inherent error in every measurement that we make in science, except for counting a small number of objects – like how many books are on a shelf. For example, if you want to measure the height of a table, the time a ball is in the air, or the mass of a textbook, you won't be able to measure these quantities *exactly*. For one, the measuring device itself has limited precision; you can't measure anything to an infinite number of significant figures. Secondly, even if you had the best possible equipment – as well as you can plausibly conceive of making, with an abundance of money and time to develop it – you will discover that there are other types of errors. For example, at some point when you proceed to measure the height of a table, you will see that it has an uneven height, that the height actually varies slightly in space and time, and that the height is not even well-defined at a very microscopic level (e.g. the electrons are in motion in a probabilistic electron cloud). In addition, there are errors that result from assumptions that we make when we develop a theoretical model, such as neglecting air resistance or the earth's rotation. A good experimentalist records and describes a couple of the most significant sources of error when making measurements, such as unaccounted for effects of air resistance or the limited precision of a protractor.

> **Important Distinction**. Note that the terms 'error' and 'mistake' are not interchangeable. A mistake is something that you can plausibly avoid during an experiment, like using equipment incorrectly or bumping the table while making a measurement. An error is an inherent problem that is challenging to minimize or account for quantiatively, and could not be plausibly avoided.

**Uncertainty**: When making a quantitative measurement, a good experiment also quantitatively determines the uncertainty in that measurement. For example, if you measure the height of a table to be 86.34 cm, you don't know how precise the measurement is unless you know what the uncertainty is. If the uncertainty is 0.03 cm, the measurement is good to better than one part in a thousand, but if the uncertainty is 0.14 cm, the table's actual height can actually be off by more than a millimeter. It's very important to establish the level of uncertainty when publishing experimental results so that the scientific community knows how precise the results are.

We use the lowercase Greek symbol sigma ($\sigma_x$) with a subscript to represent the uncertainty in the quantity designated in the subscript. For example, $\sigma_g$ would represent the uncertainty in gravitational acceleration and $\sigma_\mu$ would represent the uncertainty in the coefficient of friction.

There are four useful methods for establishing the level of uncertainty:
- Standard deviation: Use this when you measure the same quantity repeatedly, and expect the results to all be approximately the same.
- Linear regression: Use this when you measure one quantity as a function of another quantity, and expect a linear relationship (or if you can linearize the data). See Appendix B.
- Propagation of errors: Use this when you have measured some quantities and know their uncertainties, and want to use them in a formula to calculate a new quantity.

Appendix A: Propagation of Errors

- Estimate: Use this when you measure a quantity just once. It's not a blind guess, but an estimate based upon the precision of the instruments and a little qualitative judgment. If several scientists independently estimated the uncertainty, they may not agree exactly, but their estimates should all be in the same ballpark, and they should all be able to justify their numbers qualitatively.

**Notation.** The uncertainty in a measured quantity, $x$, is denoted as $\sigma_x$.

**Note.** Uncertainties have units, too. Unfortunately, students often do not realize this, and so lose points for not including units with their uncertainties in lab. If the uncertainty in a height is 0.23, it makes a huge difference if the uncertainty is 0.23 cm or 0.23 mm, so don't neglect the units.

**Significant figures**: Once you have obtained a quantitative measure of the uncertainty, there is a simple rule for recording significant figures. This rule is much better than the rules for significant figures described on page 12. The rules on page 12 are useful for solving homework problems; the rule we are about to state is much more tedious. Imagine if we gave you the uncertainties along with every quantity in the problems of this book, and then after solving each problem you also had to propagate errors to determine how many significant figures to keep! Since that would be very tedious, we use the rules from page 12 when solving physics homework problems, and reserve the better rule below for the analysis of experimental work.

Given numerical values for a measured quantity, $x$, and its corresponding uncertainty, $\sigma_x$, following are the rules for determining how many significant figures of each quantity to record:
- First look at the uncertainty, $\sigma_x$. If the leading digit is a 1 or a 2, the uncertainty should have two significant figures. Otherwise, the uncertainty should have one significant figure.[340] Round the uncertainty to one or two significant figures, depending on the leading digit.
- Next compare the uncertainty – which is now expressed with the proper number of significant figures, after using the step described above – to the measured quantity, $x$. Expressed in the same units, they must match in final decimal position. For example, if the uncertainty's final decimal position is the hundredths place, round the measured quantity to the hundredths place, also.

**Example.** A monkey measures gravitational acceleration near earth's surface to be 9.7453912 m/s$^2$ with an uncertainty of 0.0361598 m/s$^2$. Express these with appropriate use of significant figures.

The uncertainty does not begin with a 1 or a 2, so it should be rounded to one significant figure as 0.04 m/s$^2$. The uncertainty now ends in the hundredths place, so we round the measured quantity (gravitational acceleration) to the hundredths place as 9.75 m/s$^2$. It is customary to state the measurement and its uncertainty together in the form 9.75 $\pm$ 0.04 m/s$^2$.

---

[340] Trailing zeros count as significant figures, but leading zeros do not. For example, 0.00140 has three significant figures (the last three digits).

> **Example.** A monkey measures the height of a table to be 69.52643 cm with an uncertainty of 0.14397 cm. Express these with appropriate use of significant figures.
> 
> The uncertainty begins with a 1 or a 2, so it should be rounded to two significant figures as 0.14 cm. The uncertainty now ends in the hundredths place, so we round the measured quantity to the hundredths place as 69.53 cm. It is customary to state the measurement and its uncertainty together in the form $69.53 \pm 0.14$ cm.

**Standard deviation**: When you measure the same physical quantity, $x$, multiple times, each time expecting approximately the same result, you can compute its uncertainty, $\sigma_x$, using the formula for the standard deviation:[341]

$$\sigma_x = \sqrt{\frac{(\bar{x} - x_1)^2 + (\bar{x} - x_2)^2 + \cdots + (\bar{x} - x_N)^2}{N - 1}}$$

where there are $N$ measurements all together (some of which may be the same) and where $\bar{x}$ is the average value of the measurements.

> **Example.** A monkey drops a banana from rest from the same height three different times and records the following data for the time of descent: 0.642 s, 0.617 s, and 0.623 s. What values should the monkey use for the time of descent and its uncertainty?
>
> The best value to use for the time of descent is the average value:
>
> $$\bar{t} = \frac{t_1 + t_2 + t_3}{3} = \frac{0.642 + 0.617 + 0.623}{3} = 0.62733 \text{ s}$$
>
> We will keep a couple of extra digits temporarily; once we know the uncertainty, then we will be able to determine how many significant figures are appropriate. The standard deviation applies since repeated measurements are made and we expect to measure approximately the same value each time:
>
> $$\sigma_t = \sqrt{\frac{(\bar{t} - t_1)^2 + (\bar{t} - t_2)^2 + (\bar{t} - t_3)^2}{3 - 1}}$$
>
> $$= \sqrt{\frac{(0.62733 - 0.642)^2 + (0.62733 - 0.617)^2 + (0.62733 - 0.623)^2}{2}} = 0.01305 \text{ s}$$
>
> According to the rules for significant figures, $t = 0.627 \pm 0.013$ s.

---

[341] Conceptually, you can understand why we divide by $N - 1$ rather than $N$ by looking at the limit of making a small number of measurements. In the extreme, when you make a single measurement, if you were to divide by 1, you would find that there is no uncertainty in making a single measurement, which is absurd; whereas if you divide by $1 - 1 = 0$, you find that the uncertainty is indeterminate if you only make one measurement, according to this mathematical prescription.

# Appendix A: Propagation of Errors

**Propagation of errors**: Suppose that we have measured $x_1$ and its uncertainty, $\sigma_{x_1}$, as well as $x_2$ and its uncertainty, $\sigma_{x_2}$, up to $x_N$ and its uncertainty, $\sigma_{x_N}$, and that we are calculating $y$, which is a function of the measured quantities: $y = y(x_1, x_2, ..., x_N)$. The prescription for determining the uncertainty in the calculated quantity, $\sigma_y$, is called propagation of errors, and is a formula involving partial derivatives (the technique of partial differentiation is explained in Sec. 5.5):

$$\sigma_y = \sqrt{\left(\frac{\partial y}{\partial x_1}\right)^2 \sigma_{x_1}^2 + \left(\frac{\partial y}{\partial x_2}\right)^2 \sigma_{x_2}^2 + \cdots + \left(\frac{\partial y}{\partial x_N}\right)^2 \sigma_{x_N}^2}$$

---

**Example.** A monkey measures the length and width of a sheet of paper, with uncertainties, to be:

$$L = 12.42 \pm 0.12 \text{ cm} \quad , \quad W = 7.35 \pm 0.08 \text{ cm}$$

Derive a symbolic formula for the uncertainty in the perimeter of the sheet of paper and use it to establish the perimeter and its uncertainty.

The measured quantities $L$ and $W$ correspond to $x_1$ and $x_2$ in the formula for propagating errors, and the perimeter, $P$, corresponds to the calculated quantity, $y$. The formula for the perimeter of the sheet of paper is $P = 2L + 2W$. Using the error propagation formula,

$$\sigma_P = \sqrt{\left(\frac{\partial P}{\partial L}\right)^2 \sigma_L^2 + \left(\frac{\partial P}{\partial W}\right)^2 \sigma_W^2} = \sqrt{4\sigma_L^2 + 4\sigma_W^2}$$

The best value to use for the perimeter is

$$P = 2L + 2W = 2(12.42) + 2(7.35) = 39.54 \text{ cm}$$

The uncertainty in the perimeter is

$$\sigma_P = \sqrt{4\sigma_L^2 + 4\sigma_W^2} = \sqrt{4(0.12)^2 + 4(0.08)^2} = 0.29 \text{ cm}$$

Using the rules for significant figures, we find that $P = 39.54 \pm 0.29$ cm.

# Practice Problems

The hints and answers to all of the practice problems in this appendix can be found, separately, toward the back of the book.

### Significant Figures

1. A monkey measures the mass of a banana to be 463.295 kg with an uncertainty of 0.472 kg. Express these together in $\pm$ form with appropriate use of significant figures.
2. A monkey measures the speed of a banana to be 12.7385 m/s with an uncertainty of 0.2384 m/s. Express these together in $\pm$ form with appropriate use of significant figures.
3. A monkey measures the thickness of a banana to be 0.0325917 m with an uncertainty of 1.1965238 mm. Express these together in $\pm$ form with appropriate use of significant figures.

### Standard Deviation

4. A monkey measures the length of a banana five different times and records the following data: 14.517 cm, 14.465 cm, 14.538 cm, 14.444 cm, and 14.491 cm. Express the best value to use for the length of the banana together with its uncertainty in $\pm$ form with appropriate use of significant figures.
5. A monkey measures his weight every day for 30 days. On 12 of the days, the scale reads 83 pounds; on 8 of the days, the scale reads 85 pounds; on 6 of the days, the scale reads 82 pounds; and on 4 of the days, the scale reads 86 pounds. Express the best value to use for the monkey's weight (in pounds) together with its uncertainty in $\pm$ form with appropriate use of significant figures.

### Propagation of Errors

6. A monkey measures the length and width of a sheet of paper to be $L = 12.42 \pm 0.12$ cm and $W = 7.35 \pm 0.08$ cm, respectively. Derive a symbolic formula for the uncertainty in the area of the sheet of paper and use it to establish the area and its uncertainty.
7. A monkey measures the mass and volume of a banana to be $M = 367.1 \pm 0.6$ g and $V = 383.56 \pm 0.22$ cm$^3$, respectively. Derive a symbolic formula for the uncertainty in the density of the banana and use it to establish the density of the banana and its uncertainty. Note that the formula for density, $\rho$ (rho), is $\rho = M/V$, assuming that the density is uniform throughout (so maybe it has no peel).
8. A monkey measures $a = 4.83 \pm 0.04$ cm and $b = 0.315 \pm 0.008$ rad, respectively. Derive a symbolic formula for the uncertainty in $c = a \sin b$ and use it to establish $c \pm \sigma_c$.

# Appendix B: Linear Regression

**Graphical average**: Suppose that a monkey drops a banana from rest from various heights and measures its height above the ground as a function of time, such that the monkey has two columns of data – one for height, $h$, and one for time, $t$. We predict $h = gt^2/2$, where $g$ is gravitational acceleration. We can use the data for $h$ and $t$ to calculate $g$.

There is more than one way to compute $g$ from $h$ and $t$. The naïve thing to do is to plug each pair of data points into the formula, calculating $g$ separately for each case, and then averaging these to determine $g$. Statistically, it turns out to be better to determine $g$ directly from the slope of a linearized plot. The slope is a graphical average which is better than the naïve average that we have just described.

**Linearizing data**: Note that $h$ is not a linear function of $t$ in the equation $h = gt^2/2$; rather, $h$ is quadratic in $t$. If we plot $h$ as a function of $t$, the result will be a parabola. A parabola has a constantly changing slope. If instead we plot $h$ as a function of $t^2$, the graph will be linear. It is preferable to linearize the data in this way whenever possible, since a line has a single, constant slope.

In order to linearize the data, we take the theoretical equation – in this case, $h = gt^2/2$ – and make a change of variables to cast the equation in the form $y = mx + b$. If we set $y = h$ and $x = t^2$, we find that $y = gx/2$, which is a straight line. We predict the slope to equal $g/2$ and the $y$-intercept to be zero. The monkey should plot $h$ as a function of $t^2$, and interpret the slope as one-half of gravitational acceleration.

> **Example.** The period of a spring, $T$, is related to the suspended mass, $m$, and spring constant, $k$, according to the equation $T = 2\pi\sqrt{m/k}$. A monkey measures $T$ as a function of $m$. What should the monkey plot in order to obtain a straight line?
>
> We want to cast the equation in the form $y = mx + b$. First, let us square the given equation to obtain $T^2 = 4\pi^2 m/k$. If we set $y = T^2$ and $x = m$, we will obtain a straight line, $y = 4\pi^2 x/k$, with a slope of $4\pi^2/k$. The monkey could plot $T^2$ as a function of $m$.

**Linear regression**: We found that the monkey should make a linearized plot to determine gravitational acceleration from his data. But what should the monkey use for the uncertainty in gravitational acceleration? The monkey needs some measure of the amount of scatter in the data points. The technique for quantitatively determining the uncertainty in the slope of linearized data is called linear regression. We will discuss how to perform a linear regression using the method of least squares.

**Method of least-squares**: Linear regression is a useful technique for analyzing data that are expected to obey a linear relationship,

$$y = mx + b$$

where $x$ and $y$ are physical quantities. A plot of $y$ as a function of $x$ should result in a straight line with slope $m$ and $y$-intercept $b$. The problem is that $y$ and $x$ depend on measurements, and therefore have inherent uncertainties. Thus, the graph will not be perfectly linear. The questions are (1) how to determine if the relationship is indeed linear and (2) what values to use for $m$ and $b$ and their uncertainties. This problem can be approached by applying the principle of maximum likelihood.

We will use the symbols $x$ and $y$ to represent the equation of the best-fit line, whereas we will use the symbols $x_i$ and $y_i$ to represent actual data points. Due to uncertainties, the actual data points, $(x_i, y_i)$, may not lie exactly on the best-fit line. We will call $N$ the total number of data points.

If the relationship were perfectly linear, then $y - mx - b$ would be zero for each pair of points $(x_i, y_i)$. The method of least squares sums $(y_i - mx_i - b)^2$ and determines which values of $m$ and $b$ minimize this sum.

Let us define the uppercase Greek symbol lambda ($\Lambda$) as[342]

$$\Lambda \equiv \sum_{i=1}^{N} (y_i - mx_i - b)^2$$

We want to find the values of $m$ and $b$ that minimize $\Lambda$. Therefore, we take partial derivatives of $\Lambda$ with respect to $m$ and $b$, and set them equal to zero:[343]

$$\frac{\partial \Lambda}{\partial m} = 2 \sum_{i=1}^{N} (-x_i)(y_i - mx_i - b) = 0 \quad , \quad \frac{\partial \Lambda}{\partial b} = 2 \sum_{i=1}^{N} (-1)(y_i - mx_i - b) = 0$$

$$-\sum_{i=1}^{N} x_i y_i + m \sum_{i=1}^{N} x_i^2 + b \sum_{i=1}^{N} x_i = 0 \quad , \quad -\sum_{i=1}^{N} y_i + m \sum_{i=1}^{N} x_i + bN = 0$$

since $\sum_{i=1}^{N} b = b + b + \cdots + b = bN$ and $\sum_{i=1}^{N} mx_i^2 = m \sum_{i=1}^{N} x_i^2$, for example. We have two equations in two unknowns; the unknowns are $m$ and $b$. We will eliminate $b$ to solve for $m$:

$$b = \frac{1}{N} \sum_{i=1}^{N} y_i - \frac{m}{N} \sum_{i=1}^{N} x_i$$

$$-\sum_{i=1}^{N} x_i y_i + m \sum_{i=1}^{N} x_i^2 + \left( \frac{1}{N} \sum_{i=1}^{N} y_i - \frac{m}{N} \sum_{i=1}^{N} x_i \right) \sum_{i=1}^{N} x_i = 0$$

$$-\sum_{i=1}^{N} x_i y_i + m \sum_{i=1}^{N} x_i^2 + \frac{1}{N} \sum_{i=1}^{N} x_i \sum_{i=1}^{N} y_i - \frac{m}{N} \left( \sum_{i=1}^{N} x_i \right)^2 = 0$$

$$m = \frac{N \sum_{i=1}^{N} x_i y_i - \sum_{i=1}^{N} x_i \sum_{i=1}^{N} y_i}{\Delta}$$

---

[342] Summation notation is introduced in Sec. 1.2. It is also used in Chapters 6 and 7. This particular sum means $(y_1 - mx_1 - b)^2 + (y_2 - mx_2 - b)^2 + \cdots + (y_N - mx_N - b)^2$.

[343] You can also take a second derivative to verify that the final solution does indeed minimize $\Lambda$.

where the uppercase Greek delta ($\Delta$) is defined as

$$\Delta \equiv N \sum_{i=1}^{N} x_i^2 - \left(\sum_{i=1}^{N} x_i\right)^2$$

Plugging $m$ into one of the original equations, we find that

$$b = \frac{\sum_{i=1}^{N} x_i^2 \sum_{i=1}^{N} y_i - \sum_{i=1}^{N} x_i \sum_{i=1}^{N} x_i y_i}{\Delta}$$

The best fit line is then $y = mx + b$.

The uncertainties in $m$ and $b$ are given by:

$$\sigma_m^2 = \frac{N \sigma_y^2}{\Delta} \quad , \quad \sigma_b^2 = \frac{\sigma_y^2}{\Delta} \sum_{i=1}^{N} x_i^2$$

where[344]

$$\sigma_y^2 \equiv \frac{\Lambda}{N-2}$$

The coefficient of linear correlation is

$$r = \frac{\sum_{i=1}^{N}[(x_i - \bar{x})(y_i - \bar{y})]}{\sqrt{\sum_{i=1}^{N}(x_i - \bar{x})^2 \sum_{i=1}^{N}(y_i - \bar{y})^2}}$$

where $\bar{x}$ and $\bar{y}$ represent the average values of $x_i$ and $y_i$. This coefficient is useful for establishing whether or not $x$ and $y$ obey a linear relationship, or if they are even correlated. If the relationship is perfectly linear, $r = \pm 1$, and if $x$ and $y$ are uncorrelated, $r$ tends toward zero.

**Important Distinction.** The quantity $\sigma_m$ is the uncertainty in the slope, whereas the quantity $r^2$ is the coefficient of linear correlation: $r^2$ tells you to what extent the data form a linear relationship, whereas $\sigma_m$ quantifies the scatter in the data. These two quantities are not interchangeable.

---

[344] Like the $N-1$ in the formula for standard deviation (Appendix A), the reason for the $N-2$ here can be understood conceptually: You need to have three or more data points for a line to have any scatter; you can't find uncertainty in the slope of a line that only has two data points, so the formula must be indeterminate in the extreme limit that $N$ equals two.

**Example.** A monkey records the length of a simple pendulum as a function of period. The data are tabulated below. We predict that $T = 2\pi\sqrt{L/g}$. Use the summation formulas to perform a linear regression for this data set.

| L (m) | T (s) |
|---|---|
| 0.20 | 0.89 |
| 0.40 | 1.23 |
| 0.60 | 1.58 |
| 0.80 | 1.76 |
| 1.00 | 1.99 |

First, we must linearize the data. We can choose $L$ to be the $y$-axis variable and $T^2$ as the $x$-axis variable, such that the slope equals $g/(4\pi^2)$, as in the previous example (but note that we have swapped our choice of axes in this example). So let us square $T$ and work with the following table as $(x_i, y_i)$:

| $x$ (s$^2$) | $y$ (m) |
|---|---|
| 0.7921 | 0.20 |
| 1.5129 | 0.40 |
| 2.4964 | 0.60 |
| 3.0976 | 0.80 |
| 3.9601 | 1.00 |

We now perform a linear regression on these columns of $x \equiv T^2$ and $y \equiv L$. It's rather tedious work for a calculator; it's much more efficient to use a spreadsheet program like Microsoft Excel or a programming language like C++. Many graphing programs, like Microsoft Excel, provide $r^2$, but do not provide $\sigma_m$. However, if you want to establish the uncertainty in the slope, you need to perform a linear regression to compute $\sigma_m$. You can do this yourself using the summation formulas. If you write your own program, you can check that it works by entering the data in the table above and comparing it with the results in the table below. More figures are shown below than are significant for the purpose of helping you check the accuracy of your program.

| Quantity | Numerical Result |
|---|---|
| $\Delta$ | 31.49084494 s$^4$ |
| $m$ | 0.251523896 m/s$^2$ |
| $b$ | 0.003430594 m |
| $r^2$ | 0.996122661 |
| $\sigma_y^2$ | 0.000516979 m$^2$ |
| $\sigma_m$ | 0.009060019 m/s$^2$ |
| $\sigma_b$ | 0.023773125 m |

# Practice Problems

The hints and answers to all of the practice problems in this appendix can be found, separately, toward the back of the book.

## Linearizing Data

1. A monkey measures $F$ as a function of $z$. The monkey predicts that $F = k(z - z_e)$, where $k$ and $z_e$ are constants. What should the monkey plot on each axis in order to obtain a straight line? What will the slope and vertical intercept be?

2. A monkey predicts that $Q = \frac{H}{K}$. (A) If the monkey measures $Q$ as a function of $H$, with $K$ fixed, what should the monkey plot on each axis in order to obtain a straight line, and what will the slope be? (B) If the monkey measures $Q$ as a function of $K$, with $H$ fixed, what should the monkey plot on each axis in order to obtain a straight line, and what will the slope be? (C) If the monkey measures $H$ as a function of $K$, with $Q$ fixed, what should the monkey plot on each axis in order to obtain a straight line, and what will the slope be?

3. A monkey measures $q$ as a function of $u$. The monkey predicts that $q^4 = a^2 u^5$, where $a$ is a constant. What should the monkey plot on each axis in order to obtain a straight line? What will the slope and vertical intercept be?

4. A monkey measures $d$ as a function of $e$. The monkey predicts that $d^3 e^2 f^4 = 9$, where $f$ is a constant. What should the monkey plot on each axis in order to obtain a straight line? What will the slope and vertical intercept be?

5. A monkey measures $D$ as a function of $J$. The monkey predicts that $D^2 = m^2 J^3 + n^5$, where $m$ and $n$ are constants. What should the monkey plot on each axis in order to obtain a straight line? What will the slope and vertical intercept be?

6. A monkey measures $\beta$ (beta) as a function of $\alpha$ (alpha). The monkey predicts that $\beta^3 = 8\sin(4\,\alpha^2) + 9$. What should the monkey plot on each axis in order to obtain a straight line? What will the slope and vertical intercept be?

7. A monkey measures $\gamma$ (gamma) as a function of $\lambda$ (lambda). The monkey predicts that $\gamma^2 \sin(2\lambda) = 3\gamma^4 \cos(2\lambda)$. What should the monkey plot on each axis in order to obtain a straight line? What will the slope and vertical intercept be?

# Hints to Conceptual Questions and Practice Problems

## Units, Dimensions, and Significant Figures – Hints to Practice Problems

1. Convert from miles to yards to inches to centimeters to kilometers.
2. (A) Convert from kiloseconds to seconds to minutes. (B) Convert from milliyears to years to days to hours. (C) Convert from seconds to hours to days to microdays.
3. Remember to square each conversion factor.
4. Convert miles to meters and hours to seconds, remembering that time is in the denominator.
5. Remember to square the time conversion factor.
6. The argument of the sine function must be dimensionless, and the sine function itself is dimensionless. Therefore, the amplitude of the sine function must have the same dimensions as speed, and the coefficient of time must have the inverse dimensions of time.
7. Solve for $\kappa$. Plug in the dimensions of the other quantities and simplify.
8. Each term must have the same dimensions as $\rho$. Set the dimensions of each term equal to the dimensions of $\rho$, plug in the dimensions of $\delta$, and solve for the dimensions of the unknown.
9. Use the significant figures rule for multiplication/division.
10. Use the significant figures rule for addition/subtraction. Assume that the 2's of the given perimeter formula have infinite precision (since they are not measured quantities).
11. First use the significant figures rule for multiplication/division and then use the significant figures rule for addition/subtraction.

## Chapter 1 – Hints to Conceptual Questions

**Note**: You will get more out of the conceptual questions if you first try to answer and understand the question on your own, then check the hints to see if your approach was correct, and then check the answer. If you rely on the hints before attempting to reason out the answer, you miss out on the valuable learning that comes from trying to think the solution through on your own as well as learning from mistakes that you make.

1. If the object is following a predetermined course, such that the only coordinates that the object can control are forward and backward, the motion is effectively 1D. If the object can move left/right in addition to forward/backward on a surface, the motion is effectively 2D. If the object can move left/right, forward/backward, and up/down, the motion is fully 3D.
2. For the total distance traveled, you have to add up each distance that the ball travels up and down. For the net displacement, you just need to figure out where it is in the final position compared to the initial position (how it got to that position is irrelevant).

3. The definition of average speed involves the total distance traveled, while the definition of average velocity involves the net displacement.

4. Look up the definition of average acceleration in 1D: You just need to know the initial and final velocities and the total time.

5. The sign of the velocity tells you whether an object is moving forward or backward. For each graph, first figure out how you can determine velocity. You want to find the points where velocity is momentarily zero and where it is changing sign.

6. Make a sketch of slope as a function of time. Then sketch the slope of the velocity graph as a function of time.

7. For tangential acceleration, sketch the slope as a function of time. For tangential displacement, you have to work backwards: The slopes of the tangential displacement must agree with the values of the tangential velocity. There is an example of 'backwards sketching' in Sec. 1.5. When you finish your graph of tangential displacement as a function of time, it's easy to check: Sketch its slope as a function of time and see if it agrees with the given velocity graph. If it doesn't, you made a mistake.

8. You have to work backwards to make a sketch whose slopes match the values of the given acceleration graph in order to make the velocity graph, and then you must similarly work backwards from the velocity graph in order to make the tangential displacement graph. As suggested in the previous hint, study the example of this 'backwards sketching' in Sec 1.5, and take time to sketch velocity and acceleration from your displacement graph to check your solution.

9. Use the chain rule to show that $a_T = v_T \frac{dv_T}{ds_T}$. This will help you interpret the slope of the graph conceptually, and should help you see how to make the specified acceleration graph.

10. Interpret the acceleration conceptually: The train loses 8.0 m/s of speed each second.

11. Interpret the acceleration conceptually: The banana loses about 10 m/s of speed each second going upward until it runs out of speed, and then it gains about 10 m/s of speed each second coming back down.

12. Start with the acceleration graph, which equals a negative constant. The velocity graph starts out negative, and has a constant slope equal to the acceleration. Make the displacement graph last.

## Chapter 1 – Hints to Practice Problems

**Note**: You will get more out of solving the problems if you first try to solve the problem on your own, then check the hints to see if your approach was correct, and then check the answer. If you rely on the hints before attempting the problem, you miss out on the valuable learning that comes from trying to think the solution through on your own as well as learning from mistakes that you make.

1. It is incorrect to find the two speeds (20 km/hr and 12 km/hr) and average them together because more time is spent traveling at the lower speed – which requires that the average speed be less than 16 km/hr. The correct approach is to use the definition of average speed – total distance traveled over the total time.

2. It is incorrect to find the three speeds and average them together because different durations of time are spent traveling at each speed. The correct approach is to use the definition of average speed – total distance traveled over the total time.

3. (A) Figure out how far, and in which direction, the monkey winds up from where he started. (B) Add all of the distances together. (C) Use the total distance traveled. (D) Use the net displacement.

4. (A) Figure out the time for the second part of the trip, and use this to determine the time for the first part of the trip. Then you can find the speed for the first part. (B) How far, and in which direction, does the monkey wind up from where he started? (C) First find the total distance traveled. (D) Use the net displacement.

5. Define a symbol for the distance between the monkey's dorm and class. Express the time for each part of the trip in terms of this distance. Use the definition of average speed. The distance, which is unknown, will cancel out in the algebra.

6. You just need to know the initial and final velocities for the entire trip. Since the speeds have hours and the time is in minutes, you need to make some conversion(s) to put all of the time units on an equal footing.

7. (A) Read the initial and final values directly from the graph. (B) First find the total distance traveled by figuring out how far the monkey moves backward, then forward, and then backward again – by reading values directly from the graph. (C) Since velocity equals the slope of this graph, draw a tangent line at the indicated time and find its slope. (D) Since velocity equals the slope of this graph, find the initial slope. Then speed is the absolute value of velocity in 1D. (E) Read this position directly from the graph. (F) When is the velocity negative? How do you find velocity for this graph? (G) When is the velocity zero? How do you find velocity for this graph? (H) You need to find the initial and final velocities. (I) Since velocity equals the slope for this graph, find the steepest slope (ignoring the sign, since you really want speed). (J) Since velocity equals the slope for this graph, sketch the plot's slope as a function of time.

8. (A) Since distance equals the area for a velocity graph, find the area between the curve and time axis and add the areas together in absolute values. (B) First find the net displacement. Since distance equals the area for a velocity graph, find the area between the curve and time axis – calling areas above the time axis positive and areas below the time axis negative. This time direction matters, so don't use absolute values. (C) Read this directly from the graph. (D) Since acceleration equals the slope of a velocity graph, find the slope of the tangent line at the initial position. (E) Find the net displacement for the first eight seconds, as described for part (B). (F) Since acceleration equals the slope of a velocity graph, draw a tangent line at the specified time and find the slope of the tangent line. (G) Read the graph directly to see when velocity is negative. (H) To determine when the velocity is zero, read the graph directly. (I) First find the initial and final velocities. (J) Find the maximum absolute value of velocity from this graph by finding when the curve is furthest from the time axis. (K) Since acceleration equals the slope for this graph, sketch the plot's slope as a function of time. (L) You have to work backwards: The slopes of the tangential displacement must agree with the values of the tangential velocity. There is an example of 'backwards sketching' in Sec. 1.5. When you finish your graph of tangential displacement as a function of time, it's easy to check: Sketch its slope as a function of time and see if it agrees with the given velocity graph. If it doesn't, you made a mistake.

9. (A) Since the area of this graph equals the change in velocity, find the area between the curve and time axis up to twelve seconds. When the curve lies above the time axis, area is positive; and when the curve lies below the time axis, area is negative. The total area for the first twelve seconds (including correct signs when you 'add' the areas together) equals the final velocity minus the initial velocity, which was specified in the problem. Plug the initial velocity in to solve for the final velocity. (B) Read this value directly from the graph. (C) Determine the velocity at eight seconds following the prescription

Hints

described for part (A), and then take the absolute value to find the speed. (D) Read this value directly from the graph. (E) When is the velocity negative? Follow the prescription described for part (A) to determine the velocity at various points – or, since you only need the sign and not the value, you can just make a sketch, as described for part (I). (F) When is the velocity zero? Follow the prescription described for part (A) to determine the velocity at various points or make a sketch as described for part (I). (G) You already know the initial velocity. Find the final velocity following the prescription described for part (A). (H) After finding the velocity at each point, you just have to take the absolute values. (I) You have to work backwards: The slopes of the tangential velocity must agree with the values of the tangential acceleration. There is an example of 'backwards sketching' in Sec. 1.5. When you finish your graph of tangential velocity as a function of time, it's easy to check: Sketch its slope as a function of time and see if it agrees with the given acceleration graph. If it doesn't, you made a mistake.

10. Use the three equations of 1DUA. Follow the strategy described in Sec. 1.6, and study the examples. Which three symbols (of the five symbols that appear in the three equations) are known? Look at the units to help determine the known quantities. Signs are very important.

11. Use the three equations of 1DUA. Follow the strategy described in Sec. 1.6, and study the examples. Which three symbols (of the five symbols that appear in the three equations) are known? Look at the units to help determine the known quantities. Signs are very important. You may need to solve a quadratic equation.

12. Use the three equations of 1DUA. Follow the strategy described in Sec. 1.6, and study the examples. Which three symbols (of the five symbols that appear in the three equations) are known? Look at the units to help determine the known quantities. Signs are very important. Notice that the bananamobile is stopped at the final position. You may need to solve two equations in two unknowns.

13. Use the three equations of 1DUA. Follow the strategy described in Sec. 1.6, and study the examples. Which three symbols (of the five symbols that appear in the three equations) are known? Look at the units to help determine the known quantities. Signs are very important. Solve for the net displacement, then divide by the height of each step to determine the number of steps. You may need to solve two equations in two unknowns.

14. Use the three equations of 1DUA. Follow the strategy described in Sec. 1.6, and study the examples. Which three symbols (of the five symbols that appear in the three equations) are known? Look at the units to help determine the known quantities. Signs are very important. Notice that the monkey begins from rest. (Note that since the monkey is sliding down an incline, the monkey is not in free fall.)

15. Use the three equations of 1DUA. Follow the strategy described in Sec. 1.6, and study the examples. Which three symbols (of the five symbols that appear in the three equations) are known? Look at the units to help determine the known quantities. Signs are very important. Notice that the bananamobile is stopped in the final position. You may need to solve two equations in two unknowns.

16. Use the three equations of 1DUA. Follow the strategy described in Sec. 1.6, and study the examples. Which three symbols (of the five symbols that appear in the three equations) are known? Look at the units to help determine the known quantities. Signs are very important. The word 'drops' is used to imply that the banana is released from rest, as opposed to 'throws.'

17. Use the three equations of 1DUA. Follow the strategy described in Sec. 1.6, and study the examples. Which three symbols (of the five symbols that appear in the three equations) are known? Look at the units to help determine the known quantities. Signs are very important. Either work with

the trip up or the trip down, but not the roundtrip. Make sure that the three knowns that you write down apply specifically to the half-trip that you choose to work with.

18. Use the 1DUA strategy symbolically. (A) Either work with the trip up or the trip down, but not the roundtrip. Something will be zero for the half-trip that you choose to work with. Solve for the net displacement as a function of gravitational acceleration and the initial speed only. Write one equation for $\Delta y_e$ in terms of $g_e$ and rewrite it as an equation for $\Delta y_m$ in terms of $g_m$. Divide one equation by the other; $v_{y0}$ will cancel out. You should know an approximate numerical value to use for $g_e/g_m$. (B) Solve for time of the half-trip in part (A). Double the time for the roundtrip. Write the equation twice and divide, as described for part (A).

19. Use the three equations of 1DUA. Follow the strategy described in Sec. 1.6, and study the examples. Which three symbols (of the five symbols that appear in the three equations) are known? Look at the units to help determine the known quantities. Signs are very important. In part (B), you need to determine whether the firecracker is moving upward or downward (which should be naturally interpreted from the sign of the vertical component of the velocity) since velocity is a vector — i.e. it includes a direction in addition to a (positive) magnitude. In part (C), you need to redefine either initial or final; some of the knowns from parts (A) and (B) may have different values in part (C).

20. Use the three equations of 1DUA. Follow the strategy described in Sec. 1.6, and study the examples. Which three symbols (of the five symbols that appear in the three equations) are known? Look at the units to help determine the known quantities. Signs are very important. (A) Make sure that you use the equations of 1DUA. Many students have made the mistake of making things up for part (A), and always get the wrong answer by doing so. Trust the equations. The equations can work, but making things up in physics just doesn't work out. Although you are solving for the vertical component of acceleration, which may be negative, the acceleration of gravity is the magnitude of the free-fall acceleration, which is positive. (B) It's a new problem, but since you're still on Planet Fyzx, you should now know what gravitational acceleration is (but bear in mind that in the equations of 1DUA, the vertical component of acceleration may be negative, unlike the acceleration of gravity). (C) You're still on Planet Fyzx. Two of these answers you should already know, conceptually — i.e. you don't need to solve for them. Note that velocity and acceleration both include magnitude and direction.

21. Use the three equations of 1DUA. Follow the strategy described in Sec. 1.6, and study the examples. Which three symbols (of the five symbols that appear in the three equations) are known? Look at the units to help determine the known quantities. Signs are very important. (A) You may need to solve two equations in two unknowns. Think through these signs carefully, as many students tend to get incorrect answers due to sign mistakes when solving this problem. Work with the roundtrip — it's more efficient. (B) This should be easy once you solve part (A). (C) You should already know the net displacement and total time. (D) You need to split the trip up into two parts in order to find the total distance traveled.

22. Use the three equations of 1DUA. Follow the strategy described in Sec. 1.6, and study the examples. Which three symbols (of the five symbols that appear in the three equations) are known? Look at the units to help determine the known quantities. Signs are very important. (A) You may need to solve a quadratic equation. Think through these signs carefully, as many students tend to get incorrect answers due to sign mistakes when solving this problem. Read the problem carefully to determine the net displacement. (B) You need to redefine either initial or final; some of the knowns from part (A) may have different values in part (C). Remember to consider how high the ball was above the ground when it was thrown.

Hints

23. Use the three equations of 1DUA. Follow the strategy described in Sec. 1.6, and study the examples. Which three symbols (of the five symbols that appear in the three equations) are known? Look at the units to help determine the known quantities. Signs are very important. (A) Think through these signs carefully, as many students tend to get incorrect answers due to sign mistakes when solving this problem. The initial position is ten meters above the ground – this is just the initial position, not the net displacement. Read this problem carefully, think about it, and draw and label a diagram. Think through these signs carefully, as many students tend to get incorrect answers due to sign mistakes when solving this problem. You may need to solve two equations in two unknowns. (B) Note how the initial and final positions are defined. You need to split the trip up into two parts to find the average speed, but this is not necessary to find the average velocity.

24. Use the 1DUA strategy; the answer is not one-half of the initial speed. A straightforward way to solve this problem is to first figure out the maximum height of the banana, and then solve a new problem using one-half of this height.

25. Of course, the mass doesn't matter. Use the 1DUA strategy; the wrench is not thrown with twice the given speed. A straightforward way to approach this problem is to first figure out how high the wrench will rise starting with the specified speed, and then solve new problems using twice this height. This will readily give you the answers to parts (B) and (C); to get the answer for part (A), double the time it takes to rise upward (or downward) between the ground and its maximum height.

26. Write down an equation for each monkey using subscripts for any symbols that may be different for each monkey. Follow the multiple-moving object strategy described in Sec. 1.9, and study the examples. In part (A), how much distance do the two monkeys travel all together?

27. Write down an equation for each object using subscripts for any symbols that may be different for each object. Follow the multiple-moving object strategy described in Sec. 1.9, and study the examples. How far do they travel all together up to the time that they meet?

28. Write down an equation for each monkey using subscripts for any symbols that may be different for each monkey. Follow the multiple-moving object strategy described in Sec. 1.9, and study the examples. One way to account for the headstart is to add/subtract a $t_0$ to/from the time for one monkey; but you have to think this through carefully and add/subtract it to/from the correct monkey.

29. Write down an equation for each banana using subscripts for any symbols that may be different for each banana. Follow the multiple-moving object strategy described in Sec. 1.9, and study the examples. How far do the two bananas travel all together up to the time that they collide? When you express this mathematically, keep in mind that one net displacement is up and the other is down.

30. Write down an equation for each flask using subscripts for any symbols that may be different for each flask. Follow the multiple-moving object strategy described in Sec. 1.9, and study the examples. One way to account for the headstart is to add/subtract a $t_0$ to/from the time for one flask; but you have to think this through carefully and add/subtract it to/from the correct flask.

31. Write down an equation for each monkey using subscripts for any symbols that may be different for each monkey. Follow the multiple-moving object strategy described in Sec. 1.9, and study the examples – especially, the minimum headstart example. One way to account for the headstart is to add/subtract a $t_0$ to/from the time for one monkey; but you have to think this through carefully and add/subtract it to/from the correct monkey. Solve for the headstart time, $t_0$. In order to find the maximum headstart, take a derivative of the headstart time with respect to time, $t$.

32. Write down an equation for each monkey using subscripts for any symbols that may be different for each monkey. Follow the multiple-moving object strategy described in Sec. 1.9, and study the examples – especially, the minimum headstart example. One way to account for the headstart is to add/subtract a $t_0$ to/from the time for one monkey; but you have to think this through carefully and add/subtract it to/from the correct monkey. Instead of applying calculus, think conceptually: If the uncle waits just long enough, how fast will the monkey be running when the uncle catches him?

33. Write down an equation for each monkey using subscripts for any symbols that may be different for each monkey. Follow the multiple-moving object strategy described in Sec. 1.9, and study the examples – especially, the minimum headstart example. One way to account for the headstart is to add/subtract an $x_0$ to/from the time for one monkey; but you have to think this through carefully and add/subtract it to/from the correct monkey. Solve for the headstart distance, $x_0$. In order to find the minimum headstart, take a derivative of the headstart distance with respect to time, $t$.

34. Write down an equation for each object using subscripts for any symbols that may be different for each object. Follow the multiple-moving object strategy described in Sec. 1.9, and study the examples – especially, the minimum headstart example. One way to account for the headstart is to add/subtract an $x_0$ to/from the time for one object; but you have to think this through carefully and add/subtract it to/from the correct object. Solve for the headstart distance, $x_0$. In order to find the minimum headstart, take a derivative of the headstart distance with respect to time, $t$.

35. (A) Take successive derivatives of position in order to find velocity and acceleration. Note that $v_{x0}$ implies $v_x$ evaluated at $t = 0$. (B) Take successive derivatives of position in order to find velocity and acceleration. In order to find average velocity, first find the initial and final positions to get the net displacement. Since you are not given a specific time interval, find the average velocity as a function of time by using $t$ as the total time. In order to find average speed, first find the total distance traveled. Set velocity equal to zero and solve for time to see if the object ever changes direction. If so, you need to split the trip up into pieces in order to find the total distance traveled. (C) Take a derivative of velocity in order to find acceleration, and perform a definite integral to find the net displacement. (D) Perform a definite integral of acceleration in order to find the change in velocity, then plug in the initial velocity. Perform a definite integral of velocity in order to find the net displacement. (E) Take a derivative of position in order to find velocity. Use this to solve for time. (F) Set the derivative of a function equal to zero to find its relative extreme values. This gives you the times when it is a relative extreme value. Take a second derivative to see if the extreme value is a relative minimum, relative maximum, or a point of inflection. Also, check the endpoints to see if any minimum that you found is an absolute minimum. Remember to plug the time that minimizes the velocity back into the function in order to find the minimum velocity.

36. (A) Each term has to have the SI units of velocity. This readily gives you the units of gamma. Notice that beta must have the same SI units as time, since you can't subtract oranges from apples. Lastly, set the dimensions of the remaining term equal to length over time, and plug in the dimensions of beta and time to solve for the dimensions of alpha. (Although $\beta = 1/\alpha$ numerically, this is not true regarding their units.) (B) Plug in the given initial velocity along with $t = 0$ to solve for gamma. (C) Find the relative extreme values of the velocity by setting its derivative equal to zero. This gives you the times when it is a relative extreme value. Take a second derivative to see if the extreme value is a relative minimum, relative maximum, or a point of inflection. Also, check the endpoints to see if any extreme value that you found is an absolute extreme. Remember to plug the critical time(s) back into the function in order to find the extreme velocity. Since the question states 'speed' and not velocity, you're

not looking for the maximum value of the velocity, but the maximum value of the absolute value of the velocity in 1D. If the minimum is more negative than the maximum, then this will be important. (D) Take a derivative of velocity in order to find acceleration. (E) Set the velocity equal to zero to find the times when the monkey may have changed direction (i.e. changed sign). Perform a definite integral of velocity in order to find the net displacement. If the monkey changed direction one or more times, you have to perform a definite integral for each trip separately, then add them together with absolute values. (F) Perform a definite integral of velocity in order to find the net displacement. You just need one definite integral for the entire trip.

37. (A) Solve for beta. Plug in the SI units of everything else. (B) Perform a definite integral of acceleration in order to find the change in velocity. Plug in the given initial velocity in order to find the final velocity. (C) First, perform a definite integral of velocity in order to find the net displacement.

38. (A) Take a derivative of velocity in order to find acceleration. (B) Perform a definite integral of velocity in order to find the net displacement. (C) Just use the given velocity function, realizing that this speed could be a positive or negative velocity, in principle.

39. (A) Perform a definite integral of velocity in order to find the net displacement. (B) Take a derivative of velocity in order to find acceleration.

40. Perform a definite integral of acceleration in order to find the change in velocity, which is $v_x - v_{x0}$. Add $v_{x0}$ to both sides to solve for $v_x$. Use the given acceleration equation to substitute $t^2$ for $a_x$. Perform a definite integral of velocity (using $t$'s, but not $a_x$) in order to find the net displacement. When you finish, use the given acceleration equation to substitute $t^2$ for $a_x$. Divide the net displacement by time in order to find the average velocity. Use the acceleration and/or velocity equations to eliminate all of the symbols except for $v_{x0}$ and $v_x$.

## Chapter 2 – Hints to Conceptual Questions

1. Determine whether or not each measurement involves just a magnitude, or both a magnitude and a direction. Note that the derivative of a vector quantity will also be a vector.
2. The magnitude of a vector can't be negative. The components of a vector can be negative.
3. Join the tail of one to the tip of the other. You can move one or both arrows around, provided that you don't change their length or direction.
4. Negate the right vector, then join the arrows tip-to-tail.
5. The arrow will be three times longer, but will still point in the same direction.
6. Negate the second arrow, adjust the length of each arrow according to the scale factors, and then join the new arrows tip-to-tail.
7. The extremes arise when the two vectors lie in the same or opposite directions.
8. The four vectors shown plus the missing fifth vector have a resultant of zero. Use this to solve for the resultant of the four vectors shown.
9. (A) For the net displacement, look at where the monkey finished compared to where he started. For the total distance traveled, add all of the distances together. (B) Consider points near the South Pole, where the latitude has a circumference of one mile, half a mile, etc.
10. Average speed involves the total distance traveled, whereas average velocity involves the net displacement. What are the total distance traveled and net displacement for one complete orbit?

11. The definition of average acceleration requires working with the initial and final velocities. Note that they do not cancel out in the subtraction.

12. It's the product rule: One term arises from taking a derivative of the tangential component of velocity, while the second term arises from taking a derivative of the unit tangent vector. The first term is nonzero if the object is changing speed, and the second term is nonzero if the object is changing direction (so the direction of the unit tangent is changing in time).

13. Make a table of $r$ and $\theta$ for angles ranging from 0 to (at least) $2\pi$ rad. Plot each point by going $r$ units outward from the origin in a direction $\theta$, which is counterclockwise from the $+x$-axis. If you need help, you can use the equations to change from 2D polar coordinates to Cartesian coordinates.

14. Make a table of $r$ and $\theta$ for angles ranging from 0 to (at least) $2\pi$ rad. Plot each point by going $r$ units outward from the origin in a direction $\theta$, which is counterclockwise from the $+x$-axis. If you need help, try rewriting the given equation in terms of Cartesian coordinates using substitutions.

15. Make a table of $r_c$ and $z$. These points lie on a curved surface (not a curved path), since the angle $\theta$ is unrestricted. Note that $r_c$ is the distance from the $z$-axis, and not the distance from the origin.

16. Imagine varying $r$ and $\varphi$ for a fixed value of $\theta$. In this problem, $r$ is the distance from the origin.

17. The cannonball has inertia. What does this mean? It's the same path as a projectile follows.

18. The food package has inertia. What does this mean? It's the same path as a projectile follows; in this case, starting horizontally.

19. The basic idea is to realize that the $y$-component of velocity of each object is affected the same way by gravity. You can construct the proof formally by combining the $y$-equation of the falling monkey together with the $x$- and $y$-equations of the banana, in such a way as to show that they have the same height at the same time. Realize that they have the same final height, but not the same displacement – one went up, while the other went down. So you need to think about how to relate the final heights to the displacements.

20. Launching the golf balls away from the ship's velocity, due to inertia the golf balls will travel a shorter distance relative to earth than if the ship were at rest. (Note that the golf balls travel the same distance relative to the ship whether the ship is at rest or travels with constant speed in a straight line.) A launch angle above 45° shortens the horizontal range relative to earth compared to the range of a golf ball launched from a ship at rest due to the extra time spent in the air. On the other hand, a launch angle below 45° shortens the time spent in the air, minimizing the effect of the ship's velocity. Therefore, the launch angle that maximizes the range of the golf ball relative to the earth will be less than 45°. (What if the golf balls were being launched along the direction of the ship's velocity, rather than against it?)[345]

21. This is a vector addition problem: The resultant velocity, $\vec{v}_{net}$, is the vector addition of the boat's velocity, $\vec{v}_b$, plus the river current's effect on its velocity, $\vec{v}_c$: $\vec{v}_{net} = \vec{v}_b + \vec{v}_{rc}$. In each case, first determine how you want $\vec{v}_{net}$ to be oriented, and use this to figure out how to orient $\vec{v}_b$.

22. (A) The slope of each graph gives you one component of velocity. Combine both slopes together to get the speed according to $v = \sqrt{v_x^2 + v_y^2}$. (B) Make plots of slope as functions of time. The slopes of these graphs give you the components of the acceleration. Then use $a = \sqrt{a_x^2 + a_y^2}$.

---

[345] Of course, if you launch golf balls off of a real cruise ship, air resistance is also an important factor, and may easily be more significant than what we have considered here.

23. Make the sets of $x$- and $y$-plots independently. Start with the accelerations, and use the sketching techniques of Chapter 1 for each set of plots. Use 'backward sketching' to make graphs of velocity from acceleration, and position from velocity (see Sec. 1.5). Make sure that your graphs agree with the specified signs of the initial values and also of the components of the acceleration.

24. Tangential speed will increase going downhill. Tangential acceleration is proportional to the slope of the tangent line. The hill must curve the right way to make the tangential acceleration change as prescribed.

25. Tangential speed will increase going downhill. Tangential acceleration is proportional to the slope of the tangent line. The hill must curve the right way to make the tangential acceleration change as prescribed.

## Chapter 2 – Hints to Practice Problems

1. (A) Use the equations to solve for the components of a vector in terms of its magnitude and direction. (B) Use the equations to solve for the magnitude and direction of a vector in terms of its components. Note that the angle is measured counterclockwise from the $+x$-axis; when taking the inverse tangent, add $180°$ if the $x$-component of the vector is negative. (C) Follow the vector addition strategy described in Sec. 2.2, and study the examples. Note that the angles are measured counterclockwise from the $+x$-axis; when taking an inverse tangent, add $180°$ if the $x$-component of the vector is negative.

2. Follow the vector addition strategy described in Sec. 2.2, and study the examples. Note that the angles are measured counterclockwise from the $+x$-axis; when taking an inverse tangent, add $180°$ if the $x$-component of the vector is negative.

3. (A) There are two possible answers in two of the four possible quadrants. Draw them. Use an inverse trig function to solve for the angle. The calculator will give one of the two answers; you have to reason the second answer out geometrically/conceptually from your diagram. (B) You have to reason this out geometrically/conceptually. How must you orient the three arrows – of equal length – so that when they are arranged tip-to-tail, the tip of the last joins to the tail of the first (that's what it means, geometrically, to have a resultant vector of zero).

4. Follow the vector addition strategy described in Sec. 2.2, and study the examples. The vector subtraction of this problem is very similar to vector addition. One way to solve this problem is to subtract the components in the second step instead of adding them (and leave everything else the same as vector addition). Note that the angles are measured counterclockwise from the $+x$-axis; when taking an inverse tangent, add $180°$ if the $x$-component of the vector is negative.

5. Realize that finding the net force is simply the vector addition of the given forces. (A) You need to draw the vectors tip-to-tail. Notice that they are instead drawn tail-to-tail. You must move one of the arrows – without changing its length or orientation – in order to join them tip-to-tail. The resultant extends from the tail of the first to the tip of the last. (B) Follow the vector addition strategy described in Sec. 2.2, and study the examples. Note that the angles are measured counterclockwise from the $+x$-axis: Notice that this is not how the given angles are measured, so you will first have to find the Quadrant II and III angles of these vectors before you proceed with the vector addition. When taking an inverse tangent, add $180°$ if the $x$-component of the vector is negative.

6. Realize that this is just a vector addition problem. Follow the vector addition strategy described in Sec. 2.2, and study the examples. Note that the angles are measured counterclockwise from the $+x$-axis; when taking an inverse tangent, add $180°$ if the $x$-component of the vector is negative.

7. This problem is very similar to vector addition. Follow the vector addition strategy described in Sec. 2.2, and study the examples. The first and third steps are just like vector addition, but in the second step, write the vector equation in terms of components instead of adding the components together (e.g. $F_x = 3B_x - 4M_x$). Note that the angles are measured counterclockwise from the $+x$-axis; when taking an inverse tangent, add $180°$ if the $x$-component of the vector is negative.

8. (A) You are given the magnitude and direction of $\vec{R}$. The comma separates the magnitude from the direction (the comma is not part of a four-digit number). Read the note in the problem. Use the equations to solve for the components of a vector in terms of its magnitude and direction, then express the vector in terms of its components using the unit vector notation described in Sec. 2.1. (B) Perform the definite integral from $x = 0$ to $2\pi$ first. The definite integral of the cosine function (in the third term) will give you two nonzero terms over these limits – i.e. don't forget the lower limit when evaluating the integral. Use the equations to solve for the magnitude and direction of a vector in terms of its components. Note that the angle is measured counterclockwise from the $+x$-axis; when taking the inverse tangent, add $180°$ if the $x$-component of the vector is negative. (C) First, express $\vec{P}$ and $\vec{Q}$ in terms of unit vectors. In part (C), leave $\vec{P}$ as a function of $x$ – unlike part (B) where you plugged in $2\pi$ (i.e. use $x$ for the upper limit instead of $2\pi$). Vector addition is simpler than normal when the vectors are already expressed in terms of Cartesian unit vectors: You only need to do the second step of the vector addition strategy. Remember to account for the $2$ – e.g. $C_x = P_x + 2Q_x$. (D) First evaluate $\vec{P}$ and $\vec{Q}$ at $x = \pi/2$ and express each vector in terms of unit vectors. Vector subtraction is simpler than normal when the vectors are already expressed in terms of Cartesian unit vectors: You only need to do the second step of the vector addition strategy. Remember to account for the coefficients in the vector subtraction formula – e.g. $D_x = 3Q_x - 2P_x$. Finally, use the relationships between Cartesian and spherical coordinates to determine the magnitude and direction of $3\vec{Q} - 2\vec{P}$.

9. (A) Use the equations to solve for the components of a vector in terms of its magnitude and direction. Evaluate the components at the specified angle. (B) Use the equations to solve for the components of a vector in terms of its magnitude and direction. Don't evaluate the vector at the specified angle yet. Express the vector in terms of its components using the unit vector notation described in Sec. 2.1. Take the indicated derivative of the vector. After taking the derivative, evaluate both $\vec{S}$ and $d\vec{S}/d\theta$ at the specified angle. Now perform the indicated vector subtraction. Vector subtraction is simpler than normal when the vectors are already expressed in terms of Cartesian unit vectors: Start with the second step of the vector addition strategy. Remember to account for the coefficients in the vector subtraction formula – e.g. $L_x = 2S_x - \sqrt{3}\left(\frac{dS}{d\theta}\right)_x$. Notice that $\left(\frac{dS}{d\theta}\right)_x$ means something different from $\frac{dS_x}{d\theta}$. Finally, use the equations to solve for the magnitude and direction of a vector in terms of its components. Note that the angle is measured counterclockwise from the $+x$-axis; when taking the inverse tangent, add $180°$ if the $x$-component of the vector is negative. **Note**: Several students have managed, through pure coincidence, to obtain the correct answers to this problem using a totally incorrect strategy. It may be worth checking that your solution follows the prescription outlined here.

# Hints

10. Use the equations to solve for the components of each vector in terms of its magnitude and direction. Note that the labeled angles are not measured counterclockwise from the $+x$-axis. You must conceptually/geometrically reason out the correct angles to use for these Quadrant I and II vectors. This problem is very similar to vector addition. Follow the vector addition strategy described in Sec. 2.2, and study the examples. The first and third steps are just like vector addition, but in the second step, write the vector equation in terms of components instead of adding the components together (e.g. $F_x = \sqrt{3}B_x - 2\frac{dM_x}{dt}$). Note that the angles are measured counterclockwise from the $+x$-axis; when taking an inverse tangent, add $180°$ if the $x$-component of the vector is negative. Do not plug in the specified time until after taking the derivative.

11. (A) Figure out where the final position is relative to the initial position. The net displacement is a vector extending from the initial position to the final position. (B) Find the distance traveled for each section and add these together. (C) Use the total distance traveled. (D) Use the net displacement.

12. (A) You need a geometric formula. (B) Figure out where the final position is relative to the initial position. The net displacement is a vector extending from the initial position to the final position. (C) Figure out how long the first part of the trip took. Use the time for the first trip and the total trip time to figure out how long the second part of the trip took. (D) First find the total distance traveled for the whole lap. (E) First find the net displacement for the whole lap.

13. (A) Integrate over the differential arc length, $ds$, following an example similar to this in Sec. 2.3. (B) Integrate over the differential displacement vector, $d\vec{s}$. The integral for part (B) is much simpler than the integral for part (A), as it is path-independent – depending only on the initial and final positions. **Note**: The answer to part (A) is not simply the magnitude of the answer to part (B).

14. (A) Plug in the given time to find the coordinates of the monkey's position, and then determine how far the monkey is from the origin. (B) Evaluate the vector at each of the endpoints and perform the vector subtraction. (C) Take a derivative of the position vector in order to find velocity. (D) Take a derivative of velocity in order to find acceleration.

15. (A) Each term must have the same SI units as velocity. (B) Take a derivative of velocity in order to find acceleration. (C) First perform a definite integral of velocity in order to find the net displacement.

16. (A) Perform a definite integral over acceleration to find the change in velocity. Plug in the given instantaneous velocity to find the initial velocity. (B) Set the derivative of acceleration equal to zero to find its relative extrema. Remember to plug the times into the acceleration function. Take a second derivative to determine whether any relative extrema are minima, maxima, or points of inflection. Also, check the limits of the function over the interval $0 \leq t < \infty$ to see if any relative minimum is an absolute minimum. (C) Perform a definite integral of velocity in order to find the net displacement.

17. Use the relations between 2D polar coordinates and Cartesian coordinates in Sec. 2.4.
18. Use the relations between 2D polar coordinates and Cartesian coordinates in Sec. 2.4.
19. Use the relations between cylindrical and Cartesian coordinates in Sec. 2.4.
20. Use the relations between spherical and Cartesian coordinates in Sec. 2.5.
21. Use the relations between spherical and Cartesian coordinates in Sec. 2.5.
22. Treat the monkeystick as a vector in 3D space. Given its Cartesian components, find its magnitude. See Sec. 2.5.
23. Use the relations between spherical and Cartesian coordinates in Sec. 2.5.

24. Write $\vec{A}$ in terms of Cartesian components and unit vectors. Then write $\vec{A}$ in terms of 2D polar components and unit vectors. Rewrite the 2D polar unit vectors in terms of Cartesian unit vectors. Compare the expressions for $\vec{A}$ in order to relate the Cartesian and 2D polar components. One angle involves the Cartesian coordinates, $\tan^{-1}\left(\frac{y}{x}\right)$, while the other involves the Cartesian components, $\tan^{-1}\left(\frac{A_y}{A_x}\right)$.

25. Simply find the $x$-component of the initial velocity, which is constant throughout the motion. At the top of the trajectory, the $y$-component of the velocity is zero.

26. Use the equations of projectile motion. Follow the strategy described in Sec. 2.6, and study the examples. Which four symbols (of the seven symbols that appear in the four equations) are known? Look at the units to help determine the known quantities. Signs are very important. For the maximum height, you will have to redefine the initial and final positions. One of these answers you should know, conceptually; it's not something that you need to solve for. The net displacement is a vector that extends from the initial position to the final position; you should be able to find its components readily. For the speed at the top of the trajectory, simply find the $x$-component of the initial velocity, which is constant throughout the motion (since the $y$-component of the velocity is zero at the top of the trajectory).

27. Use the equations of projectile motion. Follow the strategy described in Sec. 2.6, and study the examples. Which four symbols (of the seven symbols that appear in the four equations) are known? Look at the units to help determine the known quantities. Signs are very important. First solve for the $y$-component of the final velocity, and use this together with $v_x$ (which equals $v_{x0}$) to solve for the magnitude and direction of the final velocity. Pay special note to the signs of $v_y$ and $\theta$. It's possible to prove, algebraically, that the answer for the magnitude (but not the direction) of the final velocity is exactly the same for each part.

28. Use the equations of projectile motion. Follow the strategy described in Sec. 2.6, and study the examples. Which four symbols (of the seven symbols that appear in the four equations) are known? Look at the units to help determine the known quantities. Signs are very important – especially, $v_{y0}$. Note that 10 m is only the height of the piñata above the ground, and not the net displacement. For the final speed, first solve for the $y$-component of the final velocity, and use this together with $v_x$ (which equals $v_{x0}$) to solve for the magnitude of the final velocity.

29. Use the equations of projectile motion. Follow the strategy described in Sec. 2.6, and study the examples. Which four symbols (of the seven symbols that appear in the four equations) are known? Look at the units to help determine the known quantities. Signs are very important – especially, $v_{y0}$.

30. (A) Make a right triangle and use trig. (B) Use the equations of projectile motion. Follow the strategy described in Sec. 2.6, and study the examples. Which four symbols (of the seven symbols that appear in the four equations) are known? Look at the units to help determine the known quantities. Signs are very important. (C) Use the three equations of 1DUA. Follow the strategy described in Sec. 1.6, and study the examples. Which three symbols (of the five symbols that appear in the three equations) are known? Look at the units to help determine the known quantities. Signs are very important. (D) Compare your answers to parts (B) and (C). A similar problem was posed in Conceptual Question 19.

Hints

31. Use the equations of projectile motion. Follow the strategy described in Sec. 2.6, and study the examples. Which four symbols (of the seven symbols that appear in the four equations) are known? Look at the units to help determine the known quantities. Signs are very important. In this problem, you are solving for the magnitude of the initial velocity, $v_0$, so you won't know numbers for $v_{x0}$ and $v_{y0}$ at the beginning of the problem. However, you still need to substitute expressions for $v_{x0}$ and $v_{y0}$ in terms of $v_0$ into the four main equations. Combine equations together to solve for $v_0$.

32. Since the pilot has inertia, this is actually a projectile motion problem where the initial velocity is horizontal. Use the equations of projectile motion. Follow the strategy described in Sec. 2.6, and study the examples. Which four symbols (of the seven symbols that appear in the four equations) are known? Look at the units to help determine the known quantities. Signs are very important. You will need to perform conversions in order to put all of the units on an equal footing.

33. Use the equations of projectile motion. Follow the strategy described in Sec. 2.6, and study the examples. Which four symbols (of the seven symbols that appear in the four equations) are known? Look at the units to help determine the known quantities. Signs are very important. Work with one-half of the trip in parts (A) and (B). You can double the values needed for parts (C) and (D), which relate to the complete trip. (Always make sure that the path is symmetric – i.e. the initial and final positions are at the same height – before you halve or double any values to relate a sub-trip to a whole trip.)

34. Use the equations of projectile motion. Follow the strategy described in Sec. 2.6, and study the examples. Which four symbols (of the seven symbols that appear in the four equations) are known? Look at the units to help determine the known quantities. Signs are very important. Figure out where the outfielder lands.

35. Use the equations of projectile motion. Follow the strategy described in Sec. 2.6, and study the examples. Which four symbols (of the seven symbols that appear in the four equations) are known? Look at the units to help determine the known quantities. Signs are very important. One way to solve this problem is to use the equations of projectile motion to solve for $\Delta y$ in terms of $\Delta x$, and to separately use trig to relate $\Delta y$ to $\Delta x$ (using the angle of the incline). You can then solve these two equations in two unknowns to determine both $\Delta x$ and $\Delta y$. Note that $\Delta x$ itself is not the answer, but you need to combine both $\Delta x$ and $\Delta y$ together to find the magnitude of the net displacement. (If you want to instead solve the problem with a rotated coordinate system, you must navigate some potential conceptual pitfalls. For example, note that the acceleration will have two components in such a coordinate system.)

36. Use the equations of projectile motion. Follow the strategy described in Sec. 2.6, and study the examples. Which four symbols (of the seven symbols that appear in the four equations) are known? Look at the units to help determine the known quantities. Signs are very important. Obtain the $x$-component of the initial velocity relative to the ground using the relative velocity formula, as in the last example in Sec. 2.6.

37. Use the equations of projectile motion. Follow the strategy described in Sec. 2.6, and study the examples. Which four symbols (of the seven symbols that appear in the four equations) are known? Look at the units to help determine the known quantities. Signs are very important. Obtain the $x$-component of the initial velocity relative to the ground using the relative velocity formula, as in the last example in Sec. 2.6. After solving for the range of the cannonball relative to the ground, use the hangtime to account for the distance that the ships move while the cannonball is in the air.

38. Use the equations of projectile motion. Follow the strategy described in Sec. 2.6, and study the examples. Which four symbols (of the seven symbols that appear in the four equations) are known? Look at the units to help determine the known quantities. Signs are very important. You must work symbolically, the way that we derived an equation for the horizontal range of a projectile where the final and initial positions have the same height in Sec. 2.6. Obviously, your solution will be much different, but the underlying strategy will be very similar.

39. (A) Use the equations to solve for the magnitude and direction of a vector in terms of its components. Note that the angle is measured counterclockwise from the $+x$-axis; when taking the inverse tangent, add $180°$ if the $x$-component of the vector is negative. (B)/(C) Treat these as two separate 1DUA problems – one for each component.

40. Use the equations of projectile motion. Follow the strategy described in Sec. 2.6, and study the examples. Which four symbols (of the seven symbols that appear in the four equations) are known? Look at the units to help determine the known quantities. Signs are very important. Obtain the $x$-component of the initial velocity relative to the ground using the relative velocity formula, as in the last example in Sec. 2.6. You must work symbolically, the way that we derived an equation for the horizontal range of a projectile where the final and initial positions have the same height in Sec. 2.6. Obviously, your solution will be different, but the underlying strategy will be very similar.

41. Use the relative velocity formula toward the end of Sec. 2.6, along with the strategy for projectile motion described in Sec. 2.6. Which four symbols (of the seven symbols that appear in the four equations) are known? Look at the units to help determine the known quantities. Signs are very important.

42. Use the relative velocity formula toward the end of Sec. 2.6. This is a vector addition (or subtraction) problem.

43. Treat each graph as a 1D motion plot (see Sec. 1.5), and then combine information analyzed/interpreted from each graph to answer the questions. (A) Determine the $x$- and $y$-components of the net displacement from each graph by reading the initial and final positions directly from each graph. Use the equations to solve for the magnitude and direction of a vector in terms of its components. Note that the angle is measured counterclockwise from the $+x$-axis; when taking the inverse tangent, add $180°$ if the $x$-component of the vector is negative. (B) Since velocity is the slope of a position graph, determine $x$- and $y$-components of the instantaneous velocity by finding the slope of the tangent line for each graph at the specified moment. Use the equation to solve for the magnitude of a vector in terms of its components. (C) Use the equation to solve for the direction of a vector in terms of its components. Note that the angle is measured counterclockwise from the $+x$-axis; when taking the inverse tangent, add $180°$ if the $x$-component of the vector is negative. (D) Simply look at the signs of $x$ and $y$. (E) For each position graph, make a sketch of slope as a function of time. Then sketch the slope of each component of velocity as a function of time. (F) Determine the total distance traveled along each coordinate first. Find the total distance traveled along each component by figuring out how far the monkey moves forward, then backward, then forward, etc. – by reading values directly from each graph.

# Chapter 3 – Hints to Conceptual Questions

1. Look up the definition of inertia.
2. Is it a natural tendency or is it a force? Do you feel a natural tendency or do you feel a force?
3. Figure out whether or not each quantity could be measured in Newtons.
4. The two forces add like vectors. The net force will be an extremum when the two forces are parallel or anti-parallel. Use Newton's second law.
5. If the car accelerates – i.e. changes velocity – then the necklace will not be vertical. The direction of the necklace's lean will be dictated by its inertia – i.e. the necklace wants to travel in a straight line with constant speed.
6. In each case, what causes the monkey to accelerate – i.e. to change velocity? What is pushing against the monkey in each case?
7. The answer relates to inertia.
8. For the ratio of the forces, use Newton's third law. For the ratio of the accelerations, use Newton's second law in addition to Newton's third law.
9. The answer relates to inertia.
10. The answer relates to inertia.
11. The answer relates to Newton's third law.
12. The answers relate to Newton's third law.
13. The answers relate to inertia. In the last question, the tablecloth will be slanted in one case.
14. The answer relates to Newton's second law and the definition of mass.
15. The two different results both relate to inertia. In the two cases, the initial motion is different, which explains the two different results.
16. The answer relates to Newton's third law.
17. Write the object's weight in a sentence as, "Object 1 exerts a force on object 2." Once you figure out what the two objects are, it should be very easy to identify the reaction.
18. The answer relates to Newton's third law. Therefore, you must split the shortening of the rope equally among both monkeys.
19. (A) How much does each box weigh on the moon? (B) Is the kick horizontal or vertical? Is the foot overcoming the bowling ball's mass or weight? Is this quantity the same or different on the moon? Comparing the pain, consider Newton's third law. Would the monkey need to overcome the locomotive's mass or weight?
20. Read about apparent weight and study the last example in Sec. 3.3. The elevator must be accelerating – i.e. changing velocity – in order for the scale to read anything other than the monkey's actual weight.
21. The friction force is proportional to the normal force, and the normal force depends on the slope. Draw a FBD for an inclined plane and sum the components of the forces perpendicular to the incline to determine whether the normal force increases or decreases as the slope increases.
22. The answer relates to the inequality in the equation that relates the static friction force to the normal force.

23. Does the friction force depend on the area of contact between the two surfaces? See Sec. 3.5. Does the friction force depend on the mass of the block? If so, you must also consider that a more massive block also has more inertia. Draw a FBD and apply Newton's second law to derive a symbolic equation for the acceleration of the block in order to determine whether or not mass cancels out.

24. The sand has inertia.

25. Examine Hooke's law to see how the spring constant relates to force.

26. Will the weight of the banana have the form, $F_r = -kx$? See if you can write the weight of the banana in this form, lumping all of the constants together and calling them $k$.

27. In which case will the combination be stiffer – i.e. more difficult to stretch or compress the springs – and in which case will the combination be looser – i.e. easier to stretch or compress the springs?

28. Draw a FBD for each object and apply Newton's second law. Beware of mutual surface forces (Sec. 3.7).

29. Draw a FBD for each object and apply Newton's second law. Beware of mutual surface forces (Sec. 3.7).

30. Draw a FBD for each case and apply Newton's second law. Unlike free fall, the times are not equal because the acceleration is different in the two cases.

31. Draw a FBD for the object and apply Newton's second law. Look at the form of the force of air resistance in Sec. 3.8.

32. Draw a FBD for the passenger in each case and apply Newton's second law. Look at the form of the force of air resistance in Sec. 3.8. How does the time that the passenger has been falling affect any of the answers? Remember, velocity means how fast and which way the passenger is moving, while acceleration describes how the velocity is changing.

33. This is the one place where the gravitational fields of the earth and moon are equal in magnitude and opposite in direction. If you think about the math, you will see that there is a quadratic equation to solve, which explains two mathematical answers. The mathematical solutions relate to the two points where the magnitudes of the gravitational fields are equal, but only in one of the two positions are they also oppositely directed.

34. This relates to the fact that Newton's law of universal gravitation is an inverse-square law. Study the examples of Sec. 3.9.

35. Look at the formula and see what happens to gravitational acceleration if the mass and radius (half the diameter also means half the radius) are each reduced by a factor of two. Study the examples of Sec. 3.9.

36. Read about apparent weight and weightlessness in Sec.'s 3.3 and 3.9. For lifting and punching, determine whether it's the mass or the weight of the bowling ball that matters, and how the mass and weight of the bowling ball compare to their values on earth. What does the moon have an incredible amount of compared to the space station?

37. (A) Think about the motion of both the earth and moon in the solar system. (B) In this case, being close to the earth is important – not just how much mass the object has. We can quantify this by calculating how much force each object exerts on one kilogram of mass of ocean on each side of the earth, and subtracting the two forces. For whichever object the difference in forces (on the near ocean compared to the far ocean) is greater, that object has the greater effect on earth's ocean tides. Since this force obeys Newton's law of universal gravitation, both the mass of the object and its closeness to the earth are significant. (C)/(D) The oceans are squeezed, with high tides occurring where the moon is directly above and directly away, and low tides occurring halfway in between these two points. A point

on earth rotates through all four positions (two low and two high) in one complete rotation about earth's axis. (E) How long does it take the moon to orbit the earth?

38. Solve for $G$ in Newton's law of universal gravitation and plug in the SI units of everything else.

39. Solve for $a$ and plug in the SI units of everything else.

## Chapter 3 – Hints to Practice Problems

1. Look up a conversion factor from pounds to Newtons in order to find the dummy's weight on earth. Use the formula that relates mass to weight in order to find the dummy's mass. What changes when the dummy is transported to the moon: mass or weight? By what factor?

2. From the units, you can discern whether you were given the dummy's mass or weight. Use the 1DUA strategy to determine gravitational acceleration on the other planet. Many students have made the mistake of making things up (like randomly doing something with the numbers or making up an equation), and always get the wrong answer by doing so. Trust the equations. The equations can work, but making things up in physics just doesn't work out. What changes when the dummy is transported to the moon: mass or weight? By what factor? Use the formula that relates mass to weight.

3. (A) From the units, you can discern whether you were given the dummy's mass or weight. What changes when the dummy is transported to the moon: mass or weight? By what factor? Use the formula that relates mass to weight. (B) From the units, you can discern whether you were given the dummy's mass or weight. What changes when the dummy is transported to the moon: mass or weight? By what factor? Use the formula that relates mass to weight. Notice that in part (B), they can't be identical boxes of bananas in the two cases.

4. Read about apparent weight and study the last example in Sec. 3.3. Draw a FBD and apply Newton's second law. The elevator must be accelerating – i.e. changing velocity – in order for the scale reading to read anything other than the monkey's actual weight. (C) Determine the acceleration of the quarter relative to the elevator. Use the 1DUA strategy with this relative acceleration. (E) Use the IDUA strategy with earth's gravitational acceleration.

5. (A) Use the 1DUA strategy. (B) Use Newton's second law.

6. Draw a FBD and apply Newton's second law. Study the examples. Look at the form of the force of air resistance in Sec. 3.8. Remember, velocity means how fast and which way the passenger is moving, while acceleration describes how the velocity is changing.

7. Draw two FBD's – one for the UFO and one for the monkey – and apply Newton's second law to each object individually. Study the examples in Sec. 3.4. Which of Newton's laws tells you that the tension force exerted on the UFO and the tension force exerted on the monkey have the same magnitude, but opposite direction? Note the difference in the sign of the acceleration in parts (A) and (B). (C) What is the acceleration if the UFO travels in a straight line with constant speed? **Note**: The weight of the monkey does not appear in the FBD for the UFO; rather, the effect of the monkey's weight is indirectly communicated to the UFO through the tension force.

8. Draw three FBD's – one for the helicopter, spy, and student – and apply Newton's second law to each object individually. Study the examples in Sec. 3.4. There are two different tensions in this problem – one for each cord. Each tension acts on two different objects. **Note**: The weight of the student does not appear in the FBD's for the helicopter or spy; similarly, the weight of the spy does not appear in any

FBD other than its own. Rather, the effects of their weights are indirectly communicated to the other objects through the tension forces.

9. Draw two FBD's – one for the chest and one for the monkey – and apply Newton's second law to each object individually. Study the examples in Sec.'s 3.4-3.5. Which of Newton's laws tells you that the tension forces exerted on the chest and monkey have the same magnitude? The two objects have the same acceleration because they are connected by a cord that remains taut. Use the equation that relates the friction force to the normal force. Setup your coordinate system such that $+x$ is to the right for the chest and downward for the monkey – i.e. think of the pulley as bending the $x$-axis. This way, $a_x$ will be the same for the chest and monkey.

10. Draw three FBD's – one for the box and one for each monkey – and apply Newton's second law to each object individually. Study the examples in Sec.'s 3.4-3.5. There are two different tensions in this problem – one for each cord. Each tension acts on two different objects. The three objects have the same acceleration because they are connected by cords that remain taut. Use the equation that relates the friction force to the normal force. Setup your coordinate system such that $+x$ is upward for Chimp, to the right for the box, and downward for Lemur – i.e. think of the pulley as bending the $x$-axis. This way, $a_x$ will be the same for each object. Check your FBD for the box, which should have five different forces acting on it.

11. Draw three FBD's – one for each box – and apply Newton's second law to each box individually. Study the examples in Sec.'s 3.4-3.5. There are two different tensions in this problem – one for each cord. Each tension acts on two different objects. The three objects have the same acceleration because they are connected by cords that remain taut. Resolve the pull, $\vec{P}$, into horizontal and vertical components. Check that you have four different forces acting on the middle and right boxes in your FBD's.

12. Draw two FBD's – one for each box – and apply Newton's second law to each box individually. Study the examples in Sec.'s 3.4-3.5. Which of Newton's laws tells you that the tension forces exerted on each box have the same magnitude? The two objects have the same acceleration because they are connected a by cord that remains taut. Resolve the pull, $\vec{P}$, into horizontal and vertical components. Use the equation that relates the friction force to the normal force. Check that you have five different forces acting on the right box in your FBD.

13. Draw a FBD for the box and apply Newton's second law. Study the examples in Sec. 3.5. Resolve each of the students' pulls into horizontal and vertical components. Whichever pull has the largest horizontal component will dictate the direction of the friction force. Use the equation that relates the friction force to the normal force. Check that you have five different forces in your FBD.

14. Draw a FBD for the knot and apply Newton's second law. Study the examples in Sec. 3.4. Resolve the tensions into horizontal and vertical components. The tension in the bottom cord will equal the weight of the monkey since the knot is in static equilibrium.

15. Draw a FBD for the monkey and apply Newton's second law. Study the examples in Sec. 3.5. Choose $+x$ to be down the incline and $+y$ to be perpendicular to it. You should have three forces acting on the monkey in your FBD. Resolve each force into components. Use the equation that relates the friction force to the normal force.

16. Draw two FBD's – one for each object – and apply Newton's second law to each object individually. Study the examples in Sec.'s 3.4-3.5. Which of Newton's laws tells you that the tension forces exerted on each object have the same magnitude? The two objects have the same acceleration because they

Hints

are connected by a cord that remains taut. Use the equation that relates the friction force to the normal force. Setup your coordinate system such that $+x$ is up the incline for the passenger box and downward for the stone – i.e. think of the pulley as bending the $x$-axis – and $+y$ is perpendicular to the incline for the passenger box. This way, $a_x$ will be the same for each object. Remember to account for both the mass of the passenger and the passenger box. Set $a_x > 0$ to derive an inequality for the mass of the stone; the minimum mass of the stone corresponds to $a_x \approx 0$.

17. Draw a FBD for the Neanderthal and apply Newton's second law. Draw the push, $\vec{P}$, as a pull in the FBD (whereas it was drawn as a push, instead of a pull, in the problem) – i.e. attach the tail of $\vec{P}$ to the Neanderthal, instead of the tip of $\vec{P}$. Study the examples in Sec. 3.5. Choose $+x$ to be down the incline and $+y$ to be perpendicular to it. You should have four forces acting on the Neanderthal in your FBD. Resolve each force into components. Use the equation that relates the friction force to the normal force. What is the acceleration of an object that travels in a straight line with constant speed?

18. Draw a FBD for the chest and apply Newton's second law. Study the examples in Sec.'s 3.4-3.5. Choose $+x$ to be down the incline and $+y$ to be perpendicular to it. You should have four forces acting on the chest in your FBD. Use the equation that relates the friction force to the normal force. Resolve each force into components. In part (A), the force of static friction can act in two possible directions, and also includes an inequality. Find the range of tension forces by separately treating the case of static friction acting in both possible directions with its maximum possible value. In part (C), use the 1DUA strategy.

19. Draw two FBD's – one for the box and one for the monkey – and apply Newton's second law to each object individually. Study the examples in Sec.'s 3.4-3.5. Which of Newton's laws tells you that the tension forces exerted on the box and monkey have the same magnitude? The box and monkey have the same acceleration because they are connected by a cord that remains taut. Setup your coordinate system such that $+x$ is up the incline for the box and downward for the monkey – i.e. think of the pulley as bending the $x$-axis – and $+y$ is perpendicular to the incline for the box. This way, $a_x$ will be the same for each object. The direction of the friction force depends on the direction of the instantaneous velocity. (There is only one direction that the velocity could naturally have if released from rest, but if the system were given an initial push in the opposite direction – which is no longer acting on the system – then the system could be temporarily moving in the other direction.) You should have four forces acting on the box in your FBD. Resolve each force into components. Use the equation that relates the friction force to the normal force.

20. Treat this problem like an object resting on an incline. Draw a FBD for the chest and apply Newton's second law. Study the examples in Sec. 3.5. Choose $+x$ to be down the incline and $+y$ to be perpendicular to it. You should have three forces acting on the chest in your FBD. Resolve each force into components. Use the equation that relates the friction force to the normal force. What value do you want the acceleration to have? Which coefficient is probably static and which is probably kinetic? Which coefficient do you need?

21. Draw two FBD's – one for each box – and apply Newton's second law to each box individually. Study the examples in Sec.'s 3.4-3.5. Which of Newton's laws tells you that the tension forces exerted on each box have the same magnitude? The two boxes have the same acceleration – except in part (C) – because they are connected a by cord that remains taut. You should have five forces acting on the upper box in your FBD. Choose $+x$ to be down the incline and $+y$ to be perpendicular to it. Use the equation that relates the friction force to the normal force. Resolve each force into components.

22. Draw two FBD's – one for each monkey – and apply Newton's second law to each monkey individually. Study the examples in Sec.'s 3.4-3.5. Which of Newton's laws tells you that the tension forces exerted on each monkey have the same magnitude? The two monkeys have the same acceleration because they are connected a by cord that remains taut. You should have four forces acting on each monkey in your FBD's. Choose $+x$ to be up the incline for the left monkey and down the incline for the right monkey, and $+y$ to be perpendicular to the incline. Use the equation that relates the friction force to the normal force. Resolve each force into components.

23. Draw two FBD's – one for each monkey – and apply Newton's second law to each monkey individually. Study the examples in Sec.'s 3.4-3.5. Which of Newton's laws tells you that the tension forces exerted on each monkey have the same magnitude? The monkeys have the same acceleration because they are connected by a cord that remains taut. Setup your coordinate system such that $+x$ is down the incline for Lemur and downward for Chimp – i.e. think of the pulley as bending the $x$-axis – and $+y$ is perpendicular to the incline for Lemur. This way, $a_x$ will be the same for each monkey. You should have four forces acting on the box in your FBD. Resolve each force into components. Use the equation that relates the friction force to the normal force. Which coefficient is probably static and which is probably kinetic? Which coefficient do you need? For part (D), use the 1DUA strategy.

24. Draw a FBD for the coconut and apply Newton's second law. Study the examples in Sec. 3.4. Choose $+x$ to be horizontal and $+y$ to be vertical. Use Newton's first law to determine which way the necklace will lean if the car speeds up. You should have two forces acting on the coconut in your FBD. Do not include a drive force in the FBD; the effect of the car's motor is indirectly communicated to the coconut through the tension. Resolve each force into components. What is the direction of the car's acceleration?

25. Draw two FBD's – one for each chest – and apply Newton's second law to each chest individually. Study the examples in Sec.'s 3.4-3.5 and especially Sec. 3.7. Use the equation that relates the friction force to the normal force. There are two equal and opposite force pairs in this problem, and there are two different normal forces acting on the bottom box. Check your FBD's: Five different forces act on the bottom chest and four different forces act on the top chest. Do not include the weight of the top chest as a force exerted on the bottom chest; rather, the effect of the top chest's weight is communicated through the equal and opposite normal forces exert between the two chests.

26. Draw two FBD's – one for the chest and one for the monkey – and apply Newton's second law to each object individually. Study the examples in Sec. 3.5 and especially Sec. 3.7. Use the equation that relates the friction force to the normal force. There are two equal and opposite force pairs in this problem, and there are two different normal forces acting on the chest. Check your FBD's: Five different forces act on the chest and three different forces act on the monkey. Since the monkey is stationary relative to the chest, the chest and monkey have the same (horizontal) acceleration.

27. Draw two FBD's – one for the wedge and one for the monkey – and apply Newton's second law to each object individually. Study the examples in Sec. 3.5 and especially Sec. 3.7. There is one equal and opposite force pair in this problem, and there are two different normal forces acting on the wedge. Check your FBD's: Four different forces act on the wedge and two different forces act on the monkey. Choose $+x$ to be horizontal and $+y$ to be vertical. (This is different from other incline problems, since the monkey is accelerating to the left – not down the incline.) Resolve each force into components. Since the monkey is stationary relative to the wedge, the wedge and monkey have the same (horizontal) acceleration.

Hints

28. Draw a FBD for the box and apply Newton's second law. Study the examples in Sec.'s 3.5-3.6. Choose $+x$ to be down the incline and $+y$ to be perpendicular to it. You should have three forces acting on the box in your FBD. Resolve each force into components. Apply Hooke's law to the spring. Note that the restoring force pulls the box toward the natural equilibrium position of the spring. What is the acceleration of the system?

29. Draw a FBD for the box (when the box is in contact with the spring) and apply Newton's second law. Study the examples in Sec.'s 3.5-3.6. Choose $+x$ to be down the incline and $+y$ to be perpendicular to it. You should have three forces acting on the box in your FBD. Resolve each force into components. Apply Hooke's law to the spring. Note that the restoring force pushes the box toward the equilibrium position of the spring.

30. Draw a FBD for the box and apply Newton's second law. Study the examples in Sec.'s 3.5-3.6. You should have five forces acting on the box in your FBD. Use the equation that relates the friction force to the normal force. Apply Hooke's law to the springs. Determine the direction of the net restoring force in order to determine the direction of the friction force. Note that each restoring force pulls/pushes the box toward the natural equilibrium position of the spring.

31. Draw two FBD's – one for the box and spring – and apply Newton's second law to each object. Study the examples in Sec.'s 3.5-3.6. Choose $+x$ to be up the incline for the spring and $+y$ to be perpendicular to it, and $+x$ to be downward for the box. Resolve each force into components. Apply Hooke's law to the spring. Note that the restoring force pulls the box toward the equilibrium position of the spring. What is the acceleration of the system?

32. Draw a FBD for the crate and apply Newton's second law. Study the examples in Sec. 3.5. Resolve the pull into horizontal and vertical components. Use the equation that relates the friction force to the normal force. Check that you have four different forces in your FBD. What is the acceleration of the crate if it has constant velocity? Solve for the magnitude of the pull as a function of $\theta$. Set the derivative of this function equal to zero in order to find the angle, $\theta$, that minimizes the pull. Check the sign of the second derivative to see if, in fact, that is a relative minimum. Also, check the limits as $\theta$ approaches zero and ninety degrees to ensure that you find the absolute minimum. The pull isn't most effective when it is horizontal (even though the horizontal component of the pull is greater in this case) because pulling upward at an angle actually reduces the normal force, which reduces the friction force.

33. (A) Each term must have the same SI unit as force. (B) Use Newton's second law, plug in $t = 0$, and use the equations to solve for the magnitude and direction of a vector in terms of its components. Note that the angle is measured counterclockwise from the $+x$-axis; when taking the inverse tangent, add $180°$ if the $x$-component of the vector is negative. (C) This is like part (B), except this time you know the direction of the acceleration and want to solve for the time. (D) Perform a definite integral to find the change in velocity from the acceleration. Plug in the initial velocity to solve for the instantaneous velocity. It should be easy to find the momentum once you know the velocity. Use the equation to solve for the magnitude of a vector in terms of its components.

34. (A) Each term of each vector must have the same SI units as the vector. (B)/(C)/(D)/(E) Use the equations to solve for the magnitude and direction of a vector in terms of its components. Note that the angle is measured counterclockwise from the $+x$-axis; when taking the inverse tangent, add $180°$ if the $x$-component of the vector is negative. (B) Plug in $t = 0$. (C) Use Newton's second law. (D) Perform a definite integral to find the change in velocity from the acceleration. Plug in the initial velocity to solve for the instantaneous velocity. It should be easy to find the momentum once you know the velocity. (E) Perform a definite integral to find the net displacement from velocity.

35. (A) What's the definition of momentum? This should be a piece of cake. (B) Write the momentum (as a function of time) using unit vector notation by using the equations to solve for the components of a vector in terms of its magnitude and direction. Take a derivative of momentum in order to find the net force (see the general form of Newton's second law – i.e. involving momentum). (C) When is the $y$-component of the velocity negative?

36. (A) Each term must have the same SI units as momentum. (B) Plug in $t = 0$. It should be easy to find the velocity once you know the momentum. (C)/(D) Express velocity as a function of time. Perform a definite integral to find the net displacement from velocity. Use the equations to solve for the magnitude and direction of a vector in terms of its components. Note that the angle is measured counterclockwise from the $+x$-axis; when taking the inverse tangent, add $180°$ if the $x$-component of the vector is negative. (E)/(F) Take a derivative of momentum in order to find the net force (see the general form of Newton's second law – i.e. involving momentum).

37. (A) Use the strategy for adding vectors from Sec. 2.2. (B) Perform a definite integral of the net force (as a function of time) in order to find the change in momentum (see the general form of Newton's second law – i.e. involving momentum). Plug in the initial momentum (which should be easy to find) in order to find the instantaneous momentum. (C) Use Newton's second law. (D) This should be easy to find from the momentum. (E) First perform a definite integral of velocity in order to find the net displacement.

38. (A) These are just the initial components of the position vector. (B)-(F) Use the equations to solve for the magnitude and direction of a vector in terms of its components. Note that the angle is measured counterclockwise from the $+x$-axis; when taking the inverse tangent, add $180°$ if the $x$-component of the vector is negative. (B) Take a derivative of the position vector in order to find velocity. (C) It should be easy to find the momentum once you know the velocity. (D)/(F) Take a derivative of velocity in order to find acceleration. (E) Use Newton's second law.

39. Draw a FBD and apply Newton's second law. Look at the form of the force of air resistance in Sec. 3.8. Study the examples in Sec. 3.8. Remember, velocity means how fast and which way the passenger is moving, while acceleration describes how the velocity is changing.

40. Draw a FBD and apply Newton's second law. Look at the form of the force of air resistance in Sec. 3.8. Study the examples in Sec. 3.5 and especially Sec. 3.8. Remember, velocity means how fast and which way the passenger is moving, while acceleration describes how the velocity is changing.

41. Apply Newton's second law and integrate the resulting second-order differential equation. See the technique for this in one of the examples of Sec. 3.8, for example (but note that this problem is simpler).

42. Apply Newton's second law, using Hooke's law, and integrate the resulting second-order differential equation. See the technique for this in one of the examples of Sec. 3.8, for example. It may be useful to apply the chain rule (see the end of Sec. 1.8.).

43. See Sec. 3.9. In this Information Age, it should be easy to look up the needed astronomical data. (A) Use the equation for gravitational acceleration due to a spherically symmetric distribution of mass. (B) Use Newton's law of universal gravitation. (C) What does Newton's third law say about this?

44. See Sec. 3.9. In this Information Age, it should be easy to look up the needed astronomical data. Use the equation for gravitational acceleration due to a spherically symmetric distribution of mass. Note that $R$ is the distance from the center of the earth to the field point (not necessarily the radius of the earth). Note that altitude means the height above earth's surface.

45. Use Newton's law of universal gravitation. See Sec. 3.9.

46. See Sec. 3.9. Use the equation for gravitational acceleration due to a spherically symmetric distribution of mass and Newton's law of universal gravitation.

47. See Sec. 3.9, especially the example that shows you how to take a ratio so that some of the information (which the problem prohibits you from looking up) will cancel out. Use the equation for gravitational acceleration due to a spherically symmetric distribution of mass and the definition of mass. Note that altitude means the height above earth's surface.

48. See Sec. 3.9. Use the equation for gravitational acceleration due to a spherically symmetric distribution of mass and the definition of mass. Which is the same and which changes when you travel to Planet Mnqy: mass or weight? See the hints to Conceptual Question 18.

49. See the hints to Conceptual Question 33. Read Sec. 3.9 and study the examples. Apply Newton's second law to the bananas. What will the acceleration of the bananas be? The two forces acting on the bananas must therefore be equal and opposite. Everything will cancel out except for two unknowns. Note that the $R$'s are not the radii of the earth and moon, but the distances from the center of the earth or moon to the field point. Look up the average earth-moon distance to write down a second equation in terms of the two unknowns. Solve the system of two equations in two unknowns. There are two mathematical solutions to this quadratic equation, but only one answer makes sense physically (as mentioned in the hints to Conceptual Question 33).

50. (A) The mass is proportional to the volume. Find the ratio of the volume of the original moon to its present volume, and use this to find the ratio of their masses. Density will cancel out. Note that $m_p$ was the moon's mass before mining (i.e. when it was a solid sphere). (B) Use Newton's law of universal gravitation. The equation given in the note is not relevant for part (B), since no object is inside of another. (C) Use the equation given in the note to find the gravitational field (vector) at the field point due to the original solid moon and due to the sphere that has been mined. You will need to determine the mass of the mined sphere as in part (A), and use its mass and radius. Find the net gravitational field (i.e. due to the current shape of the moon) through vector subtraction.

## Chapter 4 – Hints to Conceptual Questions

1. This relates to the equation $\Delta s_T = R_0 \Delta\theta$, and the distinction between net displacement and net tangential displacement in UCM.
2. Look at the formulas that relate angular speed and tangential speed to the period in UCM.
3. Look at the formula that relates tangential speed to angular speed in UCM. You may need to look up relevant astronomical distances.
4. Look at the formula that relates tangential speed to angular speed in UCM, and the formula for centripetal acceleration. It may be useful to express centripetal acceleration in terms of angular speed. Compare the definitions of angular, tangential, and centripetal acceleration. You may need to look up relevant astronomical data.
5. Review the precise definition of 'the' acceleration vector, and the distinction between velocity and speed. Compare the definitions of angular, tangential, centripetal, and 'the' acceleration.
6. Review the definitions in Sec. 4.1.
7. Review Sec. 4.3. Review the definitions of inertia, centripetal acceleration, centripetal force, and centrifugal force.

8. Each point on the circle has coordinates $(x, y)$ that can be expressed in terms of the radius of the circle, $R_0$, and the angle $\theta$. For example, the initial position is located at $(0, R_0)$. Plot the $x$-coordinate as a function of time. On the time axis, mark units in terms of the period, $T$ (in multiples of $T/16$ up to $T$, for example). A sketch of the slope of the $x$-coordinate graph will give the $v_x$ graph, and a sketch of the slope of the $v_x$ graph will give the $a_x$ graph. You can also determine the functions for all three graphs by writing $x = R_0 \cos(\omega_0 t + \varphi)$ and taking successive derivatives, where $\varphi$ is an appropriate phase angle that shifts the graph horizontally so that the initial coordinates match the initial position specified in the problem.

9. You can visualize this by finding some sort of a disc and slowly rolling it by hand.

10. These questions relate to the relationships between tangential displacement and angular displacement and between tangential velocity and angular velocity in UCM.

11. Use the equations of UCM (tabulated in Sec. 4.2) to express each quantity in terms of the speed and radius to see what would happen if the speed were doubled (with constant radius).

12. Use the equations of UCM (tabulated in Sec. 4.2) to express each quantity in terms of the speed and radius to see what would happen if the radius were doubled (with constant speed). (Of course, doubling the diameter also means doubling the radius.)

13. The equations of UCM are tabulated in Sec. 4.2.

14. Draw a FBD and apply Newton's second law. Read Sec.'s 4.3-4.4, and study the examples of Sec. 4.4. How does the tension depend on the speed?

15. Draw a FBD and apply Newton's second law. Read Sec.'s 4.3-4.4, and study the examples of Sec. 4.4. How does the angle with the horizontal (not vertical) depend on the speed?

16. Draw a FBD and apply Newton's second law for various positions as the Ferris wheel rotates. The illustration for Practice Problem 13 may help you to visualize some representative positions of a seat on a Ferris wheel (even though that problem features a monster truck instead of a wheel). Read Sec.'s 4.3-4.4, and study the examples of Sec. 4.4. At each position, the net support force must point in a direction that insures that the inward components of the forces have a positive sum, while the tangential components of the forces cancel out. There are just two forces: One force always has the same magnitude and direction, while the magnitude and direction of the net support force varies as the Ferris wheel rotates in such a way as to cancel the tangential components of the net force and provide the needed centripetal acceleration.

17. See the illustration for Practice Problem 11. The force of static (not kinetic!) friction can act up or down the incline, and has a range of possible values (since the formula involves an inequality). Imagine driving in a horizontal circle (with a banked curve that runs full circle, so the bank is part of a large cone) with the 'magic' speed that makes the static friction force zero. If you drive a little faster than this speed, the car tries to go up the incline, so static friction pushes down the incline to resist this; whereas if you drive a little slower than the 'magic' speed, the car tries to go down the incline, so static friction pushes up the incline to resist this.

18. What is the moon's orbital period? How does the period of a satellite relate to its distance from earth? Note that if an earth satellite had the same period as the moon, it would therefore have the same average orbital radius as the moon (since the moon is an earth satellite).

19. Read Sec. 4.6 and study the equations that we derived for satellites in UCM and apply the relevant equations to the two satellites.

20. Read Sec. 4.6 and study the equations that we derived for satellites in UCM and apply the relevant equations to the two satellites.

## Chapter 4 – Hints to Practice Problems

1. (A) You should know the period of each of these hands. Convert the period to SI units. Use the equation that relates period to angular speed. (B) Draw the hour hand in two positions that are seven hours apart. This is just a simple geometry problem.

2. Use the 1DUA strategy to determine the maximum height of the center of mass of the baton: You must work with one-half of the trip to determine this. Next, determine how many revolutions the baton completes for the entire trip based on its total time in the air and its frequency. Finally, use the total distance that the center of mass of the baton travels (double the maximum height that you found) and the number of revolutions completed to find the length of the baton (which is the diameter of the circle it makes when it rotates) – use the arc length formula.

3. (A) You should know the period of the earth's rotation about its axis. Convert the period to SI units. Use the equation that relates period to angular speed. (B)/(C) Use one of the equations in Sec. 4.2. (C) Use trig to determine the radius of the circle that the object travels in as the earth rotates about its axis. (D) Use one of the equations in Sec. 4.2. (E) Use the equations in Sec. 4.2. Look at the combined effect of the object's weight and the centripetal force.

4. (A) You should know the period of the earth's revolution around the sun. Convert the period to SI units. Use the equation that relates period to angular speed. (B)/(C)/(D) Use the equations in Sec. 4.2. (E) Use Newton's law of universal gravitation. Compare the numbers.

5. Use the equations and strategy described in Sec. 4.2. (F) Draw and label the initial and final positions. The net displacement is a vector that extends from the initial position to the final position. (H) Since you already know the mass and acceleration, this should be very easy: Use Newton's second law. (I) Draw a FBD and apply Newton's second law. See Sec. 4.5.

6. Study the basic swinging problem discussed in Sec. 4.4, including the examples. Use the equations and strategy described in Sec. 4.2. Examine the units of the given rate to determine which quantity was given in the problem.

7. Study the basic swinging problem discussed in Sec. 4.4, including the examples. Use the equations and strategy described in Sec. 4.2.

8. Study the basic swinging problem discussed in Sec. 4.4, including the examples. Use the equations and strategy described in Sec. 4.2.

9. Study the basic swinging problem discussed in Sec. 4.4, including the examples. Use the equations and strategy described in Sec. 4.2. You should have three forces in your FBD. Draw a side view with the passenger at either the left or right side of the centrifuge.

10. Study the basic swinging problem discussed in Sec. 4.4, including the examples. Use the equations and strategy described in Sec. 4.2. The passengers travel in a circle with a radius that is larger than the radius of the disc: You must use trig to determine what fraction of the length of the chains must be added to the radius of the disc.

11. Study the basic driving in a circle problem discussed in Sec. 4.5, including the examples. Use the equations and strategy described in Sec. 4.2. Draw a FBD for the racecar in the position shown. (A) You should have two forces in your FBD. (B) You should have three forces in your FBD. The direction of one of the forces is discussed in the hints to Conceptual Question 17.

12. (A) Use one of the equations in Sec. 4.2. (B) Draw a FBD for the truck in the position shown and apply Newton's second law. Label the inward and tangential directions. Resolve each force into inward and tangential components. In order for the truck to travel with constant speed, there must be a drive force in order to make the sum of the tangential components of the forces zero; but it's the net inward force that you need to use to solve for the normal force in this position. (C) Draw a FBD for the truck when it is at the top of the hill and apply Newton's second law. Label the inward and tangential directions. Resolve each force into inward and tangential components. The drive/resistive forces evidently cancel out in this position. The normal force must be nonzero in order that the truck not lose contact with the ground; set the normal force equal to zero in order to solve for the critical speed.

13. (A) Use the equations in Sec. 4.2. (B) Draw a FBD for the truck when it is in each position and apply Newton's second law. At each position, the net support force must point in a direction that ensures that the inward components of the forces have a positive sum, while the tangential components of the forces cancel out. There are just two forces: One force always has the same magnitude and direction, while the magnitude and direction of the net support force varies as the monster truck moves in such a way as to cancel the tangential components of the net force and provide the needed centripetal acceleration. (C) Set the normal force equal to zero and solve for the speed. The result varies depending on the monster truck's position; you want to find the position that requires the greatest minimum speed.

14. Study the basic swinging problem discussed in Sec. 4.4, including the examples. Use the equations and strategy described in Sec. 4.2. Read the hints to Problems 12 and 13. (E) The monkey's pull is not directly along the arm shown – i.e. it's not toward the center of the circle. Since the weight pulls straight downward, the weight has a tangential component in the position shown. Therefore, in order that the angular speed be constant (as required in the statement of the problem), the monkey's pull must have a tangential component as well as an inward component. So draw the monkey's pull toward the monkey's head (above the radius shown). (G) You will be able to derive the specified equation by setting the sum of the tangential components of the forces equal to zero. (H) Now draw a FBD for the top position. Whereas you drew a FBD for the bucket in part (F), now you are drawing a FBD for the water – so you should have a normal force (from the bucket) instead of the monkey's pull. At the top position, the normal force will be straight down, toward the center of the circle – unlike part (F), since weight does not have a tangential component at the top.

15. Study the satellite strategy in Sec. 4.6 – especially, the derivations for a satellite in UCM and for Kepler's third law in UCM. Note that 'geosynchronous' means that the satellite's period is synchronized with the earth's rotation about its axis. Therefore, you should know the satellite's period. The altitude – i.e. its height above earth's surface – can readily be found after solving for the radius of the satellite's orbit. You may need to look up astronomical data for the earth.

16. Study the satellite strategy in Sec. 4.6 – especially, the derivations for a satellite in UCM and for Kepler's third law in UCM. The altitude is the height of the satellite above earth's surface. You may need to look up astronomical data for the earth.

17. Be sure to use the appropriate masses and radii in each part of this problem. (A) See Sec. 3.9. Use the equation for gravitational acceleration due to a spherically symmetric distribution of mass. (B) How is mass related to weight? (C) Use a formula to determine Lemur's mass, then use the same formula to determine Lemur's weight on earth. (D) See Sec. 3.9. Use Newton's law of universal gravitation. (E) Use the orbital period of the planet's revolution. (F) Use one of the equations in Sec. 4.2. (G) The acceleration is centripetal. Use one of the equations in Sec. 4.2. (H) Use Newton's second law. (I) Study

the satellite strategy in Sec. 4.6 – especially, the derivations for a satellite in UCM and for Kepler's third law in UCM. The altitude is the height of the satellite above earth's surface.

18. Be sure to use the appropriate masses and radii in each part of this problem. (A) See Sec. 3.9. Use the equation for gravitational acceleration due to a spherically symmetric distribution of mass. (B) See Sec. 3.9. Use Newton's law of universal gravitation. (C)/(D) Study the satellite strategy in Sec. 4.6 – especially, the derivations for a satellite in UCM and for Kepler's third law in UCM. (E) What basic physical property of a material may help you to identify what the material is?

19. (A) Solve for $H$ and plug in the SI units of everything else. (B)/(C) Study the satellite strategy in Sec. 4.6 – especially, the derivations for a satellite in UCM and for Kepler's third law in UCM.[346]

20. A satellite in a circular orbit travels with constant speed. Its inertia is its tendency to travel in a straight line with constant speed, while the gravitational force pulling inward causes it to change direction. In this problem, we are looking at the deviation of a satellite from its tendency to go off on a tangent as a function of time. Draw a tangent. Mark and number a few points on the tangent and in the circular orbit that correspond – keeping in mind that in either case, the object travels with constant speed (so the intervals must be constant and have the same spacing in both cases). Write down a formula for the distance connecting two of these corresponding points as a function of time – the only variable should be time, but there may also be constants (like the constant speed of the satellite). Check that your formula agrees with cases that are easy to calculate, like every ninety degrees. Take subsequent derivatives to find the acceleration, and compare with the formula for centripetal acceleration.

## Chapter 5 – Hints to Conceptual Questions

1. Review Sec. 5.1. Use the form of the scalar product that involves the angle between the vectors.
2. Review Sec. 5.1. Use the form of the scalar product that involves the components of the vectors.
3. Review Sec. 5.1. Combine both forms of the scalar product.
4. Review the definition of work in Sec. 5.2 and the scalar product in Sec. 5.1. What angle between the force and displacement would make the work done by the force equal zero? Draw a FBD in each case and see examine the angle between the tension and the displacement.
5. Review the definition of work in Sec. 5.2 and the scalar product in Sec. 5.1. What angle between the force and displacement would make the work done by the force equal zero? What is the direction of the centripetal force? What is the instantaneous direction of the displacement of a monkey traveling with UCM?
6. Review the definition of work in Sec. 5.2 and the scalar product in Sec. 5.1. What angles between the force and displacement would make the work done by the force negative, zero, and positive? In each case, draw and label the direction of the specified force and the direction of the displacement in order to determine the angle between them.

---

[346] This force law is not intended to be realistic. Much deeper analysis is needed just to see if a circular orbit is actually possible with this force law. However, this contrived problem does meet the purpose of testing your understanding of the satellite strategy and your ability to apply it.

7. Review the definition of the work done by gravity in Sec. 5.2 and the scalar product in Sec. 5.1. What angles between the force and displacement would make the work done by the force negative, zero, and positive? In each case, draw and label the direction of the specified force and the direction of the displacement in order to determine the angle between them.

8. Review the definition of the work done by a spring in Sec. 5.2 and the scalar product in Sec. 5.1. What angles between the force and displacement would make the work done by the force negative, zero, and positive? In each case, draw and label the direction of the specified force and the direction of the displacement in order to determine the angle between them.

9. Is gravity conservative or nonconservative? Is the work done by such a force path-dependent or path-independent? Also, which component of motion is important for the work done by gravity? Is the net displacement or total distance traveled relevant here? Compare the relevant quantities for the two cases.

10. Review the definition of net work in Sec. 5.2. Which force(s) do you use to find the net work? Apply Newton's second law and separate the force into tangential and centripetal components. Review the derivation of the kinetic energy formula in Sec. 5.3.

11. Which force have we learned about that sometimes has both constant magnitude and direction? Is that force conservative or nonconservative? Is the work done by such a force path-dependent or path-independent? Also, which component of motion is important for the work done by that force? Is the net displacement or total distance traveled relevant here? Compare the relevant quantities for the two cases.

12. Is friction conservative or nonconservative? Is the work done by such a force path-dependent or path-independent? Also, which component of motion is important for the work done by friction? Is the net displacement or total distance traveled relevant here? Compare the relevant quantities for the two cases.

13. Review these definitions in Sec. 2.3.

14. Review the motion quantities defined in Sec. 2.3. You should recognize what this integral equals.

15. Review the motion quantities defined in Sec. 2.3. You should recognize what this integral equals.

16. Is gravity conservative or nonconservative? Is the work done by such a force path-dependent or path-independent? Also, which component of motion is important for the work done by gravity? Is the net displacement or total distance traveled relevant here? Compare the relevant quantities for the two cases.

17. Is friction conservative or nonconservative? Is the work done by such a force path-dependent or path-independent? Also, which component of motion is important for the work done by friction? Is the net displacement or total distance traveled relevant here? Compare the relevant quantities for the two cases.

18. Look at the SI units of various quantities from Chapter 5. Convert kWh to SI units.

19. Review the definition of power and the special cases considered in Sec. 5.2. Which force have we learned about that sometimes has both constant magnitude and direction? Is that force conservative or nonconservative? Is the work done by such a force path-dependent or path-independent? Also, which component of motion is important for the work done by that force? Is the net displacement or total distance traveled relevant here? Compare the relevant quantities for the two cases.

20. Review the definition of power and the special cases considered in Sec. 5.2. Which force have we learned about that is sometimes constant in magnitude and is always tangential to the path? Is that force conservative or nonconservative? Is the work done by such a force path-dependent or path-

Hints

independent? Also, which component of motion is important for the work done by that force? Is the net displacement or total distance traveled relevant here? Compare the relevant quantities for the two cases.

21. For each energy source, consider whether or not, in some way, the energy provided to earth by the sun was largely responsible, ultimately, for how its energy became stored. What causes wind? Is the sun needed for this? How were radioactive isotopes that we find on earth ultimately produced? Can you trace their origins to energy supplied by the sun? What causes tides on earth's oceans? Is the sun primarily responsible for this? Can you trace the origins of coal and/or natural gas found on earth to energy supplied by the sun?

22. In each case, write an equation for the kinetic energy for each monkey using subscripts for any quantities that are not the same. Take a ratio of the kinetic energies – anything that is the same will cancel out. For quantities that are different, write one in terms of the other – e.g. $v_1 = 2v_2$ – and substitute this equation into the ratio. Solve for $K_2$ in terms of $K_1$. See an example in Sec. 3.9 where we applied a similar ratio strategy.

23. Start with the equation for kinetic energy. We saw the result briefly in Chapter 3, and will see it much more in Chapter 6.

24. Look up the equations for kinetic energy and momentum. Make a substitution to eliminate the speed.

25. It's multiple choice: potential energy, kinetic energy, or nonconservative work. Look very closely at the definition of potential energy in Sec. 5.3 and how we derived the equation for conservation of energy from the work-energy theorem in Sec. 5.4.

26. Look at the formula for kinetic energy and the relevant formula for this type of potential energy in Sec. 5.3. Also review Sec. 5.4.

27. Look at the formula for kinetic energy and the relevant formula for this type of potential energy in Sec. 5.3. Also review Sec. 5.4.

28. Look at the formula for kinetic energy and the relevant formula for this type of potential energy in Sec. 5.3. Also review Sec. 5.4.

29. Look at the formula for kinetic energy and the relevant formula for this type of potential energy in Sec. 5.3.

30. Look at the formula for kinetic energy and the relevant formula for each type of potential energy in Sec. 5.3. Also review Sec. 5.4. Transformation of energy means transformations between potential energy, kinetic energy, and/or nonconservative work (which could include something more specific, like mechanical energy lost in the form of heat or sound waves). The question is asking which type of energy (such as potential energy) is being converted into which type of energy (such as kinetic energy).

31. Look at the relevant formula for this type of potential energy in Sec. 5.3. The question is related to the two different equations for gravitational potential energy and the two different ways of defining a reference point.

32. Is the work done by the specified force path-dependent or path-independent? See Sec.'s 5.2, 5.3, 5.5, and 5.6.

33. Review the definition of gravitational potential energy in Sec. 5.3, the definition of the gradient in Sec. 5.5, and equilibrium positions and their stability in Sec. 5.5.

## Chapter 5 – Hints to Practice Problems

1. Review Sec. 5.1. (A) Use the form of the scalar product that involves the components of the vectors. (B)/(C) Use the equations to solve for the magnitude and direction of a vector in terms of its components. Note that the angle is measured counterclockwise from the $+x$-axis; when taking the inverse tangent, add $180°$ if the $x$-component of the vector is negative. (D) Looking at the two angles, this should be easy. (E) Use the form of the scalar product that involves the angle between the vectors.
2. Three of the X-lines (not the same as edges!) meet at each corner, and all three of the X-lines meeting at a corner are parts of different X's (since they lie on different faces). The answer is not $90°$. Setup a coordinate system and express two of the lines from different X's that meet at one corner as vectors using unit vector notation. Apply both formulas for the scalar product – in Sec. 5.1 – in order to solve for the angle between these two vectors.
3. Review Sec. 5.1. Use the form of the scalar product that involves the components of the vectors. It may help to first express each of the 2D polar unit vectors in terms of Cartesian unit vectors. See Sec. 2.4.
4. Review Sec. 5.1. Use the form of the scalar product that involves the components of the vectors. It may help to first express each of the spherical or cylindrical unit vectors in terms of Cartesian unit vectors. See Sec.'s 2.4-2.5. In $\hat{\mathbf{r}} \cdot \hat{\mathbf{r}}_c$, you must realize that $\theta_c = \varphi_s$. That is, the angle $\theta$ of cylindrical coordinates is defined the same way as the angle $\varphi$ of spherical coordinates. When mixing the two coordinate systems, as in $\hat{\mathbf{r}} \cdot \hat{\mathbf{r}}_c$, don't make the mistake of thinking that the two $\theta$'s are the same. Instead, set $\theta_c = \varphi_s$ and distinguish between $\theta_c$ and $\theta_s$.
5. Review the definition of work in Sec. 5.2 and the scalar product in Sec. 5.1. (A)(i)-(iii) In each case, draw and label the direction of the specified force and the direction of the displacement in order to determine the angle between them. (iv) It's not a force if its SI unit is not a Newton. (v) A certain combination of forces is implied when a specific force is not mentioned. Read Sec. 5.2. carefully. (B) (i) Use the form of the scalar product that involves the components of the vectors. (ii)/(iii) Use the equations to solve for the magnitude and direction of a vector in terms of its components. Note that the angle is measured counterclockwise from the $+x$-axis; when taking the inverse tangent, add $180°$ if the $x$-component of the vector is negative. (iv) Review the definition of average power in Sec. 5.2.
6. Review the definition of work in Sec. 5.2 and the scalar product in Sec. 5.1. (A) Use the form of the scalar product that involves the components of the vectors. (B) Study the work integral in Sec. 5.2, including the derivations and examples that apply it. Note that the argument of the trig funtion must be in radians. (C) Review the definition of instantaneous power in Sec. 5.2.
7. Review the definition of work in Sec. 5.2. (A)/(B)/(D) The work integral should be simple for these constant forces. However, finding the displacement requires first applying Newton's second law to find the acceleration and then using the 1DUA strategy. When using the work integral, be sure to use the specified force and the correct angle. Draw and label the direction of the specified force and the direction of the displacement in order to determine the angle between them. (C) This question is simpler since you are given the displacement. Which force should you use? Draw and label the direction of the specified force and the direction of the displacement in order to determine the angle between them. (E) Review the definition of average power in Sec. 5.2. Draw and label the direction of the specified force and the direction of the displacement in order to determine the angle between them.

Hints

8. Review the definition of work in Sec. 5.2. In each case, draw and label the direction of the specified force and the direction of the displacement in order to determine the angle between them. Remember to negate the work to find the fun. (C) This question is asking about the net fun, corresponding to the net work. Which force do you use to find the net work? (D) Review the definition of average power in Sec. 5.2. It may be necessary to apply Newton's second law to find the acceleration and then use the 1DUA strategy.

9. Review the definition of work in Sec. 5.2 and the scalar product in Sec. 5.1. (A) Use the form of the scalar product that involves the components of the vectors. (B) Study the work integral in Sec. 5.2, including the derivations and examples that apply it. (C) Review the definitions of instantaneous and average power in Sec. 5.2.

10. Review the definition of work in Sec. 5.2. In each case, draw and label the direction of the specified force and the direction of the displacement in order to determine the angle between them. (D) This question is asking about the net play, corresponding to the net work. Which force do you use to find the net work? (E) Review the definition of average power in Sec. 5.2.

11. Review the definition of work in Sec. 5.2 and the scalar product in Sec. 5.1. (A) Use the form of the scalar product that involves the components of the vectors. (B)/(C)/(D) In each case, draw and label the direction of the specified force and the direction of the displacement in order to determine the angle between them.

12. Review the definition of work in Sec. 5.2. In each case, draw and label the direction of the specified force and the direction of the displacement in order to determine the angle between them. (C) Which force is nonconservative? (D) Review the definition of average power in Sec. 5.2. It may be necessary to apply Newton's second law to find the acceleration and then use the 1DUA strategy.

13. Review the definition of work in Sec. 5.2 and the scalar product in Sec. 5.1. (A) Each term must have the same SI units as force. (B) Study the work integral in Sec. 5.2, including the derivations and examples that apply it. (C) Is the work done by the specified force path-dependent or path-independent? See Sec.'s 5.2, 5.3, 5.5, and 5.6.

14. (A) Study the work integral in Sec. 5.2, including the derivations and examples that apply it. (B) Review the definition of instantaneous power in Sec. 5.2.

15. (A) Review the definition of instantaneous power in Sec. 5.2. Convert your answer to horsepower. (B) Study the work integral in Sec. 5.2, including the derivations and examples that apply it.

16. You could use the work-energy theorem to solve this problem efficiently. However, many students who apply the work-energy theorem to other problems where potential energy is changing usually make major conceptual and/or strategic mistakes in their solutions. For this reason, it's a good habit to apply the strategy for conservation of energy described in Sec. 5.4 whenever it applies.

17. (A) Choose a reference point. Use the relevant formula for this type of potential energy in Sec. 5.3. (B) Use the formula for kinetic energy in Sec. 5.3. (C) Write down the general starting equation for conservation of energy in Sec. 5.4 (in terms of potential energies, kinetic energies, and nonconservative work). Determine which of these are zero (if any), conceptually, and solve for the final kinetic energy. (D) Use your answer to part (B) and the formula for kinetic energy in Sec. 5.3. (E) Note that this is related to the change in potential energy in part (A). (F) Review the definition of work in Sec. 5.2. (G) Review the definition of average power in Sec. 5.3.

18. (A) Use the strategy for conservation of energy. Study the examples in Sec. 5.4. Choose a reference point from which to measure the heights. Determine which types of energy (i.e. potential or kinetic) there are initially and finally, and whether or not any nonconservative work is done. Use the correct

formula(s) for potential energy. (B) Recall that $v_x$ is constant throughout the motion of a projectile (and equal to $v_{x0}$). What does $v_y$ equal at the top? This is actually a very simple problem, once you understand the projectile motion details, conceptually: You just need to use the equation that relates $v_{x0}$ to the initial speed.

19. (A) Find the nonconservative work, which in this problem is the work done by the friction force. Review the definition of work in Sec. 5.2. Draw and label the direction of the specified force and the direction of the displacement in order to determine the angle between them. (B) Use the strategy for conservation of energy. Study the examples in Sec. 5.4. Choose a reference point from which to measure the heights. Determine which types of energy (i.e. potential or kinetic) there are initially and finally, and whether or not any nonconservative work is done. Use the correct formula(s) for potential energy.

20. Use the strategy for conservation of energy. Study the examples in Sec. 5.4. Measure the heights relative to the specified reference point. Determine which types of energy (i.e. potential or kinetic) there are initially and finally. Use the correct formula(s) for potential energy. Express the nonconservative work as a percentage of the total initial energy.

21. Use the strategy for conservation of energy. Study the examples in Sec. 5.4. Measure the heights relative to the specified reference point. Determine which types of energy (i.e. potential or kinetic) there are initially and finally. Use the correct formula(s) for potential energy. Express the nonconservative work as a percentage of the total initial energy.

22. (A) Use the strategy for conservation of energy. Study the examples in Sec. 5.4. Choose a reference point from which to measure the heights. Determine which types of energy (i.e. potential or kinetic) there are initially and finally, and whether or not any nonconservative work is done. Use the correct formula(s) for potential energy. (B) Write down the general starting equation for conservation of energy in Sec. 5.4 (in terms of potential energies, kinetic energies, and nonconservative work). Determine which of these are zero (if any), conceptually, and solve for the nonconservative work. Note that either the initial or final position (or both) must change between parts (A) and (B). (C) Use your answer to part (B) and review the definition of the work done by friction in Sec. 5.2. Draw and label the direction of the specified force and the direction of the displacement in order to determine the angle between them.

23. (A) Conceptually, at which point is the speed maximum? Use the strategy for conservation of energy. Study the examples in Sec. 5.4. Choose a reference point from which to measure the heights. Determine which types of energy (i.e. potential or kinetic) there are initially and finally, and whether or not any nonconservative work is done. Use the correct formula(s) for potential energy. (B) Write down the general starting equation for conservation of energy in Sec. 5.4 (in terms of potential energies, kinetic energies, and nonconservative work). Determine which of these are zero (if any), conceptually, and solve for the nonconservative work. Note that either the initial or final position (or both) must change between parts (A) and (B). Also find the nonconservative work by direct integration. Study the work integral in Sec. 5.2, including the derivations and examples that apply it. Combine the two methods of finding the nonconservative work together.

24. Except for part (G), your answers will be symbolic. (A) Use the correct formula for potential energy from Sec. 5.3. (B) Use the formula for kinetic energy from Sec. 5.3. (C) Use the correct formula for potential energy from Sec. 5.3. Express its height above the ground in terms of $R_0$ and $\theta$, using trig. (D) Use the strategy for conservation of energy. Study the examples in Sec. 5.4. Choose a reference point from which to measure the heights. Determine which types of energy (i.e. potential or kinetic) there are initially and finally, and whether or not any nonconservative work is done. (E) Draw a FBD and apply

Hints

Newton's second law. Work with tangential and centripetal components. Recall the formula for centripetal acceleration from Sec. 4.2. Use your answer to part (D) to eliminate the speed. (F) This is just like part (E), except that the normal force is now zero and you are solving for the speed. (G) The normal force equals zero when the monkey loses contact with the igloo. Therefore, you can substitute your speed from part (F) into conservation of energy – part (C) – in order to solve for the critical angle.

25. Except for part (E), your answers will be symbolic. (A) One of these should be very easy; the other involves drawing a diagram and applying trig. There is a similar problem in the examples of Sec. 5.4 (B) One of these should be very easy; the other involves applying the strategy for conservation of energy. Study the examples in Sec. 5.4. Choose a reference point from which to measure the heights. Determine which types of energy (i.e. potential or kinetic) there are initially and finally, and whether or not any nonconservative work is done. Since conservation of energy gives you speed, whereas you are solving for velocity, you will need to include an appropriate unit vector to get velocity from speed. (C) Find the tangential and centripetal components of acceleration before finding the total acceleration. Draw a FBD in each position, label the tangential and centripetal directions, and apply Newton's second law. Recall the formula for centripetal acceleration from Sec. 4.2. One or more of the initial or final components may be zero. If one of the components involves speed, use your result from part (B) to eliminate it. Include an appropriate unit vector with the total acceleration. (D) Draw a FBD in each position, label the tangential and centripetal directions, and apply Newton's second law. Recall the formula for centripetal acceleration from Sec. 4.2. If one of your answers involves a nonzero acceleration, use your result from part (C) to eliminate it. Include an appropriate unit vector with your final answers, since tension is a vector.

26. Use the strategy for conservation of energy. Study the examples in Sec. 5.4. Choose a reference point from which to measure the heights; also setup a coordinate system for the spring. Determine which types of energy (i.e. potential or kinetic) there are initially and finally, and whether or not any nonconservative work is done. Use the correct formula(s) for potential energy. One of the three answers is conceptual: One answer you should know; two you will solve for.

27. Use the strategy for conservation of energy. Study the examples in Sec. 5.4. Choose a reference point from which to measure the heights; also setup a coordinate system for the spring. Determine which types of energy (i.e. potential or kinetic) there are initially and finally, and whether or not any nonconservative work is done. Use the correct formula(s) for potential energy. (B) Draw a FBD, label the tangential and centripetal directions, and apply Newton's second law. Recall the formula for centripetal acceleration from Sec. 4.2. Use the strategy for conservation of energy to determine the speed of the box at the top of the loop.

28. Use the strategy for conservation of energy. Study the examples in Sec. 5.4. Choose a reference point from which to measure the heights; also setup a coordinate system for the spring. Determine which types of energy (i.e. potential or kinetic) there are initially and finally, and whether or not any nonconservative work is done. Use the correct formula(s) for potential energy. (A) Use the formula for potential energy in Sec. 5.3. (C) Draw a FBD, label the tangential and centripetal directions, and apply Newton's second law. Recall the formula for centripetal acceleration from Sec. 4.2. Use the strategy for conservation of energy to determine the speed of the box at the top of the loop. (D) The normal force will be zero if the box falls away from the track. Draw a FBD, label the tangential and centripetal directions, and apply Newton's second law. Recall the formula for centripetal acceleration from Sec. 4.2. Set the normal force equal to zero and solve for the speed. Use the strategy for conservation of energy along with the speed of the box at the top of the loop in order to determine what the total final

energy of the box needs to be. Compare with the total initial energy that the box actually has in the problem.

29. Use the strategy for conservation of energy. Study the examples in Sec. 5.4. Choose a reference point from which to measure the heights; also setup a coordinate system for the spring. Determine which types of energy (i.e. potential or kinetic) there are initially and finally, and whether or not any nonconservative work is done. Use the correct formula(s) for potential energy. (B) Realize that you must account for the unknown distance traveled from equilibrium to fully compressed in the gravitational potential energy, in addition to the spring potential energy. Both contributions to potential energy from this distance can be expressed in terms of the same variable, $x$. You will end up with a quadratic equation in the variable $x$.

30. Use the strategy for conservation of energy. Study the examples in Sec. 5.4. Choose a reference point from which to measure the heights; also setup a coordinate system for the spring. Determine which types of energy (i.e. potential or kinetic) there are initially and finally. Use the correct formula(s) for potential energy. Express the nonconservative work as a percentage of the initial spring energy.

31. Use the strategy for conservation of energy. Study the examples in Sec. 5.4. Choose a reference point from which to measure the heights; also setup a coordinate system for the spring. Determine which types of energy (i.e. potential or kinetic) there are initially and finally, and whether or not any nonconservative work is done. Use the correct formula(s) for potential energy. (C) Draw a FBD and apply Newton's second law while it is sliding back and forth. You should have four forces in your FBD (since the cord has been cut, there is no tension at this point). The formula for static friction has an inequality, which is why you are solving for a minimum coefficient.

32. (A) Draw a FBD and apply Newton's second law. You should have four forces in your FBD. Treat the compression/stretch of each spring separately. (B) Use the strategy for conservation of energy. Study the examples in Sec. 5.4. Choose a reference point from which to measure the heights; also setup a coordinate system for the spring. Determine which types of energy (i.e. potential or kinetic) there are initially and finally, and whether or not any nonconservative work is done. Use the correct formula(s) for potential energy. Treat the compression/stretch of each spring separately.

33. (A) Use the satellite strategy of Sec. 4.6. It should be easy to determine the period of the satellite, since it is Banana-syncrhonous. Solve for the orbital radius of the satellite and the speed of the satellite in the initial orbit. Double the radius and solve for the speed of the satellite in the final orbit (still applying the satellite strategy). (B) Use the strategy for conservation of energy. Study the examples in Sec. 5.4. Which point do you need to use as the reference point? Determine which types of energy (i.e. potential or kinetic) there are initially and finally, and whether or not any nonconservative work is done. Use the correct formula(s) for potential energy. In this problem, you know all of the potential and kinetic energies, and you're solving for the non-conservative work that must be done. (C) Use the strategy for conservation of energy. Study the examples in Sec. 5.4. Which point do you need to use as the reference point? Determine which types of energy (i.e. potential or kinetic) there are initially and finally, and whether or not any nonconservative work is done. Use the correct formula(s) for potential energy. Where is the final position? Review the concept of escape speed in Sec. 5.4.

34. Use the strategy for conservation of energy. Study the examples in Sec. 5.4. Which point do you need to use as the reference point? Determine which types of energy (i.e. potential or kinetic) there are initially and finally, and whether or not any nonconservative work is done. Use the correct formula(s) for potential energy. Where is the final position? Review the concept of escape speed in Sec. 5.4. You

Hints

may need to look up relevant astronomical data. Note that there are two initial and final potential energy terms – one for the earth and one for the moon.

35. (A) Review the definition of work in Sec. 5.2 and the scalar product in Sec. 5.1. (B) Review the relationship between potential energy and conservative forces in Sec. 5.5. (C) Review the definition of instantaneous power in Sec. 5.2. (D) Review equilibrium positions and their stability in Sec. 5.5. (E) Review Sec.'s 5.5-5.6.

36. Review Sec.'s 5.5-5.6.

37. Review Sec.'s 5.5-5.6.

38. Review Sec.'s 5.5.-5.6. There is a simple test that you can perform to determine whether or not a force is conservative. However, it may be useful to first rewrite the force in Cartesian coordinate (see Sec. 2.4).

39. Review Sec.'s 5.5-5.6.

40. Review Sec.'s 5.5-5.6.

41. Review Sec.'s 5.5-5.6.

42. Review Sec.'s 5.5-5.6.

## Chapter 6 – Hints to Conceptual Questions

1. (A) Review the definitions of these averages in Sec. 2.3. Which quantity is integrated over in these definitions? (B) Review the examples in Sec. 1.2.

2. Consider, for example, uppercase letters A, B, C… You don't have to go too far to find a letter with a shape where the center of mass clearly lies outside of the material in which the shape is made.

3. Review the definitions of center of mass and center of gravity in Sec. 6.1. Observe that electric charge and electric field are analogous to mass and gravitational field.

4. Does the ice cube sink or float? Is the solid state usually more dense or less dense than the liquid state of a substance? Does water expand or contract when it freezes? What problem might this have for water in pipes when it is so cold that water freezes?

5. The second variable will have full range, while the upper and/or lower limit of the first variable may be a function of the second variable. Draw a picture in each case, and draw the range of the first variable for a representative value. Review the technique of double integrals in Sec. 6.1 and study the examples.

6. If you put the rectangular cookie with the circular cutout and the circular cookie that was cutout together, you make the original rectangular cookie. You can write an equation for the center of mass of the original rectangular cookie in terms of the center of mass of the two pieces that make it, and then solve for the center of mass of the rectangular cookie with the circular cutout. Try working through the algebra. First, setup a coordinate system. Review the strategy for finding the center of mass for a discrete system in Sec. 6.1. You will have to find the ratio of the masses (of the three shapes) by finding the ratio of the areas.

7. (A) The mass of each shape is proportional to the area of the shape. (B) Subtract.

8. Review the definition of impulse in Sec. 6.2 and study the examples.

9. Review the conditions for which the total momentum of a system is conserved in Sec. 6.3 and study the examples. Draw FBD's for the system to help determine whether or not the net external force acting on the system is zero. If the net external force is nonzero, are there any components for which the sum of those components is zero?

10. Review the conditions for which the total momentum of a system is conserved in Sec. 6.3 and study the examples. Draw FBD's for the system to help determine whether or not the net external force acting on the system is zero. If the net external force is nonzero, are there any components for which the sum of those components is zero?

11. Review the conditions for which the total momentum of a system is conserved in Sec. 6.3 and study the examples. Draw FBD's for the system to help determine whether or not the net external force acting on the system is zero. If the net external force is nonzero, are there any components for which the sum of those components is zero? Review Newton's three laws of motion in Sec. 3.1.

12. Review the strategy for conservation of momentum in Sec. 6.5. Although it may not seem like a collision, the answer to this question actually relates to one of the types of collisions described in Sec. 6.5. (You can also determine what the monkey needs to do using one of Newton's laws of motion, but this question specifically asks you to apply the principle of conservation of momentum instead.)

13. Review the strategy for conservation of momentum in Sec. 6.5. Although it may not seem like a collision, the answer to this question actually relates to one of the types of collisions described in Sec. 6.5. Also, review the motion of the center of mass of a system of objects in Sec. 6.4.

14. Review the strategy for 1D collisions in Sec. 6.5, including the definitions of different types of collisions. Study the examples. Also, review the motion of the center of mass of a system of objects in Sec. 6.4.

15. Review the strategy for 1D collisions in Sec. 6.5 and study the examples.

16. Review the strategy for 2D collisions in Sec. 6.6 and study the examples. Remember to account for the system's initial momentum.

17. Review the strategy for 2D collisions in Sec. 6.6 and study the examples. Note that throwing the watermelon south will not produce the desired result; you must account for the system's initial momentum.

18. Use the relative velocity formula toward the end of Sec. 2.6. This is a vector addition (or subtraction) problem.

19. Review the strategy for 1D collisions in Sec. 6.5 and study the examples. What happens in a 1D elastic collision between two objects of equal mass?[347]

20. Review the ballistic pendulum in Sec. 6.7 and study the derivations. Also, review the conditions for which the total momentum of a system is conserved in Sec. 6.3 and study the examples. Draw FBD's for the system to help determine whether or not the net external force acting on the system is zero. What type of collision occurs at the bottom? Review the definitions of different types of collisions in Sec. 6.5. What problem-solving strategy applies to the upward swing of the ballistic pendulum after the collision?

21. Why did most of the alpha particles pass straight through, virtually undeflected? What evidently happens on the rare occasion where the alpha particle deflects with a significant scattering angle? Which types of objects would an alpha particle interact with? Put these ideas together to determine where the positive charge and most of the mass of the metal foil must reside.

---

[347] You might search for 'Newton's cradle' with your favorite search engine.

Hints

22. According to the formula, do longer or shorter wavelengths of light scatter more? Which colors scatter more? Which colors take a more direct route, on average, through earth's atmosphere, and which colors scatter more, on average, in earth's atmosphere? The answers to this question can explain both of the questions asked in the problem. Draw a picture to help see what is significantly different about the path that sunlight takes through the atmosphere at sunrise/sunset versus noon. (It has to do with distance, not angle.)
23. The trailer has inertia. Are the truck and trailer affected the same way by gravity?
24. (A) The dirt has inertia. (B) Use the principle of conservation of momentum in Sec. 6.3. Alternatively, you can use one of Newton's laws of motion.
25. Review the strategies for 1D and 2D collisions in Sec.'s 6.5-6.6 and study the examples. Review Conceptual Questions 15-17, including the hints.
26. Review rocket motion in Sec. 6.9 and study the examples. Review the definition of payload and the math that follows.
27. Review rocket motion in Sec. 6.9 and study the examples. Review the definition of thrust. Draw a FBD and apply Newton's second law.
28. Review rocket motion in Sec. 6.9 and study the examples. Review the definitions of payload and thrust. Draw a FBD and apply Newton's second law.

## Chapter 6 – Hints to Practice Problems

1. Review the formula for the center of mass for a discrete system. Study the examples in Sec. 6.1. Setup a coordinate system.
2. Review the formula for the center of mass for a discrete system. Study the examples in Sec. 6.1.
3. Review the formula for the center of mass for a discrete system. Study the examples in Sec. 6.1.
4. Review the formula for the center of mass for a discrete system. Study the examples in Sec. 6.1. First determine the coordinates of the center of mass of each cookie.
5. Review the formula for the center of mass for a discrete system. Study the examples in Sec. 6.1. Setup a coordinate system. Setup two equations in two unknowns – using the information given about the combined mass – and solve the system.
6. Review the formula for the center of mass for a discrete system. Study the examples in Sec. 6.1. Setup a coordinate system. Treat the weight of the plank as if all of its weight acts effectively at the center of the plank.
7. Review the formula for the center of mass for a discrete system. Study the examples in Sec. 6.1. Setup a coordinate system. Treat the weight of the platform as if all of its weight acts effectively at the center of the platform.
8. Review the formula for the center of mass for a discrete system. Study the examples in Sec. 6.1. First divide the shape up into squares or rectangles. Determine the coordinates of the center of mass of each square or rectangle. The masses of the squares or rectangles are proportional to their areas.
9. Review the formula for the center of mass for a discrete system. Study the examples in Sec. 6.1. First divide the shape up into squares or rectangles. Determine the coordinates of the center of mass of each square or rectangle. The masses of the squares or rectangles are proportional to their areas.

10. Review the formula for the center of mass for a discrete system. Study the examples in Sec. 6.1. First, determine the coordinates of the center of mass of each shape. The masses of the shapes are proportional to their areas.

11. Review the formula for the center of mass for a discrete system. Study the examples in Sec. 6.1. It is convenient to divide the mallet up into three pieces (not two) as follows: the portion of the handle to the left of the fulcrum, the portion of the handle to the right of the fulcrum, and the head. Determine the coordinates of the center of mass of each piece. Write down one equation for the center of mass, and plug in all of the knowns. Write down an equation for the sum of the three masses. Lastly, write down an equation that relates the mass of the left portion of the handle to the right portion of the handle. Solve this system of three equations in three unknowns. This problem is reconsidered, conceptually, in Sec. 7.6.

12. Review the formula for the center of mass for a discrete system. Study the examples in Sec. 6.1. If you add the funny shape of the pie to the circular pie that was cutout together, you make the original circular pie. Write down an equation for the center of mass of the original circular pie in terms of the center of mass of the two pieces that make it up, and then solve for the center of mass of the funny shape (which is on the right side of the equation; you should know the center of mass of the original circular pie, on the left side, numerically). You will have to find the ratio of the masses (of the three shapes) by finding the ratio of the areas.

13. This is like the previous question, except that when you add the given funny shape together with the cutout, it doesn't make the original large circle – instead, it makes the original large circle plus the small circle combined together.

14. Review the formula for the center of mass integral. Study the integrals in Sec. 6.1. Use the appropriate density. Eliminate the constant $a$ from the center of mass by solving for the total mass by integrating over $dm$.

15. Review the formula for the center of mass integral. Study the integrals in Sec. 6.1. Use the appropriate density. Find the total mass by integrating over $dm$. Draw the curves. The region is 2D. You will need to perform a double integral. The second variable will have full range, while the upper and/or lower limit of the first variable may be a function of the second variable. Draw the range of the first variable for a representative value.

16. Review the formula for the center of mass integral. Study the integrals in Sec. 6.1. Use the appropriate density. Work with 2D polar coordinates; write the needed differential element appropriately for this choice of coordinates. Eliminate the constant $\beta$ from the center of mass by solving for the total mass by integrating over $dm$. Draw and label a diagram. The region is 2D. You will need to perform a double integral. The second variable will have full range, while the upper and/or lower limit of the first variable may be a function of the second variable. Draw the range of the first variable for a representative value.

17. Review the formula for the center of mass integral. Study the integrals in Sec. 6.1. Use the appropriate density. Work with 2D polar coordinates; write the needed differential element appropriately for this choice of coordinates. Eliminate the constant $\beta$ from the center of mass by solving for the total mass by integrating over $dm$. The region is 2D. You will need to perform a double integral. The second variable will have full range, while the upper and/or lower limit of the first variable may be a function of the second variable. Draw the range of the first variable for a representative value.

Hints

18. Review the formula for the center of mass integral. Study the integrals in Sec. 6.1. Use the appropriate density. Work with 2D polar coordinates; write the needed differential element appropriately for this choice of coordinates. Eliminate the constant $\beta$ from the center of mass by solving for the total mass by integrating over $dm$. Draw and label a diagram. The region is 1D.

19. Review the formula for the center of mass integral. Study the integrals in Sec. 6.1. Use the appropriate density. The region is 2D. You will need to perform a double integral. The second variable will have full range, while the upper and/or lower limit of the first variable may be a function of the second variable. Draw the range of the first variable for a representative value. The mass is proportional to the area; you can work out this ratio once you find the center of mass.

20. (A)/(B) Use the strategy for conservation of energy. Study the examples in Sec. 5.4. Choose a reference point from which to measure the heights. Determine which types of energy (i.e. potential or kinetic) there are initially and finally, and whether or not any nonconservative work is done. Use the correct formula(s) for potential energy. Review the definition of momentum in Sec. 6.2. In part (A), both the initial and final positions are before the collision, whereas in part (B), both the initial and final positions are after the collision. (C) Review the definition of impulse in Sec. 6.2 and study the examples. Signs are important. (D) Now conserve energy with the initial position prior to the collision and the final position after the collision, and solve for the nonconservative work.

21. (A) Review the conditions for which the total momentum of a system is conserved in Sec. 6.3 and study the examples. Draw FBD's for the system to help determine whether or not the net external force acting on the system is zero. (B) Review the strategy for 1D collisions in Sec. 6.5, including the definitions of different types of collisions. Study the examples. Signs are important. Which type of collision is this, and which quantities are conserved or not conserved for such a collision? (C) Calculate the total initial and total final kinetic energy of the system.

22. (A) Review the strategy for 1D collisions in Sec. 6.5 and study the examples. Signs are important. Which type of collision is this, and which quantities are conserved or not conserved for such a collision? (B) Review the strategy for 2D collisions in Sec. 6.5 and study the examples. Signs are important. Which type of collision is this, and which quantities are conserved or not conserved for such a collision?

23. Review the strategy for 1D collisions in Sec. 6.5 and study the examples. Signs are important. Which type of collision is this, and which quantities are conserved or not conserved for such a collision? Is there a special equation that applies to this collision?

24. Review the strategy for 1D collisions in Sec. 6.5 and study the examples. Signs are important. You don't know what type of collision this is, so you should only write down a conservation law for the one quantity that is conserved for all types of collisions. After you solve parts (A) and (B), then you can see whether the other quantity is conserved for this collision, in order to determine which type of collision it is. (D) Review the definition of impulse in Sec. 6.2 and study the examples. Signs are important.

25. (A) Review the strategy for 1D collisions in Sec. 6.5 and study the examples. Signs are important. Which type of collision is this, and which quantities are conserved or not conserved for such a collision? Is there a special equation that applies to this collision? (B) Review the strategy for 1D collisions in Sec. 6.5 and study the examples. Signs are important. Which type of collision is this, and which quantities are conserved or not conserved for such a collision? Calculate the total initial and total final kinetic energy of the system.

26. (A) Review the strategy for 1D collisions in Sec. 6.5 and study the examples. Signs are important. Which type of collision is this, and which quantities are conserved or not conserved for such a collision? Is there a special equation that applies to this collision? (B) There could be a third collision if the middle

orangutan travels to the left (after the second collision) faster than the first orangutan travels to the left (after the first collision). (C) Review the strategy for 1D collisions in Sec. 6.5 and study the examples. Signs are important. Which type of collision is this, and which quantities are conserved or not conserved for such a collision? Calculate the total initial and total final kinetic energy of the system.

27. Review the strategy for 2D collisions in Sec. 6.6 and study the examples. Signs are important. Which type of collision is this, and which quantities are conserved or not conserved for such a collision? For the conservation of any vector quantity, you must write down two equations – one for each component.

28. Review the strategy for 1D collisions in Sec. 6.5 and study the examples. Signs are important. Which type of collision is this, and which quantities are conserved or not conserved for such a collision? Is there a special equation that applies to this collision?

29. (A)/(B) Review the strategy for 1D collisions in Sec. 6.5, including the definitions of different types of collisions. Study the examples. Signs are important. Which type of collision is this, and which quantities are conserved or not conserved for such a collision? Note that the monkey's shoes are part of the monkey's original mass. (C) Review the definition of impulse in Sec. 6.2 and study the examples. Signs are important.

30. (A) Review the strategy for 1D collisions in Sec. 6.5, including the definitions of different types of collisions. Study the examples. Signs are important. Which type of collision is this, and which quantities are conserved or not conserved for such a collision? (B) Review the strategy for 1D collisions in Sec. 6.5 and study the examples. Signs are important. Which type of collision is this, and which quantities are conserved or not conserved for such a collision? Is there a special equation that applies to this collision?

31. Review the strategy for 2D collisions in Sec. 6.6, including the definitions of different types of collisions. Study the examples. Signs are important. Which type of collision is this, and which quantities are conserved or not conserved for such a collision? For the conservation of any vector quantity, you must write down two equations – one for each component.

32. Review the strategy for 2D collisions in Sec. 6.6, including the definitions of different types of collisions. Study the examples. Signs are important. Which type of collision is this, and which quantities are conserved or not conserved for such a collision? For the conservation of any vector quantity, you must write down two equations – one for each component.

33. (A) Review the strategy for 2D collisions in Sec. 6.6, including the definitions of different types of collisions. Study the examples. Signs are important. Which type of collision is this, and which quantities are conserved or not conserved for such a collision? For the conservation of any vector quantity, you must write down two equations – one for each component. (B) Calculate the total initial and total final kinetic energy of the system. (C) Review the definition of impulse in Sec. 6.2 and study the examples. Signs are important.

34. The underlying strategy is very similar to the ballistic pendulum in Sec. 6.7. What is conserved for the swing downward? What type of collision occurs at the bottom? What is conserved for the collision? What is conserved for the swing upward? Write down a conservation law for the swing downward, the collision, and the swing upward. Be sure to distinguish between the speed just before the collision and the speed just after the collision. Combine the equations together algebraically.

35. The underlying strategy is very similar to the ballistic pendulum in Sec. 6.7. What is conserved for the swing downward? What type of collision occurs at the bottom? What is conserved for the collision? What is conserved for the swing upward? Write down a conservation law for the swing downward for

each monkey, the collision, and the swing upward. Be sure to distinguish between the speed just before the collision and the speed just after the collision. Combine the equations together algebraically.

36. Review the strategy for 1D collisions in Sec. 6.5, including the definitions of different types of collisions. Study the examples. Signs are important. Which type of collision is this, and which quantities are conserved or not conserved for such a collision? Review the definitions of the lab frame and CM frame in Sec. 6.6.

37. Review the strategy for 2D collisions in Sec. 6.6, including the definitions of different types of collisions. Study the examples. Signs are important. Which type of collision is this, and which quantities are conserved or not conserved for such a collision? For the conservation of any vector quantity, you must write down two equations – one for each component. Review the definitions of the lab frame, the CM frame, and the impact parameter. Although the collision is elastic, it is not 1D, so the velocity equation does not apply.

38. Review the strategy for 2D collisions in Sec. 6.6, including the definitions of different types of collisions. Study the examples. Signs are important. Which type of collision is this, and which quantities are conserved or not conserved for such a collision? For the conservation of any vector quantity, you must write down two equations – one for each component. Although the collision is elastic, it is not 1D, so the velocity equation does not apply. Use the equations given in the problem to express the momentum and energy of the incident and scattered photons in terms of wavelength. Use the equation given in the problem to relate the energy and momentum of the electron. Combine equations (from the conservation laws) together algebraically to derive the specified equation.

39. Review the strategy for 1D collisions in Sec. 6.5, including the definitions of different types of collisions. Study the examples. Signs are important. Which type of collision is this, and which quantities are conserved or not conserved for such a collision? Study the first example of Sec. 6.9.

40. Review the rocket propulsion strategy of Sec. 6.9 and study the examples.

41. Review the rocket propulsion strategy of Sec. 6.9 and study the examples.

## Chapter 7 – Hints to Conceptual Questions

1. Review these definitions in Sec.'s 4.1, 4.2, and 7.1. Which of these quantities depend on the radius, and, of those that do, are they directly or inversely dependent upon the radius? Make sure that you express each quantity in terms of the radius and/or quantities that are the same for both fleas.

2. Review the definitions of these three types of acceleration in the prelude to Chapter 7.

3. Review the definition of moment of inertia in Sec. 7.2, paying special attention to how to interpret moment of inertia conceptually. Study the conceptual examples. Think about how far the mass is from the axis of rotation, on average, in each case.

4. Review the definition of moment of inertia in Sec. 7.2, paying special attention to how to interpret moment of inertia conceptually. Study the conceptual examples. Think about how far the mass is from the axis of rotation, on average, in each case.

5. Review the definition of moment of inertia in Sec. 7.2, paying special attention to how to interpret moment of inertia conceptually. Study the conceptual examples. Think about how far the mass is from the axis of rotation, on average, in each case.

6. Review the definitions of moment of inertia and radius of gyration in Sec. 7.2, paying special attention to how to interpret moment of inertia conceptually. Study the examples. Think about how far the mass is from the axis of rotation, on average, in each case.

7. Review the definition of the vector product in Sec. 7.3 – especially, the derivation of the area of a parallelogram. Review the definition of the scalar product in Sec. 5.1. Draw a parallelepiped. The vector product is related to the area of one of the faces. The scalar product relates to the height – i.e. the direction perpendicular to one of the faces. The vector product and scalar product must both be combined together to make the volume.

8. Study the right-hand rule in Sec. 7.3, including the conceptual examples. Find the footnote in Sec. 2.4 regarding right-handed coordinate systems.

9. Study the right-hand rule and vector products in Sec. 7.3, including the conceptual examples. There you can also learn what the arrows $\odot$ and $\otimes$ mean. You also need to find the relevant formulas for the vector products in Sec.'s 7.4, 7.5, and 7.8; the order of the vectors is very important.

10. See how each of these unit vectors is defined in an illustration in Sec. 2.5. Study the right-hand rule and vector products in Sec. 7.3, including the conceptual examples.

11. Study the tensor notation in Sec. 7.3, including the derivations and examples.

12. Review the definition of torque in Sec. 7.4. Study the examples. Pay special attention to how the distance $r_\perp$ and the angle $\theta$ are defined in the equation for the magnitude of the torque. Draw the direction of the specified force and draw the vector $\vec{r}_\perp$. When the force is the weight of an object, treat the weight as if all of the weight effectively acts at the center of mass the object.

13. Which screwdriver(s) provide a larger $r_\perp$ in the formula for torque? Identify the axis of rotation and measure outward from this axis to determine $r_\perp$. Review the definition of torque in Sec. 7.4. Study the examples. Pay special attention to how the distance $r_\perp$ is defined in the equation for the magnitude of the torque.

14. Note that the torques are balanced, not the weights. Review the definition of torque in Sec. 7.4. Study the examples. A similar example is considered in Sec. 7.6.

15. Unlike the previous question, this time the weights are balanced, but the torques are not. Review the definition of torque in Sec. 7.4. Study the examples. A similar, yet significantly different, example is considered in Sec. 7.6. Compare this question to the previous question.

16. See the note on inverting the vector product calculation in Sec. 7.3.

17. Study the strategy for conserving energy for a rigid body that rolls without slipping in Sec. 7.7, including the derivations and examples. Does the radius of the solid disc affect the final speed, or does it cancel out? Review the definition of moment of inertia in Sec. 7.2.

18. Study the strategy for conserving energy for a rigid body that rolls without slipping in Sec. 7.7, including the derivations and examples. Due to the different densities, the solid discs have the same radius, but different mass. Does the mass of the solid disc affect the final speed, or does it cancel out? Review the definition of moment of inertia in Sec. 7.2.

19. Study the strategy for conserving energy for a rigid body that rolls without slipping in Sec. 7.7, including the derivations and examples. How does the coefficient in the moment of inertia formula affect the distance that the rigid bodies roll up the incline? Since they are rolling up an incline, we're investigating which object loses speed less rapidly (unlike the case of rolling down an incline, where we're investigating which object gains speed more rapidly). Review the definition of moment of inertia in Sec. 7.2.

Hints

20. Study the strategy for conserving energy for a rigid body that rolls without slipping in Sec. 7.7, including the derivations and examples. How does the coefficient in the moment of inertia formula affect the final speed of the rigid bodies at the bottom of the incline? Review the definition of moment of inertia in Sec. 7.2.

21. Study the strategy for conserving energy for a rigid body that rolls without slipping in Sec. 7.7, including the derivations and examples. There are similar examples of filled and unfilled containers rolling down an incline in Sec. 7.7. Since they are rolling up an incline, we're investigating which object loses speed less rapidly (unlike the case of rolling down an incline, where we're investigating which object gains speed more rapidly). Review the definition of moment of inertia in Sec. 7.2.

22. Consult the table of angular analogies in Sec. 7.8.

23. You can find the correct expression in Sec. 7.5.

24. Review the conditions for the conservation of angular momentum in Sec. 7.9 and read the conceptual examples. Review the definition of torque in Sec. 7.4.

25. Review the conditions for the conservation of angular momentum in Sec. 7.9 and read the conceptual examples. Review the definition of torque in Sec. 7.4.

26. Review the conditions for the conservation of angular momentum in Sec. 7.9 and read the conceptual examples. Review the definition of torque in Sec. 7.4. Sum the various components of torque with a suitable choice of coordinates.

27. Review the strategy for conserving angular momentum in Sec. 7.9 and study the examples. Review the definition of moment of inertia in Sec. 7.2.

28. Review the conditions for the conservation of angular momentum in Sec. 7.9 and read the conceptual examples. Review the definition of torque in Sec. 7.4. Review the strategy for conserving angular momentum in Sec. 7.9 and study the examples. Review the definition of moment of inertia in Sec. 7.2.

29. Review the conditions for the conservation of angular momentum in Sec. 7.9 and read the conceptual examples. Review the definition of torque in Sec. 7.4. Review the conditions for conservation of momentum in Sec. 6.3. Contrast this rolling collision with the collisions of Sec. 6.5. Is there anything significantly different here which may affect which quantities are conserved?

30. Review the strategy for conserving angular momentum in Sec. 7.9 and study the examples.

31. Review the strategy for conserving angular momentum in Sec. 7.9 and study the examples.

## Chapter 7 – Hints to Practice Problems

1. Follow the strategy for uniform angular acceleration. Study the examples in Sec. 7.1. You know another quantity in addition to the three numbers clearly listed in the problem: What is the angular speed of the tire just as it topples over? (Why does it topple over?) Which symbol is 60 rev? What SI unit can you convert revolutions into? Which symbol do you need to solve for in part (A)? You may need to solve two equations in two unknowns.

2. Follow the strategy for uniform angular acceleration. Study the examples in Sec. 7.1. Which symbol is 36 rev? What SI unit can you convert revolutions into?

3. Follow the strategy for uniform angular acceleration. Study the examples in Sec. 7.1. Which symbol is 90 rev? What SI unit can you convert revolutions into?

4. Follow the strategy for uniform angular acceleration. Study the examples in Sec. 7.1. First, use the initial and final periods of the earth's rotation in order to find the initial and final angular speeds. Which symbol do you need to solve for in part (B)? What SI unit can be converted into revolutions?

5. Follow the strategy for uniform angular acceleration. Study the examples in Sec. 7.1. Note that the given quantities include both angular and tangential quantities; you can use the radius to get a corresponding tangential or angular quantity. Which symbol do you need to solve for in part (A)? What SI unit can be converted into revolutions? In part (D), first find the centripetal (see Sec. 4.2) and tangential components of the acceleration, and then use the formulas to find the magnitude and direction of a vector from its components. Note that the angle will come out relative to the tangent; when taking the inverse tangent, you may need to add 180° in order to get the vector into the right quadrant. Setup a coordinate system so that your angle makes sense.

6. Follow the strategy for uniform angular acceleration. Study the examples in Sec. 7.1. Which symbol do you need to solve for in part (E)? What SI unit can be converted into revolutions? For part (F), review centripetal acceleration in Sec. 4.2.

7. Follow the strategy for uniform angular acceleration. Study the examples in Sec. 7.1. Be careful when identifying the three unknowns. The minute hand is working normally when the exam starts, so it's initial angular speed is not zero! What is the usual angular speed of a minute hand? The time is unknown – if you knew that, you wouldn't be solving for it! The third known is the angular displacement. Draw a picture of the initial and final positions of the minute hand in order to determine what the angular displacement of the minute hand is. You can work in revolutions and hours (you don't have to convert to radians or seconds – and so you can thereby avoid many of the mistakes that students often make in these conversions).

8. Read the note and hints stated in the problem. Follow the strategy for uniform angular acceleration. Study the examples in Sec. 7.1. (A) Remember to use a number from part (ii) as well as numbers from part (i). (B) First find the centripetal (see Sec. 4.2) and tangential components of the acceleration – noting that the time is different than it was in part (A) – and then use the formulas to find the magnitude and direction of a vector from its components. Note that the angle will come out relative to the tangent; when taking the inverse tangent, you may need to add 180° in order to get the vector into the right quadrant. Setup a coordinate system so that your angle makes sense. (C) Which symbol do you need to solve for? What SI unit can be converted into revolutions? (D) Which symbol do you need to solve for? Note that part (ii) is UCM (so zero angular acceleration – not uniform angular acceleration). (E) One component of acceleration is zero, but one component is not. Find the nonzero component. Remember, part (ii) is UCM. (F)/(G) Remember to use a number from part (ii) in addition to numbers from part (iii). (H) First find the angular displacement for all three parts combined together. You can find the definition of average angular speed in Sec. 7.1.

9. The monkeyball spins about its axis with constant angular speed, while the center of mass of the monkeyball is in free fall. Use the equation for constant angular speed (see Sec. 7.1) to determine how long the monkeyball is the air. Use the three equations of 1DUA. Follow the strategy described in Sec. 1.6, and study the examples. Which three symbols (of the five symbols that appear in the three equations) are known? Look at the units to help determine the known quantities. Signs are very important. Either work with the trip up or the trip down, but not the roundtrip. Make sure that the three knowns that you write down apply specifically to the half-trip that you choose to work with.

# Hints

10. (A) First find the initial and final $z$-components of the angular velocity, which you can read directly from the graph. Review the definition of average angular acceleration in Sec. 7.1. (B) Since angular acceleration equals the slope of an angular velocity graph, find the slope of the tangent line at the initial point. (C) Since angular acceleration equals the slope of an angular velocity graph, find the slope of the tangent line at the given time. (D) Find the maximum absolute value of the $z$-component of the angular velocity from this graph by finding when the curve is furthest from the time axis. (E) Use your previous answer along with the equation that relates tangential speed to angular speed. (F) First find the total angle covered, and then use the arc length formula to get the total distance traveled. Since angular displacement equals the area for an angular velocity graph, find the area between the curve and time axis. Add the areas together in absolute values since you ultimately want the total distance traveled – not the net displacement.

11. Review these definitions in Sec.'s 4.2 and 7.1. (E) Use the formulas to find the magnitude and direction of a vector from its components. When taking the inverse tangent, you may need to add $180°$ in order to get the vector into the right quadrant. Note that the angles are measured counterclockwise from the $+x$-axis; when taking an inverse tangent, add $180°$ if the $x$-component of the vector is negative. (F) How does the total distance traveled relate to the speed when speed is constant?

12. Write down an equation for each monkey using subscripts for any symbols that may be different for each monkey. Follow the multiple-moving object strategy described in Sec. 1.9, and study the examples, but use the angular equations of Sec.'s 4.2 and 7.1. How much distance do the two monkeys travel all together?

13. Use the parallel-axis theorem along with the table of moments of inertia for common geometries about common axes of rotation in Sec. 7.2.

14. (A) Use the table of moments of inertia for common geometries about common axes of rotation in Sec. 7.2. The rod's moment of inertia is approximately zero. (B) Use the parallel-axis theorem (for all three rigid bodies) along with the table of moments of inertia for common geometries about common axes of rotation in Sec. 7.2. The rod's moment of inertia is significant in this part.

15. Review the definition of the moment of inertia of a discrete system of pointlike objects in Sec. 7.2. Determine the distance between each object and the specified axis of rotation. In part (D), first determine the location of the center of mass of the system (see Sec. 6.1).

16. Review the definition of the moment of inertia of a discrete system of pointlike objects in Sec. 7.2. Determine the distance between each object and the specified axis of rotation.

17. Use the parallel-axis theorem along with the table of moments of inertia for common geometries about common axes of rotation in Sec. 7.2. In part (B), first determine the location of the center of mass of the system (see Sec. 6.1).

18. Use the parallel-axis theorem along with the table of moments of inertia for common geometries about common axes of rotation in Sec. 7.2.

19. Use the parallel-axis theorem along with the table of moments of inertia for common geometries about common axes of rotation in Sec. 7.2. If you add this funny shape to the solid disc that was cutout together, you make the original solid disc. Write down an equation for the moment of inertia of the original solid disc about an axis perpendicular to the disc and passing through its center. Also, write down an equation for the moment of inertia of the solid disc that was cut out, noting that the axis of rotation does not pass through its center (so use the parallel-axis theorem). Notice that the masses and $r_\perp$'s are not the same for the two solid discs: You will have to find the ratio of the masses (of the three shapes) by finding the ratio of the areas. Subtract the moments of inertia to find the moment of inertia

of the actual funny shape about the specified axis. Express your answer in terms of the mass of the funny shape.

20. Review the technique for carrying out the moment of inertia integral in Sec. 7.2 and study the examples. Draw a diagram of a planar blob in the $xy$ plane. Label a representative $dm$. Label $r_\perp$ for the moments of inertia about all three axes. Write down the formula for the moment of inertia integral about each axis. Rewrite $r_\perp$ in terms of Cartesian coordinates in each case: How far is a general $dm$ from each axis in terms of $x$, $y$, and $z$? Remember that $r_\perp$ is squared in the integrand. Don't carry out the integrals, and don't make any substitutions for $dm$. Just compare your expressions, and show that two of the formulas added together equal the third.

21. Review the technique for carrying out the moment of inertia integral in Sec. 7.2 and study the examples. Use the appropriate density. Draw a diagram, showing a representative $dm$. Draw the axis of rotation and label $r_\perp$ for the $dm$ that you drew. Write an expression for $r_\perp$ in terms of your integration coordinate(s) that applies to every $dm$ in the rigid body.

22. Review the technique for carrying out the moment of inertia integral in Sec. 7.2 and study the examples. Use the appropriate density. Draw a diagram, showing a representative $dm$. Draw the axis of rotation and label $r_\perp$ for the $dm$ that you drew. Write an expression for $r_\perp$ in terms of your integration coordinate(s) that applies to every $dm$ in the rigid body. Eliminate the constant $a$ from the center of mass by solving for the total mass by integrating over $dm$.

23. Review the technique for carrying out the moment of inertia integral in Sec. 7.2 and study the examples. Use the appropriate density. Draw a diagram, showing a representative $dm$. Draw the axis of rotation and label $r_\perp$ for the $dm$ that you drew. Write an expression for $r_\perp$ in terms of your integration coordinate(s) that applies to every $dm$ in the rigid body. Work with 2D polar coordinates; write the needed differential element appropriately for this choice of coordinates. The region is 2D. You will need to perform a double integral. Both integration variables will have full freedom in 2D polar coordinates. Note that there are two problems: You need to do a different integral for each axis of rotation.

24. Review the technique for carrying out the moment of inertia integral in Sec. 7.2 and study the examples. Use the appropriate density. Draw a diagram, showing a representative $dm$. Draw the axis of rotation and label $r_\perp$ for the $dm$ that you drew. Write an expression for $r_\perp$ in terms of your integration coordinate(s) that applies to every $dm$ in the rigid body. Work with 2D polar coordinates; write the needed differential element appropriately for this choice of coordinates. The region is 1D. Note that there are two problems: You need to do a different integral for each axis of rotation.

25. Review the technique for carrying out the moment of inertia integral in Sec. 7.2 and study the examples. Use the appropriate density. Draw a diagram, showing a representative $dm$. Draw the axis of rotation and label $r_\perp$ for the $dm$ that you drew. Write an expression for $r_\perp$ in terms of your integration coordinate(s) that applies to every $dm$ in the rigid body. The region is 2D. You will need to perform a double integral. Both integration variables will have full freedom in Cartesian coordinates. Note that there are two problems: You need to do a different integral for each axis of rotation.

26. Review the technique for carrying out the moment of inertia integral in Sec. 7.2 and study the examples. Use the appropriate density. Draw a diagram, showing a representative $dm$. Draw the axis of rotation and label $r_\perp$ for the $dm$ that you drew. Write an expression for $r_\perp$ in terms of your integration coordinate(s) that applies to every $dm$ in the rigid body. Work with spherical coordinates; write the needed differential element appropriately for this choice of coordinates. The region is 2D.

Hints

You will need to perform a double integral. Both integration variables (which are both angles; $r$ is a constant – but $r_\perp$ is not!) will have full freedom in spherical coordinates.

27. Review the technique for carrying out the moment of inertia integral in Sec. 7.2 and study the examples. Use the appropriate density. Draw a diagram, showing a representative $dm$. Draw the axis of rotation and label $r_\perp$ for the $dm$ that you drew. Write an expression for $r_\perp$ in terms of your integration coordinate(s) that applies to every $dm$ in the rigid body. Work with cylindrical coordinates; write the needed differential element appropriately for this choice of coordinates. The region is 3D. You will need to perform a triple integral. The angle $\theta$ will have its full range. If you let $z$ vary over the entire height of the cone, the upper limit of $r_c$ will be a function of $z$; write an equation for a straight line tangent to the cone's surface in order to express $r_c$ as a function of $z$. Draw the range of $r_c$ for a representative value of $z$.

28. Review the technique for carrying out the moment of inertia integral in Sec. 7.2 and study the examples. Use the appropriate density. Draw a diagram, showing a representative $dm$. Draw the axis of rotation and label $r_\perp$ for the $dm$ that you drew. Write an expression for $r_\perp$ in terms of your integration coordinate(s) that applies to every $dm$ in the rigid body. Work with cylindrical coordinates with the $z$-axis as the rolling axis; write the needed differential element appropriately for this choice of coordinates. The region is 3D for a solid torus. You will need to perform a triple integral. The angle $\theta$ will have its full range. If you let $z$ vary over the entire height of the torus, the upper limit of $r_c$ will be a function of $z$; write an equation for the circular cross section of the torus in order to express $r_c$ as a function of $z$. Note that you will have to consider both positive and negative roots: You can halve the range of one variable and multiply the integral by two to account for this. Draw the range of $r_c$ for a representative value of $z$. (Instead of using $r_c$ and $z$, you can develop a toroidal coordinate system by trading them for $\rho$ and $\psi$, the same way that we trade $x$ and $y$ for $r$ and $\theta$ from Cartesian to 2D polar coordinates. You still have the same third variable, $\theta$, but you get a different form for the differential volume element, and different – but more convenient – limits for the integration variables. If you try using this coordinate system, first just try a basic triple integral over $\int dV$ and look up the volume of a single-holed ring torus to verify that you understand this coordinate system before you attempt the moment of inertia integral.)

29. Review the technique for carrying out the moment of inertia integral in Sec. 7.2 and study the examples. Use the appropriate density. Draw a diagram, showing a representative $dm$. Use the $x$-axis as the axis of rotation (which is parallel to the one you ultimately want). Draw the axis of rotation and label $r_\perp$ for the $dm$ that you drew. Write an expression for $r_\perp$ in terms of your integration coordinate(s) that applies to every $dm$ in the rigid body. Work with 2D polar coordinates; write the needed differential element appropriately for this choice of coordinates. The region is 1D. Note that there are two problems: You need to do a different integral for each axis of rotation. When you finish the integral, use the parallel-axis theorem to change from the $x$-axis to the specified axis.

30. Review the technique for carrying out the moment of inertia integral in Sec. 7.2 and study the examples. Use the appropriate density. Draw a diagram, showing a representative $dm$. Draw the axis of rotation and label $r_\perp$ for the $dm$ that you drew. Write an expression for $r_\perp$ in terms of your integration coordinate(s) that applies to every $dm$ in the rigid body. Work with 2D polar coordinates; write the needed differential element appropriately for this choice of coordinates. The region is 2D. You will need to perform a double integral. Both integration variables will have full freedom in 2D polar coordinates. Eliminate the constant $\beta$ from the center of mass by solving for the total mass by integrating over $dm$.

31. Review the technique for carrying out the moment of inertia integral in Sec. 7.2 and study the examples. Use the appropriate density. Draw a diagram, showing a representative $dm$. Draw the axis of rotation and label $r_\perp$ for the $dm$ that you drew. Write an expression for $r_\perp$ in terms of your integration coordinate(s) that applies to every $dm$ in the rigid body. Work with spherical coordinates; write the needed differential element appropriately for this choice of coordinates. The region is 3D. You will need to perform a triple integral. All three integration variables will have full freedom in spherical coordinates. Eliminate the constant $\beta$ from the center of mass by solving for the total mass by integrating over $dm$.

32. Review the technique for carrying out the moment of inertia integral in Sec. 7.2 and study the examples. Use the appropriate density. Treat each rod separately. Note that the vertical rods are effectively pointlike; but you will need to first determine the mass of the horizontal rods and subtract them from the total mass in order determine what the mass of each vertical rod is. Draw a diagram for a horizontal rod, showing a representative $dm$. Draw the axis of rotation and label $r_\perp$ for the $dm$ that you drew. Write an expression for $r_\perp$ in terms of your integration coordinate(s) that applies to every $dm$ in the horizontal rod. The region is 1D. Eliminate the constant $\beta$ from the center of mass by solving for the total mass of one horizontal rod by integrating over $dm$. Add all of the moments of inertia together to get the total moment of inertia of your brain.

33. Find the prescription for taking the determinant of a $2 \times 2$ matrix in Sec. 7.3. Study the examples.

34. Find the prescription for taking the determinant of a $3 \times 3$ matrix in Sec. 7.3. Study the examples.

35. Find the prescription for taking the determinant of a $3 \times 3$ matrix in Sec. 7.3. Study the examples.

36. Study the prescription for taking the determinant of a $3 \times 3$ matrix using the method of cofactors in Sec. 7.3. The method of cofactors (but not the shortcut method) generalizes naturally to a $4 \times 4$ determinant.

37. Start out with $2 \times 2$ and $3 \times 3$ matrices. Write the matrix with symbols, using the specifications. For example, in the first part, use the same set of symbols for the first and second rows and show that the determinant is zero. As you work through the math, you begin to understand why the determinant is zero in these cases, which will help you build up a more general proof.

38. Review the definition of the vector product in Sec. 7.3. Study the examples. The first three problems can be answered by using the right-hand rule, but they can also be answered using a determinant. For example, for $\hat{k} \times \hat{i}$, the first vector is $0\hat{i} + 0\hat{j} + 1\hat{k}$, so the middle row of the determinant will have elements (0,0,1). (G) First find $\vec{A} \times \vec{B}$, then cross that into $\vec{C}$. (H) First find $\vec{B} \times \vec{C}$, then cross $\vec{A}$ into that.

39. (A) Review the definition of the scalar product in Sec. 6.1 and study the examples. Use the component formula for the scalar product. (B) Review the definition of the vector product in Sec. 7.3 and study the examples. (C) Use your result from part (A) and the formula for the scalar product that involves the angle between the vectors.

40. Review the definitions of the vector product in Sec. 7.3 and torque in Sec. 7.4. Study the examples.

41. (A) Review the definitions of the vector product in Sec. 7.3 and torque in Sec. 7.4. Study the examples. (B) Use the formula that relates the magnitude of the torque to the angle between $\vec{r}_\perp$ and $\vec{F}$, along with your answer to part (A).

42. Review the definitions of the vector product in Sec. 7.3. Study the examples. First write the cylindrical unit vectors in terms of Cartesian unit vectors. See Sec. 2.4.

Hints

43. Review the definitions of the vector product in Sec. 7.3. Study the examples. First write the cylindrical or spherical unit vectors in terms of Cartesian unit vectors. See Sec.'s 2.4-2.5. In $\hat{\mathbf{r}} \times \hat{\mathbf{r}}_c$, you must realize that $\theta_c = \varphi_s$. That is, the angle $\theta$ of cylindrical coordinates is defined the same way as the angle $\varphi$ of spherical coordinates. When mixing the two coordinate systems, as in $\hat{\mathbf{r}} \times \hat{\mathbf{r}}_c$, don't make the mistake of thinking that the two $\theta$'s are the same. Instead, set $\theta_c = \varphi_s$ and distinguish between $\theta_c$ and $\theta_s$.

44. Review the definitions of the scalar product in Sec. 6.1 the vector product in Sec. 7.3 and study the examples. Find the scalar product between $\vec{\mathbf{A}} \times \vec{\mathbf{B}}$ and $\vec{\mathbf{A}}$ and also between $\vec{\mathbf{A}} \times \vec{\mathbf{B}}$ and $\vec{\mathbf{B}}$. What does the scalar product equal when two vectors are perpendicular?

45. Study the tensor notation in Sec. 7.3, including the derivations and examples. This sum has three terms.

46. Study the tensor notation in Sec. 7.3, including the derivations and examples. This double sum has nine terms, but only three are nonzero.

47. Study the tensor notation in Sec. 7.3, including the derivations and examples. This double sum has nine terms. The index $k$ is free: It is not summed over, and could be any of the three possible values. Can you find any terms that could be nonzero, depending on the value of $k$?

48. Study the tensor notation in Sec. 7.3, including the derivations and examples. This triple sum has 27 terms, but you only need to figure out which ones are nonzero and deal with those.

49. Study the tensor notation in Sec. 7.3, including the derivations and examples. This triple sum has 27 terms, but you only need to figure out which ones are nonzero and deal with those.

50. Study the tensor notation in Sec. 7.3, including the derivations and examples. Pay special note to how the Kronecker delta can be used to contract indices, and apply this to simplify this sum.

51. Study the tensor notation in Sec. 7.3, including the derivations and examples. Write $\vec{\mathbf{A}} \times \vec{\mathbf{B}}$ in tensor notation. You can interchange the indices of the components of each vector if you also interchange the corresponding indices of the Levi-Civita symbol. What happens when you swap two indices of a Levi-Civita symbol? Identify the resulting vector product.

52. Study the tensor notation in Sec. 7.3, including the derivations and examples. Write each term in tensor notation to show that this equation is valid.

53. Study the tensor notation in Sec. 7.3, including the derivations and examples. See the note called the "triple vector product" to see what the BAC-CAB rule is. Write the triple vector product in tensor notation, noting that there are two separate vector products to express. Use the identity that applies to the product of Levi-Civita symbols. Contract indices according to the Kronecker deltas. One of the derivations in Sec. 7.3 uses a similar trick of trading a product of Levi-Civita symbols for Kronecker deltas, and then contracting indices. Study that example.

54. Review the definition of torque in Sec. 7.4. Study the examples. Pay special attention to how the distance $r_\perp$ and the angle $\theta$ are defined in the equation for the magnitude of the torque. Draw the direction of the specified force and draw the vector $\vec{\mathbf{r}}_\perp$. When the force is the weight of an object, treat the weight as if all of the weight effectively acts at the center of mass the object. Setup a coordinate system where either clockwise or counterclockwise is positive, and base the sign of your torques on this. (E) How about the torque due to the weight of the rod? (F) Sum the torques, bearing in mind that some will be positive, but others will be negative.

55. Review the definition of torque in Sec. 7.4. Study the examples. Pay special attention to how the distance $r_\perp$ and the angle $\theta$ are defined in the equation for the magnitude of the torque. Draw the direction of the specified force and draw the vector $\vec{r}_\perp$. When the force is the weight of an object, treat the weight as if all of the weight effectively acts at the center of mass the object. Setup a coordinate system where either clockwise or counterclockwise is positive, and base the sign of your torques on this. (B) Set the sum of the torques equal to zero; some of the torques will be negative. Include the torque due to the weight of the rod and the torque due to $\vec{P}$ in addition to the torques previously found.

56. Review the definition of torque in Sec. 7.4. Study the examples. Pay special attention to how the distance $r_\perp$ and the angle $\theta$ are defined in the equation for the magnitude of the torque. Draw the direction of the specified force and draw the vector $\vec{r}_\perp$. When the force is the weight of an object, treat the weight as if all of the weight effectively acts at the center of mass the object. Setup a coordinate system where either clockwise or counterclockwise is positive, and base the sign of your torques on this. Note that $\theta$ is not 90°; it is less than this for the left monkey, and more than this for the right monkey (and yet the sines of the angles may still cancel out in the sum of the torques), but in neither case does $\theta$ equal 30°. If you draw each $\vec{r}_\perp$ and the corresponding weight, it should be easy to determine $\theta$.

57. Review the definition of torque in Sec. 7.4. Study the examples. Pay special attention to how the distance $r_\perp$ and the angle $\theta$ are defined in the equation for the magnitude of the torque. Setup a coordinate system where either clockwise or counterclockwise is positive, and base the sign of your torques on this. Set the sum of the torques equal to zero, as in Sec. 7.6. Note that $r_{1\perp}$ and $r_{2\perp}$ are both unknown, but you can write an equation for $r_{1\perp}$ and $r_{2\perp}$ and solve two equations in two unknowns.

58. Review the definition of torque in Sec. 7.4. Study the examples. Pay special attention to how the distance $r_\perp$ and the angle $\theta$ are defined in the equation for the magnitude of the torque. Setup a coordinate system where either clockwise or counterclockwise is positive, and base the sign of your torques on this. Set the sum of the torques equal to zero, as in Sec. 7.6. Write down an equation for the sum of the masses and solve two equations in two unknowns.

59. Review the definition of torque in Sec. 7.4. Study the examples. Pay special attention to how the distance $r_\perp$ and the angle $\theta$ are defined in the equation for the magnitude of the torque. Setup a coordinate system where either clockwise or counterclockwise is positive, and base the sign of your torques on this. Set the sum of the torques equal to zero, as in Sec. 7.6. It is convenient to divide the mallet up into three pieces (not two) as follows: the portion of the handle to the left of the fulcrum, the portion of the handle to the right of the fulcrum, and the head. Write down an equation for the sum of the torques, and plug in all of the knowns. Write down an equation for the sum of the three masses. Lastly, write down an equation that relates the mass of the left portion of the handle to the right portion of the handle. Solve this system of three equations in three unknowns. This problem is also considered conceptually in an example in Sec. 7.6, and Conceptual Question 14 is a similar conceptual exercise. The same problem appears as Practice Problem 11 in Chapter 6, as it can alternatively be solved in terms of center of mass.

60. (C)/(D)/(E)/(F)/(G)/(J) Draw a separate FBD for the monkey and box of bananas, and draw an extended FBD for the pulley. Apply Newton's second law separately to the monkey and box of bananas, and sum the torques for the pulley. Follow the strategy outlined in Sec. 7.5 and study the examples. (A) Use the equation that relates arc length to angle and convert from radians to revolutions. (B) Use the equation for the friction force in Sec. 3.5. (H) Follow the 1DUA strategy in Sec. 1.6. (I) Use the equation that relates speed to angular speed in Sec. 4.1.

# Hints

61. Draw a separate FBD for the monkey and box of bananas, and draw an extended FBD for the pulley. Apply Newton's second law separately to the monkey and box of bananas, and sum the torques for the pulley. Follow the strategy outlined in Sec. 7.5 and study the examples. Use the equation for the friction force in Sec. 3.5.

62. Draw a FBD for the knot and apply Newton's second law. Study the examples in Sec.'s 3.4 and 7.6. Resolve the tensions into horizontal and vertical components. The tension in the bottom cord will equal the weight of the monkey since the knot is in static equilibrium. (This problem does not involve torque, but does involve the concept of static equilibrium.)

63. Follow the strategy of summing the torques and summing the horizontal and vertical components of the forces following the strategy for static equilibrium. Study the examples in Sec. 7.6, especially involving a boom and hingepin. (C) Use the equation that relates the net external torque to the angular acceleration in Sec. 7.5. Use the parallel-axis theorem along with the table of moments of inertia for common geometries about common axes of rotation in Sec. 7.2, and treat the bananas like a pointlike object.

64. Follow the strategy of summing the torques and summing the horizontal and vertical components of the forces following the strategy for static equilibrium. Study the examples in Sec. 7.6, especially involving a boom and hingepin.

65. Review the criteria for stable equilibrium in Sec. 7.6. Study the conceptual example with a stack of bricks. Review the definition of center of mass in Sec. 6.1. Start with a stack of 2 bricks, then a stack of 3 bricks, and so on to deduce the pattern.

66. Review the criteria for stable equilibrium in Sec. 7.6. Study the conceptual example with a stack of bricks. Review the definition of center of mass in Sec. 6.1.

67. Draw an extended FBD for the hollow sphere in each case. In each case, sum the components of the forces along the incline and perpendicular to the incline, and sum the torques with a sense of rotation that is consistent with your other coordinates. Follow the strategy applied in a similar example in Sec. 7.6.

68. Use the strategy for conservation of energy. Study the rolling without slipping examples in Sec. 7.7. Choose a reference point from which to measure the heights. Determine which types of energy (i.e. potential or kinetic) there are initially and finally. Use the correct formula(s) for potential energy. There are two kinds of kinetic energy involved in the problem – translational and rotational. Although there is friction in the problem, there isn't any nonconservative work done. Multiple unknowns will cancel out in the algebra – see the examples in Sec. 7.7. Use trig to relate the height to the length down the incline.

69. (A) Use the strategy for conservation of energy. Study the rolling without slipping examples in Sec. 7.7. Choose a reference point from which to measure the heights. Determine which types of energy (i.e. potential or kinetic) there are initially and finally. Use the correct formula(s) for potential energy. There are two kinds of kinetic energy involved in the problem – translational and rotational. Although there is friction in the problem, there isn't any nonconservative work done. Multiple unknowns will cancel out in the algebra – see the examples in Sec. 7.7. Use trig to relate the height to the length down the incline. (B) Use the 1DUA strategy in Sec. 1.6.

70. Use the strategy for conservation of energy. Study the rolling without slipping examples in Sec. 7.7. Choose a reference point from which to measure the heights. Determine which types of energy (i.e. potential or kinetic) there are initially and finally. Use the correct formula(s) for potential energy. There are two kinds of kinetic energy involved in the problem – translational and rotational – except for the

block of ice. Although there is friction in the problem, there isn't any nonconservative work done. Multiple unknowns will cancel out in the algebra – see the examples in Sec. 7.7. Use trig to relate the height to the length down the incline.

71. Use the strategy for conservation of energy. Study the rolling without slipping examples in Sec. 7.7. Choose a reference point from which to measure the heights. Determine which types of energy (i.e. potential or kinetic) there are initially and finally. Use the correct formula(s) for potential energy. There are two kinds of kinetic energy involved in the problem – translational and rotational. Although there is friction in the problem, there isn't any nonconservative work done. Multiple unknowns will cancel out in the algebra – see the examples in Sec. 7.7.

72. Use the strategy for conservation of energy. Study the rolling without slipping examples in Sec. 7.7. Choose a reference point from which to measure the heights. Determine which types of energy (i.e. potential or kinetic) there are initially and finally. Use the correct formula(s) for potential energy. There are two kinds of kinetic energy involved in the problem – translational and rotational. Although there is friction in the problem, there isn't any nonconservative work done. Multiple unknowns will cancel out in the algebra – see the examples in Sec. 7.7. Use trig to relate the height to the length up the incline.

73. (A)/(D)/(E) Use the strategy for conservation of energy. Study the rolling without slipping examples in Sec. 7.7. Choose a reference point from which to measure the heights. Determine which types of energy (i.e. potential or kinetic) there are initially and finally. Use the correct formula(s) for potential energy. There are two kinds of kinetic energy involved in the problem – translational and rotational. Although there is friction in the problem, there isn't any nonconservative work done. Use trig to relate the height to the length down the incline. The 50-kg mass of the unicycle includes the 20-kg wheel. Note that only the 20-kg wheel has rotational kinetic energy, while the total mass of 90 kg has translational kinetic energy (the total mass is not 110 kg because the wheel's mass is part of the unicycle's mass). (B)/(C) Draw an extended FBD for the unicycle. Sum the components of the forces along the incline and perpendicular to the incline, and sum the torques with a sense of rotation that is consistent with your other coordinates. Follow the strategy applied in a similar example in Sec. 7.6.

74. Use the strategy for conservation of energy. Study the rolling without slipping examples in Sec. 7.7. Choose a reference point from which to measure the heights. Determine which types of energy (i.e. potential or kinetic) there are initially and finally. Use the correct formula(s) for potential energy. There are two kinds of kinetic energy involved in the problem – translational and rotational. Although there is friction in the problem, there isn't any nonconservative work done. There are similar conceptual and quantitative examples with containers in Sec. 7.7, but this problem up the incline – instead of down – has a different result, explained with different reasoning. This problem is also considered in Conceptual Question 21. The subtleties for the bottle of liquid can be found by studying the quantitative example in Sec. 7.7. There is also a subtlety for the bottle of frozen juice: You can't treat it like a solid disc, but must instead treat it like a hollow ring plus a solid disc of different mass, since the container and frozen juice are different materials.

75. Neglect the thickness of the cord. (A) Separately find the moments of inertia of the two discs and hollow cylinder in rolling mode, using the table of moments of inertia of common geometries about common axes of rotation in Sec. 7.2, and then add them together. At some point you will need to convert these to SI units. (B)/(D) Draw an extended FBD for the yo-yo. Sum the components of the forces along the incline and perpendicular to the incline, and sum the torques with a sense of rotation that is consistent with your other coordinates. Follow the strategy applied in a similar example in Sec. 7.6. Note carefully which radius relates the tangential acceleration to the angular acceleration, and that

Hints

this radius may not cancel out for a yo-yo. Mass, similarly, may not cancel out. (C) Use the strategy for conservation of energy. Study the rolling without slipping examples in Sec. 7.7. Choose a reference point from which to measure the heights. Determine which types of energy (i.e. potential or kinetic) there are initially and finally. Use the correct formula(s) for potential energy. There are two kinds of kinetic energy involved in the problem – translational and rotational. Note carefully which radius relates the tangential speed to the angular speed, and that this radius may not cancel out for a yo-yo. Mass, similarly, may not cancel out. (E) Use the 1DUA strategy in Sec. 1.6.

76. (A)/(B) Use the 1DUA strategy in Sec. 1.6. (C) Use the equation that relates tangential acceleration to angular acceleration in Sec. 7.1. (D) Use the equation that relates arc length to angle. (E) Draw an extended FBD for the car. Sum the components of the forces along the incline and perpendicular to the incline, and sum the torques with a sense of rotation that is consistent with your other coordinates. Follow the strategy applied in a similar example in Sec. 7.6. Note that mass and radius may not cancel out since only the wheels of the car rotate. Also use the definition of torque in Sec. 7.4. (F) Review the definition of the work done by gravity in Sec. 5.2, or find the change in gravitational potential energy (see Sec. 5.3).

77. Draw a V and a sphere inside it, making contact at two points. Label $d$ and $R_0$. Use the strategy for conservation of energy. Study the rolling without slipping examples in Sec. 7.7. Choose a reference point from which to measure the heights. Determine which types of energy (i.e. potential or kinetic) there are initially and finally. Use the correct formula(s) for potential energy. There are two kinds of kinetic energy involved in the problem – translational and rotational. Note carefully which radius relates the tangential speed to the angular speed, and that this radius may not cancel out. Use trig to relate the height to the length down the incline.

78. Follow the strategy for conservation of angular momentum. Study the conceptual and quantitative examples in Sec. 7.9. Note that her angular speed is constant before and after the collision, so you can use the equation that relates period to angular speed to determine her initial angular speed. Similarly, after conserving angular momentum to find her final angular speed, you can use the equation that relates period to angular speed to determine the final number of revolutions completed. Note that units given for – and asked to find – in the problem are not SI units.

79. Follow the strategy for conservation of angular momentum. Study the conceptual and quantitative examples in Sec. 7.9. Note that the angular speed of the earth is constant before and after the contraction, so you can use the equation that relates period to angular speed to determine the initial angular speed. Similarly, after conserving angular momentum to find the final angular speed, you can use the equation that relates period to angular speed to determine the final period. Pretend that the earth is a solid sphere.

80. (A) Follow the strategy for conservation of angular momentum. Study the conceptual and quantitative examples in Sec. 7.9. Treat the monkey as a pointlike object and the merry-go-round as a solid disc. (B) Review the definition of angular acceleration in Sec. 7.1. (C) Assume that the motion is approximately uniform angular acceleration. Use the strategy for uniform angular acceleration in Sec. 7.1.

81. Follow the strategy for conservation of angular momentum. Study the conceptual and quantitative examples in Sec. 7.9. Treat the hamster as a pointlike object. What shape is the record? Use the equation that relates tangential speed to angular speed in order to find the hamster's initial angular speed. Note that the numbers given do not have SI units.

82. (A) Review the definition of momentum in Sec. 6.2. (B) Since there are no potential energy changes in this problem, just find the initial kinetic energy. See Sec. 5.3. (C) Treat the silly putty as a pointlike object, as in Sec. 7.8. (D) Will the direction of the total momentum of the system change? Since momentum is a vector, if the direction before and after the collision clearly change, that would make it clear that momentum is not conserved. If the net external force acting on the system is nonzero, then the total momentum of the system is not conserved. What type of collision is this? Is mechanical energy conserved for such a collision? If the net external torque acting on the system is zero, then angular momentum is conserved. Note that gravity pulls the silly putty downward, yet the door exerts a force supporting it against gravity. Any force exerted at the hinges results in zero torque; so a hinge force doesn't contribute toward a net torque, but may contribute toward a net force. The impact forces will result in equal and opposite torques. (E) Follow the strategy for conservation of angular momentum. Study the conceptual and quantitative examples in Sec. 7.9. Treat the silly putty as a pointlike object. What shape is the door, and where is the axis of rotation? (F) Calculate the total final rotational kinetic energy of the system and compare your answer with what you found in part (B). See Sec. 7.7.

83. Note that there is friction in the problem, since the spheres roll without slipping. Conserve angular momentum, following the strategy of Sec. 7.9 and noting any relative signs, and mechanical energy together, closely following the derivation of the velocity equation in Sec. 6.5. Express the angular velocity in terms of the velocity of the center of mass using the equation that relates these two quantities (see Sec.'s 4.1 and 7.1). (C) Calculate the initial and final momentum of each sphere, noting any relative signs.

84. Setup a coordinate system and include appropriate unit vectors with your answers to any quantities that are vectors. (A)/(H) Use the equation that relates tangential and angular velocity in Sec.'s 4.1 and 7.1. (B)/(J) Treat the monkey as a pointlike object. (C)/(K) It's centripetal acceleration. See Sec. 4.2. (D)/(L) Draw a FBD for the turkey, which is not moving vertically at this stage of the problem. Apply Newton's second law, as in Sec. 3.4. (E) Since the monkey is pointlike, see Sec. 7.8. (F) Note that the tension points toward the axis of rotation, and the torques due to weight and normal force cancel out. (G) Note that the tension contributes toward the net force, but not the net torque, and that there is nonconservative work. (I) Follow the strategy for conservation of angular momentum. Study the conceptual and quantitative examples in Sec. 7.9.

85. Follow the strategy for conservation of angular momentum. Study the conceptual and quantitative examples in Sec. 7.9. Note that the problem gives you numbers that do not have SI units. Also, note that you are given the diameter of one object, but the radius of the other. (B)/(C) Calculate the initial and final total rotational kinetic energy of the system. See Sec. 7.7.

## Appendix A – Hints to Practice Problems

1. Follow the two rules for significant figures and study the examples.
2. Follow the two rules for significant figures and study the examples.
3. Follow the two rules for significant figures and study the examples. First, express both the result and its uncertainty in the same units.

4. Use the formula for standard deviation and follow the rules for significant figures. Study the examples.

5. Use the formula for standard deviation and follow the rules for significant figures. Study the examples.

6. Apply the formula for propagation of errors and follow the rules for significant figures. Study the examples. Write an equation for area as a function of length and width. Take the needed partial derivatives.

7. Apply the formula for propagation of errors and follow the rules for significant figures. Study the examples. Take the needed partial derivatives.

8. Apply the formula for propagation of errors and follow the rules for significant figures. Study the examples. Take the needed partial derivatives.

## Appendix B – Hints to Practice Problems

1. Read the note on linearizing data and study the example. You want to define two variables, $x$ and $y$, in terms of the old variables, $F$ and $z$, such that after making the substitutions, the equation has the form $y = mx + b$.

2. (A) Read the note on linearizing data and study the example. You want to define two variables, $x$ and $y$, in terms of the old variables, $Q$ and $H$, such that after making the substitutions, the equation has the form $y = mx + b$. (B)/(C) These involve the same idea, but in terms of different variables.

3. Read the note on linearizing data and study the example. You want to define two variables, $x$ and $y$, in terms of the old variables, $q$ and $u$, such that after making the substitutions, the equation has the form $y = mx + b$.

4. Read the note on linearizing data and study the example. You want to define two variables, $x$ and $y$, in terms of the old variables, $d$ and $e$, such that after making the substitutions, the equation has the form $y = mx + b$.

5. Read the note on linearizing data and study the example. You want to define two variables, $x$ and $y$, in terms of the old variables, $D$ and $J$, such that after making the substitutions, the equation has the form $y = mx + b$.

6. Read the note on linearizing data and study the example. You want to define two variables, $x$ and $y$, in terms of the old variables, $\beta$ and $\alpha$, such that after making the substitutions, the equation has the form $y = mx + b$.

7. Read the note on linearizing data and study the example. You want to define two variables, $x$ and $y$, in terms of the old variables, $\gamma$ and $\lambda$, such that after making the substitutions, the equation has the form $y = mx + b$. First, simplify the given equation.

# Answers to Conceptual Questions and Selected Practice Problems

## Units, Dimensions, and Significant Figures – Answers to Practice Problems

1. 3.9 km    2. (A) 16.7 min. (B) 8.76 hr (C) 0.0864 s    3. (A) 0.00068 ft.$^2$ (B) 0.0000057 m$^3$
4. 31 m/s    5. $18.5 \times 10^3$ mi./hr$^2$    6. rad/s, m    7. m/s$^2$    8. m/s, m/s$^2$
9. 0.53 kg/m$^3$
10. 46.4 m if you follow the rules exactly (since 2 times 8.54 equals 17.1 according to the rules). However, if you first factor out the 2, you get 46.34 m. To avoid compounding round-off errors, it's a good idea not to round until you get the final answer, and then go back and see how many significant figures are prescribed for the final answer according to the rules. Remember, the rules are an easy guide, while the ultimate rule involves propagating errors (Appendix A).
11. $1.80 \times 10^3$ m$^3$

## Chapter 1 – Answers to Conceptual Questions

**Note**: You will get more out of the conceptual questions if you first try to answer and understand the question on your own, then check the hints to see if your approach was correct, and then check the answer. If you rely on the hints before attempting to reason out the answer, you miss out on the valuable learning that comes from trying to think the solution through on your own as well as learning from mistakes that you make.

1. 1D: racecar, hiker, elevator, bead, guard. 2D: ant, bowling ball, mountain climber. 3D: bumblebee, whale.
2. 2.75 m, −1.00 m (downward).    3. 2.0 m/s, −0.40 m/s (west).    4. Zero.
5. (A) Slope changes sign. (B) Crosses time axis. (C) Net area changes sign (or make a sketch of tangential velocity as a function of time and see when tangential velocity crosses the time axis).
6.

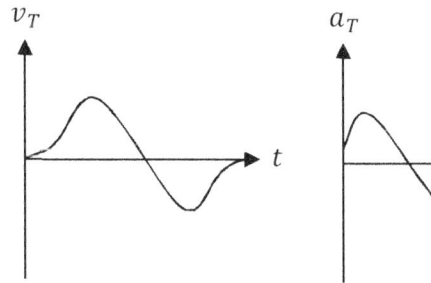

7.

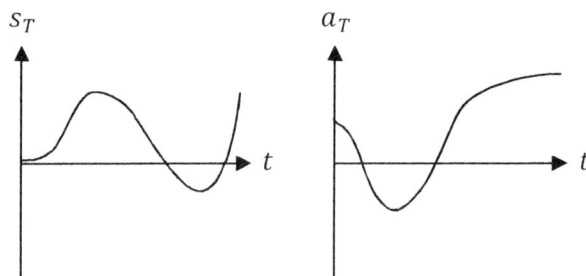

8.

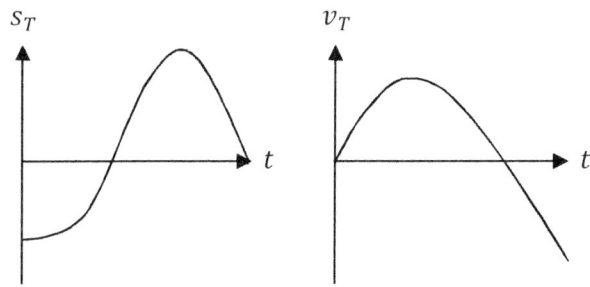

9. Multiply values of the tangential velocity by the corresponding slopes.
10. (A) 24 m/s (north). (B) 7.0 s.    11. (A) 3.5 s. (B) −15 m/s (downward). (C) 7.0 s. (D) 35 m/s.
12.

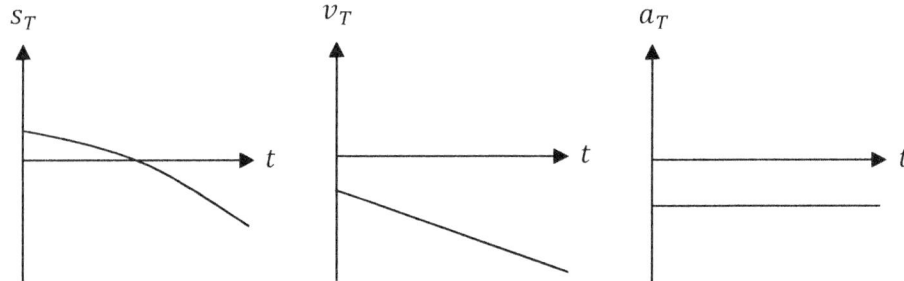

## Chapter 1 – Answers to Practice Problems

**Note**: You will get more out of solving the problems if you first try to solve the problem on your own, then check the hints to see if your approach was correct, and then check the answer. If you rely on the hints before attempting the problem, you miss out on the valuable learning that comes from trying to think the solution through on your own as well as learning from mistakes that you make.

**Note**: A metric prefix is included with the answers where it provides a convenient means of expressing the answer with the appropriate number of significant figures without using powers of ten. For example, suppose that the answer to a problem is 124 N, but the answer should only have two significant figures. In this case, the answer will be expressed as 0.12 kN.

1. (A) 15 km/hr  (B) More time is spent riding at the lower speed.
3. (A) 450 m, north  (B) 750 m  (C) 2.5 m/s  (D) 1.5 m/s, north        6. 360 km/hr$^2$
7. (A) $-20$ m  (B) 4.3 m/s  (C) 8.0 m/s  (D) 8.0 m/s  (E) $-24$ m  (F) $0 \leq t \leq 7.0$ s, $18.0$ s $\leq t \leq 27.0$ s
(G) 7.0 s, 18.0 s  (H) .026 m/s$^2$  (I) 8.0 m/s
**Note**: Your answers to the graphing questions may differ, within reason, due to interpolation errors or differences in drawing tangents, for example. It's more important that you can correctly show/explain how you get your answer than to get precisely the same value.
8. (A) 129 m  (B) $-0.50$ m/s  (C) 8.0 m/s  (D) 0.78 m/s$^2$  (E) 56 m behind his starting position
(F) 1.2 m/s$^2$  (G) $0 \leq t \leq 6.0$ s, $16.4$ s $\leq t \leq 18.0$ s  (H) 6.0 s, 16.4 s  (I) 0.56 m/s$^2$  (J) 15 m/s
**Note**: Your answers to the graphing questions may differ, within reason, due to interpolation errors or differences in drawing tangents, for example. It's more important that you can correctly show/explain how you get your answer than to get precisely the same value.
9. (A) 18 m/s  (B) $-3.8$ m/s$^2$  (C) 15 m/s  (D) 1.9 m/s$^2$  (E) $18 \leq t \leq 36$ s  (F) 18 s  (G) $-3.5$ m/s$^2$
(H) 90 m/s
**Note**: Your answers to the graphing questions may differ, within reason, due to interpolation errors or differences in drawing tangents, for example. It's more important that you can correctly show/explain how you get your answer than to get precisely the same value.
10. (A) 16 m/s$^2$  (B) 200 m        12. 5.0 s            13. 100 stairs (to 2 significant figures)
15. (A) 80 m/s  (B) $-16$ m/s$^2$    16. (A) 30.7 m  (B) 24.5 m/s
18. $6\times, 6\times$ (Perhaps the monkey can jump with a greater initial speed on the moon, which would improve these factors; but the question states that the initial speed is the same as on earth.)
20. (A) 8.0 m/s$^2$  (B) $-45$ m/s (downward)  (C) (i) 36 m  (ii) 0  (iii) $-8.0$ m/s$^2$ (downward)
21. (A) 86 m  (B) 28 m/s  (C) $-11$ m/s (downward)  (D) 21 m/s
23. (A) 28 m  (B) (i) 4.6 m/s (upward) (ii) 10 m/s           24. 21 m/s
26. (A) 1'17"  (B) 24 laps  (C) 180 m         27. (A) 0.11 km  (B) 9.3 s  (C) 23 m/s
28. (A) 5.5 s  (B) 75 m  (C) 27 m/s          30. (A) 78 m, 4.9 s              33. 0.23 m/s$^2$
35. (A) 0, 50 m/s$^2$  (B) $\frac{3}{2}$ m/s$^2$, $-\frac{21}{16}$ m/s  (C) 2.0 m/s$^2$, 40 m  (D) $\frac{22}{3}$ m/s, $\frac{248}{15}$ m  (E) 0, $\frac{12}{5}$ s
(F) $-\frac{1}{3}$ m/s
36. (A) m/s$^3$, s  (B) 12 m/s  (C) 12 m/s  (D) 2.0 m/s$^2$  (E) 62 m  (F) $-7.3$ m/s  (G) 8.1 m/s
37. (A) m/s$^{5/2}$  (B) 40 m/s  (C) $-8.0$ m/s

## Chapter 2 – Answers to Conceptual Questions

1. Scalars: volume, electric charge, density, wavelength, temperature, memory. Vectors: jerk, magnetic field, wind readings, strong nuclear force, angular momentum, relative star position.
2. Can be negative: $v_z, a_T, \Delta s_T, F_y$. Positive only: $v, \|\vec{F}\|, \Delta s, t$ (unless you defined $t_0$ to be negative).
3.                                           4.

Answers

5.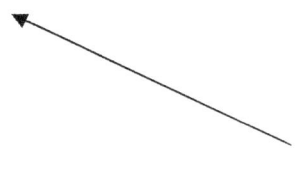
6.

7. 3.0 N, 13.0 N.   8. $-5.0$ m $\hat{\jmath}$.
9. (A) 1.0 mi. $\hat{\jmath}$ (north), 3.0 mi. (B) The set of points near the South Pole that are one mile north of the latitude that has a circumference of one mile, half a mile, a third of a mile, a quarter mile, etc.
10. 15 km/s, 0.   11. Zero.
12. Product rule: $\vec{a} = \frac{d\vec{v}}{dt} = \frac{d}{dt}(v_T \hat{t}) = \hat{t}\frac{dv_T}{dt} + v_T \frac{d\hat{t}}{dt} = a_T \hat{t} + a_c \hat{n}_i$. The object has a normal component when the direction of the velocity changes – i.e. the direction of the unit tangent, $\hat{t}$, is time-dependent.
13.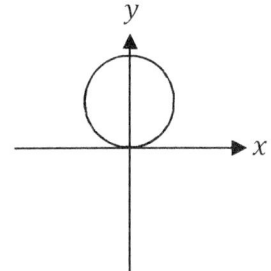
14.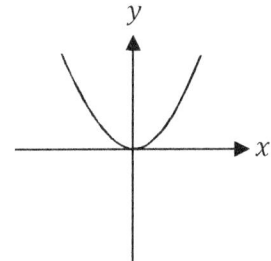

15. A double cone symmetric about the $z$-axis with the vertex at the origin.
16. A cone symmetric about the $z$-axis, with the vertex at the origin, with a vertex angle of $2\pi/3$ rad, and extending in the positive $z$-direction.
17. In the cannon; straight up and down relative to the ship, parabola relative to earth.
18.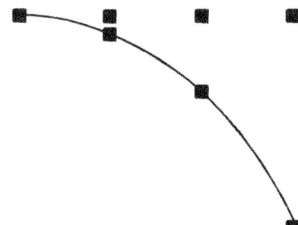

19. $\Delta y_b = \Delta x_b \tan \theta_0 - \frac{gt^2}{2}$, $\Delta y_m = -\frac{gt^2}{2}$. $\Delta y_b = \Delta x_b \tan \theta_0$ is where the banana was aimed. The banana winds up $\frac{gt^2}{2}$ below its target, while the monkey winds up $\frac{gt^2}{2}$ below its initial position.
20. $< 45°$.

548

21. (A) Angle the boat directly across so that the boat's velocity has the greatest component across the river. (B) Angle the boat partly against the river current so that the resultant velocity is directly across the river. (C) Angle the boat with the river current. Directly downriver maximizes the boat speed, but won't get the monkey across, so it has to angle a little bit.

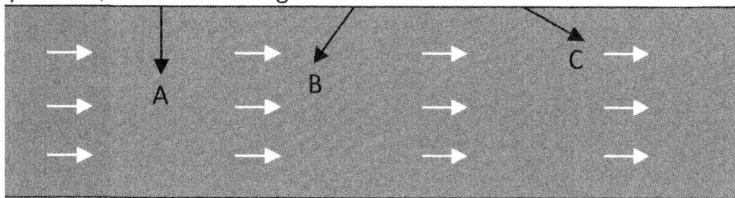

22. $v = \sqrt{(x \text{ slope})^2 + (y \text{ slope})^2}$. Sketch the slopes of the graphs. $a = \sqrt{(v_x \text{ slope})^2 + (v_y \text{ slope})^2}$.

23.

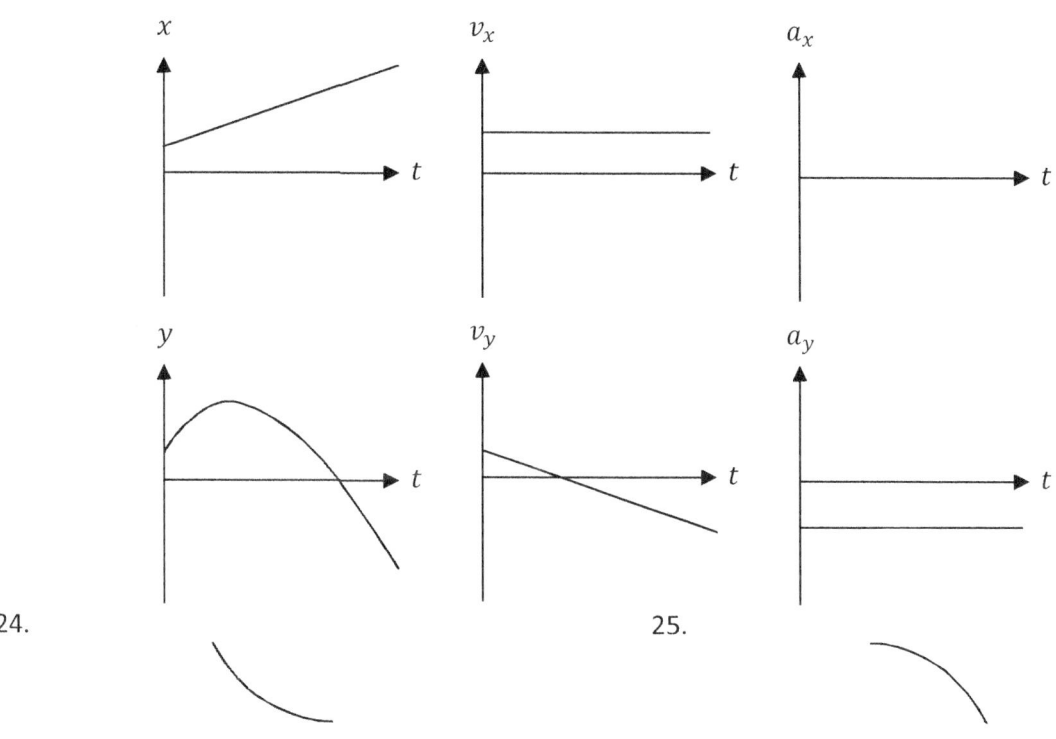

24.                                                       25.

## Chapter 2 – Answers to Practice Problems

1. (A) 0.25 km, 0.17 km  (B) 0.20 km, 79°  (C) 0.47 km, 53°    3. (A) 233°, 307°  (B) 0°, 120°, 240°
4. 0.37 'kilobananas,' −85°        8. (A) −1.84$\hat{\imath}$ − 1.54$\hat{\jmath}$ (B) 249, 5.79°, 90°
(C) $(x^3 + 12x)\hat{\imath} + 4x\hat{\jmath} - \frac{21\cos(2x)+3}{2}\hat{k}$  (D) 34.0, −31.5°, 45.1°        9. (B) 4.0, −30°
12. (A) 0.16 km (B) 0.10 km (C) 4.1 m/s (D) 6.4 m/s (E) 0
13. (A) 0.541[756185]m (B) 0.534[155744] m, 33.6[9006753]°
14. (A) 31 m (B) 25 m, 76° (C) 12 m/s, 76° (D) 4.0 m/s², 90°        17. $y = 2x^2$ (parabola)

18. $r^2 = \frac{1}{1-\sin(2\theta)}$  19. $r_c = 1, z = \theta$ (helix)  20. $x^2 + y^2 = \frac{z\sqrt{3}}{3}$ (cone)
21. $y = \frac{x\sqrt{3}}{3}$ (plane)  22. 45 cm  23. 8.0, 60°, 30°
26. (A) 64 m (B) 81 m (C) 8.5 m/s (D) $-9.81$ m/s$^2$ $\hat{\jmath}$ (E) 81 m (F) $-52°$
27. (A) 69 m/s, $-67°$ (B) 69 m/s, $-67°$ (C) 69 m/s, $-64°$ (D) same
30. (A) 52° (B) 12 m (C) 12 m (D) yes   31. 17 m/s   32. 4.4 km
33. (A) 96 m (B) 25 m/s (C) 0.22 km (D) 8.8 s   35. 68 m

## Chapter 3 – Answers to Conceptual Questions

1. Rest: not if already moving. Motion: imprecise, acceleration is a kind of motion, for example. Speed: imprecise, want to travel in a straight line, too. Velocity: imprecise, not true for rockets where mass changes. "Objects have a natural tendency to maintain constant momentum."
2. You feel forces. Inertia is not a force. Inertia is what you would do if there were no forces acting on you (or, equivalently, if all of the forces acting on you cancel out).
3. Forces: tension, weight, $\vec{\mathbf{N}}$, push, pull.   4. 5.0 m/s$^2$, 25 m/s$^2$.
5. Rest: straight down. Gains speed: backward. Circle: outward. Uphill: backward (relative to car). Constant velocity: straight down. Downhill: forward (relative to car). Deceleration: forward. Air: Straight down.
6. (A) Seat back pushes forward. (B) Seatbelt pushes backward.   7. Directly above.
8. Equal in magnitude, opposite in direction. Approximately zero acceleration for car, very large acceleration for mosquito.
9.                                                   10.

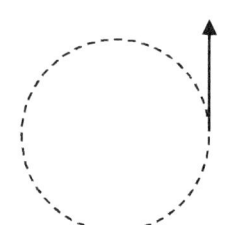
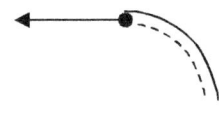

11. Throw the wrench directly away from the ship.
12. Earth spins faster; back to normal; earth spins slower during return, back to normal when they get home.
13. (A) Inertia. (B) Dinnerware moves. (C) Slightly downward: Dinnerware remains on table. Slightly upward: Dinnerware moves.
14. More mass implies less acceleration (second law).
15. Inertia: Banana is at rest, resists acceleration. Inertia: Banana is moving slowly, wants to continue moving with constant velocity.
16. No (third law).   17. The object exerts an upward force on the earth.
18. 4.5 m, 3.5 m (third law).   19. (A) 6. (B) Same. No.
20. Rest: $=150$ lbs. Upward acceleration: $>150$ lbs. Upward constant velocity: $=150$ lbs. Downward acceleration: $<150$ lbs. Downward constant velocity: $=150$ lbs. Free fall: 0.

21.

22. The force of static friction can act up or down the incline, with a magnitude $\leq \mu_s N$.
23. Same.    24. Same.    25. Larger.    26. Yes.    27. Parallel.
28. Although $\vec{P}$ does not act directly on the block, a larger magnitude, $\|\vec{P}\|$, does result in a greater normal force between the wedge and block. This normal force is upward at an angle. When the magnitude, $\|\vec{P}\|$, is small, the normal force does not have enough of a vertical component to balance the downward pull of gravity, but a large enough magnitude, $\|\vec{P}\|$, will result in a normal force with enough of a vertical component to accelerate the block up the incline. You can also explain this in terms of inertia. The block wants to have constant velocity. The wedge is accelerated to the right by the force $\vec{P}$. The wedge gains speed, while the block resists the increase in speed. The center of mass of the wedge accelerates past the center of mass of the block, as the wedge drives the block up the incline.
29. Although $\vec{P}$ does not act directly on the small block, a larger magnitude, $\|\vec{P}\|$, does result in a greater normal force between the two blocks. This normal force acts to the right. The large block also exerts an upward force of static friction on the small block, which has a magnitude $f_s \leq \mu_s N$. When the magnitude, $\|\vec{P}\|$, is small, the normal force is small, and the maximum friction force is too small to balance the downward pull of gravity, but a large enough magnitude, $\|\vec{P}\|$, will result in a large enough normal force such that the friction force is able to balance the weight of the small block.
30. Compared to the case of free fall, the banana will not rise as high in the air, and therefore will return to the monkey's hand with less speed than that with which it was thrown. On the way up, both the weight of the banana and air resistance act to slow it down. On the way down, the weight of the banana acts to speed the banana up, while air resistance resists its acceleration. Going up, the banana reaches its maximum height sooner than it would in vacuum. Falling down, the banana takes more time to reach the monkey's hand than it would if dropped from the same height in vacuum. The banana spends more time returning back down than it does rising upward.
31. The force of air resistance, directed upward, is proportional to the speed of the object. The object gains speed due to the downward pull of gravity. As the object gains speed, air resistance becomes stronger. Eventually, the speed is fast enough that the force of air resistance equals the weight of the object, at which point the object falls with constant (terminal) speed.
32. (A) Initially: Passenger is moving 100 m/s horizontally; downward component of acceleration is 9.8 m/s², horizontal component of acceleration (due to air resistance) opposes the velocity. Falling with unopened parachute: Horizontal component of velocity is less than 100 m/s, there is now a downward component of velocity; downward component of acceleration is less than 9.8 m/s² (air resistance has an upward component now), horizontal component of acceleration is decreasing (since horizontal component of velocity is decreasing due to air resistance). Long time with unopened parachute: Horizontal component of velocity approaches zero, vertical component of velocity approaches terminal speed; both components of acceleration approach zero. When parachute opens: Velocity is mainly downward, but decreases for a brief period (since there is a sudden, new upward force); acceleration is upward for a brief period. Long time with parachute: Velocity approaches a new, smaller downward terminal velocity; acceleration approaches zero. (B) One before and one after opening the parachute. (C) When the parachute is opened; downward.

33. (A) Earth (has more mass). (B) Since $R_m = R_{em} - R_e$ (where these are the distance between the field point and the moon, the earth-moon distance, and the distance between the field point and the earth, respectively; $R_m$ and $R_e$ are the two unknowns), when we equate $Gm_e/R_e^2$ to $Gm_m/R_m^2$ and plug in $R_m = R_{em} - R_e$, we get one quadratic equation with one unknown ($R_e$). The quadratic equation has two possible solutions for $R_e$. Physically, only one is viable. The incorrect solution is the one where the magnitudes are equal and the directions are collinear (instead of opposite).

34. 1/16.  35. 19.6 m/s$^2$.

36. (A) No. Yes (just put your hand 'under' it). No (inertia). (B) Unlike the space station, the moon has an astronomical amount of mass and so has a significant gravitational field of its own.

37. (A) Sun. Earth orbits the sun, not the moon. (B) Moon. Calculate the force that each object exerts on a test mass on both sides of the earth and subtract the forces. The moon's closer proximity to the earth outweighs the sun's stronger gravitational pull. (C) See below. (D) Points are marked H (high) and L (low) in the diagram below. Two of each per day. (E) See below. S = spring, N = neap. Two of each per 29.5 days (period of the moon's revolution).

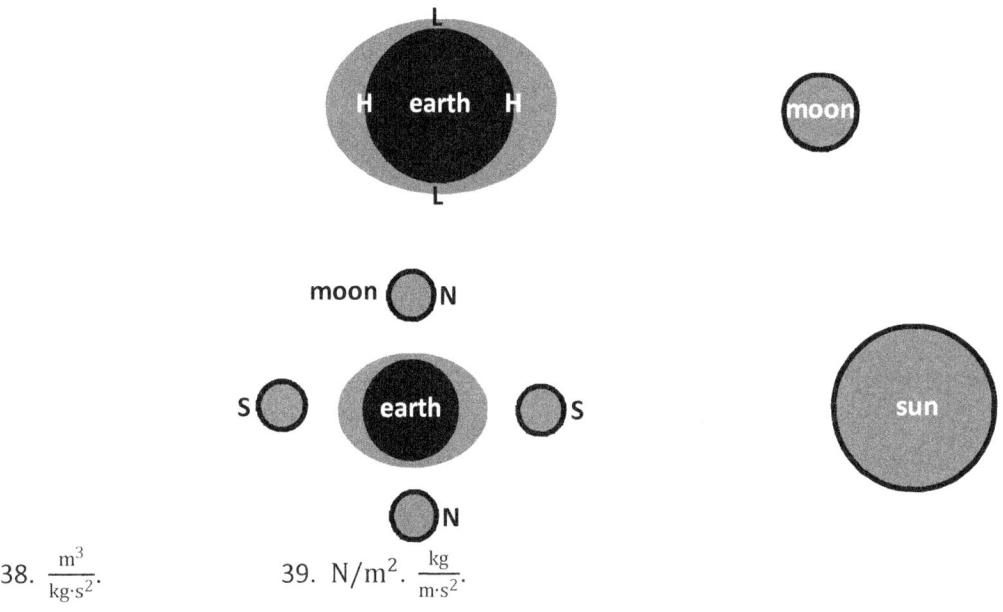

38. $\dfrac{m^3}{kg \cdot s^2}$.    39. N/m$^2$. $\dfrac{kg}{m \cdot s^2}$.

## Chapter 3 – Answers to Practice Problems

1. 81.6 kg, 801 N, 81.6 kg, 133 N    2. 123 kg, 1.5 kN
4. (A) (i) 0.49 kN (ii) 0.34 kN (B) 6.1 m/s$^2$ (upward)    5. (A) $-1.3 \times 10^6$ m/s$^2$ (B) 32 kN
6. (A) $-9.81$ m/s$^2$ (downward) (B) $-5.31$ m/s$^2$ (downward) (C) 2.69 m/s$^2$ (upward)
(D) 0 (F) 1.96 kN    8. (A) 5.1 m/s$^2$ (B) 0.89 kN, 2.1 kN
9. (A) 1.6 m/s$^2$ (B) 1.6 m/s$^2$ (C) 0.16 kN    11. (A) 0.51 m/s$^2$ (B) 0.26 kN, 0.77 kN
13. 0.54 m/s$^2$ (right)    15. $a_x = g(\sin\theta - \mu\cos\theta)$    16. 290 kg
17. 238 N    18. (A) 0 to 0.16 kN down the incline, or 0 to 2.1 kN up the incline
(B) 0.81 m/s$^2$ (C) 7.0 m/s    20. 21°    24. $a_x = g\tan\theta$

25. (A) 12 m/s² (B) 0.20 kN    27. $P = (m_1 + m_2)g \tan \theta$
28. 17 cm    29. (A) −4.9 m/s² (downward) (B) 35 m/s² (upward) (C) 75 m/s² (upward)
32. (A) 22°    34. (A) m/s³, m/s², m/s (B) −4.0 m/s², 270° (C) 0.20 kN, −24°
(D) 4.40 × 10² kg·m/s, −33° (E) 34 m, −32°    37. (A) (20 m, 0) (B) 8.0 m/s, 90°
(C) 16 kg·m/s, −159° (D) 3.2 m/s², 249° (E) 6.4 N, 249° (F) 14 s + 16 n s (n = integer)
39. $v_t = \frac{2mg}{\rho AD}$  40. $v_x = v_{x0} e^{-\frac{b\Delta t}{m}}, \Delta x = \frac{mv_{x0}}{b}\left(1 - e^{-\frac{b\Delta t}{m}}\right); v_x = \frac{mv_{x0}}{m+bv_{x0}\Delta t}, \Delta x = \frac{m}{b} \ln\left(\frac{m+bv_{x0}\Delta t}{m}\right)$
41. $v_x = v_{x0} + \frac{ct^2}{2m}, \Delta x = v_{x0}t + \frac{ct^3}{6m}$    42. $x = x_m \cos\left(\sqrt{\frac{k}{m}}t\right), v_x = -x_m\sqrt{\frac{k}{m}}\sin\left(\sqrt{\frac{k}{m}}t\right)$
43. (A) 1.62 m/s² (B) 1.99 × 10²⁰ N (C) 1.99 × 10²⁰ N    44. (B) 2.64 × 10⁶ m
45. (A) 1.6 nN (B) 1.2 Gg    47. (A) 5.45 m/s² (B) 18.3 kg, 82.6 N
49. It's 90% of the average earth-moon distance, closer to the moon. However, you should state your answer as a distance, not as a percentage.
50. (A) $7m_p/8$ (B) $\vec{F} = -\frac{41Gm_pm_t}{450R_0^2}\hat{\imath}$

## Chapter 4 – Answers to Conceptual Questions

1. The net angular displacement and net tangential displacement keep track of how far the object has traveled, in addition to whether the object is displaced in a counterclockwise or clockwise sense. In 2D polar coordinates, $\pi$ rad and $-\pi$ rad correspond to the same point on the circle, but in one case the angular and tangential displacement are positive and in the other case they are negative, which distinguishes between which way the object traveled to get there. Similarly, $2\pi$ rad and $6\pi$ rad are the same point in 2D polar coordinates, but the angular and tangential displacements differ, as they indicate whether the object has traveled around just once or three times. Note that the net tangential displacement is still more like net displacement than like total distance traveled. For example, if you travel $\pi$ rad clockwise and then $\pi$ rad counterclockwise, the net tangential displacement is zero (but if you travel $3\pi$ rad clockwise and then $\pi$ rad counterclockwise, the net tangential displacement is $2\pi$ rad clockwise, whereas the net displacement is zero).
2. (A) Same for all. (B) Points on the equator.
3. (A) Center: zero. One-third from end: 3.3 m/s. One-fourth from end: 5.0 m/s. (B) Same.
4. (A) Greatest: ends. Zero: center. 2 times. (B) Zero; zero.
5. He changes direction. Centripetal. Tangential/angular.
6. (A) Frequency (but not in SI units). (B) Period.
7. (A) Lean: left. Car's acceleration: right. Necklace's acceleration: right. Net force on car: right. Net force on necklace: right. Feels pushed: left. Push relative to inertial reference frame: right. (B) The monkey has inertia, and wants to maintain constant velocity – i.e. to go off on a tangent. This outward tendency is really inertia – not a force at all. (C) The accelerating monkey perceives a centrifugal (outward) force. (D) From an inertial perspective, the monkey is really pushed inward (the net force is centripetal, not centrifugal). This centripetal (inward) force pulls the monkey off its natural tendency, causing the monkey to instead travel in a circle. (E) A centripetal force.

# Answers

8.

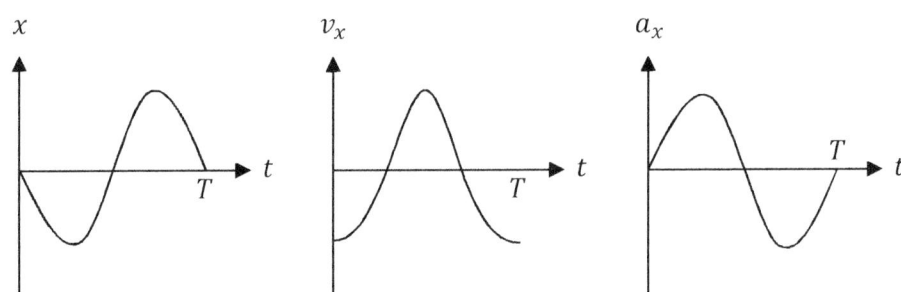

9. A sine wave.   10. (A) $v_{cm} = R_0\omega$. (B) $\Delta x = R_0 \Delta\theta$.

11. Angular speed doubles. Period halves. Frequency doubles. Acceleration quadruples. Net force quadruples.

12. Angular speed halves. Period doubles. Frequency halves. Acceleration halves. Net force halves.

13. (A)/(B) Number of revolutions (which may not be an integer).

14. Greater speed results in more tension; an increase in speed will eventually result in more tension than the cord can withstand.

15. Decreases (i.e. the cord becomes more horizontal).

16. (A) Greatest at the bottom, least at the top. (B) Yes: At the top, if the Ferris wheel is fast enough that $a_c \geq g$. The monkey could fall off.

17. Friction could act up or down the incline, and the static friction force has a magnitude $f_s \leq \mu_s N$. Friction acting down the incline corresponds to the greatest speed.

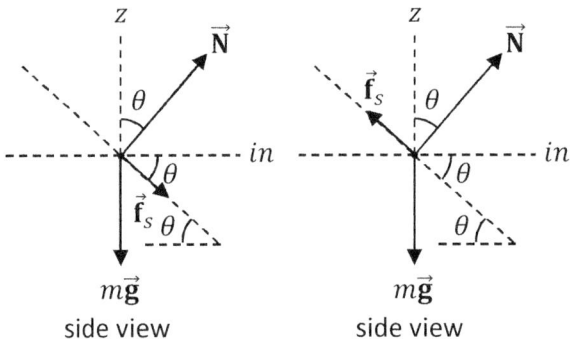

18. The moon is a satellite of earth with a period of about one month. Since orbital period depends only on the distance of the satellite from earth – and not on the mass of the satellite – any satellite of earth with a period of about a month would have an orbital radius of about the same size as the moon's orbital radius.

19. Speed is double. Angular speed is double. Acceleration is quadruple. Period is half. Frequency is double.

20. Radius quadruples. Angular speed is one-fourth. Acceleration is one-fourth. Period quadruples. Frequency is one-fourth.

## Chapter 4 – Answers to Practice Problems

1. (A) 0.105 rad/s, 0.00175 rad/s, 0.000145 rad/s  (B) 3.67 rad     2. 39 cm
3. (A) $7.27 \times 10^5$ rad/s  (B) 463 m/s  (C) 401 m/s  (D) 0.0337 m/s²  (E) 1.41 hr, weightless (normal force would equal zero)
4. (A) 0.199 μrad/s (note the metric prefix μ)  (B) 29.9 km/s  (C) 5.95 mm/s²  (D)/(E) $3.56 \times 10^{22}$ N
5. (A) 6'40"  (B) 0.0025 Hz  (C) 0.016 rad/s  (D) 3.1 m/s  (E) 0.94 km  (F) 0.28 km  (G) 0.049 m/s²
(H) 0.99 N  (I) 0.0050     6. (A) 5.24 rad/s  (B) 7.85 m/s  (C) 13.4°  (D) 1.54 (but <u>not</u> 1.50 m!)
(E) 41.1 m/s²  (F) centripetal  (G) inward (horizontally toward the center of the circle – <u>not</u> along the tail!)  (H) 8.46 N (but <u>not</u> 8.22 N!)     9. $T = 2\pi\sqrt{\frac{\mu_s R_0}{g}}$     10. 16 m/s ($R_0$ is <u>not</u> 10 m!)
11. (A) 19 m/s (maybe this is the Granny Gran Prix...)  (B) 30 m/s
12. (A) 0.25 rad/s  (B) .87 kN  (C) 28 m/s     14. (A) $R_0 \omega_0$  (B) $2\pi/\omega_0$  (C) $\omega_0/(2\pi)$
(D) $R_0 \omega_0^2$  (G) $mg \tan\theta$  (H) $\sqrt{R_0 g}$     15. (A) 3.08 km/s  (B) $3.59 \times 10^4$ km
17. (A) $2.93 \times 10^3$ m/s²  (B) 0.119 kg  (C) 0.669 N  (D) $5.28 \times 10^{20}$ N  (E) $2.95 \times 10^{-7}$ rad/s
(F) 1.94 km/s  (G) $5.72 \times 10^{-4}$ m/s²  (H) $5.28 \times 10^{20}$ N  (I) 162 s
19. (A) $\frac{\text{N·m}^4}{\text{kg}^5}$  (B) $v = m_1 m_2 \sqrt{\frac{H}{r^3}}$  (C) $T = \frac{2\pi}{m_1 m_2}\sqrt{\frac{r^5}{H}}$

## Chapter 5 – Answers to Conceptual Questions

1. Product: parallel. Zero: perpendicular. Negative product: anti-parallel. Negative: more than 90°. Positive: Less than 90°.
2. Zero.
3. Divide the scalar product by the product of their magnitudes, then take an inverse cosine.
4. Gains speed: yes, tension has a horizontal component. Constant velocity: no, tension is vertical. Loses speed: negative work is done, tension has a horizontal component (pulling backward).
5. No. It is always perpendicular to the instantaneous velocity.
6. Climbing/gravity: negative. Downhill/friction: negative. Uphill/normal force: zero. Downhill/gravity: positive. Uphill/friction: negative. Level/gravity: zero.
7. Rocket blast: negative. Circular motion: zero. Elliptical receding: negative. Meteor falls: positive.
8. From equilibrium: negative. Toward equilibrium: positive.
9. Same (same change in height).
10. Both are zero (net work is zero because net tangential force is zero; centripetal force does no work; tangential force affects only the speed, centripetal force affects only the direction).
11. Same (electric field, like gravity, is conservative – it is path-independent).
12. Semicircular path (greater distance traveled). In Question 11, the force is uniform, whereas the force changes direction in Question 12 for the semicircular path. Question 11 involves net displacement, while Question 12 involves total distance traveled (look at the scalar product in the work integral).

Answers

13. The differential element $d\vec{s}$ has components $dx, dy$, and $dz$, and a magnitude equal to $ds = \sqrt{dx^2 + dy^2 + dz^2}$. For these infinitesimal distances, there is no distinction between a curve and a right triangle. When integrating over $d\vec{s}$ or $ds$, a finite distance results, for which there is, in general, a significant distinction between the net displacement (forming a right triangle in 2D, and the diagonal of a rectangular box in 3D – with components along the edges and magnitude along the net displacement) and arc length (the total distance along the curved path joining the initial and final positions). Only for a straight path (with no reversal) is the magnitude of the net displacement equal to the arc length.

14. Path-independent. It's the net displacement: $d\vec{s} = \vec{v}dt$ since $\vec{v} = d\vec{s}/dt$.

15. Path-dependent. It's the total distance traveled: $ds = v\,dt$ since $v = ds/dt$.

16. Same (gravity is conservative – the work is path-independent).

17. Semicircular path (friction is nonconservative; in particular, a greater distance traveled results in more negative nonconservative work done by friction).

18. (A) Work (<u>not</u> power!): A Watt is a Joule per second, so a Watt times an hour is 3600 J.  (B) 3,600,000 J.

19. Straight line. Average power equals $\vec{F} \cdot \bar{\vec{v}}$ in each case, where $\bar{\vec{v}}$ is the average velocity (net displacement over total time). In both cases, the average velocity has the same direction and the net displacement is the same in both cases, but the motor takes more time for the semicircular path.

20. Same. Average power equals $F\bar{v}\cos 0°$ in each case, where $\bar{v}$ is the average speed (total distance over total time). Average speed is the same in each case. Question 19 involves average velocity, while Question 20 involves average speed.

21. Wind energy, energy from coal, and energy from natural gas have *our* sun as their ultimate energy source. The sun has a major influence on wind, through its impact on a significant temperature variation between the equator and poles. Coal comes from dead plants, which depend on the sun for their energy (through photosynthesis). Similarly, natural gas came from dead plants and animals, which depended on the sun for their energy. The radioactive isotopes for nuclear energy ultimately came from exploding stars (not *our* sun, but a sun nonetheless). The energy of tides is primarily due to the moon; the sun has a smaller influence on earth's tides than the moon.

22. (A) 4:1. (B) 2:1. (C) 8:1.  23. Momentum.  24. $K = \frac{p^2}{2m}$.

25. If the force is nonconservative, it does not have a potential energy. If a monkey suddenly enters the problem and does work, we treat this as nonconservative work. If a conservative force enters the problem (like an exploding star), we treat this by including the corresponding potential energy. Potential energy does not refer to the potential for disaster (or the potential for an unexpected influence, for example), but work that is actually done as a result of an object in the system changing position. If some event was unexpected, to account for it we either include it with potential energy or nonconservative work in the mathematics, depending upon whether the force is conservative.

26. Potential energy increases. Kinetic energy decreases. Total mechanical energy is constant.

27. Potential energy decreases. Kinetic energy increases. Total mechanical energy is constant.

28. Potential energy: minimum at bottom, maximum at turning points. Kinetic energy: zero at turning points, maximum at bottom. Total mechanical energy is constant.

29. Potential energy: zero at equilibrium, maximum when fully compressed/stretched. Kinetic energy: zero when fully compressed/stretched, maximum at equilibrium. Total mechanical energy is constant.

30. Rising: $KE \to PE$. Toward equilibrium: $PE \to KE$. Downward: $PE \to KE$. Level, slowing down: $KE \to W_{nc}$(heat). Fan: $PE$ (electrical) $\to KE + W_{nc}$(heat). Burning: $PE$ (chemical) $\to W_{nc}$(heat) $\to KE + W_{nc}$(exhaust). Monkey accelerates: $PE$ (metabolic) $\to KE + W_{nc}$(heat).

31. Gaining. Yes, because $U_g = -Gm_pm/R$: As $R$ increases, potential energy increases (becoming less negative), approaching zero as $R$ approaches infinity. (Note that $mgh$ is in appropriate for a rocket leaving earth.)

32. Spring: conservative. Earth: conservative. Friction: nonconservative. Monkey: nonconservative. Motor: nonconservative.

33. (A) Slope: The direction of the gradient tells you which way is steepest. (B) It's greater where the density of contour lines is greater. (C) The direction you would travel to get to the next equipotential (contour 'line' – which is generally a curve, even though we call it a line) in the shortest distance. (D) They lie at the centers of sets of closed equipotentials (peaks and wells). (E) Wells are stable, where potential energy is a relative minimum; peaks are unstable, where potential energy is a relative maximum. A point of inflection is neutral.

## Chapter 5 – Answers to Practice Problems

1. (A) $-96$ J (B) 6.4 N, 39° (C) 36 m, 153° (D) 115° (E) $-96$ J Of course, the answer to part (E) is *exactly* the same as part (A). You will obtain the same answer to two significant figures if you keep extra digits throughout the calculation and only round at the end of the calculation.

2. 60°   3. $\cos\theta, \cos\theta, 1, 0, 1$   4. $\cos\varphi\sin\theta, \cos\varphi, -\sin\theta, 1, 0, 1, \sin\theta$

5. (A) (i) 60° (ii) 90° (iii) $-180°$ (iv) $m\vec{g}$ (v) $\sum \vec{F}_{ext}$ (B) (i) 96 J (ii) 4.0 N (iii) 330° (iv) 14 W

6. (A) 8.0 J (B) 13 J (C) 80 W   8. (A) 2.4 kJ (B) 0 (C) $-3.2$ kJ (D) 4.7 hp

9. (A) $-7.0$ J (B) 72 J (C) 24 W, 6.0 W   13. (A) N/m², N/m (B) 1.5 J (C) yes

15. (A) 50 hp (B) 0.094 N/m²   16. $-23$ kJ

17. (A) $-0.12$ kJ (B) 40 J (C) 0.16 kJ (D) 40 m/s (E) 0.12 kJ (F) 0 (G) 25 W

19. (A) 0.24 kJ (B) 29 m/s   20. 31 m/s   22. (A) 40 m/s (B) $-0.32$ MJ (C) 2.0

**Note**: The coefficient of friction is so often less than one in problems that sometimes students develop the naïve impression that it can't be greater than one, but that's not true – materials have been made where the coefficient of friction between two of them actually exceeds one.

23. (A) 40 m/s (B) 15 N/m²   24. (A) $mgR_0$ (B) 0 (C) $mgR_0 \cos\theta$ (D) $\sqrt{2gR_0(1-\cos\theta)}$ (E) $mg(3\cos\theta - 2)$ (F) $\sqrt{gR_0\cos\theta}$ (G) $\cos^{-1}(2/3)$

25. (A) $L(1-\cos\theta), 0$ (B) $0, \pm\hat{\boldsymbol{\theta}}\sqrt{2gL(1-\cos\theta)}$ (C) $\pm\hat{\boldsymbol{\theta}}g\sin\theta, -2g\hat{r}(1-\cos\theta)$ (D) $-mg\hat{r}\cos\theta, -mg\hat{r}(3-2\cos\theta)$ (E) $\cos^{-1}\left(\frac{2+\sqrt{3}}{4}\right)$   26. $0, 39.5$ cm/s, $38.7$ cm/s

27. (A) 30 m/s (B) 0.83 kN (C) 0.50 kN/m   29. (A) 15 m/s (B) 5.0 m (but *not* 4.5 m!)

33. (A) $1.11 \times 10^7$ m, 2.36 km/s, $2.22 \times 10^7$ m, 1.67 km/s (B) $-4.16$ GJ (C) 29.2 km/s

35. (A) 13 J (B) $U_2 = -\frac{2x^3}{3} - \frac{3x^2}{2} + U_{ref}$ (C) 7.0 W (D) 2.1 m (recall that $r$ can't be negative in spherical coordinates) (E) yes, yes

36. (A) 21 J (B) 8.0 J (C) no   38. non-conservative   39. 4.0 J, no

40. 0, yes   41. 0, yes   42. $32\pi$ J, no

# Chapter 6 – Answers to Conceptual Questions

1. (A) Time. (B) A monkey walks 1.0 m/s for one minute and 3.0 m/s for one hour. The average speed is very nearly 3.0 m/s, and not 2.0 m/s, because almost all of the time was spent traveling 3.0 m/s.
2. The object could have a cavity or be concave, as in the examples below.

3.
$$\vec{r}_q = \frac{\sum_{i=1}^{N} q_i \vec{r}_i}{\sum_{i=1}^{N} q_i} \quad , \quad \vec{r}_E = \frac{\sum_{i=1}^{N} q_i \vec{E}_i \vec{r}_i}{\sum_{i=1}^{N} q_i \vec{E}_i}$$

4. Water. It floats. Contrary. It expands. Pipes may burst.
5. (A) $-\sqrt{R_0^2 - y^2} \leq x \leq \sqrt{R_0^2 - y^2}$, $0 \leq y \leq R_0$. (B) $0 \leq y \leq \sqrt{R_0^2 - x^2}$, $-R_0 \leq x \leq R_0$. (C) $0 \leq \theta \leq \pi$ rad, $0 \leq r \leq R_0$.
6. $\vec{r}_{CM}^{s}$ = CM of original cookie, $A_s$ = area of original cookie, $\vec{r}_{CM}^{c}$ = CM of circular cutout, $A_c$ = area of circular cutout, $\vec{r}_{CM}^{r}$ = CM of remainder of cookie, $A_r = A_s - A_c$ area of remainder of cookie. Setup coordinates with original square cookie at origin: $\vec{r}_{CM}^{s} = 0$.
$$\vec{r}_{CM}^{s} = 0 = \frac{A_c \vec{r}_{CM}^{c} + A_r \vec{r}_{CM}^{r}}{A_s}$$
Everything is known except for $\vec{r}_{CM}^{r}$; solve for $\vec{r}_{CM}^{r}$.
7. (A) $\pi : 4$. (B) $\pi : (4 - \pi)$.
8. Large tree. Same impulse, so a shorter duration increases the average collision force.
9. No. Yes; normal components (but this is not useful – the net normal component of the forces acting on each box of bananas is individually zero before, during, and after the collision).
10. No. Yes; horizontal components. Right (assuming the system starts from rest). Starting from rest, the system has no momentum to begin with. The monkey gains momentum to the left, so the wedge must gain an equal amount of momentum to the right in order to conserve the horizontal component of momentum (it's conserved since the net horizontal force is zero). (The vertical component of momentum is not conserved since there is a net vertical force. The monkey gains momentum downward, while the wedge does not move vertically.)
11. No. Yes. The banana moves upward. Released from rest, the initial momentum of the system is zero, so when the banana gains momentum downward, the earth gains momentum upward. The banana pulls the earth upward with the same magnitude of force as the earth pulls it downward. The earth has an enormous mass compared to the banana, so its acceleration is much too insignificant to notice.
12. The astronaut should throw the wrench directly away from the ship so that the opposite momentum of the astronaut is directed toward the ship.
13. (A) The plank moves north. (B) The plank stops. (C) The center of mass of the plank moves west, and the plank rotates as it moves. The center of mass is stationary throughout.
14. (A) Momentum only, completely inelastic. (B) Momentum only if inelastic; momentum and mechanical energy if elastic. The center of mass of the system has constant velocity.

15. (A) The monkey and plank will slow down – and if the watermelon has a large enough momentum, they could even reverse direction. (B) The monkey and plank will speed up.

16. The monkey and plank will develop a component of velocity to the south, while retaining their component of velocity to the east. The result will be southeasterly (but probably not perfectly southeast).

17. The monkey should throw the watermelon southeasterly. The eastern component of the watermelon's momentum is needed for the monkey and plank to stop traveling east, and the southern component of the watermelon's momentum gives the monkey and plank momentum to the north.

18. The monkey's uncle heads northwesterly relative to the monkey's aunt. The monkey's aunt heads southeasterly relative to the monkey's uncle. (In both cases, the precise angle depends on the speeds.)

19. (A) They swap velocities. (B) The cue ball's momentum is transferred to the last ball: The cue ball stops, while the last ball takes off with the cue ball's initial velocity.

20. No. The system loses mechanical energy to heat and sound during the completely inelastic collision at the bottom. No (the final momentum of the system is zero – all of the momentum is lost on the swing upward). Include the earth as part of the system. The earth must go just the slightest amount left and down. The amount is totally imperceptible as the earth has an astronomical amount of mass compared to the bullet and block.

21. Most of the mass and all of the positive charge are concentrated in very tiny spaces (nuclei). The alpha particles interacted electrically with this positive charge, deflecting significantly only in those rare instances where the alpha particle happened to approach very close to a nucleus.

22. On average, red photons take a more direct route, violet photons scatter more, and all of the other colors scatter varying amounts in between, in order according to ROY G. BIV. When we look at the sun, we see a greater percentage of reds, a smaller percentage of oranges, and so on, with the smallest percentage of violets. All of these different photons reach our eye at once, and what we see usually looks red or yellow – a shift toward longer wavelength, since longer wavelengths scatter less on average. When we look at the sky, but not toward the sun, we see photons which have scattered one or more times. In this case, we see a greater percentage of violets and blues, and the least percentage of oranges and reds. All of these different photons appear blue when they reach our eye together – a shift toward shorter wavelength, since shorter wavelengths scatter more on average.

At sunrise or sunset, sunlight travels a greater distance through earth's atmosphere than it does at noon (try drawing a diagram to see this, geometrically – and note the key word, 'atmosphere'). Therefore, more scattering occurs at sunrise and sunset than at noon. Hence, we see a longer wavelength (red) at sunrise or sunset than we do at noon (yellow) when we look directly at the sun.[348]

23. The truck and boat/trailer accelerate at the same rate (since the acceleration of an object down an incline is independent of mass).

24. (A) Same acceleration as if the dirt didn't fall out. (B) Increased acceleration compared to not shoveling the dirt out.

---

[348] Remember, if you value your retinas, do not look toward the sun.

# Answers

## Chapter 6 – Answers to Practice Problems

1. 5.0 m from the 90-kg monkey
2. $(0, -1.6 \text{ m})$
4. $(1.5L, 0.77L)$
6. 75 kg
8. $(1.7 \text{ m}, 1.7 \text{ m})$
9. $(1.7 \text{ m}, 4.0 \text{ m})$
10. $(22 \text{ km}, 21 \text{ km})$
11. 118 N, 275 N
12. $\left(-\frac{R_0}{6}, 0\right)$
15. (F) $\frac{32}{3}\sigma$ (H) 0 (I) $\frac{12}{5}$ m
17. (A) $\frac{3m}{7\pi R_0^3}$ (B) $\frac{45 R_0}{14\pi}\hat{\imath}$
18. $\left(\frac{2R_0}{\pi}, \frac{2R_0}{\pi}\right)$
19. (A) $\left(\frac{L}{2}, \frac{L\sqrt{3}}{6}\right)$ (B) 5:4
20. (A) $-0.70\hat{\jmath}$ kg·m/s (B) $0.63\hat{\jmath}$ kg·m/s (C) $44\hat{\jmath}$ N (D) 0.98 J
21. (B) 0.67 m/s, south (C) 0.13 kJ
22. (A) 5.3 m/s, west (B) 3.0 m/s, 37° north of east
23. 22 m/s, south; 0.12 km/s, north
24. (A) 6.0 m/s (B) north (C) no (D) 700 N
25. (A) 5.0 m/s, south; 7.0 m/s, north (B) 97%
27. $\frac{6\sqrt{7}}{7}mv_0$, 79°
28. $\frac{11v_0}{3}$, south; $\frac{4v_0}{3}$, north
29. (A) 2.0 m/s, north (B) 24 m/s, north (if thrown together)
31. 5.4 m/s, 13° south of east
32. 25° north of east or 25° north of west
33. (A) 1.1 m/s, 18° (B) 0.82 kJ (lost) (C) $1.6 \times 10^2$ kg·m/s, 175° (remember, impulse is a vector)
34. (A) $\theta = \cos^{-1}\left(\frac{5+4\cos\theta_0}{9}\right)$ (B) 56° (C) 46°
40. (A) 31% (B) 37 s
41. (A) 34% (B) $1.7 \times 10^2$ kg/s (C) 0.34 MN (D) 28 m/s² (E) 39 m/s²

## Chapter 7 – Answers to Conceptual Questions

1. (A) 1:1. (B) 2:1. (C) 2:1. (D) 2:1.
2. (A) Only tangential acceleration is nonzero. (B) Only centripetal acceleration is nonzero. (C) All are zero. (D) All are zero. (E) None are zero. (At the turning points, centripetal acceleration is momentarily zero, and at the bottom, angular and tangential acceleration are momentarily zero.)
3. Rolling, spinning in air, held at one end (based on how far the mass is, on average, from each axis of rotation).
4. Bisector (through centroid and in the plane of the triangle), perpendicular (through centroid, but perpendicular to triangle), edge (not through centroid). Answer is based on how far the mass is, on average, from each axis of rotation (and keep in mind that $r_\perp$ is squared in the integral). Passing through the centroid has a significant advantage, as you can see from the parallel-axis theorem.
5. The sphere where density increases from the center (more mass is further from the axis of rotation on average).
6. $\frac{R_0\sqrt{2}}{2}$ from the center. $R_0$ (about this axis; compare to rolling mode for a disc, where the axis passes through the center, where the radius of gyration is $\frac{R_0\sqrt{2}}{2}$). The furthest point on the disc is a distance $2R_0 - \frac{R_0\sqrt{2}}{2}$ from the axis of rotation (if instead the disc were in rolling mode, the furthest point on the disc would be $\frac{R_0}{2}$ from the axis, which would be smaller).

7. $V = \vec{a} \cdot (\vec{b} \times \vec{c}) = \vec{b} \cdot (\vec{c} \times \vec{a}) = \vec{c} \cdot (\vec{a} \times \vec{b}) = a(bc \sin \theta_{bc}) \cos \theta_{ah}$, where $\vec{a}$, $\vec{b}$, and $\vec{c}$ are vectors along edges of the parallelepiped that meet at a corner, $\theta_{bc}$ is the angle between $\vec{b}$ and $\vec{c}$, and $\theta_{ah}$ is the angle between $\vec{a}$ and the height of the parallelepiped above the base (formed by $\vec{b}$ and $\vec{c}$). Draw a picture of the parallelepiped, and draw and label these three vectors, two angles, and one altitude to see this.

8. Only the coordinate system on the left is not.

9. (A) $\otimes$. (B) $\otimes$. Note the order of the vectors. (C) ←. (D) ↓. (E) Zero ($\theta = 180°$). (F) $\otimes$. (G) ↗. (H) $\odot$. (I) ↓. (J) $\otimes$. Note which vector comes first (look up the formula). (K) ↑. Note which vector comes first (look up the formula). (L) ↖. Note which vector comes first (look up the formula). (M) $\otimes$, ↑. Look up the formulas to see which vector comes first in each case.

10. $\hat{r} \times \hat{r} = 0$. $\hat{r} \times \hat{\theta} = \hat{\varphi}$. $\hat{\theta} \times \hat{r} = -\hat{\varphi}$. $\hat{\theta} \times \hat{\theta} = 0$. $\hat{r} \times \hat{\varphi} = -\hat{\theta}$. $\hat{\varphi} \times \hat{r} = \hat{\theta}$. $\hat{\varphi} \times \hat{\varphi} = 0$. $\hat{\theta} \times \hat{\varphi} = \hat{r}$. $\hat{\varphi} \times \hat{\theta} = -\hat{r}$.

11. Assume all of the components of the vectors to be nonzero. (A) Left: 3. Right: 1. All nonzero. (B) Left: 9 (3 nonzero). Right: 3 (all nonzero). (C) Left: 1 (nonzero). Right: 9 (2 nonzero). (D) Left: 3. Right: 2. Can't tell how many are zero. (E) Left: 1 (nonzero). Right: 729 (6 nonzero).

12. Top left: $r_\perp = 10$ cm, $\theta = 90°$, $\otimes$. Top center: $r_\perp = 12$ cm, $\theta = 30°$, $\otimes$. Top right: $r_\perp = 6$ cm, $\theta = 90°$, $\otimes$. Bottom left: $r_\perp = R_0$, $\theta = 90°$, $\otimes$. Bottom center: $r_\perp = 5$ cm, $\theta = 120°$, $\otimes$. Bottom right: $r_\perp = 12$ cm, $\theta = 60°$, $\otimes$.

13. The screwdriver with the blue handle can apply the most torque; the other two are equal.

14. The right end weighs more (it's closer, on average, to the fulcrum, so it must have more weight in order that the two torques have equal magnitudes).

15. It rotates counterclockwise (falling to the left). The fulcrum is to the right of the balancing point. (It's balanced where the torques are equal in magnitude, not where the weights are equal.)

16. In $\vec{C} = \vec{A} \times \vec{B}$, $\vec{C}$ is perpendicular to the plane that contains $\vec{A}$ and $\vec{B}$, and $C = AB \sin \theta$. When solving for $\vec{B}$ given $\vec{A}$ and $\vec{C}$, there are an infinite number of directions that $\vec{B}$ could have – a whole plane where $\vec{A}$ and $\vec{B}$ will be perpendicular to $\vec{C}$ and satisfy $C = AB \sin \theta$. Algebraically, you get an expression that is indeterminate.

17. No. The final speed (and hence tangential acceleration) are independent of mass and radius, which cancel out in the algebra.

18. Same. The final speed is independent of mass, which cancels out in the algebra.

19. The solid cylinder rolls further up the incline since its mass is further from the axis of rotation, on average. The solid cylinder has a greater coefficient in its moment of inertia formula, and so it does not decelerate as quickly as the solid sphere.

20. The solid cylinder reaches the bottom first since its mass is closer to the axis of rotation, on average. (Note that one is solid, while the other is hollow.) The solid cylinder has a smaller coefficient in its moment of inertia formula, and so it has greater acceleration than the hollow sphere.

21. The hollow ball rolls furthest up the incline, while the ball filled with liquid water rolls the shortest distance up the incline. The hollow ball has more of its mass further from the axis of rotation compared to the ball with frozen water, so the hollow ball does not decelerate as quickly as the ball with frozen water. The ball with liquid water has the greatest deceleration, since only a fraction of its mass is rotating: Most of the mass only needs to have its translational inertia overcome, only a fraction of the mass also needs to have rotational inertia overcome.

Answers

22. $\omega_z = \sqrt{2\alpha_z \Delta\theta}$, $L^2/2I$, $\int \omega dI$, and $d\tau/dt$.

23. Although we often use the formula $v = R_0\omega$, which relates the magnitude of the velocity to the magnitude of the angular velocity, we can't write the same equation by simply replacing the magnitudes with the vectors because $\vec{v}$ and $\vec{\omega}$ don't point in the same direction. The correct expression is $\vec{v} = \vec{\omega} \times \vec{r}_\perp$.

24. Tension does not exert a torque, weight does. The net external torque is nonzero, so the total angular momentum is not conserved.

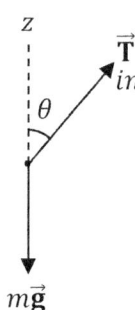

25. Tension and weight both exert torques. (Note that the axis of rotation is different in this case than it was in Question 24). The net external torque is zero, so the total angular momentum of the system is conserved.

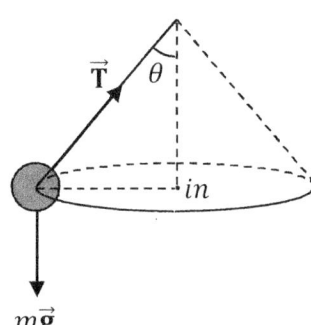

26. Her moment of inertia increases, her angular speed decreases, and her angular momentum remains constant.

27. The gravitational pull of the star is the only force, and it does not exert a torque on the comet. The net external torque is zero, so the angular momentum of the comet is conserved. The tangential speed of the comet increases.

28. Angular momentum and mechanical energy are conserved for this rotational elastic collision. The net external torque is zero.

29. The angular speed of earth's rotation decreases while they walk in order to conserve angular momentum, but returns to normal when they stop walking.

30. The train should travel east in order to slow earth's rotation through conservation of angular momentum.

## Chapter 7 – Answers to Practice Problems

1. (A) 0.13 km  (B) 50 rad/s
2. (A) 86 rad/s  (B) 4.5 s
3. (A) $60\pi$ rad/s  (B) 6.0 s
4. (A) $-9.4 \times 10^{-12}$ rad/s$^2$  (B) 25 rev
7. 5.0 min.
8. (A) $\frac{3\pi}{25}$ rad/s$^2$  (B) 9.2 m/s$^2$ (You can get a close answer with an incorrect solution. Check that you have $a_T = \frac{12\pi}{25}$ m/s$^2$ and $a_c = \frac{576\pi^2}{625}$ m/s$^2$.)  (C) $\frac{3}{4}$ rev  (D) $48\pi$ m  (E) $\frac{36\pi^2}{25}$ m/s$^2$  (F) $\frac{3\pi}{125}$ m/s$^2$  (G) 100 s  (H) $\frac{87\pi}{250}$ rad/s
10. (A) 0.36 km/s  (B) 0  (C) 2.9 km/s$^2$  (D) 0  (E) $-2.9\,\hat{\mathbf{r}}$ km/s$^2$  (F) 3.2 km
13. $\frac{7ML^2}{48}$
14. (A) $1.2 \times 10^3$ kg·m$^2$  (B) $33 \times 10^3$ kg·m$^2$
15. (A) $24\,ML^2$  (B) $100\,ML^2$  (D) $124\,ML^2$  (D) $92\,ML^2$
16. (A) 40 kg·m$^2$  (B) $1.0 \times 10^2$ kg·m$^2$
18. $\frac{3MR_0^2}{4}$
19. $\frac{13MR_0^2}{24}$
21. $\frac{7ML^2}{48}$
22. $\frac{b+1}{b+3}ML^2$
33. $-26$
34. 26
35. 0
36. $-120$
38. (A) $\hat{\mathbf{j}}$  (B) $-\hat{\mathbf{j}}$  (C) 0  (D) $-\hat{\mathbf{k}}$  (E) $5\hat{\mathbf{i}} + 2\hat{\mathbf{j}} + 4\hat{\mathbf{k}}$  (F) 0  (G) $-4\hat{\mathbf{i}} - 4\hat{\mathbf{k}}$  (H) $-\hat{\mathbf{i}} + 2\hat{\mathbf{j}} - 3\hat{\mathbf{k}}$
39. (A) $8\sqrt{3}$  (B) $-4\hat{\mathbf{i}} - 4\sqrt{3}\hat{\mathbf{j}}$  (C) $30°$
40. (A) $-\hat{\mathbf{i}}$  (B) 0  (C) $-\hat{\mathbf{i}} + \hat{\mathbf{j}}$  (D) $5\hat{\mathbf{i}} + 2\hat{\mathbf{j}} + 4\hat{\mathbf{k}}$  (E) 0
41. (A) $-208\hat{\mathbf{i}} - 390\hat{\mathbf{j}} + 104\hat{\mathbf{k}}$  (B) $82°$
42. $\hat{\mathbf{k}}\cos\theta$, $\hat{\mathbf{k}}\cos\theta$, $\hat{\mathbf{i}}\sin\theta - \hat{\mathbf{j}}\cos\theta$, $\hat{\mathbf{k}}$, $-\hat{\mathbf{j}}\sin\theta - \hat{\mathbf{i}}\cos\theta$
43. $\hat{\mathbf{j}}\cos\theta - \hat{\mathbf{k}}\sin\varphi\sin\theta$, $-\hat{\mathbf{k}}\sin\varphi$, $-\hat{\mathbf{i}}\sin\varphi\cos\theta + \hat{\mathbf{j}}\cos\varphi\cos\theta$, 0, $-\hat{\boldsymbol{\theta}}$, 0, $\hat{\boldsymbol{\varphi}}\cos\theta$
45. 3
46. 3
47. 0
48. 6
49. $2\vec{\mathbf{A}} \cdot \vec{\mathbf{B}}$
50. $A^2B^2$
54. (A) $1.2 \times 10^3$ Nm, $\odot$  (B) 0  (C) $5.0 \times 10^2$ Nm, $\odot$  (D) $1.6 \times 10^3$ Nm, $\otimes$  (E) $3.0 \times 10^3$ Nm, $\otimes$  (F) $2.9 \times 10^3$ Nm, $\otimes$
55. (A) $1.2 \times 10^4$ Nm, $\odot$; 0; $1.2 \times 10^4$ Nm, $\otimes$; 0  (B) 0.30 kN
56. (A) $3.1 \times 10^3$ Nm, $\otimes$  (B) $1.4 \times 10^3$ Nm, $\odot$  (C) 5.0 m to the left of the fulcrum
57. at the 58-cm mark
58. $1.3 \times 10^2$ kg
59. 118 N, 275 N
60. (A) 5.3 rev  (B) 0.20 kN  (D) 23 kg·m$^2$  (F) 4.0 m/s$^2$  (G) 2.7 rad/s$^2$, $\otimes$  (H) 20 m/s  (I) 13 rad/s  (J) 0.40 kN, 0.44 kN
62. 0.17 kN, 0.21 kN, 0.20 kN
63. (A) 2.6 kg  (B) 26 N, right; 70 N, up  (C) 0.56 rad/s$^2$, $\otimes$
65. The $M^{\text{th}}$ brick from the top can overhang as much as $\frac{L}{2M}$ beyond the edge of the brick below it.
66. $L/4$
67. (A) $a_x = -\frac{3}{5}g\sin\theta$ ($x$ is up the incline)  (B) $a_x = \frac{3}{5}g\sin\theta$ ($x$ is down the incline)  (C) $\frac{2}{5}\tan\theta$  (D) $\frac{2}{5}\tan\theta$
68. (A) 9.2 m/s, 8.4 m/s  (B) The solid sphere wins the race because gravity has more rotational inertia to overcome in order to accelerate the hollow sphere down the incline, since the hollow sphere has a greater percentage of its mass further from the axis of rotation compared to the solid sphere.
71. (A) (i) 37 m/s  (ii) 6.9 m/s  (B) 35 m/s  (C) $0.23\,MR_0^2$
73. (A) 28 m/s (The masses are tricky. You should have $\sqrt{2gh\frac{m}{m+m_w}}$, where $m$ is the total mass and $m_w$ is just the mass of the wheel.)  (B) 4.0 m/s$^2$  (C) 80 N  (D) 22 m/s  (E) 31 m/s
75. (A) 0.000027 kg·m$^2$  (B) 0.59 m/s$^2$  (C) 1.5 m/s  (D) 0.65 N  (E) 2.6 s
78. 6.0
79. 2.7 hr, $3.3 \times 10^3$ days
80. (A) 0.063 rev/s  (B) $-0.013$ rev/s$^2$ (deceleration)  (C) 0.24 rev

82.  (A) $1.0 \times 10^2$ kg·m/s  (B) 0.25 kJ  (C) 40 kg·m$^2$/s  (D) angular momentum (net external torque is zero)  (E) 4.4 rad/s  (F) 65% (lost)

84.  (A) 13 $\hat{\mathbf{k}}$ rad/s  (B) 3.2 kg·m$^2$  (C) $-63\,\hat{\mathbf{r}}_c$ m/s$^2$  (D) 1.3 kN  (E) 40 $\hat{\mathbf{k}}$ kg·m$^2$/s  (F) 0  (G) angular momentum (net external torque is zero)  (H) 20 $\hat{\boldsymbol{\theta}}$ m/s  (I) $2.0 \times 10^2$ $\hat{\mathbf{k}}$ rad/s  (J) 0.20 kg·m$^2$  (K) $-4.0\,\hat{\mathbf{r}}_c$ km/s$^2$  (L) 80 kN

## Appendix A – Answers to Practice Problems

1. $463.3 \pm 0.5$ kg
2. $12.74 \pm 0.24$ m/s
3. $0.0326 \pm 0.0012$ m
4. $14.49 \pm 0.04$ cm
5. $83.7 \pm 1.4$ lbs.
6. $\sigma_A = \sqrt{W^2 \sigma_L^2 + L^2 \sigma_W^2}$, $A = 91.3 \pm 1.3$ cm$^2$
7. $\sigma_\rho = \rho \sqrt{\left(\frac{\sigma_M}{M}\right)^2 + \left(\frac{\sigma_V}{V}\right)^2}$, $\rho = 0.9571 \pm 0.0017$ g/cm$^3$
8. $\sigma_c = \sqrt{\sigma_a^2 \sin^2 b + \sigma_b^2 a^2 \cos^2 b}$, $c = 1.50 \pm 0.04$ cm

## Appendix B – Answers to Practice Problems

1. $F$ (vertical), $z$ (horizontal); slope = $k$, vertical intercept = $-kz_e$
2. (A) $Q$ (vertical), $H$ (horizontal); slope = $1/K$, vertical intercept = 0
   (B) $Q$ (vertical), $1/K$ (horizontal); slope = $H$, vertical intercept = 0
   (C) $H$ (vertical), $K$ (horizontal); slope = $Q$, vertical intercept = 0
3. $q^4$ (vertical), $u^5$ (horizontal); slope = $a^2$, vertical intercept = 0
4. $d^3$ (vertical), $1/e^2$ (horizontal); slope = $9/f^4$, vertical intercept = 0
5. $D^2$ (vertical), $J^3$ (horizontal); slope = $m^2$, vertical intercept = $n^5$
6. $\beta^3$ (vertical), $\sin(4\alpha^2)$ (horizontal); slope = 8, vertical intercept = 9
7. $\gamma^2$ (vertical), $\tan(2\lambda)$ (horizontal); slope = $1/3$, vertical intercept = 0

# Index

| acceleration | |
|---|---|
| angular | 215,379-87,425,454 |
| average | 27,104-5,382 |
| center of mass | 330 |
| centripetal | 106,216,218-22,379, 425 |
| graphs | 33-43,382 |
| gravitational | 52-58,193-9 |
| instantaneous | 30-2 |
| non-uniform | 58-62,129-30,386 |
| one-dimensional | 21-2,27-32,43-62 |
| tangential | 21,29-32,43-58,106-7,216 379-80,425,454 |
| total | 104,106-7,217,219,379,425 |
| uniform | 43-58,382-6 |
| vector | 104,106-7,217,219,379,425 |
| accelerometer | 162,208p |
| action | 147 |
| additive inverse | 94 |
| air | |
| drag | 185-7 |
| resistance | 144,152,185-90 |
| angular | |
| acceleration | 215,379-87,425,454 |
| analogies | 454t |
| displacement | 215,379,454 |
| frequency | 217-8 |
| graphs | 382 |
| momentum | 426-7,451-60 |
| position | 214 |
| speed | 215,379,382,454 |
| velocity | 215,379-80,382,425,453-4 |
| apparent weight | 157-8,185,198,240 |
| arc length | |
| differential | 97-101,108,113,116,251, 310 |
| one-dimensional | 22 |
| total distance traveled | 23,25,31-2,97-101,215,283, 379,381 |

| area | |
|---|---|
| differential elements | 310-1 |
| parallelogram | 407-8 |
| under a curve | 33-43 |
| arithmetic mean | 306-7 |
| arrows in 3D | 410 |
| associative property | 248,408 |
| Atwood's machine | 163-4,430-1 |
| average | |
| acceleration | 27,104-5,382 |
| angular acceleration | 382 |
| angular speed | 382 |
| angular velocity | 382 |
| force | 321-4 |
| power | 261-2 |
| speed | 24-5,103-4,382 |
| tangential acceleration | 27 |
| tangential velocity | 26 |
| value | 23-4,306-7 |
| velocity | 26,103-4,382 |
| weighted | 307 |
| axis of rotation | 390-1,403 |
| azimuthal angle | 113-4 |
| ballistic pendulum | 352-5 |
| banked curve | 235,243p |
| banking angle | 235,243p |
| billiards | 328,343,350,355,369p |
| boom | 439 |
| buoyancy | 185 |
| burn rate | 363-4 |
| Cartesian coordinates | 82,85-6,109,114-5,310-1 |
| center | |
| of gravity | 308,319-20 |
| of mass | 306-20,329-34 |
| of mass reference frame | 333-4,347-348,351-2 |
| centrifugal force | 222-3 |
| centrifuge | 243p |

| | | | | |
|---|---|---|---|---|
| centripetal | | | curve sketching | 41-3 |
| | acceleration | 106,216,218-22,379,425 | cylinder | 399 |
| | force | 222-5 | deflection angle | 347 |
| chain rule | | 62,197,219,268 | degrees of freedom | 19,76 |
| circular motion | | 214-40,378-81,383 | density | 309 |
| closed integral | | 281,289-90 | derivative | 28-9,284 |
| coefficient of friction | | 168 | determinant | 405-6 |
| cofactors | | 405 | differential | |
| collisions | | | | arc length | 97-101,108,113,116,251, 310 |
| | ballistic pendulum | 352-5 | | |
| | completely inelastic | 336-7,341,344 | | area element | 310-1 |
| | elastic | 335-9,341,343-4,349-52 | | cross section | 358-9 |
| | forces | 321-3 | | displacement vector | 97-101,251 |
| | inelastic | 335-7,341,344 | | volume element | 311 |
| | inverse | 336-7,341,344 | dimensionality | 19,76 |
| | one-dimensional | 335-43 | dimensions | 14-5,17p |
| | scattering | 335-9 | direction | 78,83-5,117 |
| | short-lived | 328 | disc | 398 |
| | two-dimensional | 344-67 | discrete system | 307-8 |
| comet | | 463p | displacement | |
| commutative property | | 248,408,470p | | displacement vector | 97-101,251 |
| completely inelastic | | 336-7,341,344 | | net | 23,30-2,97-101,107-8,283, 379-81,454 |
| components | | 81-85 | | |
| Compton effect | | 376p | | tangential | 21,30-2,215,379-80,454 |
| cone | | 401 | distance | |
| conservation | | | | differential arc length | 97-101,108,113,116,251, 310 |
| | laws | 459 | | |
| | of angular momentum | 455-60 | | one-dimensional | 22 |
| | of energy | 246-7,269-80,335-9,341, 344,353,442-51 | | total distance traveled | 23,25,31-2,97-101,215,283, 379,381 |
| | of momentum | 324-67 | distributive property | 248,408,470p |
| conservative forces | | 258,281-8 | dot product | 247-50,407 |
| constraints | | 63 | double integral | 312-9 |
| contracting indices | | 413 | double summation | 413 |
| conversions | | 13,16 | drag coefficient | 186-7 |
| coordinate systems | | | drive force | 152,233-4,434,476p |
| | Cartesian | 82,85-6,109,114-5,310-1 | driving | 231-6,434,476p |
| | cylindrical | 108-13,115,310-1 | dumbbell | 466p |
| | polar | 108-13,310-1 | dummy indices | 414 |
| | spherical | 113-7,310-1 | dynamic equilibrium | 150,435 |
| crane | | 440 | earth | 53 |
| critical points | | 66-7 | elastic collisions | 335-9,341,343-4,349-52 |
| cross | | | ellipse | 238,401 |
| | product | 405-19,424-5 | | |
| | section | 356,358-9 | | |
| cube | | 400 | | |

| | | |
|---|---|---|
| energy | | |
|   conservation | 246-7,269-80,335-9,341,344,353,442-51 | |
|   definition | 262 | |
|   heat | 262-3 | |
|   internal | 263 | |
|   kinetic | 264,267-70,388-90,425-6,454 | |
|   mechanical | 262,329,354 | |
|   potential | 263-7,270,285-8 | |
|   rotational kinetic | 388-90,425-6,442-51,454 | |
|   translational kinetic | 267-8,442,454 | |
| equations of motion | 58-9,129-30,187,197,246,280-1 | |
| equilibrium | | |
|   conditions of | 287-8,435-42,441 | |
|   dynamic | 150,435 | |
|   neutral | 288,441 | |
|   stable | 288,441 | |
|   static | 150,435-42 | |
|   unstable | 288,441 | |
| errors | 479,482 | |
| escape speed | 279 | |
| extreme values | 66-7 | |
| Ferris wheel | 242p | |
| fields | 194-9,281-8 | |
| flipping mode | 401 | |
| fluids | 185 | |
| fluid resistance | 185-6 | |
| forces | | |
|   air resistance | 144,152,185-90 | |
|   average | 321-4 | |
|   centripetal | 222-5 | |
|   centrifugal | 222-3 | |
|   collision | 321-3 | |
|   conservative | 258,281-8 | |
|   drive | 152,233-4,434,476p | |
|   fluid resistance | 185-6 | |
|   friction | 144,152,167-71,182-4,255-6,291-2,432-4,446 | |
|   gravitational | 144,152,154-8,190-8,237-40,253-4,264-6,279,283,363-7,423 | |
|   hingepin | 439-40 | |
|   internal | 147 | |
|   mutual surface | 180-4 | |
|   nonconservative | 255-6,258,271,289-94,446 |
|   normal | 144,152,165-71,181-4,255 |
|   resistive | 144,152,167-71,182-90,232,291-2,432-4,446 |
|   restoring | 144,152,172-80,256-8,266-7,276-7 |
|   tension | 144,152,158-64 |
|   weight | 144,152,154-8,190-8 |
| force fields | 194-9,281-8 |
| free | |
|   body diagram (FBD) | 151 |
|   fall | 52-8,197-8 |
|   indices | 414 |
| frequency | 217 |
| friction | 144,152,167-71,182-4,255-6,291-2,432-4,446 |
| geosynchronous satellite | 239,244p,304p |
| gradient | 284-5 |
| graphs, motion | 33-43 |
| gravitational | |
|   acceleration | 52-58,193-9 |
|   constant | 192 |
|   field | 193-9,363-7 |
|   force | 144,152,154-8,190-8,237-40,253-4,264-6,279,283,363-7,423 |
|   potential energy | 264-6 |
|   work | 253-4,283 |
| gravity | 52-58,144,152,154-8,190-8,237-40,253-4,264-6,279,283,363-7,423 |
| Greek physics | 140-1 |
| hard spheres | 351 |
| heat | 262-3 |
| hingepin | 439-40 |
| hingepin force | 439-40 |
| hollow cylinder | 399 |
| hollow sphere | 400 |
| Hooke's law | 172-80,256-8,266-7,276-7 |
| horizontal range | 127 |
| horsepower | 262 |
| ice skater | 385,457,463p,476p |
| identity matrix | 454fn |

| | | | | |
|---|---|---|---|---|
| impact | | | magnification of a vector | 95 |
|    area | 356 | | magnitude | 77,83-5,116 |
|    parameter | 350-1 | | mallet | 372p,436,472p |
|    ring | 356 | | mass | 146,154-8,194,314,449,454 |
| impulse | 321,328 | | matrix | 405-6,453-4 |
| indices | 413-9 | | maximum value | 66-7 |
| inelastic collisions | 335-7,341,344 | | mean value | 306-7 |
| inertia | 120-1,142-6,148,223,392-3 | | mean-value theorem | 23-4 |
| inertial reference frame | 223,361 | | mechanical energy | 262,329,354 |
| instantaneous | | | merry-go-round | 378,457-8,464p,476p |
|    acceleration | 30-2 | | method of least squares | 484-7 |
|    power | 259-61 | | metric prefixes | 13-4 |
|    speed | 29-32 | | minimum | |
|    tangential acceleration | 29-32 | |    headstart | 67 |
|    tangential velocity | 29-32 | |    value | 66-7 |
|    velocity | 29-32 | | moment of inertia | 388-404,398-401t,452-4 |
| internal | | | momentum | 142,146,320-34,426,451-60 |
|    energy | 263 | | monkey gun | 132p,137p |
|    forces | 147 | | moon | 53 |
| inverse | | | motion | |
|    additive | 94 | |    circular | 214-40,378-81,383 |
|    completely inelastic | 336-7,341,344 | |    graphs | 33-43,382 |
| Jacobi identity | 412,419 | |    multi-dimensional | 76-130 |
| jerk | 130p | |    non-uniform | 58-62,129-30,386 |
| Kepler's third law | 238-9 | |    of the center of mass | 329-34 |
| kinetic | | |    one-dimensional | 19-68 |
|    energy | 264,267-70,388-90,425-6,454 | |    three-dimensional | 76 |
|    energy, rotational | 388-90,425-6,442-51,454 | |    two-dimensional | 76-130 |
|    energy, translational | 267-8,442,454 | |    uniformly accelerated | 43-58,382-6 |
|    friction | 168-9 | | multi-component motion | 76-130 |
| Kronecker delta | 413-4 | | multiple | |
| lab frame | 334,347,349-50,355 | |    integral | 312-9 |
| latitude | 114 | |    moving objects | 62-8,387 |
| law of cosines | 92,249 | | mutual surface forces | 180-4 |
| least squares | 484-7 | | natural tendency | 144 |
| lever arm | 424 | | net | |
| Levi-Civita symbol | 415-7 | |    displacement | 23,30-2,97-101,107-8,283,379-81,454 |
| line integral | 282 | |    displacement vector | 97-101,251 |
| linear | | |    tangential displacement | 21,30-2,215,379-80,454 |
|    drag | 186 | |    work | 258 |
|    regression | 484-7 | | neutral equilibrium | 288,441 |
| linearizing data | 484 | | | |
| load | 439 | | | |
| longitude | 114 | | | |
| loop-the-loop | 277-9 | | | |

| | | | | |
|---|---|---|---|---|
| Newton's | | | vector | 96-101,110-13,115,216-7, 219,329,379 |
|   first law | 145,148,426-7 | | | |
|   law of gravitation | 190-8,237-40,253-4 | | potential energy | 263-7,270,285-8 |
|   laws of motion | 140-98 | | power | 259-62,454 |
|   second law | 145-6,149-54,224-5,324-9, 330,361-3,426 | | precession | 459-60 |
| | | | problem-solving strategy | |
|   third law | 147-9,321-2,325-6 | |   angular acceleration | 383 |
| Newtonian physics | 141 | |   angular momentum | 456 |
| Noether's theorem | 459 | |   center of mass integrals | 315 |
| non-uniform | | |   collisions in 1D | 340 |
|   acceleration | 58-62,129-30,386 | |   collisions in 2D | 345 |
|   angular acceleration | 386 | |   conservation of energy | 272,445 |
| nonconservative | | |   moment of inertia | 394 |
|   forces | 255-6,258,271,289-94,446 | |   multiple moving objects | 63 |
|   work | 255-6,258,271,289-94,446 | |   Newton's second law | 153 |
| normal | | |   non-uniform motion | 60,129 |
|   components | 82-3,86,105-7,111 | |   projectile motion | 122 |
|   force | 144,152,165-71,181-4,255 | |   rotational kinetic energy | 445 |
| null vector | 86 | |   summing the torques | 427 |
| one-dimensional | | |   uniform acceleration | 48 |
|   collisions | 335-43 | |   uniform circular motion | 221 |
|   motion | 19-68 | |   vector addition | 91,96 |
|   uniform acceleration | 43-58 | |   vector subtraction | 94,96 |
| oscillatory systems | 172-80,256-8,266-7,276-7 | | projectile motion | 117-129 |
| parallel-axis theorem | 403,443-4 | | projectiles | 117,236 |
| parallelogram | 89,408 | | projection | 81,248 |
| parallelepiped | 89,461p | | propagation of errors | 479-82 |
| partial derivative | 284 | | pulleys | 162-4,429-31 |
| path | | | Pythagorean theorem | 83,92,116 |
|   dependence | 23,126,281-94,449 | | quadratic | |
|   integral | 281-2 | |   drag | 186-7 |
| payload | 365 | |   equation | 50 |
| pendulum | | | radial coordinate | 108-9,112-3 |
|   ballistic | 352-5 | | radians | 214 |
|   simple | 276-7,302p,463p | | radius of gyration | 402 |
| period | 217 | | range | 127 |
| physics | 18 | | Rayleigh scattering | 369p |
| points of inflection | 66-7 | | reaction | 147 |
| polar | | | rectangle | 400-1 |
|   angle | 108-9,113-4 | | reference frame | |
|   coordinates | 108-13,310-1 | |   center of mass | 333-4,347-348,351-2 |
| position | | |   definition | 332 |
|   center of mass | 306-20,329 | |   inertial | 223,361 |
|   coordinate | 21,23,30-2,96-101,214, 216-7,219,329,379 | |   lab | 334,347,349-50,355 |
| | | | reference height | 264-6 |
|   graphs | 33-43,382 | | | |

| | | | | |
|---|---|---|---|---|
| relative | | | solid | |
|   maxima | 66-7 | |   angle | 357 |
|   minima | 66-7 | |   cone | 401 |
|   velocity | 127-9,332-3 | |   cube | 400 |
| Renaissance physics | 141 | |   cylinder | 399 |
| resistive forces | 144,152,167-71,182-90, 232,291-2,432-4,446 | |   disc | 398 |
| | | |   sphere | 400 |
| restoring force | 144,152,172-80,256-8, 266-7,276-7 | | speed | |
| | | |   angular | 215,379,382,454 |
| resultant | 87 | |   magnitude of velocity | 22,24-32,102-5,107-8,216 |
| revolutions | 214 | |   tangential | 216,379 |
| revolving point-mass | 424-5,452-3 | |   terminal | 187,190 |
| right- | | | sphere | 400 |
|   hand rule | 410-2 | | spherical coordinates | 113-7,310-1 |
|   handed coordinates | 111fn | | spring | |
| rigid body | 388-9,393-4,398-401,404, 425-7,441-2,444 | |   constant | 173 |
| | | |   potential energy | 260-7 |
| ring | 399 | | springs | 172-80,256-8,266-7,276-7 |
| rockets | 306,360-7 | | stable equilibrium | 288,441 |
| rod | 398 | | stability | 287-8,441 |
| roller coaster loop | 277-9 | | standard deviation | 481 |
| rolling | | | static | |
|   mode | 401 | |   equilibrium | 150,435-42 |
|   without slipping | 231-6,432-4,446-51 | |   friction | 167-8 |
| rotation | 378-460 | | statics | 435-42 |
| rotational | | | strategy, problem-solving | (see problem-solving...) |
|   inertia | 392-3,452 | | summation | |
|   kinetic energy | 388-90,425-6,442-51,454 | |   notation | 23,412-9 |
|   power | 454 | |   symbol | 23,412-9 |
|   work | 454 | | summing the torques | 424-42,452,455 |
| rounding a turn | 231-6 | | swinging | 225-231 |
| Rutherford scattering | 359,369p | | system of objects | 306-67,389-90,404 |
| satellites | 236-40 | | tangent | 33-43 |
| scalar product | 247-50,407 | | tangential | |
| scalars | 77-80 | |   acceleration | 21,29-32,43-58,106-7,216, 379-80,425,454 |
| scattering | | |   components | 82-3,86,105-7,111,224,379, 425 |
|   angle | 357 | | | |
|   cone | 357 | | | |
|   cross section | 356,358-9 | |   displacement | 21,30-2,215,379-80,454 |
|   definition | 355 | |   speed | 216,379 |
|   ring | 357 | |   velocity | 21,26,28-32,105-6,216, 379-80,454 |
| significant figures | 15,17p,480,483p | | | |
| simple harmonic motion | 179 | | tension | 144,152,158-64 |
| size, effect on rolling | 449 | | tensor notation | 413-19 |
| sketching graphs | 41-3 | | tensors | 413-19,453 |
| slope | 33-43 | | terminal speed | 187,190 |

| | |
|---|---|
| test mass | 194 |
| tetherball | 457,463p |
| three-dimensional motion | 76 |
| thrust | 363 |
| time | 20 |
| tip-to-tail | 87-90 |
| top | 459-60 |
| torque | 420-42,452,454-5 |
| torus | 401 |
| total | |
|    acceleration | 104,106-7,217,219,379,425 |
|    cross section | 358-9 |
|    distance traveled | 23,25,31-2,97-101,215,283,379,381 |
|    kinetic energy | 442-3 |
| trajectory | 126 |
| translation | 80 |
| translational kinetic energy | 267-8,442,454 |
| triple | |
|    integral | 312-9 |
|    vector product | 412,470p |
| two-dimensional | |
|    collisions | 344-67 |
|    motion | 76-130 |
| uncertainty | 479-80 |
| unicycle | 475p |
| uniform | |
|    acceleration | 43-58,382-6 |
|    angular acceleration | 382-6 |
|    angular velocity | 382 |
|    circular motion (UCM) | 214-40,382 |
|    speed | 26,104 |
|    velocity | 26,104,382 |
| unit vectors | 85-6,110-3,115,249,409 |
| units | 12-16 |
| unstable equilibrium | 288,441 |
| vacuum | 52 |
| variable mass | 360-7 |
| vector | |
|    acceleration | 215,379-80,382,425,453-4 |
|    addition | 87-96,127-9,453 |
|    direction of | 78,83-5,117 |
|    components | 81-85 |
|    field | 194 |
|    magnitude of | 77,83-5,116 |
| motion quantities | 96-108,216-7,219,329-34,379 |
|    notation | 79-80 |
|    position | 96-101,110-13,115,216-7,219,329,379 |
|    product | 405-19,424-5 |
|    subtraction | 94-6,127-9 |
|    unit | 85-6,110-3,115,249,409 |
|    velocity | 102-5,107-8,127-9,216-7,219,329,379-80,425,453-4 |
|    visualizing | 80 |
| velocity | |
|    angular | 215,379-80,382,425,453-4 |
|    center of mass | 329 |
|    equation | 338-9 |
|    graphs | 33-43,382 |
|    tangential | 21,26,28-32,105-6,216,379-80,454 |
|    terminal | 187,190 |
|    vector | 102-5,107-8,127-9,216-7,219,329,379-80,425,453-4 |
| volume | 311 |
| weight | 144,152,154-8,190-8 |
| weighted average | 307 |
| weightlessness | 157-8,198,240 |
| work | |
|    by friction | 255-6 |
|    by normal force | 255 |
|    by restoring force | 256-7 |
|    conservative | 258,281-8 |
|    definition | 250-3 |
|    energy theorem | 269,425-6 |
|    gravitational | 253-4,283 |
|    integral | 250-60 |
|    net | 258 |
|    nonconservative | 255-6,258,271,289-94,446 |
|    relation to energy | 246 |
|    rotational | 454 |
| yo-yo | 431-2,476p |
| zero gravity | 364-5 |

# About the Author

Chris McMullen is a physics instructor at Northwestern State University of Louisiana. He earned his Ph.D. in physics at Oklahoma State University in phenomenological high-energy physics (particle physics), and his M.S. in physics at California State University, Northridge in theoretical, experimental, and computational work on electron spin resonance. His doctoral dissertation at OSU was on the collider phenomenology of superstring-inspired large extra dimensions, a field in which he has coauthored several papers.

Dr. McMullen previously taught a unique calculus-based physics course for eight years to advanced students who had already completed Calculus II prior to enrolling in the course. These students were very bright, highly motivated, and enthusiastically curious. His years engaging these advanced students showed him that there are many students who are capable of learning at a deeper level and more accelerated pace than typical university courses. This textbook is his effort to give other advanced, independent learners an opportunity to learn physics from a textbook that was not geared toward a large audience, but was instead specifically developed for independent students who have a strong background in mathematics and a passion for learning.

Dr. McMullen presently teaches calculus-based and other physics courses at Northwestern State University of Louisiana. This calculus-based physics course is also unique in one regard: Almost all of the students are either math or math education majors. Although some of these students are just learning calculus as part of the course, they do have strong backgrounds and interests in mathematics. They catch onto the mathematical component of the course readily, and are quite capable of learning new math skills as part of the course.

Many of the examples and problems in this textbook feature monkeys. Dr. McMullen has used monkeys in his creative examples and problems for several years to help stimulate interest in physics. A stressed-out physics student might very well be told to go throw some bananas at his or her physics textbook to help relieve some stress.

# Notes

# Notes

Printed in Great Britain
by Amazon